PROBLEM SOLVER
in
AUTOMATIC CONTROL
SYSTEMS / ROBOTICS

Staff of Research and Education Association,

Dr. M. Fogiel, Director

Research and Education Association
505 Eighth Avenue
New York, N. Y. 10018

PROBLEM SOLVER IN AUTOMATIC CONTROL SYSTEMS

Printed in the United States of America

Library of Congress Catalog Card Number 82 - 61485

International Standard Book Number 0 - 87891 - 542 - 7

WHAT THIS BOOK IS FOR

Students have generally found automatic control systems a difficult subject to understand and learn. Despite the publication of hundreds of textbooks in this field, each one intended to provide an improvement over previous textbooks, students continue to remain perplexed as a result of the numerous conditions that must often be remembered and correlated in solving a problem. Various possible interpretations of terms used in automatic control systems have also contributed to much of the difficulties experienced by students.

In a study of the problem, REA found the following basic reasons underlying students' difficulties with automatic control systems taught in schools:

(a) No systematic rules of analysis have been developed which students may follow in a step-by-step manner to solve the usual problems encountered. This results from the fact that the numerous different conditions and principles which may be involved in a problem, lead to many possible different methods of solution. To prescribe a set of rules to be followed for each of the possible variations, would involve an enormous number of rules and steps to be searched through by students, and this task would perhaps be more burdensome than solving the problem directly with some accompanying trial and error to find the correct solution route.

(b) Textbooks currently available will usually explain a given principle in a few pages written by a professional who has an insight in the subject matter that is not shared by students. The explanations are often written in an abstract manner which leaves the students confused as to the application of the principle. The explanations given are not sufficiently detailed and extensive to make the student aware of the wide range of applications and different aspects of the principle being studied. The numerous possible variations of principles and their applications are usually not discussed, and it is left for the students to discover these for themselves while doing exercises. Accordingly, the average student is expected to rediscover that which has been long known and practiced, but not published or explained extensively.

iii

(c) The examples usually following the explanation of a topic are too few in number and too simple to enable the student to obtain a thorough grasp of the principles involved. The explanations do not provide sufficient basis to enable a student to solve problems that may be subsequently assigned for homework or given on examinations.

The examples are presented in abbreviated form which leaves out much material between steps, and requires that students derive the omitted material themselves. As a result, students find the examples difficult to understand--contrary to the purpose of the examples.

Examples are, furthermore, often worded in a confusing manner. They do not state the problem and then present the solution. Instead, they pass through a general discussion, never revealing what is to be solved for.

Examples, also, do not always include diagrams/graphs, wherever appropriate, and students do not obtain the training to draw diagrams or graphs to simplify and organize their thinking.

(d) Students can learn the subject only by doing exercises themselves and reviewing them in class, to obtain experience in applying the principles with their different ramifications.

In doing the exercises by themselves, students find that they are required to devote considerably more time to automatic control systems than to other subjects of comparable credits, because they are uncertain with regard to the selection and application of the theorems and principles involved. It is also often necessary for students to discover those "tricks" not revealed in their texts (or review books), that make it possible to solve problems easily. Students must usually resort to methods of trial-and-error to discover these "tricks," and as a result they find that they may sometimes spend several hours to solve a single problem.

(e) When reviewing the exercises in classrooms, instructors usually request students to take turns in writing solutions on the boards and explaining them to the class. Students often find it difficult to explain in a manner that holds the interest of the class, and enables the remaining students to follow the material written on the boards. The remaining students seated in the class are, furthermore, too occupied with copying the material from the boards, to listen to the oral explanations and concentrate on the methods of solution.

This book is intended to aid students in automatic control systems to overcome the difficulties described, by supplying detailed illustrations of the solution methods which are usually not apparent to students. The solution methods are illustrated by problems selected from those that are most often assigned for class work and given on examinations. The problems are arranged in order of complexity to enable students to learn and understand a particular topic by reviewing the problems in sequence. The problems are illustrated with detailed step-by-step explanations, to save the students the large amount of time that is often needed to fill in the gaps that are usually found between steps of illustrations in textbooks or review/outline books.

The staff of REA considers automatic control systems a subject that is best learned by allowing students to view the methods of analysis and solution techniques themselves. This approach to learning the subject matter is similar to that practiced in various scientific laboratories, particularly in the medical fields.

In using this book, students may review and study the illustrated problems at their own pace; they are not limited to the time allowed for explaining problems on the board in class.

When students want to look up a particular type of problem and solution, they can readily locate it in the book by referring to the index which has been extensively prepared. It is also possible to locate a particular type of problem by glancing at just the material within the boxed portions. To facilitate rapid scanning of the problems, each problem has a heavy border around it. Furthermore, each problem is identified with a number immediately above the problem at the right-hand margin.

To obtain maximum benefit from the book, students should familiarize themselves with the section, "How To Use This Book," located in the front pages.

To meet the objectives of this book, staff members of REA have selected problems usually encountered in assignments and examinations, and have solved each problem meticulously to illustrate the steps which are usually difficult for students to comprehend. Special gratitude is expressed to them for their efforts in this area, as well as to the numerous contributors who devoted brief periods of time to this work.

Gratitude is also expressed to the many persons involved in the difficult task of typing the manuscript with its endless changes, and to the REA art staff who prepared the numerous detailed illustrations together with the layout and physical features of the book.

The difficult task of coordinating the efforts of all persons was carried out by Carl Fuchs. His conscientious work deserves much appreciation. He also trained and supervised art and production personnel in the preparation of the book for printing.

Finally, special thanks are due to Helen Kaufmann for her unique talents to render those difficult border-line decisions and constructive suggestions related to the design and organization of the book.

Max Fogiel, Ph. D.
Program Director

HOW TO USE THIS BOOK

This book can be an invaluable aid to students in automatic control systems as a supplement to their textbooks. The book is subdivided into 17 chapters, each dealing with a separate topic. The subject matter is developed beginning with modelling and transforms and extending through transfer functions, time analysis, frequency analysis, Nyquist diagrams, Root locus, Bode diagrams, and state-space representations. Included also are control stability, phase plane analysis, nonlinear systems, optimization, and digital control systems. An extensive number of applications have been included, since these appear to be most troublesome to students.

TO LEARN AND UNDERSTAND A TOPIC THOROUGHLY

1. Refer to your class text and read the section pertaining to the topic. You should become acquainted with the principles discussed there. These principles, however, may not be clear to you at that time.

2. Then locate the topic you are looking for by referring to the "Table of Contents" in front of this book, "Automatic Control Systems."

3. Turn to the page where the topic begins and review the problems under each topic, in the order given. For each topic, the problems are arranged in order of complexity, from the simplest to the more difficult. Some problems may appear similar to others, but each problem has been selected to illustrate a different point or solution method.

To learn and understand a topic thoroughly and retain its contents, it will be generally necessary for students to review the problems several times. Repeated review is essential in order to gain experience in recognizing the principles that should be applied, and to select the best solution technique.

TO FIND A PARTICULAR PROBLEM

To locate one or more problems related to a particular subject matter, refer to the index. In using the index, be certain to note that the numbers given there refer to problem numbers, not to page numbers. This arrangement of the index is intended to facilitate finding a problem more rapidly, since two or more problems may appear on a page.

If a particular type of problem cannot be found readily, it is recommended that the student refer to the "Table of Contents" in the front pages, and then turn to the chapter which is applicable to the problem being sought. By scanning or glancing at the material that is boxed, it will generally be possible to find problems related to the one being sought, without consuming considerable time. After the problems have been located, the solutions can be reviewed and studied in detail. For this purpose of locating problems rapidly, students should acquaint themselves with the organization of the book as found in the "Table of Contents. "

In preparing for an exam, it is useful to find the topics to be covered in the exam from the "Table of Contents, " and then review the problems under those topics several times. This should equip the student with what might be needed for the exam.

CONTENTS

CHAPTER 1

MODELLING

BLOCK DIAGRAM

● **PROBLEM** 1-1

Block diagram of a system is shown in Fig. 1.   Represent
the system in the form given in Fig. 2 and Fig. 3.

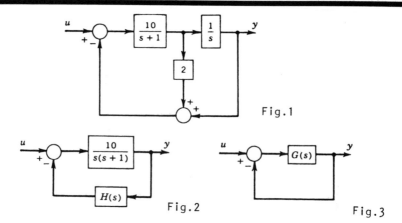

Fig.1

Fig.2

Fig.3

Solution: Let us take the summing point x of Fig. 1.
For this point we have

$$x = u - \left(2 \cdot \frac{10}{s + 1} x + \frac{10}{s(s + 1)} x\right)$$

or

$$x \left[1 + \frac{20}{s + 1} + \frac{10}{s(s + 1)}\right] = u$$

Solving for x we get

$$x = \frac{s(s + 1)}{s^2 + 21s + 10} u$$

From the diagram and the above equation we obtain

$$y = \frac{10}{s(s+1)} \quad x = \frac{10u}{s^2 + 21s + 10}$$

For the systems shown in Fig. 2 and Fig. 3 we have to get the following relationship between y and u

$$y = \frac{10u}{s^2 + 21s + 10} \qquad (*)$$

In Fig. 2 we have for the summing point x

$$x = u - \frac{10x}{s(s+1)} H$$

$$y = \frac{10x}{s(s+1)}$$

Combining the two equations we get

$$y = \frac{10u}{s^2 + s + 10H}$$

Comparing with (*) we have

$$H(s) = 2s + 1.$$

In Fig. 3 we have

$$y = \frac{G(s)}{1+G(s)} u$$

Equating to (*) we get

$$\frac{G}{1+G} = \frac{10}{s^2 + 21s + 10}$$

thus

$$G(s) = \frac{10}{s(s+21)}$$

● **PROBLEM 1-2**

The feedback control system is shown in Fig. 1.

Find the $G_{eq}(s)$ and $H_{eq}(s)$, shown in Fig. 2, of the system.

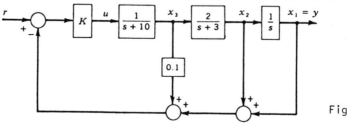

Fig.1

2

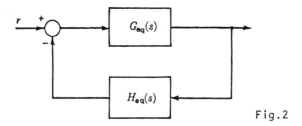

Fig.2

Solution: From Fig. 1 we have for the feedback of the system

$$H(s) = 0.1 \, x_3 + x_2 + x_1$$

and

$$x_3 = \frac{s + 3}{2} \, x_2$$

$$x_2 = sx_1$$

Substituting into the equation for the feedback

$$H(s) = (0.1 \, \frac{s + 3}{2} \, s + s + 1) \, x_1 = \frac{0.1s^2 + 2.3s + 2}{2} \, x_1$$

Since $y = x_1$, the feedback transfer function is

$$H_{eq}(s) = \frac{0.1s^2 + 2.3s + 2}{2}$$

From the block diagram (Fig. 1) the forward gain is

$$G_{eq}(s) = \frac{2K}{(s+10)(s+3)s}$$

● **PROBLEM** 1-3

An engineering organizational system consists of an intercon-
nected set of groups, which perform the whole operation. The
major groups of the system are: management, research and
development, preliminary design, experiments, product design,
fabrication and testing. The system can be analyzed by reducing it to the most elementary components. The dynamic characteristics of each component are represented by a set of
simple equations. The dynamic performance of the whole system can be determined from the relation between time and
progressive accomplishment.

Draw a functional block diagram of an engineering organizational system.

Solution: In the diagram the blocks will represent the functional activities and the lines the flow of information or product output of the system. One of the possible block diagrams is shown below.

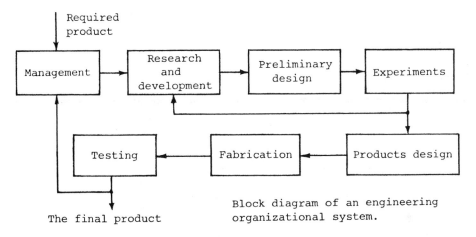

Block diagram of an engineering organizational system.

● PROBLEM 1-4

The liquid-level control system maintains the required level of the liquid in the container. The automatic controller compares the actual level with the desired level and corrects the differences by adjusting the opening of the pneumatic valve. A schematic diagram of the control system is shown below. Draw the corresponding block diagram of this system.

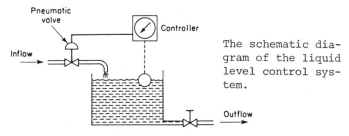

The schematic diagram of the liquid level control system.

Solution: The input of the system is the desired level of the liquid, the output--the actual level. The controller compares both levels. Thus we get the following block diagram of the system.

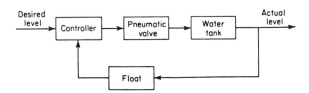

For the system shown below find the relationship between u and y, i.e., express

$$\frac{y(s)}{u(s)}$$

as a function of H_1, H_2, G_1, G_2 and G_3.

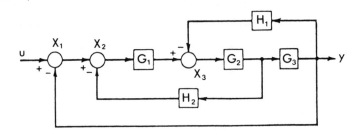

Solution: From the diagram we have the following equations:

$$x_1 = u-y \qquad (1)$$

$$x_2 = x_1 - G_2 H_2 x_3 \qquad (2)$$

$$x_3 = G_1 x_2 - H_1 y \qquad (3)$$

$$y = G_2 G_3 x_3 \qquad (4)$$

From eq. 3 and eq. 4 we compute x_2

$$x_2 = \frac{x_3 + H_1 y}{G_1} = \frac{1 + G_2 G_3 H_1}{G_1 G_2 G_3} y \qquad (5)$$

Substituting eq. (5) into eq. (2) we have

$$x_1 = x_2 + G_2 H_2 x_3 = \left(\frac{1 + G_2 G_3 H_1}{G_1 G_2 G_3} + \frac{H_2}{G_3} \right) y$$

$$= \left(\frac{1 + G_2 G_3 H_1 + G_1 G_2 H_2}{G_1 G_2 G_3} \right) y \qquad (6)$$

Substituting eq. 6 into eq. 1

$$\left(\frac{1 + G_2 G_3 H_1 + G_1 G_2 H_2}{G_1 G_2 G_3} \right) y = u - y \qquad (7)$$

Thus

$$\frac{y}{u} = \frac{G_1 G_2 G_3}{1 + G_1 G_2 H_2 + G_2 G_3 H_1 + G_1 G_2 G_3} \qquad (8)$$

5

The block diagram of the system is shown.

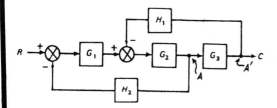

Determine the transfer function of the system and simplify the block diagram.

Solution: The system has two feedback paths. We shall re-arrange them and replace by a single feedback path. To move point A of the lower feedback path to point A' we have to replace H_2 by

$$\frac{H_2}{G_3} .$$

To move point B of the upper path to B' we have to replace H_1 by

$$\frac{H_1}{G_1} .$$

An equivalent block diagram is thus obtained.

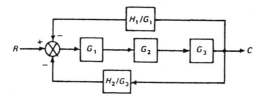

Two feedback paths shown above can be replaced by one, whose gain is

$$\frac{H_1}{G_1} + \frac{H_2}{G_3} .$$

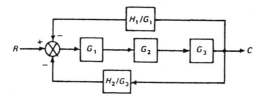

From the last block diagram the transfer function can be easily obtained. We have

$$\frac{C}{R} = \frac{G_1 G_2 G_3}{1 + G_1 G_2 G_3 \left(\frac{H_1}{G_1} + \frac{H_2}{G_3} \right)} = \frac{G_1 G_2 G_3}{1 + G_2 G_3 H_1 + G_1 G_2 H_2}$$

Reduce the block diagram of a multiple-loop feedback control system to a single block.

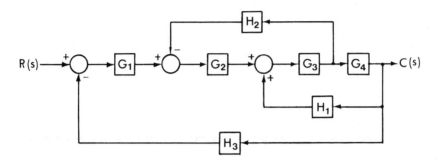

Solution: To reduce a block diagram we shall apply the following rules:

1. Cascaded blocks are multiplied together.

2. Lines may be moved across blocks, the gains must be adjusted.

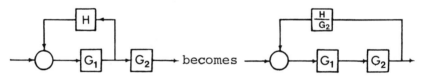

The forward gains and the loop do not change.

3. Feedback loops can be replaced by

Note that the sign changes.

Using rule 2 we shall move H_2 behind the block G_4. Using rule 3 we shall eliminate loop $G_3 G_4 H_1$. An equivalent diagram can be drawn as follows.

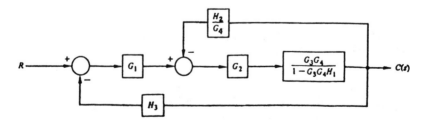

Eliminating loop

$$\frac{H_2}{G_4}$$

we obtain

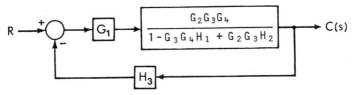

Finally, reducing the cascade (rule 1) and eliminating loop H_3 we obtain

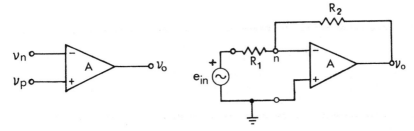

The model symbol of an operational amplifier is shown.

We can assume that $A > 10^4$.

Determine the gain $\dfrac{V_o}{e_{in}}$ of the inverting amplifier shown below.

The amplifier input current is negligibly small, because of the high input impedance of the amplifier.

Solution: Since no current enters the amplifier, the current in R_1 equals the current in R_2. At node n we may therefore write the current equation as

$$\frac{e_{in} - V_n}{R_1} + \frac{V_o - V_n}{R_2} = 0$$

Since the gain of the amplifier is A, we have

$$V_o = AV_n$$

Combining both equations we get

8

$$\frac{e_{in}}{R_1} - \frac{V_o}{AR_1} + \frac{V_o}{R_2} - \frac{V_o}{AR_2} = 0$$

or

$$V_o = \frac{A \cdot \frac{R_2}{R_1} e_{in}}{\frac{R_2}{R_1} - A}$$

We can write the last equation in the form

$$\frac{V_o}{e_{in}} = \frac{A}{1 - A \cdot \frac{R_1}{R_2}} = \frac{A}{1 - Ap}$$

where

$$p = \frac{R_1}{R_2}$$

The gain of the amplifier is $A > 10^4$. Thus

$$\frac{V_o}{e_{in}} = \frac{A}{1 - A \frac{R_1}{R_2}} = \frac{1}{\frac{1}{A} - \frac{R_1}{R_2}} \approx - \frac{R_2}{R_1}$$

The signal flow graph of inverting amplifier is

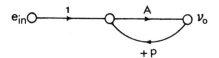

● **PROBLEM** 1-9

Draw the analog computer diagram that solves this equation:

$$A \frac{d^2x}{dt^2} + B \frac{dx}{dt} + Cx = g(t).$$

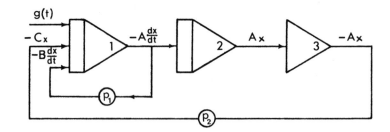

Solution: The highest order derivative is

$$A \frac{d^2x}{dt^2} = g(t) - B \frac{dx}{dt} - Cx$$

Since there are three terms on the right side of the equation, three inputs are connected to an integrator: $g(t)$, $-B \frac{dx}{dt}$, $-Cx$. The output of the integrator is $-A \frac{dx}{dt}$. This value goes to integrator 2, the output is Ax, as shown in the diagram.

The output of integrator 2 is the desired solution. Note that $g(t)$ comes from an external source.

● **PROBLEM** 1-10

The electrical network consisting of resistors and capacitors is shown in Fig. 1. The block diagram of the network is shown in Fig. 2. Find all the indicated transfer functions G_1, ... G_6 and reduce the block diagram of Fig. 2 to the one shown in Fig. 3.

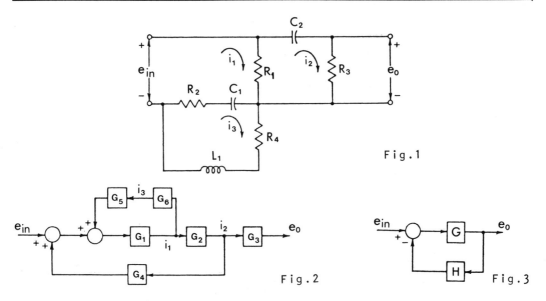

Fig.1

Fig.2

Fig.3

Solution: In the matrix form the KVL loop equations for the circuit shown in Fig. 1 are

$$
\begin{bmatrix}
(R_1 + R_2 + \frac{1}{C_1 S}) & -R_1 & -(R_2 + \frac{1}{C_1 S}) \\
-R_1 & (R_1 + R_3 + \frac{1}{C_2 S}) & 0 \\
-(R_2 + \frac{1}{C_1 S}) & 0 & (R_2 + R_4 + \frac{1}{C_1 S} + L_1 S)
\end{bmatrix}
$$

$$
\begin{bmatrix} i_1 \\ i_2 \\ i_3 \end{bmatrix} = \begin{bmatrix} e_{in} \\ 0 \\ 0 \end{bmatrix}
$$

and

$$e_o = R_3 i_2$$

From Fig. 2 we have the following equation

$$
\begin{bmatrix} 0 & G_4 & G_1 G_5 \\ G_2 & 0 & 0 \\ G_6 & 0 & 0 \end{bmatrix}
\begin{bmatrix} i_1 \\ i_2 \\ i_3 \end{bmatrix}
+
\begin{bmatrix} G_1 \\ 0 \\ 0 \end{bmatrix} e_{in}
=
\begin{bmatrix} i_1 \\ i_2 \\ i_3 \end{bmatrix}
$$

and $e_o = G_3 i_2$, thus $G_3 = R_3$

Multiplying and comparing the elements of the matrices we get

$$\frac{i_2}{i_1} = G_2 = \frac{R_1}{R_1 + R_3 + \dfrac{1}{C_2 S}} = \frac{R_1 C_2 S}{C_2 S (R_1 + R_3) + 1}$$

$$\frac{i_3}{i_1} = G_6 = \frac{R_2 + \dfrac{1}{C_1 S}}{R_2 + R_4 + \dfrac{1}{C_1 S} + L_1 S} = \frac{R_2 C_1 S + 1}{C_1 S (R_2 + R_4 + L_1 S) + 1}$$

$$i_1 = G_4 i_2 + G_1 G_5 i_3 + G_1 e_{in} = \frac{1}{R_1 + R_2 + \dfrac{1}{C_1 S}} \left[R_1 i_2 + \left(R_2 + \frac{1}{C_1 S} \right) i_3 + e_{in} \right]$$

Equating the coefficient of the above equation

$$G_4 = \frac{R_1}{R_1 + R_2 + \dfrac{1}{C_1 S}} = \frac{R_1 C_1 S}{C_1 S (R_1 + R_2) + 1}$$

$$G_1 G_5 = \frac{R_2 + \dfrac{1}{C_1 S}}{R_1 + R_2 + \dfrac{1}{C_1 S}} = \frac{R_2 C_1 S + 1}{C_1 S (R_1 + R_2) + 1}$$

$$G_1 = \frac{1}{R_1 + R_2 + \dfrac{1}{C_1 S}} = \frac{C_1 S}{(R_1 + R_2) C_1 S + 1}$$

Thus

$$G_5 = \frac{R_2 C_1 S + 1}{C_1 S}$$

Shifting the loops of the diagram shown in Fig. 2 we find

$$G = G_1 G_2 G_3$$

$$H = -\frac{G_6 G_5}{G_2 G_3} - \frac{G_4}{G_3}$$

11

Draw a block diagram of the system that obeys the following set of component equations:

$$y = z + z_1$$

$$z_2 = \begin{cases} 0 & z_1 \leq 0 \\ 4 & z_1 > 0 \end{cases}$$

$$z = z_2 z_3$$

$$z = \frac{dz_3}{dt} + 4.2\,z_3$$

Solution: The block diagram of the system can be drawn as follows.

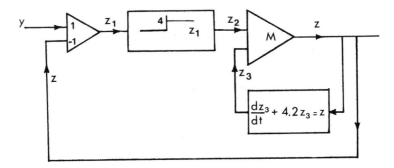

● **PROBLEM** 1-12

Design an analog computer setup to realize the transfer function

$$G(s) = \frac{E_2(s)}{E_1(s)} = \frac{1 + Ks}{1 + \alpha Ks}$$

Solution: We shall rewrite the transfer function as follows

$$(1 + \alpha Ks)E_2(s) = (1 + Ks)E_1(s)$$

Taking the inverse Laplace transform yields

$$e_2 + \alpha K\dot{e}_2 = e_1 + K\dot{e}_1$$

Let us denote $D \equiv \dfrac{d}{dt}$

thus $e_2 + \alpha KDe_2 = e_1 + KDe_1$

or

$$\alpha KDe_2 = e_1 + KDe - e_2$$

After integrating we get

$$e_2 = \frac{e_1}{\alpha} + \frac{e_1 - e_2}{\alpha KD}$$

D is an operator, not a variable. We have the following computer setup.

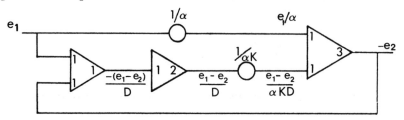

TRANSFER FUNCTION

● **PROBLEM** 1-13

From the graph below find the transfer function describing the frequency response.

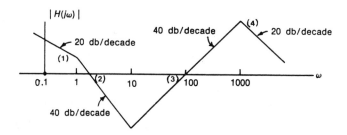

Solution: Let us denote the elements of the curve (1), (2), (3), (4).

For (1), where $0.1 \leq \omega \leq 1$, the line slopes $-20 \frac{db}{decade}$, that can be described by $\frac{1}{s}$. Since the line does not cross the ω axis at $\omega = 1$ there must be a gain factor K. Thus for (1) we have $\frac{K}{s}$. For (2) $1 \leq \omega \leq 10$, the slope changes to $-40 \frac{db}{decade}$, thus $\frac{1}{s+1}$ is the factor. For (3) $10 \leq \omega \leq 1000$ we have $+40 \frac{db}{decade}$, thus $(s + 10)^4$ is needed. For (4) $1000 \leq \omega$ we have $-20 \frac{db}{decade}$, and the factor is

$$\frac{1}{(s + 1000)^3}.$$

13

For the transfer function we obtain

$$H(S) = \frac{K(s + 10)^4}{s(s + 1)(s + 1000)^3}$$

This function is not unique and the gain K can not be determined.

● **PROBLEM** 1-14

For the armature controlled DC motor shown, find the transfer function.

Assume the following numerical values

$$e_1 = 22 \text{ V}$$

$$f = 5 \times 10^{-3} \text{ oz-in/ rad/ sec}$$

$$J = 4.27 \times 10^{-4} \text{ oz-in.-sec}^2$$

$$L_1 \simeq 0$$

$$T_s = 10 \text{ oz-in.} \qquad \text{- the stall torque}$$

$$\omega_0 = 480 \frac{\text{rad}}{\text{sec}} \qquad \text{- no - load speed}$$

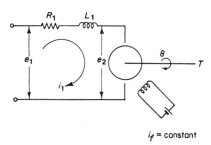

i_f = constant

Solution: The following system of equations describes the system.

$$T = Ki_1$$

$$e_2 = K_2 \frac{d\Theta}{dt}$$

$$L_1 \frac{di_1}{dt} + R_1 i_1 + e_2 = e_1$$

$$J \frac{d^2\Theta}{dt^2} + f \frac{d\Theta}{dt} = T$$

14

Taking the Laplace transforms of the above equations gives:

$$sK_2 \; \text{H}(s) = E_2(s)$$

$$I_1(s) \; [L_1 s + R_1] + E_2(s) = E_1(s)$$

$$\text{H}(s) \; [Js^2 + fs] = KI_1(s)$$

Solving for $\dfrac{\text{H}(s)}{E_1(s)}$ gives

$$\frac{\text{H}(s)}{E_1(s)} = \frac{K}{S[L_1 Js^2 + s(L_1 f + R_1 J) + R_1 f + KK_2]} \qquad (*)$$

Let us denote

$$T_m = \frac{R_1 J}{R_1 f + KK_2}$$

$$K_m = \frac{K}{R_1 f + KK_2}$$

Taking into account the fact that $L_1 \simeq 0$, equation $(*)$ becomes

$$\frac{\text{H}(s)}{E_1(s)} = \frac{K}{s[sR_1 J + R_1 f + KK_2]} = \frac{K_m}{s(T_m s + 1)}$$

where

$$K = \frac{T_s R_1}{e_1}$$

and

$$K_2 = \frac{e_1}{\omega_o}$$

The mechanical power $T\dot{\theta}$ must equal electrical power $(i_1 e_2)$ that is developed by the current, we get

$$i_1 e_2 = K_2 \dot{\theta} i_1 \; (\text{watts}) = K_2 \dot{\theta} i_1 \cdot \frac{1}{746} \; (\text{HP})$$

$$T\dot{\theta} = Ki_1 \dot{\theta} \, (\text{ft-lb/sec}) = Ki_1 \dot{\theta} \cdot \frac{1}{550} \; (\text{HP})$$

We have

$$K_2 = 1.356 \; \text{watts-sec/ft-lb} \cdot K = 7.06 \cdot 10^{-3} \; \text{watts-sec/in-oz} \cdot K$$

Using the numerical data we obtain

$$K_2 = \frac{22}{480} = 5.41 \cdot 10^{-2} \text{ volt/rad/sec}$$

$$K = \frac{K_2}{7.06 \cdot 10^{-3}} = \frac{5.41 \cdot 10^{-2}}{7.06 \cdot 10^{-3}} = 7.66 \text{ in.-oz/amp}$$

$$R_1 = \frac{Ke_1}{T_s} = \frac{7.66 \cdot 22}{10} = 16.85 \text{ ohms}$$

The transfer function is

$$\frac{\textcircled{H}(s)}{E_1(s)} = \frac{7.66}{s[s(16.85 \cdot 4.27 \cdot 10^{-4}) + 16.85 \cdot 5 \cdot 10^{-3} + 7.66 \cdot 5.41 \cdot 10^{-2}]}$$

$$= \frac{766}{s[0.719s + 49.86]} = \frac{15.36}{s(0.0144s + 1)}$$

● **PROBLEM** 1-15

The multiple-loop system and its signal flow graph are shown in Fig. 1 and Fig. 2 respectively.

Determine the closed-loop transfer function $\frac{C}{R}$ by use of Mason's gain formula.

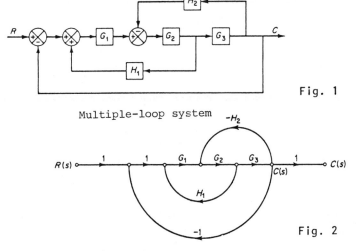

Fig. 1

Multiple-loop system

Fig. 2

Signal flow graph for the system

Solution: The system has one forward path between the input R and the output C, the forward path gain is

$$P_1 = G_1 G_2 G_3$$

There are three individual loops, whose gains are

16

$$L_1 = G_1 G_2 H_1$$

$$L_2 = - G_2 G_3 H_2$$

$$L_3 = - G_1 G_2 G_3$$

There are no nontouching loops, all three loops have a common branch, hence the determinant Δ is

$$\Delta = 1 - (L_1 + L_2 + L_3)$$

$$= 1 - G_1 G_2 H_1 + G_2 G_3 H_2 + G_1 G_2 G_3$$

Since path P_1 touches all three loops, the cofactor Δ_1 is

$$\Delta_1 = 1$$

Thus the closed-loop transfer function is

$$\frac{C(s)}{R(s)} = P = \frac{P_1 \Delta_1}{\Delta} = \frac{G_1 G_2 G_3}{1 - G_1 G_2 H_1 + G_2 G_3 H_2 + G_1 G_2 G_3}$$

We see that without a reduction of the graph Mason's gain formula gives the overall gain $\frac{C(s)}{R(s)}$.

● **PROBLEM** 1-16

The RLC network is shown below.

Let us define the current $i(t)$ and the voltage $e_c(t)$ as the dependent variables. Draw the signal flow graph equivalent of the circuit.

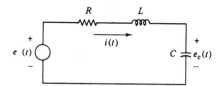

Solution: The system is described by the following differential equations of the first order.

$$L \frac{di(t)}{dt} = e(t) - Ri(t) - e_c(t)$$

$$C \frac{de_c(t)}{dt} = i(t)$$

Let us divide the first equation by L and the second by C

and take the Laplace transform. We have

$$sI(s) = i(0+) + \frac{1}{L} E(s) - \frac{R}{L} I(s) - \frac{1}{L}E_c(s)$$

$$sE_c(s) = e_c(0+) + \frac{1}{C}I(s)$$

The initial conditions are

$i(0+)$ = the initial current at $t = 0+$

$e_c(0+)$ = the initial voltage at $t = 0+$

Solving the equations for $I(s)$ and $E_c(s)$ we get

$$I(s) = \frac{1}{s + \frac{R}{L}} i(0+) + \frac{1}{L(s + \frac{R}{L})} E(s) - \frac{1}{L(s + \frac{R}{L})} E_c(s)$$

$$E_c(s) = \frac{1}{s}e_c(0+) + \frac{1}{Cs} I(s)$$

The input variables are $E(s)$ and $e_c(0+)$ and $i(0+)$.

The signal flow graph using the last two equations is

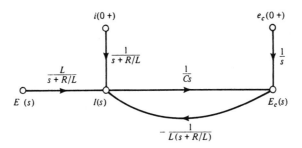

● PROBLEM 1-17

For the flow graph shown, find the output C. Sampling occurs at Y_1.

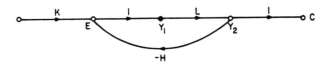

Solution: From the graph we obtain the following information:

        Type - 1 Forward Elementary Paths: none

        Type - 2 Forward Elementary Paths:

$$u_1^{(2)} = (1, E, Y_1, C)$$

Path Gain: $P_1^{(2)} = K*L$

Type - 1 Elementary Loops: none

thus $\Delta^{(1)} = 1$

Type - 2 Elementary Loops:

$$v_1^{(2)} = (Y_1, C, E, Y_1)$$

Loop gain:

$$L_1^{(2)} = - (LH)*$$

thus $\Delta^{(2)} = 1 - L_1^{(2)} = 1 + (LH)*$

Using the gain formula we obtain

$$C = \frac{\Delta_k^{(1)}}{\Delta^{(1)}} \otimes \frac{P_1^{(2)} \Delta_1^{(2)}}{\Delta^{(2)}}$$

Note that $u_1^{(2)}$ is type-2 connected to $v_1^{(2)}$, $\Delta_1^{(2)} = 1$, thus

$$C = \frac{\Delta_R^{(1)}}{\Delta^{(1)}} \otimes \frac{K * L}{1 + (LH)*}$$

In the above expression K* and L* are gains of individual segments and $\otimes$ operation must be performed on these quantities.

Since $\Delta^{(1)} = 1$ and $\Delta_k^{(1)} = 1$ for all values of k we have

$$C = \frac{K*L}{1 + (LH)*}$$

Taking the z-transformation on both sides we have

$$C(z) = \frac{L(z)}{1 + LH(z)} K(z)$$

● **PROBLEM** 1-18

Use the general gain formula to establish the relation between E_o and E_{in}. The signal flow graph is shown.

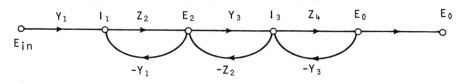

<u>Solution</u>: The forward path gain is

$$G_1 = Y_1 Z_2 Y_3 Z_4$$

As can be seen from the graph there is only one forward path. The loop gains are

$$P_1 = - Z_2 Y_1$$

$$P_2 = - Z_2 Y_3$$

$$P_3 = - Z_4 Y_3$$

There is one pair of nontouching loops, the product of their gains is

$$P_4 = (- Z_2 Y_1) (- Z_4 Y_3) = Z_2 Z_4 Y_1 Y_3 .$$

There is no higher order (that is three, four, etc.) of nontouching loops.

Thus the determinant is

$$\Delta = 1 - (P_1 + P_2 + P_3) + P_4$$

$$= 1 + Z_2 Y_1 + Z_2 Y_3 + Z_4 Y_3 + Z_2 Z_4 Y_1 Y_3 .$$

All three feedback loops touch the forward paths, thus

$$\Delta_1 = 1$$

We have

$$\frac{E_o}{E_{in}} = \frac{\Sigma G_i \Delta_i}{\Delta} = \frac{Y_1 Y_3 Z_2 Z_4}{1 + Z_2 (Y_1 + Y_3) + Z_4 Y_3 (1 + Z_2 Y_1)}$$

● **PROBLEM** 1-19

A system with several feedback loops and feedforward paths is shown. Using Mason's rule find the transfer function of the system.

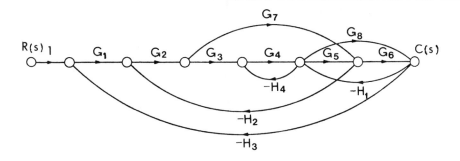

20

<u>Solution</u>: The forward paths are

$$P_1 = G_1 G_2 G_3 G_4 G_5 G_6$$

$$P_2 = G_1 G_2 G_7 G_6$$

$$P_3 = G_1 G_2 G_3 G_4 G_8$$

The feedback loops are

$$L_1 = G_2 G_3 G_4 G_5 H_2$$

$$L_2 = - G_5 G_6 H_1$$

$$L_3 = - G_8 H_1$$

$$L_4 = - G_7 H_2 G_2$$

$$L_5 = - G_4 H_4$$

$$L_6 = - G_1 G_2 G_3 G_4 G_5 G_6 H_3$$

$$L_7 = - G_1 G_2 G_7 G_6 H_3$$

$$L_8 = - G_1 G_2 G_3 G_4 G_8 H_3$$

Loop 5 does not touch loop L_4 and L_7.

L_3 does not touch L_4.

The determinant Δ is

$$\Delta = 1 - \Sigma L_i + \Sigma L_i L_j - . . .$$

$$= 1 - (L_1 + L_2 + L_3 + L_4 + L_5 + L_6 + L_7 + L_8)$$

$$+ (L_5 L_7 + L_5 L_4 + L_3 L_4).$$

The cofactors are

$$\Delta_1 = \Delta_3 = 1$$

$$\Delta_2 = 1 - L_5 = 1 + G_4 H_4$$

The transfer function is given by

$$T(s) = \frac{C(s)}{R(s)} = \frac{P_1 + P_2 \Delta_2 + P_3}{\Delta}$$

21

For the system shown obtain the closed-loop transfer function
$\dfrac{H(s)}{R(s)}$.

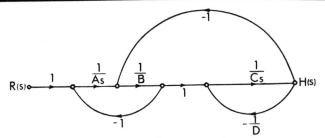

Signal flow graph of a control system

Solution: The system has only one forward path that connects the input R(s) and the output H(s). Thus

$$P_1 = \frac{1}{As} \frac{1}{B} \frac{1}{Cs}$$

The individual loops are

$$L_1 = -\frac{1}{As} \frac{1}{B}$$

$$L_2 = -\frac{1}{Cs} \frac{1}{D}$$

$$L_3 = -\frac{1}{B} \frac{1}{Cs}$$

Loop L_1 does not touch loop L_2, L_1 touches L_3 and L_2 touches L_3. Thus the determinant Δ is given by

$$\Delta = 1 - (L_1 + L_2 + L_3) + (L_1 L_2)$$

$$= 1 + \frac{1}{ABs} + \frac{1}{CDs} + \frac{1}{BCs} + \frac{1}{ABCDs^2}$$

All three loops touch the forward path P_1, thus we remove L_1, L_2 and L_3 from Δ and compute the cofactor Δ_1.

$$\Delta_1 = 1$$

The closed-loop transfer function is

$$\frac{H(s)}{R(s)} = \frac{P_1 \Delta_1}{\Delta} = \frac{\dfrac{1}{ABCs^2}}{1 + \dfrac{1}{ABs} + \dfrac{1}{CDs} + \dfrac{1}{BCs} + \dfrac{1}{ABCDs^2}}$$

$$= \frac{D}{ABCDs^2 + (CD + AB + AD)s + 1}$$

22

A pollution control system is shown below. The inputs a_1 through a_6 represent sources of impurities.

Originally the system consisted of processes G_1, G_3, G_4, and G_6, processes G_2 and G_5 were added later. For each site 1, 2, 3, 4, 5, 6 write the set of equations.

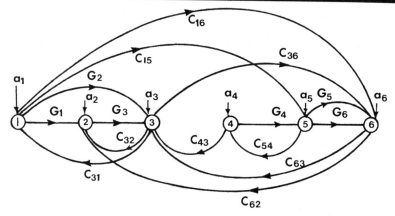

Solution: Let us denote the nodes of the system by X_1, X_2, X_3, X_4, X_5, and X_6. We have the following set of equations

$$X_1 = \bar{a}_1 + C_{31}X_3$$

$$X_2 = a_2 + X_1G_1 + C_{32}X_3 + C_{62}X_6$$

$$X_3 = a_3 + G_3X_2 + \bar{C}_2X_1 + C_{43}X_4 + C_{63}X_6$$

$$X_4 = a_4 + C_{54}X_5$$

$$X_5 = a_5 + G_4X_4 + C_{15}X_1$$

$$X_6 = a_6 + C_{16}X_1 + C_{36}X_3 + G_5X_5 + G_6X_5$$

For the signal-flow diagram shown below find the transmittance.

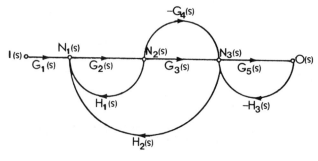

Solution: From the diagram we see that there are two forward paths, thus we have $F_1(s)$ and $F_2(s)$.

$$F_1(s) = G_1(s)G_2(s)G_3(s)G_5(s)$$

$$F_2(s) = G_1(s)G_2(s)[-G_4(s)]G_5(s) = -G_1(s)G_2(s)G_4(s)G_5(s)$$

There are four closed loops. From the diagram again we have

$$L_1 = G_2(s)H_1(s); \text{ nodes } N_1(s), N_2(s)$$

$$L_2 = G_2(s)G_3(s)H_3(s); \text{ nodes } N_1(s), N_2(s), N_3(s)$$

$$L_3 = -G_2(s)G_4(s)H_2(s); \text{ nodes } N_1(s), N_2(s), N_3(s)$$

$$L_4 = G_5(s)[-H_3(s)] = -G_5(s)H_3(s); \text{ nodes } N_3(s), 0(s)$$

The only non-touching loops are L_1 and L_4, thus

$$\Delta(s) = 1 - (L_1 + L_2 + L_3 + L_4) + L_1 L_4$$

$$= 1 - G_2(s)H_1(s) - G_2(s)G_3(s)H_2(s) + G_2(s)G_4(s)H_2(s)$$

$$+ G_5(s)H_3(s) - G_2(s)G_5(s)H_1(s)H_3(s).$$

Since all the closed loop nodes are on both forward paths

$$\Delta_1(s) = 1$$

and

$$\Delta_2(s) = 1$$

Therefore

$$\frac{0(s)}{I(s)} = \frac{F_1(s)\Delta_1(s) + F_2(s)\Delta_2(s)}{\Delta(s)}$$

$$= \frac{G_1(s)G_2(s)G_3(s)G_5(s) - G_1(s)G_2(s)G_4(s) G_5(s)}{1-G_2(s)H_1(s)-G_2(s)G_3(s)H_2(s)+G_2(s)G_4(s)H_2(s)+G_5(s)H_3(s)-}$$

$$\overline{G_2(s)G_5(s)H_1(s)H_3(s)}$$

● **PROBLEM 1-23**

Using the direct signal flow graph method determine the input-output relation of the digital system. The signal flow graph of the system is shown.

Black nodes represent the digital operations

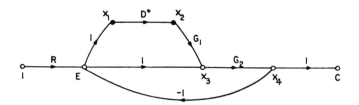

Solution: From the signal flow graph we have five segments, two forward paths and two elementary loops.

Segments:

$$\sigma_1 = (1, E, X_3, X_4, C) \qquad\qquad S_1 = RG_2$$

$$\sigma_2 = (1, E, X_1) \qquad\qquad S_2 = R*$$

$$\sigma_3 = (X_1, X_2) \qquad\qquad S_3 = D*$$

$$\sigma_4 = (X_2, X_3, X_4, C) \qquad\qquad S_4 = G_1 G_2$$

$$\sigma_5 = (X_2, X_3, X_4, E, X_1) \qquad\qquad S_5 = -(G_1 G_2)*$$

Elementary forward pahts (between nodes 1 and C)

Type 1: $u_1^{(1)} = (1, E, X_3, X_4, C)$ $\qquad$ $P_1^{(1)} = RG_2$

Type 2: $u_1^{(2)} = (1, E, X_1, X_2, X_3, X_4, C)$ $P_1^{(2)} = R*D*G_1G_2$

There is only one type-2 elementary path between nodes 1 and C. The path $(1, E, X_1, X_2, X_3, X_4, E, X_3, X_4, C)$ is not type-2 since it does not meet the condition that a path must contain distinct segments.

Type-1 Elementary Loop:

$$v_1^{(1)} = (E, X_3, X_4, E) \qquad\qquad L_1^{(1)} = -G_2$$

Type-2 Elementary Loop:

$$v_1^{(2)} = (X_1, X_2, X_3, X_4, E, X_1) \qquad\qquad L_1^{(2)} = -D*(G_1G_2)*$$

The first and second determinants are

$$\Delta^{(1)} = 1 - L_1^{(1)} = 1 + G_2$$

$$\Delta^{(2)} = 1 - L_1^{(2)} = 1 + D*(G_1G_2)*$$

25

For $u_1^{(1)}$ and $u_1^{(2)}$ we have

$$\Delta_1^{(2)} = 1 + D*(G_1G_2)*$$

$$\Delta_2^{(2)} = 1$$

The input-output transfer relation is

$$C = \frac{\Delta_k^{(1)}}{\Delta^{(1)}} \otimes \frac{P_1^{(1)}\Delta_1^{(2)} + P_1^{(2)}\Delta_2^{(2)}}{\Delta^{(2)}}$$

$$= \frac{\Delta_k^{(1)}}{\Delta^{(1)}} \otimes \frac{RG_2[1 + D*(G_1G_2)*] + R*D*G_1G_2}{1 + D*(G_1G_2)*}$$

$$= \frac{\Delta_k^{(1)}}{\Delta^{(1)}} \otimes \frac{S_1(1 - S_3S_5) + S_2S_3S_4}{1 - S_3S_5}$$

where S_i denotes the gain of the ith segment.

The determinants $\Delta_k^{(1)}$ for $k = 1, 2, \ldots, 5$ are found from $\Delta^{(1)}$ of the part of the flow graph which is not connected to the kth segment. Thus, we get:

$$\sigma_k: \quad \Delta_k^{(1)} = 1 \qquad \text{for } k = 1, 2, 4, 5$$

$$\sigma_3: \quad \Delta_3^{(1)} = \Delta^{(1)} = 1 + G_2$$

Multiplication of the segment gains by the corresponding $\frac{\Delta_k^{(1)}}{\Delta^{(1)}}$ gives

$$C = \frac{RG_2}{1+G_2} + \frac{(R/1 + G_2)* D*(G_1G_2/1 + G_2)}{1 + D*(G_1G_2/1 + G_2)*}$$

● **PROBLEM** 1-24

Using Mason's gain formula obtain the closed-loop transfer function of the system, whose signal flow graph is shown below.

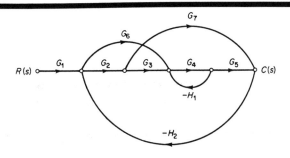

<u>Solution</u>: Between the input R(s) and the output C(s) there are three forward paths

$$P_1 = G_1 G_2 G_3 G_4 G_5$$

$$P_2 = G_1 G_6 G_4 G_5$$

$$P_3 = G_1 G_2 G_7$$

The gains of the individual loops are

$$L_1 = - G_4 H_1$$

$$L_2 = - G_2 G_7 H_2$$

$$L_3 = - G_6 G_4 G_5 H_2$$

$$L_4 = - G_2 G_3 G_4 G_5 H_2$$

Since L_1 does not touch L_2, the determinant Δ is

$$\Delta = 1 - (L_1 + L_2 + L_3 + L_4) + L_1 L_2$$

By removing the loops that touch path P_1, i.e. L_1, L_2, L_3, L_4 and $L_1 L_2$ we obtain the cofactor Δ_1 from Δ. Thus

$$\Delta_1 = 1$$

and

$$\Delta_2 = 1$$

$$\Delta_3 = 1 - L_1$$

The closed-loop transfer function is

$$\frac{C(s)}{R(s)} = P = \frac{1}{\Delta}(P_1 \Delta_1 + P_2 \Delta_2 + P_3 \Delta_3)$$

$$= \frac{G_1 G_2 G_3 G_4 G_5 + G_1 G_6 G_4 G_5 + G_1 G_2 G_7 (1 + G_4 H_1)}{1 + G_4 H_1 + G_2 G_7 H_2 + G_6 G_4 G_5 H_2 + G_4 H_1 G_2 G_7 H_2 + G_2 G_3 G_4 G_5 H_2}$$

● **PROBLEM 1-25**

The digital control system is shown in Fig. 1. Obtain the transfer function by means of a signal flow graph.

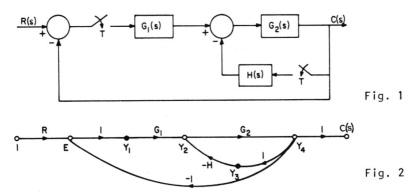

Fig. 1

Fig. 2

Solution: From Fig. 1 we can draw the signal flow graph as shown below.

From the graph we obtain

Segments: $\sigma_1 = (1, E, Y_1)$ $S_1 = R*$

$\sigma_2 = (Y_1, Y_2, Y_4, C)$ $S_2 = G_1 G_2$

$\sigma_3 = (Y_3, Y_2, Y_4, C)$ $S_3 = -G_2 H$

$\sigma_4 = (Y_1, Y_2, Y_4, Y_3)$ $S_4 = (G_1 G_2)*$

$\sigma_5 = (Y_3, Y_2, Y_4, E, Y_1)$ $S_5 = (G_2 H)*$

Elementary forward paths:

Type-1: none

$$P_k^{(1)} = 0 \qquad \text{for all k}$$

Type-2:

$$u_1^{(2)} = (1, E, Y_1, Y_2, Y_4, C) \qquad P_1^{(2)} = R*G_1 G_2$$

$$u_2^{(2)} = (1, E, Y_1, Y_2, Y_4, Y_3, Y_2, Y_4, C) \quad P_2^{(2)} = R*(G_1 G_2)*(-G_2 H)$$

Elementary loops:

Type-1: None

$$v_k^{(1)} = 0 \text{ for all k}$$

Type-2:

$$v_1^{(2)} = (E, Y_1, Y_2, Y_4, E) \qquad L_1^{(2)} = -(G_1 G_2)*$$

$$v_2^{(2)} = (Y_2, Y_4, Y_3, Y_2) \qquad L_2^{(2)} = -(G_2 H)*$$

28

$$v_3^{(2)} = (Y_1,\ Y_2,\ Y_4,\ Y_3,\ Y_2,\ Y_4,\ E,\ Y_1) \qquad L_3^{(2)} = (G_1G_2)*(G_2H)*$$

Both determinants are

$$\Delta^{(1)} = 1$$

$$\Delta^{(2)} = 1 - L_1^{(2)} + L_2^{(2)} + L_3^{(2)} + L_1^{(2)}L_2^{(2)}$$

Since $\Delta^{(1)} = 1$, $\Delta_k^{(1)} = 1$ for all values of k.

The input-output formula is obtained from

$$C = \frac{\Delta_k^{(1)}}{\Delta^{(1)}} \ \otimes \ \frac{\sum_{k=1}^{n} P_k \Delta_k^{(2)}}{\Delta^{(2)}}$$

where P_k is a path gain of kth elementary forward path, type 1 or 2.

$\Delta^{(1)}$ is the first determinant of the signal flow graph.

$$\Delta^{(1)} = 1 - \Sigma L_k^{(1)} + \Sigma L_k^{(1)}L_i^{(1)} - \Sigma L_k^{(1)}L_i^{(1)}L_j^{(1)} + \ .\ .\ .$$

$\Delta^{(2)}$ is the second determinant.

$$\Delta^{(2)} = 1 - \Sigma_k^{(2)} + \Sigma L_k^{(2)}L_i^{(2)} - \Sigma L_k^{(2)}L_i^{(2)}L_j^{(2)} + \ .\ .\ .$$

$\otimes$ is the operation of multiplying $\dfrac{\Delta_k^{(1)}}{\Delta^{(1)}}$ by the gain of the kth segment for all the segments in

$$\sum_{k=1}^{n} \frac{P_k \Delta_k^{(2)}}{\Delta^{(2)}}$$

We have

$$C = \frac{\Delta_k^{(1)}}{\Delta^{(1)}} \ \otimes \ \frac{P_1^{(2)}\Delta_1^{(2)} + P_2^{(2)}\Delta_2^{(2)}}{\Delta^{(2)}} =$$

$$\frac{R*G_1G_2[1 + (G_2H)*] - R*(G_1G_2)*G_2H}{1 + (G_1G_2)* + (G_2H)*}$$

Taking the Z-transformation on both sides we get the z-transform of the output.

$$C(z) = \frac{G_1G_2(z)[1 + G_2H(z)] - G_1G_2(z)G_2H(z)}{1 + G_1G_2(z) + G_2H(z)} R(z)$$

A block diagram of a system and its signal flow graph are shown. Use Mason's rule to find the transmittance

$$T = \frac{C}{R}.$$

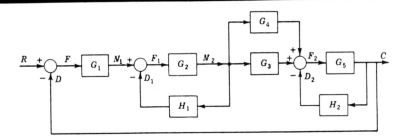

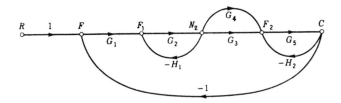

Solution: Mason's rule states that

$$T = \frac{\Sigma P_k \Delta_k}{\Delta}$$

where

$$\Delta = 1 - \Sigma L_i + \Sigma L_i L_j - \Sigma L_i L_j L_k + \ldots$$

In the expression for Δ, L_i are loop gains of loops on the signal-flow graph, $\Sigma L_i L_j$ is the sum of all the products of loop gains of non-touching loops, that is the loops that have no nodes in common. P_k is a forward path gain; the path between C and R, Δ_k is Δ less any terms that contain the loops that touch the path P_k.

There are four loops on the signal-flow graph.

$$L_1 \{F_1 N_2\} = -G_2 H_1$$

$$L_2 \{F_2 C\} = - G_5 H_2$$

$$L_3 \{FF_1 N_2 F_2 C\} = - G_1 G_2 G_3 G_5 \text{ (through } G_3)$$

$$L_4 \{FF_1 N_2 F_2 C\} = - G_1 G_2 G_4 G_5 \text{ (through } G_4)$$

In the expression for Δ we have

$$\Sigma L_i = - G_2 H_1 - G_5 H_2 - G_1 G_2 G_3 G_5 - G_1 G_2 G_4 G_5$$

There are only two non-touching loops L_1 and L_2, thus

$$\Sigma L_i L_j = (-G_2 H_1)(-G_5 H_2) = G_2 G_5 H_1 H_2$$

All the higher products $\Sigma L_i L_j L_k$ etc. are equal to zero.

The determinant Δ is given by

$$\Delta = 1 + G_2 H_1 + G_5 H_2 + G_1 G_2 G_3 G_5 + G_1 G_2 G_4 G_5 + G_2 G_5 H_1 H_2.$$

There are two forward paths, with the gains

$$P_1 \{ \text{through } G_3 \} = G_1 G_2 G_3 G_5$$

$$P_2 \{ \text{through } G_4 \} = G_1 G_2 G_4 G_5$$

From the graph we have

$$\Delta_1 = 1$$

$$\Delta_2 = 1$$

Thus

$$T = \frac{P_1 \Delta_1 + P_2 \Delta_2}{\Delta}$$

$$= \frac{G_1 G_2 G_3 G_5 + G_1 G_2 G_4 G_5}{1 + G_2 H_1 + G_5 H_2 + G_1 G_2 G_5 (G_3 + G_4) + G_2 G_5 H_1 H_2}$$

● **PROBLEM 1-27**

Find the discrete and continuous data outputs of the digital
system shown below by the sampled signal flow graph method.

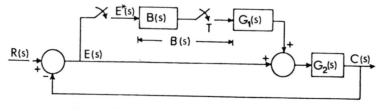

Block diagram of digital control system.

Solution: The signal flow graph of the system is

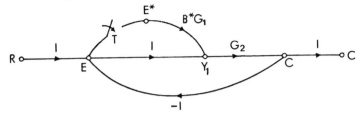

From the signal flow graph, using Mason's gain formula, we get the following equations:

$$E = \frac{R}{1 + G_2} - \frac{B*G_1 G_2}{1 + G_2} E*$$

$$Y_1 = \frac{R}{1 + G_2} + \frac{B*G_1}{1 + G_2} E*$$

$$C = \frac{RG_2}{1 + G_2} + \frac{B*G_1 G_2}{1 + G_2} E*$$

Taking the pulse transform on both sides of the equations we obtain

$$E* = \left[\frac{R}{1 + G_2}\right]^* - B*\left[\frac{G_1 G_2}{1 + G_2}\right]^* E*$$

$$Y_1^* = \left[\frac{R}{1 + G_2}\right]^* + B*\left[\frac{G_1}{1 + G_2}\right]^* E*$$

$$C* = \left[\frac{RG_2}{1 + G_2}\right]^* + B*\left[\frac{G_1 G_2}{1 + G_2}\right]^* E*$$

Using the last six equations we draw the composite signal flow graph of the system.

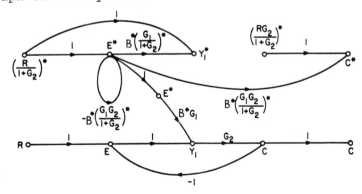

Using Mason's gain formula to the graph the continuous-data and discrete outputs of the system can be computed.

$$C^* = \left[\frac{RG_2}{1 + G_2}\right]^* * + \frac{B^* \left[\dfrac{G_1G_2}{1 + G_2}\right]^*}{1 + B^* \left[\dfrac{G_1G_2}{1 + G_2}\right]^*} \left[\frac{R}{1 + G_2}\right]^*$$

Since $\left[\dfrac{RG_2}{1 + G_2}\right]^* + \left[\dfrac{R}{1 + G_2}\right]^* = R^*$

the equation for C^* can be simplified.

$$C^* = \frac{\left[\dfrac{RG_2}{1 + G_2}\right]^* + R^*B^* \left[\dfrac{G_1G_2}{1 + G_2}\right]^*}{1 + B^* \left[\dfrac{G_1G_2}{1 + G_2}\right]^*}$$

Considering the continuous output $C(s)$ as the output node on the composite signal flow graph, we obtain

$$C = \frac{RG_2 \left[1 + B^* \left(\dfrac{G_1G_2}{1 + G_2}\right)^*\right] + \left[\dfrac{R}{1 + G_2}\right]^* G_1G_2B^*}{1 + G_2 + B^* \left[\dfrac{G_1G_2}{1 + G_2}\right]^* + G_2B^* \left[\dfrac{G_1G_2}{1 + G_2}\right]^*}$$

● **PROBLEM** 1-28

A servo potentiometer shown in Fig. 1 is driving an amplifier. Determine the output voltage as a function of excitation and a shaft position. What is the loading error?

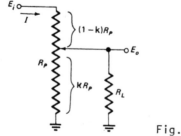

Fig. 1

Solution: From Fig. 1, we compute the transfer ratio of the potentiometer. Voltage E_i is divided between the top and the parallel combination of the load and the kR part.

The output voltage is given by

$$E_o = E_i \frac{(kR_p) || R_L}{(1 - k)R_p + (kR_p) || R_L} =$$

33

$$\frac{\dfrac{kR_p R_L}{kR_p + R_L} E_i}{(1-k)R_p + \dfrac{kR_p R_L}{kR_p + R_L}} = \frac{kR_p R_L E_i}{(1-k)R_p(kR_p + R_L) + kR_p R_L} = \frac{kE_i}{1 + k(1-k)\dfrac{R_p}{R_L}} \quad (*)$$

For the numerical values

$$k = 0.5$$

$$E_i = 20 \text{ V}$$

$$R_p = 20 \text{ k}\Omega$$

$$R_L = 100 \text{ k}\Omega$$

we have

$$E_o = 20 \text{ V} \cdot \frac{0.5}{1 + 0.5 \times 0.5 \dfrac{20}{100}} = 9.52 \text{ V}$$

The error due to loading, for a given shaft position, is the difference between the ideal and actual outputs. For ideal output

$$R_L \to \infty$$

and equation (*) reduces to

$$E_o = \frac{kE_i}{1 + 0} = kE_i$$

Thus

$$\text{Loading error} = kE_i - \frac{kE_i}{1 + k(1-k)\dfrac{R_p}{R_L}}$$

$$= E_i \frac{k^2(1-k)}{k(1-k) + \dfrac{R_L}{R_p}}$$

Substituting the numerical values we obtain

$$\text{Error} = 20 \text{ V} \cdot \frac{0.5 \cdot 0.5 \cdot 0.5}{0.5 \cdot 0.5 + 5} = 0.476 \text{ V}$$

● PROBLEM 1-29

The torsional system is shown below.

The inertia of the cylinder is

34

$$J = \frac{1}{2}Mr^2$$

where M = 6 lb and r = 5 in.

Write the differential equation for the rotation of the mass in response to an input rotation θ_i.

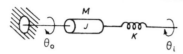

$$\theta_o \qquad \theta_i$$

Solution: Substituting values for the inertia, we obtain

$$J = \frac{1}{2} \cdot \frac{6}{32} \cdot \left(\frac{5}{12}\right)^2 = 0.016 \text{ slug-ft}^2$$

The torque is

$$T = K(\theta_i - \theta_o)$$

We have

$$K(\theta_i - \theta_o) - B\frac{d\theta_o}{dt} = J\frac{d^2\theta_o}{dt^2}$$

or

$$\frac{d^2\theta_o}{dt^2} + \frac{B}{J}\frac{d\theta_o}{dt} + \frac{K}{J}\theta_o = \frac{K}{J}\theta_i$$

For K = 1 lb-ft/rad

B = 0.3 lb-ft/(rad/s)

we obtain

$$\frac{d^2\theta_o}{dt^2} + 18.75\frac{d\theta_o}{dt} + 62.5\,\theta_o = 62.5\,\theta_i$$

● **PROBLEM 1-30**

Find a linear approximation for the torque of a pendulum oscillator when θ is small.

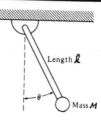

Length ℓ

θ

Mass M

Solution: The weight of the mass is Mg. The length of the moment arm is $\ell \sin \theta$, thus the torque is

$$T = Mg \; \ell\sin \theta$$

The equilibrium point for the mass M is $\theta = 0°$.

The first derivative evaluated at equilibrium gives the linear approximation, thus

$$T = Mg\ell \left. \frac{\partial \sin\theta}{\partial \theta}\right|_{\theta=0°} (\theta - 0°)$$

$$= Mg\ell \cos 0°. \; \theta = Mg\ell \; \theta$$

For $|\theta| \leq \frac{\pi}{4}$ this approximation is fairly accurate.

● PROBLEM 1-31

Write the state equations for the spring-damper mechanical system shown below.

The input f(t) is the force applied to the free end of the spring.

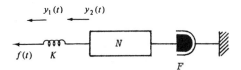

Solution: When the force f(t) is applied to the end of the spring, the spring changes its length. We thus have to define two displacements $y_1(t)$ and $y_2(t)$ for two ends of the spring.

The free body diagram of the system is shown

The force equations of the system are

$$f(t) = K[y_1(t) - y_2(t)]$$

$$K[y_1(t) - y_2(t)] = N \frac{d^2 y_2(t)}{dt^2} + F \frac{dy_2(t)}{dt}$$

Substituting the first equation into the second we obtain

$$f(t) = N \frac{d^2 y_2(t)}{dt^2} + F \frac{dy_2(t)}{dt}$$

36

Let us define
$$x_1(t) = y_2(t)$$

$$x_2(t) = \frac{dy_2(t)}{dt}$$

We get the following system of equations
$$\frac{dx_1(t)}{dt} = x_2(t)$$

$$\frac{dx_2(t)}{dt} = -\frac{F}{N}x_2(t) + \frac{1}{N}f(t)$$

● PROBLEM 1-32

For the mechanical system shown below write the equations of motion.

Then draw the state diagrams and derive transfer functions of the system.

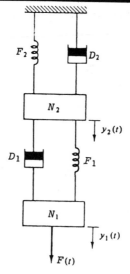

Solution: The free-body diagrams for the two masses can be drawn as follows

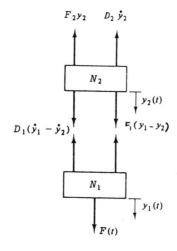

where $\dot{y} = \dfrac{dy(t)}{dt}$

The equations of motion are:

$$F(t) = N_1 \ddot{y}_1 + D_1 [\dot{y}_1 - \dot{y}_2] + F_1 (y_1 - y_2)$$

$$0 = - D_1 (\dot{y}_1 - \dot{y}_2) - F_1 (y_1 - y_2) + N_2 \ddot{y}_2 + D_2 \dot{y}_2 + F_2 y_2$$

where $y_1 = y_1(t)$ and $y_2 = y_2(t)$.

We shall replace the second order differential equations with the system of the first order differential equations.

$$x_1 = y_1$$

$$x_2 = \dot{y}_1 = \dot{x}_1$$

$$x_3 = y_2$$

$$x_4 = \dot{y}_2 = \dot{x}_3$$

We obtain

$$\dot{x}_1 = x_2$$

$$\dot{x}_2 = - \frac{F_1}{N_1} (x_1 - x_3) - \frac{D_1}{N_1} (x_2 - x_4) + \frac{1}{N_1} F(t).$$

$$\dot{x}_3 = x_4$$

$$\dot{x}_4 = \frac{F_1}{N_2} x_1 + \frac{D_1}{N_2} x_2 - \frac{F_1 + F_2}{N_2} x_3 - \frac{1}{N_2} (D_1 + D_2) x_4$$

For the displacements y_1 and y_2 we have

$$y_1(t) = x_1(t)$$

$$y_2(t) = x_3(t)$$

From the above system of equations we can draw the state diagram.

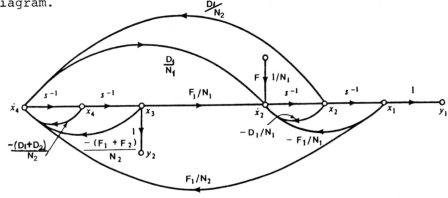

38

The transfer functions are

$$\frac{Y_1(s)}{F(s)} = \frac{N_2 s^2 + (D_1 + D_2)s + (F_1 + F_2)}{\Delta}$$

$$\frac{Y_2(s)}{F(s)} = \frac{D_1 s + F_1}{\Delta}$$

where

$$\Delta = F_1 F_2 + s[F_1 D_1 + D_1(F_1 + F_2)] + s^2[N_1(F_1 + F_2) + N_2 F_1 + D_1(D_1 + D_2) - D_1 F_1]$$

$$+ s^3[N_1(D_1 + D_2) + D_1 N_2 - D_1^2] + s^4 N_1 N_2$$

The crosscurrent extraction system with immiscible solvents of a single solute is shown below.

The system consists of three equilibrium extraction stages. The physical variables describing the system are:

$r(k)$ = solvent (extract) flow rate for k = 1, 2, 3

$x(k)$ = solute concentration leaving for k = 1, 2, 3
k-stage in raffinate

$y(k)$ = solute concentration in extract

q = solvent flow rate

$y(k) = \psi[x(k)]$ - equilibrium relationship between raffinate and extract.

Obtain a model of the system and maximize the performance index.

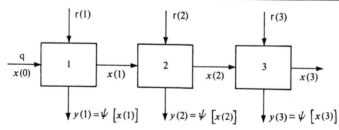

Solution: The equation describing the material balance is

$$q[x(k) - x(k - 1)] = -r(k)y(k)$$

Since the solvent flow rate q is constant we can write

$$x(k) = x(k - 1) - v(k)y(k)$$

where $$v(k) = \frac{r(k)}{q}$$

v(k) is the control variable of the discrete-time system.

For a given x(0) we choose v(1), v(2), v(3) that extremize the performance index

$$I[x(0), 3] = q[x(0) - x(3)] - \lambda[r(1) + r(2) + r(3)]$$

or

$$I^*[x(0),3] = \frac{I[x(0),3]}{q} = x(0) - x(3) - \lambda \sum_{1}^{3} v(k)$$

In this case λ can be interpreted as either a cost of solvent or a Lagrange multiplier.

Let us note that

$$\sum_{k=1}^{3} [x(k) - x(k-1)] = x(3) - x(0)$$

thus

$$\sum_{k=1}^{3} [x(k) - x(k-1)] = \sum_{k=1}^{3} v(k)y(k) = x(3) - x(0)$$

The above equation we obtain by summing the material balance equation.

We have

$$I^*[x(0),3] = \sum_{k=1}^{3} v(k)[y(k)-\lambda]$$

Let us define a new variable

$$u(k) = u(k-1) + v(k)[y(k) - \lambda]$$

$$u(0) = 0$$

It's easy to check that

$$u(3) = I^*[x(0),3].$$

Thus we have to maximize the final value u(3).

Let us write

$$x(k) = x(k-1) + g(k)$$

where

$$x(k) = \begin{bmatrix} x(k) \\ u(k) \end{bmatrix}, \quad x(k-1) = \begin{bmatrix} x(k-1) \\ u(k-1) \end{bmatrix}, \quad g(k) = \begin{bmatrix} -v(k)y(k) \\ v(k)[y(k)-\lambda] \end{bmatrix}$$

with the initial conditions

$$x(0) = \begin{bmatrix} x(0) \\ 0 \end{bmatrix}$$

We have to select v(1), v(2) and v(3) so as to maximize
u(3). The equilibrium conditions of the system are
y(k) = ψ[x(k)]

Write the state equations for the circuit shown.

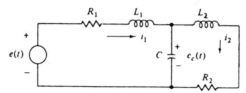

Solution: We have the following set of equations for the
voltages across the inductors and the currents in the
capacitor.

$$L_1 \frac{di_1}{dt} = -R_1 i_1 - e_c + e$$

$$L_2 \frac{di_2}{dt} = -R_2 i_2 + e_c$$

$$C \frac{de_c}{dt} = i_1 - i_2$$

where $i_1 = i_1(t)$, $i_2 = i_2(t)$, $e_c = e_c(t)$, $e = e(t)$

We can write the above equations in the following canonical
form:

$$\begin{bmatrix} \dfrac{di_1(t)}{dt} \\[2ex] \dfrac{di_2(t)}{dt} \\[2ex] \dfrac{de_c(t)}{dt} \end{bmatrix} = \begin{bmatrix} -\dfrac{R_1}{L_1} & 0 & -\dfrac{1}{L_1} \\[2ex] 0 & -\dfrac{R_2}{L_2} & \dfrac{1}{L_2} \\[2ex] \dfrac{1}{C} & -\dfrac{1}{C} & 0 \end{bmatrix} \begin{bmatrix} i_1(t) \\[2ex] i_2(t) \\[2ex] e_c(t) \end{bmatrix} + \begin{bmatrix} 1 \\[2ex] \dfrac{1}{L_1} & 0 \\[2ex] 0 \end{bmatrix} e(t)$$

● **PROBLEM** 1-35

Assuming that the capacitor is initially uncharged,
write the equation of the output of the circuit shown
below.

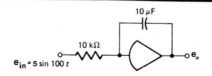

Solution: For an ideal operational amplifier, the transfer
function is

$$\frac{e_o}{e_{in}} = - \frac{R_f}{R_i}$$

(1)

R_f corresponds to the capacitor. If we use Laplace
transforms, eq. 1 becomes

$$e_o = - \frac{1}{RC} \frac{e_{in}}{s}$$

Dividing by s corresponds to integration in the time domain,
thus

$$e_o(t) = - \frac{1}{RC} \int e_{in}(t) dt$$

We obtain

$$e_o = - \frac{1}{10^4 \cdot 10^{-5}} \int (5 \sin 100t) \, dt$$

$$= - \frac{5}{10^{-1}} \frac{(-\cos 100t)}{100} = \frac{1}{2} \cos 100t \text{ V.}$$

● **PROBLEM** 1-36

The amplifier gain is 50,000. In the circuit shown compare
the actual output voltage with the output voltage of an
ideal amplifier (amplifier gain is infinitely large).

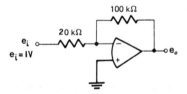

Solution: An op-amp has a very high input impedance, so no
current enters. Thus the current in each resistor is the
same. Let us denote the voltage at the negative input to
the op-amp by V. We have

$$\frac{V - e_i}{20} = \frac{e_o - e_i}{120}$$

For the amplifier

$$e_o = - Kv$$

where K is the amplifier gain.

$$e_o = \frac{-100e_i K}{20K+120} = \frac{-100}{20} e_i \left[\frac{1}{1 + \frac{1 + \frac{100}{20}}{K}} \right] \qquad (*)$$

For K = 50,000 we obtain

$$e_o = \frac{-100}{20} \cdot 1V \left[\frac{1}{1 + \frac{1 + \frac{100}{20}}{50,000}} \right] = -5 \cdot 0.99988 \ V$$

$$= -4.9994 \ V$$

When the amplifier is ideal, i.e. K → ∞, (*) becomes

$$e_o = -\frac{100}{20} e_i = -5V.$$

The error is extremely small.

● **PROBLEM** 1-37

An elastic shaft K connects a motor with a fluid pump.

F = viscous damping of the fluid

J_1 = mass of the motor (rotating part)

J_2 = mass of the pump

T_2 = oscillatory torque

T_1 = the input torque

The motor and the pump represent a two mass system.

Find $\theta_1(s)$ and $\theta_2(s)$.

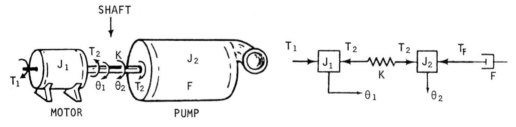

<u>Solution</u>: We have the following equations for the sum of
the torques on each mass

$$\Sigma T_1 = J_1 s^2 \theta_1 = T_1 - T_2 = T_1 - K(\theta_1 - \theta_2)$$

$$\Sigma T_2 = J_2 s^2 \theta_2 = T_2 - T_f = K(\theta_1 - \theta_2) - FS(\theta_2 - 0)$$

43

The torque T_2 is applied in a negative direction to J_1 and in a positive direction to J_2. Separating the variables we obtain

$$\left.\begin{array}{l}(J_1S^2 + K)\theta_1 - K\theta_2 = T_1(S) \\[2em] -K\theta_1 + (J_2S^2 + FS + K)\theta_2 = 0\end{array}\right\}$$

Using Cramer's rule we get

$$\theta_1(s) = \frac{\begin{vmatrix} T_1(S) & -K \\[1.5em] 0 & J_2S^2 + FS + K \end{vmatrix}}{\begin{vmatrix} J_1S^2+K & -K \\[1.5em] -K & J_2S^2+FS+K \end{vmatrix}}$$

$$\theta_2(s) = \frac{\begin{vmatrix} J_1S^2+K & T_1(S) \\[1.5em] -K & 0 \end{vmatrix}}{\begin{vmatrix} J_1S^2+K & -K \\[1.5em] -K & J_2S^2+FS+K \end{vmatrix}}$$

● **PROBLEM** 1-38

A special damper is used to reduce the undesired vibrations of the machines with long shafts and heavy masses. A wheel is placed within a wheel, between them a viscous fluid. When oscillation becomes excessive, the relative motion of the two wheels creates damping. When the system is rotating without vibrations, there is no relative motion and no damping occurs. For the system shown, find $\theta_1(s)$ and $\theta_2(s)$.

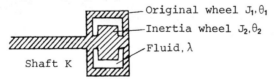

Original wheel J_1, θ_1
Inertia wheel J_2, θ_2
Fluid, λ
Shaft K

Solution: Free body diagram of the system is

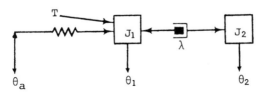

44

We have the following system of equations for the sum of torques on masses J_1 and J_2.

$$J_1\ddot{\theta}_1 = T + K(\theta_a - \theta_1) - \lambda(\dot{\theta}_1 - \dot{\theta}_2)$$

$$J_2\ddot{\theta}_2 = \lambda(\dot{\theta}_1 - \dot{\theta}_2)$$

Taking the Laplace transform of both equations, we obtain

$$K\theta_a(s) + T(s) = (J_1S^2 + K + \lambda S)\theta_1 - \lambda S\theta_2$$

$$0 = (J_2S^2 + \lambda S)\theta_2 - \lambda S\theta_1$$

Solving the above system of equations for θ_1 and θ_2 we get

$$\theta_1 = \frac{(K\theta_a + T)(J_2S + \lambda)}{J_1J_2S^3 + \lambda(J_1 + J_2)S^2 + J_2KS + \lambda K}$$

$$\theta_2 = \frac{\lambda(K\theta_a + T)}{J_1J_2S^3 + \lambda(J_1 + J_2)S^2 + J_2KS + \lambda K}$$

● **PROBLEM** 1-39

A displacement servo and its block diagram are shown below.

A servo, which uses hydraulic elements, consists of a spool valve and a piston type hydraulic actuator. Write the transfer function of the system, where displacement X_{in} of the spool valve is the input and displacement X_{out} of the piston is the output.

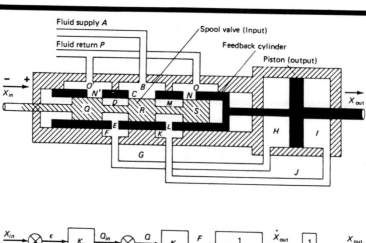

Solution: The system is shown in the off position, no fluid flows since all spool plugs Q, R and S stop the flow. Let us assume that the spool moves in the positive direction by X_{in}. Q blocks N', while C and N are open. Through the fluid supply pipe A, fluid flows to B, to C, to D. Then to the feedback port E, to F through the pipe G into the chamber H, the pressure moves the piston to the right. That causes the fluid to flow from I through pipe J and then K, L, M, N, O and P. When the piston and feedback cylinder have moved a distance X_{out}, the system will have reached a new steady state condition.

From the block diagram, we have for the inner feedback loop

$$\frac{\dot{X}_{out}}{Q_{in}} = \frac{\dfrac{K_m}{Ms + D}}{1 + \dfrac{AK_m}{Ms + D}} = \frac{K_m}{Ms + AK_m + D}$$

Then

$$\frac{X_{out}}{X_{in}\,(\text{open-loop})} = \frac{K_v}{s}\,\frac{\dot{X}_{out}}{Q_{in}} = \frac{K_m K_v}{Ms^2 + s(AK_m + D)}$$

and

$$\frac{X_{out}}{X_{in}\,(\text{closed-loop})} = \frac{\dfrac{X_{out}}{X_{in}\,(\text{open-loop})}}{1 + \dfrac{X_{out}}{X_{in}\,(\text{open-loop})}}$$

$$= \frac{K_m K_v}{Ms^2 + (AK_m + D)s + K_m K_v}$$

● **PROBLEM 1-40**

A field-controlled dc motor driving frictional and inertial load is shown below.

The input is the applied field voltage e(t) and the output is the shaft position φ (t).

Find the transfer function of the system.

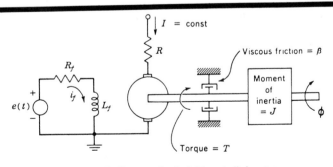

Schematic diagram of a dc field-controlled motor.

Solution: The system can be described by the following equations:

the Kirchhoff equation

$$L_f \dot{i}_f(t) + R_f i_f(t) = e(t)$$

the Newton equation

$$J\ddot{\phi}(t) + \beta\dot{\phi}(t) = T(t)$$

the torque field-current relation

$$T(t) = K i_f(t)$$

where K = torque constant

Transforming three equations we get

$$i_f(s)[sL_f + R_f] = e(s)$$

$$\phi(s)[s^2 J + \beta s] = T(s)$$

$$T(s) = K i_f(s)$$

We compute the transfer function $\dfrac{\phi(s)}{e(s)}$

$$\frac{\phi(s)}{e(s)} = \frac{K}{(Js^2 + \beta s)(L_f s + R_f)}$$

or

$$\frac{\phi(s)}{e(s)} = \frac{K}{JL_f s^3 + s^2(L_f\beta + R_f J) + \beta R_f s}$$

The corresponding differential equation is

$$JL_f \dddot{\phi}(t) + \ddot{\phi}(t)(L_f\beta + R_f J) + \dot{\phi}R_f\beta = Ke(t)$$

and the block diagram is

$$e(s) \longrightarrow \boxed{\dfrac{K}{JL_f s^3 + s^2(L_f\beta + R_f J) + \beta R_f s}} \longrightarrow \phi(s)$$

CHAPTER 2

MATRICES

RANK, ANALYSIS OF INVERSE MATRICES

Determine the rank of the following matrices:

$$A = \begin{bmatrix} 2 & 1 & 5 & 1 \\ 0 & 2 & 0 & 0 \\ 0 & 3 & 1 & 2 \\ 0 & 2 & 0 & 1 \end{bmatrix}$$

$$B = \begin{bmatrix} 1 & 1 & 0 \\ 2 & 1 & 1 \end{bmatrix}$$

<u>Solution</u>: To find the largest N × N matrix in A whose determinant is not zero let us compute

$$|2| \neq 0$$

$$\begin{vmatrix} 2 & 1 \\ 0 & 2 \end{vmatrix} \neq 0$$

$$\begin{vmatrix} 2 & 0 & 0 \\ 3 & 1 & 2 \\ 2 & 0 & 1 \end{vmatrix} \neq 0$$

$$\det A = \begin{vmatrix} 2 & 1 & 5 & 1 \\ 0 & 2 & 0 & 0 \\ 0 & 3 & 1 & 2 \\ 0 & 2 & 0 & 1 \end{vmatrix} = 2 \begin{vmatrix} 2 & 0 & 0 \\ 3 & 1 & 2 \\ 2 & 0 & 1 \end{vmatrix} \neq 0$$

Hence the rank of A is 4. For the matrix B we have

$$|1| \neq 0$$

$$\begin{vmatrix} 1 & 1 \\ 2 & 1 \end{vmatrix} \neq 0$$

Hence the rank is 2.

● **PROBLEM 2-2**

Show that the following quadratic form is positive definite:

$$V(X) = 6x_1^2 + 4x_2^2 + 2x_3^2 + 4x_1x_2 - 4x_2x_3 - 2x_1x_3$$

Solution: The quadratic form $V(X)$ can be written in the matrix form

$$V(X) = x^T P x = \begin{pmatrix} x_1 & x_2 & x_3 \end{pmatrix} \begin{bmatrix} 6 & 2 & -1 \\ 2 & 4 & -2 \\ -1 & -2 & 2 \end{bmatrix} \begin{pmatrix} x_1 \\ x_2 \\ x_3 \end{pmatrix}$$

The application of the Sylvester's Criterion gives

$$6 > 0$$

$$\begin{vmatrix} 6 & 2 \\ 2 & 4 \end{vmatrix} > 0$$

$$\begin{vmatrix} 6 & 2 & -1 \\ 2 & 4 & -2 \\ -1 & -2 & 2 \end{vmatrix} > 0$$

All the principal minors of the matrix P are positive, thus $V(X)$ is positive definite.

Let us assume that the sequence of transformations T_1, T_2, . . . T_k reduces a square matrix A to the identity matrix. Show that the same sequence of operations performed on the identity matrix produces A^{-1}.

Solution: Transformations T_1, T_2, . . . T_n reduce A to I, then

$$(T_n \cdot T_{n-1} \cdot \cdot \cdot T_2 T_1)(A) = I$$

Post multiplying both sides of the above equation by A^{-1} we get

$$(T_n \cdots T_2 T_1)(AA^{-1}) = I \, A^{-1}$$

or

$$(T_n \cdots T_2 T_1) I = A^{-1}$$

Determine whether or not the following quadratic form is negative definite:

$$Q = -2x_1^2 -5x_2^2 -10x_3^2 +4x_1x_2 -2x_1x_3 -6x_2x_3$$

Solution: The quadratic form Q can be written in the matrix form

$$Q = \begin{pmatrix} x_1 & x_2 & x_3 \end{pmatrix} \begin{bmatrix} -2 & 2 & -1 \\ 2 & -5 & -3 \\ -1 & -3 & -10 \end{bmatrix} \begin{bmatrix} x_1 \\ x_2 \\ x_3 \end{bmatrix}$$

Applying Sylvester's criterion we get

$$-2 < 0, \quad \begin{vmatrix} -2 & 2 \\ 2 & -5 \end{vmatrix} > 0, \quad \begin{vmatrix} -2 & 2 & -1 \\ 2 & -5 & -3 \\ -1 & -3 & -10 \end{vmatrix} < 0$$

The quadratic form is negative definite.

Determine the inverses of the following matrices:

a) $\begin{bmatrix} 2 & 3 \\ 4 & 5 \end{bmatrix}$
b) $\begin{bmatrix} 1 & 0 & 3 \\ 2 & 0 & 4 \\ 1 & 2 & 1 \end{bmatrix}$

c) $\begin{bmatrix} p & p+1 \\ 2 & p+2 \end{bmatrix}$

Solution: The inverse of a matrix A exists when:

1. A is a square matrix

2. A is nonsingular, i.e. det A ≠ 0. A^{-1} is given by:

$$A^{-1} = \frac{adj\ A}{|A|} \tag{1}$$

where adj(A) = [cofactor (a_{ij})]^T

a)   det A = 10 - 12 = - 2

Thus A^{-1} exists.

From (1) we get

$$A^{-1} = \frac{adj\ (A)}{|A|} = -\frac{1}{2}\begin{vmatrix} 5 & -3 \\ -4 & 2 \end{vmatrix} = \begin{bmatrix} -\frac{5}{2} & \frac{3}{2} \\ 2 & -1 \end{bmatrix}$$

b)   det A = 4

$$A^{-1} = \frac{1}{4}\begin{bmatrix} -8 & 6 & 0 \\ 2 & -2 & 2 \\ 4 & -2 & 0 \end{bmatrix} = \begin{bmatrix} -2 & 1.5 & 0 \\ 0.5 & -0.5 & 0.5 \\ 1 & -0.5 & 0 \end{bmatrix}$$

c) det $\begin{bmatrix} p & p+1 \\ 2 & p+2 \end{bmatrix}$ = p(p+2) - 2(p+1) = p² -2

The inverse matrix of A exists when p ≠ $\sqrt{2}$ and p ≠ $-\sqrt{2}$

From (1) we get

$$A^{-1} = \frac{adj(A)}{|A|} = \frac{1}{p^2-2} \begin{bmatrix} p+2 & -p-1 \\ -2 & p \end{bmatrix}$$

$$= \begin{bmatrix} \dfrac{p+2}{p^2-2} & \dfrac{-p-1}{p^2-2} \\ \dfrac{-2}{p^2-2} & \dfrac{p}{p^2-2} \end{bmatrix}$$

Determine the inverse of the matrix A

$$A = \begin{bmatrix} 1 & 2 & 1 \\ 0 & 3 & 1 \\ 0 & 1 & 2 \end{bmatrix}$$

Solution: The determinant of A is

$$det \begin{bmatrix} 1 & 2 & 1 \\ 0 & 3 & 1 \\ 0 & 1 & 2 \end{bmatrix} = 5 \neq 0$$

Thus matrix A is nonsingular and its inverse A^{-1} exists.

A^{-1} is given by

$$A^{-1} = \frac{adj(A)}{det(A)}$$

where adj (A) = $[cof(a_{ij})]^T$

For the matrix A we get

$$cof(a_{11}) = (-1)^{1+1} det \begin{bmatrix} 1 & 2 & 1 \\ 0 & 3 & 1 \\ 0 & 1 & 2 \end{bmatrix} = (-1)^{1+1} det \begin{bmatrix} 3 & 1 \\ 1 & 2 \end{bmatrix} = 5$$

$$cof(a_{12}) = (-1)^{1+2} det \begin{bmatrix} 0 & 1 \\ 0 & 2 \end{bmatrix} = 0$$

etc.

$$[\text{cof } (a_{ij})] = \begin{pmatrix} 5 & 0 & 0 \\ -3 & 2 & -1 \\ -1 & -1 & 3 \end{pmatrix}$$

$$\text{adj } (A) = [\text{cof } (a_{ij})]^T = \begin{pmatrix} 5 & -3 & -1 \\ 0 & 2 & -1 \\ 0 & -1 & 3 \end{pmatrix}$$

The inverse matrix A^{-1} is computed from

$$A^{-1} = \frac{\text{adj}(A)}{\det(A)} = \frac{1}{5} \begin{pmatrix} 5 & -3 & -1 \\ 0 & 2 & -1 \\ 0 & -1 & 3 \end{pmatrix}$$

$$= \begin{pmatrix} 1 & -\frac{3}{5} & -\frac{1}{5} \\ 0 & \frac{2}{5} & -\frac{1}{5} \\ 0 & -\frac{1}{5} & \frac{3}{5} \end{pmatrix}$$

● **PROBLEM 2-7**

Using a sequence of elementary operations show that the following matrix A can be reduced to the identity matrix I.

$$A = \begin{bmatrix} 1 & -4 \\ 1 & 1 \end{bmatrix}$$

Show that the same sequence of elementary operations applied to the identity matrix I produces A^{-1}.

Solution: In the left column we will show a sequence of transformed matrices. In the right column, the same sequence of elementary operations performed on I is shown.

$$A = \begin{bmatrix} 1 & -4 \\ 1 & 1 \end{bmatrix} \qquad I = \begin{bmatrix} 1 & 0 \\ 0 & 1 \end{bmatrix}$$

$$\begin{bmatrix} 0 & -5 \\ 1 & 1 \end{bmatrix} \qquad \begin{bmatrix} 1 & -1 \\ 0 & 1 \end{bmatrix}$$

53

$$\begin{bmatrix} 0 & 1 \\ 1 & 1 \end{bmatrix} \qquad\qquad \begin{bmatrix} -\dfrac{1}{5} & \dfrac{1}{5} \\ 0 & 1 \end{bmatrix}$$

$$\begin{bmatrix} 0 & 1 \\ 1 & 0 \end{bmatrix} \qquad\qquad \begin{bmatrix} -\dfrac{1}{5} & \dfrac{1}{5} \\ \dfrac{1}{5} & \dfrac{4}{5} \end{bmatrix}$$

$$I = \begin{bmatrix} 1 & 0 \\ 0 & 1 \end{bmatrix} \qquad\qquad A^{-1} = \begin{bmatrix} \dfrac{1}{5} & \dfrac{4}{5} \\ -\dfrac{1}{5} & \dfrac{1}{5} \end{bmatrix}$$

To verify the results let us multiply

$$\begin{bmatrix} 1 & -4 \\ 1 & 1 \end{bmatrix} \cdot \begin{bmatrix} \dfrac{1}{5} & \dfrac{4}{5} \\ -\dfrac{1}{5} & \dfrac{1}{5} \end{bmatrix} = \begin{bmatrix} 1 & 0 \\ 0 & 1 \end{bmatrix}$$

Thus $\begin{bmatrix} \dfrac{1}{5} & \dfrac{4}{5} \\ -\dfrac{1}{5} & \dfrac{1}{5} \end{bmatrix}$ is an inverse matrix of A

● **PROBLEM** 2-8

For the system described by

$$x(k+1) = A \cdot x(k) + B\, u(k)$$

the state transition matrix is defined by

$$\phi(N) = A^N = \underbrace{A \cdot A \cdot \ \ldots \ \cdot A}_{N \text{ times}}$$

Use the Cayley-Hamilton theorem to obtain $\phi(N)$ where

$$A = \begin{bmatrix} 2 & 1 \\ 1 & 3 \end{bmatrix}$$

Solution: The characteristic equation of A is

$$|\lambda I - A| = (\lambda - 2)(\lambda - 3) - 1 = \lambda^2 - 5\lambda + 5 = 0$$

The Cayley-Hamilton theorem states that every square matrix satisfies its own characteristic equation.

We have the following matrix equation

$$A^2 - 5A + 5I = 0$$

or $A^2 = 5A - 5I$

A^2 is expressed in terms of A. From the Cayley-Hamilton theorem one concludes that A^N can be expressed as an algebraic sum of A, A^2, ... A^{N-1} and by repeatedly applying the theorem, A^N can be expressed in terms of A.

$$A^2 = 5A - 5I$$

$$A^3 = A \cdot A^2 = A \cdot (5A-5I) = 5A^2 - 5A$$

$$= 5 \cdot (5A - 5I) - 5A = 20A - 25I$$

$$A^4 = A \cdot A^3 = A \cdot (20A - 25I) = 20 A^2 - 25A$$

$$= 20(5A - 5I) - 25A = 100A - 100I - 25A$$

$$= 75A - 100I$$

● **PROBLEM** 2-9

Obtain A^{-1} where A is given by

$$A = \begin{bmatrix} 1 & 1 & 0 & 0 \\ 0 & \alpha & 0 & 0 \\ 0 & \alpha^2 & \beta & 0 \\ 0 & \alpha^3 & 0 & 1 \end{bmatrix}$$

Solution: Matrix A^{-1} exists when det A ≠ 0, for the matrix A

$$\text{det } A = \alpha\beta \qquad A^{-1} \text{ exists for } \alpha, \beta \neq 0.$$

From the formula

$$A^{-1} = \frac{\text{adj}(A)}{\text{det}(A)}$$

we obtain

$$A^{-1} = \frac{1}{\alpha\beta} \begin{bmatrix} \alpha\beta & -\beta & 0 & 0 \\ 0 & \beta & 0 & 0 \\ 0 & -\alpha^2 & \alpha & 0 \\ 0 & -\alpha^3\beta & 0 & \alpha\beta \end{bmatrix} = \begin{bmatrix} 1 & -\frac{1}{\alpha} & 0 & 0 \\ 0 & \frac{1}{\alpha} & 0 & 0 \\ 0 & -\frac{\alpha}{\beta} & \frac{1}{\beta} & 0 \\ 0 & -\alpha^2 & 0 & 1 \end{bmatrix}$$

Find the inverse of the matrix P, where

$$P = \begin{bmatrix} 1 & 1 & 1 & 1 \\ -A & -A^2 & 0 & 0 \\ 0 & 0 & -B & -B^2 \\ 0 & 0 & 0 & 1 \end{bmatrix}$$

Solution: The determinant of P is

$$\det P = \begin{vmatrix} -A^2 & 0 & 0 \\ 0 & -B & -B^2 \\ 0 & 0 & 1 \end{vmatrix} + A \begin{vmatrix} 1 & 1 & 1 \\ 0 & -B & -B^2 \\ 0 & 0 & 1 \end{vmatrix} = A^2 B - AB = AB(A - 1)$$

The inverse of the matrix P exists when $A \neq 0$, $B \neq 0$ and $A \neq 1$.

Let us find adj (P) where adj (P) = $[\text{cof } (p_{ij})]^T$

We obtain

$$\left[\begin{matrix}
\begin{pmatrix} -A^2 & 0 & 0 \\ 0 & -B & -B^2 \\ 0 & 0 & 1 \end{pmatrix}, & -\begin{pmatrix} 1 & 1 & 1 \\ 0 & -B & -B^2 \\ 0 & 0 & 1 \end{pmatrix}, & \begin{pmatrix} 1 & 1 & 1 \\ -A^2 & 0 & 0 \\ 0 & 0 & 1 \end{pmatrix}, & -\begin{pmatrix} 1 & 1 & 1 \\ -A^2 & 0 & 0 \\ 0 & -B & -B^2 \end{pmatrix} \\[2em]
-\begin{pmatrix} -A & 0 & 0 \\ 0 & -B & -B^2 \\ 0 & 0 & 1 \end{pmatrix}, & \begin{pmatrix} 1 & 1 & 1 \\ 0 & -B & -B^2 \\ 0 & 0 & 1 \end{pmatrix}, & -\begin{pmatrix} 1 & 1 & 1 \\ -A & 0 & 0 \\ 0 & 0 & 1 \end{pmatrix}, & \begin{pmatrix} 1 & 1 & 1 \\ -A & 0 & 0 \\ 0 & -B & -B^2 \end{pmatrix} \\[2em]
\begin{pmatrix} -A & -A^2 & 0 \\ 0 & 0 & -B^2 \\ 0 & 0 & 1 \end{pmatrix}, & -\begin{pmatrix} 1 & 1 & 1 \\ 0 & 0 & -B^2 \\ 0 & 0 & 1 \end{pmatrix}, & \begin{pmatrix} 1 & 1 & 1 \\ -A & -A^2 & 0 \\ 0 & 0 & 1 \end{pmatrix}, & -\begin{pmatrix} 1 & 1 & 1 \\ -A & -A^2 & 0 \\ 0 & 0 & -B^2 \end{pmatrix} \\[2em]
-\begin{pmatrix} -A & -A^2 & 0 \\ 0 & 0 & -B \\ 0 & 0 & 0 \end{pmatrix}, & \begin{pmatrix} 1 & 1 & 1 \\ 0 & 0 & -B \\ 0 & 0 & 0 \end{pmatrix}, & -\begin{pmatrix} 1 & 1 & 1 \\ -A & -A^2 & 0 \\ 0 & 0 & 0 \end{pmatrix}, & \begin{pmatrix} 1 & 1 & 1 \\ -A & -A^2 & 0 \\ 0 & 0 & -B \end{pmatrix}
\end{matrix}\right]$$

=adj(P)

For the inverse matrix we have:

$$P^{-1} = \frac{1}{AB(A-1)} \begin{bmatrix} A^2B & B & A^2 & -A^2B + A^2B^2 \\ -AB & -B & -A & AB - AB^2 \\ 0 & 0 & A-A^2 & -A^2B^2 + AB^2 \\ 0 & 0 & 0 & A^2B - AB \end{bmatrix}$$

EIGENVECTORS AND DIAGONALIZATION

• **PROBLEM** 2-11

Find the eigenvectors of the matrix A, where

$$A = \begin{bmatrix} 1 & 2 \\ -2 & -3 \end{bmatrix}$$

Solution: Let us find the eigenvalues of A. The characteristic equation of A is

$$|A - \lambda I| = \begin{vmatrix} 1-\lambda & 2 \\ -2 & -3-\lambda \end{vmatrix} = \lambda^2 + 2\lambda + 1 = (\lambda + 1)^2 = 0$$

The eigenvalues of A are $\lambda_1 = \lambda_2 = -1$. The eigenvalue $\lambda_1 = -1$ has a multiplicity of two. From the equation

$$|\lambda_1 I - A| = 0$$

we check the degeneracy of the matrix $\lambda_1 I - A$, thus

$$\lambda_1 I - A = \begin{bmatrix} \lambda_1-1 & -2 \\ 2 & \lambda_1+3 \end{bmatrix} = \begin{bmatrix} -2 & -2 \\ 2 & 2 \end{bmatrix}$$

We see that $\lambda_1 I - A$ has a rank of one, thus the degeneracy of $\lambda_1 I - A$ is one. This means that we can find only one independent eigenvector for $\lambda_1 = -1$.

$$adj(\lambda_1 I - A) = \begin{bmatrix} 2 & 2 \\ -2 & -2 \end{bmatrix}$$

57

Thus, the eigenvector of $\lambda_1 = -1$ is

$$k_1 = \begin{bmatrix} 1 \\ -1 \end{bmatrix}$$

We find the generalized eigenvector from

$$(\lambda_1 I - A) k_2 = -k_1 = \begin{bmatrix} -1 \\ 1 \end{bmatrix}$$

or

$$\begin{bmatrix} -2 & -2 \\ 2 & 2 \end{bmatrix} k_2 = \begin{bmatrix} -1 \\ 1 \end{bmatrix}$$

$$-2 k_{21} - 2k_{22} = -1$$

$$2 k_{21} + 2k_{22} = 1$$

For k_2 we get

$$k_2 = \begin{bmatrix} 0 \\ \dfrac{1}{2} \end{bmatrix}$$

● **PROBLEM 2-12**

Obtain eigenvectors of the following matrix:

$$A = \begin{bmatrix} 0 & 2 & 0 & 0 \\ 0 & 0 & 1 & 0 \\ 0 & 0 & 0 & 1 \\ 1 & 0 & 0 & 0 \end{bmatrix}$$

Solution: From the characteristic equation we shall compute the eigenvalues of the matrix A.

The characteristic equation is

$$|A - \lambda I| = \begin{vmatrix} -\lambda & 2 & 0 & 0 \\ 0 & -\lambda & 1 & 0 \\ 0 & 0 & -\lambda & 1 \\ 1 & 0 & 0 & -\lambda \end{vmatrix} = \lambda^4 - 2 = 0$$

The fourth order equation has four roots

$$\lambda_1 = \sqrt[4]{2}, \qquad \lambda_2 = -\sqrt[4]{2}, \qquad \lambda_3 = j\sqrt[4]{2}, \qquad \lambda_4 = -j\sqrt[4]{2}$$

Let us denote the eigenvector for λ_i by

$$X_i = \begin{bmatrix} x_{1i} \\ x_{2i} \\ x_{3i} \\ x_{4i} \end{bmatrix}$$

Then from the definition of the eigenvector

$$\begin{bmatrix} 0 & 2 & 0 & 0 \\ 0 & 0 & 1 & 0 \\ 0 & 0 & 0 & 1 \\ 1 & 0 & 0 & 0 \end{bmatrix} \begin{bmatrix} x_{1i} \\ x_{2i} \\ x_{3i} \\ x_{4i} \end{bmatrix} = \lambda_i \begin{bmatrix} x_{1i} \\ x_{2i} \\ x_{3i} \\ x_{4i} \end{bmatrix}$$

multiplication gives

$$\begin{bmatrix} 2x_{2i} \\ x_{3i} \\ x_{4i} \\ x_{1i} \end{bmatrix} = \begin{bmatrix} \lambda_i \ x_{1i} \\ \lambda_i \ x_{2i} \\ \lambda_i \ x_{3i} \\ \lambda_i \ x_{4i} \end{bmatrix}$$

Hence the eigenvector for λ_i is

$$\begin{bmatrix} x_{1i} \\ x_{2i} \\ x_{3i} \\ x_{4i} \end{bmatrix} = k_i \begin{bmatrix} 1 \\ \dfrac{\lambda_i}{2} \\ \dfrac{\lambda_i^2}{2} \\ \dfrac{\lambda_i^3}{2} \end{bmatrix}$$

where k_i is an arbitrary constant $k_i \neq 0$. Let T be a matrix whose columns are four independent eigenvectors of A.

59

$$T = \begin{bmatrix} 1 & 1 & 1 & 1 \\ \dfrac{\lambda_1}{2} & \dfrac{\lambda_2}{2} & \dfrac{\lambda_3}{2} & \dfrac{\lambda_4}{2} \\ \dfrac{\lambda_1^2}{2} & \dfrac{\lambda_2^2}{2} & \dfrac{\lambda_3^2}{2} & \dfrac{\lambda_4^2}{2} \\ \dfrac{\lambda_1^3}{2} & \dfrac{\lambda_2^3}{2} & \dfrac{\lambda_3^3}{2} & \dfrac{\lambda_4^3}{2} \end{bmatrix} = \begin{bmatrix} 1 & 1 & 1 & 1 \\ \dfrac{\sqrt[4]{2}}{2} & -\dfrac{\sqrt[4]{2}}{2} & \dfrac{j\sqrt[4]{2}}{2} & -\dfrac{j\sqrt[4]{2}}{2} \\ \dfrac{\sqrt{2}}{2} & \dfrac{\sqrt{2}}{2} & \dfrac{-\sqrt{2}}{2} & \dfrac{-\sqrt{2}}{2} \\ \dfrac{\sqrt[4]{2^3}}{2} & -\dfrac{\sqrt[4]{2^3}}{2} & \dfrac{-j\sqrt[4]{2^3}}{2} & \dfrac{j\sqrt[4]{2^3}}{2} \end{bmatrix}$$

● **PROBLEM** 2-13

The system is described by the differential equation of the third order,

$$\dddot{x} + a_1\ddot{x} + a_2\dot{x} + a_3 x = 0$$

Find its state space representation. Assuming that the eigenvalues are

$$\lambda_1 = \lambda_2 \neq \lambda_3$$

show that the transformation

$$\begin{bmatrix} x_1 \\ x_2 \\ x_3 \end{bmatrix} = \begin{bmatrix} 0 & 1 & 1 \\ 1 & \lambda_1 & \lambda_3 \\ 2\lambda_1 & \lambda_1^2 & \lambda_3^2 \end{bmatrix} \begin{bmatrix} y_1 \\ y_2 \\ y_3 \end{bmatrix}$$

will transform the original equation into

$$\begin{bmatrix} \dot{y}_1 \\ \dot{y}_2 \\ \dot{y}_3 \end{bmatrix} = \begin{bmatrix} \lambda_1 & 0 & 0 \\ 1 & \lambda_1 & 0 \\ 0 & 0 & \lambda_3 \end{bmatrix} \begin{bmatrix} y_1 \\ y_2 \\ y_3 \end{bmatrix}$$

Solution: The state space representation is

$$x_1 = x$$

60

$$x_2 = \dot{x}_1$$

$$x_3 = \dot{x}_2$$

$$\dot{x}_3 = -a_1 x_3 - a_2 x_2 - a_3 x_1$$

or

$$
\begin{bmatrix} \dot{x}_1 \\ \dot{x}_2 \\ \dot{x}_3 \end{bmatrix}
=
\begin{bmatrix} 0 & 1 & 0 \\ 0 & 0 & 1 \\ -a_3 & -a_2 & -a_1 \end{bmatrix}
\begin{bmatrix} x_1 \\ x_2 \\ x_3 \end{bmatrix}
$$

Let us denote

$$
A = \begin{bmatrix} 0 & 1 & 0 \\ 0 & 0 & 1 \\ -a_3 & -a_2 & -a_1 \end{bmatrix}
$$

Transformation $\quad x = AQy$

where

$$
Q = \begin{bmatrix} 0 & 1 & 1 \\ 1 & \lambda_1 & \lambda_3 \\ 2\lambda_1 & \lambda_1^2 & \lambda_3^2 \end{bmatrix}
$$

gives

$$\dot{Q}y = AQy$$

$$
AQ = \begin{bmatrix} 0 & 1 & 0 \\ 0 & 0 & 1 \\ -a_3 & -a_2 & -a_1 \end{bmatrix}
\begin{bmatrix} 0 & 1 & 1 \\ 1 & \lambda_1 & \lambda_3 \\ 2\lambda_1 & \lambda_1^2 & \lambda_3^2 \end{bmatrix}
$$

$$
= \begin{bmatrix} 1 & \lambda_1 & \lambda_3 \\ 2\lambda_1 & \lambda_1^2 & \lambda_3^2 \\ -a_2-2a_1\lambda_1 & -a_3-a_2\lambda_1-a_1\lambda_1^2 & -a_3-a_2\lambda_3-a_1\lambda_3^2 \end{bmatrix}
$$

The characteristic equation is

$$(\lambda-\lambda_1)(\lambda-\lambda_1)(\lambda-\lambda_3) = \lambda^3 + (-2\lambda_1-\lambda_3)\lambda^2 + (\lambda_1^2 + 2\lambda_1\lambda_3)\lambda - \lambda_1^2\lambda_3 = 0$$

Thus
$$a_1 = -2\lambda_1 - \lambda_3$$

$$a_2 = \lambda_1^2 + 2\lambda_1\lambda_3$$

$$a_3 = -\lambda_1^2\lambda_3$$

AQ can be written

$$AQ = \begin{bmatrix} 1 & \lambda_1 & \lambda_3 \\ 2\lambda_1 & \lambda_1^2 & \lambda_3^2 \\ 3\lambda_1^2 & \lambda_1^3 & \lambda_3^3 \end{bmatrix}$$

$$Q^{-1} = \frac{1}{\lambda_1^2 - 2\lambda_1\lambda_3 + \lambda_3^2} \begin{bmatrix} \lambda_1^2\lambda_3 - \lambda_3^2\lambda_1 & \lambda_3^2 - \lambda_1^2 & \lambda_1 - \lambda_3 \\ \lambda_3^2 - 2\lambda_1\lambda_3 & 2\lambda_1 & -1 \\ \lambda_1^2 & -2\lambda_1 & 1 \end{bmatrix}$$

We have

$$Q^{-1}AQ = \begin{bmatrix} \lambda_1 & 0 & 0 \\ 1 & \lambda_1 & 0 \\ 0 & 0 & \lambda_3 \end{bmatrix}$$

● PROBLEM 2-14

Matrix A is given in the form

$$A = \begin{bmatrix} 0 & 1 & 0 \\ 3 & 0 & 2 \\ -12 & -7 & -6 \end{bmatrix}$$

Find the eigenvalues of A, λ_1, λ_2, λ_3 and the transformation matrix P such that

$$P^{-1}AP = \begin{bmatrix} \lambda_1 & 0 & 0 \\ 0 & \lambda_2 & 0 \\ 0 & 0 & \lambda_3 \end{bmatrix}$$

Solution: The characteristic equation of A is

$$|A-\lambda I| = \begin{vmatrix} -\lambda & 1 & 0 \\ 3 & -\lambda & 2 \\ -12 & -7 & -6-\lambda \end{vmatrix} = -(\lambda+1)(\lambda+2)(\lambda+3) = 0$$

The eigenvalues of A are

$$\lambda_1 = -1, \qquad \lambda_2 = -2, \qquad \lambda_3 = -3$$

The matrix P is given by

$$P = \begin{bmatrix} \begin{vmatrix} -\lambda_1 & 2 \\ -7 & -6-\lambda_1 \end{vmatrix} & \begin{vmatrix} -\lambda_2 & 2 \\ -7 & -6-\lambda_2 \end{vmatrix} & \begin{vmatrix} -\lambda_3 & 2 \\ -7 & -6-\lambda_3 \end{vmatrix} \\ -\begin{vmatrix} 3 & 2 \\ -12 & -6-\lambda_1 \end{vmatrix} & -\begin{vmatrix} 3 & 2 \\ -12 & -6-\lambda_2 \end{vmatrix} & -\begin{vmatrix} 3 & 2 \\ -12 & -6-\lambda_3 \end{vmatrix} \\ \begin{vmatrix} 3 & -\lambda_1 \\ -12 & -7 \end{vmatrix} & \begin{vmatrix} 3 & -\lambda_2 \\ -12 & -7 \end{vmatrix} & \begin{vmatrix} 3 & -\lambda_3 \\ -12 & -7 \end{vmatrix} \end{bmatrix}$$

$$= \begin{bmatrix} 9 & 6 & 5 \\ -9 & -12 & -15 \\ -9 & 3 & 15 \end{bmatrix}$$

We can divide and multiply each column of P by a constant, thus P can be written

$$P = \begin{bmatrix} 1 & 2 & 1 \\ -1 & -4 & -3 \\ -1 & 1 & 3 \end{bmatrix}$$

The inverse of P is

$$P^{-1} = \begin{bmatrix} \frac{9}{2} & \frac{5}{2} & 1 \\ -3 & -2 & -1 \\ \frac{5}{2} & \frac{3}{2} & 1 \end{bmatrix}$$

It is easy to check that

$$P^{-1}AP = \begin{bmatrix} -1 & 0 & 0 \\ 0 & -2 & 0 \\ 0 & 0 & -3 \end{bmatrix}$$

Diagonalize matrix A of the state space equation $\dot{x} = Ax$.

$$A = \begin{bmatrix} 1 & 1 + j\sqrt{2} & 0 \\ 1 - j\sqrt{2} & 2 & j\sqrt{3} \\ 0 & -j\sqrt{3} & 1 \end{bmatrix}$$

Solution: A is a Hermitian matrix, therefore it can be diagonalized by use of the transformation matrix whose columns
consist of the normalized eigenvectors of A. From the characteristic equation

$$|A - \lambda I| = \begin{vmatrix} 1 - \lambda & 1 + j\sqrt{2} & 0 \\ 1 - j\sqrt{2} & 2 - \lambda & j\sqrt{3} \\ 0 & -j\sqrt{3} & 1 - \lambda \end{vmatrix} = 0$$

we find the eigenvalues of A

$$\lambda_1 = 1 \qquad \lambda_2 = -1 \qquad \lambda_3 = 4$$

For each eigenvalue we find the eigenvector corresponding
to that value. For $\lambda_1 = 1$ we have

$$(A - \lambda_1 I) P_1 = 0$$

$$\begin{bmatrix} 0 & 1 + j\sqrt{2} & 0 \\ 1 - j\sqrt{2} & 1 & j\sqrt{3} \\ 0 & -j\sqrt{3} & 0 \end{bmatrix} \begin{bmatrix} p_{11} \\ p_{21} \\ p_{31} \end{bmatrix} = 0$$

$$(1 + j\sqrt{2}) p_{21} = 0$$

$$(1 - j\sqrt{2}) p_{11} + p_{21} + j\sqrt{3} p_{31} = 0$$

$$-j\sqrt{3} p_{21} = 0$$

$$P_1 = \begin{bmatrix} -j\sqrt{3}a \\ 0 \\ (1 - j\sqrt{2})a \end{bmatrix} \qquad a = \text{nonzero constant}$$

For $\lambda_2 = -1$

$$P_2 = \begin{bmatrix} (1 + j\sqrt{2})b \\ -2b \\ -j\sqrt{3}b \end{bmatrix} \qquad b = \text{nonzero constant}$$

For $\lambda_3 = 4$

$$P_3 = \begin{bmatrix} (1 + j\sqrt{2})c \\ 3c \\ -j\sqrt{3}c \end{bmatrix} \qquad c = \text{nonzero constant}$$

The eigenvectors P_1, P_2, P_3 are the columns of the matrix P.

$$P = \begin{bmatrix} -j\sqrt{3}a & (1 + j\sqrt{2})b & (1 + j\sqrt{2})c \\ 0 & -2b & 3c \\ (1 - j\sqrt{2})a & -j\sqrt{3}b & -j\sqrt{3}c \end{bmatrix}$$

Normalizing each column we obtain the unitary matrix U:

$$U = \begin{bmatrix} \dfrac{-j\sqrt{3}}{\sqrt{6}} & \dfrac{1 + j\sqrt{2}}{\sqrt{10}} & \dfrac{1 + j\sqrt{2}}{\sqrt{15}} \\[2mm] 0 & \dfrac{-2}{\sqrt{10}} & \dfrac{3}{\sqrt{15}} \\[2mm] \dfrac{1 - j\sqrt{2}}{\sqrt{6}} & \dfrac{-j\sqrt{3}}{\sqrt{10}} & \dfrac{-j\sqrt{3}}{\sqrt{15}} \end{bmatrix}$$

The unitary transformation $x = Ux^1$ will transform $\dot{x} = Ax$ into

$$\dot{x}^1 = Kx^1$$

where K is a diagonal matrix

$$K = U^{-1}AU = U*AU$$

$$K = \begin{bmatrix} 1 & 0 & 0 \\ 0 & -1 & 0 \\ 0 & 0 & 4 \end{bmatrix}$$

● **PROBLEM 2-16**

Diagonalize the system described by

$$\dot{x} = Ax + Bu$$

where

$$A = \begin{bmatrix} 2 & 0 \\ -1 & 1 \end{bmatrix} \qquad \text{and} \qquad B = \begin{bmatrix} 1 \\ -1 \end{bmatrix}$$

Solution: The characteristic equation of the matrix A is

$$|A - \lambda I| = \begin{vmatrix} 2 - \lambda & 0 \\ -1 & 1 - \lambda \end{vmatrix} = (2 - \lambda)(1 - \lambda) = 0$$

65

Thus the eigenvalues of A are

$$\lambda_1 = 2, \quad \lambda_2 = 1$$

The eigenvectors are found from

$$Ak_1 = \lambda_1 k_1 = 2k_1$$

$$Ak_2 = \lambda_2 k_2 = k_2$$

Solving the equations we get

$$k_1 = \begin{bmatrix} a \\ -a \end{bmatrix} \qquad k_2 = \begin{bmatrix} 0 \\ b \end{bmatrix}$$

normalizing k_1 and k_2 we obtain

$$k_1 = \begin{bmatrix} \dfrac{1}{\sqrt{2}} \\ -\dfrac{1}{\sqrt{2}} \end{bmatrix} \qquad k_2 = \begin{bmatrix} 0 \\ 1 \end{bmatrix}$$

Thus

$$K = \begin{bmatrix} \dfrac{1}{\sqrt{2}} & 0 \\ -\dfrac{1}{\sqrt{2}} & 1 \end{bmatrix} \quad \text{and} \quad K^{-1} = \begin{bmatrix} \sqrt{2} & 0 \\ 1 & 1 \end{bmatrix}$$

Multiplication gives

$$A^* = K^{-1} A K = \begin{bmatrix} 2 & 0 \\ 0 & 1 \end{bmatrix}$$

and

$$B^* = K^{-1} B = \begin{bmatrix} \sqrt{2} \\ 0 \end{bmatrix}$$

System

$$\dot{x} = \begin{bmatrix} 2 & 0 \\ -1 & 1 \end{bmatrix} \begin{bmatrix} x_1 \\ x_2 \end{bmatrix} + \begin{bmatrix} 1 \\ -1 \end{bmatrix} u$$

after the canonical transformation becomes

$$\begin{bmatrix} \dot{x}_1{}^* \\ \dot{x}_2{}^* \end{bmatrix} = \begin{bmatrix} 2 & 0 \\ 0 & 1 \end{bmatrix} \begin{bmatrix} x_1{}^* \\ x_2{}^* \end{bmatrix} + \begin{bmatrix} \sqrt{2} \\ 0 \end{bmatrix} u$$

Diagonalize matrix A, where

$$A = \begin{bmatrix} 0 & 1 & 0 \\ 0 & 0 & 1 \\ -6 & -11 & -6 \end{bmatrix}$$

<u>Solution</u>: The characteristic equation of A is

$$|A - \lambda I| = \begin{vmatrix} -\lambda & 1 & 0 \\ 0 & -\lambda & 1 \\ -6 & -11 & -6-\lambda \end{vmatrix}$$

$$= \lambda^3 + 6\lambda^2 + 11\lambda + 6 = (\lambda + 1)(\lambda + 2)(\lambda + 3)$$

$$= 0$$

The eigenvalues are

$$\lambda_1 = -1, \ \lambda_2 = -2, \ \lambda_3 = -3$$

To each eigenvalue there corresponds one eigenvector.

For $\lambda_1 = -1$ we have:

$$(A - \lambda_1 I)P_1 = 0$$

or

$$\begin{bmatrix} 1 & 1 & 0 \\ 0 & 1 & 1 \\ -6 & -11 & -5 \end{bmatrix} \begin{bmatrix} p_{11} \\ p_{21} \\ p_{31} \end{bmatrix} = 0$$

$$p_{11} + p_{21} = 0$$

$$p_{21} + p_{31} = 0$$

$$-6p_{11} - 11p_{21} - 5p_{31} = 0$$

We can assume $p_{11} = 1$, thus

$$P_1 = \begin{bmatrix} 1 \\ -1 \\ 1 \end{bmatrix}$$

For $\lambda_2 = -2$ we have

$$\begin{bmatrix} 2 & 1 & 0 \\ 0 & 2 & 1 \\ -6 & -11 & -4 \end{bmatrix} \begin{bmatrix} p_{12} \\ p_{22} \\ p_{32} \end{bmatrix} = 0$$

$$2p_{12} + p_{22} = 0$$

$$2p_{22} + p_{32} = 0$$

$$-6p_{12} - 11p_{22} - 4p_{32} = 0$$

Let $p_{12} = 1$, then we obtain

$$P_2 = \begin{bmatrix} 1 \\ -2 \\ 4 \end{bmatrix}$$

For $\lambda_3 = -3$ we get

$$P_3 = \begin{bmatrix} 1 \\ -3 \\ 9 \end{bmatrix}$$

The diagonalizing matrix P is

$$P = \begin{bmatrix} 1 & 1 & 1 \\ -1 & -2 & -3 \\ 1 & 4 & 9 \end{bmatrix}$$

The diagonal form of A can be obtained from

$$P^{-1}AP = \begin{bmatrix} 3 & \frac{5}{2} & \frac{1}{2} \\ -3 & 4 & -1 \\ 1 & \frac{3}{2} & \frac{1}{2} \end{bmatrix} \begin{bmatrix} 0 & 1 & 0 \\ 0 & 0 & 1 \\ -6 & -11 & -6 \end{bmatrix} \begin{bmatrix} 1 & 1 & 1 \\ -1 & -2 & -3 \\ 1 & 4 & 9 \end{bmatrix}$$

$$\Lambda = \begin{bmatrix} -1 & 0 & 0 \\ 0 & -2 & 0 \\ 0 & 0 & -3 \end{bmatrix}$$

● **PROBLEM** 2-18

Diagonalize the matrix A, where

$$A = \begin{bmatrix} 0 & 6 & -5 \\ 1 & 0 & 2 \\ 3 & 2 & 4 \end{bmatrix}$$

Find the transformation matrix P that transforms A into Jordan canonical form.

Solution: The characteristic equation of the matrix A is

$$|\lambda I - A| = \begin{vmatrix} \lambda & -6 & 5 \\ -1 & \lambda & -2 \\ -3 & -2 & \lambda-4 \end{vmatrix} = \lambda^3 - 4\lambda^2 + 5\lambda - 2$$

$$= (\lambda - 2)(\lambda - 1)^2 = 0$$

Matrix A has one simple eigenvalue $\lambda_1 = 2$ and one double eigenvalue $\lambda_2 = 1$.

We shall find the eigenvectors of the matrix A for each eigenvalue.

68

For $\lambda_1 = 2$ we have

$$(\lambda_1 I - A) P_1 = 0$$

or

$$\begin{bmatrix} 2 & -6 & 5 \\ -1 & 2 & -2 \\ -3 & -2 & -2 \end{bmatrix} \begin{bmatrix} p_{11} \\ p_{21} \\ p_{31} \end{bmatrix} = 0$$

Solving the system of equations and setting $p_{11} = 2$ we obtain

$$P_1 = \begin{bmatrix} 2 \\ -1 \\ -2 \end{bmatrix}$$

For the two eigenvectors P_2 and P_3 we have

$$(\lambda_2 I - A) P_2 = 0$$

$$(\lambda_2 I - A) P_3 = -P_2$$

The first equation yields

$$\begin{bmatrix} 1 & -6 & 5 \\ -1 & 1 & -2 \\ -3 & -2 & -3 \end{bmatrix} \begin{bmatrix} p_{12} \\ p_{22} \\ p_{32} \end{bmatrix} = 0$$

Setting $p_{12} = 1$ we get

$$P_2 = \begin{bmatrix} 1 \\ -\dfrac{3}{7} \\ -\dfrac{5}{7} \end{bmatrix}$$

P_3 can be found from the equation

$$\begin{bmatrix} 1 & -6 & 5 \\ -1 & 1 & -2 \\ -3 & -2 & -3 \end{bmatrix} \begin{bmatrix} p_{13} \\ p_{23} \\ p_{33} \end{bmatrix} = \begin{bmatrix} -1 \\ \dfrac{3}{7} \\ \dfrac{5}{7} \end{bmatrix}$$

Thus

$$P_3 = \begin{bmatrix} 1 \\ -\dfrac{22}{49} \\ -\dfrac{46}{49} \end{bmatrix}$$

and

$$P = \begin{bmatrix} 2 & 1 & 1 \\ -1 & -\dfrac{3}{7} & -\dfrac{22}{49} \\ -2 & -\dfrac{5}{7} & -\dfrac{46}{49} \end{bmatrix}$$

The Jordan canonical form is

$$\Lambda = P^{-1}AP = \begin{bmatrix} 2 & 0 & 0 \\ 0 & 1 & 1 \\ 0 & 0 & 1 \end{bmatrix}$$

● **PROBLEM 2-19**

For the matrices A and J where

$$A = \begin{bmatrix} 4 & 1 & -2 \\ 1 & 0 & 2 \\ 1 & -1 & 3 \end{bmatrix}, \quad J = \begin{bmatrix} 3 & 1 & 0 \\ 0 & 3 & 0 \\ 0 & 0 & 1 \end{bmatrix}$$

find the transformation matrix S such that

$$S^{-1}AS = J$$

Solution: The characteristic equation of A is

$$|A - \lambda I| = \begin{vmatrix} 4-\lambda & 1 & -2 \\ 1 & -\lambda & 2 \\ 1 & -1 & 3-\lambda \end{vmatrix} = -(\lambda - 3)(\lambda - 3)(\lambda - 1) = 0$$

The eigenvalues of A are

$$\lambda_1 = 3, \lambda_2 = 3, \lambda_3 = 1.$$

Since one eigenvalue is double, there exist two linearly independent eigenvectors.

Matrix S is given by the formula

$$S = \begin{bmatrix} \begin{vmatrix} -\lambda_1 & 2 \\ -1 & 3-\lambda_1 \end{vmatrix} & 2\lambda_1-3 & \begin{vmatrix} -\lambda_3 & 2 \\ -1 & 3-\lambda_3 \end{vmatrix} \\ -\begin{vmatrix} 1 & 2 \\ 1 & 3-\lambda_1 \end{vmatrix} & 1 & -\begin{vmatrix} 1 & 2 \\ 1 & 3-\lambda_3 \end{vmatrix} \\ \begin{vmatrix} 1 & -\lambda_1 \\ 1 & -1 \end{vmatrix} & 1 & \begin{vmatrix} 1 & -\lambda_3 \\ 1 & -1 \end{vmatrix} \end{bmatrix}$$

For $\lambda_1 = 3$ and $\lambda_3 = 1$ we have

$$S = \begin{bmatrix} 2 & 3 & 0 \\ 2 & 1 & 0 \\ 2 & 1 & 0 \end{bmatrix}$$

Since the elements of the third column are zeros, we shall replace the third column with A_{321}, A_{322}, A_{323} where

$$A_{321} = -\begin{vmatrix} 1 & -2 \\ -1 & 3-\lambda_3 \end{vmatrix}, \quad A_{322} = \begin{vmatrix} 4-\lambda_3 & -2 \\ 1 & 3-\lambda_3 \end{vmatrix}$$

$$A_{323} = -\begin{vmatrix} 4-\lambda_3 & 1 \\ 1 & -1 \end{vmatrix}$$

Note that A_{32i} are the cofactors of the second row of

$$A - \lambda_3 I$$

We get

$$S' = \begin{bmatrix} 2 & 3 & 0 \\ 2 & 1 & 8 \\ 2 & 1 & 4 \end{bmatrix}$$

It's easy to check that

$$S'^{-1}A\,S' = J$$

For a matrix A obtain a transformation matrix T such that $T^{-1}AT$ is diagonal. Assume that A has distinct eigenvalues, A is given by

$$A = \begin{bmatrix} a_{11} & a_{12} & a_{13} & a_{14} \\ a_{21} & a_{22} & a_{23} & a_{24} \\ a_{31} & a_{32} & a_{33} & a_{34} \\ a_{41} & a_{42} & a_{43} & a_{44} \end{bmatrix}$$

<u>Solution:</u> The columns of the matrix T are eigenvectors of the matrix A. Let us denote four distinct eigenvalues as λ_1, λ_2, λ_3 and λ_4.

The characteristic equation is

$$\left| A - \lambda I \right| = (\lambda - \lambda_1)(\lambda - \lambda_2)(\lambda - \lambda_3)(\lambda - \lambda_4)$$

Using the definiton of the eigenvector and solving the system of equations we obtain for the elements of the ith column of T

$$T_i = \begin{bmatrix} \begin{vmatrix} a_{22}-\lambda_i & a_{23} & a_{24} \\ a_{32} & a_{33}-\lambda_i & a_{34} \\ a_{42} & a_{43} & a_{44}-\lambda_i \end{vmatrix} \\[6pt] - \begin{vmatrix} a_{21} & a_{23} & a_{24} \\ a_{31} & a_{33}-\lambda_i & a_{34} \\ a_{41} & a_{43} & a_{44}-\lambda_i \end{vmatrix} \\[6pt] \begin{vmatrix} a_{21} & a_{22}-\lambda_i & a_{24} \\ a_{31} & a_{32} & a_{34} \\ a_{41} & a_{42} & a_{44}-\lambda_i \end{vmatrix} \\[6pt] - \begin{vmatrix} a_{21} & a_{22}-\lambda_i & a_{23} \\ a_{31} & a_{32} & a_{33}-\lambda_i \\ a_{41} & a_{42} & a_{43} \end{vmatrix} \end{bmatrix}$$

For four different values of λ_i, $i = 1, 2, 3, 4$ we get four columns of T.

Show that for the matrix A

$$A = \begin{bmatrix} 1 & 1 \\ 0 & 1 \end{bmatrix}$$

there does not exist any nonsingular matrix P such that

$$P^{-1}AP = I$$

Solution: From the characteristic equation we find the eigenvalues of A.

$$|A - \lambda I| = \begin{vmatrix} 1-\lambda & 1 \\ 0 & 1-\lambda \end{vmatrix} = (1 - \lambda)^2 = 0.$$

Thus $\lambda_1 = \lambda_2 = 1$

Let us assume that there exists a matrix P such that

$$\det P \neq 0$$

$$P^{-1}AP = I$$

Multiplying by P

$$AP = P$$

$$\begin{bmatrix} 1 & 1 \\ 0 & 0 \end{bmatrix} \begin{bmatrix} P_{11} & P_{12} \\ P_{21} & P_{22} \end{bmatrix} = \begin{bmatrix} P_{11} & P_{12} \\ P_{21} & P_{22} \end{bmatrix}$$

or

$$P_{11} + P_{21} = P_{11}$$

$$P_{12} + P_{22} = P_{12}$$

$$0 = P_{21}$$

$$0 = P_{22}$$

thus

$$P = \begin{bmatrix} P_{11} & P_{12} \\ 0 & 0 \end{bmatrix}$$

P is singular. This is a contradiction since P is assumed to be nonsingular, therefore P does not exist.

The characteristic equation of a system

$$
\begin{bmatrix} \dot{x}_1 \\ \dot{x}_2 \\ \vdots \\ \dot{x}_{n-1} \\ \dot{x}_n \end{bmatrix} = \begin{bmatrix} 0 & 1 & 0 & \cdots & 0 & 0 \\ 0 & 0 & 1 & & \vdots & \vdots \\ \vdots & \vdots & \vdots & & 1 & \\ 0 & 0 & 0 & \cdots & 0 & 1 \\ -k_n & -k_{n-1} & -k_{n-2} & \cdots & -k_2 & -k_1 \end{bmatrix} \begin{bmatrix} x_1 \\ x_2 \\ \vdots \\ x_{n-1} \\ x_n \end{bmatrix}
$$

has a pair of double roots.

For the above system find a transformation matrix S such that after the coordinates are transformed the equation is in the Jordan canonical form.

Assuming that the characteristic equation involves roots of multiplicity four find a transformation matrix that will transform the system equation into the Jordan canonical form.

Solution: Let us assume that the roots of the characteristic equation are different, thus λ_1, $\lambda_1 + \Delta\lambda_1$, λ_3, $\ldots \lambda_n$, and obtain the transformation matrix S. The Vandermonde transformation matrix is

$$
P = \begin{bmatrix} 1 & 1 & 1 & \cdots & 1 \\ \lambda_1 & \lambda_1 + \Delta\lambda_1 & \lambda_3 & & \lambda_n \\ \vdots & \vdots & \vdots & & \\ \lambda_1^{n-1} & (\lambda_1+\Delta\lambda_1)^{n-1} & \lambda_3^{n-1} & \cdots & \lambda_n^{n-1} \end{bmatrix}
$$

The transformation x = Py transforms the initial system equation into

$$\dot{y}_1 = \lambda_1 y_1$$

$$\dot{y}_2 = (\lambda_1 + \Delta\lambda_1) y_2$$

$$\dot{y}_3 = \lambda_3 y_3$$

$$\dot{y}_n = \lambda_n y_n$$

Let us define the second transformation as y = Tz, where

$$T = \begin{bmatrix} 0 & -\dfrac{1}{\Delta\lambda_1} & 0 & . & . & 0 \\[2ex] 1 & \dfrac{1}{\Delta\lambda_1} & 0 & & & 0 \\[2ex] 0 & 0 & 1 & & & 0 \\[1ex] . & . & . & & & . \\ . & . & . & & & . \\[1ex] 0 & 0 & 0 & . & . & 1 \end{bmatrix}$$

We obtain the following equations:

$$\dot{z}_1 = (\lambda_1 + \Delta\lambda_1)z_1 + z_2$$

$$\dot{z}_2 = \lambda_1 z_2$$

$$\dot{z}_3 = \lambda_3 z_3$$

$$\vdots \qquad \vdots$$

$$\dot{z}_n = \lambda_n z_n$$

It is easy to check that $\lim\limits_{\Delta\lambda_1 \to 0} \det P = 0$

and

$$\lim\limits_{\Delta\lambda_1 \to 0} \det PT \neq 0 .$$

For $\Delta\lambda_1 \to 0$ we obtain

$$\dot{z}_1 = \lambda_1 z_1 + z_2$$

$$\dot{z}_2 = \lambda_1 z_2 \tag{1}$$

$$\dot{z}_3 = \lambda_3 z_3$$

$$\vdots$$

$$\dot{z}_n = \lambda_n z_n$$

The above system yields the Jordan canonical form when the characteristic equation of the initial system involves double eigenvalues $\lambda_1 = \lambda_2$. The transformation matrix S that transforms the initial equation into (1) can be obtained from the formula:

$$S = \lim\limits_{\Delta\lambda_1 \to 0} PT$$

$$= \lim_{\Delta\lambda_1 \to 0} \begin{bmatrix} 1 & 0 & 1 & \cdots & 1 \\ \lambda_1+\Delta\lambda_1 & 1 & \lambda_3 & & \lambda_4 \\ (\lambda_1+\Delta\lambda_1)^2 & 2\lambda_1+\Delta\lambda_1 & \lambda_3^2 & & \lambda_4^2 \\ (\lambda_1+\Delta\lambda_1)^3 & 3\lambda_1^2+3\lambda_1\Delta\lambda_1+\Delta\lambda_1^2 & \lambda_3^3 & & \lambda_4^3 \\ (\lambda_1+\Delta\lambda_1)^4 & 4\lambda_1^3+6\lambda_1^2\Delta\lambda_1+4\lambda_1\Delta\lambda_1^2+\Delta\lambda_1^3 & \lambda_3^4 & & \lambda_4^4 \\ \vdots & \vdots & \vdots & & \vdots \\ (\lambda_1+\Delta\lambda_1)^{n-1} & \cdot & \lambda_3^{n-1} & \cdots & \lambda_n^{n-1} \end{bmatrix}$$

$$= \begin{bmatrix} 1 & 0 & 1 & \cdots & 1 \\ \lambda_1 & 1 & \lambda_3 & \cdots & \lambda_n \\ \lambda_1^2 & 2\lambda_1 & \lambda_3^2 & & \lambda_n^2 \\ \lambda_1^3 & 3\lambda_1^2 & \lambda_3^3 & & \lambda_n^3 \\ \vdots & \vdots & \vdots & & \vdots \\ \lambda_1^{n-1} & (n-1)\lambda_1^{n-2} & \lambda_3^{n-1} & \cdots & \lambda_n^{n-1} \end{bmatrix} \qquad (2)$$

We obtain the following equation in Z coordinates

$$\begin{bmatrix} \dot{z}_1 \\ \dot{z}_2 \\ \cdot \\ \cdot \\ \dot{z}_n \end{bmatrix} = \begin{bmatrix} \lambda_1 & 1 & 0 & \cdot & \cdot & 0 \\ 0 & \lambda_1 & 0 & & & \cdot \\ \cdot & 0 & \lambda_3 & & & \cdot \\ \cdot & & \cdot & \cdot & & \\ 0 & 0 & 0 & \cdot & \cdot & \lambda_n \end{bmatrix} \begin{bmatrix} z_1 \\ z_2 \\ \cdot \\ \cdot \\ z_n \end{bmatrix}$$

In case, when the characteristic equation involves the roots of multiplicity four we shall obtain the transformation matrix S' from (2). The second column is the first derivative of the first column, the third column is the second derivative of the first column divided by 2!, the fourth column is the third derivative of the first column divided by 3!. If multiplicity of the roots is higher we can continue this procedure.

Thus for S' we have

$$S' = \begin{bmatrix} 1 & 0 & 0 & 0 & 1 & \cdots & 1 \\ \lambda_1 & 1 & 0 & 0 & \lambda_5 & \cdots & \lambda_n \\ \lambda_1^2 & 2\lambda_1 & 1 & 0 & \lambda_5^2 & & \lambda_n^2 \\ \lambda_1^3 & 3\lambda_1^2 & 3\lambda_1 & 1 & \lambda_5^3 & & \lambda_n^3 \\ \cdot & \cdot & \cdot & \cdot & \cdot & & \cdot \\ \cdot & \cdot & \cdot & \cdot & \cdot & & \cdot \\ \lambda_1^{n-1} & \frac{d}{d\lambda_1}\left(\lambda_1^{n-1}\right) & \frac{1}{2!}\frac{d^2}{d\lambda_1^2}\left(\lambda_1^{n-1}\right) & \frac{1}{3!}\frac{d^3}{d\lambda_1^3}\left(\lambda_1^{n-1}\right) & \lambda_5^{n-1} & \cdots & \lambda_n^{n-1} \end{bmatrix}$$

Matrix S' transforms the initial equation into

$$\begin{bmatrix} \dot{z}_1 \\ \dot{z}_2 \\ \cdot \\ \cdot \\ \dot{z}_n \end{bmatrix} = \begin{bmatrix} \lambda_1 & 1 & 0 & 0 & & & 0 \\ 0 & \lambda_1 & 1 & 0 & & & \\ 0 & 0 & \lambda_1 & 1 & & & \\ 0 & 0 & 0 & \lambda_1 & & & \\ \hline & & & & \lambda_5 & & 0 \\ & & & & & \lambda_6 & \\ & & & & & & \ddots \\ 0 & & & & 0 & & \lambda_n \end{bmatrix} \begin{bmatrix} z_1 \\ z_2 \\ \cdot \\ \cdot \\ z_n \end{bmatrix}$$

● **PROBLEM 2-23**

Obtain a diagonal matrix similar to the following real symmetric matrix A:

$$A = \begin{bmatrix} 3 & 0 & 2 \\ 0 & 3 & -2 \\ 2 & -2 & 1 \end{bmatrix}$$

Solution: Since matrix A is real and symmetric it can be diagonalized by the use of the transformation matrix P whose columns consist of the normalized eigenvectors of A. These eigenvectors are orthonormal to each other and are linearly independent. From the characteristic equation

$$|A-\lambda I| = \begin{vmatrix} 3-\lambda & 0 & 2 \\ 0 & 3-\lambda & -2 \\ 2 & -2 & 1-\lambda \end{vmatrix}$$

76

$$= (1-\lambda)(3-\lambda)(3-\lambda) - 4(3-\lambda) - 4(3-\lambda)$$

$$= -\lambda^3 + 7\lambda^2 - 7\lambda - 15 = (5-\lambda)(3-\lambda)(-1-\lambda) = 0.$$

The eigenvalues of A are

$$\lambda_1 = -1 \qquad \lambda_2 = 3 \qquad \lambda_3 = 5$$

The eigenvectors corresponding to λ_1, λ_2 and λ_3 are obtained from

$$(A - \lambda_i I)P_i = 0 \qquad \text{for } i = 1,2,3$$

For $\lambda_1 = -1$ we have

$$\begin{bmatrix} 4 & 0 & 2 \\ 0 & 4 & -2 \\ 2 & -2 & 2 \end{bmatrix} \begin{bmatrix} p_{11} \\ p_{21} \\ p_{31} \end{bmatrix} = 0$$

$4p_{11} + 2p_{31} = 0$

$4p_{21} - 2p_{31} = 0$

$2p_{11} - 2p_{21} + 2p_{31} = 0$

Thus

$$P_1 = \begin{bmatrix} a \\ -a \\ -2a \end{bmatrix} \xrightarrow{\text{normalized}} \begin{bmatrix} \dfrac{1}{\sqrt{6}} \\ -\dfrac{1}{\sqrt{6}} \\ -\dfrac{2}{\sqrt{6}} \end{bmatrix}$$

where a is a nonzero constant.

For $\lambda_2 = 3$ we have

$$\begin{bmatrix} 0 & 0 & 2 \\ 0 & 0 & -2 \\ 2 & -2 & -2 \end{bmatrix} \begin{bmatrix} p_{12} \\ p_{22} \\ p_{32} \end{bmatrix} = 0$$

$p_{32} = 0$

$2p_{12} - 2p_{22} = 0$

Thus

$$P_2 = \begin{bmatrix} b \\ b \\ 0 \end{bmatrix} \xrightarrow{\text{normalized}} \begin{bmatrix} \dfrac{1}{\sqrt{2}} \\ \dfrac{1}{\sqrt{2}} \\ 0 \end{bmatrix}$$

For $\quad \lambda_3 = 5 \quad$ we obtain

$$\begin{bmatrix} -2 & 0 & 2 \\ 0 & -2 & -2 \\ 2 & -2 & -4 \end{bmatrix} \begin{bmatrix} p_{13} \\ p_{23} \\ p_{33} \end{bmatrix} = 0$$

$-2p_{13} + 2p_{33} = 0$.

$-2p_{23} - 2p_{33} = 0$.

$2p_{13} - 2p_{23} - 4p_{33} = 0$.

$$P_3 = \begin{bmatrix} c \\ -c \\ c \end{bmatrix} \xrightarrow{\text{normalized}} \begin{bmatrix} \dfrac{1}{\sqrt{3}} \\ -\dfrac{1}{\sqrt{3}} \\ \dfrac{1}{\sqrt{3}} \end{bmatrix}$$

Hence

$$P = \begin{bmatrix} a & b & c \\ -a & b & -c \\ -2a & 0 & c \end{bmatrix}$$

or using normalized vectors

$$K = \begin{bmatrix} \dfrac{1}{\sqrt{6}} & \dfrac{1}{\sqrt{2}} & \dfrac{1}{\sqrt{3}} \\ -\dfrac{1}{\sqrt{6}} & \dfrac{1}{\sqrt{2}} & -\dfrac{1}{\sqrt{3}} \\ -\dfrac{2}{\sqrt{6}} & 0 & \dfrac{1}{\sqrt{3}} \end{bmatrix}$$

The orthogonal matrix K will transform A into diagonal form as follows:

$$K^{-1} A K = K^T A K =$$

$$\begin{bmatrix} \dfrac{1}{\sqrt{6}} & -\dfrac{1}{\sqrt{6}} & -\dfrac{2}{\sqrt{6}} \\[2mm] \dfrac{1}{\sqrt{2}} & \dfrac{1}{\sqrt{2}} & 0 \\[2mm] \dfrac{1}{\sqrt{3}} & -\dfrac{1}{\sqrt{3}} & \dfrac{1}{\sqrt{3}} \end{bmatrix} \begin{bmatrix} 3 & 0 & 2 \\ 0 & 3 & -2 \\ 2 & -2 & 1 \end{bmatrix} \begin{bmatrix} \dfrac{1}{\sqrt{6}} & \dfrac{1}{\sqrt{2}} & \dfrac{1}{\sqrt{3}} \\[2mm] -\dfrac{1}{\sqrt{6}} & \dfrac{1}{\sqrt{2}} & -\dfrac{1}{\sqrt{3}} \\[2mm] -\dfrac{2}{\sqrt{6}} & 0 & \dfrac{1}{\sqrt{3}} \end{bmatrix} =$$

$$\begin{bmatrix} -1 & 0 & 0 \\ 0 & 3 & 0 \\ 0 & 0 & 5 \end{bmatrix}$$

● **PROBLEM 2-24**

Find the matrix S that will transform the matrix A into Jordan canonical form.

$$A = \begin{bmatrix} 1 & 2 & 0 \\ 0 & 1 & 0 \\ -3 & 3 & 5 \end{bmatrix}$$

Solution: The characteristic equation is

$$|A - \gamma I| = \begin{vmatrix} 1-\gamma & 2 & 0 \\ 0 & 1-\gamma & 0 \\ -3 & 3 & 5-\gamma \end{vmatrix} = (1-\gamma)(1-\gamma)(5-\gamma) = 0$$

The eigenvalues of A are

$$\gamma_1 = 1, \ \gamma_2 = 1, \ \gamma_3 = 5$$

For the multiple eigenvalue $\gamma_1 = \gamma_2 = 1$, the eigenvectors are

$$(A - I) X_1 = 0$$

or

$$\begin{bmatrix} 0 & 2 & 0 \\ 0 & 0 & 0 \\ -3 & 3 & 4 \end{bmatrix} \begin{bmatrix} x_{11} \\ x_{21} \\ x_{31} \end{bmatrix} = 0$$

79

$$2x_{21} = 0$$

$$-3x_{11} + 3x_{21} + 4x_{31} = 0$$

we get

$$\begin{bmatrix} a \\ 0 \\ \frac{3}{4}a \end{bmatrix} \qquad a = \text{nonzero constant}$$

For $\gamma_3 = 5$ we have

$$\begin{bmatrix} -4 & 2 & 0 \\ 0 & -4 & 0 \\ -3 & 3 & 0 \end{bmatrix} \begin{bmatrix} x_{13} \\ x_{23} \\ x_{33} \end{bmatrix} = 0$$

$$-4x_{13} + 2x_{23} = 0$$

$$-4x_{23} = 0$$

$$-3x_{13} + 3x_{23} = 0$$

$$\begin{bmatrix} 0 \\ 0 \\ b \end{bmatrix} \qquad b = \text{nonzero constant}$$

The matrix A has two linearly independent eigenvectors.

Thus the Jordan canonical form is $\qquad J = \begin{bmatrix} 1 & 1 & 0 \\ 0 & 1 & 0 \\ 0 & 0 & 5 \end{bmatrix}$

We shall find the transformation matrix S such that

$$S^{-1} A S = J$$

For a 3x3 system with one distinct and one double eigenvalue

$(\gamma_1 = \gamma_2 \neq \gamma_3)$ we have

$$A = \begin{bmatrix} a & b & c \\ d & e & f \\ g & h & i \end{bmatrix}$$

$$S = \begin{bmatrix} \begin{vmatrix} e-\gamma_1 & f \\ h & i-\gamma_1 \end{vmatrix} & 2\gamma_1 - (i+e) & \begin{vmatrix} e-\gamma_3 & f \\ h & i-\gamma_3 \end{vmatrix} \\[20pt] -\begin{vmatrix} d & f \\ g & i-\gamma_1 \end{vmatrix} & d & -\begin{vmatrix} d & f \\ g & i-\gamma_3 \end{vmatrix} \\[20pt] \begin{vmatrix} d & e-\gamma_1 \\ g & h \end{vmatrix} & g & \begin{vmatrix} d & e-\gamma_3 \\ g & h \end{vmatrix} \end{bmatrix}$$

where $\gamma_1 = \gamma_2 \neq \gamma_3$ are the eigenvalues of A. The first column of S is

$$
\begin{bmatrix}
\begin{vmatrix} e-\gamma_1 & f \\ h & i-\gamma_1 \end{vmatrix} \\[2mm]
-\begin{vmatrix} d & f \\ g & i-\gamma_1 \end{vmatrix} \\[2mm]
\begin{vmatrix} d & e-\gamma_1 \\ g & h \end{vmatrix}
\end{bmatrix}
=
\begin{bmatrix}
\begin{vmatrix} 0 & 0 \\ 3 & 4 \end{vmatrix} \\[2mm]
-\begin{vmatrix} 0 & 0 \\ -3 & 4 \end{vmatrix} \\[2mm]
\begin{vmatrix} 0 & 0 \\ -3 & 3 \end{vmatrix}
\end{bmatrix}
=
\begin{pmatrix} 0 \\ 0 \\ 0 \end{pmatrix}
$$

Since the first column of S is zero we have to modify S as follows:

$$A_{ijk} = \text{cofactor } |A - \gamma_i I|$$

$$
S =
\begin{bmatrix}
A_{121}(\gamma_1) & \dfrac{d}{d\gamma_1} A_{121}(\gamma_1) & A_{311}(\gamma_3) \\[3mm]
A_{122}(\gamma_1) & \dfrac{d}{d\gamma_1} A_{122}(\gamma_1) & A_{312}(\gamma_3) \\[3mm]
A_{123}(\gamma_1) & \dfrac{d}{d\gamma_1} A_{123}(\gamma_1) & A_{313}(\gamma_3)
\end{bmatrix}
$$

$$
=
\begin{bmatrix}
-\begin{vmatrix} b & c \\ h & i-\gamma_1 \end{vmatrix} & b & \begin{vmatrix} e-\gamma_3 & f \\ h & i-\gamma_3 \end{vmatrix} \\[3mm]
\begin{vmatrix} a-\gamma_1 & c \\ g & i-\gamma_1 \end{vmatrix} & 2\gamma_1 - (a+i) & -\begin{vmatrix} d & f \\ g & i-\gamma_3 \end{vmatrix} \\[3mm]
-\begin{vmatrix} a-\gamma_1 & b \\ g & h \end{vmatrix} & h & \begin{vmatrix} d & e-\gamma_3 \\ g & h \end{vmatrix}
\end{bmatrix}
$$

$$
=
\begin{bmatrix}
-\begin{vmatrix} 2 & 0 \\ 3 & 4 \end{vmatrix} & 2 & \begin{vmatrix} -4 & 0 \\ 3 & 0 \end{vmatrix} \\[3mm]
\begin{vmatrix} 0 & 0 \\ -3 & 4 \end{vmatrix} & 2-6 & -\begin{vmatrix} 0 & 0 \\ -3 & 0 \end{vmatrix} \\[3mm]
-\begin{vmatrix} 0 & 2 \\ -3 & 3 \end{vmatrix} & 3 & \begin{vmatrix} 0 & -4 \\ -3 & 3 \end{vmatrix}
\end{bmatrix}
=
\begin{bmatrix}
-8 & 2 & 0 \\
0 & -4 & 0 \\
-6 & 3 & -12
\end{bmatrix}
$$

det S = -32·12

$$S^{-1} = \frac{1}{32} \begin{bmatrix} -4 & -2 & 0 \\ 0 & -8 & 0 \\ 2 & -1 & -\frac{8}{3} \end{bmatrix}$$

We have

$$S^{-1}AS =$$

$$\frac{1}{32} \begin{bmatrix} -4 & -2 & 0 \\ 0 & -8 & 0 \\ 2 & -1 & -\frac{8}{3} \end{bmatrix} \begin{bmatrix} 1 & 2 & 0 \\ 0 & 1 & 0 \\ -3 & 3 & 5 \end{bmatrix} \begin{bmatrix} -8 & 2 & 0 \\ 0 & -4 & 0 \\ -6 & 3 & -12 \end{bmatrix} = \begin{bmatrix} 1 & 1 & 0 \\ 0 & 1 & 0 \\ 0 & 0 & 5 \end{bmatrix}$$

● **PROBLEM 2-25**

The state space equation is given in the form

$$\dot{x} = \begin{bmatrix} 6 & 1 & 0 \\ -12 & 0 & 1 \\ 8 & 0 & 0 \end{bmatrix} x$$

Obtain a transformation matrix T which transforms the above equation into the Jordan canonical form.

Solution: The characteristic equation of A is

$$|A - \gamma I| = \begin{vmatrix} 6-\gamma & 1 & 0 \\ -12 & -\gamma & 1 \\ 8 & 0 & -\gamma \end{vmatrix} = -\gamma^3 + 6\gamma^2 - 12\gamma + 8 = -(\gamma-2)^3 = 0$$

For the system with multiple eigenvalues the transformation matrix T is given by

$$T = \begin{bmatrix} 1 & 0 & 0 \\ \gamma_1 + a_1 & 1 & 0 \\ \gamma_1^2 + a_1\gamma_1 + a_2 & \gamma_1 \cdot 2 + a_1 & 1 \end{bmatrix} = \begin{bmatrix} 1 & 0 & 0 \\ -4 & 1 & 0 \\ 4 & -2 & 1 \end{bmatrix}$$

82

The state space equation is

$$\dot{x} = Ax$$

where

$$A = \begin{bmatrix} -k_1 & 1 & 0 & 0 & 0 \\ -k_2 & 0 & 1 & 0 & 0 \\ -k_3 & 0 & 0 & 1 & 0 \\ -k_4 & 0 & 0 & 0 & 1 \\ -k_5 & 0 & 0 & 0 & 0 \end{bmatrix}$$

Assuming that the eigenvalues of A are distinct, $\gamma_1 \neq \gamma_2$ $\neq \gamma_3 \neq \gamma_4 \neq \gamma_5$, obtain the transformation matrix P that will transform A into diagonal form.

Solution: Let us denote the (n,m) th element of P as A_{nlm}. In the characteristic equation we shall substitute $\gamma = \gamma_i$.

$$|A - \gamma_i I| = \begin{vmatrix} -k_1 - \gamma_i & 1 & 0 & 0 & 0 \\ -k_2 & -\gamma_i & 1 & 0 & 0 \\ -k_3 & 0 & -\gamma_i & 1 & 0 \\ -k_4 & 0 & 0 & -\gamma_i & 1 \\ -k_5 & 0 & 0 & 0 & -\gamma_i \end{vmatrix} = 0$$

The sum of the elements of any row and the cofactors of corresponding elements of a different row is zero. Taking the elements of the second row and cofactors of the first row and then the third, the fourth and the fifth row and cofactors of the first row we obtain

$$A_{i12} = \left(-\frac{k_2}{\gamma_i} - \frac{k_3}{\gamma_i^2} - \frac{k_4}{\gamma_i^3} - \frac{k_5}{\gamma_i^4} \right) A_{i11}$$

$$A_{i13} = \left(-\frac{k_3}{\gamma_i} - \frac{k_4}{\gamma_i^2} - \frac{k_5}{\gamma_i^3} \right) A_{i11} \tag{1}$$

$$A_{i14} = \left(-\frac{k_4}{\gamma_i} - \frac{k_5}{\gamma_i^2} \right) A_{i11}$$

83

$$A_{i15} = -\frac{k_5}{\gamma_i} A_{i11}$$

For A_{i11} we get

$$A_{i11} = \gamma_i^4$$

Substituting into (1) we obtain

$$A_{i11} = \gamma_i^4$$

$$A_{i12} = -(k_2\gamma_i^3 + k_3\gamma_i^2 + k_4\gamma_i + k_5)$$

$$A_{i13} = -(k_3\gamma_i^3 + k_4\gamma_i^2 + k_5\gamma_i)$$

$$A_{i14} = -(k_4\gamma_i^3 + k_5\gamma_i^2)$$

$$A_{i15} = -k_5\gamma_i^3$$

Thus P can be written in the form

$$P = \begin{bmatrix} \gamma_1^4 & & \cdots & & \gamma_5^4 \\ -(k_2\gamma_1^3 + k_3\gamma_1^2 + k_4\gamma_1 + k_5) & & & & -(k_2\gamma_5^3 + k_3\gamma_5^2 + k_4\gamma_5 + k_5) \\ -(k_3\gamma_1^3 + k_4\gamma_1^2 + k_5\gamma_1) & & & & -(k_3\gamma_5^3 + k_4\gamma_5^2 + k_5\gamma_5) \\ -(k_4\gamma_1^3 + k_5\gamma_1^2) & & & & -(k_4\gamma_5^3 + k_5\gamma_5^2) \\ -k_5\gamma_1^3 & & & & -k_5\gamma_5^3 \end{bmatrix}$$

Since $\gamma_i^5 + k_1\gamma_i^4 + k_2\gamma_i^3 + k_3\gamma_i^2 + k_4\gamma_i + k_5 = 0$

We have

$$-(k_2\gamma_i^3 + k_3\gamma_i^2 + k_4\gamma_i + k_5) = \gamma_i^5 + k_1\gamma_i^4$$

$$-(k_3\gamma_i^3 + k_4\gamma_i^2 + k_5\gamma_i) = \gamma_i^6 + k_1\gamma_i^5 + k_2\gamma_i^4$$

$$-(k_4\gamma_i^3 + k_5\gamma_i^2) = \gamma_i^7 + k_1\gamma_i^6 + k_2\gamma_i^5 + k_3\gamma_i^4$$

$$-k_5\gamma_i^3 = \gamma_i^8 + k_1\gamma_i^7 + k_2\gamma_i^6 + k_3\gamma_i^5 + k_4\gamma_i^4$$

Dividing each column of P by γ_i^4 we get the transformation matrix P'.

$$
P' = \begin{bmatrix}
1 & \cdots\cdots & 1 \\
\gamma_1 + k_1 & \cdots\cdots & \gamma_5 + k_1 \\
\gamma_1^2 + k_1\gamma_1 + k_2 & \cdots\cdots & \gamma_5^2 + k_1\gamma_5 + k_2 \\
\gamma_1^3 + k_1\gamma_1^2 + k_2\gamma_1 + k_3 & \cdots & \gamma_5^3 + k_1\gamma_5^2 + k_2\gamma_5 + k_3 \\
\gamma_1^4 + k_1\gamma_1^3 + k_2\gamma_1^2 + k_3\gamma_1 + k_4 & \cdots & \gamma_5^4 + k_1\gamma_5^3 + k_2\gamma_5^2 + k_3\gamma_5 + k_4
\end{bmatrix}
$$

● PROBLEM 2-27

The state space equation is given in the form

$$
\dot{x} = \begin{bmatrix}
0 & 1 & 0 & 0 \\
0 & 0 & 1 & 0 \\
0 & 0 & 0 & 1 \\
-k_1 & -k_2 & -k_3 & -k_4
\end{bmatrix} x
$$

We assume that the characteristic equation

$$
\gamma^4 + k_4\gamma^3 + k_3\gamma^2 + k_2\gamma + k_4 = 0
$$

has two identical roots $\gamma_1 = \gamma_2$ and one pair of complex conjugate roots

$$
\gamma_3 = \sigma + jw, \quad \gamma_4 = \sigma - jw.
$$

Simplify the state space equation.

Solution: Let us define the following transformation

$$
x = Sy
$$

where

$$
S = \begin{bmatrix}
1 & 0 & 1 & 1 \\
\gamma_1 & 1 & \sigma+jw & \sigma-jw \\
\gamma_1^2 & 2\gamma_1 & (\sigma+jw)^2 & (\sigma-jw)^2 \\
\gamma_1^3 & 3\gamma_1^2 & (\sigma+jw)^3 & (\sigma-jw)^2
\end{bmatrix}
$$

which transforms the state space equation into

85

$$y = \begin{bmatrix} \gamma_1 & 1 & 0 & 0 \\ 0 & \gamma_1 & 0 & 0 \\ 0 & 0 & \sigma+jw & 0 \\ 0 & 0 & 0 & \sigma-jw \end{bmatrix} y$$

The second transformation is defined by

$$y = Tz$$
where

$$T = \begin{bmatrix} 1 & 0 & 0 & 0 \\ 0 & 1 & 0 & 0 \\ 0 & 0 & \frac{1}{2} & \frac{j}{2} \\ 0 & 0 & \frac{1}{2} & \frac{j}{2} \end{bmatrix}$$

The transformed state space equation becomes

$$\dot{z} = Az$$
where

$$A = \begin{bmatrix} \gamma_1 & 1 & 0 & 0 \\ 0 & \gamma_1 & 0 & 0 \\ 0 & 0 & \sigma & w \\ 0 & 0 & -w & \sigma \end{bmatrix}$$

● **PROBLEM 2-28**

The state space equation is

$$\dot{x} = Ax$$

where

$$x = \begin{bmatrix} x_1 \\ x_2 \\ x_3 \end{bmatrix}, \qquad A = \begin{bmatrix} -a_1 & 1 & 0 \\ -a_2 & 0 & 1 \\ -a_3 & 0 & 0 \end{bmatrix}$$

The transformation matrix K

$$K = \begin{bmatrix} 1 & 1 & 1 \\ \gamma_1 + a_1 & \gamma_2 + a_1 & \gamma_3 + a_1 \\ \gamma_1^2 + a_1\gamma_1 + a_2 & \gamma_2^2 + a_1\gamma_2 + a_2 & \gamma_3^2 + a_1\gamma_3 + a_2 \end{bmatrix}$$

86

transforms the matrix A into the diagonal form. Show that if $\gamma_1 = \gamma_2 \neq \gamma_3$, the matrix S will transform the coefficient matrix into the Jordan canonical form

$$S = \begin{bmatrix} 1 & 0 & 1 \\ \gamma_1 + a_1 & 1 & \gamma_3 + a_1 \\ \gamma_1^2 + a_1\gamma_1 + a_2 & 2\gamma_1 + a_1 & \gamma_3^2 + a_1\gamma_3 + a_2 \end{bmatrix}$$

When $\gamma_1 = \gamma_2 = \gamma_3$, matrix S is replaced by

$$P = \begin{bmatrix} 1 & 0 & 0 \\ \gamma_1 + a_1 & 1 & 0 \\ \gamma_1^2 + a_1\gamma_1 + a_2 & 2\gamma_1 + a_1 & 1 \end{bmatrix}$$

<u>Solution:</u> We shall start with the case when $\gamma_1 = \gamma_2 \neq \gamma_3$. Let us define the following transformation

$$x = K'x'$$

where

$$K' = \begin{bmatrix} 1 & 1 \\ \gamma_1 + a_1 & \gamma_1 + \Delta\gamma_1 + a_1 \\ \gamma_1^2 + a_1\gamma_1 + a_2 & (\gamma_1 + \Delta\gamma_1)^2 + a_1(\gamma_1 + \Delta\gamma_1) + a_2 \end{bmatrix}$$

$$\begin{bmatrix} 1 \\ \gamma_3 + a_1 \\ \gamma_3^2 + a_1\gamma_3 + a_2 \end{bmatrix}$$

The above transformation will transform the state space equation

$$\dot{x} = Ax$$

into

$$\begin{bmatrix} \dot{x}_1^1 \\ \dot{x}_2^1 \\ \dot{x}_3^1 \end{bmatrix} = \begin{bmatrix} \gamma_1 & 0 & 0 \\ 0 & \gamma_1 + \Delta\gamma_1 & 0 \\ 0 & 0 & \gamma_3 \end{bmatrix} \begin{bmatrix} x_1^1 \\ x_2^1 \\ x_3^1 \end{bmatrix}$$

The second transformation

$$x^1 = K^{11}x^{11}$$

where

$$K^{11} = \begin{bmatrix} 0 & -\dfrac{1}{\Delta\gamma_1} & 0 \\ 1 & \dfrac{1}{\Delta\gamma_1} & 0 \\ 0 & 0 & 1 \end{bmatrix}$$

we obtain

$$\dot{x}^{11} = \begin{bmatrix} \gamma_1 + \Delta\gamma_1 & 1 & 0 \\ 0 & \gamma_1 & 0 \\ 0 & 0 & \gamma_3 \end{bmatrix} x^{11}$$

$K^1 K^{11}$ is nonsingular as $\Delta\gamma_1 \to 0$, thus the transformation

$$x = Sx^{11} = \lim_{\Delta\gamma_1 \to 0} K^1 K^{11} x^{11}$$

transforms the coefficient matrix A into the Jordan canonical form.

S is given by

$$S = \lim_{\Delta\gamma_1 \to 0} K^1 K^{11} = \begin{bmatrix} 1 & 0 & 1 \\ \gamma_1 + a_1 & 1 & \gamma_3 + a_1 \\ \gamma_1^2 + a_1\gamma_1 + a_2 & 2\gamma_1 + a_1 & \gamma_3^2 + a_1\gamma_3 + a_2 \end{bmatrix}$$

For the case where $\gamma_1 = \gamma_2 = \gamma_3$ we shall also define two transformations L^1 and L^{11}

$$x = L^1 x^1$$

where

$$L^1 = \begin{bmatrix} 1 & 1 \\ \gamma_1 + a_1 & \gamma_1 + \Delta\gamma_1 + a_1 \\ \gamma_1^2 + a_1\gamma_1 + a_2 & (\gamma_1 + \Delta\gamma_1)^2 + a_1(\gamma_1 + \Delta\gamma_1) + a_2 \end{bmatrix}$$

$$\begin{bmatrix} 1 \\ \gamma_1 + 2\Delta\gamma_1 + a_1 \\ (\gamma_1 + 2\Delta\gamma_1)^2 + a_1(\gamma_1 + 2\Delta\gamma_1) + a_2 \end{bmatrix}$$

The equation x = Ax becomes

$$\dot{x}^1 = \begin{bmatrix} \gamma_1 & 0 & 0 \\ 0 & \gamma_1 + \Delta\gamma_1 & 0 \\ 0 & 0 & \gamma_1 + 2\Delta\gamma_1 \end{bmatrix} x^1$$

The second transformation

$$x^1 = L^{11}x^{11}$$

where

$$L^{11} = \begin{bmatrix} 0 & 0 & \dfrac{1}{2\Delta\gamma_1^2} \\[2ex] 0 & \dfrac{-1}{\Delta\gamma_1} & -\dfrac{1}{\Delta\gamma_1^2} \\[2ex] 1 & \dfrac{1}{\Delta\gamma_1} & \dfrac{1}{2\Delta\gamma_1^2} \end{bmatrix}$$

gives

$$\dot{x}^{11} = \begin{bmatrix} \gamma_1 + 2\Delta\gamma_1 & 1 & 0 \\ 0 & \gamma_1 + \Delta\gamma_1 & 1 \\ 0 & 0 & \gamma_1 \end{bmatrix} x^{11}$$

We can check that $L^1 L^{11}$ is nonsingular as $\Delta\gamma_1 \to 0$. The transformation

$$x = Px^{11} = \lim_{\Delta\gamma_1 \to 0} L^1 L^{11} x^{11}$$

transforms the coefficient matrix A into the Jordan canonical form, where

$$P = \lim_{\Delta\gamma_1 \to 0} L^1 L^{11} = \begin{bmatrix} 1 & 0 & 0 \\ \gamma_1 + a_1 & 1 & 0 \\ \gamma_1^2 + a_1\gamma_1 + a_2 & 2\gamma_1 + a_1 & 1 \end{bmatrix}$$

89

The state equation is given by

$$\dot{x} = Ax + Br \qquad (1)$$

where

$$A = \begin{bmatrix} 0 & 1 \\ -2 & -2 \end{bmatrix} \qquad B = \begin{bmatrix} 0 \\ 1 \end{bmatrix} \qquad (2)$$

Transform equation 1 into modal form and find $\phi(t)$.

Solution: The eigenvalues of A are

$$|\lambda I - A| = \begin{vmatrix} \lambda & -1 \\ 2 & \lambda+2 \end{vmatrix} = \lambda(\lambda+2) + 2 = 0.$$

$$\lambda_1 = -1+j \qquad \lambda_2 = -1-j$$

From

$$(\lambda_i I - A)q_i = 0$$

we find eigenvectors q_1 and q_2.

$$\begin{bmatrix} -1+j & -1 \\ 2 & 1+j \end{bmatrix} \begin{bmatrix} q_{11} \\ q_{21} \end{bmatrix} = 0$$

$$q_1 = \begin{bmatrix} 1 \\ -1+j \end{bmatrix} = \alpha_1 + j\beta_1 = \begin{bmatrix} 1 \\ -1 \end{bmatrix} + j\begin{bmatrix} 0 \\ 1 \end{bmatrix}$$

For q_2 we have

$$\begin{bmatrix} -1-j & -1 \\ 2 & 1-j \end{bmatrix} \begin{bmatrix} q_{12} \\ q_{22} \end{bmatrix} = 0$$

$$q_2 = \begin{bmatrix} 1 \\ -1-j \end{bmatrix} = \alpha_2 + j\beta_2 = \begin{bmatrix} 1 \\ -1 \end{bmatrix} + j\begin{bmatrix} 0 \\ -1 \end{bmatrix}$$

Thus

$$P = [\alpha_1 \beta_1] = \begin{bmatrix} 1 & 0 \\ -1 & 1 \end{bmatrix}$$

$$\Lambda = P^{-1} AP = \begin{bmatrix} -1 & 1 \\ -1 & -1 \end{bmatrix}$$

$$\Gamma = P^{-1} B = \begin{pmatrix} 0 \\ 1 \end{pmatrix}$$

The state equation (1) is transformed to

$$\dot{y} = \Lambda y + \Gamma r$$

The state transition matrix is

$$\phi(t) = L^{-1}\{[sI-\Lambda]^{-1}\}$$

$$= L^{-1}\left\{ \begin{bmatrix} s+1 & -1 \\ 1 & s+1 \end{bmatrix}^{-1} \right\}$$

$$= L^{-1}\left\{ \frac{1}{(s+1)^2 +1} \begin{bmatrix} s+1 & 1 \\ -1 & s+1 \end{bmatrix} \right\}$$

$$= L^{-1}\left\{ \begin{bmatrix} \dfrac{s+1}{(s+1)^2 +1} & \dfrac{1}{(s+1)^2 +1} \\[3mm] \dfrac{-1}{(s+1)^2 +1} & \dfrac{(s+1)}{(s+1)^2 +1} \end{bmatrix} \right\} = \begin{bmatrix} e^{-t} \cos t & e^{-t} \sin t \\[2mm] -e^{-t} \sin t & e^{-t} \cos t \end{bmatrix}$$

$$= e^{-t} \begin{bmatrix} \cos t & \sin t \\ -\sin t & \cos t \end{bmatrix}$$

● **PROBLEM** 2-30

For the matrix A, where

$$A = \begin{bmatrix} -1 & 1 & 2 \\ 0 & -2 & 0 \\ 0 & 0 & -3 \end{bmatrix}$$

find a non-singular matrix K that will transform A into a diagonal matrix Λ, such that

$$\Lambda = K^{-1} AK$$

Solution: The characteristic equation of A is

91

$$|A - \lambda I| = \begin{vmatrix} -1-\lambda & 1 & 2 \\ 0 & -2-\lambda & 0 \\ 0 & 0 & -3-\lambda \end{vmatrix} = (-\lambda-1)(-\lambda-2)(-\lambda-3) = 0$$

Thus, the eigenvalues of A are

$$\lambda_1 = -1, \qquad \lambda_2 = -2, \qquad \lambda_3 = -3$$

Let the eigenvector associated with $\lambda_1 = -1$ be represented by

$$K_1 = \begin{bmatrix} k_{11} \\ k_{21} \\ k_{31} \end{bmatrix}$$

Then K_1 must satisfy

$$(\lambda_1 I - A) K_1 = 0$$

or

$$\left(\begin{bmatrix} -1 & 0 & 0 \\ 0 & -1 & 0 \\ 0 & 0 & -1 \end{bmatrix} - \begin{bmatrix} -1 & 1 & 2 \\ 0 & -2 & 0 \\ 0 & 0 & -3 \end{bmatrix} \right) \begin{pmatrix} k_{11} \\ k_{21} \\ k_{31} \end{pmatrix} = 0$$

or

$$\begin{bmatrix} 0 & -1 & -2 \\ 0 & 1 & 0 \\ 0 & 0 & 2 \end{bmatrix} \begin{bmatrix} k_{11} \\ k_{21} \\ k_{31} \end{bmatrix} = 0$$

Multiplying and solving the system we get

$$K_1 = \begin{pmatrix} 1 \\ 0 \\ 0 \end{pmatrix}$$

For $\lambda_2 = -2$ we have

$$\begin{bmatrix} -1 & -1 & -2 \\ 0 & 0 & 0 \\ 0 & 0 & 1 \end{bmatrix} \begin{bmatrix} k_{12} \\ k_{22} \\ k_{32} \end{bmatrix} = 0$$

or

92

$$-k_{12} - k_{22} - 2k_{32} = 0$$

$$k_{32} = 0$$

Let $k_{12} = 1$, we thus have

$$K_2 = \begin{bmatrix} 1 \\ -1 \\ 0 \end{bmatrix}$$

For $\lambda_3 = -3$ we get

$$\begin{bmatrix} -2 & -1 & -2 \\ 0 & -1 & 0 \\ 0 & 0 & 0 \end{bmatrix} \begin{bmatrix} k_{13} \\ k_{23} \\ k_{33} \end{bmatrix} = 0$$

or

$$-2k_{13} - k_{23} - 2k_{33} = 0$$

$$k_{23} = 0$$

For $k_{13} = 1$ we have

$$K_3 = \begin{bmatrix} 1 \\ 0 \\ -1 \end{bmatrix}$$

The matrix K is given by

$$K = [K_1 \; K_2 \; K_3] = \begin{bmatrix} 1 & 1 & 1 \\ 0 & -1 & 0 \\ 0 & 0 & -1 \end{bmatrix}$$

det K = 1 ≠ 0 thus the matrix K^{-1} exists.

$$K^{-1} = \begin{bmatrix} 1 & 1 & 1 \\ 0 & -1 & 0 \\ 0 & 0 & -1 \end{bmatrix}$$

It is easy to show that

$$\Lambda = K^{-1} AK =$$

$$\begin{bmatrix} 1 & 1 & 1 \\ 0 & -1 & 0 \\ 0 & 0 & -1 \end{bmatrix} \begin{bmatrix} -1 & 1 & 2 \\ 0 & -2 & 0 \\ 0 & 0 & -3 \end{bmatrix} \begin{bmatrix} 1 & 1 & 1 \\ 0 & -1 & 0 \\ 0 & 0 & -1 \end{bmatrix} = \begin{bmatrix} -1 & 0 & 0 \\ 0 & -2 & 0 \\ 0 & 0 & -3 \end{bmatrix}$$

CHAPTER 3

LAPLACE TRANSFORMS

LAPLACE TRANSFORMS AND THEOREMS

Find the Laplace transform of the following functions.

a) $f(t) = K$

b) $f(t) = Kt$

c) $f(t) = K \sin \omega t$

Solution:

a) $L[K] = \int_0^\infty e^{-st} K \, dt = K \left[\dfrac{e^{-st}}{-s} \right]_0^\infty = \dfrac{K}{s}$

b) $L[Kt] = \int_0^\infty e^{-st} Kt \, dt = K \int_0^\infty t e^{-st} \, dt$

Using the integration by parts identity

$$\int f \, dg = fg - \int g \, df$$

we have for

$$f = t$$

and

$$dg = e^{-st} \, dt$$

$$\int t \, e^{-st} \, dt = \frac{t e^{-st}}{-s} - \int \frac{e^{-st}}{-s} dt$$

$$\int_0^\infty t \, e^{-st} \, dt = \left[\frac{t e^{-st}}{-s} - \frac{e^{-st}}{s^2} \right]_0^\infty = \frac{1}{s^2}$$

Thus

$$L[Kt] = \frac{K}{s^2}$$

94

c) $L[K \sin \omega t] = K \int_0^\infty \sin \omega t \, e^{-st} \, dt$

$$= K \int_0^\infty \left(\frac{e^{j\omega t} - e^{-j\omega t}}{2j}\right) e^{-st} \, dt = \frac{K}{2j}\left[-\frac{e^{-(s-j\omega)t}}{s - j\omega} + \frac{e^{-(s+j\omega)t}}{s + j\omega}\right]_0^\infty$$

$$= \frac{K}{2j}\left(\frac{1}{s - j\omega} - \frac{1}{s + j\omega}\right) = \frac{K\omega}{s^2 + \omega^2}$$

● **PROBLEM 3-2**

Find the Laplace transform of $g(t) = \cos \omega t$.

Solution: Using the definition of the Laplace transform we write

$$L[g(t)] = \int_0^\infty e^{-st} \cos \omega t \, dt$$

Using Euler's identities

$$\cos \omega t = \frac{e^{j\omega t} + e^{-j\omega t}}{2}$$

we obtain

$$L[g(t)] = \int_0^\infty e^{-st}\left(\frac{e^{j\omega t} + e^{-j\omega t}}{2}\right) dt$$

$$= \frac{1}{2} \int_0^\infty \left(e^{-(s-j\omega)t} + e^{-(s+j\omega)t}\right) dt$$

$$= \frac{1}{2}\left[\frac{-e^{-(s-j\omega)t}}{s - j\omega} - \frac{e^{-(s+j\omega)t}}{s + j\omega}\right]_0^\infty$$

$$= \frac{1}{2}\left[\frac{1}{s - j\omega} + \frac{1}{s + j\omega}\right] = \frac{s}{s^2 + \omega^2}$$

Thus

$$L[\cos \omega t] = \frac{s}{s^2 + \omega^2}$$

● **PROBLEM 3-3**

Using the formula

$$L[\sin \omega t] = \frac{\omega}{s^2 + \omega^2}$$

and the differentiation theorem find the Laplace transform of

$$f(t) = A \cos \omega t$$

Solution: The differentiation theorem states that:

If f(t) is the time function such that

$$L[f(t)] = F(s)$$

then

$$L\left[\frac{df(t)}{dt}\right] = sF(s) - \lim_{t \to 0^+} f(t)$$

We shall use the above theorem and the equation

$$A \cos \omega t = \frac{d}{dt} \frac{A}{\omega} \sin \omega t$$

to obtain

$$L[A \cos \omega t] = L\left[\frac{d}{dt} \frac{A}{\omega} \sin \omega t\right]$$

$$= \frac{A}{\omega} [s \, L[\sin \omega t] - \sin \omega t \Big|_{t=0}] = \frac{A}{\omega}\left[\frac{s\omega}{s^2 + \omega^2} - 0\right]$$

$$= \frac{As}{s^2 + \omega^2}$$

● **PROBLEM 3-4**

Find the Laplace transform of the following functions.

a) $f(t) = e^{-at}$, $t \geq 0$ where a is a constant

b) $g(t) = Ae^{-at}$, $t \geq 0$ where a, A are constants.

Solution:

a)   From the definition of the Laplace transform we have

$$F(s) = L[f(t)] = \int_0^\infty e^{-st} e^{-at} \, dt = \int_0^\infty e^{-(s+a)t} \, dt$$

$$= \frac{e^{-(s+a)t}}{-(s+a)}\Big|_0^\infty = \frac{1}{s+a}$$

b)   Using the results of a) we have

$$L[g(t)] = A \, L[e^{-at}] = A \cdot \frac{1}{s+a}$$

● **PROBLEM 3-5**

Obtain the Laplace transform of the impulse function defined by

$$f(t) = \begin{cases} 0 & \text{for } t < 0, \ t_0 < t \\ \lim_{t_0 \to 0} \dfrac{A}{t_0} & \text{for } 0 < t < t_0 \end{cases}$$

Solution: The size of an impulse is measured by its area. In our problem the area is equal to A, since the height of the impulse function is $\frac{A}{t_0}$ and the duration is t_0,

$\frac{A}{t_0} \cdot t_0 = A$. The Laplace transform can be found by observing that f(t) is the difference of two step functions, one at t = 0 and one at t = t_0. Thus

$$f(t) = \lim_{t \to 0} \frac{A}{t_0}[u_s(t) - u_s(t - t_0)]$$

We have

$$L[f(t)] = \lim_{t_0 \to 0} \frac{A}{t_0 s}(1 - e^{-st_0})$$

Using l'Hôpital's rule for limits we find

$$\lim_{t_0 \to 0} \frac{A(1 - e^{-st_0})}{t_0 s} = \lim_{t_0 \to 0} \frac{\frac{d}{dt_0}[A(1 - e^{-st_0})]}{\frac{d}{dt_0}(t_0 s)}$$

$$= \frac{As}{s} = A$$

We conclude that the Laplace transform of the impulse function is equal to the area under the impulse.

● PROBLEM 3-6

Find the Laplace transforms of the following functions.

a) $f(t) = 5 \cos[(\omega_1 + \omega_2)tu(t)]$

b) $g(t) = \sin(\omega t + \theta)$

Solution:

a)   u(t) is a step function defined by

$$u(t) = \begin{cases} 0 & \text{for} \quad t < 0 \\ 1 & \text{for} \quad t > 1 \end{cases}$$

We have

$$L[f(t)] = L[5 \cos[(\omega_1 + \omega_2)tu(t)]]$$

$$= 5L[\cos(\omega_1 + \omega_2)t] = 5 \cdot \frac{s}{s^2 + (\omega_1 + \omega_2)^2}$$

We used the formula

$$L[\cos \omega t] = \frac{s}{s^2 + \omega^2}$$

97

b) To compute the Laplace transform of the function $\sin(\omega t + \theta)$ we shall use the identity

$$\sin(\alpha + \beta) = \sin \alpha \cos \beta + \cos \alpha \sin \beta$$

thus

$$\sin(\omega t + \theta) = \sin \omega t \cos \theta + \cos \omega t \sin \theta$$

From the previous results we have

$$L[\cos \omega t] = \frac{s}{s^2 + \omega^2}$$

$$L[\sin \omega t] = \frac{\omega}{s^2 + \omega^2}$$

Remembering that $\sin \theta$ and $\cos \theta$ are constants we compute

$$L[\sin(\omega t + \theta)] = L[\sin \omega t \cos \theta + \cos \omega t \sin \theta]$$

$$= L[\sin \omega t \cos \theta] + L[\cos \omega t \sin \theta]$$

$$= \cos \theta \cdot \frac{\omega}{s^2 + \omega^2} + \sin \theta \frac{s}{s^2 + \omega^2}$$

$$= \frac{\omega \cos \theta + s \sin \theta}{s^2 + \omega^2}$$

● **PROBLEM 3-7**

Let $f_p(t)$ be a periodic function with period T, thus

$$f_p(t) = f_p(t + T)$$

And let $f(t)$ be defined as

$$f(t) = \begin{cases} f_p(t) & \text{for} \quad 0 < t \le T \\ 0 & \text{for} \quad T < t \end{cases}$$

Express the Laplace transform of $f_p(t)$ in terms of the Laplace transform of $f(t)$.

Solution: It is easy to check that

$$f_p(t) = f(t) + f(t - T) + f(t - 2T) + \ldots$$

We shall define the Laplace transform of $f(t)$ to be $F(s)$ where

$$L[f(t)] = F(s).$$

Using the theorem

98

$$L[f(t - T)u_s(t - T)] = e^{-Ts} F(s)$$

where $u_s(t - T)$ is the unit step function, we obtain

$$L[f_p(t)] = F_p(s) = L[f(t)] + L[f(t - T)u_s(t - T)] +$$

$$L[f(t - 2T)u_s(t - 2T)] + \ldots$$

$$= F(s) + e^{-Ts}F(s) + e^{-2Ts}F(s) + e^{-3Ts}F(s) + \ldots$$

$$= \frac{F(s)}{1 - e^{-Ts}}$$

Thus

$$L[f_p(t)] = \frac{1}{1 - e^{-Ts}} L[f(t)]$$

● **PROBLEM 3-8**

Find the steady-state value of f(t). The Laplace transform of f(t) is

$$L[f(t)] = F(s) = \frac{3}{s(s^2 + s + 2)}$$

Solution: The function $sF(s) = \dfrac{3}{s^2 + s + 2}$ is analytic on the imaginary axis and in the right half of the s-plane, therefore we can apply the final-value theorem.

$$\lim_{t\to\infty} f(t) = \lim_{s\to 0} sF(s) = \lim_{s\to 0} \frac{3}{s^2 + s + 2} = \frac{3}{2}$$

● **PROBLEM 3-9**

What is the steady-state value of f(t)? The Laplace transform of f(t) is

$$L[f(t)] = F(s) = \frac{\omega}{s^2 + \omega^2}$$

Solution: The final-value theorem gives

$$\lim_{t\to\infty} f(t) = \lim_{s\to 0} sF(s) = \lim_{s\to 0} \frac{\omega s}{\omega^2 + s^2} = 0$$

On the other hand we know, that

$$L[\sin \omega t] = \frac{\omega}{s^2 + \omega^2}$$

99

and the function sin ωt has no final value.

In the above example the final-value theorem can not be used.

The function sF(s) has two poles on the imaginary axis and the final-value theorem can only be applied if sF(s) is analytic (has no poles) on the jω-axis and in the right-hand plane.

● **PROBLEM 3-10**

Find f'(0⁺) given

$$F(s) = \frac{3}{s + 0.2}$$

Solution: The following theorems will be used to solve this problem

$$g'(0^+) = \lim_{t \to 0^+} \frac{dg(t)}{dt} = \lim_{s \to \infty} s[sG(s) - g(0^+)] \qquad (1)$$

and

$$g(0^+) = \lim_{t \to 0^+} g(t) = \lim_{s \to \infty} sG(s). \qquad (2)$$

f(0⁺) can be calculated from (2)

$$f(0^+) = \lim_{s \to \infty} sF(s) = \lim_{s \to \infty} \frac{3s}{s + 0.2} = 3$$

and f'(0⁺) from (1)

$$f'(0^+) = \lim_{s \to \infty} s[sF(s) - f(0^+)]$$

$$= \lim_{s \to \infty} s\left[\frac{3s}{s + 0.2} - 3\right] = \lim_{s \to \infty} s\left[\frac{3s - 3s - 0.6}{s + 0.2}\right]$$

$$= \lim_{s \to \infty} s\left[\frac{-0.6}{s + 0.2}\right] = \lim_{s \to \infty}\left[-\frac{0.6}{1 + \frac{0.2}{s}}\right] = -0.6$$

The other way of solving the above problem is to find f'(t) and then evaluate this at t = 0⁺.

For f(t) we have

$$f(t) = L^{-1}\left[\frac{3}{s + 0.2}\right] = 3e^{-0.2t}u(t)$$

The first derivative of f(t) is

$$f'(t) = 3(-0.2)e^{-0.2t}u(t) + 3e^{-0.2t}\frac{du(t)}{dt}$$

$$f'(t) = -0.6e^{-0.2t}u(t) + 3e^{-0.2t}\delta(t)$$

100

To evaluate $f'(0^+)$ let us observe that

$$\delta(t) = \begin{cases} 0 & \text{for} \quad t \neq 0 \\ \infty & \text{for} \quad t = 0 \end{cases}$$

Then

$$3e^{-0.2t} \, \delta(t) \Big|_{t = 0^+} = 3e^{-0.2(0^+)} \delta(0^+) = 0$$

since

$$\delta(0^+) = 0$$

Thus

$$f'(t) = -0.6e^{-0.2t}u(t) \quad \text{for } t > 0$$

and

$$f'(0^+) = \lim_{t \to 0^+} f'(t) = -0.6e^{-0.2(0^+)}u(0^+) = -0.6$$

The above result agrees with the previous answer.

● **PROBLEM** 3-11

Find $\lim_{t \to \infty} g(t)$, where

$$L[g(t)] = G(s) = \frac{\omega}{s^2 + \omega^2}$$

Solution: Let us apply the final value theorem

$$\lim_{t \to \infty} g(t) = \lim_{s \to 0} sG(s) = \lim_{s \to 0} \frac{s\omega}{s^2 + \omega^2} = 0$$

On the other hand we know that

$$g(t) = L^{-1}\left[\frac{\omega}{s^2 + \omega^2}\right] = \sin \omega t$$

which is a periodic function and does not have any limit as $t \to \infty$. The general conclusion is that the final value theorem does not apply to periodic functions.

The final-value theorem can not be applied to any function of s with poles in the right-hand plane or on the $j\omega$-axis. Since

$$G(s) = \frac{\omega}{(s + j\omega)(s - j\omega)} \quad \text{has poles } \pm j\omega, \text{ the}$$

final value theorem will fail.

101

Find the Laplace transforms of the following functions:

a) $u_{(t)} = \begin{cases} 0 & t < 0 \\ 1 & t \geq 0 \end{cases}$

b) $g(t) = \begin{cases} A & 0 < t < t_0 \\ 0 & t < 0, \ t_0 < t \end{cases}$

Solution:

a)   The Laplace transform of a unit step function $u(t)$ is

$$U(s) = L[u(t)] = \int_0^\infty u(t) e^{-st} dt = \left. \frac{e^{-st}}{-s} \right|_0^\infty = \frac{1}{s}$$

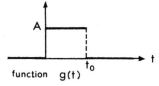

function g(t)

b)   From the definition of the Laplace transform we have

$$G(s) = L[g(t)] = \int_0^\infty e^{-st} g(t) dt$$

$$= \int_0^{t_0} e^{-st} A \, dt + \int_{t_0}^\infty 0 \cdot e^{-st} dt$$

$$= A \left(\left. \frac{e^{-st}}{-s} \right| \begin{matrix} t = t_0 \\ t = 0 \end{matrix} \right)$$

$$= \frac{A}{s} (1 - e^{-st_0})$$

Find the Laplace transform of a shifted sine wave.

a) $f_1(t) = 5 \sin \omega (t - \alpha)$

b)   The function described in a) has been multiplied by a unit-step beginning at $t = \alpha$.   Find the Laplace transform of the function

$$f_2(t) = 5 \sin \omega (t - \alpha) \cdot u(t - \alpha)$$

Solution:

a)   To find the Laplace transform of

$$f_1(t) = 5 \sin \omega(t - \alpha)$$

we shall use the equation

$$\sin(\alpha - \beta) = \sin \alpha \cos \beta - \sin \beta \cos \alpha$$

Thus

$$L[5 \sin \omega(t - \alpha)] = 5L[\sin \omega t \cos \omega \alpha - \sin \omega \alpha \cos \omega t]$$

$$= 5\left(\frac{\omega}{s^2 + \omega^2} \cos \omega \alpha - \sin \omega \alpha \frac{s}{s^2 + \omega^2}\right)$$

$$= \frac{5}{s^2 + \omega^2}(\omega \cos \omega \alpha - s \sin \omega \alpha).$$

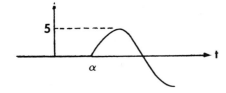 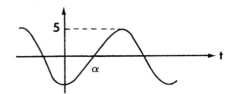

b)   We know that

$$L[f(t - a)u(t - a)] = e^{-as}F(s)$$

where

$$F(s) = L[f(t)u(t)]$$

$$f(t - \alpha) = 5 \sin \omega(t - \alpha)$$

thus

$$f(t) = 5 \sin \omega t$$

We obtain

$$F(s) = L[f(t)u(t)] = 5\frac{\omega}{s^2 + \omega^2}$$

and

$$L[5 \sin \omega(t - \alpha) \cdot u(t - \alpha)] = \frac{5\omega}{s^2 + \omega^2} \cdot e^{-as}$$

● **PROBLEM 3-14**

Obtain the algebraic expression and then compute the Laplace transform of the function shown in the figure.

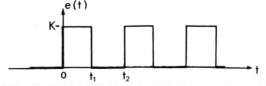

Solution: First we shall consider the time interval $[0, t_2]$; $e_1(t)$ for this period is

$$e_1(t) = Ku(t) - Ku(t - t_1)$$

103

The Laplace transform of $e_1(t)$ is

$$E_1(s) = L[e_1(t)] = KL[u(t)] - KL[u(t - t_1)]$$

$$= \frac{K}{s} - \frac{K}{s}e^{-t_1 s} = \frac{K}{s}\left(1 - e^{-t_1 s}\right)$$

The following equation represents a periodic function

$$f(t) = f_1(t)u(t) + f_1(t - T)u(t - T) + \ldots$$

$$+ f_1(t - nT)u(t - nT)$$

If

$$F_1(s) = L[f_1(t)]$$

we can write

$$L[f(t)u(t)] = F_1(s) \sum_{k=0}^{n} e^{-kTs}$$

The Laplace transform of $e(t)$ is

$$L[e(t)] = E(s) = E_1(s) \sum_{k=0}^{n} e^{-Kt_2 s}$$

$$= \frac{K}{s}(1 - e^{-t_1 s}) \sum_{k=0}^{n} e^{-kt_2 s}$$

Since

$$\lim_{n \to \infty} \sum_{k=0}^{n} e^{-ka} = \frac{1}{1 - e^{-a}}$$

we obtain

$$E(s) = \frac{K}{s} \frac{1 - e^{-t_1 s}}{1 - e^{-t_2 s}}$$

● **PROBLEM 3-15**

Find the Laplace transform of the signal $e(t)$ shown in the figure.

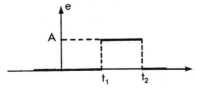

Solution: We shall solve this problem twice. First by applying the definition of the Laplace transform

$$E(s) = L[e(t)] = \int_{0}^{\infty} e^{-st} \cdot e(t)dt$$

$$= \int_{0}^{t_1} 0 \cdot e^{-st}dt + \int_{t_1}^{t_2} Ae^{-st}dt + \int_{t_2}^{\infty} 0 \cdot e^{-st}dt$$

$$= \int_{t_1}^{t_2} Ae^{-st}dt = A\left(\frac{e^{-st}}{-s}\bigg|_{t_1}^{t_2}\right) = \frac{A}{s}\left[e^{-st_1} - e^{-st_2}\right]$$

We will get the same result using another method. Let us define

$$e_1(t) = Au(t - t_1)$$

$$e_2(t) = -Au(t - t_2)$$

where $u(t) = \begin{cases} 0 & \text{for } t < 0 \\ 1 & \text{for } t \geq 0 \end{cases}$

Note that

$$e(t) = e_1(t) + e_2(t)$$

$$E(s) = L[e(t)] = L[e_1(t) + e_2(t)] = L[e_1(t)] + L[e_2(t)]$$

$$= A\left[\frac{e^{-t_1 s}}{s} - \frac{e^{-t_2 s}}{s}\right] = \frac{A}{s}\left(e^{-t_1 s} - e^{-t_2 s}\right)$$

● **PROBLEM 3-16**

Obtain an algebraic expression for the signal shown in the figure and find the Laplace transform of the expression.

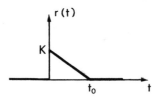

Solution: The signal $r(t)$ can be expressed as a linear combination of two step functions

$$u(t) \text{ and } u(t - t_0)$$

Let us verify that

$$r(t) = \left[-\frac{K}{t_0}t + K\right]u(t) + \left[\frac{K}{t_0}(t - t_0)u(t - t_0)\right]$$

indeed for

$$0 < t < t_0; \; u(t) = 1 \text{ and } u(t - t_0) = 0$$

thus

$$r(t) = -\frac{K}{t_0}t + K$$

and for $t_0 < t$; $u(t) = u(t - t_0) = 1$, therefore

$$r(t) = -\frac{K}{t_0}t + K + \frac{K}{t_0}t - \frac{K}{t_0}t_0 = 0$$

We have

$$R(s) = L[r(t)] = -\frac{K}{t_0}\frac{1}{s^2} + \frac{K}{s} + \frac{K}{t_0}\frac{e^{-t_0 s}}{s^2}$$

$$= -\frac{K}{t_0 s^2}(1 - e^{-t_0 s}) + \frac{K}{s} .$$

105

Compute the Laplace transform of the function

$$g(t) = \begin{cases} 0 & 0 < t < 1 \\ (t - 1)^2 & t > 1 \end{cases}$$

whose graph is shown in the figure.

Use the following property of the Laplace transform.

If

$$h(t) = \begin{cases} 0 & 0 < t < a \\ f(t - a), & t > a \end{cases}$$

(1)

then

$$H(s) = e^{-as}F(s), \quad s > a$$

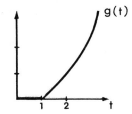

g(t)

Solution: The Laplace transform of t^n, n = positive integer is

$$L[t^n] = \frac{n!}{s^{n+1}}$$

Thus in our case

$$h(t) = t^2, \text{ then } L[h(t)] = \frac{2}{s^3}$$

It follows from property (1) that

$$L[g(t)] = e^{-s} L[h(t)] = \frac{2}{s^3}e^{-s}$$

Find the Laplace transform of the function shown in the figure.

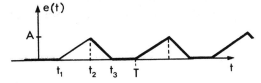

e(t)

Solution: First we shall find the algebraic expression of function e(t) shown in the figure. Let us assume for simplicity that

106

$$t_3 - t_2 = t_2 - t_1 = \Delta t$$

The basic equation of a straight-line is

$$e(t) = \pm mt$$

where

$$m = \frac{A}{t_3 - t_2} = \frac{A}{t_2 - t_1} = \frac{A}{\Delta t}$$

For $0 < t \leq t_2$, we have

$$e_1(t) = \frac{A}{\Delta t}(t - t_1)u(t - t_1)$$

For $t_2 < t < t_3$, we have

$$e_2(t) = -\frac{A}{\Delta t}(t - t_2)u(t - t_2)$$

But for $t > t_2$

$$e_1(t) + e_2(t) = \frac{A}{\Delta t}(t_2 - t_1) = A$$

which is the amplitude of $e(t)$ for $t = t_2$. Since $e_1(t) + e_2(t)$ equals a constant for $t > t_2$, to get zero we have to add function $e_3(t)$ where

$$e_3(t) = -\frac{A}{\Delta t}(t - t_2)u(t - t_2).$$

To cancel the effects of $e_3(t)$ for $t > t_3$ we add

$$e_4(t) = \frac{A}{\Delta t}(t - t_3)u(t - t_3)$$

For the first period we obtain

$$e_{1,p}(t) = e_1(t) + e_2(t) + e_3(t) + e_4(t)$$

$$= \frac{A}{\Delta t}(t - t_1)u(t - t_1) - \frac{A}{\Delta t}(t - t_2)u(t - t_2)$$

$$- \frac{A}{\Delta t}(t - t_2)u(t - t_2) + \frac{A}{\Delta t}(t - t_3)u(t - t_3)$$

$$= \frac{A}{\Delta t}[(t - t_1)u(t - t_1) - 2(t - t_2)u(t - t_2)$$

$$+ (t - t_3)u(t - t_3)]$$

To check if the above result is correct we shall compare it with the graph for $0 \leq t \leq T$.

For $t < t_1$

$$u(t - t_1) = u(t - t_2) = u(t - t_3) = 0$$

thus

$$e_{1,p}(t) = 0$$

For $t_1 \leq t \leq t_2$
_____

$\quad\quad u(t - t_1) = 1; \quad u(t - t_2) = u(t - t_3) = 0$

and

$\quad\quad e_{1,p}(t) = \frac{A}{\Delta t}(t - t_1)$, which is correct

for $\quad t = t_1 \quad\quad e_{1,p}(t_1) = 0$

for $\quad t = t_2 \quad\quad e_{1,p}(t_2) = A$

For $t_2 < t < t_3$

$\quad\quad u(t - t_1) = u(t - t_2) = 1 \quad$ and $\quad u(t - t_3) = 0$

therefore

$$e_{1,p}(t) = \frac{A}{\Delta t}\ [(t - t_1) - 2(t - t_2)]$$

$$= -\frac{A}{t_2 - t_1}t + \frac{2t_2 - t_1}{t_2 - t_1}A$$

$$e_{1,p}(t)\Big|_{t=t_2} = \frac{-At_2 + 2At_2 - At_1}{t_2 - t_1} = A$$

Comparing with the graph we see that the result is correct

$$e_{1,p}(t)\Big|_{t=t_3} = \frac{-At_3}{t_2 - t_1} + \frac{A(2t_2 - t_1)}{t_2 - t_1}$$

but

$\quad\quad t_3 - t_2 = t_2 - t_1$

thus

$\quad\quad t_3 = 2t_2 - t_1$

and

$$e_{1,p}(t)\Big|_{t=t_3} = 0$$

For $t_3 < t < T \quad$ and $\quad t > T$

$\quad\quad u(t - t_1) = u(t - t_2) = u(t - t_3) = 1$

$$e_{1,p}(t) = \frac{A}{\Delta t}(t - t_1) - \frac{2A}{\Delta t}(t - t_2) + \frac{A}{\Delta t}(t - t_3)$$

$$= \frac{-At_1}{\Delta t} + \frac{2At_2}{\Delta t} - \frac{At_3}{\Delta t} = \frac{-At_3}{\Delta t} + \frac{A(2t_2 - t_1)}{\Delta t} = 0$$

We conclude that the equation is correct. Using

$$L[f(t)u(t)] = F_1(s) \sum_{n=0}^{m} e^{-nTs}$$

108

we obtain

$$L[e(t)] = E(s) = E_1(s) \sum_{n=0}^{k} e^{-nTs}$$

$$E_1(s) = L[e_{1,p}(t)] = \frac{A}{\Delta t} L[(t - t_1)u(t - t_1)]$$

$$- \frac{2A}{\Delta t} L[(t - t_2)u(t - t_2)] + \frac{A}{\Delta t} L[(t - t_3)u(t - t_3)]$$

$$= \frac{A}{\Delta t} \frac{e^{-t_1 s}}{s^2} - \frac{2A}{\Delta t} \frac{e^{-t_2 s}}{s^2} + \frac{A}{\Delta t} \frac{e^{-t_3 s}}{s^2}$$

Hence

$$L[e(t)u(t)] = \frac{A}{\Delta t s^2}[e^{-t_1 s} - 2e^{-t_2 s} + e^{-t_3 s}] \sum_{n=0}^{k} e^{-nTs}$$

Summing the series, for m periods

$$L[e(t)u(t)] = \frac{A}{\Delta t \cdot s^2}[e^{-t_1 s} - 2e^{-t_2 s} + e^{-t_3 s}] \frac{1 - e^{-(m+1)Ts}}{1 - e^{-Ts}}$$

If the function is periodic and does not vanish, we obtain for $m \to \infty$

$$E(s) = \frac{A}{\Delta t \cdot s^2}[e^{-t_1 s} - 2e^{-t_2 s} + e^{-t_3 s}] \frac{1}{1 - e^{-Ts}}$$

● **PROBLEM** 3-19

Using the theorem

$$L[f'(t)] = sF(s) - f(0), \text{ where } L[f(t)] = F(s)$$

for the Laplace transform of the derivative f' solve the following initial value problem.

$$x'(t) + 2x(t) = e^{-t} \qquad x(0) = 2$$

Solution: Let us assume that both functions x and x' possess Laplace transforms.

$$L[e^{-t}] = \frac{1}{s + 1},$$

thus

$$L[x'(t) + 2x(t)] = L[e^{-t}] = \frac{1}{s + 1}$$

or

$$L[x'(t)] + 2L[x(t)] = \frac{1}{s + 1}$$

Using the theorem for the derivative

$$L[x'(t)] = sX(s) - x(0)$$

we obtain

$$sX(s) - x(0) + 2X(s) = \frac{1}{s + 1}$$

Solving the above equation for $X(s)$ we have

$$X(s) = \frac{2s + 3}{(s + 1)(s + 2)}$$

or using partial fractions

$$X(s) = \frac{1}{s + 1} + \frac{1}{s + 2}$$

From the equation

$$L[e^{at}] = \frac{1}{s - a} \text{ for } s > 0$$

we obtain

$x(t) = e^{-t} + e^{-2t}$, which is the solution of the equation $x'(t) + 2x(t) = e^{-t}$ and $x(0) = e^0 + e^0 = 2$.

INVERSE LAPLACE TRANSFORMS AND SOLUTIONS OF DIFFERENTIAL EQUATIONS

● **PROBLEM 3-20**

The Laplace transform of $f(t)$ is given by

$$F(s) = \frac{(s + 1)e^{-7s}}{s^2(s + 5)}$$

Find $f(t)$.

Solution: $F(s)$ can be written

$$F(s) = e^{-7s} \frac{s + 1}{s^2(s + 5)}$$

Let

$$G(s) = \frac{s + 1}{s^2(s + 5)}$$

Using partial fraction expansion

$$G(s) = \frac{K_1}{s} + \frac{K_2}{s^2} + \frac{K_3}{s + 5}$$

The coefficients K_1, K_2, K_3 are calculated from

$$K_1 = \left\{ \frac{d}{ds}\left[\frac{s + 1}{s + 5} \right] \right\}_{s = 0} = \frac{4}{25}$$

$$K_2 = \left[\frac{s + 1}{s + 5} \right]_{s = 0} = \frac{1}{5}$$

110

$$K_3 = \left[\frac{s + 1}{s^2} \right]_{s = -5} = -\frac{4}{25}$$

$G(s)$ can be written

$$G(s) = \frac{4}{25s} + \frac{1}{5s^2} - \frac{4}{25(s + 5)}$$

The inverse Laplace transform of $G(s)$ is

$$g(t) = \left(\frac{4}{25} + \frac{1}{5}t - \frac{4}{25}e^{-5t} \right) u(t)$$

We shall apply the following theorem.

If $h(t) = e^{ct}f(t)$, then $H(s) = F(s - c)$

or $L[g(t - a)u(t - a)] = e^{-as}G(s)$

In our example

$$F(s) = e^{-7s}G(s) = L[g(t - 7)u(t - 7)]$$

and

$$f(t) = L^{-1}[F(s)] = g(t - 7)u(t - 7)$$

For $g(t - 7)$ we get

$$g(t - 7) = \left[\frac{4}{25} + \frac{1}{5}(t - 7) - \frac{4}{25}e^{-5(t - 7)} \right] u(t - 7)$$

Since

$$u(t - a) \cdot u(t - a) = u(t - a)$$

we get

$$g(t - 7)u(t - 7) = \left[\frac{4}{25} + \frac{1}{5}(t - 7) - \frac{4}{25}e^{-5(t - 7)} \right] u(t - 7)$$

Thus

$$f(t) = \left[\frac{4}{25} + \frac{1}{5}(t - 7) - \frac{4}{25}e^{-5(t - 7)} \right] u(t - 7)$$

● **PROBLEM 3-21**

Find the inverse Laplace transform of the function $F(s)$ where

$$F(s) = \frac{3}{s^2(s^2 + \omega_n^2)}$$

ω_n is a constant.

Solution: The convolution method states that:

if

111

$$L[f_1(t)] = F_1(s)$$

and

$$L[f_2(t)] = F_2(s)$$

then

$$L[f_1(t) * f_2(t)] = F_1(s) \cdot F_2(s)$$

$$= L[\int_0^t f_1(\tau) f_2(t - \tau) d\tau]$$

$$= L[\int_0^t f_1(t - \tau) f_2(\tau) d\tau]$$

The function $F(s)$ can be written

$$F(s) = \frac{3}{s^2(s^2 + \omega_n^2)} = \frac{3}{s} \cdot \frac{1}{s(s^2 + \omega_n^2)} = F_1(s) \cdot F_2(s)$$

The convolution method gives

$$f(t) = L^{-1}[F_1(s) F_2(s)] = f_1(t) * f_2(t) = \int_0^t f_2(\tau) f_1(t - \tau) d\tau$$

In our case

$$F_1(s) = \frac{3}{5}$$

and

$$F_2(s) = \frac{1}{s(s^2 + \omega_n^2)}$$

The inverse Laplace transforms are

$$f|_1(x) = 3u(x)$$

$$f_2(x) = \frac{1}{\omega_n^2}(1 - \cos \omega_n x) u(x)$$

Then

$$\text{for } x = t - \tau, \quad f_1(t - \tau) = 3u(t - \tau)$$

and

$$f_2(\tau) = \frac{1}{\omega_n^2}(1 - \cos \omega_n \tau) u(\tau) \text{ for } x = \tau.$$

$$\int_0^t f_2(\tau) f_1(t - \tau) d\tau = \int_0^t \frac{1}{\omega_n^2}(1 - \cos \omega_n \tau) u(\tau) \cdot 3u(t - \tau) d\tau$$

$$= \frac{3}{\omega_n^2}\int_0^t (1 - \cos \omega_n \tau) d\tau$$

In calculating the above integral we used

$$u(\tau) u(t - \tau) = \begin{cases} 0 & \text{for } \tau < 0 \text{ and } \tau > t \\ 1 & 0 < \tau < t \end{cases}$$

Thus, $f(t)$ is equal

$$f(t) = \frac{3}{\omega_n^2}\int_0^t (1 - \cos \omega_n \tau) d\tau = \frac{3}{\omega_n^2}[\int_0^t d\tau - \int_0^t \cos \omega_n \tau d\tau]$$

112

$$= \frac{3}{\omega_n^2}[\tau \big|_0^t - \frac{1}{\omega_n} \sin \omega_n \tau \big|_0^t] = \frac{3}{\omega_n^2}[t - \frac{1}{\omega_n} \sin \omega_n t]u(t).$$

For practical purposes our choice of functions $F_1(s)$ and $F_2(s)$ is the best.

Some other possible choices, like

$$F_1(s) = \frac{3}{s^2} \quad \text{and} \quad F_2 = \frac{1}{s^2 + \omega_n^2}$$

lead to longer calculations.

● **PROBLEM** 3-22

Given the function

$$X(s) = \frac{\omega_n^2}{s(s^2 + 2\mu\omega_n s + \omega_n^2)}$$

where the values of μ and ω_n are such that the nonzero poles of $X(s)$ are complex numbers, find $x(t)$.

<u>Solution:</u> Using partial fraction expansion we obtain

$$X(s) = \frac{K_1}{s} + \frac{K_{-\alpha + j\omega}}{s + \alpha - j\omega} + \frac{K_{-\alpha - j\omega}}{s + \alpha + j\omega}$$

where

$$\omega = \omega_n\sqrt{1 - \mu^2}$$

and

$$\alpha = \mu\omega_n$$

The coefficients K are determined from

$$K_1 = sX(s)\big|_{s = 0} = 1$$

$$K_{-\alpha + j\omega} = (s + \alpha - j\omega)X(s)\big|_{s = -\alpha + j\omega}$$

$$= \frac{\omega_n^2}{2j\omega(-\alpha + j\omega)} = \frac{\omega_n}{2\omega}e^{-j(\theta + \frac{\pi}{2})}$$

where

$$\theta = \tan^{-1}[-\frac{\omega}{\alpha}] = \tan^{-1}[-\sqrt{\frac{1}{\mu^2} - 1}]$$

From the equation

$$K_{-\alpha - j\omega} = K^*_{-\alpha + j\omega}$$

we obtain

113

$$K_{-\alpha - j\omega} = \frac{\omega_n}{2\omega}e^{j\left(\theta + \frac{\pi}{2}\right)}$$

The function $X(s)$ can be written

$$X(s) = \frac{1}{s} + \frac{\omega_n}{2\omega}\left[\frac{e^{-j\left(\theta + \frac{\pi}{2}\right)}}{s + \alpha - j\omega} + \frac{e^{j\left(\theta + \frac{\pi}{2}\right)}}{s + \alpha + j\omega}\right]$$

Taking the inverse Laplace transform we get

$$L^{-1}[X(s)] = x(t) = 1 + \frac{\omega_n}{2\omega}(e^{-j\left(\theta + \frac{\pi}{2}\right)}e^{-(\alpha - j\omega)t}$$

$$+ e^{j\left(\theta + \frac{\pi}{2}\right)}e^{-(\alpha + j\omega)t}) = 1 + \frac{\omega_n}{\omega}e^{-\alpha t}\sin(\omega t - \theta)$$

Or, in terms of μ and ω_n

$$x(t) = 1 + \frac{1}{\sqrt{1 - \mu^2}}e^{-\mu\omega_n t}\sin(\omega_n\sqrt{1 - \mu^2} \cdot t - \theta)$$

● **PROBLEM 3-23**

Given the Laplace transform $X(s)$ where

$$X(s) = \frac{3s^3 + 17s^2 + 33s + 15}{s^3 + 6s^2 + 11s + 6}$$

find $x(t)$.

Solution: First, let us make $X(s)$ a proper fraction, hence

$$\begin{array}{r} 3 \\ 3s^3 + 17s^2 + 33s + 15 \,:\quad s^3 + 6s^2 + 11s + 6 \\ \underline{-(3s^3 + 18s^2 + 33s + 18)} \\ -s^2 - 3 \end{array}$$

We can write $X(s)$ in the following form

$$X(s) = 3 - \frac{s^2 + 3}{s^3 + 6s^2 + 11s + 6}$$

We shall write a proper fraction $\dfrac{s^2 + 3}{s^3 + 6s^2 + 11s + 6}$ in partial fractional form.

Let us investigate the denominator of the fraction. It is easy to check that for $s = -3$

$$s^3 + 6s^2 + 11s + 6 = 0$$

Thus

$$\frac{s^3 + 6s^2 + 11s + 6}{s + 3} = s^2 + 3s + 2$$

The two remaining roots are $s = -1$, $s = -2$ thus

114

$$s^3 + 6s^2 + 11s + 6 = (s + 1)(s + 2)(s + 3)$$

To find the partial fractional form, we set

$$\frac{s^2 + 3}{s^3 + 6s^2 + 11s + 6} = \frac{K_1}{s + 1} + \frac{K_2}{s + 2} + \frac{K_3}{s + 3} = P(s)$$

$$K_1 = [(s + 1)P(s)] \Big|_{s=-1} = 2$$

$$K_2 = [(s + 2)P(s)] \Big|_{s=-2} = -7$$

$$K_3 = [(s + 3)P(s)] \Big|_{s=-3} = 6$$

We obtained $X(s)$ in form

$$X(s) = 3 - \frac{2}{s + 1} + \frac{7}{s + 2} - \frac{6}{s + 3}$$

Taking $L^{-1}[X(s)]$ we get

$$L^{-1}[X(s)] = L^{-1}[3 - \frac{2}{s + 1} + \frac{7}{s + 2} - \frac{6}{s + 3}]$$

and

$$x(t) = 3\delta(t) - 2e^{-t}u(t) + 7e^{-2t}u(t) - 6e^{-3t}u(t)$$

$$= 3\delta(t) - (2e^{-t} - 7e^{-2t} + 6e^{-3t})u(t).$$

● **PROBLEM 3-24**

Through the convolution integral technique find the inverse Laplace transform of

$$X(s) = \frac{1}{s(s + 2)^2}$$

Solution: We shall write $X(s)$ as a product of simple fractions

$$X(s) = \frac{1}{s} \cdot \frac{1}{s + 2} \cdot \frac{1}{s + 2} = F_1(s)F_2(s)F_3(s)$$

In the convolution integral method the following equation is applied

$$L^{-1}[F_1(s) \cdot F_2(s)] = \int_0^t f_1(t - \tau)f_2(\tau)d\tau$$

$$= \int_0^t f_1(\tau)f_2(t - \tau)d\tau$$

where

$$L^{-1}[F_1(s)] = f_1(t)$$

115

$$L^{-1}[F_2(s)] = f_2(t)$$

Thus for $F_3(s)$ and $F_2(s)$ we have

$$F_2(s) = \frac{1}{s + 2}$$

$$F_3(s) = \frac{1}{s + 2}$$

$$f_2(t) = e^{-2t}$$

$$f_3(t) = e^{-2t}$$

and

$$L^{-1}\left[\frac{1}{(s + 2)^2}\right] = \int_0^t e^{-2(t - \tau)} \cdot e^{-2\tau} d\tau = e^{-2t}\int_0^t d\tau = te^{-2t}$$

For the function $X(s)$ we have

$$L^{-1}[X(s)] = L^{-1}[\frac{1}{s(s + 2)^2}] = \int_0^t \tau e^{-2\tau} d\tau$$

$$= \tau \frac{e^{-2\tau}}{-2}\Big|_0^t - \int_0^t \frac{e^{-2\tau}}{-2} d\tau = t\frac{e^{-2t}}{-2} + \frac{1}{2}(\frac{e^{-2\tau}}{-2})\Big|_0^t$$

$$= -\frac{1}{2}te^{-2t} - \frac{1}{4}e^{-2t} + \frac{1}{4}$$

● **PROBLEM 3-25**

The Laplace transform of a function $x(t)$ is $X(s)$ where

$$X(s) = \frac{100}{s(s + 2)(s + 3)^2(s^2 + 8s + 25)}$$

Find $x(t)$.

Solution: We shall express $X(s)$ as a sum of simpler fractions. Let us note that $s^2 + 8s + 25 = (s + 4 - 3j)(s + 4 + 3j)$, thus

$$X(s) = \frac{A_1}{s} + \frac{A_2}{s + 2} + \frac{A_3}{(s + 3)^2} + \frac{A_4}{s + 3} + \frac{A_5}{s + 4 - 3j} + \frac{A_6}{s + 4 + 3j}$$

We evaluate the constants from the equations:

$$A_1 = [sX(s)]_{s=0} = 0.22$$

$$A_2 = [(s + 2)X(s)]_{s=-2} = -3.85$$

$$A_3 = [(s + 3)^2 X(s)]_{s=-3} = 3.33$$

116

$$A_4 = \frac{d}{ds}[(s + 3)^2 X(s)] \bigg|_{s=-3} = 3.78$$

$$A_5 = [(s + 4 - 3j)X(s)] \bigg|_{s=-4+3j} = 0.0923 \lfloor -214°$$

$$A_6 = A_5^* = 0.0923 \lfloor 214°$$

The inverse transform of $X(s)$ gives $x(t) = 0.22 - 3.85e^{-2t}$ $+ 3.33te^{-3t} + 3.78e^{-3t} + 0.185e^{-4t} \sin(3t - 124°)$

Calculating the above equation we used the formula

$$L^{-1}\left[\frac{A}{s + a - bj} + \frac{A^*}{s + a + bj}\right] = 2|A|e^{-at}\sin(bt + \lfloor A)$$

● PROBLEM 3-26

Find the inverse Laplace transform of the function

$$F(s) = \frac{3s}{s^2 + 4s + 5}$$

<u>Solution</u>: Note that the denominator of $F(s)$ can be written

$$s^2 + 4s + 5 = (s + 2)^2 + 1$$

We can write $F(s)$ as follows

$$F(s) = \frac{3s}{(s + 2)^2 + 1} = \frac{3(s + 2)}{(s + 2)^2 + 1} - \frac{6}{(s + 2)^2 + 1}$$

We know that

$$L[\sin t] = \frac{1}{s^2 + 1}$$

and

$$L[\cos t] = \frac{s}{s^2 + 1}$$

We shall use the following property of the Laplace transform. If $g(t) = e^{at}f(t)$,

then $G(s) = F(s - a)$, $s > a$
We obtain

$$L^{-1}[F(s)] = L^{-1}\left[\frac{3(s + 2)}{(s + 2)^2 + 1} - \frac{6}{(s + 2)^2 + 1}\right]$$

$$= 3L^{-1}\left[\frac{s + 2}{(s + 2)^2 + 1}\right] - 6L^{-1}\left[\frac{1}{(s + 2)^2 + 1}\right]$$

$$= e^{-2t} \cdot 3 \cos t - e^{-2t} 6 \sin t$$

$$= e^{-2t}(3 \cos t - 6 \sin t)$$

Find the inverse Laplace transform of

$$F(s) = \frac{s + 1}{s(s^2 + 2s + 2)}$$

Solution: We expand $F(s)$ as follows

$$\frac{s + 1}{s(s^2 + 2s + 2)} = \frac{a_1 s + a_2}{s^2 + 2s + 2} + \frac{a_3}{s} \qquad (1)$$

It is easy to verify that

$$s^2 + 2s + 2 = (s + 1 - j)(s + 1 + j)$$

We multiply both sides of (1) by $s^2 + 2s + 2$

$$\frac{s + 1}{s} = a_1 s + a_2 + \frac{a_3}{s}(s^2 + 2s + 2)$$

Let $s = -1 - j$, then

$$\frac{-1 - j + 1}{-1 - j} = a_1(-1 - j) + a_2$$

which can be simplified as follows

$$-j = a_1(1 + j)(1 + j) + a_2(-1 - j)$$

Comparing the real and imaginary parts of the equation we get
$$a_1 = -0.5, \quad a_2 = 0$$

To determine a_3, let us multiply (1) by s and let $s = 0$

$$a_3 = \left[\frac{s + 1}{s^2 + 2s + 2}\right]_{s=0} = \frac{1}{2}$$

We have

$$F(s) = \frac{-0.5s}{s^2 + 2s + 2} + \frac{0.5}{s} = \frac{0.5}{s} - \frac{0.5s}{(s + 1)^2 + 1}$$

$$= 0.5\left[\frac{1}{s} - \frac{s + 1}{(s + 1)^2 + 1} + \frac{1}{(s + 1)^2 + 1}\right]$$

The inverse Laplace transform
$$f(t) = L^{-1}[F(s)] = 0.5[1 - e^{-t}\cos t + e^{-t}\sin t]$$

Find the inverse Laplace transform of

$$G(s) = \frac{s^3 + 8s^2 + 20s + 17}{s^2 + 4s + 3}$$

Solution: Dividing the numerator by the denominator, we obtain

$$G(s) = \frac{s^3 + 8s^2 + 20s + 17}{s^2 + 4s + 3} = s + 4 + \frac{s + 5}{s^2 + 4s + 3}$$

Next, we expand the fraction

$$\frac{s + 5}{s^2 + 4s + 3} = \frac{s + 5}{(s + 1)(s + 3)} = \frac{A}{s + 1} + \frac{B}{s + 3}$$

Multiplying both sides by $(s + 1)$ we get

$$\frac{s + 5}{s + 3} = A + \frac{B(s + 1)}{s + 3}$$

Let $s = -1$, then $A = \frac{-1 + 5}{-1 + 3} = \frac{4}{2} = 2$

Multiplying the fraction by $(s + 3)$ we obtain

$$\frac{s + 5}{s + 1} = \frac{A(s + 3)}{s + 1} + B$$

For $s = -3$ we get $B = \frac{-3 + 5}{-3 + 1} = \frac{2}{-2} = -1$

We can rewrite $G(s)$ as follows

$$G(s) = s + 4 + \frac{2}{s + 1} - \frac{1}{s + 3}$$

$$g(t) = L^{-1}[G(s)] = L^{-1}[s] + L^{-1}[4] + L^{-1}[\frac{2}{s + 1}]$$

$$+ L^{-1}[\frac{-1}{s + 3}]$$

Note that if $\delta(t)$ is a unit-impulse function

$$L[\delta(t)] = \int_0^\infty \delta(t)e^{-st}dt = 1$$

and

$$L[\frac{d}{dt}\delta(t)] = s$$

Thus we have

$$g(t) = \frac{d}{dt}\delta(t) + 4\delta(t) + 2e^{-t} - e^{-3t} \quad \text{for } t \geq 0$$

● PROBLEM 3-29

Find the inverse Laplace transform of the following function
$F(s)$;

$$F(s) = \frac{s^2 + 2s + 5}{(s + 1)^3}$$

Solution: Expanding $F(s) = \frac{A(s)}{B(s)}$ into partial fractions,
we obtain

$$F(s) = \frac{A(s)}{B(s)} = \frac{a_3}{(s + 1)^3} + \frac{a_2}{(s + 1)^2} + \frac{a_1}{s + 1}$$

119

From the following equations we find a_3, a_2 and a_1

$$a_3 = \left[\frac{A(s)}{B(s)}(s + 1)^3\right]_{s=-1} = (s^2 + 2s + 5)_{s=-1} = 4$$

$$a_2 = \left\{\frac{d}{ds}\left[\frac{A(s)}{B(s)}(s + 1)^3\right]\right\}_{s=-1} = \left[\frac{d}{ds}(s^2 + 2s + 5)\right]_{s=-1}$$

$$= (2s + 2)_{s=-1} = 0$$

$$a_1 = \frac{1}{(3 - 1)!}\left\{\frac{d^2}{ds^2}\left[\frac{A(s)}{B(s)}(s + 1)^3\right]\right\}_{s=-1}$$

$$= \frac{1}{2!}\left[\frac{d^2}{ds^2}(s^2 + 2s + 5)\right]_{s=-1} = \frac{1}{2} \cdot 2 = 1$$

Thus, we obtain

$$f(t) = L^{-1}[F(s)] = L^{-1}\left[\frac{4}{(s + 1)^3} + \frac{1}{s + 1}\right]$$

$$= L^{-1}\left[\frac{4}{(s + 1)^3}\right] + L^{-1}\left[\frac{1}{s + 1}\right] = 2t^2 e^{-t} + e^{-t}$$

$$= e^{-t}(2t^2 + 1) \quad \text{for} \quad t \geq 0$$

● **PROBLEM** 3-30

Using the convolution formula find the inverse transform of the function

$$H(s) = \frac{1}{s^2(s^2 + 1)}$$

Solution: Since

$$H(s) = H_1(s) \cdot H_2(s)$$

where

$$H_1(s) = \frac{1}{s^2} \quad \text{and} \quad H_2(s) = \frac{1}{s^2 + 1}$$

we have

$$h = h_1 * h_2$$

We can compute or find from the tables that

$$h_1(t) = t$$

$$h_2(t) = \sin t,$$

thus using the definition of convolution

$$f * g(t) = \int_0^t f(t - \tau)g(\tau)d\tau$$

120

we obtain

$$h(t) = h_1 * h_2(t) = \int_0^t (t - \tau) \sin \tau \, d\tau$$

$$= [-(t - \tau)\cos \tau - \sin \tau] \Big|_{\tau = 0}^{\tau = t} = t - \sin t.$$

Find the inverse Laplace transform of

$$G(s) = \frac{5se^{-4s}}{(s + 2)^2(s + 3)}$$

Solution: From the equation

$$L[Kf(t - a)u(t - a)] = Ke^{-as}F(s)$$

we obtain

$$g(t) = L^{-1}[G(s)] = Kf(t - a)u(t - a) = L^{-1}[Ke^{-as}F(s)]$$

and

$$G(s) = Ke^{-as}F(s)$$

In our case K = 5 and a = 4

$$G(s) = 5e^{-4s}\frac{s}{(s + 2)^2(s + 3)}$$

Using partial fractions we obtain

$$F(s) = \frac{s}{(s + 2)^2(s + 3)} = \frac{a_1}{s + 3} + \frac{a_2}{s + 2} + \frac{a_3}{(s + 2)^2}$$

$$a_1 = \left[\frac{s}{(s + 2)^2}\right]\Bigg|_{s=-3} = -3$$

$$a_2 = \left\{\frac{d}{ds}\left[\frac{s}{s + 3}\right]\right\}\Bigg|_{s=-2} = \frac{(s + 3) - s}{(s + 3)^2} = 3$$

$$a_3 = \left[\frac{s}{s + 3}\right]\Bigg|_{s=-2} = -2$$

Then

$$F(s) = \frac{-3}{s + 3} + \frac{3}{s + 2} - \frac{2}{(s + 2)^2}$$

Using

$$L[K e^{-at}u(t)] = \frac{K}{s + a}$$

and

121

$$L\left[\frac{K}{(n-1)!}t^{n-1}\ e^{-at}u(t)\right] = \frac{K}{(s+a)^n}$$

we get

$$f(t) = (3e^{-2t} - 2te^{-2t} - 3e^{-3t})u(t)$$

For f(t - a) we have the following expression

$$f(t-a)u(t-a) = f(t-4)u(t-4)$$

$$= [3e^{-2(t-4)} - 2(t-4)e^{-2(t-4)} - 3e^{-3(t-4)}]u(t-4)$$

Thus

$$g(t) = Kf(t-4)u(t-4) = 5[3e^{-2(t-4)} - 2(t-4)e^{-2(t-4)}$$

$$+ 3e^{-3(t-4)}]u(t-4) = 5[(11-2t)e^8 \cdot e^{-2t}$$

$$- 3e^{12}e^{-3t}]u(t-4)$$

● **PROBLEM** 3-32

Investigate the circuit shown in the figure. Assume that
the switch is closed at time t = 0. Use the Laplace transform method to determine the current i in the circuit, apply
the final-value and initial-value theorems.

E, R, L are constants.

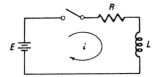

Solution: The differential equation describing the current
in the circuit is

$$Ri + L\frac{di}{dt} = E$$

The Laplace transform of the above equation is

$$L[Ri(t) + L\frac{di}{dt}] = L[E]$$

Using the superposition property and the initial-value
theorem we obtain

$$L[Ri(t) + L\frac{di}{dt}] = L[Ri(t)] + L[L\frac{di}{dt}]$$

$$= R\ I(s) + L[sI(s) - i(0)]$$

$$= L[E] = \frac{E}{s}$$

I(s) denotes the current as a function of the Laplace variable s.

Since the switch is closed at time t = 0 we have i(0) = 0, and

$$(R + Ls) \ I(s) = \frac{E}{s}$$

Solving for I(s) we have

$$I(s) = \frac{E}{s(R + Ls)} = \frac{E}{L}\left[\frac{1}{s(s+\frac{R}{L})}\right]$$

or

$$I(s) = \frac{E}{R}\frac{1}{s} - \frac{E}{R}\left[\frac{1}{s + \frac{R}{L}}\right]$$

Taking the inverse transform

$$i(t) = \frac{E}{R} - \frac{E}{R}e^{-\frac{Rt}{L}} = \frac{E}{R}(1 - e^{-\frac{R}{L}t})$$

The final-value theorem states that

$$\lim_{t\to\infty} f(t) = \lim_{s\to 0} sF(s)$$

thus

$$\lim_{t\to\infty} i(t) = \lim_{s\to 0} s\left[\frac{E}{s(R + Ls)}\right] = \frac{E}{R}$$

The initial-value theorem is

$$\lim_{t\to 0} f(t) = \lim_{s\to\infty} sF(s)$$

thus

$$\lim_{t\to 0} i(t) = \frac{E}{R}(1 - 1) = 0$$

and

$$\lim_{s\to\infty} sI(s) = \lim_{s\to\infty} \frac{E}{R + Ls} = 0$$

The result agrees with the initial condition i(0) = 0 for the current.

● **PROBLEM 3-33**

Consider the linear differential equation of the second order

$$\frac{d^2x(t)}{dt^2} + 15\frac{dx(t)}{dt} + 100x(t) = 100u_s(t) \qquad (1)$$

where $u_s(t)$ is the unit step function

$$u_s(t) = \begin{cases} 0 & t < 0 \\ 1 & t \geq 0 \end{cases}$$

Using the Laplace transform find the solution of equation (1) for the initial values

$$x(0) = 0, \quad \frac{dx(0)}{dt} = 0$$

Solution: The Laplace transform of (1) for $x(0) = 0$ and

$$\frac{dx(0)}{dt} = 0$$

is

$$s^2 X(s) + 15sX(s) + 100X(s) = \frac{100}{s} \qquad (2)$$

From the last equation we have

$$X(s) = \frac{100}{s(s^2 + 15s + 100)}$$

The poles of $X(s)$ are at

$$s = 0, \quad s = -7.5 + j6.61, \quad s = -7.5 - j6.61$$

$X(s)$ can be written

$$X(s) = \frac{100}{s(s + 7.5 - j6.61)(s + 7.5 + j6.61)}$$

The partial-fraction expansion of $X(s)$ gives

$$X(s) = \frac{K_0}{s} + \frac{K_1}{s + 7.5 - j6.61} + \frac{K_1^*}{s + 7.5 + j6.61}$$

We obtain

$$K_0 = [s \, X(s)] \Big|_{s=0} = \frac{100}{100} = 1$$

$$K_1 = (s + 7.5 - j6.61)X(s) \Big|_{s=-7.5+j6.61}$$

$$= \frac{100}{(-7.5 + j6.61) \cdot 2j \cdot 6.61}$$

$$= \frac{\sqrt{100}}{2(6.61)} e^{-j(\theta + \frac{\pi}{2})}$$

where $\theta = \tan^{-1}\left(-\frac{6.61}{7.5}\right) = \tan^{-1}(-0.881) = -41.39°$

$$X(s) = \frac{1}{s} + \frac{10}{2 \cdot (6.61)} \left[\frac{e^{-j(\theta + \frac{\pi}{2})}}{s + 7.5 - j6.61} + \frac{e^{j(\theta + \frac{\pi}{2})}}{s + 7.5 + j6.61} \right]$$

$$\theta = -41.39°$$

The inverse Laplace transform gives

$$x(t) = 1 + \frac{10}{2 \cdot 6.61} \left[e^{-j(\theta + \frac{\pi}{2})} \cdot e^{-(7.5 - j6.61)t} \right.$$

$$+ e^{j(\theta + \frac{\pi}{2})} e^{-(7.5 + j6.61)t} \Big]$$

$$= 1 + \frac{10}{2 \cdot 6.61} e^{-7.5t} \Big[e^{-j(\theta + \frac{\pi}{2})} + j6.61t$$

$$+ e^{j(\theta + \frac{\pi}{2})} - j6.61t \Big]$$

$$= 1 + 1.51e^{-7.5t} \cos[6.61t - \theta]$$

● **PROBLEM** 3-34

Taking the inverse Laplace transform of X(s) solve the equation

$$\ddot{x} + 4\dot{x} + 8x = 0$$

with the initial conditions

$$x(0) = 0, \quad \dot{x}(0) = 4$$

Solution: The Laplace transform of the differential equation is

$$s^2 X(s) - sx(0) - \dot{x}(0) + 4sX(s) - 4x(0) + 8X(s) = 0$$

Substituting the initial conditions we have

$$s^2 X(s) - 4 + 4sX(s) + 8X(s) = 0$$

or

$$X(s) = \frac{4}{s^2 + 4s + 8} = 2\frac{2}{(s + 2)^2 + (2)^2}$$

The inverse Laplace transform of X(s) is

$$x(t) = 2e^{-2t} \sin 2t.$$

● **PROBLEM** 3-35

Using the Laplace transform solve a differential equation

$$x^{(5)} + 11x^{(4)} + 46\dddot{x} + 89\ddot{x} + 74\dot{x} + 24x = 1$$

with a set of initial conditions

$$x(0) = 0, \quad \dot{x}(0) = 0, \quad \ddot{x}(0) = 0, \quad \dddot{x}(0) = 0, \quad x^{(4)}(0) = 4$$

Solution: Taking the Laplace transform of both sides

$$s^5 X(s) - s^4 x(0) - s^3 \dot{x}(0) - s^2 \ddot{x}(0) - s\dddot{x}(0) - x^{(4)}(0)$$

$$+ 11[s^4 X(s) - s^3 x(0) - s^2 \dot{x}(0) - s\ddot{x}(0) - \dddot{x}(0)]$$

$$+ 46[s^3X(s) - s^2x(0) - s\dot{x}(0) - \ddot{x}(0)]$$

$$+ 89[s^2X(s) - sx(0) - \dot{x}(0)]$$

$$+ 74[sX(s) - x(0)] + 24X(s) = \frac{1}{s}$$

Substitution of the initial conditions gives

$$X(s)[s^5 + 11s^4 + 46s^3 + 89s^2 + 74s + 24] = \frac{1}{s} + 4$$

$$X(s) = \frac{4s + 1}{s(s^5 + 11s^4 + 46s^3 + 89s^2 + 74s + 24)}$$

It's easy to check that the denominator is equal to

$$s(s^5 + 11s^4 + 46s^3 + 89s^2 + 74s + 24)$$

$$= s(s + 1)^2(s + 2)(s + 3)(s + 4)$$

The partial fraction expansion of $X(s)$ is

$$X(s) = \frac{A_1}{s} + \frac{A_2}{s + 1} + \frac{A_3}{(s + 1)^2} + \frac{A_4}{s + 2} + \frac{A_5}{s + 3} + \frac{A_6}{s + 4}$$

$$A_1 = [X(s) \cdot s]_{s=0} = \frac{1}{24}$$

$$A_3 = [X(s)(s + 1)^2]_{s=-1} = \frac{1}{2}$$

$$A_4 = [(s + 2)X(s)]_{s=-2} = \frac{7}{4}$$

$$A_5 = [X(s)(s + 3)]_{s=-3} = -\frac{11}{12}$$

$$A_6 = [X(s)(s + 4)]_{s=-4} = \frac{5}{24}$$

$$A_2 = [\frac{d}{ds}(s + 1)^2 X(s)]_{s=-1} = [\frac{d}{ds}\frac{4s + 1}{s(s + 2)(s + 3)(s + 4)}]_{s=-1}$$

$$= [\frac{d}{ds}(\frac{4s + 1}{s^4 + 9s^3 + 26s^2 + 24s})]_{s=-1}$$

$$= \frac{4(s^4 + 9s^3 + 26s^2 + 24s) - (4s + 1)(4s^3 + 27s^2 + 52s + 24)}{(s^4 + 9s^3 + 26s^2 + 24s)^2}\bigg|_{s=-1}$$

$$= \frac{4(-6) - (-3)(-5)}{(-6)^2} = -\frac{39}{36}$$

Taking the inverse Laplace transform we obtain the solution of the differential equation

$$x(t) = \frac{1}{24} + \frac{1}{2}te^{-t} - \frac{39}{36}e^{-t} + \frac{7}{4}e^{-2t} - \frac{11}{12}e^{-3t} + \frac{5}{24}e^{-4t}$$

CHAPTER 4

Z-TRANSFORMS

Z-TRANSFORMS AND THEOREMS

● **PROBLEM** 4-1

Find the z-transform of a^k and A^k, where A is an n × n matrix.

Solution: From the definition of the z-transform we have

$$Z[a^k] = \sum_{k=0}^{\infty} a^k z^{-k} = 1 + az^{-1} + a^2 z^{-2} + \ldots$$

$$= \frac{1}{1 - az^{-1}} = \frac{z}{z - a}$$

For A^k we have

$$Z[A^k] = \sum_{k=0}^{\infty} A^k z^{-k} = I + Az^{-1} + A^2 z^{-2} + \ldots$$

$$= (I - Az^{-1})^{-1} = (zI - A)^{-1} z$$

● **PROBLEM** 4-2

Find the z-transform of $f(t) = \cos\omega t$.

Solution: By definition

$$F(z) = \sum_{k=0}^{\infty} f(kT) z^{-k}$$

Then

127

$$F(z) = \sum_{k=0}^{\infty} \cos\omega kT \cdot z^{-k}$$

Expressing $\cos\omega kT$ in terms of $e^{j\omega Tk}$ and $e^{-j\omega Tk}$

$$\cos\omega kT = \frac{e^{j\omega kT} + e^{-j\omega kT}}{2}$$

and substituting into $F(z)$ one gets

$$F(z) = \sum_{k=0}^{\infty} \frac{e^{j\omega kT} + e^{-j\omega kT}}{2} z^{-k}$$

For $|z^{-1}| < 1$ the above series is covergent, and can be written

$$F(z) = \frac{1}{2}\left[\frac{1}{1 - e^{j\omega T}z^{-1}} + \frac{1}{1 - e^{-j\omega T}z^{-1}}\right]$$

Simplifying

$$F(z) = \frac{1}{2}\left[\frac{z}{z - e^{j\omega T}} + \frac{z}{z - e^{-j\omega T}}\right]$$

$$= \frac{1}{2}\left[\frac{z(z - e^{-j\omega T}) + z(z - e^{j\omega T})}{(z - e^{j\omega T})(z - e^{-j\omega T})}\right]$$

$$= \frac{1}{2}\left[\frac{z^2 - ze^{-j\omega T} + z^2 - ze^{j\omega T}}{z^2 - z(e^{-j\omega T} + e^{j\omega T}) + 1}\right]$$

$$= \frac{z(z - \cos\omega T)}{z^2 - 2z\cos\omega T + 1}$$

● PROBLEM 4-3

Find the z-transform of $f(t) = \sin\omega t$ for $t \geq 0$.

Solution: The function $\sin\omega t$ can be expressed as

$$\sin\omega t = \frac{e^{j\omega t} - e^{-j\omega t}}{2j}$$

Thus

$$Z\{\sin\omega t\} = Z\left\{\frac{e^{j\omega t}}{2j} - \frac{e^{-j\omega t}}{2j}\right\}.$$

We know that

$$Z\{e^{at}\} = \frac{z}{z - e^{-at}}$$

Therefore

$$F(z) = \frac{1}{2j} \left(\frac{z}{z - e^{j\omega T}} - \frac{z}{z - e^{-j\omega T}} \right)$$

$$= \frac{1}{2j} \left[\frac{z(e^{j\omega T} - e^{-j\omega T})}{z^2 - z(e^{j\omega T} + e^{-j\omega T}) + 1} \right]$$

$$= \frac{z\sin\omega T}{z^2 - 2z\cos\omega T + 1}$$

● **PROBLEM 4-4**

Determine the z-transform of the exponential function $f(t) = e^{-at}$, where a is a real constant.

Solution: From the definition of the z-transform we get

$$F(z) = \sum_{k=0}^{\infty} f(kT) z^{-k} = \sum_{k=0}^{\infty} e^{-akT} z^{-k}$$

For the above series to converge the following condition must be satisfied

$$|e^{-aT} z^{-1}| < 1.$$

Let us expand F(z)

$$F(z) = 1 + e^{-aT} z^{-1} + e^{-2aT} z^{-2} + e^{-3aT} z^{-3} + \ldots \quad (1)$$

Multiplying both sides of the above equation by $e^{-aT} z^{-1}$ we get

$$e^{-aT} z^{-1} F(z) = e^{-aT} z^{-1} + e^{-2aT} z^{-2} + e^{-3aT} z^{-3} + \ldots \quad (2)$$

Subtracting (2) from (1) yields

$$(1 - e^{-aT} z^{-1}) F(z) = 1$$

Thus

$$F(z) = \frac{1}{1 - e^{-aT} z^{-1}} = \frac{z}{z - e^{-aT}}$$

● **PROBLEM 4-5**

Determine the z-transform of the ramp function $f(t) = tu_s(t)$.

Solution: From definition of the z-transform

$$F(z) = \sum_{k=0}^{\infty} kTz^{-k} = Tz^{-1} + 2Tz^{-2} + 3Tz^{-3} + \ldots \qquad (1)$$

We multiply both sides of (1) by z^{-1}

$$z^{-1}F(z) = Tz^{-2} + 2Tz^{-3} + 3Tz^{-4} + \ldots \qquad (2)$$

Subtracting (2) from (1)

$$(1 - z^{-1})F(z) = Tz^{-1} + Tz^{-2} + Tz^{-3} + \ldots$$

$$= \frac{Tz^{-1}}{1 - z^{-1}}$$

Thus

$$F(z) = \frac{Tz^{-1}}{(1 - z^{-1})^2} = \frac{Tz}{(z - 1)^2} .$$

● **PROBLEM** 4-6

Given

$$X(s) = \frac{s + 3}{(s + 1)(s + 2)}$$

obtain X(z) using both methods, partial-fraction expansion and inversion formula. Compare the results.

Solution: X(s) can be written

$$X(s) = \frac{2}{s + 1} - \frac{1}{s + 2}$$

thus

$$X(t) = 2e^{-t} - e^{-2t}$$

Taking the z-transforms we obtain

$$X(z) = \frac{2z}{z - e^{-T}} - \frac{z}{z - e^{-2T}}$$

The application of the inversion formula leads to

$$X(z) = [\text{residue of } \frac{(3+s)z}{(s+1)(s+2)(z-e^{Ts})} \text{ at } s = -1]$$

$$+ [\text{residue of } \frac{(s+3)z}{(s+1)(s+2)(z-e^{Ts})} \text{ at } s = -2]$$

$$= \left.\frac{(s+3)z}{(s+2)(z-e^{Ts})}\right|_{s=-1} + \left.\frac{(s+3)z}{(s+1)(z-e^{Ts})}\right|_{s=-2}$$

$$= \frac{2z}{z - e^{-T}} - \frac{z}{z - e^{-2T}}$$

That agrees with the result obtained previously.

Find the z-transform of a unit-step function which is delayed by one sampling period T.

<u>Solution</u>: From the definition of the z-transform we can write

$$Z\{u(t - T)\} = \sum_{k=0}^{\infty} u(kT - T) z^{-k}$$

Let $\qquad n = k - 1$, thus $k = n + 1$

The z-transform of $u(t - T)$ becomes

$$Z\{u(t - T)\} = \sum_{n=-1}^{\infty} u(nT) z^{-n-1} = z^{-1} \sum_{n=-1}^{\infty} u(nT) z^{-n}$$

From the definition of the unit step function we have

$$u(-T) = 0$$

Thus

$$Z\{u(t - T)\} = z^{-1} \sum_{n=0}^{\infty} u(nT) z^{-n} = z^{-1} Z\{u(t)\}$$

$$= z^{-1} \frac{z}{z - 1} = \frac{1}{z - 1} \; .$$

Prove the following theorems, where $X(z) = Z[x(t)]$

1. $Z[a^k x(t)] = X(\frac{z}{a})$

2. $Z[e^{-at} x(t)] = X(ze^{at})$

3. $Z[tx(t)] = -Tz\frac{d}{dz}X(z)$

4. $Z[t] = \dfrac{Tz}{(z - 1)^2}$

<u>Solution</u>: Using the definition of the z-transform we obtain

1. $Z[a^k x(t)] = \sum\limits_{k=0}^{\infty} a^k x(kT) z^{-k} = \sum\limits_{k=0}^{\infty} x(kT) (\frac{z}{a})^{-k} = X(\frac{z}{a})$

2. $Z[e^{-at} x(t)] = \sum\limits_{k=0}^{\infty} e^{-akT} x(kT) z^{-k} = \sum\limits_{k=0}^{\infty} x(kT) (ze^{aT})^{-k}$

   $= X(ze^{aT})$

3. $Z[tx(t)] = \sum\limits_{k=0}^{\infty} kT x(kT) z^{-k} = -T \sum\limits_{k=0}^{\infty} x(kT) z \frac{d}{dz} (z^{-k})$

   $= -Tz \frac{d}{dz} [\sum\limits_{k=0}^{\infty} x(kT) z^{-k}] = -Tz \frac{d}{dz} X(z)$

4. Using

   $$Z[1(t)] = \frac{z}{z - 1}$$

   and the results obtained in (3) we write

   $$Z[t] = -Tz \frac{d}{dz} \{Z[1(t)]\} = -Tz \frac{d}{dz} (\frac{z}{z - 1}) = \frac{Tz}{(z - 1)^2}$$

● **PROBLEM** 4-9

Prove the following formulae

$$Z[\sum\limits_{k=0}^{n} x(k)] = \frac{z}{z - 1} X(z)$$

$$Z[\sum\limits_{k=0}^{n-1} x(k)] = \frac{1}{z - 1} X(z)$$

$$\sum\limits_{k=0}^{\infty} x(k) = \lim\limits_{z \to 1} X(z)$$

<u>Solution</u>: The z-transform of the first forward difference is

$$Z[\Delta f(k)] = Z[f(k + 1)] - Z[f(k)]$$

$$= [zF(z) - zf(0)] - F(z)$$

$$= (z - 1)F(z) - zf(0)$$

The term x(n) can be written

132

$$x(n) = \sum_{k=0}^{n} x(k) - \sum_{k=0}^{n-1} x(k) = \Delta \sum_{k=0}^{n-1} x(k)$$

and

$$Z[\Delta \sum_{k=0}^{n-1} x(k)] = Z[x(n)] = X(z) = (z - 1)Z[\sum_{k=0}^{n-1} x(k)]$$

where

$$\sum_{k=0}^{n-1} x(k) = 0 \quad \text{for } n = 0.$$

Therefore

$$Z[\sum_{k=0}^{n-1} x(k)] = \frac{X(z)}{z - 1}$$

and

$$Z[\sum_{k=0}^{n} x(k)] = Z[\sum_{k=0}^{n-1} x(k)] + Z[x(n)] = \frac{X(z)}{z - 1} + X(z)$$

$$Z[\sum_{k=0}^{n} x(k)] = \frac{z}{z - 1} X(z)$$

That proves the first two equations.

Using the final value theorem we find

$$\lim_{n\to\infty}[\sum_{k=0}^{n} x(k)] = \lim_{z\to 1}[(z - 1)\frac{z}{z - 1} X(z)]$$

thus

$$\sum_{k=0}^{\infty} x(k) = \lim_{z\to 1} X(z)$$

● **PROBLEM** 4-10

Find the z-transform of $f(t) = e^{-at}\sin\omega t$.

Solution: The z-transform of $\sin\omega t$ is

$$Z[\sin\omega t] = \frac{z\sin\omega t}{z^2 - 2z\cos\omega t + 1}$$

We shall use the following theorem to find the z-transform of f(t)

$$Z[e^{\mp at}f(t)] = F(ze^{\pm aT})$$

Replacing z by ze^{aT} we get

$$F(z) = \frac{ze^{aT}\sin\omega T}{z^2 e^{2aT} - 2ze^{aT}\cos\omega T + 1}$$

$$= \frac{ze^{-aT}\sin\omega T}{z^2 - 2ze^{-aT}\cos\omega T + e^{-2aT}}$$

● **PROBLEM 4-11**

Using the partial-differentiation theorem determine the z-transform of

$$f(t) = te^{-at}$$

Solution: The z-transform of te^{-at} is

$$Z[f(t)] = Z[te^{-at}] = Z[-\frac{\partial}{\partial a}e^{-at}]$$

Applying the partial-differentiation theorem we get

$$Z[-\frac{\partial}{\partial a}e^{-at}] = -\frac{\partial}{\partial a}Z[e^{-at}] = -\frac{\partial}{\partial a}[\frac{z}{z-e^{-at}}]$$

$$= \frac{Tze^{-at}}{(z-e^{-at})^2}$$

● **PROBLEM 4-12**

Determine the z-transform of

$$g(t) = e^{-at} \qquad \text{for } t \geq 0$$

Solution: By definition

$$Z\{f(t)\} = \sum_{k=0}^{\infty} f(kT)z^{-k}$$

Then for e^{-at}

$$Z\{e^{-at}\} = F(z) = \sum_{k=0}^{\infty} e^{-akT}z^{-k} = \sum_{k=0}^{\infty} (ze^{aT})^{-k}$$

This is an infinite geometric series, thus

$$F(z) = \frac{1}{1 - (ze^{aT})^{-1}} = \frac{z}{z - e^{-aT}}$$

It can be shown in general that

$$Z\{e^{-at}f(t)\} = F(e^{aT}z)$$

Indeed

$$Z\{e^{-at}f(t)\} = \sum_{k=0}^{\infty} f(kT)e^{-akT}z^{-k}$$

$$= \sum_{k=0}^{\infty} f(kT)(e^{aT}z)^{-k} = F(e^{aT}z).$$

● **PROBLEM 4-13**

Given the z-transform

$$F(z) = \frac{0.387\ z^2}{(z - 1)(z^2 - 2.37z + 0.25)}$$

determine the final value of $f(kT)$ by use of the final-value theorem.

Solution: The final-value theorem states that

$$\lim_{k \to \infty} f(kT) = \lim_{z \to 1}(1-z^{-1})F(z).$$

In our case we have

$$(1-z^{-1})F(z) = (1-z^{-1})\frac{0.387\ z^2}{(z-1)(z^2-2.37z+0.25)}$$

$$= \frac{z^{-1}(z-1)}{z-1}\ \frac{0.387\ z^2}{z^2-2.37z+0.25}.$$

Since $\dfrac{0.387z}{z^2-2.37z+0.25}$ does not have poles

on or outside the unit circle $|z| = 1$, the final-value theorem may be applied.

We have

$$\lim_{k \to \infty} f(kT) = \lim_{z \to 1} \frac{0.387z}{z^2-2.37z+0.25} = -0.345$$

135

INVERSE Z-TRANSFORMS AND RESPONSE OF SYSTEMS

The z-transform function is

$$R(z) = \frac{(1 - e^{-aT})z}{(z - 1)(z - e^{-aT})}$$

Find the inverse z-transform.

Solution: We shall expand $R(z)$ into partial fractions such that

$$R(z) = \frac{Az}{z - 1} + \frac{Bz}{z - e^{-aT}}$$

In order to use the conventional partial-fraction method, we expand $\frac{R(z)}{z}$, thus

$$\frac{R(z)}{z} = \frac{1}{z - 1} - \frac{1}{z - e^{-aT}}$$

and

$$R(z) = \frac{z}{z - 1} - \frac{z}{z - e^{-aT}}$$

The inverse z-transform of $R(z)$ is:

$$r(kT) = 1 - e^{-akT}$$

$X(z)$ is given by

$$X(z) = \frac{10z}{(z-1)(z-2)}$$

Using the inversion integral find $x(kT)$.

Solution: The inversion formula is

$$X(kT) = \frac{1}{2\pi j} \oint X(z) z^{k-1} dz$$

and

$$\frac{1}{2\pi j} \oint X(z) z^{k-1} dz = \sum_{k} (\text{residues of } X(z) z^{k-1} \text{ at poles of } X(z)).$$

In our problem we have

$$X(kT) = \frac{1}{2\pi j} \oint \frac{10z}{(z-1)(z-2)} z^{k-1} dz$$

It is easy to find that

$$\frac{1}{(z-1)(z-2)} = \frac{-1}{z-1} + \frac{1}{z-2}$$

X(kT) becomes

$$X(kT) = \frac{1}{2\pi j} \oint \left(\frac{10z^k}{z-2} - \frac{10z^k}{z-1} \right) dz$$

then, using the second equation of the inversion formula

$$X(kT) = (\text{residue of } \frac{10z^k}{z-2} \text{ at pole } z = 2)$$

$$+ (\text{residue of } - \frac{10z^k}{z-1} \text{ at pole } z = 1)$$

$$= 10(-1 + 2^k)$$

where $k = 0,1,2,\ldots$

● **PROBLEM 4-16**

Find x(kT), given

$$X(z) = \frac{2z(z^2-1)}{(z^2+1)^2}$$

Solution: Expansion of $\frac{X(z)}{z}$ into partial fractions gives

$$\frac{X(z)}{z} = \frac{1}{(z+j)^2} + \frac{0}{(z+j)} + \frac{1}{(z-j)^2} + \frac{0}{z-j}$$

thus

$$\frac{z}{(z+j)^2} + \frac{z}{(z-j)^2} = X(z)$$

In the Euler's formula

$$e^{j\theta} = \cos\theta + j\sin\theta$$

let us set $\theta = \frac{\pi}{2}$, then

$$j = e^{j\frac{\pi}{2}} \quad \text{and} \quad -j = e^{-j\frac{\pi}{2}}$$

X(z) can thus be written

$$X(z) = e^{j\frac{\pi}{2}} \frac{ze^{-j\frac{\pi}{2}}}{\left(z - e^{-j\frac{\pi}{2}}\right)^2} + e^{-j\frac{\pi}{2}} \frac{ze^{j\frac{\pi}{2}}}{\left(z - e^{j\frac{\pi}{2}}\right)^2}$$

Using the transform pair

$$\frac{Tze^{-aT}}{(z - e^{-aT})^2} \iff ke^{-ak}$$

we obtain the inverse z-transform of X(z)

$$x(kT) = e^{j\frac{\pi}{2}} ke^{-j\frac{\pi}{2}} + e^{-j\frac{\pi}{2}} ke^{j\frac{\pi}{2}k}$$

$$= jk[\cos\frac{\pi k}{2} - j\sin\frac{\pi k}{2}] - jk[\cos\frac{\pi k}{2} + j\sin\frac{\pi k}{2}]$$

$$= 2k\sin\frac{\pi k}{2} \qquad \text{for } k = 0,1,2,\ldots$$

● **PROBLEM 4-17**

Find the inverse z-transform of

$$F(z) = \frac{(1 - e^{-aT})z}{z^2 - (1+e^{-aT})z + e^{-aT}}$$

Solution: The division of the numerator by the denominator yields

$$F(z) = (1-e^{-aT})z^{-1} + (1-e^{-2aT})z^{-2} + (1-e^{-3aT})z^{-3}$$

$$+ \ldots$$

We see that

$$f(kT) = 1 - e^{-akT} \qquad k = 0,1,2,\ldots$$

and

$$f^*(t) = \sum_{k=0}^{\infty} (1 - e^{-akT})\delta(t-kT)$$

To solve the above problem we can also use the partial fraction expansion, thus

$$F(z) = \frac{(1-e^{-aT})z}{(z-e^{-aT})(z-1)} = \frac{z}{z-1} - \frac{z}{z-e^{-aT}}$$

138

$$= \frac{1}{1-z^{-1}} - \frac{1}{1-e^{-aT}z^{-1}}$$

The inverse transform of F(z) yields

$$f(kT) = 1 - e^{-akT}$$

which agrees with the previous result.

Determine the inverse z-transform of

$$F(z) = \frac{(1-e^{-aT})z}{(z-1)(ze^{-aT})} \tag{1}$$

using the inversion integral

$$f(kT) = \frac{1}{2\pi j} \oint_\Gamma F(z)z^{k-1}dz \tag{2}$$

Solution: Substituting (1) in equation (2) we obtain

$$f(kT) = \frac{1}{2\pi j} \oint_\Gamma \frac{z(1-e^{-aT})}{(z-1)(z-e^{-aT})} z^{k-1}dz \tag{3}$$

where Γ is a circle, large enough to enclose both poles of F(z) at z = 1 and z = e^{-aT}. Using the residue theorem we can evaluate the contour integral

$$f(kT) = \Sigma \text{ residue of } F(z)z^{k-1} \text{ at the poles of } F(z)$$

Thus equation (3) becomes

$$f(kT) = \Sigma \text{ residues of } \frac{z(1-e^{-aT})z^{k-1}}{(z-1)(z-e^{-aT})} \text{ at } z=1, \ z=e^{-aT}$$

$$= \frac{z^k(1-e^{-aT})}{z-e^{-aT}}\bigg|_{z=1} + \frac{z^k(1-e^{-aT})}{z-1}\bigg|_{z=e^{-aT}}$$

$$= 1^k - e^{-akT} = 1 - e^{-akT} \ .$$

Find the response of the system

$$x(k+2) - 3x(k+1) + 2x(k) = u(k) \qquad (1)$$

where $u(0) = 1$

$\qquad\qquad u(k) = 0 \qquad$ for $k \neq 0$

$\qquad\qquad x(k) = 0 \qquad$ for $k \leq 0$.

Solution: For $k = -1$ we obtain

$$x(1) = 0$$

Taking the z-transform of (1) with the initial data

$$x(0) = x(1) = 0$$

we get

$$(z^2 - 3z + 2)X(z) = U(z)$$

Where $U(z)$ denotes the z-transform of $u(k)$ thus

$$U(z) = \sum_{k=0}^{\infty} u(k)z^{-k} = 1$$

$X(z)$ becomes

$$(z^2 - 3z + 2)X(z) = 1$$

$$X(z) = \frac{1}{z^2 - 3z + 2} = \frac{1}{(z-1)(z-2)} = \frac{-1}{z-1} + \frac{1}{z-2}$$

Remembering that

$$Z[x(k+1)] = zX(z) - zx(0)$$

we obtain for $x(0) = 0$

$$Z[x(k+1)] = zX(z) = -\frac{z}{z-1} + \frac{z}{z-2}$$

Since

$$Z[1^k] = \frac{z}{z-1}$$

and $\qquad Z[2^k] = \frac{z}{z-2}$

$x(k+1)$ is equal to

$$x(k+1) = -1 + 2^k \qquad \text{for } k = 0,1,2,\ldots$$

and $\qquad x(k) = -1 + 2^{k-1} \qquad \text{for } k = 1,2,3,\ldots$

Find the solution of the following difference equation in terms of $x(0)$ and $x(1)$

$$x(k+2) + Ax(k+1) + Bx(k) = 0$$

A and B are constants and $k \geq 0$.

Solution: Taking the z-transforms of $x(k+2)$, $x(k+1)$ and $x(k)$, we obtain

$$Z[x(k+2)] = z^2 X(z) - z^2 x(0) - zx(1)$$

$$Z[x(k+1)] = zX(z) - zx(0)$$

$$Z[x(k)] = X(z)$$

Substituting into the initial equation and solving for $X(z)$ we get

$$[z^2 X(z) - z^2 x(0) - zx(1)] + A[zX(z) - zx(0)]$$
$$+ BX(z) = 0$$

or $(z^2 + Az + B)X(z) = (z^2 + Az)x(0) + zx(1)$

$$X(z) = \frac{(z^2 + Az)x(0) + zx(1)}{z^2 + Az + B}$$

The roots of $z^2 + Az + B = 0$ are

$$z_1 = \frac{-A + \sqrt{A^2 - 4B}}{2}$$

$$z_2 = \frac{-A - \sqrt{A^2 - 4B}}{2}$$

thus

$$z^2 + Az + B = (z - z_1)(z - z_2)$$

The partial fractions expansion of $\frac{X(z)}{z}$ is

$$\frac{X(z)}{z} = \frac{(z_1 + A)x(0) + x(1)}{z_1 - z_2} \cdot \frac{1}{z - z_1}$$

$$+ \frac{(z_2 + A)x(0) + x(1)}{z_2 - z_1} \cdot \frac{1}{z - z_2} \quad -z_1 - z_2 = A$$

thus

$$X(z) = \frac{z_2 x(0) - x(1)}{z_2 - z_1} \cdot \frac{z}{z - z_1} + \frac{z_1 x(0) - x(1)}{z_1 - z_2}$$

$$\cdot \frac{z}{z - z_2}$$

141

The inverse z-transform of $X(z)$ is

$$x(k) = \frac{z_2 x(0) - x(1)}{z_2 - z_1} z_1^k + \frac{z_1 x(0) - x(1)}{z_1 - z_2} z_2^k$$

$$k = 0,1,2,\ldots$$

for $z_1 \neq z_2$.

The above equation is the solution of

$$x(k+2) + Ax(k+1) + Bx(k) = 0$$

where $A^2 \neq 4B$.

If $A^2 = 4B$ we get

$$z_1 = z_2 \quad \text{and} \quad A = -2z_1 \quad B = z_1^2, \quad \text{then}$$

$$X(z) = \frac{(z^2 - 2z_1 z)x(0) + zx(1)}{(z - z_1)^2}$$

expanding by partial fractions

$$\frac{X(z)}{z} = \frac{x(1) - z_1 x(0)}{(z - z_1)^2} + \frac{x(0)}{z - z_1}$$

And

$$X(z) = [x(1) - z_1 x(0)] \frac{z}{(z-z_1)^2} + x(0) \frac{z}{z-z_1}$$

Using the definition of the z-transform

$$Z\{ta^t\} = \sum_{k=0}^{\infty} kTa^{kT}z^{-k}$$

$$= T[0 + (a^T z^{-1}) + 2(a^T z^{-1})^2 + 3(a^T z^{-1})^3 + \ldots]$$

Dividing each side by $a^T z^{-1}$:

$$a^{-T} z Z\{ta^t\} = T[1 + 2(a^T z^{-1}) + 3(a^T z^{-1})^2 + 4(a^T z^{-1})^3 + \ldots]$$

Subtracting the equations, we have

$$Z\{ta^t\}(a^{-T}z - 1) = T[1 + (a^T z^{-1}) + (a^T z^{-1})^2 + (a^T z^{-1})^3 + \ldots]$$

$$= \frac{T}{1 - a^T z^{-1}} = \frac{Tz}{z - a^T}$$

Thus

$$Z\{ta^t\} = \frac{1}{za^{-T} - 1} \cdot \frac{Tz}{z - a^T} = \frac{Ta^T z}{(z - a^T)^2}$$

Substituting the above into x(z) and setting $z_1 = A^T$ we get

$$X(k) = \frac{1}{z_1 T} [x(1) - z_1 x(0)](kTz_1^k) + x(0)z_1^k$$

$$= \{[\frac{X(1)}{z_1} - x(0)]k + x(0)\}z_1^k .$$

● **PROBLEM** 4-21

For the circuit shown derive the difference equation describing the system dynamics.

The input voltage e(t) is piecewise constant

$$e(t) = e(kT) \quad \text{for } kT \leq t \leq (k+1)T$$

Solution: From the Kirchoff's law we have

$$RC\dot{x}(t) + x(t) = e(kT) \quad \text{for } kT \leq t < (k+1)T$$

The Laplace transform of the above equation for t = kT to be the initial time, is

$$£[RC\dot{x}(t) + x(t)] = £[e(kT)]$$

Using

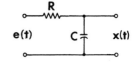

$$£[\frac{d}{dt}x(t)] = sX(s) - x(0)$$

we obtain

$$RC[sX(s) - x(0)] + X(s) = \frac{e(kT)}{s}$$

Since the initial time is t = kT

$$RC[sX(s) - x(kT)] + X(s) = \frac{e(kT)}{s}$$

Solving for X(s) gives

$$X(s) = \frac{e(kT)}{s} + \frac{x(kT) - e(kT)}{s + \frac{1}{RC}}$$

Using the transform pairs

$$F(s) = \frac{1}{s} \iff \text{unit step} = 1(t)$$

$$F(s) = \frac{k}{s+a} \iff ke^{-at} \quad k, \text{ a real constants}$$

143

we have for the inverse Laplace transform of X(s)

$$x(t) = e(kT) + [x(kT) - e(kT)]e^{-(t-kT)/RC}$$

for $kT \leq t < (k+1)T$.

Substituting $t = (k+1)T$, we obtain

$$x((k+1)T) = e^{-\frac{T}{RC}} x(kT) + (1 - e^{-\frac{T}{RC}})e(kT)$$

● **PROBLEM** 4-22

Find the pulse response of the system shown in Fig. 1.
Re-form the system as a parallel combination of two
first-order systems.

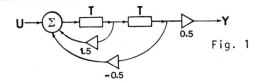

Fig. 1

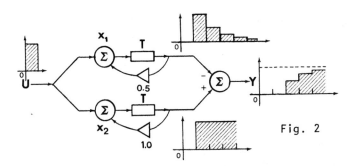

Fig. 2

Solution: We have at the summing point

$$x(k) = u(k) + 1.5x(k-1) - 0.5x(k-2) \qquad (1)$$

and the output

$$y(k) = 0.5x(k-2) \qquad (2)$$

The z-transforms of equations (1) and (2) are

$$x = u + 1.5z^{-1}x - 0.5 z^{-2}x \qquad (3)$$

$$y = 0.5z^{-2}x \qquad (4)$$

From (3) we get

$$x = \frac{u}{1 + 0.5z^{-2} - 1.5z^{-1}}$$

Substituting into (4)

$$y = \frac{0.5z^{-2}u}{0.5z^{-2} - 1.5z^{-1} + 1} = \frac{0.5u}{z^2 - 1.5z + 0.5} \qquad (5)$$

We have for a pulse $u(z) = 1$, thus

$$h(z) = \frac{0.5}{z^2 - 1.5z + 0.5} = \frac{0.5z}{z(z-0.5)(z-1)}$$

$$= 1 + \frac{z}{z - 1} - \frac{2z}{z - 0.5} \qquad (6)$$

Taking the inverse transforms

$$h(k) = \delta(k) + 1 - 2(0.5)^k \qquad (7)$$

We shall break the system shown in the Fig. 1 down to a parallel combination of first-order systems.

We have

$$x_1(k) = a_1 x_1(k-1) + u(k) \qquad (8)$$

$$x_2(k) = a_2 x_2(k-1) + u(k) \qquad (9)$$

$$y(k) = c_1 x_1(k-1) + c_2 x_2(k-1) \qquad (10)$$

Taking the z-transforms and rearranging we get

$$x_1 = \frac{zu}{z - a_1}$$

$$x_2 = \frac{zu}{z - a_2} \qquad (11)$$

$$y = \left(\frac{c_1}{z - a_1} + \frac{c_2}{z - a_2} \right) u$$

Equation (5) is of the above form,

$$y = \left(\frac{1}{z - 1} - \frac{1}{z - 0.5} \right) u$$

● **PROBLEM** 4-23

Obtain the output $C(z)_N$, of the multirate system shown below where

$$G(s) = \frac{1}{s(s+1)}$$

The input is a unit step function, and $\dot{T}_1 = 1$ sec, $T_2 = \frac{1}{3}$ sec.

<u>Solution:</u> To obtain the modified z-transform of G(s), let us expand G(s) in partial fractions.

INPUT ——→ ✕ —— → $\boxed{G(s)}$ —— → ✕ ——→ OUTPUT
T_1 T_2

$$G(s) = \frac{1}{s(s+1)} = \frac{1}{s} - \frac{1}{s+1}$$

Taking the inverse Laplace transform of G(s)

$$g(t) = \mu(t) - e^{-t}$$

The modified z-transform of g(t) gives

$$G(z,m) = z^{-1} \sum_{k=0}^{\infty} \mu(k+m) T z^{-k} - z^{-1} \sum_{k=0}^{\infty} e^{-(k+m)T} z^{-k}$$

and

$$z^{-1} \sum_{k=0}^{\infty} \mu(k+m) T z^{-k} = z^{-1} \sum_{k=0}^{\infty} T z^{-k} = G_1(z,m)$$

$$= z^{-1} [T + Tz^{-1} + Tz^{-2} + Tz^{-3} + \ldots]$$

$$= Tz^{-1} + Tz^{-2} + Tz^{-3} + \ldots$$

Then

$$z^{-1} G_1(z,m) = Tz^{-2} + Tz^{-3} + Tz^{-4} + \ldots$$

Subtracting the equations we get

$$(1 - z^{-1}) G_1(z,m) = Tz^{-1}$$

$$G_1(z,m) = \frac{Tz^{-1}}{1 - z^{-1}} = \frac{T}{z - 1}$$

For $G_2(z,m)$ we obtain

$$G_2(z,m) = z^{-1} \sum_{k-0}^{\infty} e^{-(k+m)T} z^{-k} = z^{-1} e^{-mT} \sum_{k=0}^{\infty} e^{-kT} z^{-k}$$

$$= z^{-1} e^{-mT} (1 + e^{-T} z^{-1} + e^{-2T} z^{-2} + e^{-3T} z^{-3} + \ldots)$$

$$e^{-T} z^{-1} G_2(z,m) = z^{-1} e^{-mT} (e^{-T} z^{-1} + e^{-2T} z^{-2} + \ldots)$$

Subtracting the above equations we get

$$(1 - e^{-T} z^{-1}) G_2(z,m) = z^{-1} e^{-mT}$$

$$G_2(z,m) = \frac{z^{-1} e^{-mT}}{1 - e^{-T} z^{-1}} = \frac{e^{-mT}}{z - e^{-T}}$$

146

Thus

$$G(z,m) = G_1(z,m) - G_2(z,m) = \frac{T}{z-1} - \frac{e^{-mT}}{z-e^{-T}}$$

Thus for T = 1 sec.

$$G(z,m) = \frac{1}{z-1} - \frac{e^{-m}}{z-e^{-1}}$$

Using

$$C(z)_N = R(z) \sum_{k=0}^{N-1} z^{1-\frac{k}{N}} G(z,\frac{k}{N})$$

for $T_1 = 3T_2$ and $N = 3$.

The z-transform of the output is

$$C(z)_3 = \sum_{k=0}^{2} z^{1-\frac{k}{3}} \left[\frac{1}{z-1} - \frac{e^{-\frac{k}{3}}}{z-0.368} \right] \frac{z}{z-1}$$

We should note that the above equations contains powers z, $z^{\frac{1}{3}}$ and $z^{\frac{2}{3}}$, introducing the variable z_3 where

$$z = z_3^3$$

we get $z^{\frac{1}{3}} = z_3$ and $z^{\frac{2}{3}} = z_3^2$.

Expanding $C(z)_3$ into a power series in z_3^{-1} we obtain the response of c(t) at the fast-rate sampling instants where

$$t = \frac{kT}{3} \qquad k = 0,1,2,\ldots$$

● PROBLEM 4-24

Obtain the z-transform of the output of the system C(z), shown below where

$$R(s) = \frac{1}{s} \qquad G(s) = \frac{1}{s+1}$$

Fast-slow multirate sampled system.

Solution: The inverse Laplace transforms of R(s) and G(s) are

$$r(t) = \mu(t) \qquad \text{and} \qquad g(t) = e^{-t}$$

Then

147

$$R(z)_N = \frac{z_N}{z_N - 1}$$

and
$$G(z)_N = \frac{z_N}{z_N - e^{-\frac{T}{N}}}$$

Then

$$C(z)_N = G(z)_N R(z)_N = \frac{z_N^2}{(z_N - 1)(z_N - e^{-\frac{T}{N}})}$$

From the equation

$$C(z) = \Sigma \ (\text{residues of } C(z)_N \ \cdot \ \frac{z_N^{-1}}{1 - z_N^N z^{-1}} \qquad \text{at}$$

$$\text{poles of } C(z)_N z_N^{-1})$$

we find the z-transform of the output of the system $C(z)$

$$C(z) = \Sigma \ \text{residues of} \ \frac{z_N^2}{(z_N - 1)(z_N - e^{-\frac{T}{N}})}$$

$$\cdot \ \frac{z_N^{-1}}{(1 - z_N^N z^{-1})} \qquad \text{at}$$

$$z_N = 1, \ z_N = e^{-\frac{T}{N}} = \frac{1}{1 - e^{-\frac{T}{N}}} \left[\frac{z}{z-1} - \frac{ze^{-\frac{T}{N}}}{z - e^{-T}} \right].$$

• **PROBLEM 4-25**

Determine the output of the system shown in Fig. 1, by means of the modified z-transform.

The function G(s) is

$$G(s) = \frac{1}{s + a}$$

where a is a constant. The input of the system is a unit-step function; i.e., $e(t) = u_s(t)$.

Solution: The modified z-transform of the output is

$$Z_m[C(s)] = C(z,m) = Z_m[G(s)E^*(s)] = G(z,m)E(z).$$

148

The modified z-transform of G(s) is

$$G(z,m) = z^{-1} \sum_{k=0}^{\infty} G[(k+m)T] z^{-k}$$

since

$$G(s) = \frac{1}{s+a} \quad \text{and} \quad g(t) = e^{-at}$$

then

$$G(z,m) = z^{-1} \sum_{k=0}^{\infty} e^{-a(k+m)T} z^{-k}$$

$$= z^{-1} e^{-amT} \sum_{k=0}^{\infty} e^{-akT} z^{-k} \ .$$

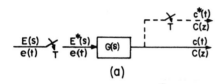

(a)

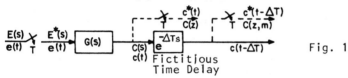

Fictitious
Time Delay

Fig. 1

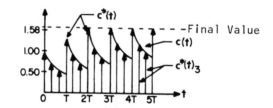

Fig. 2

We know that

$$\sum_{k=0}^{\infty} e^{-akT} z^{-k} = \frac{z}{z - e^{-aT}}$$

and G(z,m) becomes

$$G(z,m) = z^{-1} e^{-amT} \frac{z}{z - e^{-aT}} = \frac{e^{-amT}}{z - e^{-aT}}$$

The input of the system is a unit-step function

$$e(t) = u_s(t)$$

then

$$E(z) = \frac{z}{z - 1}$$

Thus

$$C(z,m) = G(z,m) E(z) = \frac{e^{-maT}}{z - e^{-aT}} \frac{z}{z - 1}$$

149

CHAPTER 5

TRANSFER FUNCTION AND BLOCK DIAGRAMS

TRANSFER FUNCTIONS FROM BLOCK DIAGRAMS

Obtain the transfer function of the system, whose block diagram is shown below.

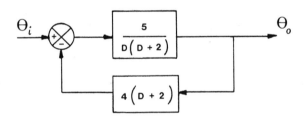

Solution: The transfer function of the system is

$$\frac{\theta_o}{\theta_i} = \frac{G}{1 + GH}$$

where

$$G = \frac{5}{D(D+2)}$$

and $H = 4(D+2)$

thus

$$\frac{\theta_o}{\theta_i} = \frac{\dfrac{5}{D(D+2)}}{1 + \dfrac{5 \cdot 4(D+2)}{D(D+2)}} = \frac{\dfrac{5}{D(D+2)}}{\dfrac{D+20}{D}} = \frac{5}{D(D+2)} \cdot \frac{D}{D+20}$$

$$= \frac{5}{(D+2)(D+20)}$$

150

Find the transfer function $\frac{C(s)}{R(s)}$ for the block diagram given in the figure.

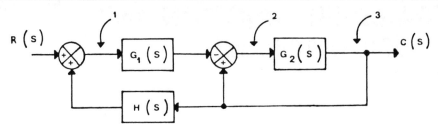

Solution: For points 1,2,3 we have

$$R(s) + H(s)C(s)$$

$$C(s) - G_1(s)[R(s) + H(s)C(s)]$$

$$G_2(s)\{C(s) - G_1(s)[R(s) + H(s)C(s)]\} = C(s)$$

Multiplying and solving the equation we find the transfer function $\frac{C(s)}{R(s)}$ of the system

$$G_2(s)C(s) - G_1(s)G_2(s)R(s) + G_2(s)G_1(s)H(s)C(s) = C(s)$$

$$C(s)[G_2(s) + G_1(s)G_2(s)H(s) - 1] = G_1(s)G_2(s)R(s)$$

$$\frac{C(s)}{R(s)} = \frac{G_1(s)G_2(s)}{G_2(s)[1 + G_1(s)H(s)] - 1}$$

● PROBLEM 5-3

Show that the input-output transfer function for both block diagrams is the same.

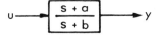

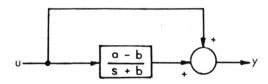

Solution: For the first diagram we have the following
relationship between u and y

$$\frac{s+a}{s+b} u = y$$

Thus the transfer function is

$$\frac{y}{u} = \frac{s+a}{s+b}$$

For the second diagram

$$y = u + u \frac{a-b}{s+b}$$

Thus the transfer function is

$$\frac{y}{u} = 1 + \frac{a-b}{s+b} = \frac{s+b+a-b}{s+b} = \frac{s+a}{s+b}$$

● **PROBLEM 5-4**

The system shown below has two input signals, one the
reference input R(s) and the other the disturbance
input K(s). Show that the characteristic equation of
the system does not change when the input R(s) is replaced
by K(s).

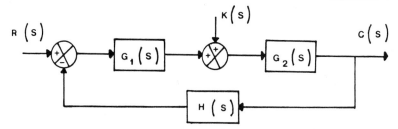

Solution: If we neglect the disturbance input K(s),
then the transfer function of the system is

$$\frac{C(s)}{R(s)} = \frac{G_1(s)G_2(s)}{1 + G_1(s)G_2(s)H(s)}$$

The transfer function relating K(s) and C(s) in the
absence of the reference input R(s) is

$$\frac{C(s)}{K(s)} = \frac{G_2}{1 + G_1 G_2 H}$$

The denominators of both equations are the same, thus the
characteristic equation for both is

$$1 + G_1(s)G_2(s)H(s) = 0$$

For a given system there is only one characteristic
equation. Regardless of which signal we choose as input
we obtain the same characteristic equation.

Find the transfer function of the system, whose block diagram is shown.

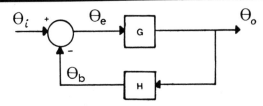

Solution: From the diagram we have the following equations:

$$\frac{\theta_o}{\theta_e} = G$$

$$\frac{\theta_b}{\theta_o} = H$$

$$\theta_i - \theta_b = \theta_e$$

Combining the equations we get

$$\theta_b = \theta_o H$$

$$\theta_i - \theta_o H = \frac{\theta_o}{G}$$

The transfer function $\frac{\theta_o(s)}{\theta_i(s)}$ can be computed from the last equation, thus

$$\theta_i = \theta_o H + \frac{\theta_o}{G} = (H + \frac{1}{G}) \theta_o$$

or

$$\frac{\theta_o}{\theta_i} = \frac{1}{H + \frac{1}{G}} = \frac{G}{1 + GH} \cdot$$

Determine the closed-loop transfer function $\frac{y(s)}{r(s)}$ for the system whose block diagram is shown below.

Solution: Let us denote on the block diagram x_1, x_2, x_3, x_4. From the diagram we see that

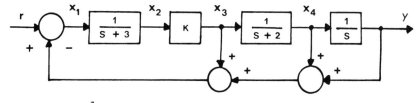

$$x_1 \cdot \frac{1}{s+3} = x_2$$

$$x_2 \cdot K = x_3$$

$$x_3 \cdot \frac{1}{s+2} = x_4$$

$$x_4 \frac{1}{s} = y$$

$$x_1 = r - (x_4 + y + x_3)$$

Combining the equations we obtain

$$\frac{1}{K} sy(s+2)(s+3) + sy + y + sy(s+2) = r$$

Thus the transfer function is

$$\frac{y(s)}{r(s)} = \frac{K}{s(s+2)(s+3) + Ks + K + Ks(s+2)}$$

$$= \frac{K}{s(s+2)(s+3+K) + K(s+1)}$$

● **PROBLEM** 5-7

Find the transfer functions $\dfrac{\theta_o}{\theta_i}$ of the systems, whose block diagrams are shown in Figs. 1 through 4.

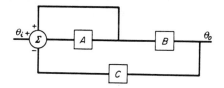

Fig. 1

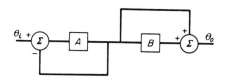

Fig. 2

154

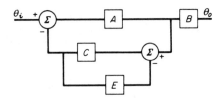

Fig. 3

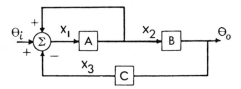

Fig. 4

Solution: Fig. 1: Let us denote x_1, x_2, x_3 as shown.

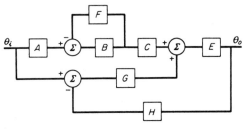

From the above diagram we have

$$\theta_o = Bx_2$$

$$x_3 = \theta_o C$$

$$x_1 = \theta_i - x_3 + x_2$$

$$x_2 = x_1 A.$$

Combining all four equations we have

$$x_1 = \theta_i - x_3 + x_2$$

$$x_1 = \frac{x_2}{A} = \frac{\theta_o}{AB}$$

$$\frac{\theta_o}{AB} = \theta_i - \theta_o C + \frac{\theta_o}{B}$$

$$\frac{\theta_o}{\theta_i} = \frac{AB}{1 - A + ABC}$$

In the same manner we find for Fig. 2

155

$$\frac{\theta_o}{\theta_i} = \frac{A(1+B)}{1+A}$$

for Fig. 3

$$\frac{\theta_o}{\theta_i} = \frac{AB(1+EC)}{1+C(A+E)}$$

for Fig. 4

$$\frac{\theta_o}{\theta_i} = \frac{ABCE + EG + BEFG}{(1+BF)(1+EGH)}$$

● **PROBLEM** 5-8

From the block diagram find the transfer function $\frac{C(s)}{R(s)}$.

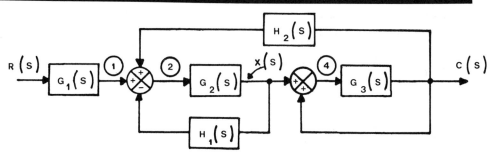

Solution: Let us define the output of $G_2(s)$ as $X(s)$.
At point (1) we have

$$R(s)G_1(s)$$

at point (2)

$$RG_1 + CH_2 - XH_1$$

and for $X(s)$

$$X = G_2 [RG_1 + CH_2 - XH_1]$$

$$X = \frac{G_1G_2R + CG_2H_2}{1 + H_1G_2}$$

at point (4)

$$X + C$$

for the output $C(s)$ we obtain

$$C = G_3(C+X)$$

$$C = \frac{G_3G_1G_2R + G_3G_2H_2C}{1 + H_1G_2} + G_3C$$

156

$$C(1 + H_1G_2) - G_3C(1 + H_1G_2) = G_1G_2G_3R + CG_2G_3H_2$$

thus

$$\frac{C(s)}{R(s)} = \frac{G_1G_2G_3}{(1 + H_1G_2)(1 - G_3) - G_2G_3H_2}$$

● **PROBLEM** 5-9

Determine the overall transfer function for the system shown.

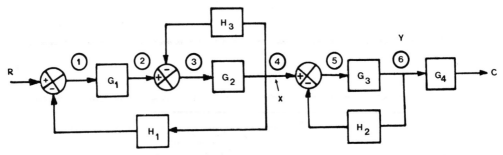

Solution: Let us denote the signal at point (4) by X. For (1) we have $R - H_1X$

for (2) $G_1[R - H_1X]$

for (3) $G_1(R - H_1X) - H_3X$

for (4) $X = G_2[G_1R - G_1H_1X - H_3X]$

thus

$$X = \frac{G_1G_2R}{1 + G_1G_2H_1 + G_2H_3}$$

Let us denote the signal at point (6) by Y. We have

for (5) $X - H_2Y$

for (6) $Y = G_3[X - H_2Y]$.

Thus

$$Y = \frac{G_3X}{1 + H_2G_3}$$

Finally
$$C = G_4Y = G_4 \cdot \frac{G_3}{1 + H_2G_3} \cdot X$$

$$= \frac{G_3G_4}{1 + H_2G_3} \cdot \frac{G_1G_2}{1 + G_1G_2H_1 + G_2H_3} \cdot R$$

The transfer function is

$$\frac{C}{R} = \frac{G_1G_2G_3G_4}{(1 + H_2G_3)(1 + G_1G_2H_1 + G_2H_3)}$$

TRANSFER FUNCTIONS OF NETWORKS AND SYSTEMS

Find the transfer functions of the linear time-invariant systems described by the following equations.

a) $\dfrac{dx}{dt} = \dfrac{dy}{dt} + 2y$

b) $\dfrac{dy}{dt} + 3y = \dfrac{dx}{dt} + 4x$

c) $3\dfrac{d^2y}{dt^2} + 5\dfrac{dy}{dt} - 2y = x$

d) $\dfrac{dy}{dt} = \dfrac{dx}{dt} + x(t-a)$ where $x(t-a) = 0$ for $t < a$

where $x(t)$ is the input variable and $y(t)$ is the output variable. Assume initial conditions are zero.

Solution: a) The Laplace transformation of the differential equation is

$$X(s) = sY(s) + 2Y(s) = Y(s)(s+2)$$

Therefore the transfer function is

$$G(s) = \frac{Y(s)}{X(s)} = \frac{1}{s+2}$$

b) The Laplace transform is

$$sY(s) + 3Y(s) = sX(s) + 4X(s)$$
or
$$Y(s)(s+3) = X(s)(s+4)$$

and the transfer function is

$$G(s) = \frac{Y(s)}{X(s)} = \frac{s+4}{s+3}$$

c) In this case the Laplace transform is

$$(3s^2 + 5s - 2)Y(s) = X(s)$$

The transfer function is

$$G(s) = \frac{Y(s)}{X(s)} = \frac{1}{3s^2 + 5s - 2} = \frac{1}{(3s-1)(s+2)}$$

d) The Laplace transform is

$$sY(s) = sX(s) + e^{-as}X(s)$$

and the transfer function is

$$G(s) = \frac{Y(s)}{X(s)} = \frac{s + e^{-as}}{s} = 1 + \frac{1}{se^{as}}$$

● **PROBLEM** 5-11

Find out if the systems whose transfer functions are given below are equivalent.

1) $\dfrac{X(s)}{Y(s)} = \dfrac{s+\gamma}{(s+\alpha)(s+\gamma)}$

2) $\dfrac{X(s)}{Y(s)} = \dfrac{s+\beta}{(s+\alpha)(s+\beta)}$ where $\gamma \neq \beta$

Solution: If we cancel the common terms in the numerator and denominator, the two systems will have the same transfer function, but their basis functions are different. The basis functions of the first system are $\frac{1}{s+\alpha}$ and $\frac{1}{s+\gamma}$ of the second $\frac{1}{s+\alpha}$ and $\frac{1}{s+\beta}$. Two systems are equivalent if their transfer functions are the same and if the basis functions of one system are linear combinations of the basis functions of the other system and vice-versa. The systems described in our problem don't meet the second condition, thus they are not equivalent.

● **PROBLEM** 5-12

Determine the transfer function of the spring-mass system shown below. The displacement x is the input and displacement y the output of the system.

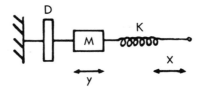

Solution: Assume that positive displacement is to the left. The spring will exert force equal to

$$F_s = K(x-y)$$

The damper will exert force

$$F_D = -D\frac{dy}{dt}$$

Using Newton's Law for the sum of the forces on M, we have

$$\Sigma F = Ma = M \frac{d^2 y}{dt^2} = K(x-y) - D \frac{dy}{dt}$$

or

$$M \frac{d^2 y}{dt^2} + D \frac{dy}{dt} + K(y-x) = 0$$

Transforming the equation and assuming zero initial conditions, we have

$$Ms^2 Y(s) + DsY(s) + KY(s) = KX(s)$$

The transfer function is

$$\frac{Y(s)}{X(s)} = \frac{K}{Ms^2 + Ds + K}$$

● **PROBLEM 5-13**

Obtain the transfer function of the system shown in the figure.

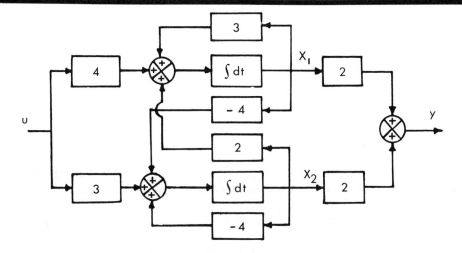

Solution: From the diagram we obtain the following equations

$$\int (4u + 3x_1 + 2x_2) dt = x_1$$

$$\int (3u - 4x_1 - 4x_2) dt = x_2$$

$$2x_1 + 2x_2 = y$$

Differentiating the first and the second equation we obtain

$$\dot{x}_1 = 3x_1 + 2x_2 + 4u$$

160

$$\dot{x}_2 = -4x_1 - 4x_2 + 3u$$

$$y = 2x_1 + 2x_2$$

In vector-matrix form, we have

$$\begin{bmatrix} \dot{x}_1 \\ \dot{x}_2 \end{bmatrix} = \begin{bmatrix} 3 & 2 \\ -4 & -4 \end{bmatrix} \begin{bmatrix} x_1 \\ x_2 \end{bmatrix} + \begin{bmatrix} 4 \\ 3 \end{bmatrix} (1)$$

$$y = [2 \quad 2] \begin{pmatrix} x_1 \\ x_2 \end{pmatrix}$$

The transfer function of the system is given by

$$G(s) = C(sI - A)^{-1}B$$

$$= [2 \quad 2] \begin{bmatrix} s-3 & -2 \\ 4 & s+4 \end{bmatrix}^{-1} \begin{bmatrix} 4 \\ 3 \end{bmatrix}$$

$$= [2 \quad 2] \frac{1}{detA} \begin{bmatrix} s+4 & 2 \\ -4 & s-3 \end{bmatrix} \begin{bmatrix} 4 \\ 3 \end{bmatrix}$$

$$= \frac{1}{detA} \cdot [2(s+4) - 8, \ 4+2(s-3)] \begin{bmatrix} 4 \\ 3 \end{bmatrix}$$

$$= \frac{1}{detA} [2s, \ 2s-2] \begin{bmatrix} 4 \\ 3 \end{bmatrix} = \frac{8s + 6s - 6}{detA}$$

$$= \frac{14s - 6}{s^2 + s - 4}$$

● **PROBLEM** 5-14

Compare the sensitivity of the output C to variations in system parameters for the open-loop and feedback control systems Fig. 1, 2.

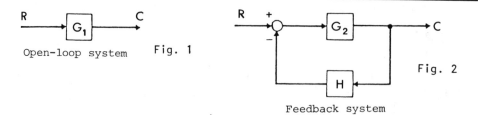

Open-loop system   Fig. 1

Fig. 2

Feedback system

Solution: Let us investigate first the open-loop case. We have

$$C = G_1R \qquad \text{and} \qquad \frac{\partial C}{\partial G_1} = R$$

161

A change δG_1 produces a change in the output

$$\delta C = R \delta G_1$$

The sensitivity S of the system is defined

$$S = \frac{\% \text{ change in } \frac{C}{R}}{\% \text{ change in the process transfer function}}$$

% indicates the percentage.

For the open-loop system

$$S = 1 = \frac{\delta G_1 / G_1}{\delta G_1 / G_1}$$

For the closed-loop feedback system

$$C = \frac{G_2 R}{1 + G_2 H}$$

and

$$\delta G_2 \frac{\partial C}{\partial G_2} = \partial G_2 \cdot \frac{R}{1 + G_2 H} - \frac{G_2 RH}{(1 + G_2 H)^2} \delta G_2 = \frac{R \delta G_2}{(1 + G_2 H)^2}$$

The sensitivity is

$$S = \frac{\delta \left(\frac{C}{R}\right)}{\frac{C}{R}} \frac{G_2}{\delta G_2} = \frac{\delta G_2}{(1 + G_2 H)^2} \frac{(1 + G_2 H)}{G_2} \frac{G_2}{\delta G_2} = \frac{1}{1 + G_2 H}$$

Note that

$$\lim_{G_2 \to \infty} S = 0$$

● **PROBLEM 5-15**

The air heating system is shown in Fig. 1. Draw the block diagram of the system considering small deviations from steady-state operations. Neglect the heat loss and the heat capacitance of the metal parts of the heater.

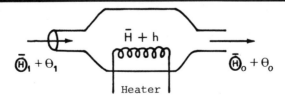

Heater

Solution: Let us assume that the heat input changes from $\bar{H}$ to $\bar{H} + h$, the inlet air temperature changes from $(\bar{H})_1$ to $(\bar{H})_1 + \theta_1$ and the outlet air temperature from $(\bar{H})_o$ to $(\bar{H})_o + \theta_o$. The following equation describes the system

$$\frac{d\theta_o}{dt} = \frac{h + Gc(\theta_1 - \theta_o)}{C}$$

where

G = flow rate of air

c = specific heat of air

R = thermal resistance

C = thermal capacitance

Using

$$RGc = 1$$

we get

$$C\frac{d\theta_o}{dt} = h + \frac{1}{R}(\theta_1 - \theta_o)$$

or

$$RC\frac{d\theta_o}{dt} + \theta_o = Rh + \theta_1$$

Taking the Laplace transform of both sides with the initial condition $\theta_o(0) = 0$, we obtain

$$\text{(H)}_o(s) = \frac{R}{RCs + 1}H(s) + \frac{1}{RCs + 1}\text{(H)}_1(s)$$

The block diagram representing the above relationship is

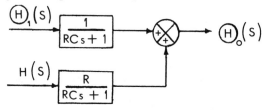

● **PROBLEM** 5-16

Find the transfer function $\frac{Y(s)}{U(s)}$ for the electrical circuit shown below.

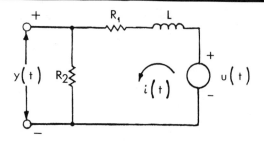

Solution: The Kirchhoff loop equation for the above circuit is

$$R_1 i(t) + R_2 i(t) + L \frac{di}{dt} = u(t)$$

For the output voltage we have

$$y(t) = R_2 i(t)$$

Combining both equations, we get

$$L\dot{y}(t) + (R_1 + R_2)y(t) = R_2 u(t)$$

Assuming zero initial conditions we take the Laplace transform of the above equation

$$LsY(s) + (R_1 + R_2)Y(s) = R_2 U(s)$$

The transfer function is

$$G(s) = \frac{Y(s)}{U(s)} = \frac{R_2}{Ls + R_1 + R_2}$$

The block diagram of the system is

$$U \longrightarrow \boxed{\frac{R_2}{LS + R_1 + R_2}} \longrightarrow Y$$

● **PROBLEM** 5-17

Obtain the transfer function of the two-phase servomotor shown below. The maximum voltages for both phases are 115 volts. The torque-speed curve is

The moment of inertia

$$J = 6 \times 10^{-4} \text{ oz-in.-sec}^2$$

The viscous-friction coefficient

$$f = 0.004 \text{ oz-in./rad/sec.}$$

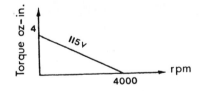

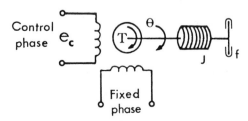

164

Solution: The transfer function of the system can be obtained from the functions

$$T = -K_n \dot{\theta} + K_c E_c$$

$$T = J\ddot{\theta} + f\dot{\theta}$$

K_c, K_n are constants

T = torque

θ = the angle of the motor shaft

E_c = control voltage

J = moment of inertia

Combining both equations

$$J\ddot{\theta} + (f + K_n)\dot{\theta} = K_c E_c$$

Taking the Laplace transform and assuming zero initial conditions

$$Js^2\theta(s) + (f + K_n)s\theta(s) = K_c E_c(s)$$

The transfer function is

$$\frac{\theta(s)}{E_c(s)} = \frac{K_c}{Js^2 + (f + K_n)s} = \frac{K_m}{s(T_m s + 1)}$$

where

$$K_m = \frac{K_c}{f + K_n}$$

$$T_m = \frac{J}{f + K_n}$$

We have

$$K_n = \frac{4}{4000} \cdot \frac{60}{2\pi} = 0.0095 \text{ oz-in/rad/sec}$$

$$K_c = \frac{4}{115} = 0.0348 \text{ oz-in./volt}$$

Thus

$$\frac{\theta(s)}{E_c(s)} = \frac{0.0348}{s[s \cdot 6 \cdot 10^{-4} + 0.004 + 0.0095]} = \frac{348}{s(6s + 135)}$$

● **PROBLEM 5-18**

Draw the feedback circuit and obtain the transfer function of the controller shown below.

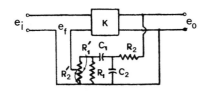 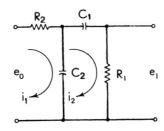

Solution: The feedback circuit of the system is shown.

The equations for the above circuit are

$$\frac{1}{C_2 s} [I_1(s) - I_2(s)] + R_2 I_1(s) = E_0(s)$$

$$\frac{1}{C_2 s} [I_2(s) - I_1(s)] + \frac{1}{C_1 s} I_2(s) + R_1 I_2(s) = 0$$

From both equations we get

$$\frac{I_2(s)}{E_0(s)} = \frac{C_1 s}{R_1 C_1 R_2 C_2 s^2 + s(R_1 C_1 + R_2 C_1 + R_2 C_2) + 1}$$

Multiplying both sides by R_1 we have

$$\frac{E_1(s)}{E_0(s)} = \frac{R_1 C_1 s}{R_1 C_1 R_2 C_2 s^2 + (R_1 C_1 + R_2 C_1 + R_2 C_2)s + 1}$$

e_i , e_f , e_1 and e_0 satisfy the following equations

$$K(e_i - e_f) = e_0 , \qquad e_f = e_1 \frac{R_2'}{R_1'}$$

Thus

$$K[E_i(s) - \frac{R_2'}{R_1'} (\frac{R_1 C_1 s E_0(s)}{R_1 C_1 R_2 C_2 s^2 + (R_1 C_1 + R_2 C_1 + R_2 C_2)s + 1})]$$

$$= E_0(s)$$

The transfer function of the system $\dfrac{E_0(s)}{E_i(s)}$ is

$$\frac{E_0(s)}{E_i(s)} = \frac{KR_1'[R_1 C_1 R_2 C_2 s^2 + s(R_1 C_1 + R_2 C_1 + R_2 C_2) + 1]}{KR_2' R_1 C_1 s + \underbrace{R_1'[C_1 R_1 R_2 C_2 s^2 + s(R_1 C_1 + R_2 C_1 + R_2 C_2) + 1]}_{N}}$$

If the loop gain is much greater than unity then we can
neglect N and the transfer function becomes

$$\frac{E_0(s)}{E_i(s)} = \frac{R_1'[R_1 C_1 R_2 C_2 s^2 + s(R_1 C_1 + R_2 C_1 + R_2 C_2) + 1]}{R_2' R_1 C_1 s}$$

Let us denote

$$K' = \frac{R_1'}{R_2'} \; ; \; T_1 = R_1 C_1 \; ; \; T_2 = R_2 C_2 \; \text{and} \; \alpha = 1 + \frac{R_2}{R_1} + \frac{T_2}{T_1}$$

then

$$\frac{E_0(s)}{E_1(s)} = K'\alpha(1 + \frac{T_2}{\alpha} s + \frac{1}{\alpha T_1 s})$$

Find e as a function of V, for the ladder network shown below. The values on the diagram represent resistances.

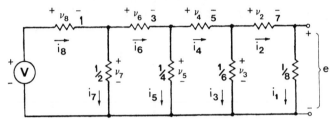

Solution: The desired result will be of the form

e = AV

since the network is linear.
To find A we can find any pair of values of e and V.
We shall work backward using KVL and KCL and Ohm's law.
The easiest value of e to work with is e = 1.   We have

$$e = 1 \quad \overset{\text{Ohm's law}}{\longrightarrow} \quad i_1 = 8$$

$$i_1 = 8 \quad \overset{\text{KCL}}{\longrightarrow} \quad i_1 = i_2 = 8$$

$$i_2 = 8 \quad \overset{\text{Ohm's law}}{\longrightarrow} \quad v_2 = 7i_2 = 56$$

$$v_2 = 56 \rightarrow v_3 = 57$$
$$v_3 = 57 \rightarrow i_3 = 342$$

$$i_4 = 350 \rightarrow v_7 = 24,541$$

$$v_4 = 1750 \rightarrow i_7 = 49,082$$

$$v_5 = 1807 \rightarrow i_8 = 56,660$$

$$i_5 = 7,228 \rightarrow v_8 = 56,660$$

$$i_6 = 7,578 \rightarrow V = 81,201.$$

We have $e = 1 = A \cdot 81,201$, $A = \dfrac{1}{81,201} = 123 \cdot 10^{-7}$

thus

$$e = 123 \cdot 10^{-7} \, V$$

Find v_0 in terms of I, V and R_1, R_2, R_3, R_4 for the network shown.

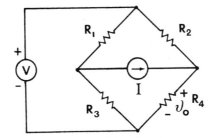

Solution: Since the network is linear the principle of superposition applies. That is, we may find v_0 due to V with no I, and V_0 due to I with no V and add the results. Let us first find $v_0 = v_{0V}$ caused by the voltage source with I = 0. That is equivalent to replacing the network with

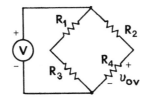

R_2 and R_4 form a voltage divider, thus

$$v_{0V} = \frac{R_4}{R_2 + R_4} V$$

To find v_{0I} we set V = 0. That means that the voltage across the voltage source is zero. For V = 0, an equivalent network is

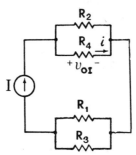

R_2 and R_4 form a current divider. Thus for the current i passing through R_4 we have

$$i = \frac{G_4}{G_2 + G_4} I$$

where $G_2 = \dfrac{1}{R_2}$, $G_4 = \dfrac{1}{R_4}$.

But

$$v_{0I} = R_4 I = \frac{I}{G_2 + G_4}$$

or

$$v_{0I} = \frac{R_2 R_4}{R_2 + R_4} I$$

Using the principle of superposition we can write

$$v_0 = v_{0V} + v_{0I}$$

$$= \frac{R_4}{R_2 + R_4} V + \frac{R_2 R_4}{R_2 + R_4} I.$$

● **PROBLEM** 5-21

Let us consider a network composed of two negative-conductance elements as shown in Fig. 1. The e(i) characteristics for elements 1 and 2 are shown in Fig. 2.

Construct the composite e(i) characteristic of the circuit. How many states can the circuit be in? Assume that only the horizontal states are stable. Represent the circuit as a state machine, and draw a state diagram showing transitions for each input to each state.

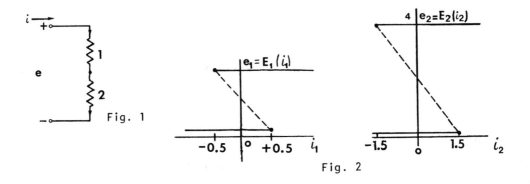

Fig. 1

Fig. 2

Solution: Since both elements are in series we have

$$e(i) = e_1(i) + e_2(i).$$

Thus the composite can be obtained by graphically adding the two characteristics shown in Fig. 2. For $|i| > 1.5$, e_1 and e_2 are single valued, so e is also single valued. For $0.5 < |i| < 1.5$, e_1 is single-valued and e_2 is triple-valued, thus e is triple-valued. For $|i| < 0.5$ both e_1 and e_2 are triple-valued, so the composite e(i) has 9 values. The composite driving-point e(i) curve is shown in Fig. 3.

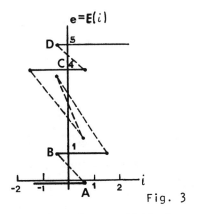

Fig. 3

It has four stable open-circuit equilibrium voltages; A, B, C, D. The network can be viewed as a sequential machine whose input can take values -2, -1, 1, 2. The machine has four possible states A, B, C, D corresponding respectively to the values e = 0, 1, 4, 5. The state diagram of the network is shown below.

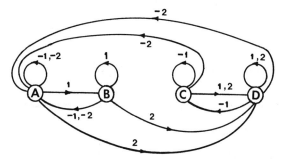

Each directed branch indicates that the state transition will take place when the input associated with that branch is applied.

● **PROBLEM 5-22**

The network shown below has one negative-resistance device. The characteristics of both elements are

Calculate the driving-point e(i)-characteristic of the network.

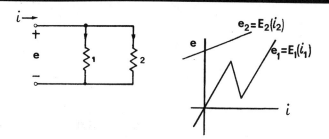

Solution: The function $e_1 = E_1(i_1)$ can not be inverted as required. It is however possible to obtain a converse,

$E_1^{-1}(e)$, by reflecting the curve across the line e = i. We do the same with the e_2 characteristic and add both graphically. We obtain

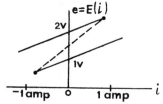

The above curve is single valued neither in e nor in i. The physical interpretation is that when a characteristic curve is multivalued then the device can be in any of the states represented by the multiple values.

● **PROBLEM** 5-23

The positional servomechanism is shown below. Obtain the closed-loop transfer function of the mechanism, assuming that the input and output of the system are the input shaft position and the output shaft position, respectively. We denote

r = angular displacement of the reference input shaft

C = angular displacement of the output shaft

θ = angular displacement of the motor shaft

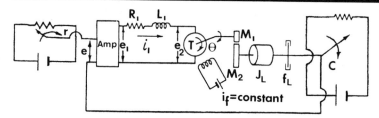

Solution: The following equations describe the system dynamics.

$$E_1(s) = K_a E(s)$$

$$E(s) = K_1 [R(s) - C(s)]$$

where K_a is amplifier gain, K_1 is gain of the potentiometric error detector.

For the armature-controlled dc motor

$$J = J_m + n^2 J_L$$

$$f = f_m + n^2 f_L$$

where J_m is moment of inertia of the motor,

J_L moment of inertia of the load,

f_m viscous-friction coefficient of the motor,

171

f_L viscous-friction coefficient of the load,

$n = \dfrac{M_1}{M_2}$ gear ratio.

We have

$$\frac{\textcircled{H}\,(s)}{E_1(s)} = \frac{K_m}{s(T_m s + 1)}$$

where

$$K_m = \frac{K}{R_1 f + KK_2}$$

$$T_m = \frac{R_1 J}{R_1 f + KK_2}$$

for $K_a = 10$

$\quad n = \dfrac{1}{10}$

$\quad K_1 = \dfrac{15}{\pi}\,\dfrac{\text{volts}}{\text{rad}}$

$\quad J_m = 1.4 \times 10^{-5}\,\text{lb-ft-sec}^2$

$\quad J_L = 3.9 \times 10^{-3}\,\text{lb-ft-sec}^2$

$\quad f_m \approx 0$

$\quad f_L = 4.2 \times 10^{-2}\,\text{lb-ft/rad/sec}$

$\quad K = 5.7 \times 10^{-5}\,\text{lb-ft/amp}$

$\quad R_1 = 0.3\;\Omega$

$\quad K_2 = 4.5 \times 10^{-2}\,\text{volts-sec/rad}$

Substituting the numerical values into the equations we obtain

$$E_1(s) = 10E(s)$$

$$E(s) = 4.77[R(s) - C(s)].$$

$J = 1.4 \times 10^{-5} + 3.9 \times 10^{-5} = 5.3 \times 10^{-5}$

$f = 4.2 \times 10^{-4}$

$$K_m = \frac{5.7 \times 10^{-5}}{(0.3)(4.2 \times 10^{-4}) + (5.7 \times 10^{-5})(4.5 \times 10^{-2})} = 0.4433$$

$$T_m = \frac{0.3 \cdot 5.3 \times 10^{-5}}{(0.3)(4.2 \times 10^{-4}) + (5.7 \times 10^{-5})(4.5 \times 10^{-2})} = 0.1237$$

Thus for

$$\frac{\text{(H)}(s)}{E_1(s)} = \frac{10C(s)}{E_1(s)} = \frac{0.4433}{s(0.1237s+1)}$$

Using the results obtained above we can draw the block diagram of the system.

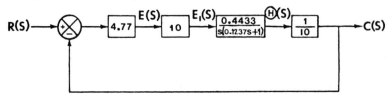

Simplifying this block diagram we obtain

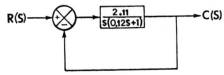

The transfer function is

$$\frac{C(s)}{R(s)} = \frac{2.11}{0.12s^2 + s + 2.11} = \frac{17.58}{s^2 + 8.33s + 17.58}$$

THE TRANSFER MATRIX AND PULSE TRANSFER FUNCTION

● **PROBLEM** 5-24

The system shown below has two inputs $N(s)$ and $R(s)$ and one output $C(s)$. Obtain the transfer matrix between the output and the inputs.

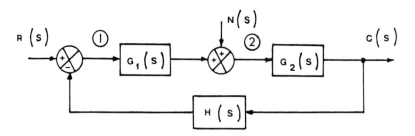

Solution: The function at point (1) is

$$R(s) - C(s)H(s)$$

at point (2)

$$G_1(R - CH) + N$$

Thus we obtain for the output $C(s)$

$$C = G_2 [N + G_1 (R - CH)] = G_2N + G_2G_1R - G_1G_2CH.$$

Thus

$$C(1 + G_1G_2H) = G_1G_2R + G_2N.$$

$$C = \frac{G_1G_2}{1 + G_1G_2H} R + \frac{G_2}{1 + G_1G_2H} N$$

In the matrix form

$$C(s) = \left[\frac{G_1(s)G_2(s)}{1 + G_1(s)G_2(s)H(s)} \quad \frac{G_2(s)}{1 + G_1(s)G_2(s)H(s)} \right] \begin{bmatrix} R(s) \\ N(s) \end{bmatrix}$$

The transfer matrix between the output and the inputs is

$$T(s) = \left[\frac{G_1(s)G_2(s)}{1 + G_1(s)G_2(s)H(s)} \quad \frac{G_2(s)}{1 + G_1(s)G_2(s)H(s)} \right]$$

● **PROBLEM** 5-25

The block diagram of a multiple-input multiple-output system is shown. Determine the transfer matrix $P_a(s)$ such that the closed-loop transfer matrix is

$$P(s) = \begin{bmatrix} \frac{1}{s+1} & 0 \\ 0 & \frac{1}{5s+1} \end{bmatrix}$$

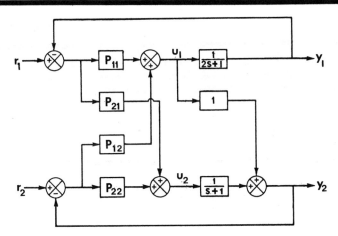

Solution: We have

$$P_0 = P(I - P)^{-1}$$

Substituting P we obtain

$$P_0 = \begin{bmatrix} \frac{1}{s+1} & 0 \\ 0 & \frac{1}{5s+1} \end{bmatrix} \begin{bmatrix} \frac{s+1}{s} & 0 \\ 0 & \frac{5s+1}{5s} \end{bmatrix} = \begin{bmatrix} \frac{1}{s} & 0 \\ 0 & \frac{1}{5s} \end{bmatrix}$$

174

From the diagram

$$\begin{bmatrix} Y_1(s) \\ Y_2(s) \end{bmatrix} = \begin{bmatrix} \dfrac{1}{2s+1} & 0 \\ 1 & \dfrac{1}{s+1} \end{bmatrix} \begin{bmatrix} U_1(s) \\ U_2(s) \end{bmatrix}$$

and

$$\begin{bmatrix} U_1(s) \\ U_2(s) \end{bmatrix} = \begin{bmatrix} P_{11}(s) & P_{12}(s) \\ P_{21}(s) & P_{22}(s) \end{bmatrix} \begin{bmatrix} R_1(s) - Y_1(s) \\ R_2(s) - Y_2(s) \end{bmatrix}$$

$$\begin{bmatrix} Y_1(s) \\ Y_2(s) \end{bmatrix} = \begin{bmatrix} \dfrac{1}{2s+1} & 0 \\ 1 & \dfrac{1}{s+1} \end{bmatrix} \begin{bmatrix} P_{11} & P_{12} \\ P_{21} & P_{22} \end{bmatrix} \begin{bmatrix} R_1 - Y_1 \\ R_2 - Y_2 \end{bmatrix}$$

$$= \begin{bmatrix} \dfrac{1}{s} & 0 \\ 0 & \dfrac{1}{5s} \end{bmatrix} \begin{bmatrix} R_1 - Y_1 \\ R_2 - Y_2 \end{bmatrix}$$

Hence

$$P_a(s) = \begin{bmatrix} P_{11} & P_{12} \\ P_{21} & P_{22} \end{bmatrix} = \begin{bmatrix} \dfrac{1}{2s+1} & 0 \\ 1 & \dfrac{1}{s+1} \end{bmatrix}^{-1} \begin{bmatrix} \dfrac{1}{s} & 0 \\ 0 & \dfrac{1}{5s} \end{bmatrix}$$

$$= \begin{bmatrix} \dfrac{2s+1}{s} & 0 \\ \dfrac{-(s+1)(2s+1)}{s} & \dfrac{s+1}{5s} \end{bmatrix}$$

● **PROBLEM 5-26**

Decompose the transfer function T(z) by parallel decomposition and determine the dynamic equations of the system. The transfer function is given by

$$T(z) = \frac{C(z)}{R(z)} = \frac{10(z^2+z+1)}{z^2(z-0.5)(z-0.8)}$$

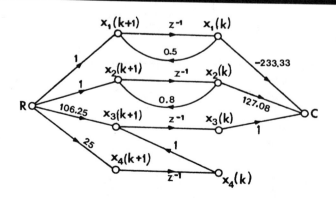

Solution: The partial fraction expansion of T(z) gives

$$T(z) = \frac{-233.33}{z-0.5} + \frac{127.08}{z-0.8} + \frac{25}{z^2} + \frac{106.25}{z} .$$

The desired transfer function is realized by the parallel connection of first-order components.

There are four time-delay units in the state diagram, because $T(z)$ is of the fourth order.

The state equations are, in Jordan canonical form

$$\begin{bmatrix} x_1(k+1) \\ x_2(k+1) \\ x_3(k+1) \\ x_4(k+1) \end{bmatrix} = \begin{bmatrix} 0.5 & 0 & 0 & 0 \\ 0 & 0.8 & 0 & 0 \\ 0 & 0 & 0 & 1 \\ 0 & 0 & 0 & 0 \end{bmatrix} \begin{bmatrix} x_1(k) \\ x_2(k) \\ x_3(k) \\ x_4(k) \end{bmatrix} + \begin{bmatrix} 1 \\ 1 \\ 106.25 \\ 25 \end{bmatrix} r(kT)$$

The output equation is

$$C(k) = [-233.33 \quad 127.08 \quad 1 \quad 0] x(k)$$

● **PROBLEM** 5-27

The system is described by the following equations

$$\dot{x}_1 = x_3$$

$$\dot{x}_2 = x_4$$

$$\dot{x}_3 = -3x_2 - 2x_3 + y_1(t)$$

$$\dot{x}_4 = -x_2 - 2x_3 + y_2(t)$$

Obtain the transfer matrix of the system.

Solution: In matrix form we have

$$\dot{x} = \begin{bmatrix} 0 & 0 & 1 & 0 \\ 0 & 0 & 0 & 1 \\ 0 & -3 & -2 & 0 \\ 0 & -1 & -2 & 0 \end{bmatrix} x + \begin{bmatrix} 0 & 0 \\ 0 & 0 \\ 1 & 0 \\ 0 & 1 \end{bmatrix} y$$

or

$$\dot{x} = Ax + By$$

Taking the Laplace transform

$$sIx = Ax + By$$

solving for x

$$x = [sI - A]^{-1}By$$

Since $x_3 = \dot{x}_1$ and $x_4 = \dot{x}_2$ we have to find x_1 and x_2. We can write

$$x = \begin{pmatrix} x_1 \\ x_2 \end{pmatrix} = \begin{pmatrix} a_{13} & a_{14} \\ a_{23} & a_{24} \end{pmatrix} \begin{pmatrix} Y_1 \\ Y_2 \end{pmatrix} = Ty$$

where $[a_{ij}]$ is a submatrix of $[sI - A]^{-1}$. Let us denote

$$x_1 = a_{13}Y_1 + a_{14}Y_2$$

$$x_2 = a_{23}Y_1 + a_{24}Y_2$$

we have

$$x_1 = \frac{\begin{vmatrix} 0 & 0 & -1 & 0 \\ 0 & s & 0 & -1 \\ Y_1 & 3 & s+2 & 0 \\ Y_2 & 1 & 2 & s \end{vmatrix}}{\begin{vmatrix} s & 0 & -1 & 0 \\ 0 & s & 0 & -1 \\ 0 & 3 & s+2 & 0 \\ 0 & 1 & 2 & s \end{vmatrix}} = \frac{Y_1(s^2 + 1) - 3Y_2}{s(s^3 + 2s^2 + s - 4)}$$

In a similar way we obtain

$$x_2 = \frac{-2sY_1 + (s+2)Y_2}{s(s^3 + 2s^2 + s - 4)}$$

Thus the elements of T are

$$\begin{bmatrix} \dfrac{s^2 + 1}{s(s^3 + 2s^2 + s - 4)} & \dfrac{-3}{s(s^3 + 2s^2 + s - 4)} \\[3ex] \dfrac{-2s}{s(s^3 + 2s^2 + s - 4)} & \dfrac{s + 2}{s(s^3 + 2s^2 + s - 4)} \end{bmatrix}$$

● **PROBLEM** 5-28

Consider a system described by

$$\dot{x} = Ax + Bu$$

$$y = Cx + Du$$

where x is a state vector, n vector
      y is an output vector, m vector
      u is a control vector, r vector

Matrices A, B, C, D are constant.

Determine the transfer matrices of this system, if

$$A = \begin{bmatrix} -5 & -4 & 2 \\ 3 & 3 & -2 \\ 0 & 2 & -2 \end{bmatrix} \qquad B = \begin{bmatrix} -1 & 0 \\ 1 & 0 \\ 0 & 2 \end{bmatrix}$$

$$C = \begin{bmatrix} 1 & 1 & 0 \end{bmatrix} \qquad D = \begin{bmatrix} 0 & 0 \end{bmatrix}.$$

Solution: Assuming zero initial conditions, the Laplace transform of the equations describing the system is

$$sX(s) = AX(s) + BU(s)$$

$$Y(s) = CX(s) + DU(s)$$

or

$$X(s) = (sI-A)^{-1}BU(s)$$

$$Y(s) = [C(sI-A)^{-1}B+D]U(s)$$

From the above equations we see that the transfer matrix between $X(s)$ and $U(s)$ is $(sI-A)^{-1}B$ and between $Y(s)$ and $U(s)$ is $C(sI-A)^{-1}B+D$.

We have

$$|sI-A| = (s+1)^2(s+2) \qquad \text{and}$$

$$(sI-A)^{-1} = \begin{bmatrix} \dfrac{s-2}{(s+1)(s+2)} & \dfrac{-4}{(s+1)(s+2)} & \dfrac{2}{(s+1)(s+2)} \\[4mm] \dfrac{3}{(s+1)^2} & \dfrac{s+5}{(s+1)^2} & \dfrac{-2}{(s+1)^2} \\[4mm] \dfrac{6}{(s+1)^2(s+2)} & \dfrac{2(s+5)}{(s+1)^2(s+2)} & \dfrac{(s+3)(s-1)}{(s+1)^2(s+2)} \end{bmatrix}$$

Thus the transfer matrix between $X(s)$ and $U(s)$ is

$$X(s) = T_{xu}(s)U(s)$$

$$T_{xu}(s) = \begin{bmatrix} \dfrac{-1}{s+1} & \dfrac{4}{(s+1)(s+2)} \\[4mm] \dfrac{s+2}{(s+1)^2} & \dfrac{-4}{(s+1)^2} \\[4mm] \dfrac{2}{(s+1)^2} & \dfrac{2(s-1)(s+3)}{(s+1)^2(s+2)} \end{bmatrix}$$

The transfer matrix T_{YU} is

$$Y(s) = [C(sI-A)^{-1}B+D]U(s) = [C\cdot T_{xu}+D]U(s).$$

Thus

$$T_{YU}(s) = C \cdot T_{xu} + D = \left(\frac{1}{(s+1)^2} \quad \frac{-4}{(s+1)^2(s+2)} \right)$$

● **PROBLEM** 5-29

Obtain the pulse transfer function of the system shown in the figure.

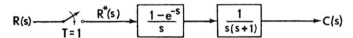

Solution: From the block diagram we have

$$G(s) = \frac{C(s)}{R^*(s)} = \frac{1-e^{-s}}{s^2(s+1)} = (1-e^{-s})(\frac{1}{s^2} - \frac{1}{s} + \frac{1}{s+1})$$

The inverse Laplace transformation gives

$$g(t) = (t-1+e^{-t})1(t) - (t-1-1+e^{-(t-1)})1(t-1)$$

For $T = 1$, we have $kT = k$ and

$$g(k) = (k-1+e^{-k}) - (k-2+e^{-(k-1)}) = e^{-k} + 1 - e^{-(k-1)}$$

$$\text{for } k = 1,2,3$$

$$g(0) = 0$$

For $G(z)$ we obtain

$$G(z) = \sum_{k=0}^{\infty} g(k) z^{-k} = \sum_{k=0}^{\infty} (e^{-k}+1-e^{-(k-1)})z^{-k}+e-2$$

$$= \frac{1-e}{1-e^{-1}z^{-1}} + \frac{1}{1-z^{-1}} + e - 2$$

$$= \frac{e^{-1}z + 1 - 2e^{-1}}{z^2 - (1+e^{-1})z + e^{-1}} \cdot$$

● **PROBLEM** 5-30

The block diagram of a discrete-time system is shown below. Obtain the pulse transfer function of the system.

$$x(t) \longrightarrow \overset{x^*(t)}{\underset{T}{\diagup}} \longrightarrow \boxed{\frac{K}{(s+\alpha)(s+\beta)}} \longrightarrow y(t)$$

Solution: The transfer function can be written

$$G(s) = \frac{K}{(s+\alpha)(s+\beta)} = \frac{K}{\beta-\alpha} \left(\frac{1}{s+\alpha} - \frac{1}{s+\beta}\right)$$

The impulse-response function is

$$g(t) = \frac{K}{\beta-\alpha} \left(e^{-\alpha t} - e^{-\beta t}\right)$$

and

$$g(kT) = \frac{K}{\beta-\alpha} \left(e^{-\alpha kT} - e^{-\beta kT}\right)$$

The function G(z) is obtained as follows

$$G(z) = \sum_{k=0}^{\infty} \frac{K}{\beta-\alpha} \left(e^{-\alpha kT} - e^{-\beta kT}\right) z^{-k}$$

and

$$\sum_{k=0}^{\infty} e^{-\alpha kT} z^{-k} = 1 + \sum_{k=1}^{\infty} e^{-\alpha kT} z^{-k}$$

$$= 1 + \sum_{k=1}^{\infty} e^{-\alpha T} z^{-1} \left(e^{-\alpha kT + \alpha T} z^{-k+1}\right)$$

$$= 1 + e^{-\alpha T} z^{-1} \sum_{k=0}^{\infty} e^{-\alpha kT} z^{-k}$$

We obtain

$$\sum_{k=0}^{\infty} e^{-\alpha kT} z^{-k} = \frac{1}{1 - e^{-\alpha T} z^{-1}}$$

Using the above formula we obtain the following pulse transfer function

$$G(z) = \frac{K}{\beta-\alpha} \left(\frac{1}{1 - e^{-\alpha T} z^{-1}} - \frac{1}{1 - e^{-\beta T} z^{-1}}\right)$$

● **PROBLEM** 5-31

For the system shown below determine the pulse transfer function of G(s) in cascade with the z.o.h. (zero order hold) when

$$R(s) = \frac{1}{s+a}$$

180

<u>Solution:</u> From $R(s) = \dfrac{1}{s+a}$ we get

$$R(z) = \frac{z}{z-e^{-aT}}$$

$z = e^{Ts}$ and $z^{-1} = e^{-Ts}$, we have

$$R^*(s) = \frac{e^{Ts}}{e^{Ts} - e^{-aT}}$$

For the output H of z.o.h. we have

$$H(s) = G_{ho}(s)R^*(s) = \frac{1 - e^{-Ts}}{s} \cdot \frac{e^{Ts}}{e^{Ts} - e^{-aT}}$$

where $G_{ho}(s) = \dfrac{1-e^{-Ts}}{s}$ is the transfer function of a zero-order hold.

Using l'Hopital's rule, we can show that as T approaches zero

$$\lim_{T \to 0} H(s) = \frac{1}{s+a} = R(s)$$

We will prove the following property

$$\lim_{T \to 0} z[G_{ho}(s)G(s)] = G(s)$$

Indeed, using the expression for $G_{ho}(s)$ we have

$$\lim_{T \to 0} z[G_{ho}(s)G(s)] = \lim_{T \to 0} z\left[\frac{1-e^{-Ts}}{s} G(s)\right]$$

Expanding e^{-Ts} into a power series and taking only the first two terms, we get

$$\lim_{T \to 0} z[G_{ho}(s)G(s)] = \lim_{T \to 0} TG(z)$$

Using the identity

$$G(z) = G^*(s)\bigg|_{z=e^{Ts}} = \frac{1}{T} \sum_{n=-\infty}^{\infty} G(s+j2\pi n/T)\bigg|_{z=e^{Ts}}$$

we obtain

$$\lim_{T \to 0} z[G_{ho}(s)G(s)] = \lim_{T \to 0} \sum_{n=-\infty}^{\infty} G\left(s + \frac{j2\pi n}{T}\right)\bigg|_{z=e^{Ts}} = G(s)$$

Obtain the pulse transfer function of the closed-loop
system shown below.

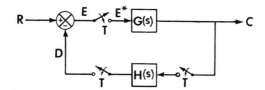

Solution: From the block diagram we have the following
equations:

$$E(s) = R(s) - D(s)$$

or $\quad\quad E(z) = R(z) - D(z)$

$$C(z) = E(z) \cdot G(z)$$

$$D(z) = C(z) \cdot H(z)$$

Thus

$$C(z) = [R(z) - H(z)C(z)]G(z)$$

The pulse transfer function is

$$\frac{C(z)}{R(z)} = \frac{G(z)}{1 + H(z)G(z)}$$

● **PROBLEM** 5-33

Obtain the pulse transfer function of the closed-loop
system shown below.

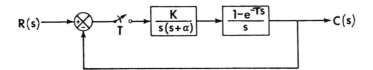

Solution: The open-loop transfer function is

$$G(s) = \frac{K}{s(s+\alpha)} \cdot \frac{1-e^{-Ts}}{s}$$

The z transform of G(s) is

$$Z\left[\frac{K(1-e^{-Ts})}{s^2(s+\alpha)}\right] = K(1-z^{-1})\, Z\left[\frac{1}{s^2(s+\alpha)}\right]$$

182

$$\frac{1}{s^2(s+\alpha)} = \frac{1}{\alpha s^2} - \frac{1}{\alpha^2 s} + \frac{1}{\alpha^2(s+\alpha)} \qquad (*)$$

We have $\mathcal{L}^{-1}\left\{\dfrac{1}{s^2}\right\} = t$. Sampled, this becomes kT, and the z-transform is

$$Z\{kT\} = \sum_{k=0}^{\infty} kTz^{-k} = T(0+z^{-1}+2z^{-2}+3z^{-3}+\ldots)$$

Multiplying the above equation by z gives

$$z \cdot z\{kT\} = T(1+2z^{-1}+3z^{-2}+4z^{-3}+\ldots)$$

Subtracting the above equations we obtain

$$z \cdot z\{kT\} - z\{kT\} = T(1+z^{-1}+z^{-2}+z^{-3}+\ldots) = \frac{Tz}{z-1}$$

Thus

$$z\{kT\} = \frac{Tz}{(z-1)^2}$$

$$\mathcal{L}\{\tfrac{1}{s}\} = 1, \quad \text{and}$$

$$Z\{1\} = \sum_{k=0}^{\infty} z^{-k} = \frac{z}{z-1} \ .$$

$\mathcal{L}\left\{\dfrac{1}{s+\alpha}\right\} = e^{-\alpha t}$, which becomes $e^{-\alpha kT}$ when sampled.

$$Z\{e^{-\alpha kT}\} = \sum_{k=0}^{\infty} e^{-\alpha kT}z^{-k} = \sum_{k=0}^{\infty} (e^{-\alpha T}z^{-1})^k$$

$$= \frac{1}{1 - e^{-\alpha T}z^{-1}} = \frac{z}{z - e^{-\alpha T}}$$

Taking the Z transform of (*) we obtain

$$Z\left(\frac{1}{s^2(s+\alpha)}\right) = \frac{Tz}{\alpha(z-1)^2} - \frac{z}{\alpha^2(z-1)} + \frac{z}{\alpha^2(z-e^{-\alpha T})}$$

Thus

$$Z\left[\frac{k(1-e^{-Ts})}{s^2(s+\alpha)}\right] = \frac{kT}{\alpha(z-1)} - \frac{k}{\alpha^2} + \frac{k(z-1)}{\alpha^2(z-e^{-\alpha T})}$$

$$= \frac{k[(\alpha T-1+e^{-\alpha T})z + (1-e^{-\alpha T}-\alpha Te^{-\alpha T})]}{\alpha^2(z-1)(z-e^{-\alpha T})} = \frac{\beta}{\gamma}$$

We have for the closed-loop pulse transfer function

$$\frac{C(z)}{R(z)} = \frac{\beta}{\gamma + \beta}$$

where $\beta = k[(\alpha T - 1 + e^{-\alpha T})z + (1 - e^{-\alpha T} - \alpha T e^{-\alpha T})]$

and $\gamma = \alpha^2(z-1)(z - e^{-\alpha T})$

● **PROBLEM** 5-34

Given the discrete system

$$x(k+1) = Ax(k)$$

where

$$A = \begin{bmatrix} 0 & 1 & 0 \\ 0 & 0 & 1 \\ -6 & -11 & -6 \end{bmatrix}$$

Use the z transform method to find the transformation matrix.

Solution: The characteristic equation of A is

$$|\lambda I - A| = \begin{vmatrix} \lambda & -1 & 0 \\ 0 & \lambda & -1 \\ 6 & 11 & \lambda+6 \end{vmatrix} = \lambda^3 + 6\lambda^2 + 11\lambda + 6$$

$$= (\lambda - 1)(\lambda - 2)(\lambda - 3) = 0.$$

The coefficients of the characteristic equation are

$$\lambda^3 + 6\lambda^2 + 11\lambda + 6 = 0$$

$$a_4 = 1 \qquad a_3 = 6 \qquad a_2 = 11 \qquad a_1 = 6.$$

Using

$$F = (zI-A)^{-1}z = \left(\sum_{j=1}^{n} z^j \sum_{i=j}^{n} a_{i+1} A^{i-j} \right) \cdot \frac{1}{|zI-A|}$$

$$= \frac{a_4 z^3 I + (a_3 I + a_2 A)z^2 + (a_2 I + a_3 A + a_4 A^2)z}{|zI-A|}$$

$$= \frac{z^3 I + \begin{bmatrix} 6 & 1 & 0 \\ 0 & 6 & 1 \\ -6 & -11 & 0 \end{bmatrix} z^2 + \begin{bmatrix} 11 & 6 & 1 \\ -6 & 0 & 0 \\ 0 & -6 & 0 \end{bmatrix} z}{(z-1)(z-2)(z-3)}$$

184

The partial-fraction expansion of the above expression gives

$$F = \frac{z}{2(z-1)}\begin{bmatrix} 18 & 7 & 1 \\ -6 & 7 & 1 \\ -6 & -17 & 1 \end{bmatrix} - \frac{z}{z-2}\begin{bmatrix} 27 & 8 & 1 \\ -6 & 16 & 2 \\ -12 & -28 & 4 \end{bmatrix}$$

$$+ \frac{z}{2(z-3)}\begin{bmatrix} 38 & 9 & 1 \\ -6 & 27 & 3 \\ -18 & -39 & 9 \end{bmatrix}$$

Taking the inverse z-transform on both sides, we obtain

$$\phi(k) = \frac{1}{2}\begin{bmatrix} 18 & 7 & 1 \\ -6 & 7 & 1 \\ -6 & -17 & 1 \end{bmatrix} - \begin{bmatrix} 27 & 8 & 1 \\ -6 & 16 & 2 \\ -12 & -28 & 4 \end{bmatrix}e^{-0.694k}$$

$$+ \frac{1}{2}\begin{bmatrix} 38 & 9 & 1 \\ -6 & 27 & 3 \\ -18 & -39 & 9 \end{bmatrix}e^{-1.1k}$$

It is easy to check that

$$\phi(0) = I.$$

● **PROBLEM** 5-35

Given the discrete system

$$x(k+1) = Ax(k) + Bu(k)$$

where

$$A = \begin{bmatrix} 3 & 2 \\ 2 & 3 \end{bmatrix}$$

find the transition matrix using Sylvester's expansion theorem.

Solution: The eigenvalues of A are

$$|\lambda I - A| = \begin{vmatrix} \lambda-3 & -2 \\ -2 & \lambda-3 \end{vmatrix} = (\lambda-3)^2 - 4 = \lambda^2 - 6\lambda + 5$$

$$= (\lambda-5)(\lambda-1)$$

185

Thus $\lambda_1 = 5$ and $\lambda_2 = 1$

From $f(\lambda_i) = e^{\lambda_i T}$ we obtain

$$f(\lambda_1) = e^{5T}$$

$$f(\lambda_2) = e^{T}$$

From

$$F(\lambda_i) = \sum_{\substack{j=1 \\ i \neq j}}^{n} \frac{A - \lambda_j I}{\lambda_i - \lambda_j} \qquad i = 1, 2, \ldots n$$

we get

$$F(\lambda_1) = \frac{A - \lambda_2 I}{\lambda_1 - \lambda_2} = \begin{bmatrix} 0.5 & 0.5 \\ 0.5 & 0.5 \end{bmatrix}$$

$$F(\lambda_2) = \frac{A - \lambda_1 I}{\lambda_2 - \lambda_1} = \begin{bmatrix} 0.5 & -0.5 \\ -0.5 & 0.5 \end{bmatrix}$$

Finally from

$$\phi(T) = \sum_{i=1}^{n} e^{\lambda_i T} F(\lambda_i) \qquad \text{we get the transformation}$$

matrix $\phi(T)$

$$\phi(T) = e^{5T} \begin{bmatrix} 0.5 & 0.5 \\ 0.5 & 0.5 \end{bmatrix} + e^{T} \begin{bmatrix} 0.5 & -0.5 \\ -0.5 & 0.5 \end{bmatrix}$$

$$= 0.5 \begin{bmatrix} e^{5T} + e^{T} & e^{5T} - e^{T} \\ e^{5T} - e^{T} & e^{5T} + e^{T} \end{bmatrix}$$

The above method of finding the transition matrix can be applied only when A has distinct eigenvalues.

CHAPTER 6

TIME ANALYSIS

RESPONSE

Find the damping ratio of a feedback system with the given open loop transfer function:

$$G(s) = \frac{1638(s^2 + 2.6s + 1.65)}{s(s^2 + 40s + 300)(s^2 + 4.65s + 0.45)}$$

<u>Solution:</u> The closed loop transfer function is

$$\frac{C(s)}{R(s)} = \frac{G(s)}{1 + G(s)} =$$

$$\frac{1638\ (s^2 + 2.6s + 1.65)}{1638(s^2+2.6s+1.65)+s(s^2+40s+300)(s^2+4.65s+0.45)} =$$

$$\frac{1638\ (s^2 + 2.6s + 1.65)}{(s+32.4)(s+0.855+j0.672)(s+0.855-j0.672)(s+5.24+j6.566)(s+5.24-j6.566)}$$

The dominant poles of the system are

$$s + 0.855 \pm j0.672$$

They are the closest to the imaginary axis.

A small change in the system may cause the root locus to cross the imaginary axis and destabilize the system.

θ is found from

$$\tan\theta = \frac{0.672}{0.855} = 0.786$$

$$\theta = 38.2^\circ$$

The damping ratio is

$$\zeta = \cos 38.2^\circ = 0.785.$$

The mechanical system shown, is initially at rest.

A unit-impulse force sets the system into motion. Find the resulting oscillations.

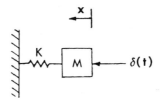

Solution: From Newton's laws, we have

$$M\ddot{x} + Kx = \delta(t)$$

Taking the Laplace transform:

$$M[s^2 X(s) - sx(0) - \dot{x}(0)] + KX(s) = 1.$$

Initially the system is at rest, thus

$$\dot{x}(0) = 0 \quad \text{and} \quad x(0) = 0$$

Solving for X(s)

$$X(s) = \frac{1}{Ms^2 + K}$$

Taking the inverse Laplace transform

$$x(t) = \frac{1}{\sqrt{MK}} \sin \sqrt{\frac{K}{M}} \cdot t$$

where $\dfrac{1}{\sqrt{MK}}$ is the amplitude of the oscillation.

The block diagram of a servomechanism is shown below.

Determine the values of K and k so that the maximum over-shoot in unit-step response is 50% and the peak time is 5 sec.

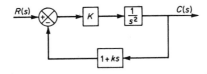

<u>Solution:</u> The maximum overshoot M_p is given by

$$M_p = e^{-\dfrac{\zeta}{\sqrt{1-\zeta^2}}\pi}$$

For the maximum overshoot 50% we have

$$0.5 = e^{-\dfrac{\zeta}{\sqrt{1-\zeta^2}}\pi}$$

$$\ln 0.5 = \dfrac{-\zeta}{\sqrt{1-\zeta^2}}\pi$$

or

$$\dfrac{\zeta\pi}{\sqrt{1-\zeta^2}} = 0.693$$

$$\zeta^2\pi^2 = 0.48(1-\zeta^2)$$

$$\zeta^2\pi^2 + 0.48\zeta^2 = 0.48$$

$$\zeta = \sqrt{\dfrac{0.48}{\pi^2+0.48}} = 0.21$$

The peak time $t_p = 5$ sec,

$$t_p = \dfrac{\pi}{\omega_d}$$

$$\omega_d = \dfrac{\pi}{t_p} = 0.63$$

The undamped natural frequency ω_n is

$$\omega_n = \dfrac{\omega_d}{\sqrt{1-\zeta^2}} = \dfrac{0.63}{\sqrt{1-(0.21)^2}} = 0.644$$

From the block diagram we have

$$\dfrac{C(s)}{R(s)} = \dfrac{K}{s^2 + Kks + K}$$

For the general system

$$\dfrac{C(s)}{R(s)} = \dfrac{\omega_n^2}{s^2 + 2\zeta\omega_n s + \omega_n^2}$$

Equating the coefficients

$$\omega_n^2 = K, \quad 2\zeta\omega_n = Kk$$

189

we have

$$K = \omega_n^2 = 0.415$$

$$k = \frac{2\zeta\omega_n}{K} = \frac{2 \cdot 0.21}{0.644} = 0.652.$$

● **PROBLEM** 6-4

The system shown below has parameters

$$\zeta = 0.4 \quad \text{and} \quad \omega_n = 5\text{rd/sec.}$$

The system is subjected to a unit step input, find the resulting rise time (t_r), peak time (t_p), maximum overshoot (M_p), and settling time (t_s).

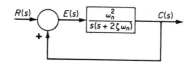

__Solution:__ Having $\zeta = 0.4$ and $\omega_n = 5 \frac{\text{rad}}{\text{sec}}$ we find

$$\omega_d = \omega_n\sqrt{1 - \zeta^2} = 4.58$$

$$\sigma = \zeta\omega_n = 2$$

Rise time:

$$t_r = \frac{\pi - \beta}{\omega_d}$$

where

$$\beta = \tan^{-1}\frac{\omega_d}{\sigma} = \tan^{-1}\frac{4.58}{2} = 1.16$$

then

$$t_r = \frac{\pi - 1.16}{4.58} = 0.43 \text{ sec.}$$

Peak time:

$$t_p = \frac{\pi}{\omega_d} = \frac{3.14}{4.58} = 0.69 \text{ sec.}$$

Maximum overshoot:

$$M_p = e^{-\frac{\sigma}{\omega_d}\pi} = e^{-\frac{2 \cdot 3.14}{4.58}} = e^{-1.37} = 0.254.$$

190

Settling time t_s:

For the 2% criterion

$$t_s = \frac{4}{\sigma} = 2 \text{ sec}$$

for 5%

$$t_s = \frac{3}{\sigma} = 1.5 \text{ sec.}$$

● **PROBLEM** 6-5

In the speed control system the output $\Omega_c(s)$ is subject to a torque disturbance $T'(s)$.

$T(s)$ is the driving torque.

$T'(s)$ is the disturbance torque.

When $T'(s) = 0$, then $\Omega_r = \Omega_c$.

Examine the response of the system to a unit step disturbance torque, assume $\Omega_r(s) = 0$.

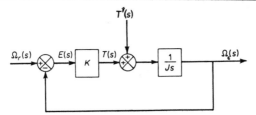

Solution: The equivalent block diagram, when

$$\Omega_r(s) = 0$$

is shown.

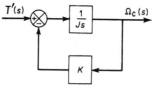

The closed-loop transfer function is

$$\frac{\Omega_c(s)}{T'(s)} = \frac{1}{Js + K} .$$

For a unit-step disturbance torque, the steady-state output velocity is

$$\omega_{T'}(\infty) = \lim_{s \to 0} s\,\Omega_c(s)$$

$$= \lim_{s \to 0} \frac{s}{Js + K} \frac{1}{s} = \frac{1}{K} .$$

The closed-loop transfer function of the system is

$$\frac{C(s)}{R(s)} = \frac{1}{s^2 + 2\zeta s + 1} \quad \text{for} \quad 0 < \zeta < 1$$

For a unit-impulse input find $\int_0^\infty c^2(t)dt$.

Solution: We have $R(s) = 1$ for a unit-impulse input. Thus

$$C(s) = \frac{1}{s^2 + 2\zeta s + 1} = \frac{1}{(s + \zeta)^2 + 1 - \zeta^2}$$

$$= \frac{1}{\sqrt{1 - \zeta^2}} \frac{\sqrt{1 - \zeta^2}}{(s + \zeta)^2 + 1 - \zeta^2}$$

The inverse Laplace transform is

$$c(t) = \frac{1}{\sqrt{1 - \zeta^2}} e^{-\zeta t} \sin\sqrt{1 - \zeta^2} \cdot t \qquad (t \geq 0)$$

$c^2(t)$ can be written in the form

$$c^2(t) = \frac{1}{1 - \zeta^2} e^{-2\zeta t} \sin^2\sqrt{1 - \zeta^2} \, t$$

From

$$\sin^2 x = \frac{1 - \cos 2x}{2}$$

we obtain

$$c^2(t) = \frac{1}{1 - \zeta^2} e^{-2\zeta t} \frac{1}{2}[1 - \cos 2\sqrt{1 - \zeta^2} \cdot t].$$

Taking the Laplace transform of $c^2(t)$

$$L[c^2(t)] = \frac{1}{2(1 - \zeta^2)}\left[\frac{1}{s + 2\zeta} - \frac{s + 2\zeta}{(s + 2\zeta)^2 + 4(1 - \zeta^2)}\right]$$

Using the final limit theorem

$$\int_0^\infty c^2(t)dt = \lim_{s\to 0} L[c^2(t)] = \frac{1}{2(1 - \zeta^2)}\left[\frac{1}{2\zeta} - \frac{2\zeta}{(2\zeta)^2 + 4(1 - \zeta^2)}\right]$$

$$= \frac{1}{4\zeta}.$$

For the following transfer function

$$\frac{y(s)}{r(s)} = \frac{3(s + 2)}{(s + 4)(s + 1)^2}$$

find the system response to a unit step input.

<u>Solution:</u> For a unit step, $r(s) = \frac{1}{s}$, then

$$y(s) = \frac{3(s + 2)}{s(s + 4)(s + 1)^2}$$

Expanding $y(s)$ in partial fraction yields

$$y(s) = \frac{A}{s} + \frac{B}{s + 4} + \frac{C}{(s + 1)^2} + \frac{D}{s + 1}$$

Then $A = \frac{3(s + 2)}{(s + 4)(s + 1)^2}\bigg|_{s=0} = \frac{6}{4} = \frac{3}{2}$

$$B = \frac{3(s + 2)}{s(s + 1)^2}\bigg|_{s=-4} = \frac{-3 \cdot 2}{(-4) \cdot 9} = \frac{1}{6}$$

$$C = \frac{3(s + 2)}{s(s + 4)}\bigg|_{s=-1} = \frac{3}{-3} = -1$$

$$D = \frac{d}{ds}\left[\frac{3(s + 2)}{s(s + 4)}\right]\bigg|_{s=-1} = \frac{3s(s + 4) - 3(s + 2)(2s + 4)}{s^2(s + 4)^2}\bigg|_{s=-1}$$

$$= -\frac{5}{3}$$

The partial-fraction form for $y(s)$ is

$$y(s) = \frac{\frac{3}{2}}{s} + \frac{\frac{1}{6}}{s + 4} + \frac{-1}{(s + 1)^2} + \frac{-\frac{5}{3}}{s + 1}$$

From the Laplace transform tables,

$$y(t) = \frac{3}{2} + \frac{1}{6}e^{-4t} - te^{-t} - \frac{5}{3}e^{-t} .$$

Consider the system

$$x^{(n)} + a_1 x^{(n-1)} + a_2 x^{(n-2)} + \ldots + a_{n-1}\dot{x} + a_n x$$

$$= \dot{u} + \alpha u$$

with the initial conditions

$$x(0) = \dot{x}(0) = \ldots = x^{(n-1)}(0) = 0$$

Prove that for u given by

$$u(t) = \begin{cases} 0 & \text{for} \quad t < 0 \\ e^{-\alpha t} & \text{for} \quad t \geq 0 \end{cases}$$

the response x(t) is identical with the impulse response function h(t), which satisfies

$$h^{(n)} + a_1 h^{(n-1)} + \ldots + a_{n-1}\dot{h} + a_n h = \delta(t)$$

Solution: u(t) can be written as u(t)f(t) where

$$f(t) = \begin{cases} 0 & \text{for} \quad t < 0 \\ 1 & \text{for} \quad t \geq 0 \end{cases} .$$

Then $u(t) = f(t)e^{-\alpha t}$. Differentiating we get

$$\dot{u}(t) = \frac{d}{dt}(f(t))e^{-\alpha t} + \frac{d}{dt}(e^{-\alpha t})f(t)$$

$\frac{d}{dt}[f(t)]$ is a unit impulse

$$\frac{d}{dt}[f(t)] = \delta(t)$$

$$\dot{u}(t) = \delta(t)e^{-\alpha t} - \alpha e^{-\alpha t}f(t)$$

Since u(t) = 0 for t < 0, we can drop f(t) in the second term, thus

$$\dot{u}(t) = \delta(t)e^{-\alpha t} - \alpha e^{-\alpha t}$$

Hence for t ≥ 0,

$$\dot{u}(t) + \alpha u(t) = \delta(t)e^{-\alpha t} - \alpha e^{-\alpha t} + \alpha e^{-\alpha t} = \delta(t)e^{-\alpha t}$$

$$= \delta(t)$$

We proved that the response of the system is given by the impulse response function h(t).

● **PROBLEM** 6-9

The system is described by

$$\frac{dx}{dt} + x = k$$

Using a) the time domain, and b) the frequency domain, find the zero-state response for a unit-step function input k(t) = S(t).

<u>Solution:</u> The transfer function is found from

$$\frac{dx}{dt} + x = k$$

$$L\left[\frac{dx}{dt} + x\right] = L[k(t)]$$

$$sX(s) + X(s) = K(s)$$

$$\frac{X(s)}{K(s)} = \frac{1}{s + 1}$$

thus

$$G(s) = \frac{1}{s + 1}$$

a) The impulse response is the inverse of $G(s)$

$$g(t - t_0) = e^{-(t - t_0)}$$

and

$$x(t) = \int_0^t e^{-(t - \theta)} d\theta = 1 - e^{-t} \qquad \text{for} \qquad t \geq 0.$$

b) The frequency domain

$$K(s) = \frac{1}{s}$$

Then

$$X(s) = K(s) \frac{1}{s + 1} = \frac{1}{s(s + 1)}$$

Let $s = j\omega$, we have

$$X(j\omega) = \frac{1}{j\omega(j\omega + 1)}$$

The partial fraction expansion yields

$$\frac{1}{j\omega(j\omega + 1)} = \frac{A}{j\omega} + \frac{B}{j\omega + 1}$$

$$\frac{1}{j\omega + 1}\bigg|_{j\omega=0} = A$$

$$A = 1$$

$$B = \frac{1}{j\omega}\bigg|_{j\omega=-1} = -1$$

Then

$$X(j\omega) = \frac{1}{j\omega} - \frac{1}{j\omega + 1}$$

Taking the inverse Laplace transform we get

$$x(t) = 1 - e^{-t}.$$

The closed-loop transfer function of the system is

$$\frac{y(s)}{r(s)} = \frac{6(s + 3)}{(s + 8)(s^2 + 4s + 8)}$$

Find the impulse response of this system.

Solution: We have for the impulse response $r(s) = 1$, and

$$y(s) = \frac{6(s + 3)}{(s + 8)(s^2 + 4s + 8)}$$

Note that

$$s^2 + 4s + 8 = (s + 2)^2 + 2^2$$

We can rewrite $y(s)$ as follows

$$y(s) = \frac{6(s + 3)}{(s + 8)(s + 2 + 2j)(s + 2 - 2j)}$$

$$= \frac{A}{s + 8} + \frac{B}{s + 2 + 2j} + \frac{\overline{B}}{s + 2 - 2j}$$

$$A = \left. \frac{6(s + 3)}{(s + 2)^2 + 2^2} \right|_{s=-8} = -\frac{3}{4}$$

$$B = \left. \frac{6(s + 3)}{(s + 8)(s + 2 - 2j)} \right|_{s=-2-2j} = \frac{3}{8} + \frac{3}{8}j$$

We have

$$y(s) = \frac{-\frac{3}{4}}{s + 8} + \frac{\frac{3}{8} + \frac{3}{8}j}{s + 2 + 2j} + \frac{\frac{3}{8} - \frac{3}{8}j}{s + 2 - 2j}$$

$y(t)$ can be found directly from the tables. Let us add the last two terms of the fraction expansion.

$$y(s) = \frac{-\frac{3}{4}}{s + 8} + \frac{0.75(s + 4)}{(s + 2)^2 + 2^2}$$

The inverse transform is

$$y(s) = -0.75e^{-8t} + \frac{0.75}{2}[(4 - 2)^2 + 2^2]^{1/2}e^{-2t}\sin(2t + \psi)$$

where

$$\psi = \text{arc tan } \frac{2}{4 - 2}$$

$$\psi = 45^{\circ}$$

$$y(s) = -0.75e^{-8t} + 1.06e^{-2t}\sin(2t + 45^{\circ}).$$

Using a system shown below, explain why the proportional control of a plant which does not have an integrating property, that is the transfer function of the plant does not include the factor $\frac{1}{s}$, suffers offset in response to step inputs.

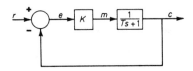

Solution: For the system shown above, if

$$c = r$$

at the steady state, then

$$e = 0$$

and

$$K \cdot 0 = 0 = m$$

and

$$c = 0$$

which contradicts the assumption. Thus, such a control system must have a nonzero offset. That is, at steady state if

$$e = \frac{r}{1 + K}$$

then

$$m = \frac{Kr}{1 + K}$$

and

$$c = \frac{Kr}{1 + K}$$

Thus the assumed error signal is

$$e = \frac{r}{1 + K}$$

and the offset is $\frac{r}{1 + K}$.

The transfer function of the system is

$$\frac{\theta_0}{\theta_i} = \frac{2}{2D^2 + 5D - 3}$$

For a step forcing function, $\theta_i = 2$, obtain the time response of the system.

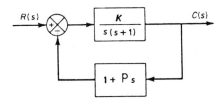

Solution: Assuming the zero initial conditions for θ_0 and θ_i, the Laplace transform is

$$\frac{\theta_0}{\theta_i}(s) = \frac{2}{2s^2 + 5s - 3} = \frac{1}{s^2 + 2.5s - 1.5}$$

$$= \frac{1}{(s + 3)(s - 0.5)}$$

For $\theta_i = 2$, $\theta_i(s) = \frac{2}{s}$ and

$$\theta_0(s) = \frac{2}{s(s + 3)(s - 0.5)} = \frac{\dfrac{2}{3 \cdot (-0.5)}}{s\left(\dfrac{s}{3} + 1\right)\left(\dfrac{s}{-0.5} + 1\right)}$$

$$= \frac{-1.33}{s(0.33s + 1)(-2s + 1)}$$

$\theta_0(s)$ is in a standard form. Using the transform pair

$$\frac{HK}{s(1 + T_1 s)(1 + T_2 s)} \Leftrightarrow HK\left[1 + \frac{1}{T_2 - T_1}\left(T_1 e^{-\frac{t}{T_1}} - T_2 e^{-\frac{t}{T_2}}\right)\right]$$

we obtain

$$\theta_0(t) = -1.33\left[1 + \frac{1}{-2 - 0.33}\left(0.33e^{-\frac{t}{0.33}} + 2e^{\frac{t}{2}}\right)\right]$$

$$= -1.33 + 0.188e^{-\frac{t}{0.33}} + 1.141e^{\frac{t}{2}} .$$

The block diagram of a control system is shown below.

Determine the values of K and P so that the maximum over-shoot in the unit step response is 0.4 and the peak time is 1 sec. Then obtain the rise time.

<u>Solution:</u> The maximum overshoot M_p is given by

$$M_p = e^{-\frac{\zeta}{\sqrt{1 - \zeta^2}}\pi}$$

Thus for $M_p = 0.4$ we obtain

$$0.4 = e^{-\frac{\zeta}{\sqrt{1 - \zeta^2}}\pi}$$

$$\ln 0.4 = -\frac{\zeta}{\sqrt{1 - \zeta^2}}\pi$$

$$\frac{\zeta^2 \pi^2}{1 - \zeta^2} = (\ln 0.4)^2 = 0.84$$

$$\zeta^2 \pi^2 = 0.84 - 0.84\zeta^2$$

$$\zeta^2 = \frac{0.84}{\pi^2 + 0.84}$$

$$\zeta = 0.28 \ .$$

The peak time is

$$t_p = \frac{\pi}{\omega_d}$$

$t_p = 1$ sec, thus $\omega_d = 3.14$.

ω_n is given by

$$\omega_n = \frac{\omega_d}{\sqrt{1 - \zeta^2}} = \frac{3.14}{\sqrt{1 - (0.28)^2}} = 3.27 \ .$$

From the block diagram we have

$$\frac{C(s)}{R(s)} = \frac{G}{1 + GH} = \frac{\frac{K}{s(s + 1)}}{1 + \frac{K(1 + Ps)}{s(s + 1)}} = \frac{K}{s^2 + (KP + 1)s + K}$$

Thus the characteristic equation is

$$s^2 + (KP + 1)s + K = s^2 + 2\zeta\omega_n s + \omega_n^2$$

Comparing the coefficients we obtain

$$K = \omega_n^2, \quad KP + 1 = 2\zeta\omega_n$$

$$K = 3.27^2 = 10.69$$

$$P = \frac{2\zeta\omega_n - 1}{K} = 0.077$$

The rise time t_r is given by

$$t_r = \frac{\pi - \beta}{\omega_d}$$

where

$$\beta = \tan^{-1}\left(\frac{\omega_d}{T}\right)$$

and

$$T = \zeta\omega_n = 0.9156$$

$$\beta = \tan^{-1}(3.43)$$

$$\beta = 1.29.$$

$$t_r = \frac{\pi - 1.29}{3.14} = 0.59 \text{ sec.}$$

● **PROBLEM** 6-14

The open-loop transfer function of the system is

$$G(j\omega) = \frac{K}{(1 + j\omega T_1)(1 + j\omega T_2)}$$

From the Bode plot determine the gain constant.

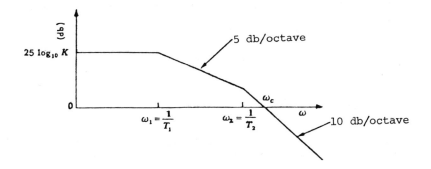

Solution: The gain constant is the very low-frequency value of $G(j\omega)$, setting $\omega = 0$ we obtain

$$G(j\omega) = \frac{K}{(1 + 0)(1 + 0)} = K$$

in decibels

$$[db] = 25 \log_{10} K$$

● **PROBLEM** 6-15

The electrical circuit is shown below.

Assuming the initial conditions are zero, find the time response of the current in the circuit to a unit voltage impulse.

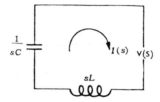

Solution: The system is described by the following equation.

$$V(s) = I(s)sL + \frac{I(s)}{sC}$$

Solving for $I(s)$ and taking a unit voltage

$$I(s) = \frac{V(s)Cs}{s^2LC + 1} = \frac{Cs}{s^2LC + 1}$$

Inverting the last equation, we obtain

$$i(t) = \frac{1}{LC} \cdot C \cdot \cos \frac{t}{\sqrt{LC}} = \frac{1}{L} \cos \frac{t}{\sqrt{LC}}$$

For $t = 0^+$, $i(0^+) = \frac{1}{L}$.

At $t = 0^+$ the capacitor is uncharged. Across the inductor, we get

$$v(t) = L\frac{di}{dt},$$

$$i_{t=0^+} = \frac{1}{L} \int_0^{0^+} v(t)dt$$

The integral represents the area under the time-voltage function, which equals unity. Thus

$$i(0^+) = \frac{1}{L}$$

This agrees with initial current found by the Laplace transform method.

● **PROBLEM** 6-16

For the system

$$\frac{\theta_0}{\theta_i} = \frac{2}{4D^2 + 3D + 1}$$

obtain and plot the response expression, when the system is subjected to the forcing function $\theta_i = 2$. θ_0 is the change in the response variable due to θ_i.

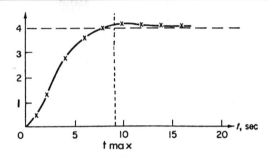

Solution: The system equation is

$$4D^2\theta_0 + 3D\theta_0 + \theta_0 = 4.$$

To find the particular solution, let us try

$$\theta_{01} = \text{const},$$

then

$$D^2\theta_{01} = D\theta_{01} = 0$$

thus

$$\theta_{01} = 4$$

The system's characteristic equation is

$$4D^2 + 3D + 1 = 0$$

and its roots are

$$D_1 = -0.375 + j0.33$$

$$D_2 = -0.375 - j0.33.$$

The complementary solution is in the form

$$\theta_{02} = A_1 e^{D_1 t} + A_2 e^{D_2 t} = Ae^{-0.375t} \sin(0.33t + \phi)$$

Thus the full solution is

$$\theta_0 = \theta_{01} + \theta_{02} = 4 + Ae^{-.375t} \sin(0.33t + \phi).$$

We shall evaluate A and ϕ from two initial conditions

1. $\theta_0 = 0$ at $t = 0$

    $4 + A \sin\phi = 0$

2. $\dot{\theta} = 0$ at $t = 0$

Thus

$$0.375A \sin \phi = 0.33A \cos \phi$$

$$\frac{\sin \phi}{\cos \phi} = \tan \phi = \frac{0.33}{0.375} = 0.88$$

$$\phi = 41.3^\circ.$$

From

$$0 = 4 + A \sin \phi$$

we obtain

$$A = \frac{-4}{\sin \phi} = -6.08$$

The response function is

$$\theta_0 = 4 - 6.08e^{-0.375t} \sin(0.33t + 41.3^\circ).$$

Note that

$$\lim_{t \to \infty} \theta_0 = 4$$

The function has only one maximum

$$\frac{d\theta_0}{dt} = 0$$

for

$$t = t_{max}$$

● **PROBLEM 6-17**

For the system shown in the figure which has one closed-loop zero and three closed-loop poles, obtain the time response of the system to a unit step input.

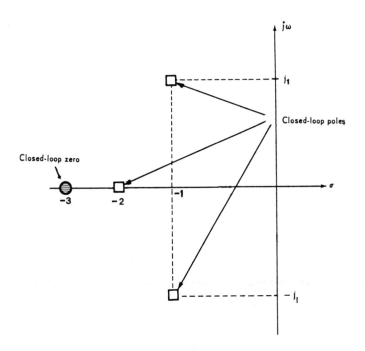

Solution: The closed loop poles are located:

$$\lambda_1 = -1 + j$$

$$\lambda_2 = -1 - j$$

$$\lambda_3 = -2$$

The response of the system to a unit step input is of the form

$$c(t) = 1 + A_1 e^{(-1 + j)t} + A_2 e^{(-1 - j)t} + A_3 e^{-2t}$$

where the amplitudes A_1, A_2 and A_3 are determined from

$$A_K = -\frac{\overset{n}{\Pi_*}(-\lambda_i)\overset{m}{\Pi}(\lambda_K - z_j)}{\overset{n}{\Pi}(-z_i)\overset{n}{\Pi_*}(\lambda_K - \lambda_j)}$$

where Π denotes product of all terms and Π_* denotes product over all j except $j = k$.

We get

$$A_1 = -\frac{(1 + j)2}{3} \cdot \frac{(-1 + j + 3)}{(-1 + j + 1 + j)(-1 + j + 2)}$$

$$= -\frac{1}{3} + \frac{2}{3}j$$

$$A_2 = -\frac{(1 - j) \cdot 2}{3} \cdot \frac{(-1 - j + 3)}{(-1 - j + 1 - j)(-1 - j + 2)}$$

204

$$= -\frac{1}{3} - \frac{2}{3}j$$

$$A_3 = \frac{-(1 - j)(1 + j)}{3} \quad \frac{(-2 + 3)}{(-2 + 1 - j)(-2 + 1 + j)}$$

$$= -\frac{1}{3}$$

The time response will be

$$c(t) = 1 + (-\frac{1}{3} + \frac{2}{3}j)e^{(-1+j)t} + (-\frac{1}{3} - \frac{2}{3}j)e^{(-1-j)t}$$

$$+ (-\frac{1}{3})e^{-2t}$$

● **PROBLEM** 6-18

Consider the nonlinear system of differential equations:

$$\dot{x}_1 = -x_1 - Bx_1x_2$$

$$\dot{x}_2 = Bx_1x_2 - x_2 + u(t) \qquad (*)$$

$$\dot{x}_3 = x_1 + x_2$$

The coupling coefficient B varies in time B = B(t) and is approximated by

$$B(t) = 1 + \sin\frac{\Pi}{2}t$$

Using discrete approximation find the system response for k = 1 and k = 2. Assume the following initial conditions:

$$x_1(0) = 1, \; x_2(0) = 0, \; x_3(0) = 0, \; u(0) = 1$$

$$u(k) = 0 \quad \text{for } k \geq 1.$$

Select the time increment T = 0.2 sec.

Solution: The discrete approximation of the derivative is

$$\dot{x}(t) = \frac{1}{T} [x(t+T) - x(t)]$$

Let   t = kT , we have

$$\dot{x}(k) = \frac{1}{T} [x(k+1) - x(k)]$$

Substituting the above equation into (*) we get

$$\frac{x_1(k+1) - x_1(k)}{T} = -x_1(k) - B(k)x_1(k)x_2(k)$$

$$\frac{x_2(k+1) - x_2(k)}{T} = B(k)x_1(k)x_2(k) - x_2(k) + u(k)$$

$$\frac{x_3(k+1) - x_3(k)}{T} = x_1(k) + x_2(k)$$

For   T = 0.2s we obtain

$x_1(k+1) = 0.8x_1(k) - 0.2B(k)x_1(k)x_2(k)$

$x_2(k+1) = 0.8x_2(k) + 0.2B(k)x_1(k)x_2(k) + 0.2u(k)$

$x_3(k+1) = x_3(k) + 0.2\ x_1(k) + 0.2\ x_2(k)$

The response for k = 0 , t = T is

   $x_1(1) = 0.8$

   $x_2(1) = 0.2$

   $x_3(1) = 0.2$

For $B(t) = 1 + \sin\frac{\pi}{2}t$ we have

   $B(k) = 1 + \sin\left(\frac{\pi}{2}\right)kT$

For k = 1 we have

   $x_1(2) = 0.598$

   $x_2(2) = 0.202$

   $x_3(2) = 0.40$

To get approximations for the higher values of k we repeat the above procedure.

● **PROBLEM 6-19**

A mechanical system shown below is originally at rest.

Find the motion $x_2 = x_2(t)$ assuming that the input x_1 is a unit step function.

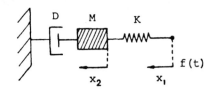

Solution: Let us establish the system of equations describing the motion of the system. The forces on the mass M are due to the damper and the spring; summing them and using Newton's law we get

$$- D\dot{x}_2 - K(x_2 - x_1) = M \ddot{x}_2$$

or

$$M \ddot{x}_2 + D\dot{x}_2 + Kx_2 = Kx_1 \tag{1}$$

We represent $x_2(t)$ as a sum of the steady-state solution and the transient solution

$$x_2(t) = x_{2,ss} + x_{2,t}$$

Note that when x_2 reaches a constant value, the velocity and acceleration become zero.

Thus

$$\ddot{x}_2 = \dot{x}_2 = 0 \ ,$$

substituting into eq. (1) we get for the steady-state value

$$x_2(\infty) = x_1$$

The characteristic equation is

$$Mm^2 + Dm + K = K \left(\frac{M}{K} m^2 + \frac{D}{K} m + 1 \right) = 0$$

or in terms of ζ and ω_n

$$\frac{1}{\omega_n^2} m^2 + \frac{2\zeta}{\omega_n} m + 1 = 0$$

where

$$\omega_n = \sqrt{\frac{K}{M}}$$

and

$$\zeta = \frac{D}{2\sqrt{KM}}$$

The roots are

$$m_1 = - \zeta\omega_n + \omega_n \sqrt{\zeta^2 - 1}$$

$$m_2 = - \zeta\omega_n - \omega_n \sqrt{\zeta^2 - 1}$$

Depending on the value of ζ we obtain the transient solution.

For $\zeta > 1$, m_1, m_2 are real and the transient response is

$$x_{2,t} = A_1 e^{(-\zeta + \sqrt{\zeta^2 - 1})\omega_n t} + A_2 e^{(-\zeta - \sqrt{\zeta^2 - 1})\omega_n t}$$

For $\zeta = 1$, the roots are real, $m_1 = m_2 = - \zeta \omega_n$

$$x_{2,t} = A_1 e^{-\zeta \omega_n t} + A_2 t e^{-\zeta \omega_n t}$$

For $\zeta < 1$, the roots are complex

$$m_{1,2} = - \zeta \omega_n \pm j \omega_n \sqrt{1 - \zeta^2}$$

and the transient solution is

$$x_{2,t} = A e^{-\zeta \omega_n t} \sin(\omega_n \sqrt{1 - \zeta^2}\, t + \phi)$$

For the case when $\zeta < 1$, the complete solution is

$$x_2(t) = 1 + A e^{-\zeta \omega_n t} \sin(\omega_n \sqrt{1 - \zeta^2} \cdot t + \phi).$$

The two constants A and ϕ are determined from the initial conditions. The system was initially at rest, thus $x_2(0) = 0$.

The second condition, $\frac{d}{dt} x_2(0) = 0$, we obtain from the fact that the velocity of a system with mass cannot change instantaneously. Differentiating the solution of $x_2(t)$ we get

$$\frac{d}{dt} x_2(t) = - \zeta \omega_n A e^{-\zeta \omega_n t} \sin(\omega_n \sqrt{1 - \zeta^2}\, t + \phi)$$

$$+ \omega_n \sqrt{1 - \zeta^2}\, A e^{-\zeta \omega_n t} \cos(\omega_n \sqrt{1 - \zeta^2}\, t + \phi).$$

Inserting the initial conditions

$$x_2(0) = 0 \quad \text{and} \quad \frac{d}{dt} x_2(0) = 0$$

we find

$$A \sin \phi + 1 = 0$$

$$- \zeta \omega_n A \sin\phi + \omega_n \sqrt{1-\zeta^2} A \cos\phi = 0$$

Solving for A and ϕ

$$A = - \frac{1}{\sqrt{1-\zeta^2}}$$

$$\phi = \tan^{-1} \frac{\sqrt{1-\zeta^2}}{\zeta} = \cos^{-1}\zeta$$

We can write the complete solution

$$x_2(t) = 1 - \frac{e^{-\zeta\omega_n t}}{\sqrt{1-\zeta^2}} \sin(\omega_n \sqrt{1-\zeta^2}\, t + \cos^{-1}\zeta)$$

where

$$\omega_n = \sqrt{\frac{K}{M}}$$

and

$$\zeta = \frac{D}{2\sqrt{KM}} \quad .$$

● **PROBLEM** 6-20

The system is described by the equations

$$\ddot{x}_1 + \dot{x}_2 + x_1^2 = y_1$$

$$\ddot{x}_2 + \dot{x}_1 + x_2^2 = y_2 \qquad (*)$$

Obtain the first four terms of the Taylor series expansion of x_1 in the neighborhood of $t_0 = 0$. The initial conditions are

$$x_1(0) = x_2(0) = \frac{dx_1(0)}{dt} = \frac{dx_2(0)}{dt} = 0$$

and the functions y_1 and y_2 are

$$y_1 = 2e^{-t}, \qquad y_2 = e^t$$

<u>Solution:</u> The Taylor series of a function $x(t)$ is

$$x(t) = \sum_{k=0}^{\infty} x^{(k)}(t_0)(t-t_0)^k \cdot \frac{1}{k!}$$

where

$$x^{(k)}(t_0) = \left.\frac{d^k x}{dt^k}\right|_{t=t_0} \quad , \quad x^{(0)} = x$$

For $t_0 = 0$ we have

$$x_1(t) = x_1^{(0)}(0)\frac{1}{0!} + \dot{x}_1(0) \cdot t \frac{1}{1!} + \ddot{x}_1(0)t^2\frac{1}{2!} + \dddot{x}_1(0)t^3\frac{1}{3!}$$

$$+ x_1^{(4)}(0)t^4\frac{1}{4!} + x_1^{(5)}(0)t^5\frac{1}{5!} + \dots$$

From the initial conditions we conclude that the first two terms are zero.

From the first equation of (*) we have

$$\left.\ddot{x}_1\right|_{t=0} = y_1(0) = 2$$

Differentiating the first equation of (*) we get

$$\left.\dddot{x}_1\right|_{t=0} = \dot{y}_1 - y_2 = -3$$

Repeating this procedure we obtain

$$\left.x_1^{(4)}\right|_{t=0} = \ddot{y}_1 - x_2^{(3)} = \ddot{y}_1 - \dot{y}_2 = 2e^{-t} - e^t = 1$$

$$\left.x_1^{(5)}\right|_{t=0} = \dddot{y}_1 - \ddot{y}_2 + 2\Big|_{t=0} = -2 - 1 + 2 = -1$$

Substituting the above results we obtain

$$x_1(t) = 2t^2 \cdot \frac{1}{2!} - 3t^3\frac{1}{3!} + t^4\frac{1}{4!} - t^5\frac{1}{5!} + \dots$$

or

$$x_1(t) = t^2 - \frac{1}{2}t^3 + \frac{1}{24}t^4 - \frac{1}{120}t^5 + \dots$$

For the hydraulic positioning device shown in Fig. 1,

where:

a = 4 in.; b = 12 in.; A = 10 in.^{2}; k_V = 20 in.^{3}/sec/in.

Examine the dynamic nature of the system by plotting the response to

(1) a step input disturbance x_i = 0.5 in.,

(2) a ramp input disturbance x_i = 0.25t in.,

(3) a sinusoidal input disturbance x_i = sin 2t in. (amplitude H = 1).

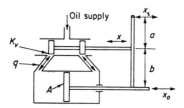

Fig. 1

Solution: First we need the system transfer function

$$\frac{x_0}{x_i} \quad .$$

From fluid dynamics we obtain

$$q = xK_V \tag{1}$$

$$\frac{q}{A} = \dot{x}_0 = Dx_0 \tag{2}$$

When a displacement x_i is introduced, the lever pivots about the lower joint by the amount

$$x_1 = x_i \left[\frac{b}{a+b}\right] \quad .$$

Then oil begins to flow down one side causing movement in the opposite direction x_0 by the amount

$$x_2 = x_0 \left[\frac{a}{a+b}\right] .$$

x_1 and x_2 are of opposite sign and the net movement is

$$x = \left[\frac{b}{a+b}\right] x_i - \left[\frac{a}{a+b}\right] x_0 \; .$$

Using eqs. (1) and (2), we eliminate x, then q, resulting in

$$\left[\left(\frac{b}{a+b}\right) x_i - \left(\frac{a}{a+b}\right) x_0\right] \frac{K_v}{AD} = x_0 \; .$$

Then the transfer function is

$$\frac{x_0}{x_i} = \frac{b/a}{1 + \left[\frac{a+b}{a}\right] \frac{AD}{K_v}}$$

Then substituting in the values of a, b, A, and K_v we get

$$\frac{x_0}{x_i} = \frac{3}{1 + 2D}$$

The transfer function can be written as the system equation

$$2Dx_0 + x_0 = 3x_i$$

(1) For a step input x_i = 0.5 in., the system equation becomes

$$2Dx_0 + x_0 = 1.5$$

which is of the form $TD\theta_0 + \theta_0 = HK$

and hence will have a solution of the form

$$\theta_0 = KH\left(1 - e^{-t/T}\right)$$

thus

$$x_0 = 3 \times 0.5\left(1 - e^{-t/2}\right) = 1.5\left(1 - e^{-0.5t}\right)$$

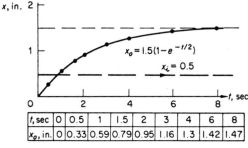

t, sec	0	0.5	1	1.5	2	3	4	6	8
x_0, in.	0	0.33	0.59	0.79	0.95	1.16	1.3	1.42	1.47

Fig. 2

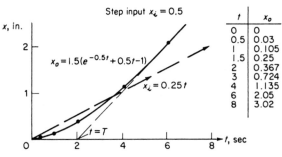

Step input $x_i = 0.5$

$x_0 = 1.5(e^{-0.5t} + 0.5t - 1)$

$x_i = 0.25t$

$t = T$

Ramp input $x_i = 0.25t$

t	x_0
0	0
0.5	0.03
1	0.105
1.5	0.25
2	0.367
3	0.724
4	1.135
6	2.05
8	3.02

Fig. 3

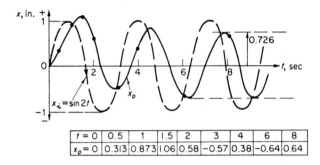

$t=0$	0.5	1	1.5	2	3	4	6	8
$x_o=0$	0.313	0.873	1.06	0.58	-0.57	0.38	-0.64	0.64

Sine input $x_i = \sin 2t$ ($H=1, \omega=2$)   Fig. 4

Figure 2 shows tabulated and plotted values of t and x_0. Note that

(a) x_0 reaches 63 per cent of its final steady-state value in $t = T = 2$ sec;

(b) x_0 reaches 98 per cent of its final steady-state value in $t = 4T = 8$ sec;

(c) the slope of the $x_0(t)$ curve at $t = 0$ is x_0 (steady-state)/T = 1.5/2 = 0.75.

(2) For a ramp input $x_i = 0.25t$, the system equation becomes

$$2Dx_0 + x_0 = 0.75t$$

which is of the form $TD\theta_0 + \theta_0 = HKt$

and hence will have a solution of the form

$$\theta_0 = HKT \left(e^{-t/T} + \frac{t}{T} - 1\right)$$

then

$$x_0 = 3 \times 0.25 \times 2 \left(e^{-t/2} + \frac{t}{2} - 1\right) = 1.5(e^{-0.5t} + 0.5t - 1)$$

Figure 3 shows tabulated and plotted values of t and x_0. Note that for $t = T = 2$ sec, the steady-state asymptote $x_{01} = 1.5(0.5t - 1)$ cuts the t axis.

(3) For a sinusoidal input $x_i = \sin 2t$, the system equation becomes

$$2Dx_0 + x_0 = 3 \sin 2t$$

which is of the form $TD\theta_0 + \theta_0 = KH \sin \omega t$

and hence will have a solution of the form

$$\theta_0 = \frac{HK}{(1 + \omega^2 T^2)^{\frac{1}{2}}} \left[\sin(\omega t - \phi) + \frac{\omega T}{(1 + \omega^2 T^2)^{\frac{1}{2}}} e^{-t/T} \right]$$

so

$$x_0 = \frac{3 \times 1}{(1 + 2^2 \times 2^2)^{\frac{1}{2}}} \left[\sin(2t - \tan^{-1} 2 \times 2) + \frac{2 \times 2}{(1 + 2^2 \times 2^2)^{\frac{1}{2}}} \right.$$

$$\left. \cdot \; e^{-t/2} \right]$$

$$= 0.726 \; [\sin(2t - \tan^{-1} 4) + 0.973 \; e^{-0.5t}]$$

Figure 4 shows tabulated and plotted values of t and x_0 .

● **PROBLEM 6-22**

Determine the current delivered by the battery in the
circuit after closure of the switch.

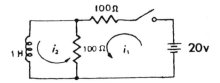

Solution: We shall use the mesh-current method summing the
voltage drops around each loop, thus

$$20 = 100i_1 + 100i_1 - 100i_2$$

$$0 = 100i_2 - 100i_1 + \frac{di_2}{dt}$$

Taking the Laplace transform of the equations we obtain

$$\frac{20}{s} = 200I_1(s) - 100I_2(s)$$

$$0 = -100I_1(s) + (100+s)I_2(s)$$

From the second equation

$$100I_1(s) = (100+s)I_2(s)$$

or

$$I_2(s) = \frac{100}{100+s} I_1(s)$$

Substituting into the first equation

$$\frac{20}{s} = 200I_1(s) - \frac{100 \cdot 100}{100+s} I_1(s)$$

Solving for $I_1(s)$

$$I_1(s) = \frac{20(s + 100)}{s[200s + 200 \cdot 100 - 100 \cdot 100]} = \frac{s + 100}{s(10s + 500)}$$

$$= \frac{s}{s(10s + 500)} + \frac{100}{s(10s + 500)} = \frac{1}{10s + 500} + \frac{10}{s(s + 50)}$$

$$= \frac{\frac{1}{10}}{s + 50} + \frac{10}{s(s + 50)}$$

Taking the inverse Laplace transform we get

$$i_1(t) = \frac{1}{10} \cdot e^{-50t} + 10 \cdot \frac{1}{50}\left(1 - e^{-50t}\right)$$

$$= \frac{1}{5} - \frac{1}{5} e^{-50t} \quad .$$

● **PROBLEM** 6-23

Consider the space-vehicle control system shown in Fig. 1. Obtain the unit step response.

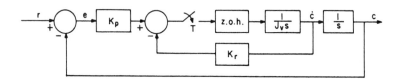

A digital space-vehicle control system.   Fig. 1

Solution: The open-loop and the closed-loop transfer functions of the system are

$$G(z) = \frac{T^2 K_p(z+1)}{2J_v z^2 + (2K_r T - 4J_v)z + 2J_v - 2K_r T}$$

and

$$\frac{C(z)}{R(z)} = \frac{T^2 K_p(z+1)}{2J_v z^2 + (2K_r T - 4J_v + T^2 K_p)z + (2J_v - 2K_r T + T^2 K_p)}$$

respectively. Let $T = 0.1$ sec, $K_r = 3.17 \times 10^5$, $J_v = 41822$, and K_p be the variable parameter, the open-loop transfer function becomes

$$G(z) = \frac{1.2 \times 10^{-7} K_p(z+1)}{(z-1)(z-0.242)}$$

215

and the characteristic equation of the closed-loop system is

$$z^2 + (1.2 \times 10^{-7} K_p - 1.242)z + 0.242 + 1.2 \times 10^{-7} K_p = 0$$

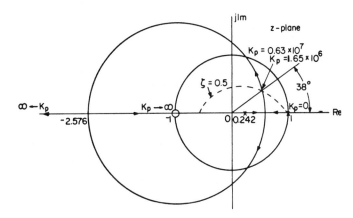

Root loci of the digital system   Fig. 2

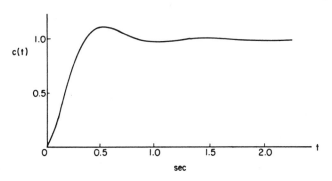

Unit-step response of the system   Fig. 3

The root locus plot of the system when K_p varies between 0 and ∞ is constructed based on the pole-zero configuration of G(z), as shown in Fig. 2. From the root locus plot we find that when the root loci cross the unit circle in the z-plane, the value of K_p is 6.32×10^6. Thus, the critical value of K_p for stability is 6.32×10^6. If it is desired to realize a relative damping ratio of 50% for the system, we sketch the constant-damping-ratio locus of $\zeta = 0.5$ in Fig. 2. The intersect between the $\zeta = 0.5$ locus and the root loci gives the desired root location, and the corresponding value of K_p which is 1.65×10^6. The frequency ω that corresponds to $K_p = 1.65 \times 10^6$ is determined from the constant-frequency locus which has an angle of 38 degrees, as shown in Fig. 2. Thus,

$$\omega T = \theta = 38 \text{ deg} = 0.66 \text{ rad}$$

The natural undamped frequency is determined as
(for T = 0.1 sec)

216

$$\omega_n = \frac{\omega}{\sqrt{1-\zeta^2}} = \frac{6.6}{\sqrt{1-0.5^2}} = 7.62 \text{ rad/sec}$$

Using the values of ζ, and ω, and the pole-zero configuration of the closed-loop transfer function, or, using a digital computer, the unit-step response of the system for $K_p = 1.65 \times 10^6$ can be computed and is plotted in Fig. 3.

● **PROBLEM** 6-24

The sampled-data system of Fig. 1 has the transfer function

$$G(s) = \frac{1}{s+1}$$

The input $r(t)$ is a unit-step function. The sampling period is 1 second. Find the response of the system at time instants of $t = kT/3$, $k = 0,1,2, \ldots$.

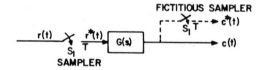

Fig. 1

Solution: From $C(z)_N = G(z)_N R(z)$, the z-transform of the system at the submultiple-sampling instants is written as

$$C(z)_3 = G(z)_3 R(z)$$

where

$$G(z)_3 = G(z)\Bigg|_{\substack{z=z^{1/3}\\T=T/3}} = \frac{z}{z-e^{-T}}\Bigg|_{\substack{z=z^{1/3}\\T=T/3}}$$

Thus,

$$G(z)_3 = \frac{z^{1/3}}{z^{1/3} - e^{-1/3}} = \frac{z^{1/3}}{z^{1/3} - 0.717}$$

The z-transform of the unit-step input is $R(z) = z/(z-1)$. The z-transform of the submultiple-sampled output is

$$C(z)_3 = \frac{z^{1/3}}{z^{1/3} - 0.717} \; \frac{z}{z-1} \tag{1}$$

However, one difficulty remains in that the last expression has fractional powers as well as integral powers of z. To overcome this difficulty, we introduce a new variable z_3, such that

$$z_3 = z^{1/3}$$

Equation (1) becomes

$$C(z)_3 = \frac{z_3}{z_3 - 0.717} \frac{z_3^3}{z_3^3 - 1} = \frac{z_3^4}{(z_3 - 0.717)(z_3^3 - 1)}$$

Expanding $C(z)_3$ into a power series in z_3^{-1}, we have

$$C(z)_3 = 1 + 0.717z_3^{-1} + 0.513z_3^{-2} + 1.368z_3^{-3} + 0.98z_3^{-4} + 0.703z_3^{-5}$$

$$+ 1.504z_3^{-6} + 1.08z_3^{-7} + 0.773z_3^{-8} + 1.55z_3^{-9} + \ldots$$

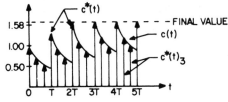

Output responses of the system   Fig. 2

The coefficients of the power-series expansion of $C(z)_3$ are the values of $c*(t)_3$ at $t = kT/3$, $k = 0,1,2, \ldots$. The response $c*(t)_3$ is shown in Fig. 2. In this case, the value of the submultiple sampling method is clearly demonstrated, since the ordinary z-transform obviously would produce a misleading result.

● PROBLEM 6-25

Consider the liquid-level control system shown in Fig. 1. Assume that the set point of the controller is fixed. Assuming a step disturbance of magnitude n_0, determine the error. Assume that n_0 is small and the variations in the variables from their respective steady-state values are also small. The controller is a proportional one.

If the controller is not a proportional one, but integral, what is the steady-state error?

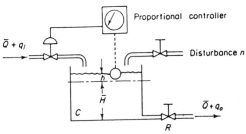

Liquid-level control system.   Fig. 1

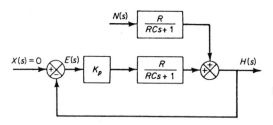

Fig. 2 Block diagram of the liquid-level control system shown in Fig. 1

Solution: Figure 2 is a block diagram of the system when the controller is proportional with gain K_p. (We assume the transfer function of the pneumatic valve to be unity.) Since the set point is fixed, the variation in the set point is zero, or $X(s) = 0$. The Laplace transform of $h(t)$ is

$$H(s) = \frac{K_p R}{RCs + 1} E(s) + \frac{R}{RCs + 1} N(s)$$

Then

$$E(s) = -H(s) = -\frac{K_p R}{RCs + 1} E(s) - \frac{R}{RCs + 1} N(s)$$

Hence

$$E(s) = -\frac{R}{RCs + 1 + K_p R} N(s)$$

Since

$$N(s) = \frac{n_0}{s}$$

we obtain

$$E(s) = -\frac{R}{RCs + 1 + K_p R} \frac{n_0}{s}$$

$$= \frac{R n_0}{1 + K_p R} \left(\frac{1}{s + \dfrac{1 + K_p R}{RC}} \right) - \frac{R n_0}{1 + K_p R} \frac{1}{s}$$

The time solution for $t > 0$ is

$$e(t) = \frac{R n_0}{1 + K_p R} \left[\exp\left(-\frac{1 + K_p R}{RC} t \right) - 1 \right]$$

Thus, the time constant is $RC/(1 + K_p R)$. (In the absence of the controller, the time constant is equal to RC.) As the gain of the controller is increased, the time constant is decreased. The steady-state error is

$$e(\infty) = -\frac{R n_0}{1 + K_p R}$$

219

As the gain K_p of the controller is increased, the steady-state error, or offset, is reduced. Thus, mathematically, the larger the gain K_p is, the smaller the offset and time constant are. In practical systems, however, if the gain K_p of the proportional controller is increased to a very large value, oscillation may result in the output since in our analysis all the small lags and small time constants which may exist in the actual control system are neglected. (If these small lags and time constants are included in the analysis, the transfer function becomes higher order and for very large values of K_p the possibility of oscillation or even instability may occur.)

If the controller is an integral one, then assuming the transfer function of the controller to be

$$G_c = \frac{K}{s}$$

we obtain

$$E(s) = - \frac{Rs}{RCs^2 + s + KR} N(s)$$

The steady-state error for a step disturbance $N(s) = n_0/s$ is

$$e(\infty) = \lim_{s \to 0} sE(s)$$

$$= \lim_{s \to 0} \frac{-Rs^2}{RCs^2 + s + KR} \frac{n_0}{s}$$

$$= 0$$

Thus, an integral controller eliminates steady-state error or offset due to the step disturbance. (The value of K must be chosen so that the transient response due to the command input and/or disturbance damps out with a reasonable speed.)

● **PROBLEM** 6-26

A windup oscillating device consists of two steel spheres on each end of a long slender rod. The entire display is hung on a thin wire which can be twisted many revolutions without breaking. The device will be wound up 4000° (about 11.1 revolutions) when the store opens. How long will it take until the motion decays down to 10°? The device is shown in Fig. 1.

Solution: The system is a torsional one. The mass J is identified as the rigid rod with spheres on either end. A torsion spring K (in the form of a thin wire) connects the mass J to the ground. The spheres, acting as air

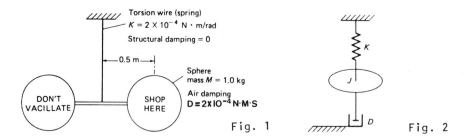

Fig. 1   Fig. 2

damper (at low velocity, the damper is approximately linear) D also connects the mass J to the ground. Hence, the free body diagram is shown in Fig. 2.

Since the system is initially wound up, it has an initial displacement θ_0. The differential equation becomes

$$J\ddot{\theta} + D\dot{\theta} + K\theta = 0$$

$$\text{IC } \theta(0) = \theta_0, \quad \dot{\theta}(0) = 0 \tag{1}$$

While eq. (1) is the system equation, it will not be necessary to solve it. We are interested only in the form of the solution.

Transforming eq. (1) into the Laplace domain,

$$J(s^2\theta - s\theta_0 - \dot{\theta}_0) + D(s\theta - \theta_0) + K\theta = 0 \tag{2}$$

Since $\dot{\theta}_0 = 0$, we may rewrite eq. (2) as

$$\theta = \frac{\left(s + \dfrac{D}{J}\right)\theta_0}{s^2 + \dfrac{D}{J}s + \dfrac{K}{J}} = \frac{(s + 2\zeta\omega_n)\theta_0}{s^2 + 2\zeta\omega_n s + \omega_n^2} \tag{3}$$

This is a decaying sinusoid with a decay envelop given by

$$\theta = \theta_0 e^{-at} \tag{4}$$

where

$$a = \zeta\omega_n \tag{5}$$

Consequently, to solve this problem, we must find J, the damping ratio ζ, and the natural frequency ω_n.

In order to find the moment of inertia, neglect the mass of the rod. Assume the rod is rigid and that the mass of each sphere is concentrated at its center of mass. Then,

$$J = 2Mr^2 = 2(1.0)(0.5)^2 = 0.5 \text{ kg} \cdot \text{m}^2 \tag{6}$$

and

$$\omega_n = \sqrt{\frac{K}{J}} = \sqrt{\frac{2 \times 10^{-4}}{0.5}} = 0.02 \text{ rad/s} \tag{7}$$

221

$$\zeta = \frac{D}{2J\omega_n} = \frac{2 \times 10^{-4}}{2(0.5)(0.02)} = 0.01 \tag{8}$$

Then,

$$a = (0.01)(0.02) = 2 \times 10^{-4} \tag{9}$$

Given that

$$\theta_0 = \text{initial displacement} = 4000° \tag{10}$$

and

$$\theta = \text{final displacement} = 10° \tag{11}$$

apply numerical values to eq. (4) and solve for t

$$t = \frac{1}{a} \ln \frac{\theta_0}{\theta} = \frac{1}{2 \times 10^{-4}} \ln \frac{4000}{10} = 3 \times 10^4 s = 8.35h \tag{12}$$

If the store stays open for 8 h, then one winding should be sufficient.

● **PROBLEM 6-27**

A scalar system is described by

$$\dot{x} = -x + e^t$$

$$x(0) = 10$$

Find the solution x(t).

Derive a discrete approximation using

$$t_{k+1} - t_k = 1$$

and solve the discrete system.

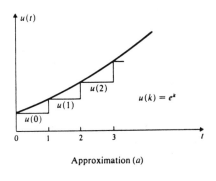

Approximation (a)

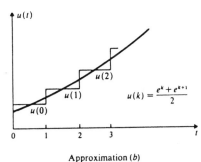

Approximation (b)

Fig. 1

Table 1

		$t = k = 0$	$t = k = 1$	$t = k = 2$	$t = k = 3$	$t = k = 4$
Continuous result: sinh t		0	1.1752	3.6269	10.018	27.290
Discrete result: $\sum\limits_{j=1}^{k} (0.368)^{k-j}[0.632 u(j-1)]$	input (a)	0	0.632	1.951	5.388	14.677
	input (b)	0	1.175	3.626	10.016	27.286

<u>Solution</u>: The general form of the equation

$$\dot{x} = Ax + Bu$$

has the solution

$$x(t) = e^{At} x(0) + \int_0^t e^{A(t-\tau)} B(\tau) u(\tau) d\tau$$

In our case

$$A = -1, \ x(0) = 10, \ B(\tau) = 1 \text{ and } u(\tau) = e^{\tau}, \text{ thus}$$

$$x(t) = 10e^{-t} + \int_0^t e^{-(t-\tau)} e^{\tau} d\tau = 10e^{-t} \sinh t.$$

Using the scalar transition matrix, we find

$$x(k+1) = \phi(k+1,k) x(k) + \int_k^{k+1} \phi(k+1,\tau) B(\tau) u(\tau) d\tau$$

where

$$\phi(t,0) = e^{At} = e^{-t}$$

and

$$\phi(k+1,k) = \phi([k+1]-k,0) = e^{-1}$$

$$\phi(k+1,\tau) = \phi(k+1-\tau) = e^{-(k+1-\tau)} ; \qquad B(\tau) = 1$$

Thus we have

$$x(k+1) = e^{-1} x(k) + e^{-k-1} \int_k^{k+1} e^{\tau} u(\tau) d\tau$$

Assuming that $u(\tau)$ is constant over the interval $[k+1,k]$, we obtain

$$x(k+1) = e^{-1} x(k) + [1 - e^{-1}] u(k) \approx 0.368 \, x(k) + 0.632u(k)$$

For the general equation

$$x(k+1) = Ax(k) + Bu(k)$$

we have

$$x(1) = Ax(0) + Bu(0)$$

$$x(2) = Ax(1) + Bu(1) = A^2 x(0) + ABu(0) + Bu(1)$$

$$x(3) = Ax(2) + Bu(2) = A^3 x(0) + A^2 Bu(0) + ABu(1) + Bu(2)$$

Thus in general

$$x(k) = A^k x(0) + \sum_{j=1}^{k} A^{k-j} Bu(j-1)$$

In our case

$$A = 0.368, \ B = 0.632, \ x(0) = 10$$

$$x(k) = 10(0.368)^k + \sum_{j=1}^{k} (0.368)^{k-j} [0.632u(j-1)]$$

● **PROBLEM** 6-28

Consider a system consisting of a spring, a mass and a damper. The system is described by the following equations

$$\begin{bmatrix} \dot{x}_1 \\ \\ \dot{x}_2 \end{bmatrix} = \begin{bmatrix} 0 & 1 \\ \\ -\dfrac{K}{M} & -\dfrac{f}{M} \end{bmatrix} \begin{bmatrix} x_1 \\ \\ x_2 \end{bmatrix} + \begin{bmatrix} 0 \\ \\ \dfrac{1}{M} \end{bmatrix} r(t) \ .$$

Obtain the system response, assuming that

$$\frac{K}{M} = 2, \qquad \frac{f}{M} = 3 , \qquad M = 1 \ .$$

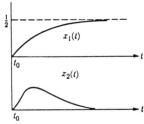

The transient response

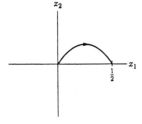

Phase plane trajectory

224

Solution: Substituting in the values for f, k, M we get

$$A = \begin{bmatrix} 0 & 1 \\ -2 & -3 \end{bmatrix}$$

The matrix $\phi(t)$ is given by

$$\phi(t) = L^{-1} \{ [sI-A]^{-1} \} \qquad\qquad (*)$$

The characteristic equation is

$\det[sI-A] = s(s+3) + 2 = 0$

The roots are

$$s_1 = -1, \qquad s_2 = -2$$

The inverse matrix is given by

$$[sI-A]^{-1} = \frac{\text{adj } [sI-A]}{|sI - A|} = \frac{1}{(s+1)(s+2)} \begin{bmatrix} s+3 & 1 \\ -2 & s \end{bmatrix}$$

From (*) we get

$$\phi(t) = \begin{bmatrix} 2e^{-t} - e^{-2t} & e^{-t} - e^{-2t} \\ -2e^{-t} + 2e^{-2t} & -e^{-t} + 2e^{-2t} \end{bmatrix}$$

where $t_0 = 0$.

Let us assume the following initial conditions

$$x(0) = \begin{bmatrix} 1 \\ 0 \end{bmatrix}$$

For x(t) we get

$$x(t) = \phi(t)x(0) = \phi(t) \begin{bmatrix} 1 \\ 0 \end{bmatrix} = \begin{bmatrix} 2e^{-t} - e^{-2t} \\ -2e^{-t} + 2e^{-2t} \end{bmatrix} = \begin{bmatrix} x_1(t) \\ x_2(t) \end{bmatrix}$$

Note that

$$\lim_{t \to \infty} x_1(t) = \lim_{t \to \infty} x_2(t) = 0$$

Let us consider the system response to the unit step input. The transient response is given by

$$x(t) = \phi(t-t_0)x(t_0) + \int_{t_0}^{t} \phi(t-\tau)B(\tau)r(\tau)d\tau$$

Assuming zero initial conditions and a unit step input, we obtain

$$x(t) = \begin{bmatrix} \int_{t_0}^{t} \phi_{12}(t-\tau)d\tau \\ \\ \int_{t_0}^{t} \phi_{22}(t-\tau)d\tau \end{bmatrix}$$

Since $B(\tau)r(\tau) = [0,1]^T$

we have

$$x_1(t) = \int_{t_0}^{t} \left(e^{-(t-\tau)} - e^{-2(t-\tau)} \right) d\tau = \frac{1}{2} - e^{-(t-t_0)} + \frac{1}{2} e^{-2(t-t_0)}$$

$$x_2(t) = \int_{t_0}^{t} \left(-e^{-(t-\tau)} + 2e^{-2(t-\tau)} \right) d\tau = e^{-(t-t_0)} - e^{-2(t-t_0)}$$

● **PROBLEM** 6-29

Find the response of the system shown.

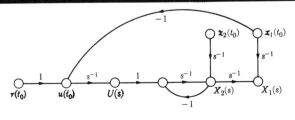

226

Solution: The following equation describes the sample and hold element:

$$u(t_0^+) = e(t_0^+) = r(t_0^+) - x_1(t_0^+)$$

or

$$u(kT) = r(kT) - x_1(kT).$$

From Mason's gain formula we obtain the state equations.

$$X_1(s) = (\frac{1}{s} - \frac{1}{s^2(s+1)})x_1(t_0) + \frac{1}{s(s+1)}x_2(t_0)$$

$$+ \frac{1}{s^2(s+1)}r(t_0)$$

$$X_2(s) = \frac{-1}{s(s+1)}x_1(t_0) + \frac{1}{(s+1)}x_2(t_0) + \frac{1}{s(s+1)}r(t_0)$$

Taking the inverse Laplace transform we obtain

$$x(t) = \begin{bmatrix} 2 - (t - t_0) - e^{-(t-t_0)} & 1 - e^{-(t-t_0)} \\ -1 + e^{-(t-t_0)} & e^{-(t-t_0)} \end{bmatrix} x(t_0)$$

$$+ \begin{bmatrix} (t - t_0) - 1 + e^{-(t-t_0)} \\ 1 - e^{-(t-t_0)} \end{bmatrix} r(t_0)$$

To obtain the discrete transition equation we set $t = (k + 1)T$ and $t_0 = kT$, thus

$$x[(k + 1)T] = \begin{bmatrix} 2 - T - e^{-T} & 1 - e^{-T} \\ -1 + e^{-T} & e^{-T} \end{bmatrix} x(kT)$$

$$+ \begin{bmatrix} T - 1 + e^{-T} \\ 1 - e^{-T} \end{bmatrix} r(kT)$$

For the sampling period $T = 1$ we get

$$x[(k + 1)T] = \begin{bmatrix} 0.6321 & 0.6321 \\ -0.6321 & 0.3679 \end{bmatrix} x(kT)$$

$$+ \begin{bmatrix} 0.3679 \\ 0.6321 \end{bmatrix} r(kT)$$

The response of the system to a unit step input at the first two sampling instants is

$$x(T) = \begin{bmatrix} 0.3679 \\ 0.6321 \end{bmatrix}$$

$$x(2T) = \phi(T)x(T) + D(T)$$

$$= \begin{bmatrix} 0.6321 & 0.6321 \\ -0.6321 & 0.3679 \end{bmatrix} \begin{bmatrix} 0.3679 \\ 0.6321 \end{bmatrix} + \begin{bmatrix} 0.3679 \\ 0.6321 \end{bmatrix} = \begin{bmatrix} 1.000 \\ 0.6321 \end{bmatrix}$$

● **PROBLEM** 6-30

Find the response of the computer control system shown below.

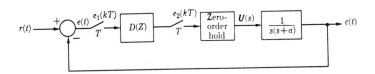

D(Z) is the first order digital compensator.

Solution: To find the discrete difference equations we shall use the signal flow graph technique.

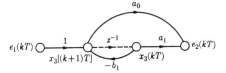

Flow graph of first-order digital compensator

Let us take the z-transform of the continuous portion of the system.

$$G(z) = Z\{G(s)\} = \left\{ Z \; \frac{1 - e^{-sT}}{s^2(s+a)} \right\} = \frac{w_1 z^{-1} + w_2 z^{-2}}{1 + u_1 z^{-1} + u_2 z^{-2}}$$

where

$$w_1 = \frac{aT - 1 + e^{-aT}}{a^2}, \quad w_2 = \frac{1 - (aT + 1)e^{-aT}}{a^2}$$

$$u_1 = -1 - e^{-aT}, \quad u_2 = e^{-aT}$$

We get the following signal flow graph representing the continuous portion of the system.

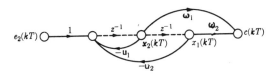

228

The difference equations that describe the plant-hold combination are

$$x[(k+1)T] = \begin{bmatrix} 0 & 1 \\ -u_2 & -u_1 \end{bmatrix} x(kT) + \begin{bmatrix} 0 \\ 1 \end{bmatrix} e_2(kT).$$

$$c(kT) = \begin{bmatrix} w_2 \\ w_1 \end{bmatrix} x(kT).$$

We can draw now the canonical flow graph of the computer control system.

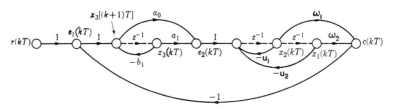

We can write now the difference equations describing the system operation.

$$x_1[(k+1)T] = x_2(kT)$$

$$x_2[(k+1)T] = (-u_2 - a_0 w_2) x_1(kT) + (-u_1 - a_0 w_1) x_2(kT)$$

$$+ (a_1 - a_0 b_1) x_3(kT) + a_0 r(kT)$$

$$x_3[(k+1)T] = -w_2 x_1(kT) - w_1 x_2(kT) - b_1 x_3(kT) + r(kT)$$

$$c(kT) = w_2 x_1(kT) + w_1 x_2(kT).$$

● PROBLEM 6-31

Open-loop sampled second-order system is shown below.

For $t_o = 0$, the transition matrix of the continuous system is

$$\phi(t) = \begin{bmatrix} 2e^{-t} - e^{-2t}, & e^{-t} - e^{-2t} \\ -2e^{-t} + 2e^{-2t}, & -e^{-t} + 2e^{-2t} \end{bmatrix}$$

Obtain the response of the system.

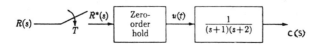

Solution:

The discrete transition matrix is

$$\phi(T) = \begin{bmatrix} 2e^{-T}-e^{-2T}, & e^{-T}-e^{-2T} \\ -2e^{-T}+2e^{-2T}, & -e^{-T}+2e^{-2T} \end{bmatrix}$$

and

$$D(T) = \int_0^T \phi(\tau)B d\tau$$

where $B = [0, 1]^T$. Thus

$$D(T) = \begin{bmatrix} \int_0^T \phi_{12}(\tau) d\tau \\ \\ \int_0^T \phi_{22}(\tau) d\tau \end{bmatrix} = \begin{bmatrix} \frac{1}{2} - e^{-T} + \frac{1}{2}e^{-2T} \\ \\ e^{-T}-e^{-2T} \end{bmatrix}$$

Taking $T = 1$, we get

$$\phi(1) = \begin{bmatrix} 0.6004 & 0.2325 \\ -0.4651 & -0.0972 \end{bmatrix} \qquad D(T) = \begin{bmatrix} 0.1998 \\ 0.2325 \end{bmatrix}$$

If the input is a unit step and the initial conditions are zero, we have

$$x(T) = D(T)r(0) = D(T) = [0.1998, 0.2325]$$

Since $\qquad r(kT) = 1$

the response is

$$x[(k + 1) T] = \phi(T) \quad x(kT) + D(T).$$

● **PROBLEM** 6-32

The symbol of a transistor, which is a three-terminal device, is shown below.

Its behavior, under certain circumstances is equivalent to the network containing linear conductances G_1, G_2 and a current source, as shown below, the encircled part of the drawing.

Find e_2 in terms of V_1.

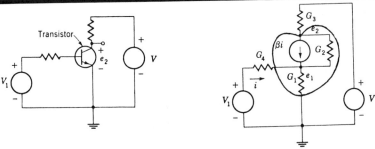

Solution: We have the following system of equations describing the network:

$$G_1 e_1 + G_2 (e_1 - e_2) + G_4 (e_1 - V_2) = \beta i \qquad (1)$$

$$G_3 (e_2 - V) + G_2 (e_2 - e_1) = -\beta i \qquad (2)$$

For the current i flowing through G_4

$$i = G_4 (V_1 - e_1)$$

$$(3)$$

Substituting eq. (3) we get the following system

$$G_1 e_1 + G_2 (e_1 - e_2) + G_4 (\beta + 1)(e_1 - V_1) = 0$$

$$G_2 (e_2 - e_1) + G_3 (e_2 - V) - \beta G_4 (e_1 - V_1) = 0$$

The values of G_1, G_2, G_3, G_4, β, e_1, V are known.

Thus from the above system we can determine e_2 and V_1.

● **PROBLEM** 6-33

Obtain the system response for the RC network shown.

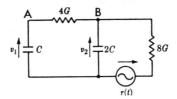

Solution: The Kirchhoff's current law for points A and
B is

$$C \frac{dv_1}{dt} = 4G(v_2 - v_1) \qquad \text{at A}$$

$$2C \frac{dv_2}{dt} = -4G(v_2 - v_1) - 8Gv_2 + 8Gr(t) \quad \text{at B}$$

In the matrix form it can be written

$$\begin{bmatrix} \dot{x}_1 \\ \\ \dot{x}_2 \end{bmatrix} = \begin{bmatrix} -\dfrac{4G}{C} & \dfrac{4G}{C} \\ \\ \dfrac{2G}{C} & \dfrac{-6G}{C} \end{bmatrix} \begin{bmatrix} x_1 \\ \\ x_2 \end{bmatrix} + \begin{bmatrix} 0 \\ \\ \dfrac{4G}{C} \end{bmatrix} r(t)$$

where $x = [x_1 , x_2]^T$.

The state variables are capacitor voltages.

Thus, $x_1 = v_1$

$$x_2 = v_2$$

Let $\dfrac{G}{C} = \alpha$ thus

231

$$(sI - A) \begin{bmatrix} s+4\alpha & -4\alpha \\ -2\alpha & s+6\alpha \end{bmatrix}$$

The characteristic equation is

$$\det (sI - A) = s^2 + 10\alpha s + 16\alpha^2 = (s + 2\alpha)(s + 8\alpha) = 0$$

The roots are -2α and -8α.

The transition matrix is

$$\phi(t) = L^{-1}\{[sI-A]^{-1}\} = L^{-1}\left\{ \begin{bmatrix} s+6\alpha & 4\alpha \\ 2\alpha & s+4\alpha \end{bmatrix} \frac{1}{(s+2\alpha)(s+8\alpha)} \right\}$$

$$= \begin{bmatrix} \frac{2}{3}e^{-2\alpha t} + \frac{1}{3}e^{-8\alpha t}, & \frac{2}{3}e^{-2\alpha t} - \frac{2}{3}e^{-8\alpha t} \\ \frac{1}{3}e^{-2\alpha t} - \frac{1}{3}e^{-8\alpha t}, & \frac{1}{3}e^{-2\alpha t} + \frac{2}{3}e^{-8\alpha t} \end{bmatrix}$$

We have

$$x(t) = \phi(t)\, x(0)$$

Let's assume that

$$x_1(0) = 1$$

$$x_2(0) = 0$$

Thus

$$x(t) = \phi(t)\, x(0) = \begin{bmatrix} \frac{2}{3}e^{-2\alpha t} + \frac{1}{3}e^{-8\alpha t} \\ \frac{1}{3}e^{-2\alpha t} - \frac{1}{3}e^{-8\alpha t} \end{bmatrix}$$

● **PROBLEM 6-34**

Find the two normal modes of operation for the circuit shown below.

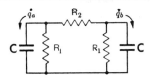

Solution: The equations describing the network are

$$\dot{q} = -R_N C_N q$$

where

$$R_N = \begin{bmatrix} R_1 + R_2 & -R_2 \\ -R_2 & R_1 + R_2 \end{bmatrix}, \qquad C_N = \begin{bmatrix} C & 0 \\ 0 & C \end{bmatrix}, \qquad q = \begin{bmatrix} q_a \\ q_b \end{bmatrix}$$

If $q_a = q_b$, then there is no current in R_2 and the charge on each capacitor decays as

$$e^{-R_1 Ct}$$

Let $\qquad \lambda_1 = -R_1 C$

Thus λ_1 is one of the eigenvalues of the matrix $-R_N C_N$, the associated normal mode is

$$Q_1(t) = \begin{bmatrix} 1 \\ 1 \end{bmatrix} e^{-R_1 Ct}$$

When $q_a = -q_b$, then we can bisect the network as shown below.

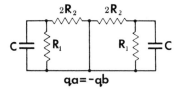

$$q_a = -q_b$$

The voltage is zero along the line of bisection. Each charge decays as

$$e^{-(R_1 + 2R_2)Ct}$$

Let $\lambda_2 = -(R_1 + 2R_2)C$, thus λ_2 is the other eigenvalue of $-R_N C_N$ with associated normal mode,

$$Q_2(t) = \begin{bmatrix} 1 \\ -1 \end{bmatrix} e^{-(R_1 + 2R_2)Ct}$$

The normal modes can be shown geometrically

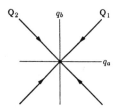

Given $Q_1(t)$ and $Q_2(t)$ we can represent any other solution in the form

$$q(t) = a_1 Q_1(t) + a_2 Q_2(t)$$

Examine the response of the system to a step input disturbance $\theta_i = 3$. The transfer function of the system is

$$\frac{\theta_0}{\theta_i} = \frac{20(\frac{5}{4}D + 1)}{(2D + 1)(2D^2 + 4D + 1)}$$

Solution: Assuming that all initial conditions of θ_0 and θ_i are zero, we obtain the Laplace transform form

$$\frac{\theta_0(s)}{\theta_i(s)} = \frac{20(\frac{5}{4}s + 1)}{(2s + 1)(2s^2 + 4s + 1)}$$

The partial fraction expansion gives:

$$\frac{20(\frac{5}{4}s+1)}{(2s+1)(2s^2+4s+1)} = \frac{A}{2s+1} + \frac{Bs}{2s^2+4s+1} + \frac{C}{2s^2+4s+1}$$

For $s = 0$

$$20 = A + C$$

Multiplying both sides by $\left(2s + 1\right)$:

$$\frac{20(\frac{5}{4}s + 1)}{2s^2+4s+1} = A + \frac{Bs(2s+1) + C(2s+1)}{2s^2+4s+1}$$

Let $2s + 1 = 0$, or

$$s = -\frac{1}{2}$$

$$A = \frac{20 \cdot (-\frac{5}{4} \cdot \frac{1}{2}+1)}{\frac{1}{2} - 2 + 1} = \frac{\frac{3}{4} \cdot 10}{-\frac{1}{2}} = -15$$

In a similar way we find

$$B = 15$$

$$C = 35$$

Thus

$$\frac{\theta_0(s)}{\theta_i(s)} = \frac{-15}{2s+1} + \frac{15s}{2s^2+4s+1} + \frac{35}{2s^2+4s+1}$$

For $\theta_i(t) = 3$, $\theta_i(s) = \frac{3}{s}$ and

$$\theta_0(s) = \frac{-45}{s(2s+1)} + \frac{45}{2s^2+4s+1} + \frac{105}{s(2s^2+4s+1)}$$

Using the Laplace transform pairs:

$$\frac{HK}{s(1 + Ts)} \Longleftrightarrow HK(1 - e^{-\frac{t}{T}})$$

$$\frac{K}{(1+T_1s)(1+T_2s)} \Longleftrightarrow \frac{K}{T_1-T_2}(e^{-\frac{t}{T_1}} - e^{-\frac{t}{T_2}})$$

$$\frac{HK}{s(1+T_1s)(1+T_2s)} \Longleftrightarrow HK\left[1 + \frac{1}{T_2-T_1}(T_1e^{-\frac{t}{T_1}} - T_2e^{-\frac{t}{T_2}})\right]$$

Note that

$$(2s^2 + 4s + 1) = (0.588s + 1)(3.4s + 1)$$

$$\theta_0(t) = -45(1 - e^{-\frac{t}{2}}) + \frac{45}{3.4 - 0.59}\left(e^{-\frac{t}{3.4}} - e^{-\frac{t}{0.59}}\right)$$

$$+ 105\left[1 + \frac{1}{3.4 - 0.59}(0.59\,e^{-\frac{t}{0.59}} - 3.4e^{-\frac{t}{3.4}})\right]$$

$$= 60 + 45\,e^{-\frac{t}{2}} - 111e^{-\frac{t}{3.4}} + 6e^{-\frac{t}{0.59}}$$

● PROBLEM 6-36

A large capacitor is used to light a 2W, 60V lamp. When the voltage falls below 30V, the capacitor has to be recharged. Determine how long the lamp will serve on a single charge.

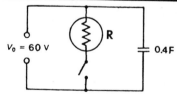

$V_0 = 60$ V R 0.4 F

Solution: The Kirchhoff's law for the system when the voltage source V_0 is removed and the switch is closed is

$$R\frac{dq}{dt} + \frac{q}{c} = 0$$

235

where $q(0) = q_0$.

$\frac{dq}{dt}$ is the current through the capacitor. Separating the variables and integrating, we get

$$\ln q = -\frac{t}{RC} + \ln Q$$

or

$$q = Qe^{-\frac{t}{RC}}$$

Since $q(0) = q_0$, we can write

$$q(t) = q_0 e^{-\frac{t}{RC}}.$$

The capacitor is described by

$$C = \frac{q}{V}, \text{ thus}$$

$$\frac{q(t)}{C} = \frac{q_0}{C} e^{-\frac{t}{RC}}$$

or

$$V(t) = V_0 e^{-\frac{t}{RC}}$$

The resistance R can be determined from

$$P = \frac{E^2}{R}$$

or

$$R = \frac{E^2}{P} = \frac{60^2}{2} = 1800\Omega.$$

Substituting the numerical values:

$$V(t) = V_0 e^{-\frac{t}{RC}}$$

$$30 = 60 \, e^{-\frac{t}{1800 \cdot 0.4}}$$

$$e^{\frac{t}{720}} = 2$$

$$t = 720 \ln 2 = 499s.$$

The series RL circuit shown below has the input
e(t) = E√2 sin ωt.

Find the output voltage $V_R(t)$ if the switch is closed at
time t = 0.

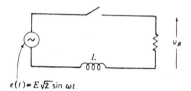

$e(t) = E\sqrt{2}\sin \omega t$

Solution: The equation for the voltages of the circuit is

$$e = \frac{L}{R} Dv_R + v_R$$

We can represent the output voltage as a sum

$$v_R = v_{R,ss} + v_{R,t}$$

where $v_{R,ss}$ is the steady-state solution,

   $v_{R,t}$ is the transient solution.

Let us convert the equation for voltages to phasor
notation, thus

$$E(j\omega) = \frac{L}{R} j\omega V_R(j\omega) + V_R(j\omega)$$

Solving for $V_R(j\omega)$ and returning to the time domain

$$V_{R,ss}(j\omega) = \frac{E(j\omega)}{1 + \frac{L}{R} j\omega} = \frac{E(j\omega)}{\sqrt{1 + (\frac{L}{R}\omega)^2}} \left| - \tan^{-1} \frac{\omega L}{R} \right.$$

and

$$v_{,ss} = \frac{E\sqrt{2}}{\sqrt{1 + (\frac{\omega L}{R})^2}} \sin (\omega t - \tan^{-1} \frac{\omega L}{R}).$$

The characteristic equation is

$$\frac{L}{R} m + 1 = 0$$

where $m = -\frac{R}{L}$ is the root of the equation. There is only
one root of the characteristic equation, therefore the
transient solution has only one exponential term,

$$v_{R,t} = Ae^{-\frac{R}{L}t}$$

Thus

$$v_R = \frac{E\sqrt{2}}{\sqrt{1 + (\frac{\omega L}{R})^2}} \sin(\omega t - \tan^{-1}\frac{\omega L}{R}) + Ae^{-\frac{R}{L}t} \quad (1)$$

To find the constant A let us use the initial conditions.
We have
at $t = 0$, $v_R = 0$

From eq. (1) we obtain

$$A = \frac{\omega RLE\sqrt{2}}{R^2 + (\omega L)^2}$$

RESPONSE-DISCRETE

● **PROBLEM** 6-38

A closed-loop discrete-time system is shown below.

Find the unit-step response of the system.

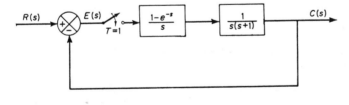

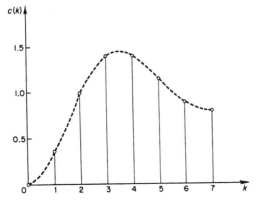

<u>Solution:</u> The pulse transfer function of this closed-loop discrete-time system is

$$\frac{C(z)}{R(z)} = \frac{G(z)}{1 + G(z)}$$

where

$$G(s) = \frac{1 - e^{-s}}{s} \cdot \frac{1}{s(s + 1)}$$

It is easy to find the partial fractions expansion of G(s). We have

$$G(s) = (1 - e^{-s})(\frac{1}{s^2} - \frac{1}{s} + \frac{1}{s + 1})$$

From the tables we find the inverse Laplace transform of G(s).

$$L^{-1}[G(s)] = g(t) = (t - 1 + e^{-t})\, 1\,(t) - (t - 2 + e^{-(t-1)})\, 1\,(t-1)$$

Let t = kT, since T = 1 we get

$$g(k) = (k - 1 + e^{-k}) - (k - 2 + e^{-(k-1)}) =$$

$$e^{-k} + 1 - e^{-(k-1)}$$

for k = 1, 2, 3 ...   and g(0) = 0

Then

$$G(z) = \sum_{k=0}^{\infty} g(k) z^{-k}$$

After some calculations we obtain

$$G(z) = \frac{e^{-1}z + 1 - 2e^{-1}}{z^2 - (1 + e^{-1})z + e^{-1}} = \frac{0.368z + 0.264}{z^2 - 1.368z + 0.368}$$

Then

$$\frac{C(z)}{R(z)} = \frac{G(z)}{1 + G(z)} = \frac{\dfrac{0.368z + 0.264}{z^2 - 1.368z + 0.368}}{1 + \dfrac{0.368z + 0.264}{z^2 - 1.368z + 0.368}}$$

$$= \frac{0.368z + 0.264}{z^2 - z + 0.632}$$

For a unit step input we have

239

$$R(z) = \frac{z}{z-1}$$

For C(z) we obtain

$$C(z) = \frac{(0.368z + 0.264)z}{(z^2-z+0.632)(z-1)} = \frac{0.368z^{-1} + 0.264z^{-2}}{1-2z^{-1}+1.632z^{-2}-0.632z^{-3}}$$

$$= 0.368z^{-1} + z^{-2} + 1.4z^{-3} + 1.4z^{-4} + 1.147z^{-5} + 0.895z^{-6} + \ldots$$

From the inverse z transform we get

$$c(0) = 0$$

$$c(1) = 0.368$$

$$c(2) = 1$$

$$c(3) = 1.4$$

$$c(4) = 1.4$$

$$c(5) = 1.147$$

$$c(6) = 0.895$$

.

.

.

We obtained information about the response for k = 1, 2, 3, ... The z transform analysis does not provide information about the response of the system between sampling instants.

● **PROBLEM 6-39**

The block diagram of the system is shown below.

Find the impulse response and draw the signal-flow graph of the system.

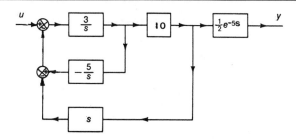

Solution: Moving the gain block ⌐10⌐ into the output and combining the two feedback paths, we have

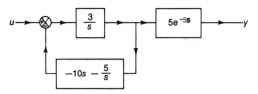

Applying the feedback rule, we get

$$H(s) = 5e^{-5s} \frac{\frac{3}{s}}{1 - \frac{3}{s}\frac{5}{s}(2s^2 + 1)} = -\frac{15}{29}e^{-5s} \frac{s}{s^2 + \frac{15}{29}}$$

From the tables we have

$$\cos \omega_0 t = L^{-1}[\frac{s}{s^2 + \omega_0^2}]$$

Thus

$$h(t) = [-\frac{15}{29} \cos\sqrt{\frac{15}{29}} (t-5)] \ 1 \ (t-5)$$

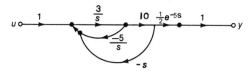

The signal-flow graph of the system is shown.

● **PROBLEM** 6-40

Consider the system shown below.

The Laguerre polynomial is defined by

$$L_n(\alpha t) = \sum_{i=0}^{n} \frac{n!}{(n-i)!} \frac{(\alpha t)^i}{(i!)^2} \ 1 \ (t)$$

Show how the impulse response of the shown system is related to some L_n.

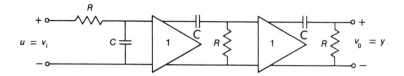

Solution: The amplifiers are used in the system for isolation and both have unity voltage gain. The transfer function of the system can be obtained from

$$H(s) = \frac{L[v_0]}{L[v_i]} = \frac{\frac{1}{sC}}{R + \frac{1}{sC}} \cdot \frac{R}{R + \frac{1}{sC}} \cdot \frac{R}{R + \frac{1}{sC}}$$

$$= \frac{\frac{s^2}{RC}}{(s + \frac{1}{RC})^3}$$

To find h(t) note that

$$L^{-1}\{sH(s)\} = \frac{d}{dt}h(t).$$

Differentiating twice $H'(s) = \dfrac{\frac{1}{RC}}{(s + \frac{1}{RC})^3}$, we get

$$h(t) = \frac{1}{RC} \frac{d^2}{dt^2} \left\{ \frac{t^2}{2} e^{-\frac{t}{RC}} 1(t) \right\} = \frac{e^{-\frac{t}{RC}}}{RC} \left\{ 1 - \frac{2t}{RC} + \frac{t^2}{2R^2C^2} \right\} 1(t)$$

From the definition of the Laguerre polynomial we have

$$L_0(\alpha t) = 1(t)$$

$$L_1(\alpha t) = [1 + \alpha t] 1(t)$$

$$L_2(\alpha t) = [1 + 2\alpha t + \frac{\alpha^2}{2}t^2] 1(t)$$

Comparing h(t) and $L_2(\alpha t)$ we get

$$h(t) = L_2(-\frac{t}{RC}) \frac{e^{-\frac{t}{RC}}}{RC}$$

To obtain the unit-step response, let us integrate h(t), thus

$$y_u(t) = \frac{t}{RC} e^{-\frac{t}{RC}}[1 - \frac{t}{2RC}] = \frac{t}{RC} e^{-\frac{t}{RC}} L_1(\frac{-t}{RC})$$

where we used the result

$$L_1(-\frac{t}{RC}) = [1 - \frac{t}{2RC}].$$

● **PROBLEM** 6-41

Given the digital space-vehicle control system shown in Fig. 1 where

$$K_p = \text{positional sensor gain}$$

$$= 1.65 \times 10^6$$

K_r = rate sensor gain

$= 3.17 \times 10^5$

J_v = moment of inertia of vehicle

$= 41822$

The unit-step responses of the system are plotted in Fig. 2 for several values of the sampling period T. Determine the maximum overshoot and the peak time of the system.

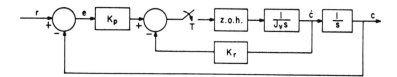

A digital space-vehicle control system.   Fig. 1

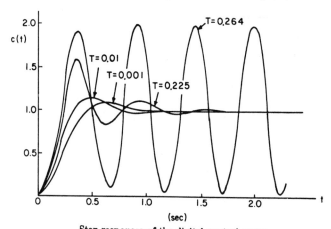

Step responses of the digital control system   Fig. 2

Solution: The method that will be used to solve this problem has the advantage that the overshoot and peak-time properties of the unit-step response can be approximated without actually computing the response c*(t) numerically.

We need to find the closed loop system transfer function
$$\frac{C(z)}{R(z)} = \frac{G(z)}{1 + G(z)}$$

The open loop transfer function G(z) is $\frac{C(z)}{E(z)}$

or
$$G(z) = \frac{K_p \, Z\left[G_h(s) \dfrac{1}{J_v s^2}\right]}{1 + K_r Z\left[G_h(s) \dfrac{1}{J_v s^2}\right]} \tag{1}$$

243

where $G_h(s)$ is the transfer function of the zero order hold (z.o.h).

$$Z\left[G_h(s)\frac{1}{J_v s^2}\right] = Z\left[\frac{1 - e^{-s}}{s}\frac{1}{J_v s^2}\right]$$

$$= (1 - z^{-1})\; Z\left[\frac{1}{J_v s^3}\right]$$

Using the transform pair

$$\frac{2}{s^3} \iff \frac{T^2 z(z + 1)}{(z - 1)^3}$$

we get

$$Z\left[G_h(s)\frac{1}{J_v s^2}\right] = (1 - z^{-1})\frac{T^2 z(z + 1)}{2J_v(z - 1)^3}$$

$$= \frac{T^2(z + 1)}{2J_v(z-1)^2} \tag{2}$$

and

$$Z[G_h(s)\frac{1}{J_v s}] = Z[\frac{1 - e^{-s}}{s}\frac{1}{J_v s}]$$

$$= (1 - z^{-1})\; Z[\frac{1}{J_v s^2}]$$

Using the transform pair

$$\frac{1}{s^2} \iff \frac{Tz}{(z - 1)^2}$$

we get

$$Z[G_h(s)\frac{1}{J_v s}] = \frac{(1 - z^{-1})Tz}{J_v(z - 1)^2}$$

$$= \frac{T}{J_v(z - 1)} \tag{3}$$

Then substituting eqs. (2) and (3) into eq. (1) yields

$$G(z) = \frac{\dfrac{K_p T^2(z + 1)}{2J_v(z - 1)^2}}{1 + \dfrac{K_r T}{J_v(z - 1)}}$$

244

$$= \frac{\dfrac{K_p T^2 (z + 1)}{2 J_v (z - 1)^2}}{\dfrac{J_v (z - 1) + K_r T}{J_v (z - 1)}}$$

$$= \frac{\dfrac{K_p T^2 (z + 1)}{2 (z - 1)}}{J_v (z - 1) + K_r T}$$

$$= \frac{K_p T^2 (z + 1)}{2 (z - 1)(J_r [z - 1] + K_r T)}$$

Thus

$$G(z) = \frac{K_p T^2 (z + 1)}{2 J_v z^2 + (2 K_r T - 4 J_v) z + (2 J_v - 2 K_r T)}$$

Then

$$\frac{C(z)}{R(z)} = \frac{G(z)}{1 + G(z)}$$

$$= \frac{\dfrac{K_p T^2 (z + 1)}{2 J_v z^2 + (2 K_r T - 4 J_v) z + 2 J_v - 2 K_r T}}{1 + \dfrac{K_p T^2 (z + 1)}{2 J_v z^2 + (2 K_r T - 4 J_v) z + 2 J_v - 2 K_r T}}$$

$$= \frac{K_p T^2 (z + 1)}{2 J_v z^2 + (2 K_r T - 4 J_v) z + 2 J_v - 2 K_r T + K_p T^2 (z + 1)}$$

$$= \frac{K_p T^2 (z + 1)}{2 J_v z^2 + (2 K_r T - 4 J_v + T^2 K_p) z + 2 J_v - 2 K_r T + T^2 K_p}$$

Substituting in the values for K_p, K_r and J_v yields

$$\frac{C(z)}{R(z)} = \frac{1.65 \times 10^6 T^2 (z + 1)}{83644 z^2 + (6.34 \times 10^5 T - 167288 + 1.65 \times 10^6 T^2) z + C}$$

where $C = 83644 - 6.34 \times 10^5 T + 1.65 \times 10^6 T^2$

For $T = 0.225$ sec, the closed-loop transfer function becomes

$$\frac{C(z)}{R(z)} = \frac{83531.25(z + 1)}{83644z^2 + 58893.25z + 24525.25}$$

The zero of $C(z)/R(z)$ is at $z = -1$, and the two poles are at $z = p_1 = -0.352 + j\,0.4114$ and $z = \bar{p}_1 = -0.352 - j\,0.4114$. The pole-zero configuration of $C(z)/R(z)$ is shown in Fig. 3.

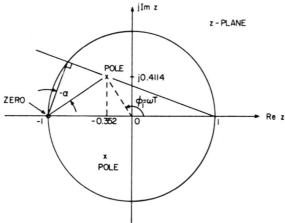

Pole-zero configuration of the closed-loop transfer function
$T = 0.225$ sec.

Fig. 3

From Fig. 3 we get

$$a = -40.66^{\circ}$$

and

$$\phi_1 = \omega T = 130.55^{\circ} = 2.278 \text{ rad}, \quad |p_1| = 0.54$$

Substituting the values of ϕ_1 (in radians) and $|p_1|$ into

$$|p_1| = e^{-\zeta\omega_n T} = \exp\left(\frac{-\zeta\phi_1}{\sqrt{1 - \zeta^2}}\right) \tag{4}$$

the damping ratio ζ is found to be $\zeta = 0.26$

Once a and ζ are found, we go to Fig. 4 to determine the maximum overshoot. In this case, the interpolated value on the curves for the maximum overshoot is approximately 50 percent. This result agrees quite well with the computed response curve in Fig. 2 which shows a maximum overshoot of slightly over 50 percent.

246

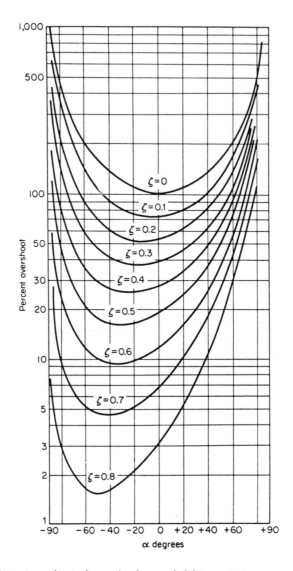

Percent overshoot of second-order sampled-data control systems.

Fig. 4

Now let us consider the case for T = 0.01 sec. The closed-loop transfer function becomes

$$\frac{C(z)}{R(z)} = \frac{165(z + 1)}{83644z^2 - 160783z + 77469}$$

The zero of this transfer fucntion is at z = - 1, and the two poles are at z = p_1 = 0.96 + j0.0494 and z = $\bar{p}_1$ = 0.96 - j0.0494. The pole-zero configuration of C(z)/R(z) is shown in Fig. 5, from which we get

$$a = - 37.56^{\circ}$$

and

$$\phi_1 = 2.95^{\circ} = 0.0514 \text{ rad}$$

$$|p_1| = 0.9613$$

Using Eq. 4, the damping ratio of the closed-loop digital control system is found to be $\zeta = 0.609$.

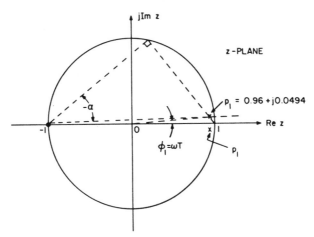

Pole-zero configuration of the closed-loop transfer function
T = 0.01 sec.

Fig. 5

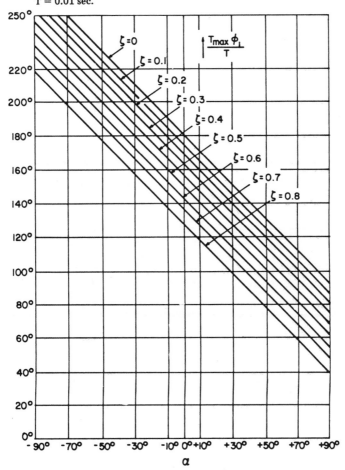

Peak time of step response of second-order sampled-data systems versus a.

Fig. 6

The maximum overshoot is now determined from Fig. 4, and the result is approximately 10 percent which agrees quite well with the exact value indicated by the response for T = 0.01 sec in Fig. 2.

The peak times for the two cases considered above can be computed directly from

$$T_{max} = \frac{T}{\phi_1} \left(\tan^{-1} \frac{-\zeta}{\sqrt{1 - \zeta^2}} \mp a + \pi \right) \qquad (5)$$

or from Fig. 6. For T = 0.225 sec, Eq. 5 gives

$$T_{max} = \frac{0.225}{130.66°} \left\{ \tan^{-1} \frac{-0.26}{\sqrt{1 - 0.26^2}} + 40.66° + 180° \right\}$$

$$= 0.354 \text{ sec}$$

For T = 0.01 sec,

$$T_{max} = \frac{0.01}{2.95°} \left\{ \tan^{-1} \frac{-0.609}{\sqrt{1 - 0.609^2}} + 37.56° + 180° \right\}$$

$$= 0.61 \text{ sec.}$$

The same results would have been obtained using Fig. 6, and these also agree very well with the results shown in Fig.2

RESPONSE-ERROR

● **PROBLEM** 6-42

The open-loop transfer function of a unity-feedback control system is

$$G(s) = \frac{K}{s(Ms + N)}$$

Find the steady-state error in unit-ramp response and investigate how it changes with the varying values of K and N.

Solution: The closed-loop transfer function of the system is

$$\frac{C(s)}{R(s)} = \frac{K}{Ms^2 + Ns + K}$$

For a unit ramp input, $R(s) = \frac{1}{s^2}$, thus

$$\frac{E(s)}{R(s)} = \frac{R(s) - C(s)}{R(s)} = \frac{Ms^2 + Ns}{Ms^2 + Ns + K}$$

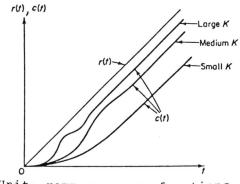

Unit- ramp response functions
of the system.

or

$$E(s) = \frac{1}{s^2} \frac{Ms^2 + Ns}{Ms^2 + Ns + K}$$

The steady-state error is

$$e_{ss} = e(\infty) = \lim_{s \to 0} sE(s) = \frac{N}{K}$$

The steady-state error e_{ss} can be reduced by increasing the gain K or decreasing N.

● **PROBLEM** 6-43

The open-loop transfer function of a unity feedback control system is

$$G(s) = \frac{K}{s + 1}$$

What are the error constants? What is the steady-state error for a step, ramp and parabolic input? Evaluate the first three terms of the error series. Find the error as a function of time for a step, ramp and parabolic input and for

$$r(t) = a_o + a_1 t + \frac{1}{2}a_2 t^2 + e^{-a_3 t}.$$

Solution: The constants are

$$K_p = \lim_{s \to 0} G(s) = K$$

$$K_v = \lim_{s \to 0} s\, G(s) = 0$$

$$K_a = \lim_{s \to 0} s^2\, G(s) = 0$$

The system is of type 0.

The steady-state errors are: for unit step input, $u_s(t)$,

$$e_{ss} = \frac{1}{1 + K_p} = \frac{1}{1 + K}$$

unit ramp input, $tu_s(t)$,

$$e_{ss} = \frac{1}{K_v} = \infty$$

unit parabolic input, $t^2 u_s(t)$,

$$e_{ss} = \frac{1}{K_a} = \infty$$

We have

$$W_e(s) = \frac{1}{1 + G(s)} = \frac{s + 1}{s + K + 1}$$

The error coefficients are

$$C_0 = \lim_{s \to 0} W_e(s) = \frac{1}{K + 1}$$

$$C_1 = \lim_{s \to 0} \frac{dW_e(s)}{ds} = \frac{K}{(1 + K)^2}$$

$$C_2 = \lim_{s \to 0} \frac{d^2 W_e(s)}{ds^2} = \frac{-2K}{(1 + K)^3}$$

The error series can be written

$$e_s(t) = \frac{C_0}{0!} r_s(t) + \frac{C_1}{1!} \dot{r}_s(t) + \frac{C_2}{2!} \ddot{r}_s(t) + \dots$$

$$= \frac{1}{K + 1} r_s(t) + \frac{K}{(1 + K)^2} \dot{r}_s(t) + \frac{-K}{(1 + K)^3} \ddot{r}_s(t)$$

$$+ \dots$$

For the basic types of inputs we have

1)   unit step, $r_s(t) = u_s(t)$

$$\dot{r}_s(t) = \ddot{r}_s(t) = 0$$

and

$$e_s(t) = \frac{1}{K + 1}$$

2) unit ramp function, $r_s(t) = tu_s(t)$

$$\dot{r}_s(t) = u_s(t) \quad , \quad \ddot{r}_s(t) = 0$$

$$e_s(t) = \left[\frac{1}{1 + K} t + \frac{K}{(1 + K)^2} \right] u_s(t)$$

3) parabolic input, $r_s(t) = \frac{t^2}{2} u_s(t)$

$$\dot{r}_s(t) = tu_s(t) \qquad \ddot{r}_s(t) = u_s(t)$$

thus

$$e_s(t) = \left[\frac{1}{1 + K} \frac{t^2}{2} + \frac{K}{(K + 1)^2} t - \frac{K}{(1 + K)^3} \right] u_s(t)$$

4) for

$$r(t) = \left[a_0 + a_1 t + \frac{a_2 t^2}{2} + e^{-a_3 t} \right] u_s(t)$$

we have

$$r_s(t) = \left[a_0 + a_1 t + \frac{a_2 t^2}{2} \right] u_s(t)$$

$$\dot{r}_s(t) = [a_1 + a_2 t] u_s(t)$$

$$\ddot{r}_s(t) = a_2 u_s(t)$$

$$e_s(t) = \left\{ \frac{1}{1 + K} [a_0 + a_1 t + \frac{a_2 t^2}{2}] + \frac{K}{(1 + K)^2}(a_1 + a_2 t) \right.$$

$$\left. - \frac{Ka_2}{(1 + K)^3} \right\} u_s(t)$$

● **PROBLEM** 6-44

For the system given by

$$\frac{C(s)}{R(s)} = \frac{\omega_n^2}{s^2 + 2\zeta\omega_n s + \omega_n^2}$$

obtain the unit-ramp response and the steady-state error.

<u>Solution:</u> We have for a unit-ramp input

$$R(s) = \frac{1}{s^2}$$

Thus $$C(s) = \frac{1}{s^2} \frac{\omega_n^2}{s^2 + 2\zeta\omega_n s + \omega_n^2}$$

We shall consider three cases

1. the overdamped case, $\zeta > 1$

2. The critically damped case, $\zeta = 1$

3. the underdamped case, $0 \le \zeta < 1$

For $\zeta > 1$, the inverse Laplace transform of $C(s)$ is

$$c(t) = t - \frac{2\zeta}{\omega_n} - \frac{2\zeta^2 - 1 - 2\zeta\sqrt{\zeta^2 - 1}}{2\omega_n\sqrt{\zeta^2 - 1}} \; e^{-(\zeta+\sqrt{\zeta^2-1})\,\omega_n t}$$

$$+ \frac{2\zeta^2 - 1 + 2\zeta\sqrt{\zeta^2 - 1}}{2\omega_n\sqrt{\zeta^2 - 1}} \; e^{-(\zeta-\sqrt{\zeta^2-1})\,\omega_n t} \qquad \text{for } t \ge 0$$

For $\zeta = 1$,

$$c(t) = t - \frac{2}{\omega_n} + \frac{2}{\omega_n} \; e^{-\omega_n t}\left(1 + \frac{\omega_n t}{2}\right) \qquad \text{for } t \ge 0$$

For $\zeta < 1$

$$c(t) = t - \frac{2\zeta}{\omega_n} + \frac{e^{-\zeta\omega_n t}}{\omega_n\sqrt{1 - \zeta^2}} \; \sin\left(\omega_k t + \tan^{-1}\frac{2\zeta\sqrt{1-\zeta^2}}{2\zeta^2 - 1}\right)$$

where $\omega_k = \omega_n\sqrt{1 - \zeta^2}$.

The error signal $e(t)$ is (for $\zeta > 0$)

$$e(t) = r(t) - c(t) = t - c(t)$$

The steady-state error, for $\zeta > 0$ is

$$e_{ss} = e(\infty) = \frac{2\zeta}{\omega_n}$$

● **PROBLEM** 6-45

Compare the errors of the unity-feedback control systems
with the following feedforward transfer functions:

$$G_1(s) = \frac{10}{s(s + 1)} \qquad , \qquad G_2(s) = \frac{10}{s(2s + 1)}$$

The input of the system is

$$r(t) = 2t^2 + 3t + 7.$$

Solution: We have

$$\frac{E_1(s)}{R(s)} = \frac{1}{1 + G_1(s)} = \frac{s^2 + s}{s^2 + s + 10}$$

$$= 0.1s + 0.09s^2 - 0.019s^3 + \ldots$$

and for $G_2(s)$

$$\frac{E_2(s)}{R(s)} = \frac{1}{1 + G_2(s)} = \frac{2s^2 + s}{2s^2 + s + 10}$$

$$= 0.1s + 0.19s^2 - 0.039s^3 + \ldots$$

Note that, if $\frac{E(s)}{R(s)}$ is given by

$$\frac{E(s)}{R(s)} = a_1 s + a_2 s^2 + a_3 s^3 + \ldots$$

then

$$E(s) = a_1 s R(s) + a_2 s^2 R(s) + a_3 s^3 R(s) + \ldots$$

Taking the inverse Laplace transform

$$e(t) = a_1 \dot{r}(t) + a_2 \ddot{r}(t) + a_3 \dddot{r}(t) + \ldots$$

For $\quad r(t) = 2t^2 + 3t + 7$

$$\dot{r}(t) = 4t + 3, \qquad \ddot{r}(t) = 4, \qquad \dddot{r} = r^{(4)} = 0$$
$$= \ldots$$

The error time functions are given by

$$\lim_{t \to \infty} e_1(t) = \lim_{t \to \infty} \left[0.1(4t + 3) + 0.09 \cdot 4 \right]$$

and

$$\lim_{t \to \infty} e_2(t) = \lim_{t \to \infty} \left[0.1(4t + 3) + 0.19 \cdot 4 \right]$$

Thus

$$\lim_{t \to \infty} e_1(t) < \lim_{t \to \infty} e_2(t)$$

Find the steady-state error in the output of a
linear control system with unity feedback. The input
is given by

$$r(t) = r_o + r_1 t + \frac{r_2}{2} t^2$$

Solution: Since the steady-state error is the sum of
the errors due to each input component, we have

$$e_{ss} = \frac{r_o}{1 + K_p} + \frac{r_1}{K_v} + \frac{r_2}{K_a}$$

e_{ss} becomes infinity unless the system is of type 2
or higher.

Determine the error constants for the general second-order
system

$$H(s) = \frac{K\omega_n^2}{s^2 + 2\zeta\omega_n s + \omega_n^2} \tag{1}$$

having all zeros at infinity.

Insert a finite pole at $s = -\alpha$ and find how the error
constants change for

$$H(s) = \frac{K\omega_n^2}{(s^2 + 2\zeta\omega_n s + \omega_n^2)(s + \alpha)} \tag{2}$$

Solution: Expanding eq. (1) we obtain

$$H(s) = K - \frac{2\zeta K s}{\omega_n} + \frac{K(4\zeta^2 - 1)}{\omega_n^2} s^2 + \ldots$$

Identifying with the error constants, we have

$$K = \frac{K_p}{1 + K_p} \qquad \text{thus} \qquad K_p = \frac{K}{1 - K}$$

$$-\frac{2\zeta K}{\omega_n} = -\frac{1}{K_v} \qquad \text{thus} \qquad K_v = \frac{\omega_n}{2\zeta K}$$

$$-\frac{K(4\zeta^2 - 1)}{\omega_n^2} = -\frac{1}{K_a} \qquad \text{thus} \qquad K_a = \frac{\omega_n^2}{K(4\zeta^2 - 1)}$$

Using the same method for eq. (2) we find

$$H(s) = K - K\left(\frac{1}{\alpha} + \frac{2\zeta}{\omega_n}\right)s + \frac{[(2\zeta\alpha + \omega_n)^2 K - (\alpha + 2\zeta\omega_n)]}{\alpha\omega_n^2}s^2$$

$$+ \dots$$

$$K_p = \frac{K}{1 - K}$$

$$K_v = \frac{\alpha\omega_n}{K(\omega_n + 2\zeta\alpha)}$$

$$K_a = \frac{-\alpha^2\omega_n^2}{[(\alpha^2 + 2\zeta\omega_n) - K(2\alpha\zeta + \omega_n)^2]}$$

● **PROBLEM** 6-48

The steady-state error to a unit-ramp input, of the system shown below is $e_{ss} = \frac{2\zeta}{\omega_n}$.

Show that the steady-state error for following a ramp input may be eliminated, if we introduce the input through a $(1 + ks)$ element and choose the proper value for k.

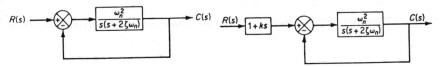

Solution: For the system with the additional element, the closed-loop transfer function is

$$\frac{C(s)}{R(s)} = \frac{(1 + ks)\omega_n^2}{s^2 + 2\zeta\omega_n s + \omega_n^2}$$

Thus

$$R(s) - C(s) = R(s) \cdot \frac{s^2 + 2\zeta\omega_n s - \omega_n^2 ks}{s^2 + 2\zeta\omega_n s + \omega_n^2}$$

When the input is a unit ramp, the steady-state error is

$$e(\infty) = r(\infty) - c(\infty) = \lim_{s \to 0} s \cdot \frac{1}{s^2} \frac{s^2 + 2\zeta\omega_n s - \omega_n^2 k s}{s^2 + 2\zeta\omega_n s + \omega_n^2}$$

$$= \frac{2\zeta\omega_n - \omega_n^2 k}{\omega_n^2}$$

From the condition

$$e_{ss} = 0$$

we obtain

$$2\zeta\omega_n - \omega_n^2 k = 0$$

$$k = \frac{2\zeta}{\omega_n}$$

For this value of k, the steady-state error for following a ramp input is equal to zero.

● **PROBLEM** 6-49

A system is described by

$$\alpha_o \frac{d\lambda}{dt} + \lambda = 2x(t)$$

Determine the sensitivity functions for the system with $\alpha_o = 0.01$.

For x = 3t determine the error from the ideal response at 0.1 sec. and at 1 second.

Solution: Let us find the impulse response, $\delta(t) = x$. Taking the Laplace transform

$$\alpha_o \cdot s \cdot \lambda + \lambda = 2$$

or

$$\lambda = \frac{2}{\alpha_o s + 1} = \frac{\frac{2}{\alpha_o}}{s + \frac{1}{\alpha_o}}$$

Thus

$$I(t) = \frac{2}{\alpha_o} e^{-\frac{t}{\alpha_o}}$$

257

For $\alpha_o = 0.01$ we have

$$I(t) = 200\ e^{-100t}$$

Using the method of convolution

$$\lambda(t) = \int_o^t I(\tau)\ \ x(t - \tau)d\tau$$

Expanding x into a Taylor series

$$x(t - \tau) = \sum_{n=0}^{\infty} (-1)^n\ x^{(n)}(\tau = 0)\ \frac{\tau^n}{n!}$$

and substituting into the integral, we get

$$\lambda(t) = \sum_{n=0}^{\infty} S_n(t) x^{(n)}(t)$$

where the sensitivity functions are

$$S_n(t) = \int_0^t (-1)^n I(\tau)\ \frac{\tau^n}{n!}\ d\tau$$

For the function $x = 3t$ we have

$$x^{(0)} = 3t, \qquad x^{(1)} = 3, \qquad x^{(2)} = x^{(3)} = \ldots = 0$$

Thus

$$\lambda(t) = S_o(t) x^{(0)}(t) + S_1(t) x^{(1)}(t)$$

We get

$$S_o = 2(1 - e^{-100t})$$

$$S_1 = 2[e^{-100t}(t + 0.01) - 0.01]$$

For $t = 0.01$, $S_o = 1.2662$, $S_1 = -0.0200$

$$\lambda(0.01) = S_o(t) \cdot 3t + 3\ S_1 = 0.0221$$

For $t = 1$, $S_o = 2$, $S_1 = -0.0200$

and $\lambda = 5.94$

For an ideal system we have

$$\lambda_I = 2(3t)\ \text{and the relative error is}$$

$$\delta = \left| \frac{\lambda_I - \lambda}{\lambda_I} \right|$$

For $t = 0.01$

$$\lambda = 0.0221, \qquad \lambda_I = 0.0600, \qquad \delta = 0.6309$$

and for $t = 1$

$$\lambda = 5.94, \qquad \lambda_I = 6, \qquad \delta = 0.01$$

We see that for $t = 0.01$ the relative error is 63.1% of the expected value, and for $t = 1$ sec., 1%.

● **PROBLEM** 6-50

Use the Lyapunov's direct method to estimate the speed of the response of the system, i.e. the dominant time constant.

Solution: We have to estimate how much time it takes for the system to return to the origin. We assume that the origin is the equilibrium point. $V(x,t)$ is a Lyapunov function and

$$\alpha = - \frac{\dot{V}(x,t)}{V(x,t)}$$

From all possible Lyapunov functions the best is the one which gives the largest value of α. Integrating both sides

$$\int_{t_o}^{t} \alpha dt = - \int_{t_o}^{t} \frac{\dot{V}}{V} dt = - \int_{V(x(t_o),t_o)}^{V(x(t),t)} \frac{dV}{V} = - \ln \left[\frac{V(x(t),t)}{V(x(t_o),t_o)} \right]$$

Solving the equation

$$- \int_{t_o}^{t} \alpha dt = \ln \frac{V(x(t),t)}{V(x(t_o),t_o)}$$

we obtain

$$V(x(t),t) = V(x(t_o),t_o) \cdot e^{-\int_{t_o}^{t} \alpha dt}$$

Let us denote α_{min} as the minimum value of α, then

$$V(x(t),t) \leq V(x(t_o),t_o) \, e^{-\alpha_{min}(t-t_o)}$$

When the system is asymptotically stable $x(t) \to 0$ as

$V(x,t) \rightarrow 0$, thus α_{min}^{-1} can be interpreted as a bound on the system time constant.

Feedforward transfer function of the unity-feedback control system is

$$G(s) = \frac{10}{s(s + 1)}$$

Find the dynamic error coefficients and the steady-state error to the input defined by

$$r(t) = a_0 + a_1 t + a_2 t^2.$$

<u>Solution:</u> We have

$$\frac{E(s)}{R(s)} = \frac{1}{1 + G(s)} = \frac{s^2 + s}{s^2 + s + 10}$$

$$= 0.1s + 0.09s^2 - 0.019s^2 + \ldots$$

or

$$E(s) = 0.1sR(s) + 0.09s^2R(s) - 0.019s^3R(s) + \ldots$$

The steady-state error in the time domain is given by

$$\lim_{t \to \infty} e(t) = \lim_{t \to \infty} [0.1\dot{r}(t) + 0.09\ddot{r}(t) - 0.019r^{(3)}(t)$$

$$+ \ldots]$$

For the dynamic error coefficients, we get

$$k_1 = \frac{1}{0} = \infty$$

$$k_2 = \frac{1}{0.1} = 10$$

$$k_3 = \frac{1}{0.09} = 11.1$$

Differentiating the input we get

$$r(t) = a_0 + a_1 t + a_2 t^2$$

$$\dot{r}(t) = a_1 + 2a_2 t$$

$$\ddot{r}(t) = 2a_2$$

$$\overset{\cdots\cdot}{r}(t) = 0$$

Thus, the steady-state error is

$$\lim_{t\to\infty} e(t) = \lim_{t\to\infty} [0.1(a_1 + 2a_2 t) + 0.09 \cdot 2a_2]$$

The steady-state error is infinite unless

$$a_2 = 0.$$

Consider a unity feedback system, whose forward gain is

$$G(s) = \frac{K}{s + 1}$$

and the input to the system is

$$r(t) = \sin \omega_o t$$

Show, that in this situation the error constant can not provide a solution to the steady-state error.

Solution: Let us compute the first four terms of the error series and approximate the steady-state error as a function of time. The derivatives of the input are:

$$r(t) = \sin \omega_o t$$

$$\overset{\cdot}{r}(t) = \omega_o \cos \omega_o t$$

$$\overset{\cdot\cdot}{r}(t) = -\omega_o^2 \sin \omega_o t$$

$$\overset{\cdot\cdot\cdot}{r}(t) = -\omega_o^3 \cos \omega_o t$$

We can write the error series

$$e_s(t) = [C_o - \frac{C_2}{2!} \omega_o^2 + \frac{C_4}{4!} \omega_o^4 - \dots] \sin \omega_o t$$

$$+ [C_1 \omega_o - \frac{C_3}{3!} \omega_o^3 + \dots] \cos \omega_o t$$

The error series is an infinite series, whose convergence depends on the value of ω_o and K. The question of convergence of the series is crucial in arriving at the answer to the steady-state error. Let us assign the values,

$$K = 100$$

$$\omega_o = 4$$

Then

$$C_o = \frac{1}{1 + K} = 0.0099$$

$$C_1 = \frac{K}{(1 + K)^2} = 0.0098$$

$$C_2 = - \frac{2K}{(1 + K)^3} = - 0.000194$$

$$C_3 = \frac{6K}{(1 + K)^5} = 5.70 \times 10^{-8}$$

The above constants were obtained from

$$\omega_e(s) = \frac{1}{1 + G(s)} = \frac{s + 1}{s + K + 1}$$

and

$$c_n = \lim_{s \to 0} \frac{d^n \omega_e(s)}{ds^n} = \lim_{s \to 0} \frac{d^n}{ds^n} \left[\frac{s + 1}{s + K + 1} \right]$$

To evaluate $e_s(t)$ we shall use the first four coefficients, then

$$e_s(t) \underset{\sim}{\sim} [0.0099 + \frac{0.000194}{2!} 16] \sin 4t$$

$$+ [0.0392 - 6.08 \cdot 10^{-7}] \cos 4t$$

$$= 0.01145 \sin 4t + 0.0392 \cos 4t$$

● **PROBLEM** 6-53

A block diagram of a velocity servomechanism is shown below.

For the input voltage 5V, determine the steady-state error in the velocity $\dot{\theta}$.

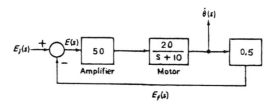

Solution: The system can be regarded as a unity-feedback system, where the tachometer voltage e_f is the controlled output. The open-loop transfer function is then

$$G(s) = \frac{E_f(s)}{E(s)} = \frac{50 \cdot 20 \cdot 0.5}{s + 10} = \frac{500}{s + 10}$$

which represents a Type 0 system. The position error constant is

$$\dot{K}_p = \lim_{s \to 0} G(s) = 50$$

and the steady-state error

$$e_{ss} = \lim_{s \to 0} \frac{sR(s)}{1 + G(s)} = \lim_{s \to 0} s \, (\frac{V}{s}) \frac{1}{1 + G(s)}$$

$$= \frac{V}{1 + K_p} = \frac{5}{51} = 0.098 \text{ V.}$$

From the block diagram we have

$$E_f(s) = 0.5 \, \dot{\theta}(s)$$

Thus the steady-state velocity error $\dot{\theta}_\varepsilon$ is

$$\dot{\theta}_\varepsilon = \frac{e_{ss}}{0.5} = 0.196 \text{ rad/sec.}$$

● **PROBLEM** 6-54

Find the error coefficients of

$$G(s) = \frac{K}{s(1 + sT)}$$

Solution: The output of the system is

$$C(s) = \frac{GR}{1 + G}$$

The error is defined as the input minus the output, therefore

$$E = R - \frac{GR}{1 + G} = \frac{R}{1 + G}$$

and

$$\frac{E}{R} = \frac{1}{1 + G(s)} = \frac{s(1 + s T)}{s(1 + sT) + K} = \frac{s + s^2 T}{K + s + s^2 T}$$

Dividing $s + s^2 T$ by $K + s + s^2 T$ we obtain

263

$$\frac{E}{R} = \frac{1}{K}s + \frac{KT - 1}{K^2}s^2 - \frac{2KT - 1}{K^3}s^3 + \dots$$

The error coefficients are

$$C_0 = 0$$

$$C_1 = \frac{1}{K}$$

$$C_2 = \frac{KT - 1}{K^2}$$

$$C_3 = \frac{-2KT + 1}{K^3}$$

The error series is

$$e(t) = C_0 r(t) + C_1 \dot{r}(t) + C_2 \ddot{r}(t) + C_3 \dddot{r}(t) + \dots$$

$$= \frac{1}{K}\dot{r}(t) + \frac{KT - 1}{K^2}\ddot{r}(t) - \frac{2KT - 1}{K^3}\dddot{r}(t) + \dots$$

● **PROBLEM** 6-55

The block diagram of the loop is shown below. The block z.o.h. is a zero-order-hold with sampling period T, thus

$$c_c(t) = c(t_k) , \qquad t_k < t \le t_{k+1}$$

$$t_k = KT.$$

Describe the performance of the loop.

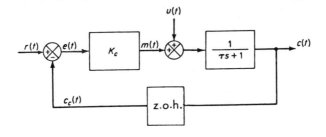

Solution: We have

$$m(t) = K_c [r(t) - c_c(t)]$$

and

$$\tau \frac{dc}{dt} + c = m(t) + u(t)$$

264

The state vector is

$$x(t) = c(t)$$

Let us set $u(t) = 0$ and investigate the set-point response. Let,

$$r(t) = S(t) \text{ for the unit-step change in set}$$
point.

The loop output signal at $t = 0$ is $c_o = x(0)$.

Thus

$$m(t) = K_c[1 - c_c(t)]$$

or

$$m(t) = K_c - K_c x(t_k) \qquad t_k < t \leq t_{k+1}$$

In state variables, the loop equation becomes

$$\frac{dx}{dt} = -\frac{x}{\tau} + \frac{K_c - K_c x(t_k)}{\tau} \quad .$$

For a sampled input $m_1(t_k)$, where

$$m_1(t_k) = K_c - K_c x(t_k)$$

we will determine the response of the system described by

$$\frac{dx}{dt} = \frac{-x}{\tau} + \frac{m_1(t)}{\tau}$$

Comparing with

$$\frac{dx}{dt} = Ax(t) + Bm(t)$$

we obtain

$$A = -\frac{1}{\tau}, \qquad B = \frac{1}{\tau}$$

From $\quad \phi(T) = e^{AT} \quad$ we get

$$\phi(T) = e^{-\frac{T}{\tau}} = \lambda$$

From

$$D(T) = \int_0^T e^{A\zeta} B d\zeta$$

we get

$$D(T) = \int_0^T e^{-\frac{\zeta}{\tau}} \left(\frac{1}{\tau}\right) d\zeta = 1 - \lambda$$

From

$$x(t_{k+1}) = \phi(T)x(t_k) + D(T)m(t_k)$$

we obtain

$$x(t_{k+1}) = \lambda x(t_k) + (1 - \lambda)[K_c - K_c x(t_k)]$$

$$= [\lambda - K_c(1 - \lambda)]x(t_k) + K_c(1 - \lambda).$$

From

$$x(t_k) = A^k x(t_o) + \sum_{k=0}^{k-1} A^{k-i-1} Bm(t_i) ,$$

where $k \geq 1$

we obtain the solution in the form

$$x(t_k) = [\lambda - K_c(1 - \lambda)]^k c_o$$

$$+ K_c(1 - \lambda) \sum_{i=0}^{k-1} [\lambda - K_c(1 - \lambda)]^{k-i-1}$$

Summing the geometric series we have

$$x(t_k) = [\lambda - K_c(1 - \lambda)]^k c_o + \frac{K_c}{1 + K_c}$$

$$(1 - [\lambda - K_c(1 - \lambda)]^k)$$

Stability condition is

$$\left| [\lambda - K_c(1 - \lambda)] \right| < 1$$

or

$$K_c < \frac{1 + \lambda}{1 - \lambda}$$

The final value is

$$\lim_{k \to \infty} x(t_k) = \frac{K_c}{1 + K_c}$$

In case of a unit-step load disturbance, we set

$$r(t) = 0 , \qquad u(t) = S(t) \qquad \text{and obtain}$$

266

$$x(t_{k+1}) = [\lambda - K_c(1 - \lambda)] \, x(t_k) + (1 - \lambda)$$

or

$$x(t_k) = [\lambda - K_c(1 - \lambda)]^k c_0$$

$$+ \, \frac{1}{1 + K_c} \left\{ 1 - [\lambda - K_c(1 - \lambda)]^k \right\}$$

with the final value

$$\frac{1}{1 + K_c}$$

● **PROBLEM** 6-56

In the elementary integrator

$$RC = 1, \qquad A = 10^4, \qquad e_i = 1$$

After what time will the error be 1%?

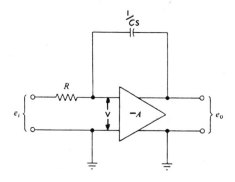

Solution: No current enters the integrator, thus

$$\frac{v - e_i}{R} = \frac{e_o - e_i}{R + \frac{1}{Cs}}$$

and

$$e_o = - \, Av.$$

Substituting in the second equation into the first

$$- (e_o + Ae_i)(RCs + 1) = ARCs(e_o - e_i)$$

$$RC = 1, \qquad \text{thus}$$

$$-(e_o + Ae_i)(s + 1) = As(e_o - e_i)$$

267

or

$$e_o = \cfrac{- \cfrac{A}{A + 1}}{s + \cfrac{1}{A + 1}} e_i$$

We have $e_i(t) = 1$, thus $e_i(s) = \frac{1}{s}$. Then

$$e_o = -\frac{A}{s} + \cfrac{A}{s + \cfrac{1}{A + 1}}$$

Thus the output is

$$e_o(t) = A[-1 + e^{-\frac{t}{A+1}}]$$

Ideally, the output should be

$$e_o(t) = -t$$

thus the percent error is

$$error = 100\% \cdot \left(\frac{-t - e_o(t)}{-t} \right)$$

For the error of 1% we have

$$0.01 = \left| \frac{t_1 - 10^4 (1 - e^{\frac{t}{10^4}})}{t_1} \right|$$

Solving the above equation we get

$$t_1 = 198.746 \text{ sec}$$

The open-loop and closed-loop control systems are shown below.

Assume that $K_2 K >> 1$.

Compare the steady-state errors for these control systems assuming a unit-step input.

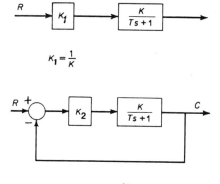

Solution: For the open-loop control system, the error
is

$$e(t) = r(t) - c(t)$$

$$E(s) = R(s) - C(s) = [1 - G_o(s)]R(s)$$

where

$$G_o(s) = K_1 \cdot \frac{K}{Ts + 1} = \frac{1}{Ts + 1}$$

The steady-state error for the unit-step input is

$$e_{ss} = \lim_{s \to 0} sE(s) = \lim_{s \to 0} s[1 - G_o(s)]\frac{1}{s}$$

$$= 1 - G_o(0)$$

When $G_o(0) = 1$ the steady-state error is zero.

For the closed-loop system

$$E(s) = R(s) - C(s) = \frac{1}{1 + G(s)} R(s)$$

where $G(s) = \dfrac{K_2 K}{Ts + 1}$

Thus, the steady-state error is

$$e_{ss} = \lim_{s \to 0} s[\frac{1}{1 + G(s)}]\frac{1}{s} = \frac{1}{1 + G(0)} = \frac{1}{1 + K K_2}$$

where the input is a unit-step function.

When $K K_2 >> 1$, the steady-state error is small.

● **PROBLEM** 6-58

Find a second-order system for which the harmonics are
shifted no more than 5° and distorted no more than 5%.

Solution: A second order system is

$$H(s) = \frac{\omega_0^2}{s^2 + 2\zeta\omega_0 s + \omega_0^2}$$

For $s = j\omega$ we have

$$H(j\omega) = \frac{\omega_0^2}{(\omega_0^2 - \omega^2) + 2j\zeta\omega\omega_0}$$

The phase shift is given by

$$\phi = \tan^{-1} \frac{2\zeta\omega_0\omega}{\omega^2 - \omega_0^2} = \frac{2\zeta v}{1 - v^2}$$

where $v = \dfrac{\omega_0}{\omega}$.

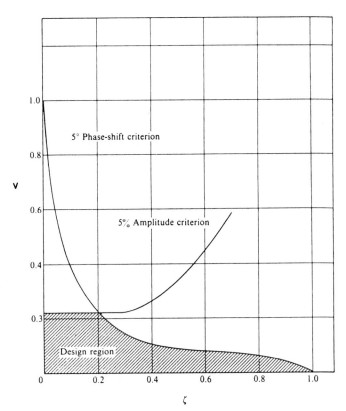

The gain of the system is

$$|H(j\omega)| = \frac{\omega_0^2}{\sqrt{(\omega^2 - \omega_0^2)^2 + 4\zeta^2\omega_0^2\omega^2}} = \frac{1}{\sqrt{(1 - v^2)^2 + (2\zeta v)^2}} \,.$$

Let

$$\frac{2\zeta v}{1 - v^2} = \tan 5° = 0.0875$$

The above equation defines a 5° phase-shift curve.

For the 5% distortion, we get

$$\frac{1}{[(1 - v^2)^2 + (2\zeta v)^2]^{\frac{1}{2}}} = \begin{cases} 1.05 \text{ for } \zeta < 0.707 \\ 0.95 \text{ for } \zeta > 0.707 \end{cases}$$

The above equation defines a 5% amplitude curve. The region of the space (ζ, v) that meets both criteria is shown.

● **PROBLEM** 6-59

The block diagram of a speed control system is shown below.

Try to eliminate the speed errors due to torque disturbances. Investigate this problem for the steady state.

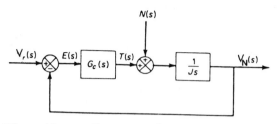

Fig. 1

Solution: When the reference input is zero, the closed-loop transfer function between the output velocity $V_N(s)$ and the disturbance torque $N(s)$ is

$$\frac{V_N(s)}{N(s)} = \frac{\dfrac{1}{Js}}{1 + \dfrac{1}{Js}G_c(s)} = \frac{1}{Js + G_c(s)} .$$

The steady-state output speed due to a unit-step disturbance torque is

$$V_N(\infty) = \lim_{s \to 0} sV_N(s) = \lim_{s \to 0} \frac{1}{s} \frac{s}{Js + G_c(s)} = \frac{1}{G_c(0)} .$$

For $V_N(\infty) = 0$, we must choose $G_c(0) = \infty$. One of possible choices is

$$G_c(s) = \frac{K}{s} .$$

The characteristic equation has two imaginary roots, thus the system is not stable. One of the methods of stabilizing the system is to choose

$$G_c(s) = \frac{K}{s} + K_p .$$

In the absence of the reference input the block diagram of the system is

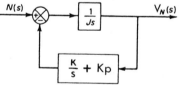

Fig. 2

and its closed-loop transfer function

$$\frac{V_N(s)}{N(s)} = \frac{s}{Js^2 + K_p s + K}$$

Thus for a unit-step disturbance torque, the steady-state speed is

$$V_N(\infty) = \lim_{s \to 0} sV_N(s) = 0 .$$

We conclude that the proportional-plus-integral controller eliminates speed error at steady state.

The transfer function of a linear process is

$$G(s) = \frac{C(s)}{E(s)} = \frac{20}{s^2(s + 1)} .$$

$E(s)$ is the system error equal to

$$E(s) = R(s) - C(s).$$

We want to add state-variable feedback to $E(s)$ from all three states in such way that the system experiences no steady-state error for a unit-step input. Additionally, two of the closed loop poles are at $-1 + j$ and $-1 - j$. Determine the feedback constants g_1, g_2, g_3.

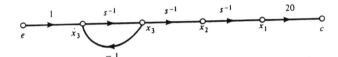

Fig. 1

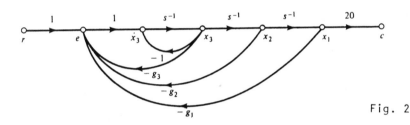

Fig. 2

Solution: Fig. 1 shows the uncompensated system and Fig. 2 shows the system with state-variable feedback.

Using Mason's gain formula, we obtain

$$\frac{C(s)}{R(s)} = \frac{20s^{-3}}{g_1 s^{-3} + g_2 s^{-2} + g_3 s^{-1} + s^{-1} + 1}$$

$$= \frac{20}{g_1 + g_2 s + s^2(g_3 + 1) + s^3}$$

For $R(s) = \frac{1}{s}$, a unit-step

$$C_{ss}(t) = \lim_{s \to 0} s \frac{C(s)}{R(s)} R(s) = \frac{20}{g_1} .$$

Since we want no error, $g_1 = 20$.

The characteristic equation of the system is

$$s^3 + (g_3 + 1)s^2 + g_2 s + 20 = (s + 1 - j)(s + 1 + j)(s + a)$$

or

272

$$s^3 + (g_3 + 1)s^2 + g_2 s + 20$$

$$= s^3 + (a + 2)s^2 + (2a + 2)s + 2a.$$

Equating the coefficients, we obtain

$$g_2 = 22$$

and

$$g_3 = 11.$$

The third pole is at s = -10.

● **PROBLEM** 6-61

The block diagram of an automobile speed control system is shown below.

Where

$$G(s) = \frac{K}{\tau s + 1}$$

$$G_1(s) = K_1 + \frac{K_2}{s}.$$

Assuming $K_2 = 0$, what is the steady-state error for a step input of height A?

What is the steady-state error for a step input of height A when $K_2 \neq 0$?

What is the steady-state error if the input is a ramp of slope A?

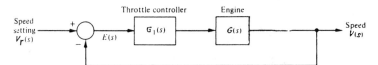

Solution: In general, the error is defined by

$$E(s) = \frac{R(s)}{1 + G(s)}$$

where G(s) is the forward gain, R(s) is the input and there is a unity negative feedback. The steady state error is

$$e_{ss}(t) = \lim_{t \to \infty} e(t) = \lim_{s \to 0} sE(s) = \lim_{s \to 0} \frac{sR(s)}{1 + G(s)}$$

In our case,

$$G(s) = (K_1 + \frac{K_2}{s})(\frac{K}{\tau s + 1}).$$

For $K_2 = 0$ and a step input $R(s) = \frac{A}{s}$ and

$$e_{ss}(t) = \lim_{s \to 0} \frac{s\frac{A}{s}}{1 + K_1\frac{K}{\tau s + 1}} = \frac{A}{1 + K_1 K}$$

When $K_2 \neq 0$, we have a type one system and

$$G_1(s) = \frac{K_1 s + K_2}{s} \; .$$

For a step input the steady-state error is zero. For the ramp input, $R(s) = \frac{A}{s^2}$, we get

$$e_{ss}(t) = \lim_{s \to 0} \frac{s\frac{A}{s^2}}{\frac{K_1 s + K_2}{s} \cdot \frac{K}{\tau s + 1}} = \lim_{s \to 0} \frac{A}{(K_1 s + K_2)\frac{K}{\tau s + 1}}$$

$$= \frac{A}{K_2 K} \; .$$

● **PROBLEM** 6-62

The signal-flow graph of a space vehicle attitude control system is shown in Fig. 1. It is desired to select the magnitude of the gain K_3 in order to minimize the effect of the disturbance $U(s)$. The disturbance in this case is equivalent to an initial attitude error. Use $K_1 = 0.5$ and $K_1 K_2 K_p = 2.5$. Find $c(t)$ for $u(t) = 1$. Find the

$$\text{ISE} = \int_0^\infty c^2(t)dt,$$ and find the value of K_3 that makes the ISE minimal.

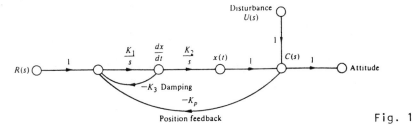

Fig. 1

Solution: We use Mason's rule between $U(s)$ and $C(s)$ in Fig. 1 to determine $C(s)/U(s)$. This gives us

$$\frac{C(s)}{U(s)} = \frac{P_1(s)\Delta_1(s)}{\Delta(s)}$$

$$= \frac{1 \cdot (1 + K_1 K_3 s^{-1})}{1 + K_1 K_3 s^{-1} + K_1 K_2 K_p s^{-2}}$$

$$= \frac{s(s + K_1 K_3)}{s^2 + K_1 K_3 s + K_1 K_2 K_p} \; . \tag{1}$$

Typical values for the constants are $K_1 = 0.5$ and $K_1 K_2 K_p = 2.5$. Then the natural frequency of the vehicle is $f_n = \sqrt{2.5}/2\pi = 0.25$ cycles/sec. For a unit step disturbance, the minimum ISE can be analytically calculated. $U(s) = 1/s$, so

$$C(s) = \frac{s + 0.5K_3}{s^2 + 0.5K_3 + 2.5} \tag{2}$$

Taking the inverse transform yields

$$c(t) = \frac{\sqrt{10}}{\beta}\left[e^{-0.25K_3 t} \sin\left(\frac{\beta}{2}t + \psi\right) \right], \tag{3}$$

where $\beta = K_3\sqrt{(K_3^2/8)} - 5$. Squaring $c(t)$ and integrating the result, we have

$$I = \int_0^\infty \frac{10}{\beta^2} e^{-0.5K_3 t} \sin^2\left(\frac{\beta}{2}t + \psi\right) dt$$

$$= \int_0^\infty \frac{10}{\beta^2} e^{-0.5K_3 t}\left(\frac{1}{2} - \frac{1}{2}\cos(\beta t + 2\psi)\right) dt$$

$$= \left(\frac{1}{K_3} + 0.1K_3\right) \tag{4}$$

Differentiating I and equating the result to zero, we obtain

$$\frac{dI}{dK_3} = -K_3^{-2} + 0.1 = 0. \tag{5}$$

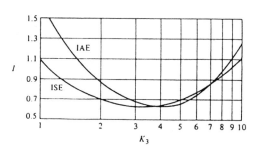

Fig. 2

Therefore, the minimum ISE is obtained when $K_3 = \sqrt{10} = 3.2$. This value of K_3 corresponds to a damping ratio ζ of 0.50. The values of ISE and IAE for this system are plotted in Fig. 2. The minimum for the IAE performance index is obtained when $K_3 = 4.2$ and $\zeta = 0.665$. While the

ISE criterion is not as selective as the IAE criterion, it is clear that it is possible to solve analytically for the minimum value of ISE. The minimum of IAE is obtained by measuring the actual value of IAE for several values of the parameter of interest.

For the feedback control system shown in Fig. 1, sketch the system response, approximate the value of

$$\int_0^\infty e(t)\,dt \qquad (e(t) \text{ is the error signal})$$

and verify that $\int_0^\infty e(t)\,dt = e_{ssr}$ (1)

where $e(t)$ is error in the unit step response and e_{ssr} is the steady state error in the unit ramp response, using numerical integration on the response curves.

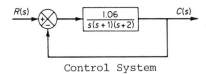

Control System   Fig.1

<u>Solution</u>: The closed-loop transfer function for the present system is given by

$$\frac{C(s)}{R(s)} = \frac{G(s)}{1 + G(s)}$$

$$= \frac{1.06}{s(s + 1)(s + 2) + 1.06}$$

$$= \frac{1.06}{(s + 2.33)(s + 0.33 + j0.58)(s + 0.33 - j0.58)}$$

For a unit-step input,

$$R(s) = \frac{1}{s}$$

The unit-step response is thus obtained as

$$c(t) = L^{-1}[C(s)]$$

$$= L^{-1}\left[\frac{1.06}{s(s + 2.33)[(s + 0.33)^2 + 0.58^2]}\right]$$

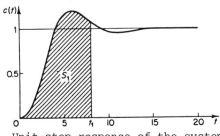

Unit-step response of the system   Fig.2

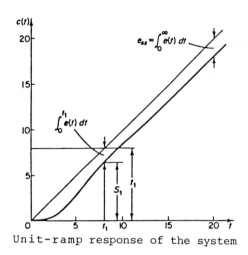

Unit-ramp response of the system   Fig.3

Referring to the formula

$$L^{-1}\left[\frac{1}{s(s+a)\left[(s+\zeta\omega_n)^2+\omega_d^2\right]}\right] = \frac{1}{a\omega_n^2} - \frac{e^{-at}}{a\left[(a-\zeta\omega_n)^2+\omega_d^2\right]}$$

$$+ \frac{e^{-\zeta\omega_n t}}{\omega_n\left[(a-\zeta\omega_n)^2+\omega_d^2\right]}\left[\left(2\zeta - \frac{a}{\omega_n}\right)\cos\omega_d t\right.$$

$$\left.+ \frac{1}{\sqrt{1-\zeta^2}}\left(2\zeta^2 - \frac{a\zeta}{\omega_n} - 1\right)\sin\omega_d t\right]$$

and noting that in the present system

$$a = 2.33, \quad \zeta = 0.5, \quad \omega_n = 0.667, \quad \omega_d = 0.58$$

we obtain c(t) as follows:

$$c(t) = 1 - 0.103e^{-2.33t} - e^{-0.33t}(0.897 \cos 0.58t$$

$$+ 0.933 \sin 0.58t) \qquad (t \geq 0) \quad (2)$$

Equation (2) gives the response to a unit-step input; see Fig. 2.

Now we shall examine the response of the system to a unit-ramp input. When all initial conditions are zero, the unit-ramp response of the system can be obtained by integrating the unit-step response curve. Figure 2 shows the unit-ramp response curve for the present system. [Note that the area S_1 under the unit-step response curve from t = 0 to t = t_1 gives the value of the output at t = t_1 with a unit-ramp input. See Fig's.2 and 3.]

In the present system, the unit-step response settles at approximately t = 20. Hence, the difference between 20 and the area under the unit-step response curve from t = 0 to t = 20 gives the steady-state error in the unit-ramp

277

response. This value of the error should be equal to the value given by 1/lim sG(s).
$$\underset{s \to 0}{}$$

From the response curve of Fig. 2, we obtain the following data:

t	Area S under Unit-Step Response Curve from 0 to t	Error Area $1 \times t - S$
0	0	0
2	0.16	1.84
4	1.50	2.50
6	3.80	2.20
8	6.10	1.90
10	8.15	1.85
12	10.10	1.90
14	12.06	1.94
16	14.07	1.93
18	16.08	1.92
20	18.08	1.92

The error area at steady state is found as 1.92.
The static velocity error coefficient for this system is obtained from the open-loop transfer function as

$$K_V = \lim_{s \to 0} sG(s)$$

$$= \frac{1.06}{2}$$

$$= 0.53$$

Hence the steady-state error in the unit-ramp response is given by
$$e_{ss} = \frac{1}{K_V} = 1.89$$

The graphically obtained value and the analytically obtained value of the steady-state error in the unit-ramp response are close to each other. Thus, we verified Eq. 1. (Note that if these two values differ considerably, then we should check to see where the discrepancy comes from, find the errors, and correct them.)

● **PROBLEM** 6-64

A pendulum is suspended from a pivot wich generates a linear torque $-C\dot\theta$ in which θ is the angular displacement.

The complete assembly is restricted to translation in a vertical plane and is accelerated with an acceleration (a)

on the same plane. Show that this device can be used to de-
tect acceleration and obtain the static sensitivity of the
system (Fig. 1).

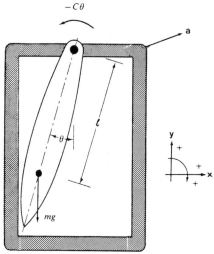

Fig. 1

Solution: To obtain the equation of motion we use Lagrange's
equation

$$\frac{d}{dt}\left(\frac{\partial T}{\partial \dot{\theta}}\right) - \frac{\partial T}{\partial \theta} = Q \tag{1}$$

in which T is the kinetic energy of the system and Q
represents the applied torques. From Figure 1 we obtain

$$T = \frac{1}{2}m(\dot{x}^2 + \dot{y}^2) + \frac{1}{2}I_0\dot{\theta}^2 \tag{2}$$

where

   I_0 = moment of inertia about center of gravity

   m = mass of pendulum

From the geometry of the system it follows that

$$\dot{x} = \dot{u} - \ell\dot{\theta} \cos \theta \tag{3}$$

$$\dot{y} = \dot{v} + \ell\dot{\theta} \sin \theta \tag{4}$$

in which u and v are the horizontal and vertical displacements of the pivot, respectively. Thus we see that

$$T = \frac{1}{2}m[(\dot{u} - \ell\dot{\theta} \cos \theta)^2 + (\dot{v} + \ell\dot{\theta} \sin \theta)^2] + \frac{1}{2}I\dot{\theta}^2 \tag{5}$$

Application of Lagrange's equation yields

$$(I_0 + m\ell^2)\ddot{\theta} + m\ell(\ddot{v} \sin \theta - \ddot{u} \cos \theta) = Q \tag{6}$$

Q is easily obtained by observation

$$Q = -C\dot{\theta} - mg\ell \sin \theta \qquad (7)$$

If we let $I = I_0 + m\ell^2$, we obtain the performance equation of the system

$$I\ddot{\theta} + C\dot{\theta} + mg\ell \sin \theta = -m\ell(\ddot{v} \sin \theta - \ddot{u} \cos \theta) \qquad (8)$$

It is obvious that the quantity $\theta(t)$ is a function of both the acceleration components $\ddot{u}$ and $\ddot{v}$.

The static sensitivity is obtained by setting $\ddot{\theta} = \dot{\theta} = 0$ and solving for $\theta = \theta_u$:

$$\tan \theta_u = \frac{\ddot{u}}{g + \ddot{v}} \qquad (9)$$

This result shows the difficulties encountered in the use of this device. θ_u depends nonlinearly on $\ddot{v}$ as well as on $\ddot{u}$. Hence the pendulum is a device for detecting one component of acceleration only if $\ddot{v} \ll g$. Let us find the static sensitivity of the pendulum to $\ddot{u}$.

$$S_u = \frac{\partial \theta}{\partial \ddot{u}} = \frac{\cos^2 \theta}{g + \ddot{v}} \qquad (10)$$

Thus since $\cos^2 \theta = 1/(1 + \tan^2 \theta) = 1/(1 + \frac{\ddot{u}^2}{(g + \ddot{v})^2})$, $\qquad (11)$

$$S_u = \frac{g + \ddot{v}}{(g + \ddot{v})^2 + \ddot{u}^2} \qquad (12)$$

The sensitivity of the system is not a constant but is a function of $\ddot{v}$ and $\ddot{u}$. This static sensitivity is plotted in Figure 2 in a dimensionless form

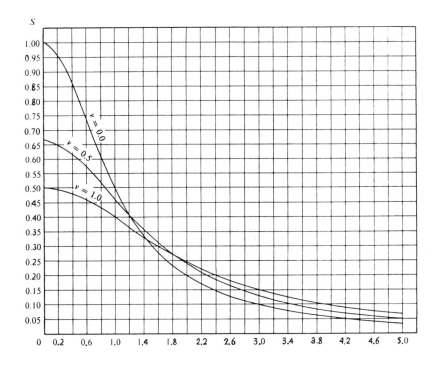

Fig. 2

$$gS_u = \frac{1 + (\ddot{v}/g)}{[1 + (\ddot{v}/g)^2] + (\ddot{u}/g)^2}$$

(13)

At low horizontal acceleration $\ddot{u}$ the response is not single valued. At high horizontal acceleration the "accelerometer" becomes insensitive to cross acceleration, but the system sensitivity goes down. We must conclude that the pendulum is not a very good accelerometer.

CHAPTER 7

FREQUENCY ANALYSIS, NYQUIST DIAGRAM, ROOT LOCUS, BODE DIAGRAM

NYQUIST DIAGRAM

● **PROBLEM** 7-1

For the integrator shown in Fig. 1 draw the Nyquist diagram, assuming K > 0.

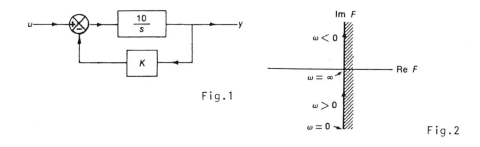

Fig.1

Fig.2

Solution: From the diagram we get

$$F(s) = \frac{10K}{s} = 10K \frac{\sigma - j\omega}{\sigma^2 + \omega^2} \tag{1}$$

The sketch of F(s) in the F(s) - plane is the Nyquist diagram. Fig. 2 shows the sketch of

$$F(j\omega) = -\frac{j10K}{\omega}$$

where we set s = jω.
For K > 0 the figure is closed to the right; it can be seen by taking s = R → ∞ , where F(R) > 0. Point -1 is not encircled, therefore the system is stable from the Nyquist criterion, for K > 0.

F(s) has a zero on the s-plane curve C at s = ∞ since the curve passes through the origin.

Assume that a polar plot of $KG(j\omega)H(j\omega)$ is given. What information is available from it?

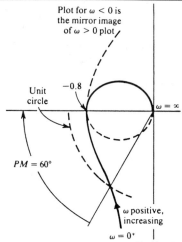

Plot for $\omega < 0$ is the mirror image of $\omega > 0$ plot

−0.8

Unit circle

$\omega = \infty$

$PM = 60°$

ω positive, increasing

$\omega = 0^+$

Solution: From Nyquist's criterion we can determine the number of closed-loop poles in the right-half plane in terms of the number of encirclements of -1. The system type is indicated assuming a minimum phase system, by the phase angle at $\omega = 0$. Type 0 systems have a finite magnitude and zero phase angle. Type 1 systems approach infinite magnitude at an angle of -90°. The plot shows the system of type 1.

Type 2 systems approach infinite magnitude at an angle of -180°. The behavior of the system as $\omega \to \infty$ indicates the excess of open-loop poles compared to zeros. For the equal number of poles and zeros the magnitude approaches a finite constant. The magnitude approaches zero along the -90° axis when there is one more pole than zero, along -180° axis when there are two more poles than zeros, etc. From the plot we can easily determine relative stability in terms of gain and phase margins.

The system shown in the plot has three more poles than zeros, it is type 1. The system is stable since the plot does not encircle the point -1, assuming no open-loop, right half-plane poles. The phase margin is 60° and the gain margin is 1.25.

Discuss the stability of the system for different values of K_2, where

$$G(s)H(s) = \frac{K_2(1 + T_4s)}{s^2(1 + T_1s)(1 + T_2s)(1 + T_3s)}$$

The Nyquist plot of $G(s)H(s)$ is shown. Assume $T_4 > T_1, T_2, T_3$.

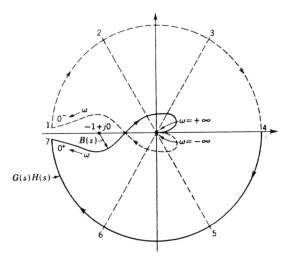

Solution: The point -1 + j0 is not encircled, thus the system is stable. By increasing K_2 sufficiently, however, the plot can be made to enclose -1 + j0 and the system will become unstable.

● **PROBLEM** 7-4

Fig. 1 shows a single-loop control system.

where

$$GH(s) = \frac{K}{s(\tau s + 1)}$$

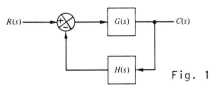

Fig. 1

Draw the Nyquist plot for the system and investigate its stability.

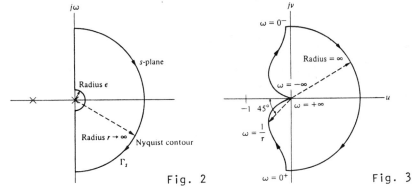

Fig. 2   Fig. 3

Solution: We first determine the contour $\Gamma_p = \Gamma_{GH}$ where P(s) = GH(s), in the GH(s)-plane. Fig. 2 shows the contour Γ_s in the s-plane. The infinitesimal detour around the pole at the origin is a semicircle of radius ε, where $\varepsilon \to 0$. Cauchy's theorem states that the contour cannot pass through the pole at the origin. Fig. 3 shows the contour Γ_{GH} .

We shall investigate each part of the Nyquist contour Γ_s and determine the corresponding portions of the GH(s) -plane contour Γ_{GH} .

1) The origin of the s-plane:

We shall represent a semicircular detour around the pole by

$$s = \varepsilon e^{j\phi}$$

where ϕ varies from $-90°$ at $\omega = 0^-$ to $+90°$ at $\omega = 0^+$. We have

$$\lim_{\varepsilon \to 0} GH(s) = \lim_{\varepsilon \to 0} \frac{K}{\varepsilon e^{j\phi}} = \lim_{\varepsilon \to 0} \left(\frac{K}{\varepsilon}\right) e^{-j\phi}$$

Thus the angle of the contour in the GH(s)-plane changes from $90°$ at $\omega = 0^-$ to $-90°$ at $\omega = 0^+$, it passes $0°$ at $\omega = 0$. In the GH(s) plane the radius of this portion of the contour is infinite as shown in Fig. 3.

2) The part from $\omega = 0^+$ to $\omega = +\infty$:

Since $\omega j = s$

and

$$GH(s)\Big|_{s=j\omega} = GH(j\omega)$$

the portion of Γ_s from $\omega = 0^+$ to $\omega = +\infty$ is mapped by the function GH(s) as the real frequency polar plot, see Fig. 3.

We have
$$\lim_{\omega \to +\infty} GH(j\omega) = \lim_{\omega \to +\infty} \frac{K}{+j\omega(j\omega\tau + 1)}$$
$$= \lim_{\omega \to \infty} \left|\frac{K}{\tau\omega^2}\right| \underline{/-\frac{\pi}{2} - \tan^{-1}\omega\tau}$$

The magnitude approaches zero at an angle of $-180°$.

3) The part from $\omega = +\infty$ to $\omega = -\infty$:

This part of Γ_s is mapped into the point zero at the origin of the GH(s)-plane, the mapping is

$$\lim_{r \to \infty} GH(s)\Big|_{s=re^{j\phi}} = \lim_{r \to \infty} \left|\frac{K}{r^2}\right| e^{-2j\phi}$$

where ϕ changes from $\phi = +90°$ at $\omega = +\infty$ to $\phi = -90°$ at $\omega = -\infty$. The contour moves from an angle of $-180°$ at $\omega = +\infty$ to an angle of $+180°$ at $\omega = -\infty$, with the magnitude constant.

4) The part from $\omega = -\infty$ to $\omega = 0^-$:

This part is mapped by the function GH(s) as

$$GH(s)\bigg|_{s=-j\omega} = GH(-j\omega)$$

We see that the plot for the part of the polar plot from $\omega = -\infty$ to $\omega = 0^-$ is symmetrical to the polar plot from $\omega = +\infty$ to $\omega = 0^+$, see Fig. 3.

To examine the stability of the system we note that there are no poles in the right-hand s-plane. Thus, for the system to be stable, we require N = Z = 0 and the contour Γ_{GH} must not encircle the (-1 + j0) point in the GH-plane. From Fig. 3 we see that the system is always stable, for all values of K and τ.

● PROBLEM 7-5

Consider the equation

$$G(s)H(s) = \frac{K_0(1+T_1s)^2}{(1+T_2s)(1+T_3s)(1+T_4s)(1+T_5s)^2}$$

where $T_s < T_1 < T_2$, T_3 and T_4, whose complete polar plot is shown in the figure for a particular value of gain. Discuss the stability of the system for other values of K_0.

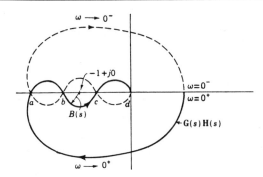

Solution: We see that increasing or decreasing the gain we can make the system unstable. When the gain is increased sufficiently the point (-1+j0) lies between c and d of the polar plot, the net clockwise rotation is equal to 2 and the system is unstable. When the gain is decreased so that the point (-1+j0) lies between a and b of the polar plot, the net clockwise rotation is 2 and the system is unstable. When the gain is further decreased the point (-1+j0) lies to the left of a of the polar plot and the system is stable. Thus the system is conditionally stable.

Determine the stability of the control system shown in Fig. 1, without and with derivative feedback (i.e., $K_2 = 0$ and $K_2 > 0$). Use the Nyquist criterion.

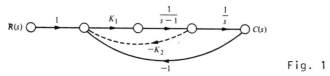

Fig. 1

Solution: First, we consider the case $K_2 = 0$. We have

$$GH(s) = \frac{K_1}{s(s-1)}$$

The open-loop transfer function has one pole in the right-hand plane, $P = 1$. The system is stable when $N = -P = -1$.

At the semicircular detour at the origin of the s-plane, we let $s = \varepsilon e^{j\phi}$ for

$$-\frac{\pi}{2} \le \phi \le \frac{\pi}{2} .$$

We have

$$\lim_{\varepsilon \to 0} GH(s) = \lim_{\varepsilon \to 0} \frac{K_1}{-\varepsilon e^{j\phi}} = \left(\lim_{\varepsilon \to 0} \left| \frac{K_1}{\varepsilon} \right| \right) \angle{-180° - \phi}$$

Thus, this part of the contour Γ_{GH} is a semicircle of infinite magnitude as shown in Fig. 2.

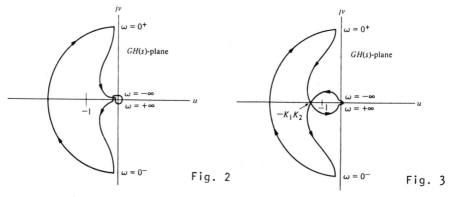

Fig. 2

Fig. 3

For $s = j\omega$, we have

$$GH(j\omega) = \frac{K_1}{j\omega(j\omega-1)} = \frac{K_1}{(\omega^2+\omega^4)^{\frac{1}{2}}} \angle{-\frac{\pi}{2} - \tan^{-1}(-\omega)}$$

$$= \frac{K_1}{(\omega^2 + \omega^4)^{\frac{1}{2}}} \bigg/ \frac{\pi}{2} + \tan^{-1} \omega$$

For the semicircle of radius r, as $r \to \infty$, we have

$$\lim_{r \to \infty} GH(s) \bigg|_{s=re^{j\phi}} = \left(\lim_{r \to \infty} \left| \frac{K_1}{r^2} \right| \right) e^{-2j\phi}$$

where ϕ varies in a clockwise direction from $\frac{\pi}{2}$ to $-\frac{\pi}{2}$. Thus the contour Γ_{GH} at the origin of the GH-plane varies 2π rad in a counterclockwise direction. The contour Γ_{GH} in the GH(s)-plane encircles the -1 point once in the clockwise direction and N = +1, thus

$$Z = N + P = 2.$$

The system is unstable, since two zeros of the characteristic equation lie in the right half of the s-plane.

We shall investigate the case when the derivative feedback is included. The open-loop transfer function is

$$GH(s) = \frac{K_1(1 + K_2 s)}{s(s-1)}$$

The part of the contour Γ_{GH} when $s = \varepsilon e^{j\phi}$ is the same as for the system without derivative feedback, see Fig. 3. For $s = re^{j\phi}$ as $r \to \infty$ we have

$$\lim_{r \to \infty} GH(s) \bigg|_{s=re^{j\phi}} = \lim_{r \to \infty} \left| \frac{K_1 K_2}{r} \right| e^{-j\phi}$$

and the Γ_{GH} -contour at the origin of the GH-plane varies π rad in a counterclockwise direction.

The frequency locus GH($j\omega$) crosses the u-axis. We have

$$GH(j\omega) = \frac{K_1(1 + K_2 j\omega)}{-\omega^2 - j\omega}$$

$$= \frac{-K_1(\omega^2 + \omega^2 K_2) + j(\omega - K_2 \omega^3)K_1}{\omega^2 + \omega^4}$$

The point of intersection of the GH($j\omega$)-locus and the u-axis is determined from

$$\omega - K_2 \omega^3 = 0$$

or

$$\omega^2 = \frac{1}{K_2}$$

The value of the real part of GH($j\omega$) at the intersection is

$$u\Big|_{\omega^2 = \frac{1}{K_2}} = \frac{-\omega^2 K_1 (1 + K_2)}{\omega^2 + \omega^4}\Big|_{\omega^2 = \frac{1}{K_2}} = -K_1 K_2$$

When $K_1 K_2 > 1$ the contour Γ_{GH} encircles the -1 point once in a counterclockwise direction, $N=-1$. The number of zeros of the system, in the right-hand plane is

$$Z = N + P = -1 + 1 = 0$$

The system is stable when

$$K_1 K_2 > 1$$

ROOT LOCUS

● PROBLEM 7-7

The open loop transfer function of the system is

$$\frac{\theta_{on}}{\theta} = \frac{K_0{}'(D + 12)}{D^2(D + 20)}$$

Sketch the root locus for the system.

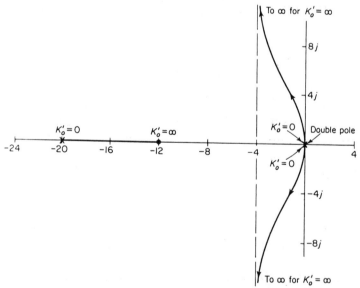

Solution: From the transfer function we have three open loop poles, one at -20 and two at 0 due to the term D^2. There is one finite zero (-12). Therefore two root locus branches start at 0 for $K_0' = 0$ and the third starts at -20 for $K_0' = 0$. Two branches finish at ∞, and one finishes at -12 for $K_0' = \infty$.

We find the infinite zero asymptotes

$$\alpha = \frac{\pm 180}{2} = \pm 90$$

$$\bar{x} = \frac{-20 - (-12)}{2} = -4$$

We see that the root locus does not cut the imaginary axis, it occupies the real axis between -20 and -12. The root locus for the system is shown in the figure.

● PROBLEM 7-8

The transfer function of a system is

$$\frac{\theta_0}{\theta_i} = \frac{K(D+2)}{D^2 + (4+K)D + 2K}$$

where K has a constant value.

Determine the relationship between the value of K and the nature of the system.

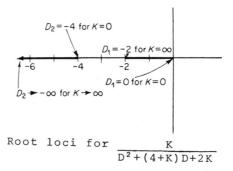

Root loci for $\dfrac{K}{D^2+(4+K)D+2K}$

Solution: The characteristic equation of the system is

$$D^2 + (4+K)D + 2K = 0$$

Solving this equation we find the roots

$$D_1 = \frac{-(4+K) + \sqrt{16 + K^2}}{2} \quad ;$$

$$D_2 = \frac{-(4+K) - \sqrt{16 + K^2}}{2}$$

Since the characteristic equation has two real roots, there are two paths on the root locus plot. For $K = 0$, $D_1 = 0$ and $D_2 = -4$ which are the starting points of the two loci. The roots D_1 and D_2 cannot be complex for any value of K because $16 + K^2 > 0$. The roots for any value of K will be real and negative, since

$$\sqrt{16 + K^2} \leq 4 + K$$

For $K \to \infty$

1) $D_1 \to -2$, thus the root locus of D_1 is a line from 0 to -2 on the real axis.

2) $D_2 \to -\infty$; thus the root locus of D_2 is a line from -4 to $-\infty$ on the real axis. Evaluating the roots we find the intermediate values of K that can be fitted onto the two loci. Thus the root locus of the system can be represented by the plot for all values of K. Since the whole root locus plot is located in the lefthand half plane the system is stable for all values of K.

● **PROBLEM** 7-9

Sketch the root locus for the system shown in Fig. 1.

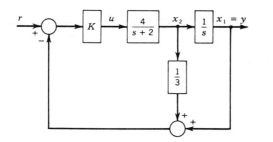

Solution: From Fig. 1 we get the following equations.

$$G_p(s) = \frac{4}{s(s+2)} \quad ,$$

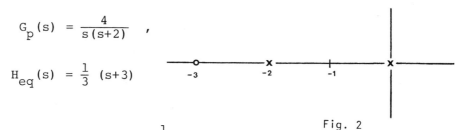

$$H_{eq}(s) = \frac{1}{3}(s+3)$$

Thus

$$K_p = 4 \qquad K_n = \frac{1}{3}$$

Fig. 2

and the loop transfer function is

$$KG_p(s) \, H_{eq}(s) = K \frac{4}{3} \frac{s+3}{s(s+2)}$$

Fig. 2 shows the pole-zero plot of the loop transfer function.

The breakaway point is the point between s = -2 and s = 0 where the gain reaches its maximum. The reentry point is located to the left of s = -3 at a point midway between two very close values of gain. We shall examine some more points in the upper half plane. Since the root locus is symmetrical about the real axis we don't have to worry about

291

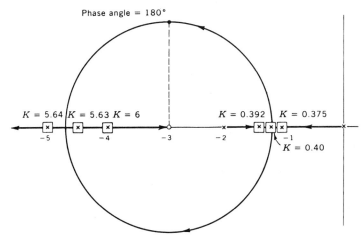

Phase angle = 180°

$K = 5.64$ $K = 5.63$ $K = 6$ $K = 0.392$ $K = 0.375$

$K = 0.40$

Fig. 3

the lower half plane. We shall find the point on the
line s = -3 for which the phase angle is 180°. The branches
of the root locus must be 90° apart at the breakaway and
reentry points. Thus the angles between the root loci
and the real axis must be ±90°. Connecting the points
found so far with a smooth curve that passes through the
point on the s = -3 line and is perpendicular to the real
axis at the point of breakaway or reentry we obtain a
semicircle.

Fig. 3 shows the complete root locus of the system.

● **PROBLEM 7-10**

Sketch the root locus of

$$KGH = \frac{64K}{s(s+4)(s+16)} \tag{1}$$

Solution: 1) There are no zeros. The poles are at s = 0,
s = -4, s = -16. Branches of loci start at open-loop
poles and terminate at open-loop zeros at infinity.

2) The root locus exists on the real axis between s = 0 and
s = -4, s = -16 and s = -∞. The root locus exists on
the real axis when an odd number of poles plus zeros is
found to the right of the point.

3) The asymptotic angles are

$$\alpha = \frac{(2k+1)\,180°}{\Sigma P - \Sigma Z} = \frac{(2k+1)\,180°}{3} = 60°(2k+1) \tag{2}$$

for   k = 0 $\alpha = 60°$

         k = 1 $\alpha = 180°$

         k = 2 $\alpha = 300°$

4) The center of gravity (C.G.) is given by

292

$$\text{C.G.} = \frac{\sum_{\text{values}} P - \sum_{\text{values}} Z}{\sum P - \sum Z} = \frac{-16-4-0+0}{3-0} = -6\frac{2}{3} \tag{3}$$

The center of gravity gives the starting point for the asymptotic lines.

5) The breakaway point is given by

$$\frac{1}{s_b} = \frac{1}{16-s_b} + \frac{1}{4-s_b} \tag{4}$$

or

$$(16-s_b)(4-s_b) = s_b(4-s_b) + s_b(16-s_b)$$

The approximate value of s_b is

$$s_b \approx -1.86$$

The breakaway angle from the real axis is $\pm 90°$.

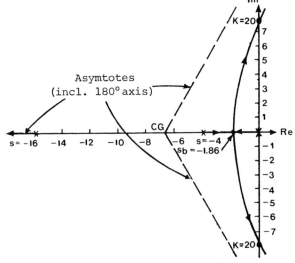

6) The maximum value of K for which the system is stable is found by substituting $s = j\omega$, thus

$$KGH(j\omega) = \frac{64K}{j\omega(j\omega+4)(j\omega+16)} \tag{5}$$

Setting $KGH(j\omega) = -1$ we have

$$\frac{64K}{j\omega(j\omega+4)(j\omega+16)} = -1 \tag{6}$$

Solving for K

$$K = \frac{20\omega^2 + j\omega(\omega^2 - 64)}{64} \tag{7}$$

For K to be a real number $\omega^2 - 64$ must be equal to zero.

Thus $\omega = \pm 8$.

From eq. (7), for $\omega = \pm 8$, we get $K = 20$.

The root locus of the system

$$KGH = \frac{64K}{s(s+4)(s+16)}$$

is shown in the figure.

● **PROBLEM** 7-11

Sketch the root locus of the system, whose transfer function is

$$KGH(s) = \frac{K}{s(s+3)(s^2+6s+64)} \qquad (1)$$

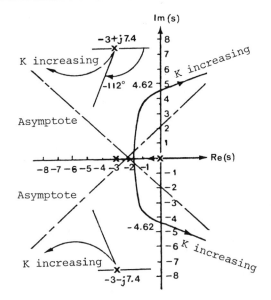

Solution: We shall follow the standard procedure to sketch the root locus of the system.

1) Find the zeros and poles of KGH. KGH(s) has no zeros, it has two real poles at $s_0 = 0$ and $s_1 = -3$, and two complex poles. To find the complex poles we solve the equation

$$s^2 + 6s + 64 = 0$$

or

$$s = \frac{-6 \pm \sqrt{36 - 4 \times 64}}{2}$$

We obtained the following poles

$$s_0 = 0$$

$$s_1 = -3$$

$$s_2 = -3 + j\sqrt{55} = -3 + j\ 7.4$$

$$s_3 = -3 - j\sqrt{55} = -3 - j\ 7.4$$

(2)

The zeros and poles of KGH(s) are the termination and starting points, respectively, of branches of loci for K between infinity and zero.

2) The root locus exists on the real axis between s = 0 and s = -3.

3) Find the asymptotic angles.

They are given by the equation

$$\frac{(2k+1)\,180°}{\Sigma P - \Sigma Z} = \frac{(2k+1)\,180°}{4} = (2k+1)\cdot 45°$$

where ΣP indicates the number of poles.   (3)

4) Find the center of gravity.

$$\text{Center of gravity} = \frac{\underset{\text{values}}{\Sigma\ P} - \underset{\text{values}}{\Sigma\ Z}}{\Sigma P - \Sigma Z}$$

$$= \frac{-0 - 3 - 3 + j\ 7.4 - 3 - j\ 7.4 + 0}{4 - 0} = -\frac{9}{4} = -2.25 \qquad (4)$$

5) The breakaway point is at s $\approx$ -1.5, half way between the real negative axis poles. The breakaway angle is $\pm 90°$.

6) Find the angle of departure from the open-loop pole s_2. We have

$$\theta_0 + \theta_1 + \theta_2 + \theta_3 = (2k+1)\,180°$$

and

$$\theta_2 = (2k+1)\,180° - (\theta_0 + \theta_1 + \theta_3) \qquad (5)$$

For k = 1

$$\theta_2 = 540° - (112° + 90° + 90°) = 248°$$

Note that for large s, this root locus approaches the k=1 asymptote.

The value of K, for which the root locus crosses the imaginary axis is given by substituting s = jω into eq. (1). We have

$$KGH(j\omega) = \frac{K}{(j\omega)^4 + 9(j\omega)^3 + 82(j\omega)^2 + 192j\omega} \qquad (6)$$

and

$$\frac{1}{(\omega^4 - 82\omega^2) + j(192\omega - 9\omega^3)} = -\frac{1}{K} \tag{7}$$

or

$$K = (82\omega^2 - \omega^4) - j(192\omega - 9\omega^3) \tag{8}$$

Since K is a real number, we have

$$192 - 9\omega^2 = 0 \tag{9}$$

$$\omega = \pm \frac{8}{\sqrt{3}} = \pm 4.6$$

Substituting the value of ω into eq.(8) we have

$$K = 82 \times \frac{64}{3} - \left(\frac{64}{3}\right)^2 = 1.27 \times 10^3$$

The root locus of the system is shown in the figure.

● **PROBLEM** 7-12

For the system

$$\frac{\theta_0}{\theta_i} = \frac{K}{D^2 + 3D + K}$$

obtain a root locus plot.

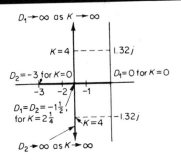

Root loci for $\frac{K}{D^2+3D+K}$

Solution: The roots of the characteristic equation are

$$D_1 = \frac{-3 + (9-4K)^{\frac{1}{2}}}{2}$$

$$D_2 = \frac{-3 - (9-4K)^{\frac{1}{2}}}{2}$$

For 9 - 4K > 0 both roots are negative real numbers. We have for K = $2\frac{1}{4}$ $\begin{cases} D_1 = -1\frac{1}{2} \\ \\ D_2 = -1\frac{1}{2} \end{cases}$

For K > $2\frac{1}{4}$ the roots are a complex pair, with the real part $-1\frac{1}{2}$ and the imaginary part

$$\pm \frac{(9-4K)^{\frac{1}{2}}}{2}$$

The imaginary part reaches $\pm$ infinity as K $\rightarrow \infty$.

$$D_1 = \frac{-3 + 2.64j}{2} = -1.5 + 1.32j$$
$$\left.\right\} \text{ for K = 4}$$
$$D_2 = \frac{-3 - 2.64j}{2} = -1.5 - 1.32j$$

For all values of K the system is stable since the root loci are located in the lefthand half plane.

● **PROBLEM** 7-13

Sketch the root locus of the following system

$$KG_p(s) H_{eq}(s) = \frac{K(s+1)}{s(s-1)(s+6)}$$

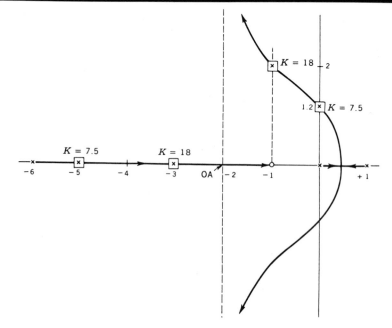

Solution: The system has a pole in the right half s-plane. This, of course, does not change the rules of construction for

the root locus. The figure shows the complete root locus.
The zero in the system "pulls" the root locus back into the
left half s-plane and stabilizes the system though for certain values of K, K > 7.5 only.

A breakaway point exists in the right-half plane between
s = 0 and s = +1. The gains at s = 0.4 and 0.5 are 1.09
and 1.08 respectively. The phase shift is 180° at approximately s = j1.2 along the imaginary axis. Here the sum
of the open-loop poles is -5, so that when the complex conjugate poles are on the jω axis, the remaining closed-loop
pole is at s = -5. At this point the gain K is

$$\frac{6 \times 5 \times 1}{4} = 7.5$$

We see that if the gain K is less than 7.5 the system is
unstable. The last point in the construction of the root
locus is the point along the line s = -1 where the phase-
angle criterion is satisfied. This condition is met at
s ≈ -1 + j2, the gain is 18 and the real pole is at
s = -3.

● **PROBLEM** 7-14

The open loop transfer function of the system is

$$\frac{\theta_{on}}{\theta} = \frac{K'}{D(D^2 + 2D + 2)}$$

Sketch the root locus of the system. On the locus branches
locate some intermediate points and find their appropriate
values of open loop sensitivity K'.

Solution: We shall follow the standard procedure to sketch
the root locus for the system.

1) The open loop transfer function is

$$\frac{\theta_{on}}{\theta} = \frac{K'}{D(D+1+j)(D+1-j)}$$

2) The open loop poles are

0, -1-j, -1+j

Hence there are three branch loci, starting from these
points for K' = 0.

3) Each branch locus finishes at ∞ , because there are no
finite open loop zeros. The asymptotes which the branch loci
approach when K' → ∞ are at the angles

$$\alpha = \pm \frac{180°}{3} = \pm 60° \quad \text{and} \quad \alpha = \pm \frac{3 \times 180°}{3} = 180°$$

298

The asymptotes intersect the real axis at

$$\bar{x} = -\frac{2}{3}.$$

4) There are no breakaway or breakin points. The root locus occupies the negative real axis. Hence one branch starts at 0 for K' = 0 and follows the negative real axis to $-\infty$ as K' $\rightarrow$ + ∞.

5) Substituting jb for D in the characteristic equation we we find where the root locus crosses the imaginary axis.

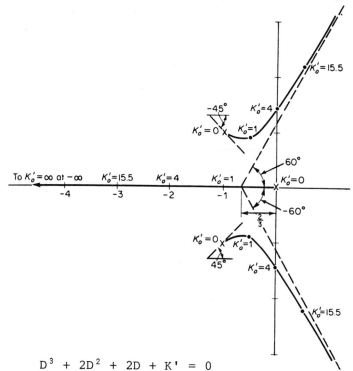

Fig. 1

$$D^3 + 2D^2 + 2D + K' = 0$$

Substituting jb = D

$$[-2b^2 + K'] + j[-b(b^2 - 2)] = 0$$

Solving the equation we find

$$b = \pm 1.414, \qquad K' = 4$$

Loci cross the imaginary axis at

$$\pm 1.414j \qquad \text{for } K' = 4.$$

6) Angles of departure are for pole -1+j

$$\Psi_1 = 0 - 135° - 90° - 180° = -45°$$

for pole -1-j

$$\Psi_2 = 45°.$$

The root locus for the system is shown in Fig. 1.

299

7) Applying the angular criterion to the point in the region $(-\frac{1}{2} + j)$ that seems to lie near the upper root locus branch (see Fig. 2) we get

$$\Sigma \phi_{z0} = 0, \quad \Sigma \phi_{po} = 120° + 75° + 0° = 195°$$

$$\Sigma \phi_{z0} - \Sigma \phi_{po} = -195° \neq 180°$$

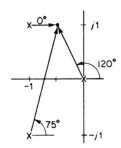

Fig. 2

Thus the point $(-\frac{1}{2} + j)$ does not lie on the root locus. We find that $(-\frac{1}{2} + j \times 0.9)$ lies on the root locus. From symmetry $(-\frac{1}{2} - j \times 0.9)$ lies on the lower branch locus. For points $(-\frac{1}{2} \pm j \times 0.9)$ $K' = 1$.

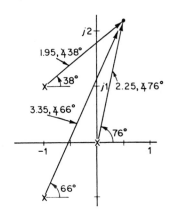

Fig. 3

8)   Points $(\frac{1}{2} \pm j \times 2.2)$ lie on the root locus.   Indeed, applying the angular criterion, we find

$$\Sigma \phi_{z0} - \Sigma \phi_{po} = 0 - 76° - 66° - 38° = -180°$$

and

$$K' = 2.25 \times 3.55 \times 1.95 = 15.5$$

Fig. 3 shows point $(\frac{1}{2} + j \times 2.2)$

300

The open loop transfer function of the system is

$$\frac{\theta_{on}}{\theta} = \frac{K_0'}{D(D+1)(D+8)}$$

Establish the values of K_0' for which the system is stable and sketch the root locus for the system.

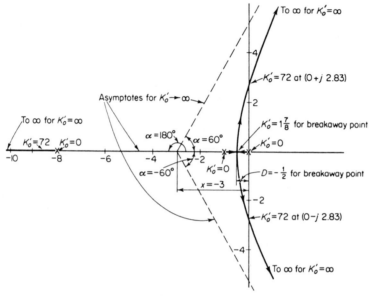

Solution: The characteristic equation of the system has three roots, thus there are three branch loci. Loci start at 0, -1, -8 and end at infinity. The asymptote angles are given by

$$\alpha° = \frac{n \times 180°}{N_p - N_z} = \frac{\pm 180°}{3-0} = \pm 60°$$

and

$$\alpha° = \frac{3 \times (\pm 180°)}{3} = 180°$$

The asymptotes intersect the real axis at

$$x = \frac{\Sigma P_0 - \Sigma Z_0}{N_p - N_z} = \frac{(-1-8) - 0}{3-0} = -3$$

A breakaway point occurs between 0 and -1. Loci exist on the real axis between 0 and -1, and between -8 and $-\infty$. The characteristic equation of the system is

$$D^3 + 9D^2 + 8D + K_0' = 0$$

or

$$K_0' = -D^3 - 9D^2 - 8D$$

To find a breakin point, we differentiate

$$\frac{dK_0'}{dD} = -3D^2 - 18D - 8 = 0$$

Solving the equation

$$3D^2 + 18D + 8 = 0$$

we get

$$D = -0.5$$

and

$$-5.5$$

Thus, $D = -0.5$ is the required breakaway point. For K_0' we have

$$K_0' = -(-0.5)^3 - 9(-0.5)^2 - 8(-0.5) = 1.875$$

at the breakaway point.

Let us substitute $jb = D$ in the characteristic equation

$$j^3b^3 + 9j^2b^2 + 8jb + K_0' = 0$$

$$(K_0' - 9b^2) + jb(8-b^2) = 0$$

We get

$$K_0' - 9b^2 = 0$$

$$jb(8-b^2) = 0$$

Solving we obtain

$$b = \pm\sqrt{8}, \qquad K_0' = 72$$

We see that root loci cross the imaginary axis at ±2.83, at this point $K_0' = 72$. There are no complex open loop poles or zeros. The sum of the roots of the characteristic eq. is -9. For $K_0' = 72$ two of the roots are 2.83 and -2.83, thus the third root must be -9. We see that $K_0' = 72$ is located at -9 on the branch locus from -8 to $-\infty$.

For $K_0' < 72$ the system is stable,

for $K_0' = 72$ the system is critically stable,

for $K_0' > 72$ the system is unstable.

Consider the system

$$G(s)H(s) = \frac{K(1 + Ts)}{s(s+1)(s+2)} \qquad (1)$$

This case represents the effect of derivative control on the root locations of the characteristic equation. Show that the presence of T helps stabilize the system for higher values of K.   Use root locus techniques.

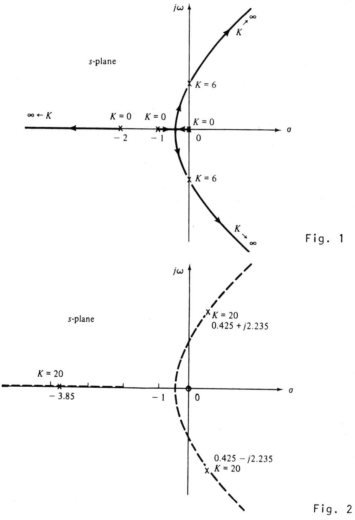

Fig. 1

Fig. 2

Solution: The characteristic equation of the system is

$$s(s+1)(s+2) + K(1+Ts) = 0 \qquad (2)$$

We shall investigate the effect of varying the parameter K.   Let $T = 0$. Eq. (2) yields

$$s(s+1)(s+2) + K = 0 \qquad (3)$$

or

$$1 + \frac{K}{s(s+1)(s+2)} = 0 \qquad (4)$$

Based on the pole-zero configuration of

$$G_1(s)H_1(s) = \frac{K}{s(s+1)(s+2)} \qquad (5)$$

we sketch the root loci of eq (3), as shown in Fig. 1.

We write eq.(2) for T varying between zero and infinity, as

$$1 + G_2(s)H_2(s) = 1 + \frac{TKs}{s(s+1)(s+2)+K} = 0 \qquad (6)$$

Note that the points corresponding to T = 0 on the root contour are at the roots of $s(s+1)(s+2) + K = 0$, see Fig. 1. Fig. 2 shows the pole-zero configuration of $G_2(s)H_2(s)$ for K = 20.

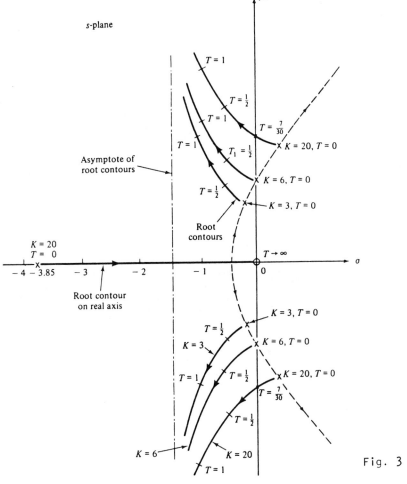

Fig. 3

Fig. 3 shows the root contours of eq.(2) for $0 \leq T < \infty$ for three values of K. The intersection of the asymptotes of the root contours is obtained from

$$\sigma_1 = \frac{\sum\limits_{value} \text{poles} - \sum\limits_{value} \text{zeros}}{n_p - n_z} = \frac{-3.85 + 0.425 + 0.425}{3 - 1}$$

$$= -1.5 \qquad (7)$$

Since the sum of the poles of $G_2(s)H_2(s)$ is always equal to -3 regardless of the value of K and the sum of the zeros of $G_2(s)H_2(s)$ is zero, the intersection of the asymptotes is always at $s = -1.5$.

From Fig. 3 we see that the derivative control generally improves the relative stability of the closed-loop system by moving the roots of the characteristic equation toward the left in the s-plane.

● **PROBLEM** 7-17

Plot the root contours for equation (1)

$$s^3 + K_2 s^2 + K_1 s + K_1 = 0 \qquad (1)$$

where K_1, K_2 are the variable parameters whose values are between 0 and ∞.

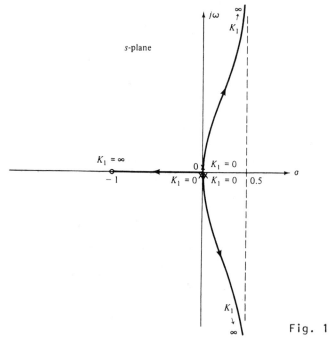

Fig. 1

Solution: Let us take $K_2 = 0$, eq.(1) becomes

$$s^3 + K_1 s + K_1 = 0 \qquad (2)$$

or

$$1 + \frac{K_1(s+1)}{s^3} = 0 \qquad (3)$$

Fig. 1 shows the root loci of eq.(2)drawn from the poles and zeros of

$$G_1(s)H_1(s) = \frac{K_1(s+1)}{s^3} \qquad (4)$$

Let us fix K_1 at a constant nonzero value and vary K_2 between 0 and ∞. We transform eq.(1) to

$$1 + \frac{K_2 s^2}{s^3 + K_1 s + K_1} = 0 \qquad (5)$$

The root contours of eq.(1)when K_2 varies is drawn from the pole-zero configuration of

$$G_2(s)H_2(s) = \frac{K_2 s^2}{s^3 + K_1 s + K_1} \qquad (6)$$

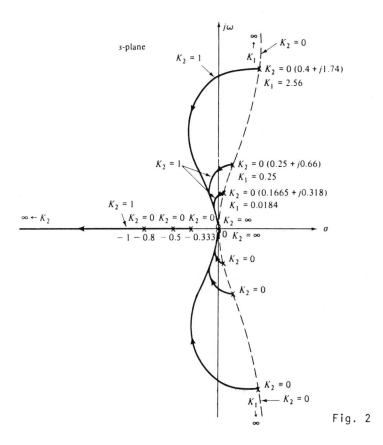

Fig. 2

The zeros of eq.(6) are at $s = 0,0$. The poles are the zeros of $1 + G_1(s)H_1(s)$, which have been found in Fig. 1. Thus for fixed K_1 the root contours when K_2 varies must result from Fig. 1 as shown in Fig. 2.

Sketch the root locus of the system whose open loop transfer function is

$$KG_p(s)H_{eq}(s) = \frac{0.0854K[(s+3.25)^2 + 1.21^2]}{s(s+1)(s+2)}$$

where the gain K = 48.

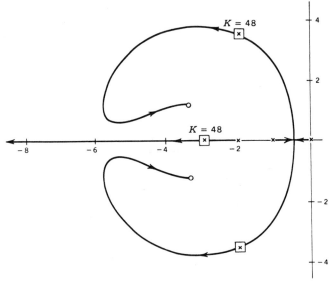

Solution: We find the breakaway point at s $\approx$ - 0.5. The
angle of arrival is 187°. Note that the two branches of
the root locus that break away from the real axis near
s =-0.5 must follow a fairly long path if they are to terminate on the complex zeros at an angle of 187°.

In the construction of the root locus it is necessary to
find points along lines s = j0.5, j1.0, j2.0, j3.0, j4.0
that satisfy the angle criterion. Connecting these points
we obtain the root locus for the system.

Sketch the root-locus for the transfer function

$$KG_p(s) H_{eq}(s) = \frac{K(s+2)}{s(s+1)(s+19)}$$

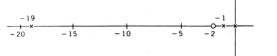

Fig. 1

Solution: The pole-zero plot is shown in Fig. 1. The
pole-zero excess is 2, thus the asymptotes are at ±90°

and the origin of the asymptotes is

$$OA = \frac{(-19-1) - (-2)}{2} = -9$$

Note that the breakaway point is located between the poles at s = -1 and 0. For s = -0.5, -0.6, -0.7 we get the following values for K: 3.08, 3.15 and 2.96. Thus the breakaway point is located approximately at s = -0.6. A branch of the root locus exists between -19 and -2. This system has one point of breakaway and one point of reentry, see Fig. 2.

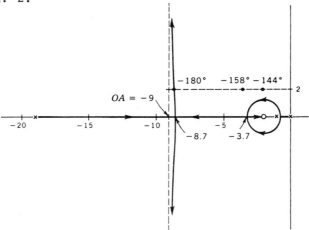

Fig. 2

Our guess is that the two branches seek the asymptotes. Since the asymptotes are far from the breakaway point we should establish another point on the branches of the root locus that are approaching the asymptote. As one possible choice we shall check the phase angle at various points along the line s = j2, looking for a point with a phase angle of 180°. Fig. 2 shows some of the points. Near the breakaway point, the phase angle is far from 180°. The phase angle reaches the desired value close to the asymptote.

We mark gain values along the segment [-19, -2] of the negative real axis, to examine the necessity of breakaway and reentry points. At s = -10 the gain is 101, at s = -6 the gain is 97.5. This rapid decrease in gain could not occur if the branch of the root locus would go directly from the pole at s = -19 to the zero at s = -2. It is easy to find that a relative maximum gain exists at s ≈ -8.7 and a relative minimum gain at s ≈ -3.7. Two branches of the root locus are involved in each case, the breakaway and reentry angles must be ±90°. To complete the root locus we don't have to check any further points.

● **PROBLEM 7-20**

Draw the root locus of the system whose open loop transfer function is

$$KG_p(s)H_{eq}(s) = \frac{K[(s+1.5)^2 + 1]}{s^2(s+0.5)(s+8)(s+9)}$$

Solution: From the transfer function we see that a branch of the locus exists on the real axis between s = -0.5 and s = -8 and for s < -9. The gain reaches the maximum value along the former segment at s = -2.4; this is the point of breakaway. The origin of the asymptotes is at (-17.3 + 3)/3 = -4.83 and the angles of the asymptotes are at ±60° and 180°.

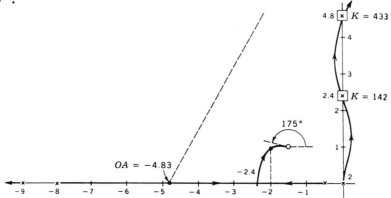

We shall determine the branch of the root locus which starts at the origin and approaches infinity along the asymptotes. Points of 180° phase shift are located on the imaginary axis at s = j2.4 and s = j4.8. The completed root locus is shown below. The system is stable for 142 ≤ K ≤ 433.

● **PROBLEM** 7-21

Show that a part of the root locus of a system with

$$G(s) = \frac{s+3}{s(s+2)} \quad , \quad H(s) = 1 \tag{1}$$

is circular.

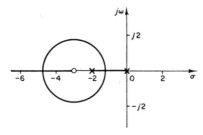

Solution: The angle condition on the root locus is

$$\angle G(s) = \angle s+3 - \angle s - \angle s+2 = 180° \tag{2}$$

Let us substitute in eq.(2)

$$s = \sigma + j\omega$$

we obtain

$$\underline{/\sigma + j\omega + 3} - \underline{/\sigma + j\omega} - \underline{/\sigma + j\omega + 2} = 180° \qquad (3)$$

or

$$\tan^{-1} \frac{\omega}{\sigma+3} - \tan^{-1} \frac{\omega}{\sigma} = 180° + \tan^{-1} \frac{\omega}{\sigma+2} \qquad (4)$$

Using the identity

$$\tan(\alpha \pm \beta) = \frac{\tan\alpha \pm \tan\beta}{1 \mp \tan\alpha\tan\beta} \qquad (5)$$

we can transform eq. (4)

$$\tan\left[\tan^{-1} \frac{\omega}{\sigma+3} - \tan^{-1} \frac{\omega}{\sigma}\right] = \frac{\frac{\omega}{\sigma+3} - \frac{\omega}{\sigma}}{1 + \frac{\omega}{\sigma+3}\frac{\omega}{\sigma}} = \frac{-3\omega}{\sigma(\sigma+3) + \omega^2} \qquad (6)$$

$$\tan(180° + \tan^{-1} \frac{\omega}{\sigma+2}) = \frac{0 + \frac{\omega}{\sigma+2}}{1 - 0 \times \frac{\omega}{\sigma+2}} = \frac{\omega}{\sigma+2} \qquad (7)$$

Thus

$$\frac{-3\omega}{\sigma(\sigma+3) + \omega^2} = \frac{\omega}{\sigma+2} \qquad (8)$$

or

$$(\sigma+3)^2 + \omega^2 = (\sqrt{3})^2 \qquad (9)$$

Equation (9) represents a circle with a center at $\sigma = -3$, $\omega = 0$ and radius $\sqrt{3}$.

The figure shows the root-locus of the system with

$$G(s) = \frac{(s+3)}{s(s+2)}, \qquad H(s) = 1$$

● **PROBLEM** 7-22

The forward transfer function of a unity feedback position control system is

$$G(s) = \frac{C(s)}{E(s)} = \frac{K_1}{s(T_m s+1)} = \frac{N_1}{D_1} \qquad (1)$$

T_m is the motor time constant, which varies over a given range. K_1 is constant. Draw a root locus plot for eq. (1).

Solution: Let $\qquad \sigma = T_m = \sigma_0 + \sigma' \qquad (2)$

where σ_0 is the reference value and σ' represents the change in T_m.

We have

$$D_1(s) = A(s) = s(\sigma s+1) = s[(\sigma_0 s+1) + \sigma' s]$$

$$= [s(\sigma_0 s+1)] + (\sigma' s^2) = U(s) + V(s) \tag{3}$$

The characteristic equation is

$$D_1(s) + K_1 = 0 \tag{4}$$

or

$$s(\sigma_0 s+1) + \sigma' s^2 + K_1 = 0 \tag{5}$$

dividing eq.(5) by $K_1 + s(\sigma_0 s+1)$ gives

$$\frac{\sigma' s^2}{s(\sigma_0 s+1) + K_1} = -1 \tag{6}$$

dividing by σ_0 gives

$$\frac{\left(\sigma'/\sigma_0\right)s^2}{s(s + \frac{1}{\sigma_0}) + \frac{K_1}{\sigma_0}} = -1 \tag{7}$$

Note that the denominator of eq.(7) is equal to the denominator of

$$\frac{C(s)}{R(s)}$$

for $\sigma' = 0$.

We have

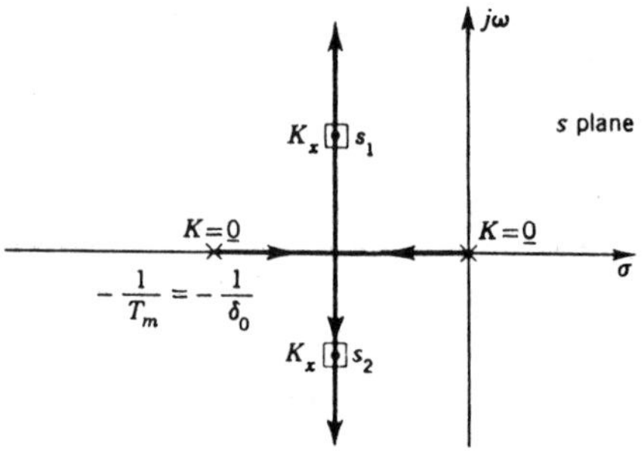

Fig. 1

$$\frac{C(s)}{R(s)} = \frac{G(s)}{1 + G(s)}\bigg|_{\sigma'=0} = \frac{\frac{K_1}{T_m}}{s(s + \frac{1}{T_m}) + \frac{K_1}{T_m}} = \frac{\frac{K_1}{\sigma_0}}{s(s + \frac{1}{\sigma_0}) + \frac{K_1}{\sigma_0}} \tag{8}$$

From the root locus we obtain the poles of

$$\frac{C(s)}{R(s)}$$

for $\sigma' = 0$, using

$$G(s) = \frac{K}{s(s + \frac{1}{T_m})} = \frac{K}{s(s + \frac{1}{\sigma_0})} = -1 \tag{9}$$

311

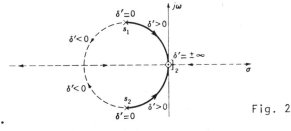

Fig. 2

where

$$K = \frac{K_1}{T_m} \quad .$$

Fig. 1 shows the root locus for the system.

The roots s_1 and s_2 are selected on the basis of performance requirements. The value of K_1 is determined by the roots s_1 and s_2. The values of s_1 and s_2 are the poles of eq.(7). The root locus for eq.(7) can be drawn as a function of σ'. Fig. 2 shows this locus, where solid lines represent the locus for $\sigma'' > 0$ and the dashed lines for $\sigma' < 0$.

● **PROBLEM** 7-23

Inspecting the following transfer functions, find their poles and finite zeros.

1) $D(D^2 + 3D + 2)$

2) $\dfrac{6D(D + 3)}{D^2 + 3D + 2}$

3) $(3D+1)(2D+3D+4)$

4) $\dfrac{10}{(3D+1)(\frac{1}{2}D+1)(D+1)}$

5) $\dfrac{K(\frac{1}{2}D + 1)}{D^2(4D+1)(3D+2D+1)}$

Solution:

1)   zeros at $D = 0$, -1, -2; no poles

2)   zeros at 0, -3; poles at -1, -2

3)   zeros at $D = -\dfrac{1}{3}$, $-\dfrac{4}{5}$

4)   poles at $D = -1$, -2, $-\dfrac{1}{3}$; no zeros

5)   zeros at $D = -2$; poles at 0, 0, $-\dfrac{1}{4}$, $-\dfrac{1}{5}$

The system with transportation lag is shown in Fig. 1

Sketch the root-locus plot for the system and find the two pairs of closed-loop poles nearest the jω axis. Obtain the unit-step response and sketch the response curve using the dominant closed-loop poles.

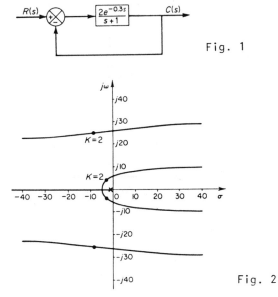

Fig. 1

Fig. 2

Solution: The characteristic equation of the system is

$$1 + \frac{2e^{-0.3s}}{s + 1} = 0$$

or in the magnitude and angle conditions

$$\left| \frac{2e^{-0.3s}}{s + 1} \right| = 1$$

$$\left/ \frac{2e^{-0.3s}}{s + 1} \right. = \pm 180° (2n+1)$$

The angle condition can be reduced to $\underline{/s+1} = \pm\pi (2n+1) - 0.3\omega$

for n=0 $\underline{/s+1} = \pm\pi - 0.3\omega = \pm 180° - 17.2°\omega$

for n=1 $\underline{/s+1} = \pm 3\pi - 0.3\omega = \pm 540° - 17.2°\omega$

Fig. 2 shows the root-locus plot for the system.

To find the closed-loop poles, let us set $s = \sigma + j\omega$ in the magnitude condition and replace 2 by K. We have

313

$$K = \frac{\sqrt{(1+\sigma)^2 + \omega^2}}{e^{-0.3\sigma}}$$

Evaluating K at different points on the root loci we find
points for which K = 2. These are closed-loop points.
The dominant pair is

$$s = -2.5 \pm j3.9$$

and the next pair is

$$s = -8.6 \pm j25.1$$

The closed-loop transfer function may be approximated using
the pair of dominant closed-loop poles.

We have

$$\frac{C(s)}{R(s)} = \frac{2e^{-0.3s}}{1 + s + 2e^{-0.3s}} = \frac{2e^{-0.3s}}{1 + s + 2(1 - 0.3s + \frac{0.09s^2}{2} + \dots)}$$

$$= \frac{2e^{-0.3s}}{3 + 0.4s + 0.09s^2 + \dots}$$

and

$$(s + 2.5 + j3.9)(s + 2.5 - j3.9) = s^2 + 5s + 21.46$$

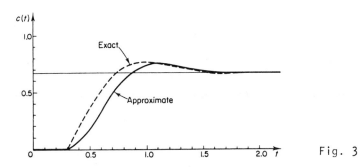

Fig. 3

We can approximate $\frac{C(s)}{R(s)}$ by

$$\frac{C(s)}{R(s)} = \frac{\frac{2}{3} \times 21.46e^{-0.3s}}{s^2 + 5s + 21.46} = \frac{14.31e^{-0.3s}}{(s+2.5)^2 + 3.9^2}$$

We obtain for a unit-step input

$$C(s) = \frac{14.31e^{-0.3s}}{s[(s+2.5)^2 + 3.9^2]}$$

After simple transformations we can write

314

$$C(s) = \frac{\frac{2}{3}}{s} e^{-0.3s} + \left[\frac{-\frac{2}{3}s - \frac{10}{3}}{(s+2.5)^2 + 3.9^2} \right] e^{-0.3s}$$

Taking the inverse Laplace transform of C(s) we obtain

$$c(t) = \frac{2}{3} [1 - e^{-2.5(t-0.3)} \cos 3.9(t-0.3)$$

$$- 0.64e^{-2.5(t-0.3)} \sin 3.9(t-0.3)] 1(t-0.3)$$

where 1(t-0.3) is the unit-step function occurring at
t = 0.3

Fig. 3 shows the approximate response curve with the
exact unit-step response curve.

● **PROBLEM** 7-25

An airplane with an autopilot in the longitudinal mode
has a simplified open-loop transfer function

$$G(s)H(s) = \frac{K(s+a)}{s(s-b)(s^2 + 2\zeta\omega_n s + \omega_n^2)}$$

where

$$a > 0, \qquad b > 0 .$$

This system, which involves an open-loop pole in the right-
half s plane may be conditionally stable.   For a=b=1,
$\zeta=0.5$, $\omega_n=4$ sketch the root-locus plot and determine the
range of gain K for stability.

Solution: For a=b=1, $\zeta=0.5$, $\omega_n=4$ the open-loop transfer
function for the system is

$$G(s)H(s) = \frac{K(s+1)}{s(s-1)(s^2+4s+16)}$$

We shall follow this procedure to sketch the root-locus:

1)   In the complex plane plot the open-loop poles and zero.
Root loci exist on the real axis between 1 and 0 and
between -1 and -∞.

2)   Find the asymptotes of the root loci.

In our case there are three asymptotes whose angles are

$$\phi = \frac{180°(2k+1)}{4-1} = 60°, -60°, 180°$$

The abscissa of the intersection of the asymptotes and the real axis is

$$-\sigma_a = -\frac{(0-1+2+j2\sqrt{3}+2-j2\sqrt{3}) - 1}{4-1} = -\frac{2}{3}$$

3) The characteristic equation is

$$\frac{K(s+1)}{s(s-1)(s^2+4s+16)} + 1 = 0$$

computing K we get

$$K = -\frac{s(s-1)(s^2+4s+16)}{s+1}$$

differentiating K with respect to s we obtain

$$\frac{dK}{ds} = \frac{3s^4 + 10s^3 + 21s^2 + 24s - 16}{(s+1)^2}$$

Factoring we have

$$3s^4 + 10s^3 + 21s^2 + 24s - 16 = 3(s + 0.79 + j2.16)$$

$$\times (s + 0.79 - j2.16)(s + 2.22)(s - 0.46).$$

We get the following breakaway and break-in points:

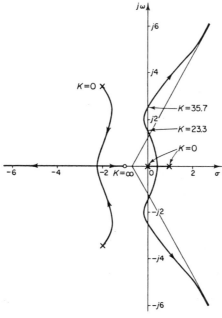

$$s = 0.46 , \qquad s = -2.22$$

4) From the Routh stability criterion we determine the value of K at which the root loci cross the imaginary axis.

The characteristic equation is

$$s^4 + 3s^3 + 12s^2 + (K-16)s + K = 0$$

and the Routh array

s^4	1	12	K
s^3	3	K-16	0
s^2	$\dfrac{52-K}{3}$	K	0
s^1	$\dfrac{-K^2 + 59K - 832}{3(52-K)}$	0	
s^0	K		

$$\frac{-K^2 + 59K - 832}{3(52-K)} = 0$$

for   K = 35.7   and   K = 23.3

The auxiliary equation obtained from the s^2 row is

$$\frac{52-K}{3} s^2 + K = 0$$

its solution is

$$s_1 = \pm j2.56 \qquad \text{for} \qquad K = 35.7$$

$$s_2 = \pm j1.56 \qquad \text{for} \qquad K = 23.3$$

s_1 and s_2 are the crossing points on the imaginary axis.

5)   For the open-loop pole at s = -2 + j2√3 the angle of departure θ is 106° - 120° - 130.5° - 90° - θ = ± 180°(2k+1)

or
$$\theta = -54.5°$$

For the open-loop pole at s = -2 - j2√3

the angle of departure is

$$\theta = 54.5°$$

6)   We apply the angle condition to the test points in the neighborhood of the jω axis and the origin, until we find one that satisfies this condition. The neighborhood we choose has to be sufficiently big to locate a number of points which satisfy the angle condition.

The root-locus plot for the system is shown in the figure.

The system is stable for 23.3 < K < 35.7. Note that the zero tends to pull the root locus toward itself while a pole tends to push it away.

Fig. 1 shows the system with an unstable feedforward transfer function.

Sketch the root-locus plot and locate the closed-loop poles.
Show that the unit-step response curve will exhibit overshoot.

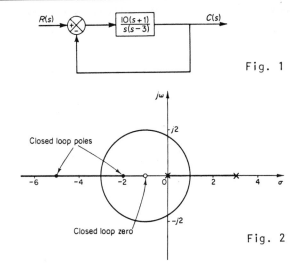

Fig. 1

Fig. 2

Solution: The closed-loop poles are located at s = -2
and s = -5. The root-locus of this system is shown in
Fig. 2.

The closed-loop transfer function of the system is

$$\frac{C(s)}{R(s)} = \frac{10(s + 1)}{s^2 + 7s + 10}$$

and the unit-step response is

$$C(s) = \frac{10(s + 1)}{s(s + 2)(s + 5)}$$

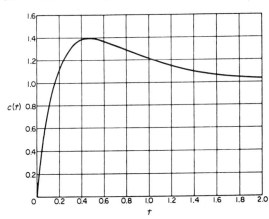

Fig. 3

Taking the inverse Laplace transform of C(s), we obtain

$$c(t) = 1 + 1.66e^{-2t} - 2.66e^{-5t} \qquad t \geq 0$$

318

Fig. 3 shows the unit-step response curve. The system has
a zero at s = -1, therefore the unit-step response curve
exhibits overshoot.

Sketch the root locus for the system shown in Fig. 1.

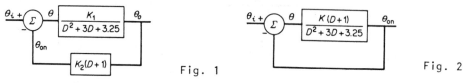

Fig. 1   Fig. 2

Solution: We shall rearrange the system block diagram to
the nondimensional form shown in Fig. 2.

The open-loop transfer is simply $G(s)H(s)$ obtained from
Fig. 1 or it can be obtained from Fig. 2 in the form

$$\frac{\theta_{on}}{\theta} = \frac{K(D + 1)}{D^2 + 3D + 3.25}$$

where $K = K_1 K_2$.

The system has a second-order characteristic equation with
two roots, there are thus two root locus branches. We find
the open loop poles from

$$D^2 + 3D + 3.25 = [D + (1.5 + j)][D + (1.5 - j)]$$

The branch loci start at

$$D = -1.5 \pm j, \text{ for which } K = 0.$$

Fig. 3 shows the root locus for the system.

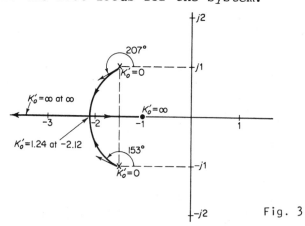

Fig. 3

The branch loci end at open loop zeros $D = -1$, and $D = \infty$
for $K = \infty$. The $D = \infty$ zero lies along the 180° real axis,

$$\alpha = \frac{\pm 180°}{N_p - N_z} = \frac{\pm 180°}{2 - 1} = \pm 180°$$

The root locus occupies the real axis between $-\infty$ and -1. The two zeros lie on the real axis and the two poles are complex. Hence there must be a breakin point on the root locus. It will lie between $-\infty$ and -1. The characteristic equation of the system

$$D^3 + (3 + K)D + (3.25 + K) = 0$$

can be written in the form

$$K = \frac{-D^2 - 3D - 3.25}{D + 1}$$

Let us differentiate K with respect to D

$$\frac{dK}{dD} = \frac{(D + 1)(-2D - 3) + D^2 + 3D + 3.25}{(D + 1)^2} = \frac{-D^2 - 2D + 0.25}{(D + 1)^2}$$

A breakin point is determined from the equation

$$\frac{dK}{dD} = 0 \qquad \text{or} \qquad D^2 + 2D - 0.25 = 0$$

Thus $D = -2.12$ is the breakin point. For $D = -2.12$, $K = 1.24$. The angles of departure of the branch loci are

$$\psi = \Sigma\phi_z - \Sigma\phi_p - n \times 180°$$

n is an odd number. We have for the pole $-1.5 + j$, $\psi = 117° - 90° - 180° = -153° = 207°$. For the pole $-1.5 - j$, $\psi = 153°$.

● PROBLEM 7-28

The block diagram of the system is shown in Fig. 1

where $K > 0$.

Sketch the root-locus plot for the system. Note that for large and small values of K the system is overdamped, and for medium values of K it is underdamped.

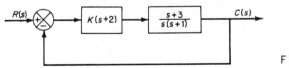

Fig. 1

Solution: Sketching the root-locus we shall follow the procedure:

1) Plot on the complex plane the open-loop poles and zeros. Root loci exist on the negative real axis between -3 and -2 and between -1 and 0.

2) There are no asymptotes in the complex region of the s plane since the number of open-loop poles and finite zeros are the same.

320

3)   From the characteristic equation of the system

$$1 + \frac{K(s + 2)(s + 3)}{s(s + 1)} = 0$$

or

$$K = - \frac{s(s + 1)}{(s + 2)(s + 3)}$$

we determine the breakaway and break-in points. Differentiating with respect to s we obtain

$$\frac{dK}{ds} = - \frac{(2s + 1)(s + 2)(s + 3) - s(s + 1)(2s + 5)}{(s + 2)^2(s + 3)^2}$$

$$= - \frac{4(s + 0.634)(s + 2.366)}{(s + 2)^2(s + 3)^2} = 0$$

Solving the equation we have

$$s = -0.634, \quad s = -2.366$$

For $s = -0.634$, the value of K is

$$K = - \frac{(-0.634)(0.366)}{1.366 \times 2.366} = 0.0718$$

and for $s = -2.366$

$$K = 14$$

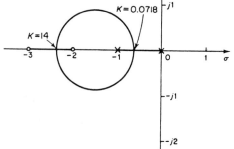

Fig. 2

The values of K in both cases are positive, these points are actual breakaway and break-in points. Point $s = -2.366$ lies between two zeros, it is therefore a break-in point, point $s = -0.634$ is a breakaway point.

4)   Fig. 2 shows the root-locus plot of the system. We found a sufficient number of points that satisfy the angle condition.

5)   We can calibrate the root loci in terms of K using the magnitude condition. For a given value of K, the closed-loop poles, which satisfy both the angle and magnitude conditions, can be found from the root-locus plot.

The system is stable for any positive value of K.

The system is overdamped for $0 < K < 0.0718$ and for $14 < K$, and is underdamped for $0.0718 < K < 14$.

321

Consider the system shown in Fig. 1.

If a = 2 and $k_2 = \frac{17}{60}$ then

$$\frac{y(s)}{r(s)} = \frac{120}{[(s + 2)^2 + 2^2](s + 15)}$$

and the root locus plot is shown in Fig. 2.

Determine the changes in the root locus plot as a and k_2 are varied.

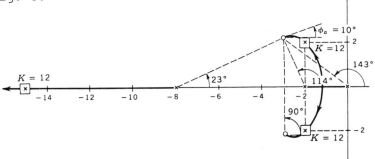

Fig. 1

Solution: Let us determine the position of the pole. We shall find the characteristic equation as a function of the parameter a and then transform it to

$$\alpha W(s) = -1 = 1 \underline{/180°} \tag{1}$$

where α is any system parameter and W(s) is any transfer function that is independent of α. To achieve it we reduce the block diagram shown in Fig. 1 to the form shown in Fig. 3.

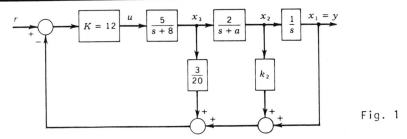

Fig. 2

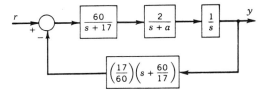

Fig. 3

The characteristic equation is now

$$1 + \frac{120}{s(s + a)(s + 17)} \left[\frac{17}{60}(s + \frac{60}{17}) \right] = 0$$

or

$$\frac{34(s + 3.53)}{s(s + a)(s + 17)} = -1 \qquad\qquad (2)$$

We transform eq. 2 into

$$-a(s^2 + 17s) = s^3 + 17s^2 + 34s + 120$$

or

$$a\frac{s^2 + 17s}{s^3 + 17s^2 + 34s + 120} = -1 \qquad\qquad (3)$$

Equation 3 is the desired form of eq.1. Before we plot the root locus versus a, the denominator has to be factored. It can be done in the following way:

From equation 3 we obtain the auxiliary equation

$$\frac{120}{s(s^2 + 17s + 34)} = -1$$

One of the methods of finding the roots of the polynomial is to draw the root locus of the equation. We get $s^3 + 17s^2 + 34s + 120 \approx (s + 15.3)[(s + 0.85)^2 + 2.7^2]$. The root locus to be plotted versus a is based upon the equation

$$a\frac{s(s + 17)}{(s + 15.3)[(s + 0.85)^2 + 2.7^2]} = -1 \qquad\qquad (4)$$

Fig. 4 shows the root locus for the eq. 4.

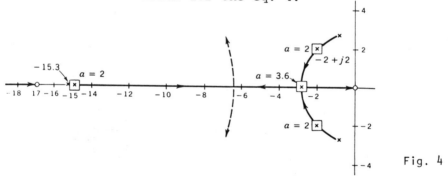

Fig. 4

For a = +2 the closed-loop poles are at s = -2 ± j2 and s = -15. Note that the dotted parts of the locus are of little consequence. In that region the values of the parameter a are vastly different from the designed value.

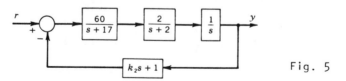

Fig. 5

Regardless of the value of a the system is always stable. To examine the behavior of the system when k_2 changes we redraw the block diagram of Fig. 1 as shown in Fig. 5.

The parameter k_2 should appear as few times as possible in the characteristic equation. From Fig. 5 we get

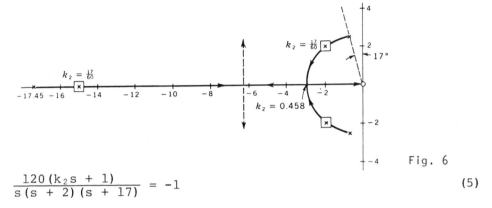

Fig. 6

$$\frac{120(k_2 s + 1)}{s(s + 2)(s + 17)} = -1 \tag{5}$$

Putting eq. 5 into the form of eq. 1 we obtain

$$k_2 \frac{120s}{s^3 + 19s^2 + 34s + 120} = -1$$

To obtain the final form of the equation we have to factor the denominator, thus

$$k_2 \frac{120s}{(s + 17.45)[(s + 0.775)^2 + 2.54^2]} = -1 \tag{6}$$

The root locus is sketched in Fig. 6.

Note that the system is stable for $k_2 = 0$.

● **PROBLEM** 7-30

Sketch the root locus plots for the systems whose poles are located as shown in Fig. 1, Fig. 2 and Fig. 3.

Note that the pole placements of the three systems are similar.

Solution: s = -4 is the origin of the asymptotes in each case, the angles of the asymptotes are

$$\frac{180°}{4}, \quad \frac{180° - 360°}{4}, \text{ etc.}$$

The root locus for Fig. 1 is shown in Fig. 4. The two breakaway points exist on the real axis at s ≈ -0.8 and s ≈ -7.2. A reentry point is located at s = -4, so that there are three places on the real axis at which branches of the root locus are separated from each other by 90°.

Fig. 5 shows the root locus for the pole-zero plot shown in Fig. 2. The breakaway points and the point of reentry are the same. The branches of the root locus follow directly the asymptotes.

Fig. 6 shows the root locus for the case shown in Fig. 3.

In this case, we have two branches of the root locus that
meet in the lower and upper half planes.

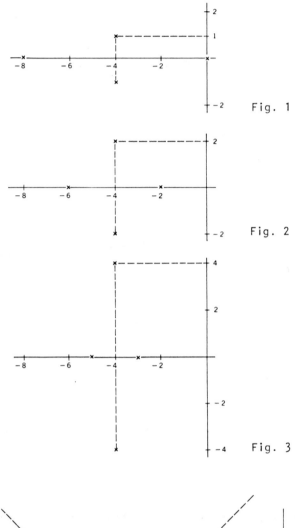

Fig. 1

Fig. 2

Fig. 3

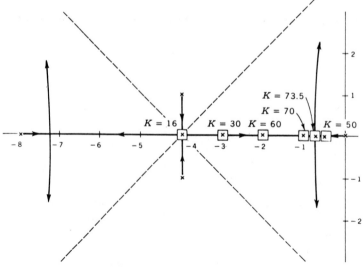

Fig. 4

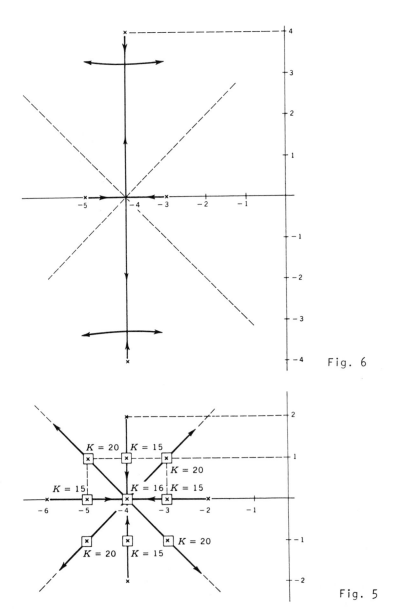

Fig. 6

Fig. 5

BODE DIAGRAM

For the transfer function

$$G(j\dot{\omega}) = \frac{1}{(1 + \frac{j\omega}{20})^2}$$

draw the frequency-response curves.

<u>Solution:</u> We shall start with plotting the magnitude

326

and phase angle of $\dfrac{1}{1 + \dfrac{j\omega}{20}}$ and then graphically doubling

the curves. The plot of magnitude versus frequency for
$\dfrac{1}{1 + \dfrac{j\omega}{20}}$ and $\dfrac{1}{(1 + \dfrac{j\omega}{20})^2}$ is shown in Fig. 1.

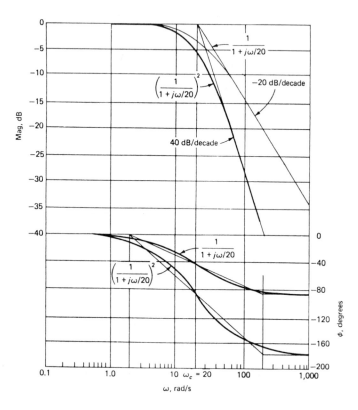

Fig.1

Note that the $\dfrac{1}{(1 + \dfrac{j\omega}{20})^2}$ curve is obtained by doubling the

decibel values at various frequencies.

At the higher frequencies the $\dfrac{1}{(1 + \dfrac{j\omega}{20})^2}$ curve follows the

-40dB per decade asymptote and not the -20dB per decade
slope. This is due to the fact that the logarithm of a
squared term is multiplied by two.

For the third power of the term in the denominator the
asymptote would be -60dB per decade, for the fourth
-80dB per decade. Fig. 1 shows the phase angle versus fre-
quency for $\dfrac{1}{(1 + \dfrac{j\omega}{20})^2}$, the actual curve and its straight-line

approximation. In drawing this plot we used the same method
as for the magnitude, i.e., we draw the response for
$\dfrac{1}{1 + \dfrac{j\omega}{20}}$ and then double the angle values.

The block diagram of the system is shown in Fig. 1.

Obtain the phase and gain margins of the system when K = 10 and K = 100.

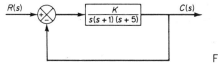

Fig. 1

Solution: We shall obtain the phase and gain margins from the logarithmic plot. Fig. 2 shows a logarithmic plot for K = 10.

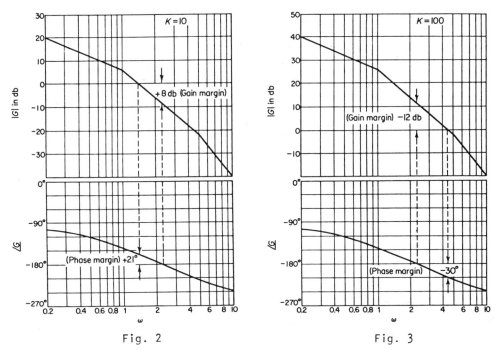

Fig. 2   Fig. 3

For K = 10 the phase and gain margins are

     Phase margin = 21°

     Gain margin = 8db

Thus, the gain of the system can be increased by 8db before instability occurs.

Fig. 3 shows the increase of the gain from K = 10 to K = 100. It shifts the 0-db axis down by 20db. We have

     Phase margin = -30°

     Gain margin = -12db

The system is unstable for K = 100, but stable for K = 10.

Obtain the Bode diagram asymptotic approximation to the magnitude ratio-frequency relationship for the system whose transfer function is

$$\frac{\theta_0}{\theta_i} = \frac{3}{1 + 2D} \; ; \; \theta_i = 2 \sin \omega t$$

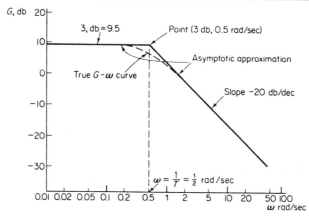

Solution: We shall follow the outlined procedure:

1) Draw the horizontal line 3

$$db = 20 \log 3 = 20 \times 0.477 = 9.54$$

2) The first and only corner frequency is $\omega_1 = \frac{1}{2} \frac{rad}{sec}$ associated with the denominator factor $(1 + 2D)$.

Locate the point $\omega = \frac{1}{T} = \frac{1}{2} \frac{rad}{sec}$ on the horizontal line.

3) Draw a line from this point at a slope of $-20\frac{db}{dec}$. The figure shows the asymptotic approximation plot.

The transfer function of the system is

$$G(j\omega) = \frac{e^{-j\omega L}}{1 + j\omega T}$$

Draw the Bode diagram of the system.

Solution: We first compute the log magnitude

$$20 \log|G(j\omega)| = 20 \log|e^{-j\omega L}| + 20 \log\left|\frac{1}{1 + j\omega T}\right|$$

$$= 0 + 20 \log\left|\frac{1}{1 + j\omega T}\right|$$

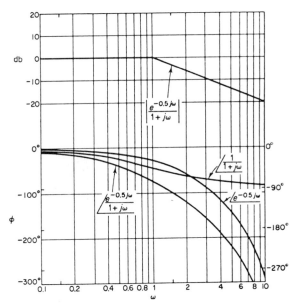

The phase angle of $G(j\omega)$ is

$$\underline{/G(j\omega)} = \underline{/e^{-j\omega L}} + \underline{/\dfrac{1}{1 + j\omega T}} = -\omega L \ -\tan^{-1}\omega T.$$

The figure shows the log-magnitude and phase-angle curves for the transfer function

$$G(j\omega) = \dfrac{e^{-0.5j\omega}}{1 + j\omega}$$

• **PROBLEM** 7-35

Consider a positional servomechanism whose open-loop transfer function is

$$G(j\omega) = \dfrac{1}{j\omega(1 + j\omega)(1 + j0.2\omega)} \qquad (1)$$

$$H(j\omega) = 1 \qquad (2)$$

Find the gain margin and phase margin and the appropriate closed-loop frequency response.

Solution: Fig. 1 shows the Bode diagram for the open-loop frequency response. At the gain-crossover frequency the phase angle is 135° and the phase margin 180° - 135° = 45°. The gain margin is 16dB, since the magnitude of GH(jω)at the phase-crossover frequency is -16dB.

Since the $(1 + j0.2\omega)^{-1}$ factor does not affect significantly the response in the vicinity of the gain-crossover frequency, a second-order approximation of the transfer function is possible. Fig. 2 shows the damping ratio vs. phase margin plot.

From the plot the damping ratio corresponding to a 45° phase margin is 0.42. From Fig. 3 we have the peak magnitude 1.3 or 2.4dB. M_m occurs at approximately the gain-crossover frequency. Fig. 1 shows the closed-loop magnitude response $\frac{C}{R}$.

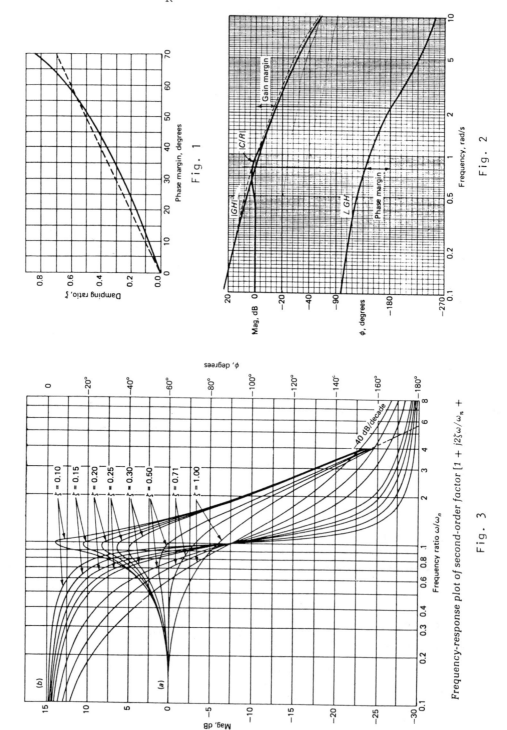

Fig. 1

Fig. 2

Fig. 3

Frequency-response plot of second-order factor $[1 + j2\zeta\omega/\omega_n +$

331

For the following transfer function

$$G(j\omega) = \frac{10(j\omega + 3)}{(j\omega)(j\omega + 2)[(j\omega)^2 + j\omega + 2]}$$

draw the Bode diagram. Make necessary corrections so that the log-magnitude curve is accurate.

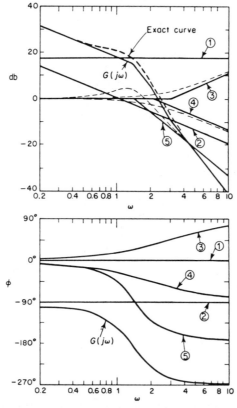

Fig. 1

Solution: We shall put G(jω) in the normalized form to avoid possible mistakes in drawing the log-magnitude curve. In this form the low-frequency asymptotes for the first-order factors and the second-order factor are the 0-db line.

$$G(j\omega) = \frac{7.5(\frac{j\omega}{3} + 1)}{(j\omega)(\frac{j\omega}{2} + 1)[\frac{(j\omega)^2}{2} + \frac{j\omega}{2} + 1]}$$

This function consists of the following factors:

$$7.5, \quad (j\omega)^{-1}, \quad 1 + \frac{j\omega}{3}, \quad (1 + \frac{j\omega}{2})^{-1}, \quad [1 + \frac{j\omega}{2} + \frac{(j\omega)^2}{2}]^{-1}$$

The corner frequencies of the third, fourth and fifth term are ω = 3, ω = 2, ω = √2.

Fig. 1 shows the separate asymptotic curves for each of the factors. The composite curve is obtained by adding algebraically the individual curves.

From the Fig. 1 we see that when the individual asymptotic

332

curves are added at each frequency, the slope of the composite curve is cumulative. The plot has the slope of $-20\frac{db}{decade}$ below $\omega = \sqrt{2}$.

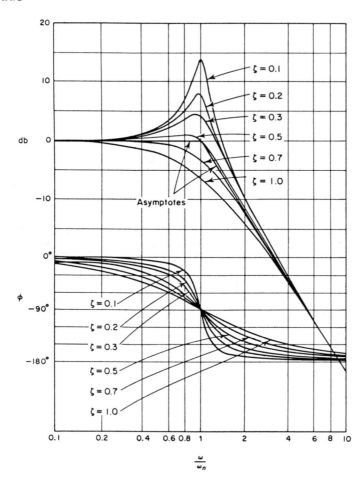

Fig. 2

At the first corner frequency $\omega = \sqrt{2}$ the slope changes to $-60\frac{db}{decade}$, at the next corner frequency $\omega = 2$ the slope changes to $-80\frac{db}{decade}$, at the last corner frequency $\omega = 2$ the slope changes to $-60\frac{db}{decade}$. From the approximate log-magnitude curve we obtain the exact curve by adding corrections at each corner frequency and at frequencies one octave below and above the corner frequencies. For first-order factors $1 + j\omega T$ and $\frac{1}{1 + j\omega T}$, the corrections are ±3db at the corner frequency and ±1db at the frequencies one octave below and above the corner frequency. From Fig. 2 we obtain the necessary corrections for the quadratic factor.

The dotted line in Fig. 1 shows the exact log-magnitude curve for $G(j\omega)$. Note that we first sketch the phase-angle curves for all factors and then take their algebraic sum to obtain the complete phase-angle curve.

Make a Bode and phase plot of

$$G(s) = \frac{10^6 s(s + 10^3)}{(s + 50)(s^2 + 12.2 \times 10^3 s + 64 \times 10^6)} \qquad (1)$$

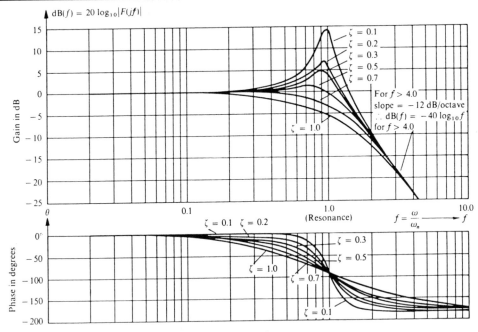

Fig. 1

<u>Solution:</u> From the expression $s^2 + 2\zeta\omega_n s + \omega_n^2$ we have

$$2\zeta\omega_n = 12.2 \times 10^3$$
$$\omega_n^2 = 64 \times 10^6 \qquad (2)$$

Solving the system of equations (2) we have

$$\omega_n = 8 \times 10^3$$
$$\zeta = 0.76 \qquad (3)$$

From Fig. 1 we estimate the curve $\zeta = 0.76$.

Writing $s^2 + 12.2 \times 10^3 s + 64 \times 10^6$ in the form of

$$G(j\omega) = \frac{1}{\omega_n^2 \left[\left(\dfrac{j\omega}{\omega_n} \right)^2 + j2\zeta\dfrac{\omega}{\omega_n} + 1 \right]}$$

we get

$$G(s) = \frac{10^6 s(s + 10^3)}{64 \times 10^6 (s + 50)\left[\left(\dfrac{s}{8 \times 10^3} \right)^2 + \dfrac{12.2 \times 10^3}{64 \times 10^6} s + 1 \right]} \qquad (4)$$

$$= \frac{s(s + 10^3)}{64(s + 50)\left[\left(\dfrac{s}{8 \times 10^3} \right)^2 + \dfrac{s}{5.25 \times 10^3} + 1 \right]}$$

Thus

$$G(j\omega) = \frac{j\omega(j\omega + 10^3)}{64(j\omega + 50)\left[\left(\dfrac{j\omega}{8 \times 10^3} \right)^2 + j\dfrac{\omega}{5.25 \times 10^3} + 1 \right]}$$

$$= \frac{j\omega(j\omega + 10^3)}{64(j\omega + 50)} F(jf) \tag{5}$$

where

$$F(jf) = \frac{1}{j2\zeta f + (1 - f^2)}$$

and

$$f = \frac{\omega}{\omega_n} = \frac{\omega}{8 \times 10^3}$$

The corner frequencies are 50 and 10^3, and the pseudocorner frequencies are $\omega = 0.1\omega_n = 800$ and $\omega = 4\omega_n = 32 \times 10^3$.

ω-ranges are

$$\omega < 50, \ 50 < \omega < 800, \ 800 < \omega < 10^3, \ 10^3 < \omega < 32 \times 10^3$$

and $32 \times 10^3 < \omega$.

For the ω-ranges we have

$\omega < 50$:

$$G(j\omega) \cong \frac{j\omega(10^3)}{64(50)} = \frac{\omega}{3.2} \underline{/90°} \text{ since } F(jf) \cong 1\underline{/0°}$$

$$dB(\omega) = 20 \log_{10}\omega - 20 \log_{10}3.2$$

$$= 20 \log_{10}\omega - 10dB \text{ at } \underline{/90°}. \tag{6}$$

$50 < \omega < 800$:

$$G(j\omega) \cong \frac{j\omega(10^3)}{64(j\omega)}1\underline{/0°} = 15.6\underline{/0°}$$

$$dB(\omega) = 20 \log_{10}15.6 = 24dB \text{ at } \underline{/0°} \tag{7}$$

$800 < \omega < 10^3$:

$$G(j\omega) \cong \frac{j\omega(10^3)}{64(j\omega)}F(jf) = 15.6|F(jf)| \text{ at } \underline{/\theta(f)}$$

$$dB(\omega) = 24 + dB(f)dB \text{ at } \underline{/\theta(f)} \tag{8}$$

$10^3 < \omega < 32 \times 10^3$:

$$G(j\omega) \cong \frac{j\omega(j\omega)}{64(j\omega)}F(jf) = \frac{\omega}{64}|F(jf)| \underline{/90° + \theta(f)} \tag{9}$$

$$dB(\omega) = 20 \log_{10}\omega - 36.4 + dB(f) \text{ at } \underline{/90° + \theta(f)}$$

$32 \times 10^3 < \omega$:

$$G(j\omega) \cong \frac{j\omega(j\omega)}{64(j\omega)}(\frac{\omega_n}{\omega})^2 \underline{/-180°} = \frac{\omega_n^2}{64\omega}\underline{/90° - 180°}$$

$$= \frac{64 \times 10^6}{64\omega}\underline{/-90°} = \frac{10^6}{\omega}\underline{/-90°}$$

$$dB(\omega) = -20 \log_{10}\omega + 120dB \text{ at } \underline{/-90°} \tag{10}$$

We shall first plot all the asymptotes for gain and phase, that is eqs. (6),(7), 24dB of eq. (8), 20 $\log_{10}\omega$ - 36.4 of eq. (9) and eq. (10).

As a next step we shall put corner frequency corrections. From Fig. 1 we get the values for $\zeta = 0.76$ for eq. (9). The expression 20 $\log_{10}\omega$ - 36.4 can be corrected by adding dB(f) to it. Fig. 2 shows the completed graph.

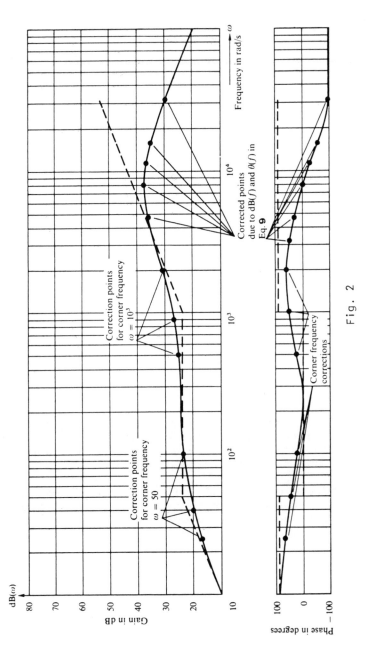

Fig. 2

336

Consider a system that measures the signals of all frequencies. Assume that the set of unwanted frequencies lies between $20\frac{rad}{s}$ and $30\frac{rad}{s}$. Find the transfer function and performance equation of a filter which attenuates the signals in the given band. The signals in the range between 24 and $26\frac{rad}{s}$ should be down -3dB.

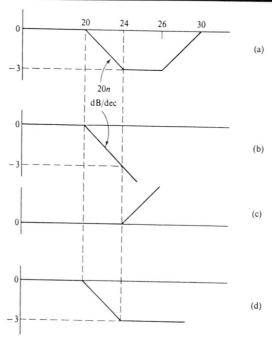

Fig. 1

<u>Solution</u>: Fig. 1a shows a straight-line approximation to the sinusoidal gain characteristics. The systems shown in Fig. 1b (breakpoint $\tau_1 = 0.05s$) and Fig. 1c (breakpoint $\tau_2 = 0.0417s$) build up the initial graph.

Fig. 1d is the sum of (b) and (c). The two slopes are selected to cancel and generate a straight-line response. To find (n) it is necessary to compute the slope. The first negative slope is equal to

$$\frac{-3}{\log 24 - \log 20} \qquad \text{(from (a) and (d))}$$

so

$$\frac{-3}{\log \frac{24}{20}} = -20n = -38$$

thus n ≈ 2. We use a slope of $40\frac{dB}{dec}$ and complete the curve by adding a positive $40\frac{dB}{dec}$ slope at the breakpoint $26\frac{rad}{s}$, and

337

a negative slope of $40\frac{dB}{dec}$ at the breakpoint $30\frac{rad}{s}$. $40\frac{dB}{dec}$ slopes we obtain by superposing two first-order systems. Thus the transfer function is given by

$$G(s) = \frac{(0.0417s + 1)^2 (0.0387s + 1)^2}{(0.050s + 1)^2 (0.0333s + 1)^2}$$

Inverting we get the performance equation

$$(0.0500s + 1)^2 (0.0333s + 1)^2 Z = K(0.0417s + 1)^2 (0.0387s + 1)^2 Y$$

● **PROBLEM** 7-39

The block diagram of a feedback control system with unity feedback is shown in Fig. 1.

Draw the magnitude and phase plots of the open-loop gain.

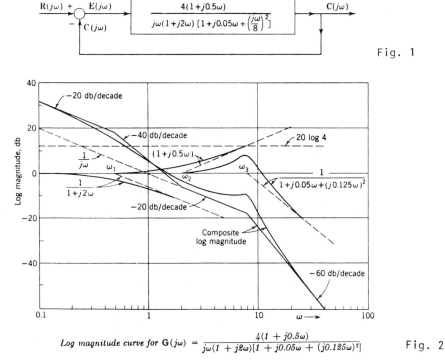

Fig. 1

Fig. 2

Log magnitude curve for $G(j\omega) = \dfrac{4(1 + j0.5\omega)}{j\omega(1 + j2\omega)[1 + j0.05\omega + (j0.125\omega)^2]}$

Solution: The important characteristics for each factor are listed in the table.

Fig. 2 shows the log magnitude asymptotes for each factor.

Fig. 3 shows the angle curves for each factor. In both cases the curves are added algebraically to obtain the composite curve.

1) For frequencies smaller than ω_1, the first corner frequency, only the factors Lm 4 and Lm $(j\omega)^{-1}$ are effective, the other factors have zero value.

338

At ω_1 the composite curve has the value of 18dB, since
Lm 4 = 12db and Lm$(j\omega_1)^{-1}$ = 6db. Below ω_1 the composite
curve has a slope of -20 db/decade.

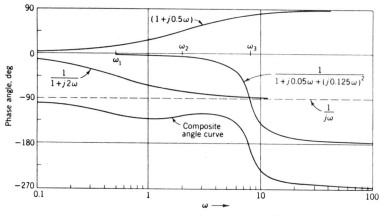

Phase angle curve for $G(j\omega) = \dfrac{4(1 + j0.5\omega)}{j\omega(1 + j2\omega)[1 + j0.05\omega + (j0.125\omega)^2]}$

Fig. 3

Characteristics of log magnitude and angle diagram for various factors

Factor	Corner frequency ω_{cf}	Log magnitude	Angle characteristics
4	None	Constant magnitude of +12 db	Constant 0°
$(j\omega)^{-1}$	None	Constant slope of −20 db/decade	Constant −90°
$(1 + j2\omega)^{-1}$	$\omega_1 = 0.5$	0 slope below the corner frequency −20 db/decade slope above the corner frequency	Varies from 0° to −90°
$1 + j0.5\omega$	$\omega_2 = 2.0$	0 slope below the corner frequency +20 db/decade slope above the corner frequency	Varies from 0° to +90°
$\{1 + j0.05\omega + [j\omega/8]^2\}^{-1}$ $\zeta = 0.2$ $\omega_n = 8$	$\omega_3 = 8.0$	0 slope below the corner frequency −40 db/decade slope above the corner frequency	Varies from 0° to −180°

2) The factor Lm$(1 + j2\omega)^{-1}$ has a slope of -20 db/decade, above ω_1 and must be added to the terms in step 1. The composite curve has a total slope of -40 db/decade for frequency $\omega_1 < \omega < \omega_2$, after the slopes are added. This bandwidth represents 2 octaves, thus the value of the composite curve at ω_2 is -6db.

3) Above ω_2, the factor Lm$(1 + j0.5\omega)$ is effective, it has a slope of +20 db/decade above ω_2 and must be added to obtain the composite curve. In the frequency band from ω_2 to ω_3, the composite curve has a total slope of -20 db/decade. The value of the composite curve at ω_3 is -18db.

4) Above ω_3 the term Lm$[1 + j0.05\omega + (\frac{j\omega}{8})^2]^{-1}$ must be added.

339

This factor has a slope of -40 db/decade, thus the total slope of the composite curve above ω_3 is -60 db/decade.

5)   The corrections can be added to the asymptotic plot of the log magnitude of $G(j\omega)$. It is usually sufficient to add corrections at each corner frequency and at an octave above and below the corner frequency. For first-order terms the corrections are ±3db at the corner frequencies and ±1db at an octave above and below the corner frequency. For the quadratic factor the corrections at the frequencies $\omega = \omega_n$ and $\omega = 0.707\ \omega_n$ can be calculated from

$$1 - \left(\frac{\omega}{\omega_n}\right)^2 + 2j\zeta\frac{\omega}{\omega_n} \ .$$

Fig. 2 shows the corrected log magnitude curves for each factor and for the composite curve.

● **PROBLEM** 7-40

Using the function

$$G(j\omega) = \frac{1}{-\omega^2 T^2 + 2\zeta j\omega T + 1} \qquad (1)$$

plot the Bode diagram of a quadratic function.

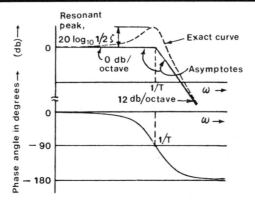

Fig. 1

Solution: We shall determine the asymptotes for low and high frequencies. At the very high frequency

$$|G(j\omega)| = \left|\frac{1}{-\omega^2 T^2}\right| \text{ with } 180° \text{ phase shift.} \qquad (2)$$

converting to db, we have

$$db = -20\ \log_{10}\omega^2 T^2 = -40\ \log_{10}\omega T$$

At low frequency $|G(j\omega)| = 1$, with zero phase shift, converting to db

$$db = 20\ \log_{10} 1 = 0 \qquad (3)$$

db = $-20 \log_{10} \omega^2 T^2 = -40 \log_{10} \omega T$ plotted on suitable coordinates, is a straight line with a negative slope of 40 db/decade or 12 db/octave with phase shift approaching 180° at very high frequencies. This straight line asymptote intersects the zero db axis at $-40 \log_{10} \omega T = 0$ or $\omega T = 1$. We see that a quadratic expression of the form of eq. (1) has two asymptotes. We shall consider now the region in the neighborhood of the corner frequency, in eq. (1) we set $\omega T = 1$. Then

$$G(j\omega) = \frac{1}{2\zeta j} \tag{4}$$

and the gain is $\frac{1}{2\zeta}$ and a phase lag is 90°.

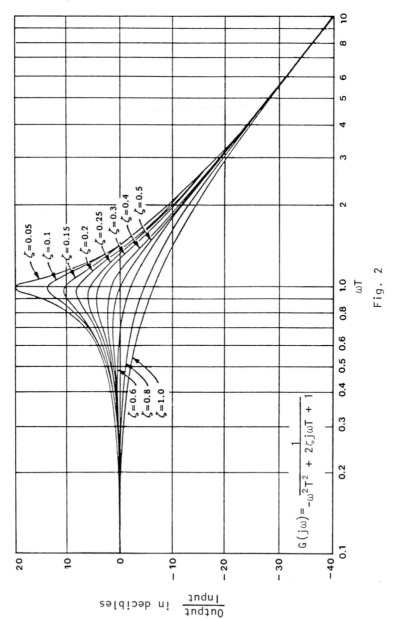

Fig. 2

341

Fig. 1 shows the approximate gain curve. We plotted the high and low frequency asymptotes, the exact corner frequency point and sketched in phase shift.

Figs.2 and 3 show amplitude and gain functions of the quadratic plotted versus dimensionless frequency ωT as a function of the parameter ζ.

Note that the damping factor ζ for the quadratic expression is exactly the same as for the single second-order servo. When the damping factor is zero, we have infinite amplitude at the corner or "resonant" frequency. The resonant peak is eliminated for heavy damping. The quadratic expression may be factored into two linear expressions when a damping factor is greater than unity.

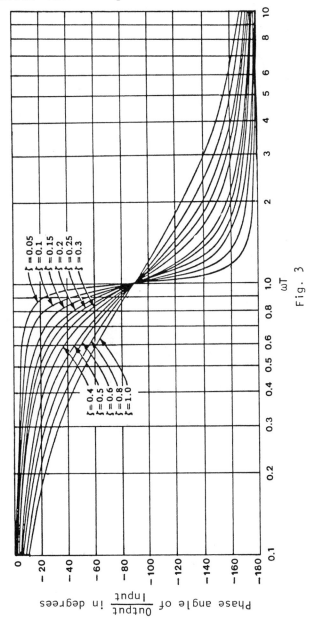

Fig. 3

FREQUENCY RESPONSE

The transfer function of the system is

$$\frac{\theta_0}{\theta_i} = \frac{1}{1 + 0.7D + 0.1D^2}$$

Draw the frequency response characteristics of this system.

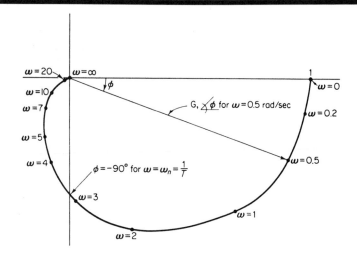

ω, rad/sec	0	0.2	0.5	1	2	3	4	5	7	10	20	100	∞
G	1	0.994	0.96	0.87	0.65	0.49	0.35	0.26	0.16	0.09	0.024	~0	0
ϕ, degrees	0	−8	−19.5	−38	−67	−87	−102	−113	−128	−142	−160	−176	−180

<u>Solution:</u> To find the steady-state frequency response let $j\omega = D$, we get

$$\left(\frac{\theta_0}{\theta_i}\right)_{j\omega} = \frac{1}{1 - 0.1\omega^2 + j \cdot 0.7\omega}$$

We compute G and ϕ

$$G = \frac{1}{[(1 - 0.1\omega^2)^2 + (0.7\omega)^2]^{\frac{1}{2}}} = \left(\frac{1}{1 + 0.29\omega^2 + 0.01\omega^4}\right)^{\frac{1}{2}}$$

$$\phi = 0 - \tan^{-1}\frac{0.7\omega}{1 - 0.1\omega^2}$$

Frequency response of the system is as shown in figure.

Evaluating $\omega = \omega_n = \frac{1}{T} = 3.16$ rad/sec and taking values of ω about this frequency we established the useful frequency range. Setting $1 - 0.1\omega^2 = 0$ we find $\phi = -90°$ for $\omega = \omega_n = \frac{1}{T}$.

A first-order system

$$1.25\frac{dz}{dt} + z = y \tag{1}$$

is subjected to a sinusoidal input

$$y(t) = 3 \sin(\omega_0 t)\text{mV}.$$

Assuming that the output of the system goes into the filter that cuts off all input signals smaller in amplitude than 0.01mV, determine the critical frequency ω_c such that for all $\omega_0 > \omega_c$ the filter will not observe the input signal.

Solution: For the equation

$$1.25 \frac{dz}{dt} + z = 3 \sin \omega_0 t \tag{2}$$

the particular solution can be found using the method of undetermined coefficients, we get

$$z_u = \frac{3}{\sqrt{1 + (1.25\omega_0)^2}} \sin (\omega_0 t + \phi) \tag{3}$$

$$\tan \phi = -1.25\omega_0$$

The amplitude A is given by

$$A = \frac{3}{\sqrt{1 + (1.25\omega_0)^2}} \tag{4}$$

$\omega_0 = \omega_c$ can be found by setting $A = 0.01$.   We have

$$\omega_c = 240 \frac{\text{rad}}{\text{s}} = 38.2\text{Hz} \tag{5}$$

The system is represented by the differential equation

$$A\ddot{x} + B\dot{x} + Cx = f(t) \tag{1}$$

Determine the sinusoidal response of the system.

Solution: Let us represent $f(t)$ as a rotating vector of unit amplitude and angular frequency ω, thus

$$f(t) = e^{j\omega t}$$

and the response as

$$x(t) = ke^{j(\omega t + \phi)}$$ (2)

where k is the relative amplitude and ϕ the phase angle with respect to the input vector.

Differentiating eq. (2) we obtain

$$\frac{dx}{dt} = j\omega ke^{j(\omega t + \phi)}, \quad \frac{d^2x}{dt^2} = (j\omega)^2 ke^{j(\omega t + \phi)}$$ (3)

Substituting eq. (3) in eq. (1) we get

$$A(j\omega)^2 ke^{j(\omega t + \phi)} + Bj\omega ke^{j(\omega t + \phi)} + Cke^{j(\omega t + \phi)}$$

$$= e^{j\omega t}$$ (4)

Dividing both sides of eq. (4) by $e^{j\omega t}$ we get

$$ke^{j\phi}[A(j\omega)^2 + B(j\omega) + C] = 1$$ (5)

or

$$ke^{j\phi} = \frac{1}{A(j\omega)^2 + B(j\omega) + C}$$ (6)

Equation (6) gives the relative amplitude k and phase shift ϕ with respect to the sinusoidal input.

● **PROBLEM 7-44**

The system is described by

$$\tau_1 \frac{dz_1}{dt} + z_1 = Ky_1$$ (1)

Show that this system tends to suppress high-frequency inputs and the system described by

$$z_2 = k(\tau_2 \frac{dy_2}{dt} + y_2)$$ (2)

amplifies high-frequency signals.

Solution: Let us represent the input in form

$$y = C \sin \omega t$$

or in phasor notation

$$y = Ce^{j\omega t}$$

and let

$$z = Ae^{j(\omega t + \phi)}$$

Equation (1) becomes

345

$$A(j\omega\tau_1 + 1)e^{j\phi_1} = KC \tag{3}$$

so

$$Ae^{j\phi_1} = \frac{KC}{j\omega\tau_1 + 1} = \frac{KC}{\sqrt{1 + \omega^2\tau_1^2}}e^{-j\tan^{-1}\omega\tau_1} \tag{4}$$

We have

$$z_1(t) = \frac{KC}{\sqrt{1 + \omega^2\tau_1^2}}\sin(\omega t + \phi_1) \tag{5}$$

From eq. (5) we conclude that

$$\lim_{\omega \to \infty} z_1(t) = 0$$

In the similar way we have

$$Z_2(t) = kC\sqrt{1 + \omega^2\tau_2^2}\ \sin(\omega t + \phi_2) \tag{6}$$

and

$$\lim_{\omega \to \infty} |z_2(t)| \to \infty$$

● **PROBLEM** 7-45

The transfer function of the system is

$$\frac{\theta_0}{\theta_i} = \frac{K(1 + T_1D)}{(1 + T_2D)(1 + T_3D)}$$

where K = 20, T_1 = 0.1 sec, T_2 = 0.2 sec, T_3 = 0.04 sec.
Find the frequency response characteristic of the system;
make Cartesian and polar plots.

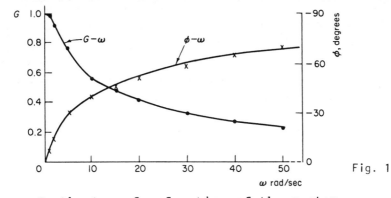

Fig. 1

Solution: In the transfer function of the system

$$\frac{\theta_0}{\theta_i} = \frac{20(1 + T_1D)}{(1 + T_2D)(1 + T_3D)}$$

we substitute $j\omega$ for D

346

$$\left(\frac{\theta_0}{K\theta_i}\right) = \frac{1 + j\omega T_1}{(1 + j\omega T_2)(1 + j\omega T_3)}$$

Let us denote

$$\overline{P} = (1 + j\omega T_1)$$

$$\overline{Q} = (1 + j\omega T_2)$$

$$\overline{R} = (1 + j\omega T_3)$$

in polar notation, the magnitude and angle can be written

$$\overline{P} = (1 + \omega^2 T_1^2)^{\frac{1}{2}}, \quad \tan^{-1}\omega T_1$$

$$\overline{Q} = (1 + \omega^2 T_2^2)^{\frac{1}{2}}, \quad \tan^{-1}\omega T_2$$

$$\overline{R} = (1 + \omega^2 T_3^2)^{\frac{1}{2}}, \quad \tan^{-1}\omega T_3$$

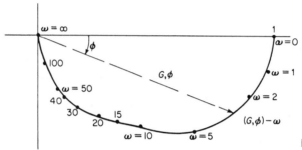

Fig. 2

The magnitude ratio is

$$G = \frac{20(1 + \omega^2 T_1^2)^{\frac{1}{2}}}{(1 + \omega_2^2 T_2)^{\frac{1}{2}}(1 + \omega^2 T_3^2)^{\frac{1}{2}}}$$

and phase $\phi = \tan^{-1}\omega T_1 - \tan^{-1}\omega T_2 - \tan^{-1}\omega T_3$. Substituting $T_1 = 0.1$ sec, $T_2 = 0.2$ sec, $T_3 = 0.04$ sec we evaluate G and ϕ for various values of ω.

ω =	0	1	2	5	10	15	20	30	40	50	100	∞
G =	20	19.7	18.4	15.6	11.6	9.75	8.6	6.7	5.42	4.54	2.4	0
$\phi°$ =	0	−8.2	−15.2	−30	−40	−46	−51	−59.5	−65	−69	−76	−90

Figs.1 and 2 show Cartesian and polar plots of G, ϕ against ω.

● **PROBLEM 7-46**

A loop transfer function of a feedback system is

$$G(j\omega) = \frac{1}{j\omega(j\omega+1)(0.2j\omega+1)} \tag{1}$$

Plot the function on the Nichols chart. What is the maximum closed-loop magnitude, and at what frequency does it occur? What is the phase angle at that point?

At what frequency is the closed-loop magnitude equal to −3dB? What is the 3dB bandwidth? What is the phase angle at that frequency?

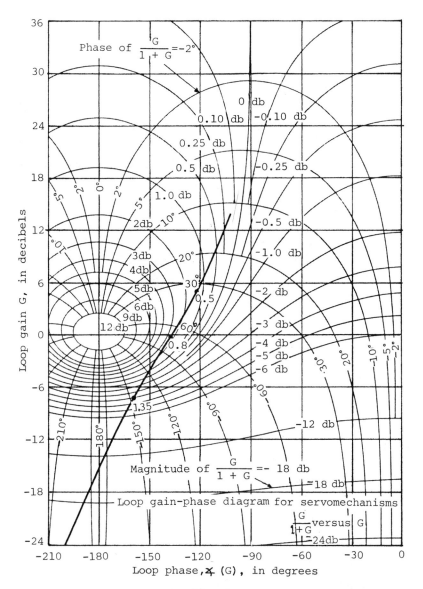

Fig. 1

Solution: Fig. 1 shows the Nichols diagram for
$$G(j\omega) = \frac{1}{j\omega(j\omega+1)(0.2j\omega+1)}$$

On a Nichols chart, a function is plotted as $|G|$dB versus $\angle G$. On the chart are contours of constant closed-loop gain
$$\left|\frac{G}{1+G}\right|$$

and contours of constant closed-loop phase
$$\angle\frac{G}{1+G} .$$

We can construct the plot using Bode diagrams or substituting values of ω into $G(j\omega)$. For $\omega = 0.8$ the closed-loop magnitude is greatest approximately 2.5dB, and the phase about 72°. The plot crosses the -3dB contour when $\omega = 1.35$. At this point the phase is about -145°.

348

The transfer function of the system is

$$G_x(j\omega) = \frac{1.47}{j\omega(1 + j0.25\omega)(1 + j0.1\omega)} \tag{1}$$

Determine the actual gain K_1 needed and the amount by which the original gain K_x must be changed to obtain $M_m = 1.3$.

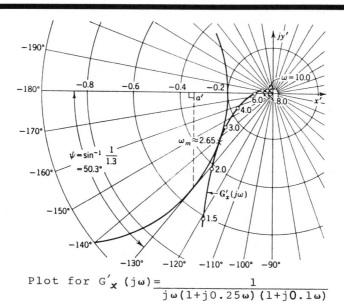

Plot for $G'_x(j\omega) = \dfrac{1}{j\omega(1+j0.25\omega)(1+j0.1\omega)}$

Solution: We draw a polar plot of $G'_x(j\omega)$, where

$$G'_x(j\omega) = \frac{1}{K_x}G_x(j\omega) = \frac{1}{1.47}G_x(j\omega) \tag{2}$$

The radius of the M_m circle is

$$r_0 = \frac{1}{K_x}\left|\frac{M_m}{M_m^2 - 1}\right| \tag{3}$$

Note that M_m is the maximum magnitude of $\dfrac{C(j\omega)}{R(j\omega)}$.

The circle is located with center on the real axis and tangent both to the plot and to the line $\psi = \sin^{-1}\dfrac{1}{M_m}$.

If the function is to have the required $M_m = 1.3$ the distance oa' must be equal to 1. Thus, we find the required value of K_1 by scaling the plot. Therefore

$$K_1 = \frac{1}{oa'} \approx \frac{1}{0.34} = 2.94 \text{ sec}^{-1} \tag{4}$$

The additional gain required is

$$\Delta K = \frac{K_1}{K_x} = \frac{2.94}{1.47} = 2.0 \tag{5}$$

To obtain $M_m = 1.3$ we must double the original gain.

An open-loop transfer function of a system is

$$G(j\omega) = \frac{0.64}{j\omega[(j\omega)^2 + j\omega + 1]} \qquad (1)$$

$\zeta = 0.5$ for the complex poles and $H(j\omega) = 1$. Plot $G(j\omega)$ on the Nichols chart. What is the phase margin for the system? The system damping ratio, for a second order system can be approximated by $\zeta = 0.01\phi^\circ_{pm}$. This approximation still holds for a higher order system dominated by a pair of complex poles. Use this to obtain a value for ζ. From the Nichols chart find M_{p_ω}, the maximum value of

$$\left|\frac{G}{1+G}\right| \ .$$

Obtain another value of ζ.

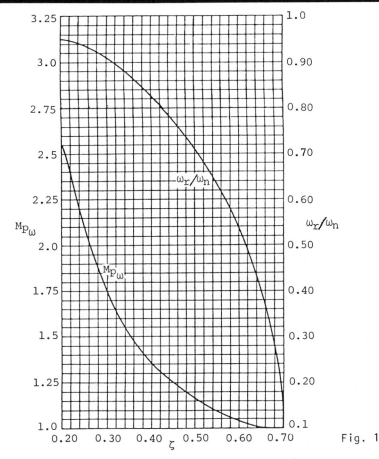

Fig. 1

Solution: The function is shown plotted, $|G|$dB vs. $\angle G$ in Fig. 1. When $|G| = 0$dB, $\angle G = -150°$. Therefore, the phase margin is $\phi_{pm} = 30°$.

The system damping ratio, based on the phase is $\zeta = 0.30$. The maximum magnitude occurs at a frequency $\omega_r = 0.88$ and is equal to +9 db. We have

$$20 \log M_{p_\omega} = 9 \text{ db} \tag{2}$$

Solving equation (2) we obtain

$$M_{p_\omega} = 2.8$$

From Fig. 1 we estimate the damping ratio $\zeta \approx 0.175$.
We obtained two different damping ratios ($\zeta \approx 0.30$ and
$\zeta \approx 0.175$) one from a peak frequency-response measure, the
other from a phase margin measure.

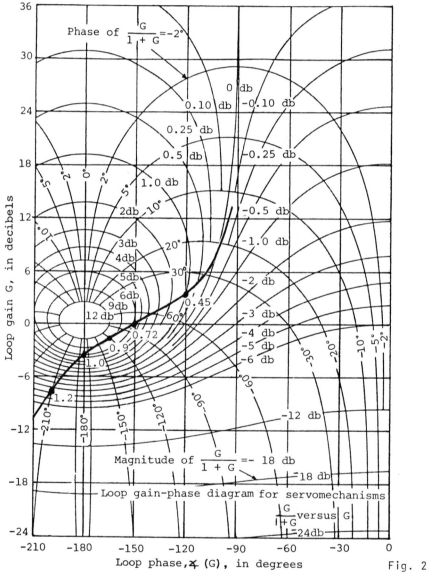

Fig. 2

There is no clear relation, in this case, between the
frequency domain and the time domain. It is due to the
form of the G(jω) locus which slopes rapidly toward the
180° line from the 0-db axis. The roots of the charac-

teristic equation for 1 + GH(s) are

$$q(s) = (s + 0.77)(s^2 + 0.225s + 0.826) = 0 \qquad (3)$$

From eq. (3) we see that the damping ratio of the complex roots is equal to 0.124. The complex roots do not dominate the response of the system and the real root adds some damping. We can estimate the damping ratio as being approximately the value determined from the M_{p_ω} index, that is $\zeta = 0.175$.

Usually it is safer to use the lower value of the damping ratio obtained from the phase margin and the M_{p_ω} relation.

● **PROBLEM** 7-49

The Fourier sine series

$$f(t) = \sum_{n=0}^{\infty} \frac{(-1)^n}{(2n+1)!} \sin(2n + 1)t \qquad (1)$$

represents a periodic function $f(t)$

$$f(t) = \begin{cases} \frac{\pi}{4} t & 0 < t < \frac{\pi}{2} \\ \\ \frac{\pi}{4}(\pi-t) & \frac{\pi}{2} < t < \pi \end{cases} \qquad (2)$$

When presented as input to the first-order system

$$a_0 \frac{dz}{dt} + K = y(t) \qquad (3)$$

the output will have the form

$$z = \sum_{k=0}^{\infty} A_k'(t) \sin[n\omega t + \psi_n(t)] \qquad (4)$$

Define

$$A_n = \lim_{t \to \infty} A_n'(t)$$

and

$$\phi_n = \lim_{t \to \infty} \psi_n(t)$$

For the first-order system $K = 1$, $a_0 = 1$ determine $A_n(t)$ and $\phi_n(t)$.

Solution: We have

352

$$\frac{dz}{dt} + z = f(t)$$

so the impulse response is $I(t) = e^{-t}$. The output, using the convolution method is

$$z(t) = \int_0^t I(t-\tau)y(\tau)d\tau \qquad (5)$$

and substituting eq. (1) for y we have

$$z(t) = e^{-t}\int_0^t e^{\tau} \sum_{n=0}^{\infty} \frac{(-1)^n}{(2n+1)!} \sin(2n+1)\tau \ d\tau$$

$$= e^{-t} \sum_{n=0}^{\infty} \frac{(-1)^n}{(2n+1)!} \int_0^t e^{\tau} \sin(2n+1)\tau \ d\tau$$

$$= e^{-t} \sum_{n=0}^{\infty} \frac{(-1)^n}{(2n+1)!} \frac{1}{1+(2n+1)^2} \{e^t \sin(2n+1)t - (2n+1)e^t$$

$$\cos(2n+1)t + (2n+1)\}$$

$$= \sum_{n=0}^{\infty} \frac{(-1)^n}{(2n+1)![1+(2n+1)^2]}$$

$$\left[\sin(2n+1)t - (2n+1)\cos(2n+1)t + (2n+1)e^{-t}\right] \qquad (6)$$

Since we want the steady-state values A_n and ϕ_n we may disregard the transient term $(2n+1)e^{-t}$ in eq. (6). Note that

$$\sin(2n+1)t - (2n+1)\cos(2n+1)t$$

$$= \sqrt{1+(2n+1)^2} \ \sin[(2n+1)t - \tan^{-1}(2n+1)] \qquad (7)$$

We can thus write eq. (6) as

$$z(t) = \sum_{n=0}^{\infty} \frac{(-1)^n}{(2n+1)!\sqrt{1+(2n+1)^2}} \sin[(2n+1)t - \tan^{-1}(2n+1)] \qquad (8)$$

and we obtain

$$A_n = \frac{(-1)^n}{(2n+1)!\sqrt{1+(2n+1)^2}} \ ; \quad \tan\phi_n = -(2n+1) \qquad (9)$$

Fig. 1 shows a level control system.

Its loop-transfer function is

$$GH(s) = G_A(s) G(s) G_f(s) e^{-sT}$$

$$= \frac{31.5}{(s+1)(30s+1)\left(\frac{s^2}{9} + \frac{s}{3} + 1\right)} e^{-sT} \tag{1}$$

Determine if the system is stable with zero time delay. Determine the stability for the time-delay T = 1 sec. Use Bode diagrams to solve the problem.

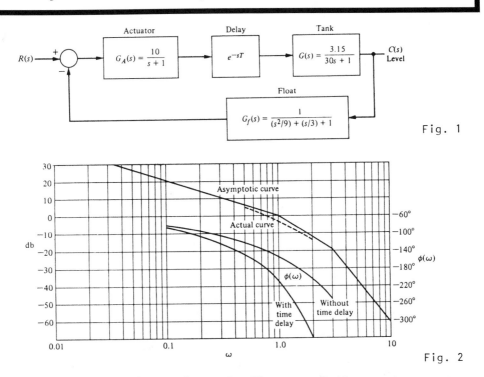

Fig. 1

Fig. 2

Solution: Fig. 2 shows the Bode diagram of the system.

Time delay causes the additional phase lag, the phase angle is shown both for the denominator factors alone and with the phase lag. The logarithmic gain curve crosses the 0-db line at $\omega = 0.8$, and the phase margin of the system without the pure time delay would be 40°. With the added time-delay the phase margin is equal to -3°, thus the system is unstable. In order to have a reasonable phase margin the system gain must be reduced. For a phase margin of 30°, the gain should be decreased by a factor of 5db to

$$K = \frac{31.5}{1.78} = 17.7.$$

A time delay, e^{-sT} in a feedback makes the system less stable. We cannot avoid pure time delays in many systems and to obtain a stable response we have to reduce the loop gain.

● **PROBLEM** 7-51

The loop transfer function of the digital control system shown in Fig. 1 is

$$G(s)H(s) = \frac{1.57}{s(s+1)}$$

The sampling frequency of the system is

$$\omega_s = 4 \; \frac{rad}{s}$$

and the sampling period is $T = \frac{\pi}{2}$ sec.

Sketch the Nyquist plot for the system.

Solution: The z-transform of $G(s)H(s)$ is obtained by replacing terms

$$\frac{1}{s+a}$$

with

$$\frac{z}{z-e^{-aT}} \quad .$$

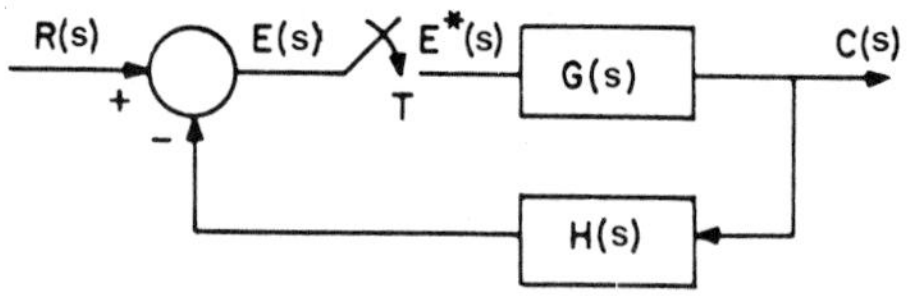

Fig. 1

That can be easily shown by

$$Z\left\{L^{-1}\left\{\frac{1}{s+a}\right\}\right\} = Z\left\{e^{-aT}\right\} = \sum_{k=0}^{\infty} z^{-k} e^{-akT}$$

$$= \sum_{k=0}^{\infty} (z^{-1} e^{-aT})^k = \frac{1}{1-z^{-1} e^{-aT}} = \frac{z}{z-e^{-aT}}$$

In our case we have

$$GH = 1.57\left(\frac{1}{s} - \frac{1}{s+1}\right) = 1.57\left(\frac{z}{z-1} - \frac{z}{z-e^{-T}}\right)$$

$$= 1.57 \; \frac{z(1-e^{-T})}{(z-1)(z-e^{-T})}$$

$$GH(z) = 1.57 \; \frac{0.792z}{(z-1)(z-0.208)}$$

355

From

$$G(z) \Big|_{z=e^{j\omega T}} = G^*(s) \Big|_{s=j\omega}$$

the frequency-response locus of GH(z) is described by

$$GH(e^{j\omega T}) = \frac{1.243 \ e^{j\omega T}}{(e^{j\omega T}-1)(e^{j\omega T}-0.208)}$$

$$= \frac{1.243(\cos\omega T + j\sin\omega T)}{(\cos\omega T + j\sin\omega T-1)(\cos\omega T + j\sin\omega T - 0.208)}$$

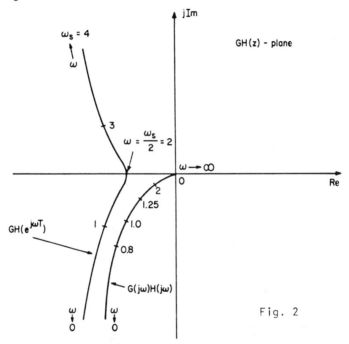

Fig. 2

Fig. 2 shows the plot of GH(z) when z takes on values $|z| = 1$.

The frequency locus of GH(z) repeats over

$$-n\omega_s|2 \le \omega \le n\omega_s|2$$

for

$$n = 1,2, \ldots$$

since the unit circle $|z| = 1$ is traversed once for every frequency interval

$$-n\omega_s/2 \le \omega \le n\omega_s/2$$

The transfer function G(s)H(s) of the system is of the

356

second-order, thus a continuous-data control system with this function as its loop transfer function will be stable. The plot $G(s)H(s)$ does not intersect the negative real axis. The plot $GH(z)$ intersects the negative real axis at -0.515 with $K = 1.57$ for the digital system. Since the point $(-1, j0)$ is to the left of -0.515 it is not enclosed by the $GH(z)$ plot, thus the digital control system is stable for

$$K = 1.57 \qquad \text{and} \qquad T = \frac{\pi}{2} \text{ sec.}$$

The digital control system is shown in Fig. 1.

Its open-loop transfer function is

$$G(z) = \frac{1.2 \times 10^{-7}}{(z-1)(z-0.242)} \tag{1}$$

with $\quad T = 0.1$ sec, $\quad K_r = 3.17 \times 10^5$,

$J_v = 41822$.

Obtain the Nyquist plot for the system.

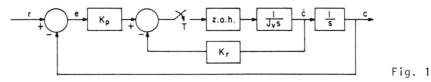

Fig. 1

Solution: Let us substitute in eq (1) $z = e^{j\omega T}$ and compute the magnitude and phase of $G(z)$. The figure shows the Nyquist plot for three values of K_p.

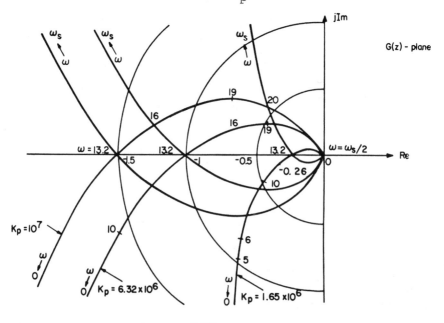

Note that for $K_p = 1.65 \times 10^6$, the $G(z)$ locus crosses the negative real axis at -0.26. The frequency corresponding to that point is $13.2 \frac{rad}{sec}$. According to the Nyquist criterion the closed loop system is asymptotically stable, since the $(-1,j0)$ point is to the left of the crossover point. Using the same criterion we find that for $K_p = 6.32 \times 10^6$ the system is not asymptotically stable and for $K_p = 10^7$ the system is unstable.

● **PROBLEM** 7-53

Fig. 1 shows a positional servomechanism that uses a synchro transmitter-transformer pair.

Assume the following values for the parameters

$$K_1 = 5.73 \; \frac{V}{rad}$$

$$K_3 = 0.5 \; \left(\frac{rad}{s}\right)\bigg/ V$$

$$K_4 = 10$$

$$\tau_m = 0.05 \; sec.$$

$$\tau_e = 0.02 \; sec.$$

where τ_m is a mechanical time constant of motor and load and τ_e electrical time constant of motor.

Determine the amplifier gain K_2 for the system to have an M_m of 1.4 or 3dB.

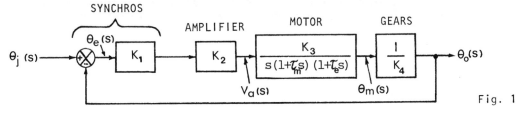

Fig. 1

Solution: From the diagram we obtain the following transfer function of the motor

$$\frac{\theta_m(s)}{V_a(s)} = \frac{K_3}{s(1+\tau_m s)(1+\tau_e s)} = \frac{0.5}{s(1+0.05s)(1+0.02s)} \tag{1}$$

The system shown in the diagram is a unity-feedback system, its open-loop transfer function is

358

$$G(s) = \frac{K_1 K_2 K_3 / K_4}{s(1+\tau_m s)(1+\tau_e s)} = \frac{5.73 \times 0.5 \times 0.1 \times K_2}{s(1+0.05s)(1+0.02s)} \qquad (2)$$

or in the frequency domain

$$G(j\omega) = \frac{K_T}{j\omega(1+j0.05\omega)(1+j0.02\omega)} \qquad (3)$$

The loop gain $K_T = 0.286 K_2$.

The Bode diagram of $G(j\omega)$ is shown in Fig. 2. To make the diagram simpler we set $K_T = 10$.

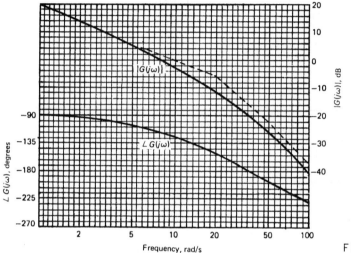

Fig. 2

Fig. 3 is the open-loop Bode diagram sketched on the Nichols chart, with the points plotted for frequencies $1,2,5,10,20 \frac{\text{rad}}{\text{s}}$. If we want the closed-loop system to have $M_m = 1.4$, the plot of $G(j\omega)$ on the Nichols chart must be tangent to the M contour of the value +3dB. The plot of $G(j\omega)$ must be moved upward by 5dB in order to be tangent to this contour (see the dashed curve). Note that the increase on the vertical scale of 5dB is equivalent to a gain increase of 1.77.

We have

$$K_T = 10 \times 1.77 = 17.7 = 0.286 \ K_2 \qquad (4)$$

The necessary amplifier gain K_2 is

$$K_2 = 61.9$$

or

$$35.8 \text{ dB}.$$

Plotting the values of the M and N contours passing through the dashed plot of $G(j\omega)$ at given frequencies gives the Bode diagram of the closed-loop response, see Fig. 4.

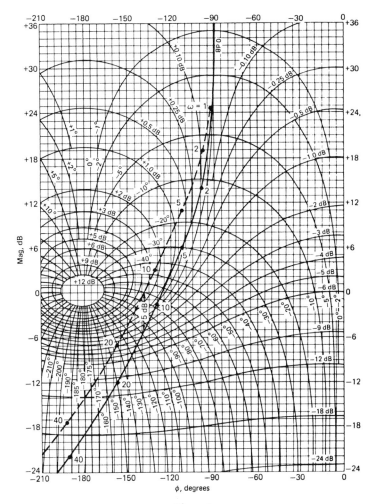

Fig. 3

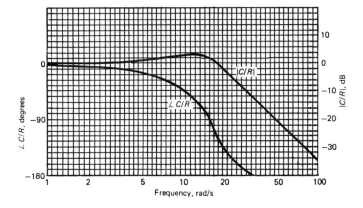

Fig. 4

Fig. 1 shows the RLC network.

Discuss the response of the system to sinusoidal inputs. What is the peak value of the magnitude response? What is the 3dB bandwidth (i.e., the distance from the peak value at which the magnitude response has decreased to

$$\frac{1}{\sqrt{2}}$$

of peak)?

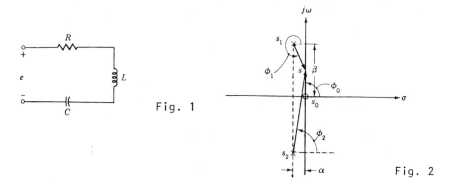

Fig. 1

Fig. 2

Solution: The driving-point impedance is

$$\frac{E}{I}(s) = Z(s) = sL + R + \frac{1}{Cs} = \frac{LCs^2 + RCs + 1}{Cs} \qquad (1)$$

The inverse of the driving-point impedance is the driving-point admittance $Y(s)$, thus

$$Y(s) = \frac{I}{E}(s) = \frac{Cs}{LCs^2 + RCs + 1} \qquad (2)$$

The admittance has one zero at $s = 0 + j0$ and two poles. Setting the denominator of $Y(s)$ equal to 0 we compute the poles.

$$s_1, s_2 = -\frac{R}{2L} \pm \sqrt{\left(\frac{R}{2L}\right)^2 - \frac{1}{LC}} \qquad (3)$$

The poles are located in the left half plane, they have negative real part.

Let us assume the following values for $R = 1$, $L = 0.5$, $C = \frac{2}{101}$

From eq. (3) the poles are

$$\bar{s}_{1,2} = -1 \pm j10 \qquad (4)$$

361

We write Y(s) in the form

$$Y(s) = \frac{\frac{1}{L}s}{s^2 + \frac{R}{L}s + \frac{1}{LC}} = \frac{2s}{(s+1-j10)(s+1+j10)} \tag{5}$$

$$= \frac{\frac{1}{L}(s-s_0)}{(s-s_1)(s-s_2)}$$

where s_0 is the zero and $\bar{s}_1, \bar{s}_2$ are two poles. For a sinusoidal excitation $s = j\omega$, we obtain

$$|Y| = \frac{\frac{1}{L}A_0}{B_1B_2} \tag{6}$$

$$\angle Y = \phi_0 - \phi_1 - \phi_2 \tag{7}$$

Fig. 2 shows the vectors $s-s_0$, $s-s_1$, $s-s_2$, their magnitudes are A_0, B_1, B_2 and angles are ϕ_0, ϕ_1, ϕ_2 respectively.

The magnitude of $Y(j\omega)$ is given by eq. (6)

$$|Y(j\omega)| = \frac{2|\omega|}{\sqrt{1+(\omega-10)^2}\sqrt{1+(\omega+10)^2}} \tag{8}$$

and the angle by eq. (7)

$$\angle Y(j\omega) = \frac{\pi}{2} - \tan^{-1}(\omega-10) - \tan^{-1}(\omega+10) \tag{9}$$

where $\omega > 0$

The general form of. eq (5) is

$$H(s) = \frac{Ks}{(s-s_1)(s-s_2)} \tag{10}$$

where

$$s_1 = s_2^* \quad \text{and} \quad \bar{s}_1 = -\alpha + j\beta, \qquad \alpha, \beta > 0$$

Fig. 3 shows H(s) for $\alpha << \beta$

We shall estimate qualitatively the magnitude and phase of H(s) for $s = \omega j$. We shall use the graphical method.

For $\omega \approx 0$ and $\omega \to \pm\infty$, $|H(j\omega)| = 0$,

for large ω $|H(j\omega)| \simeq \frac{1}{\omega}$. Fig. 4 and Fig. 5 show the magnitude and the angle of $H(j\omega)$.

From the equation

$$\omega = \beta \quad \text{then} \quad |H(j\omega)| \simeq \frac{K}{2\alpha}$$

we estimate the peak value of $|H(j\omega)|$.

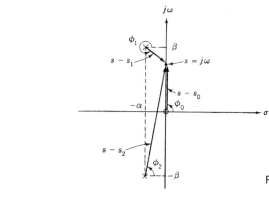

Fig. 3

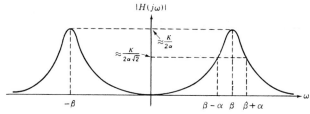

Fig. 4

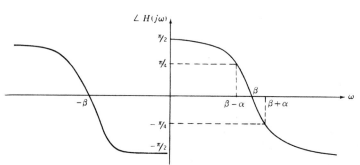

Fig. 5

Let us assume that as ω varies around $\omega = \beta$, the magnitude and the phase contributed by s and $s-s_2$ are constant.

When $\phi_1 = \frac{\pi}{4}$, the value of ω is $\alpha + \beta$, the magnitude of $s-s_1$ is $\alpha\sqrt{2}$ and

$$|H(j\omega)| \simeq \frac{K\beta}{2\beta\alpha\sqrt{2}} = \frac{K}{2\alpha\sqrt{2}}$$

and the angle is

$\angle H(j\omega) \cong -\frac{\pi}{4}$. For $\phi_1 = -\frac{\pi}{4}$ the value of ω is $\beta - \alpha$ and the magnitude of $s-s_1$ is $\alpha\sqrt{2}$ and

$$|H(j\omega)| \simeq \frac{K}{2\sqrt{2}\alpha} \quad \text{while} \quad \angle H(j\omega) \cong \frac{\pi}{4}.$$

363

The circuit is shown in Fig. 1.

Show that this circuit is a high pass filter. Investigate its use in an amplifier system.

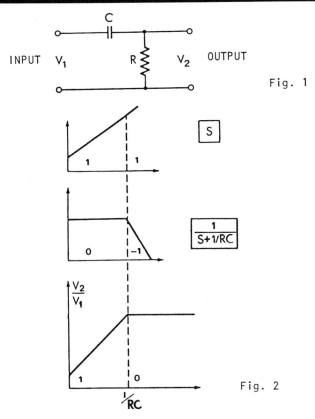

INPUT V_1 R V_2 OUTPUT

Fig. 1

S

$\dfrac{1}{S+1/RC}$

$\dfrac{V_2}{V_1}$

$\dfrac{1}{RC}$

Fig. 2

Solution: A capacitor C (see Fig. 1) acts as an open circuit to low frequencies and a short circuit to high frequencies. Therefore, at low frequencies most of V_1 will be dropped across C and little will be dropped across R, so the output is low. At high frequencies, C will drop very little voltage, so R will drop most of V_1 and the output will be high. Thus, the system appears to be a high-pass filter.

From Kirchhoff's law we have

$$V_1 = \left[R + \frac{1}{CS}\right] I = \left[\frac{RCS + 1}{CS}\right] I \tag{1}$$

and

$$V_2 = IR \tag{2}$$

Eliminating I we get

$$V_2 = \frac{RCS}{RCS + 1} V_1$$

364

or

$$\frac{V_2}{V_1}(s) = \frac{S}{S + \frac{1}{RC}}$$

(3)

Note that equation 3 is formed by the product of an exponential and a reciprocal first order factor. Fig. 2 shows the individual curves and their composite.

The circuit shown in Fig. 1 is used as a coupling between stages in the amplifier. It passes only the high frequencies so the sound is rich in high notes but lacks low notes.

CHAPTER 8

DESIGN AND COMPENSATION

DESIGN

● PROBLEM 8-1

What are the major advantages and disadvantages of the open-loop control systems?

Solution: The advantages are:

1. There is no stability problem.

2. Simple construction and easy maintenance.

3. Low cost of construction.

The disadvantages are:

1. The errors caused by disturbances and changes in calibration make the output different from that desired.

2. The systems require frequent recalibration to maintain the desired output.

● PROBLEM 8-2

The system is described by

$$\dot{x}(t) = Ax(t) + Bu(t)$$

where

$$A = \begin{bmatrix} 0 & 1 \\ -1 & 0 \end{bmatrix}, \quad B = \begin{bmatrix} 0 \\ 1 \end{bmatrix}, \quad \dot{x}(t) = \begin{bmatrix} \dot{x}_1(t) \\ \dot{x}_2(t) \end{bmatrix}$$

Using the variable feedback method, place the poles of the system at -4 and -6.

Solution: We write $u(t)$ in the form

$$u(t) = -G \begin{bmatrix} x_1(t) \\ x_2(t) \end{bmatrix}.$$

Thus

$$G = [g_1, g_2]$$

and

$$u(t) = -[g_1, g_2] \begin{bmatrix} x_1(t) \\ x_2(t) \end{bmatrix} = -[g_1 x_1(t) + g_2 x_2(t)]$$

The eigenvalues of A are

$$\lambda_1 = j, \lambda_2 = -j$$

We must introduce feedback if the obtained values are undesirable. The closed-loop system for any values of g_1 and g_2 is

$$\begin{bmatrix} \dot{x}_1(t) \\ \dot{x}_2(t) \end{bmatrix} = \begin{bmatrix} 0 & 1 \\ -1 & 0 \end{bmatrix} \begin{bmatrix} x_1(t) \\ x_2(t) \end{bmatrix} - \begin{bmatrix} 0 \\ 1 \end{bmatrix} [g_1 g_2] \begin{bmatrix} x_1(t) \\ x_2(t) \end{bmatrix}$$

$$= \begin{bmatrix} 0 & 1 \\ -1 & 0 \end{bmatrix} \begin{bmatrix} x_1(t) \\ x_2(t) \end{bmatrix} - \begin{bmatrix} 0 & 0 \\ g_1 & g_2 \end{bmatrix} \begin{bmatrix} x_1(t) \\ x_2(t) \end{bmatrix}$$

$$= \begin{bmatrix} 0 & 1 \\ -1 - g_1 & -g_2 \end{bmatrix} \begin{bmatrix} x_1(t) \\ x_2(t) \end{bmatrix}$$

$$A - BG = \begin{bmatrix} 0 & 1 \\ -1 - g_1 & -g_2 \end{bmatrix}$$

If we want the eigenvalues of the A - BG matrix to be at

$$\mu_1 = -4 \quad \text{and} \quad \mu_2 = -6$$

we shall apply the following method. Since [A,B] are controllable we can use the variable feedback gain theorem. It states: Given a system $\dot{x} = Ax + Bu$, A is any real $n \times n$ matrix, B is any real $n \times m$ matrix and [A,B] is controllable, there exists at least one $m \times n$ real state feedback gain matrix G such that the eigenvalues of A - BG are equal to the prespecified values (assuming that any complex eigenvalues come in complex conjugate pairs). The characteristic polynomial $p(\mu)$ of A - BG is

$$p(\mu) = \det \begin{bmatrix} \mu & -1 \\ 1 + g_1 & \mu + g_2 \end{bmatrix} = \mu(\mu + g_2) + 1 + g_1$$

367

$$= \mu^2 + g_2\mu + 1 + g_1$$

Since

$$p(\mu) = (\mu - \mu_1)(\mu - \mu_2) = (\mu + 4)(\mu + 6)$$

$$= \mu^2 + 10\mu + 24 \quad,$$

then g_1 and g_2 must have the values

$$g_1 = 23, \; g_2 = 10.$$

Using the state feedback, decouple the following system.

$$A = \begin{bmatrix} 1 & 1 & 0 \\ 0 & 1 & 0 \\ 0 & 0 & 1 \end{bmatrix}, \quad B = \begin{bmatrix} 0 & 1 \\ 1 & 0 \\ 1 & 0 \end{bmatrix}$$

$$C = \begin{bmatrix} 1 & 1 & -1 \\ 0 & 1 & 0 \end{bmatrix}, \quad D = [0]$$

Solution: We shall introduce the control

$$U = -K_d x + F_d v$$

where v is the input of the system. Let us denote the ith row of C by c_i.

For each row c_i we define a number d_i such that:

$$d_i = \min_{j} \; \{ j \mid c_i A^j B \neq 0 \quad j = 1, 2 \ldots n - 1 \}$$

if $c_i A^j B = 0$ for all j then $d_i = n - 1$. We form $m \times m$ matrix

$$N = \begin{bmatrix} c_1 A^{d_1} B \\ c_2 A^{d_2} B \\ \vdots \\ c_m A^{d_m} B \end{bmatrix}$$

The system can be decoupled if N is nonsingular. One set of decoupling matrices is

368

$$F_d = N^{-1}, \quad K_d = N^{-1} \begin{bmatrix} c_1 A^{d_1} + 1 \\ c_2 A^{d_2} + 1 \\ \cdot \\ \cdot \\ \cdot \\ c_m A^{d_m} + 1 \end{bmatrix}$$

The decoupled transfer matrix is

$$H(s) = C[sI - A + BK_d]^{-1} BF_d$$

In our case we have

$$c_1 B = [0 \quad 1] \neq 0, \text{ so } d_1 = 0$$

and

$$c_2 B = [1 \quad 0] \neq 0, \text{ so } d_2 = 0.$$

The matrix N is given by

$$N = \begin{bmatrix} 0 & 1 \\ 1 & 0 \end{bmatrix}$$

and is nonsingular. F_d and K_d are

$$F_d = N^{-1} = \begin{bmatrix} 0 & 1 \\ 1 & 0 \end{bmatrix}; \quad K_d = \begin{bmatrix} 0 & 1 \\ 1 & 0 \end{bmatrix} \begin{bmatrix} c_1 A \\ c_2 A \end{bmatrix}$$

$$= \begin{bmatrix} 0 & 1 & 0 \\ 1 & 2 & -1 \end{bmatrix}$$

The decoupled transfer matrix is

$$H(s) = \begin{bmatrix} \dfrac{1}{s} & 0 \\ 0 & \dfrac{1}{s} \end{bmatrix}$$

The system described in the example is not completely controllable, therefore the decoupled system is not controllable.

● **PROBLEM** 8-4

The open loop transfer function of the system is

$$\frac{\theta_{on}}{\theta} = \frac{25}{D(1 + \frac{1}{4}D)(1 + \frac{1}{16}D)}$$

We want the system's gain and phase margins to exceed 1.5 and 15° respectively. Determine if the original system meets the specification, if not compensate the system with first-order lead action. We want the open loop gain to be maintained at 25. If necessary an additional proportional amplifier can be used.

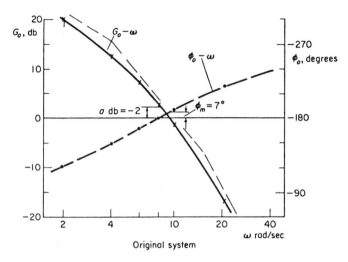

Original system   Fig.1

Solution: The open loop Bode diagram for the original system is shown in Fig. 1. The gain and phase margins are a, db = -2 and a = 0.8, ϕ_m = -6°. We see that the system is unstable.

The first-order lead action

$$\frac{\beta(1 + TD)}{1 + \beta TD} \text{ , } \beta < 1$$

reduces phase lag and open loop gain at any particular frequency. By appropriate selection of T and β we concentrate the effect of this action on the region near ϕ_0 = -180°. Let us try

$$T = \frac{1}{\omega} = 0.2 \quad \text{where} \quad \omega = 5$$

$$\beta = 0.2$$

We obtain

$$\frac{\theta_c}{\theta} = \frac{0.2(1 + 0.2D)}{(1 + 0.04D)} \text{ ;}$$

$$\left(\frac{\theta_{on}}{\theta}\right)_c = \frac{5(1 + \frac{1}{5}D)}{D(1 + \frac{1}{25}D)(1 + \frac{1}{4}D)(1 + \frac{1}{16}D)}$$

Thus, the open loop gain constant has been reduced to 5 and the original system becomes stable. Since we want to maintain K_0 = 25, a 5x amplifier must be included in the

370

forward loop (see Fig. 2). We obtain

$$\frac{\theta_{on}}{\theta} = \frac{25(1 + \frac{1}{5}D)}{D(1 + \frac{1}{25}D)(1 + \frac{1}{4}D)(1 + \frac{1}{16}D)}$$

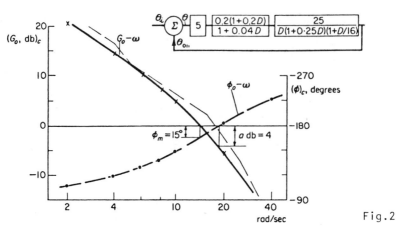

Fig.2

The open loop Bode diagram of the compensated system is shown in Fig. 2. The gain and phase margins are a, db = 4 and a = 1.6, ϕ_m = 15°.

● **PROBLEM** 8-5

A system is described by

$$\dot{x} = Ax + B$$

where

$$A = \begin{bmatrix} -2 & 1 & 0 \\ 0 & -2 & 1 \\ 0 & 0 & -2 \end{bmatrix}, \quad B = \begin{bmatrix} 0 \\ 0 \\ 1 \end{bmatrix}$$

Find a constant state feedback matrix K which gives closed-loop eigenvalues $\lambda_1 = \lambda_2 = \lambda_3 = -1$.

Solution: We form the matrice $[\lambda I - A]$ and its inverse, $\phi(\lambda)$ where

$$\phi(\lambda) = \begin{bmatrix} \dfrac{1}{\lambda + 2} & \dfrac{1}{(\lambda + 2)^2} & \dfrac{1}{(\lambda + 2)^3} \\ 0 & \dfrac{1}{\lambda + 2} & \dfrac{1}{(\lambda + 2)^2} \\ 0 & 0 & \dfrac{1}{\lambda + 2} \end{bmatrix}$$

We have

$$\psi(\lambda) = \phi(\lambda) B = \begin{bmatrix} \dfrac{1}{(\lambda + 2)^3} \\[3mm] \dfrac{1}{(\lambda + 2)^2} \\[3mm] \dfrac{1}{\lambda + 2} \end{bmatrix}$$

We have to obtain three linearly independent columns, thus

$$\frac{d\psi}{d\lambda} = \begin{bmatrix} \dfrac{-3}{(\lambda + 2)^4} \\[3mm] \dfrac{-2}{(\lambda + 2)^3} \\[3mm] \dfrac{-1}{(\lambda + 2)^2} \end{bmatrix}$$

For $\lambda = -1$ we get

$$G = \begin{bmatrix} 1 & -3 & 12 \\ 1 & -2 & 6 \\ 1 & -1 & 2 \end{bmatrix}$$

Correspondingly, we obtain

$$\mathcal{Y} = \begin{bmatrix} 1 & 0 & 0 \end{bmatrix}$$

and

$$K = -\begin{bmatrix} 1 & 0 & 0 \end{bmatrix} G^{-1} = \begin{bmatrix} -1 & 3 & -3 \end{bmatrix}.$$

● **PROBLEM** 8-6

A system is described by

$$\dot{x} = Ax + Bu$$

$$y = Cx + Du$$

where

$$A = \begin{bmatrix} 0 & 1 & 0 \\ 0 & 0 & 1 \\ 1 & 0 & 0 \end{bmatrix}, \quad B = \begin{bmatrix} 0 \\ 1 \\ 0 \end{bmatrix}, \quad C = \begin{bmatrix} 1 & 0 & 0 \\ 1 & 1 & 0 \end{bmatrix}, \quad D = \begin{bmatrix} 0 \\ 0 \end{bmatrix}$$

Find K' so that $\lambda_1 = 1$, $\lambda_2 = \varepsilon$ where ε is arbitrarily close to zero. Note that since the system is completely controllable and rank C = m = 2, two closed-loop eigenvalues can be made arbitrarily close to any specified values by using output feedback.

We form the matrices

$$\Phi(\lambda) = (\lambda I - A)^{-1} = \frac{\begin{bmatrix} \lambda^2 & \lambda & 1 \\ 1 & \lambda^2 & \lambda \\ \lambda & 1 & \lambda^2 \end{bmatrix}}{\lambda^3 - 1}$$

and

$$C\Phi(\lambda)\,B = \frac{\begin{bmatrix} \lambda \\ \lambda + \lambda^2 \end{bmatrix}}{\lambda^3 - 1} \tag{1}$$

Setting $\lambda_2 = \varepsilon$ and then using a limiting process allows λ_2 to approach 0. The value $\lambda_1 = 1$ causes unbounded components. Thus we must use another limiting process.

Let

$$\alpha = \lambda_1^3 - 1$$

and select

$$G' = \begin{bmatrix} \dfrac{1}{\alpha} & \dfrac{\varepsilon}{\varepsilon^3 - 1} \\[2ex] \dfrac{2}{\alpha} & \dfrac{\varepsilon + \varepsilon^2}{\varepsilon^3 - 1} \end{bmatrix}$$

where

$$G' = [\,c\Phi(1)B \qquad c\Phi(\varepsilon)\,]$$

Then

$$K' = [1 \quad 1]G'^{-1} = \frac{-[\alpha(1 + \varepsilon) - 2(\varepsilon^2 - \frac{1}{\varepsilon}) \qquad -\alpha + \varepsilon^2 - \frac{1}{\varepsilon}]}{\varepsilon - 1} \tag{2}$$

For $\alpha \to 0$ and $\varepsilon \to 0$ we have

$$K' \to \begin{bmatrix} \dfrac{2}{\varepsilon} & \dfrac{-1}{\varepsilon} \end{bmatrix}.$$

It indicates that $\lambda_2 = 0$ can be achieved only by using infinite feedback gains.

● **PROBLEM** 8-7

For the system shown in Fig. 1 select K in such way that the the phase margin is greater than 30° and the gain margin is greater than 10 db.

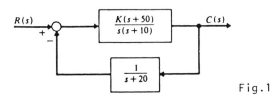

Fig.1

Solution: We have in Bode form,

$$KGH = \frac{50K}{200} \frac{(0.02s + 1)}{s(0.1s + 1)(0.05s + 1)} \qquad (1)$$

Fig. 2 shows the Bode plots with the Bode gain $K_b = \frac{K}{4}$
set to unity. At $\omega = 10$, the phase is $-150°$ and the gain
is -24 db. Thus K_b could be increased from 0 db to $+24$ db
and the phase margin specification would be satisfied. In
such case, at $\omega = 24\frac{rad}{sec}$ where the phase is $-180°$, the gain
would increase from -39db to -15db. The gain margin of
15db satisfies the specifications, so $K_b = 24$ db, that
converts to a real gain of about 15. Therefore $K = 4K_b = 60$
can be used to satisfy both specifications.

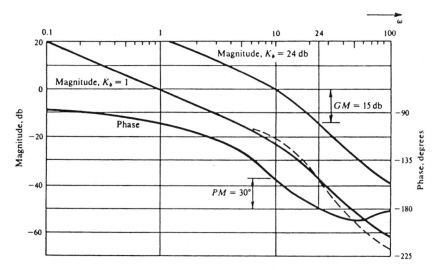

Fig.2

• **PROBLEM** 8-8

Fig. 1 shows the initial design of a closed loop system.

We want the system to have a gain margin of about 4 and a
phase margin of about 30°, while the open loop gain constant
is to be maintained at 30. In case the system does not
satisfy these specifications, use first-order lag series
compensation to achieve the desired result. Obtain the
frequency response characteristics for both the original
system and the compensated system.

374

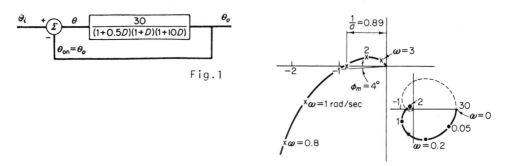

Fig.1

Fig.2

Solution: The original open loop transfer function is

$$\frac{\theta_{on}}{\theta} = \frac{30}{(1 + 0.5D)(1 + D)(1 + 10D)}$$

Fig. 2 shows the original Nyquist plot. The original
system is stable, with gain and phase margins

$$a = 1.12, \quad \phi_m = 4°$$

We see that the margins of stability do not meet the
performance specifications. The equation

$$\frac{\theta_c}{\theta} = \frac{1 + TD}{1 + \alpha TD} \ , \quad \alpha > 1$$

describes the first-order lag action which increases the
open loop lag at any frequency. This destabilizing effect
of increased open loop lag can be diminished by reducing
the open loop magnitude ratio to shrink the Nyquist plot
away from enclosing (-1,0). α and T must be chosen to
produce a strong attenuation effect on G_0. We see from
Fig. 2 that reshaping of the Nyquist plot should start
from about the $\omega = 0.5\frac{rad}{sec}$ region. Thus, select a compensator numerator time constant $T = \frac{1}{\omega} = 2$ sec ($\omega = 0.5$).

To get a pronounced attenuation effect from the compensator, we choose $\alpha = 10$. Fig. 3 shows the block diagram
of the new system with the compensator

$$\frac{1 + 2D}{1 + 20D} \ .$$

The compensated system's open loop transfer function is

$$\left(\frac{\theta_{on}}{\theta}\right)_c = \frac{30(1 + 2D)}{(1 + 20D)(1 + 0.5D)(1 + D)(1 + 10D)}$$

The Nyquist plot of the new system is shown in Fig. 4. The
gain and phase margins of the compensated system are

$$a = 4.7, \quad \phi_m = 27°$$

and the open loop gain is $K_0 = 30$.

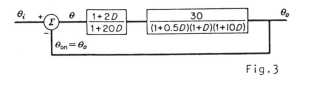

Fig.3

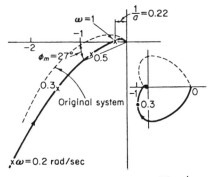

Fig.4

The closed loop transfer funcon of the original system is

$$\frac{\theta_0}{\theta_i} = \frac{0.97}{0.16D^3 + 0.5D^2 + 0.37D + 1}$$

and of the compensated system is

$$\left(\frac{\theta_0}{\theta_i}\right)_c = \frac{0.97(1 + 2D)}{0.1D^4 + 0.33D^3 + 7.3D^2 + 1.08D + 1}$$

The frequency response characteristics of both systems are shown in Fig. 5.

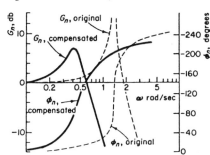

Fig.5

● **PROBLEM** 8-9

Using the series

$$\frac{\theta_c}{\theta} = \frac{(1 + 0.1D)0.1}{1 + 0.01D}$$

compensate the system, whose open loop transfer function is

$$\frac{\theta_{on}}{\theta} = \frac{40}{D(1 + \frac{1}{2}D)(1 + \frac{1}{4}D)}$$

Sketch the root loci of both systems uncompensated and compensated. Discuss the results.

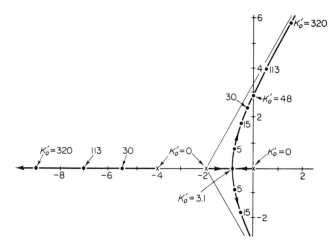

Fig. 1

Solution: The original open loop and compensator transfer functions, in pole-zero form, are

$$\frac{\theta_{on}}{\theta} = \frac{320}{D(D + 2)(D + 4)}$$

$$\frac{\theta_{c}}{\theta} = \frac{D + 10}{D + 100}$$

$$\left(\frac{\theta_{on}}{\theta}\right)_{c} = \frac{320(D + 10)}{D(D + 2)(D + 4)(D + 100)}$$

Fig. 1 shows the root locus of the original system.

Note that:

K_0' = 48 as branches cross imaginary axis;

branches start at the open loop poles 0, -2, -4;

high K_0' asymptotes have angles of

$$\alpha = \frac{n \times 180}{3 - 0} = \pm 60° \text{ and } 180°;$$

the asymptotes start at $\bar{x} = \frac{-(6 - 0)}{3 - 0} = -2.$

For K_0' = 320, the system is unstable. The root locus of the compensated system is shown in Fig. 2. The critical locations are:

branches start at 0, -2, -4, and -100;

one branch ends at -10, the other three at ∞;

the asymptotes have angles ±60°, 180°;

the asymptotes intersect the negative real axis at $\frac{106 - 10}{4 - 1} = 32.$

377

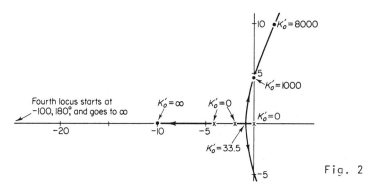

Fig. 2

From Fig. 2 we see that the lead action (zero dominant) has drawn the complex branches to the left. The compensated system is stable for $K_0' < 1002$. The result of the compensation was the stabilization of the system.

● **PROBLEM** 8-10

Design the system shown in the figure (find K_p, J, D) that satisfies the following specifications:

a) $\varepsilon_{ss} \leq 0.05$ rad for step input $\theta_{in} = 1.0$ rad.

b) $\varepsilon_{ss} \leq 0.05$ rad for ramp input $\theta_{in} = 75\dfrac{rad}{s}$.

c) $\varepsilon_{ss} \leq 0.05$ rad for step disturbance $T_d = 50N \cdot m$.

d) Settling time $T_s = 0.1s$ for 5% transient residue.

e) Phase margin PM = 45°.

The margins of stability are shown in the table.

MARGINS OF STABILITY

ζ	PM	GM
0	0	0
0.125	20°	6.3
0.25	30°	25
0.5	45°	100
0.707	60°	200
1.0	90°	400

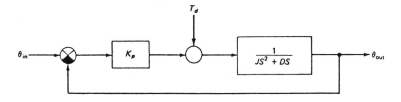

Solution: The open loop gain is

$$L(s) = \frac{K_p}{S(JS + D)} \tag{1}$$

and the closed loop transfer function is

$$\frac{\theta_{out}}{\theta_{in}}(s) = \frac{K_p}{JS^2 + DS + K_p} = \frac{\omega_n^2}{S^2 + 2aS + \omega_n^2} \tag{2}$$

where

$$\omega_n^2 = \frac{K_p}{J} \tag{3}$$

$$a = \zeta\omega_n \tag{4}$$

$$\zeta = \frac{D}{2J\omega_n} \tag{5}$$

Including T_d, the output is

$$\theta_{out} = \frac{K_p\theta_{in} + T_d}{JS^2 + DS + K_p} \tag{6}$$

The error is given by

$$\varepsilon(S) = \theta_{in}(S) - \theta_{out}(S)$$

$$= \frac{S(JS + D)\theta_{in}(S) - T_d(S)}{S(JS + D) + K_p} \tag{7}$$

The error due to a step input $\frac{U}{S}$ with $T_d = 0$ is

$$\varepsilon_{ss} = \lim_{s \to 0}\left[\frac{S^2(JS + D)U/S}{S(JS + D) + K_p}\right] = 0 \tag{8}$$

We see that the system automatically meets specification a). The error due to the ramp input $\frac{\dot{\theta}}{S^2}$ is

$$\varepsilon_{ss} = \lim_{s \to 0}\left[\frac{S^2(JS + D)\dot{\theta}/S^2}{S(JS + D) + K_p}\right] = \frac{\dot{\theta}}{K_p} \tag{9}$$

Applying condition b) to eq. (9) we find

$$K_p \geq \frac{\dot{\theta}}{\varepsilon_{ss}} = \frac{75}{0.05} = 1500 \tag{10}$$

The error due to a step disturbance $\frac{T_d}{S}$ with $\theta_{in} = 0$ is

$$\varepsilon_{ss} = \lim_{s \to 0} \left[\frac{-ST_d/S}{S(JS + D) + K_p} \right] = \frac{T_d}{K_p} \qquad (11)$$

We apply condition c) to eq. (11).

$$K_p \geq \frac{T_d}{\varepsilon_{ss}} = \frac{50}{0.05} = 1000 \qquad (12)$$

Eqs. (10) and (12) give two values for K_p. Observing the inequality, the final choice is

$$K_p = 1500 \qquad (13)$$

Adjusting the damping ratio we attain stability, the appropriate value is selected from the table. Applying condition e) to this table we find for

$$PM = 45°, \ \zeta = 0.5 \qquad (14)$$

The settling times T_s for 5% transient residue is

$$T_s = 3\tau = \frac{3}{\zeta \omega_n} \qquad (15)$$

Applying condition d) we find

$$\omega_n = \frac{3}{\zeta T_s} = \frac{3}{0.5 \times 0.1} = 60 \ \frac{rad}{s} \qquad (16)$$

Using eq. (3),

$$J = \frac{K_p}{\omega_n^2} = \frac{1500}{60^2} = 0.4166 \ kg \qquad (17)$$

Applying eqs. (14), (16) and (17) to eq. (5) we obtain

$$D = 2\zeta \omega_n J = 2 \times 0.5 \times 60 \times 0.4166 = 25N \cdot m \cdot s \qquad (18)$$

This completes the design.

● **PROBLEM** 8-11

The open-loop transfer function of a system is

$$\frac{\theta_{on}}{\theta} = \frac{25(D + 2)}{D(D + 1)(D + 1)(D + 5)}$$

We want the system to be stable, with a damping ratio ≈ 0.7 and sensitivity $K_0^\prime = 25$.

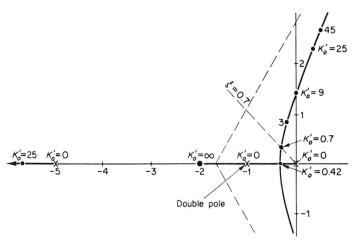

Fig. 1

Solution: The root locus of the system is shown in Fig. 1.

From the sketch we see that the system is unstable for $K_0' > 9$. To reach the damping ratio $\xi = 0.7$, the value of K_0' would decrease to $K_0' = 0.7$. Therefore we must compensate.

We need a dominant zero effect to move the complex root loci into the lefthand half plane.

The first-order lead action yields

$$\frac{\theta_c}{\theta} = \frac{K'(D + z)}{D + p}, \quad p > z$$

$$\frac{\theta_c}{\theta} = \frac{\beta(1 + TD)}{1 + \beta TD}, \quad \beta < 1$$

and

$$\frac{\theta_c}{\theta} = \frac{\beta T(D + \frac{1}{T})}{\beta T(D + \frac{1}{\beta T})} = \frac{D + z}{D + p}$$

where $z = \frac{1}{T}$ and $p = \frac{1}{\beta T}$.

Choosing $z = 1$ as the compensator zero, we cancel one of the $(D + 1)$ poles and move the original root locus to the left. For $\beta = 0.1$, we get $p = 10$ as the pole, that does not counter too strongly the effect of $z = 1$.

The open loop transfer function of the compensated system is

$$\left(\frac{\theta_{on}}{\theta}\right)_c = \frac{K_0'(D + 2)(D + 1)}{D(D + 1)^2(D + 5)(D + 10)} = \frac{K_0'(D + 2)}{D(D + 1)(D + 5)(D + 10)}$$

Fig. 2 shows the root locus of the compensated system with $K_0' = 25$ and $\xi \approx 0.7$.

381

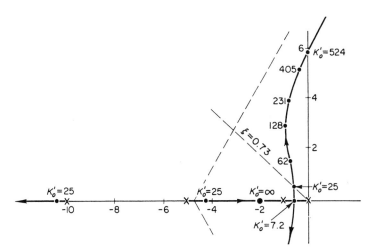

Fig. 2

For $K_0' = 25$, the damping ratio is $\xi = 0.73$, the natural frequencies of the system are $\omega_n = 0.9 \dfrac{rad}{sec}$ and $\omega_d = 0.6 \dfrac{rad}{sec}$ (damped).

The system's characteristic equation from the root locus, for $K_0' = 25$, is

$$(D + 10.4)(D + 4.3)(D + 0.65 + j0.6)(D + 0.65 - j0.6) = 0$$

or $\quad D^4 + 16D^3 + 65D^2 + 75D + 50 = 0$

The compensated system is stable for $K_0' < 524$.

FREQUENCY RESPONSE

• **PROBLEM** 8-12

Consider a closed-loop system whose closed-loop transfer function is nearly equal to the inverse of the feedback transfer function whenever the open-loop gain is much greater than unity. We can modify the open-loop charac- teristic by adding an internal feedback loop with a characteristic equal to the inverse of the desired open- loop characteristic. Assume that a unity-feedback system has the open-loop transfer function

$$G(s) = \frac{K}{(T_1 s+1)(T_2 s+1)}$$

Determine the transfer function $H(s)$ of the element in the internal feedback loop so that the inner loop becomes ineffective at both low and high frequencies.

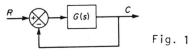

Fig. 1

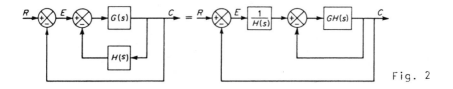

Fig. 2

Solution: The original system is shown in Fig. 1

The addition of the interval feedback loop around G(s) is shown in Fig. 2.

We have

$$\frac{C(s)}{E(s)} = \frac{G(s)}{1 + G(s)H(s)} = \frac{1}{H(s)} \cdot \frac{G(s)H(s)}{1 + G(s)H(s)}$$

When the gain around the inner loop is much larger than unity, then

$$\frac{G(s)H(s)}{1 + G(s)H(s)} \approx 1$$

and

$$\frac{C(s)}{E(s)} \approx \frac{1}{H(s)}$$

When $G(s)H(s) << 1$, then the inner loop becomes ineffective and

$$\frac{C(s)}{E(s)} \approx G(s).$$

For the inner loop to be ineffective at both the low- and high-frequency ranges, we require that

$$G(j\omega)H(j\omega) \ll 1 \quad , \quad \text{for } \omega \ll 1 \quad \text{and} \quad \omega \gg 1$$

In this problem,

$$G(j\omega) = \frac{K}{(1 + j\omega T_1)(1 + j\omega T_2)}$$

The requirement can be satisfied for

$$H(s) = ks$$

since

$$\lim_{\omega \to 0} G(j\omega)H(j\omega) = \lim_{\omega \to 0} \frac{Kk\, j\omega}{(1 + j\omega T_1)(1 + j\omega T_2)} = 0$$

383

$$\lim_{\omega \to 0} G(j\omega)H(j\omega) = \lim_{\omega \to 0} \frac{Kkj\omega}{(1 + j\omega T_1)(1 + j\omega T_2)} = 0$$

We see that for $H(s) = ks$ the inner loop becomes ineffective at both the low and high frequency regions.

BILINEAR TRANSFORM

● **PROBLEM** 8-13

For the system shown in Fig. 1 with

$$G_p(s) = \frac{K}{s(s+1)}$$

with $K = 1.57$, $T = 1.57$ sec.,

design the system so that it will have a phase margin of 45°.

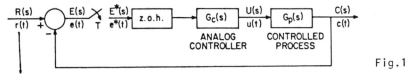

Fig.1

Solution: The z-transform of the open-loop transfer function is

$$z[G_{ho}(s)G_p(s)] = G_{ho}G_p(z) = (1-z^{-1})z\left[\frac{K}{s^2(s+1)}\right]$$

$$= \frac{1.22(z + 0.598)}{(z-1)(z-0.208)}$$

The w-transform of the last equation is

$$G_{ho}G_p(w) = G_{ho}G_p(z)\Big|_{z = \frac{1+w}{1-w}} = \frac{1.232(1 + 0.251w)(1-w)}{w(1 + 1.525w)}$$

Fig. 2 shows the Bode diagram of $G_{ho}G(w)$. The phase-crossover frequency is at $\omega_w = 1$ and the uncompensated system has very low margin of stability. To achieve a phase margin of 45° we must move the gain crossover to $\omega_w = 0.4$. At this point the magnitude of $G_{ho}G_p(w)$ is approximately 8 db. Therefore, to realize the gain crossover of $\omega_w = 0.4$, the magnitude curve must be attenuated by 8 db. in the vicinity of $\omega_w = 0.4$. We should do that without reducing the gain of $G_{ho}G_p(w)$ at $\omega_w = 0$.

We select the "phase-lag" model for the continuous-data controller $G_c'(w)$, thus

$$G_c'(w) = \frac{1 + a\tau w}{1 + \tau w} \qquad (a < 1)$$

The required attenuation is 8 db, we set

$$20 \log_{10} a = -8\text{db}.$$

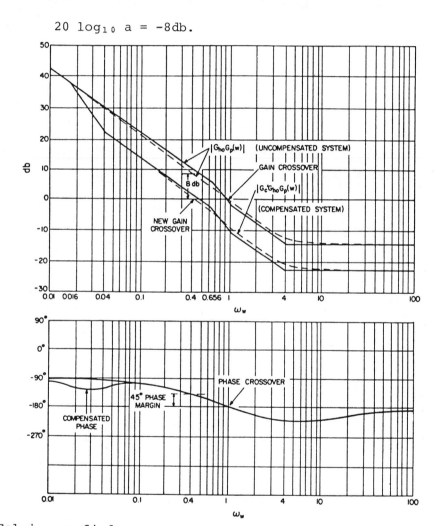

Fig.2

Solving we find

$$a = 10^{-\frac{8}{20}} = 0.398.$$

We set the value of $\frac{1}{a\tau}$ to be one tenth of the new gaincrossover frequency, then the phase lag of the controller produces negligible effect on the phase characteristics of the original system.

We have

$$\frac{1}{a\tau} = 0.04$$

or

$$\frac{1}{\tau} = 0.016$$

The transfer function of the controller is

$$G_c'(w) = \frac{1 + 25w}{1 + 62.8w}$$

thus

$$G_c'(w) \; G_{ho}G_p(w) = \frac{1.232(1+0.251w)(1-w)(1+25w)}{w(1+1.525w)(1.62.8w)} \tag{1}$$

Substituting $w = \frac{z-1}{z+1}$ into eq. (1) we find the z-transform of the open-loop transfer function of the compensated system. We have

$$G_{ho}\,G_c\,G_p(z) = G_c'(w)G_{ho}G_p(w)\Big|_{w\,=\,\frac{z-1}{z+1}}$$

$$= \frac{0.497(z-0.92)(z+0.599)}{(z-0.969)(z-1)(z-0.208)}$$

The closed-loop transfer function of the compensated system is

$$\frac{C(z)}{R(z)} = \frac{G_{ho}G_cG_p(z)}{1 + G_{ho}G_cG_p(z)} = \frac{0.497z^2-0.16z-0.274}{z^3-1.68z^2 + 1.219z-0.476}$$

Fig. 3 shows the unit-step response of the compensated system.

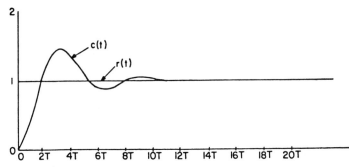

Fig.3

We have now to determine the transfer function $G_c(s)$ of the continuous-data controller. From

386

$$z\left[\frac{G_c(s)G_p(s)}{s}\right]\Bigg|_{z=\frac{1+w}{1-w}} = \frac{w+1}{2w}\,G_c'(w)\,G_{h_o}G_p(w)$$

we get

$$z\left[\frac{G_c(s)G_p(s)}{s}\right]\Bigg|_{z=\frac{1+w}{1-w}} = \frac{w+1}{2w}\,\frac{1.232(1+0.251w)(1-w)(1+25w)}{w(1+1.525w)(1+62.8w)} \quad (2)$$

We shall expand the function using the partial fraction expansion. Note that the transfer function

$$\frac{G_p(s)}{s}$$

has two poles at s=0. Table 1 shows that corresponding to $G(s) = \frac{1}{s^2}$, $G(w)$ has the term $(1+w)(1-w)$ in its numerator, thus the partial fraction expansion of

$$\frac{G_c'(w)G_{ho}G_p(w)}{1-w}$$

should preserve $(1+w)(1-w)$.
We have

$$\frac{G_c'(w)G_{ho}G_p(w)}{1-w} = \frac{1.232(1+0.251w)(1+25w)}{w(1+1.525w)(1+62.8w)}$$

$$= \frac{0.0807(w+3.984)(w+0.04)}{w(w+0.656)(w+0.016)}$$

$$= \frac{7.688}{w} + \frac{0.192}{w+0.656} - \frac{7.881}{w+0.016}$$

Substituting into eq. (2) we obtain

$$z\left[\frac{G_c(s)G_p(s)}{s}\right]\Bigg|_{z=\frac{1+w}{1-w}} = \frac{7.688(1+w)(1-w)}{2w^2} + \frac{0.293(1+w)(1-w)}{2w(1+1.525w)}$$

$$- \frac{492.56(1+w)(1-w)}{2w(1+62.5w)} \quad (3)$$

From Table 1 we find the transform pairs for the terms in eq. (3).

$$\frac{G_c(s)G_p(s)}{s} = \frac{7.688}{s^2} + \frac{0.293}{s(s+1)} - \frac{10.04}{s(s+0.0204)}$$

$$= \frac{0.1568 - 2.188s - 2.058s^2}{s^2(s+1)(s+0.0204)}$$

we get

$$G_c(s) = \frac{0.1 - 1.394s - 1.31s^2}{s + 0.0204}$$

To make $G_c(s)$ physically realizable, we add a remote pole at $s = -10$, thus

$$G_c(s) = \frac{1 - 13.94s - 13.1s^2}{(s+0.0204)(s+10)}$$

Table 1

Laplace transform $G(s)$	z-transform $G(z)$	w-transform $G(w)$
$\dfrac{1}{s}$	$\dfrac{z}{z-1}$	$\dfrac{w+1}{2w}$
$\dfrac{1}{s^2}$	$\dfrac{Tz}{(z-1)^2}$	$\dfrac{T(1+w)(1-w)}{4w^2}$
$\dfrac{1}{s^3}$	$\dfrac{T^2z(z+1)}{2(z-1)^3}$	$\dfrac{T^2(1+w)(1-w)}{8w^3}$
$\dfrac{1}{s+a}$	$\dfrac{z}{z-e^{-aT}}$	$\dfrac{1+w}{(1-e^{-aT})\left[1+\dfrac{1+e^{-aT}}{1-e^{-aT}}w\right]}$
$\dfrac{1}{(s+a)^2}$	$\dfrac{Tze^{-aT}}{(z-e^{-aT})^2}$	$\dfrac{(1+w)(1-w)Te^{-aT}}{(1-e^{-aT})^2\left[1+\dfrac{1+e^{-aT}}{1-e^{-aT}}w\right]^2}$
$\dfrac{a}{s(s+a)}$	$\dfrac{(1-e^{-aT})z}{(z-1)(z-e^{-aT})}$	$\dfrac{(1+w)(1-w)}{2w\left[1+\dfrac{1+e^{-aT}}{1-e^{-aT}}w\right]}$
$\dfrac{a}{s^2(s+a)}$	$\dfrac{Tz}{(z-1)^2} - \dfrac{1-e^{-aT}}{a(z-1)(z-e^{-aT})}$	$\dfrac{T(1-w)(1+w)}{4w^2} - \dfrac{(1-w)(1+w)}{2aw\left[1+\dfrac{1+e^{-aT}}{1-e^{-aT}}w\right]}$
$\dfrac{\omega}{s^2+\omega^2}$	$\dfrac{z\sin\omega T}{z^2-2z\cos\omega T+1}$	$\dfrac{(1-w)(1+w)\sin\omega T}{2[(1+w^2)-(1-w^2)\cos\omega T]}$
$\dfrac{\omega}{(s+a)^2+\omega^2}$	$\dfrac{ze^{-aT}\sin\omega T}{z^2-2ze^{-aT}\cos\omega T+e^{-2aT}}$	$w\dfrac{\dfrac{(1+w)(1-w)e^{-aT}\sin\omega T}{(1+w)^2-2(1+w)(1-w)}}{e^{-aT}\cos\omega T+(1-w)e^{-2aT}}$

COMPENSATION

Consider a situation of a disturbance in the performance
of a plant. Some time will pass before any effect of
this disturbance on the output of the plant can be detected.
This malfunction can be corrected sooner if we measure the
disturbance itself and not the response to the disturbance.
A block diagram of the feedforward compensation for the
disturbance is shown in the figure. Examine the limitations
of the disturbance-feedforward scheme in general and the
advantages of the scheme shown.

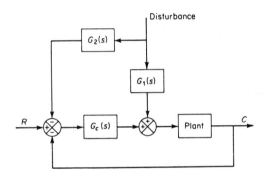

Solution: A disturbance-feedforward scheme depends on the
constancy of the parameter values since it is an open-loop
scheme. Imperfect compensation will occur whenever these
values will change. In the system shown in the figure,
open-loop and closed-loop schemes operate simultaneously.
Large errors caused by the main disturbance source can be
reduced by the open-loop compensation without a high loop
gain and the small errors caused by other disturbance sources
can be eliminated by the closed-loop control scheme. Thus we can
reduce all errors without requiring a large loop gain.

● **PROBLEM** 8-15

For the signal flow graph shown in Fig. 1:

a) Determine the unit-impulse, the unit-step and zero-
   input responses.

b) Assuming that the step response has an initial magnitude of 1/4 and a time constant of 5 secs, find the
   a and K actually used.

c) Using the results of b), we have to compensate further
   and reach the time constant of 2.5 secs.

Solution: From the signal flow graph we have

$$H(s) = \frac{-s}{s+a} \cdot \frac{\frac{1}{K}}{1 - \left(\frac{1}{K}\right)\left(\frac{-1}{s+a}\right)} = \frac{-s}{Ks + Ka + 1} = \frac{-\frac{s}{K}}{s+a + \frac{1}{K}}$$

389

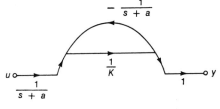

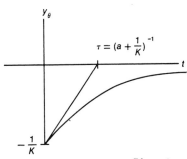

Fig. 1

Fig. 2

a) We compute the following responses:

The unit-step response,

$$H(s) \frac{1}{s} = \frac{-\frac{1}{K}}{s + \left(a + \frac{1}{K}\right)} \Rightarrow \left. Y_\theta \right|_{u=1(t)} = -\frac{1}{K} e^{-\left(a + \frac{1}{K}\right)t} 1(t);$$

The unit-impulse response

$$\left. Y_\theta \right|_{u=\delta} = \frac{a + \frac{1}{K}}{K} e^{-\left(a + \frac{1}{K}\right)t} 1(t) + -\left(\frac{1}{K}\right)\delta(t);$$

The zero-input response

$$\frac{\beta_0}{s + \left[a + \frac{1}{K}\right]} \Rightarrow Y_0 = \beta_0 e^{-\left(a + \frac{1}{K}\right)t} 1(t)$$

b) We have

$$\left. Y_\theta \right|_{u=1(t)} = -\frac{1}{K} e^{-\left(a + \frac{1}{K}\right)t} 1(t)$$

which is represented in Fig. 2.

We get

$$-\frac{1}{K} = -\frac{1}{4}, \quad K = 4$$

and

$$\frac{1}{a + \frac{1}{K}} = 5$$

Solving for a we obtain

$$a = -\frac{1}{20}$$

Note that if $K = -4$, then $a = \dfrac{9}{20} = 0.45$

c)   We can achieve the required result by

1) Cancellation compensation with no feedback.

2) Feedback with simple transfer functions.

3) Insertion of gain in a forward loop of feedback configuration.

● **PROBLEM** 8-16

Fig. 1 shows the root locus for a Type 2 system with a forward transfer function

$$G_x(s) = \frac{K}{s^2\left(s + \dfrac{1}{T_1}\right)}$$

Discuss the stability of the system. Suppose a zero were added at $s = \dfrac{-1}{T_2}$ between the origin and the pole at $-\dfrac{1}{T_1}$. Sketch the new root locus. Determine the stability of the system.

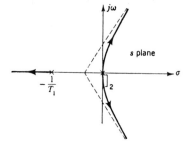

Fig. 1

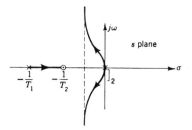

Fig. 2

Solution: Note that in Fig. 1 an entire branch of the locus is in the right half-plane. Thus for all values of K the system is unstable. With the addition of the zero at $-\dfrac{1}{T_2}$, the number of poles minus the number of zeros is two. Therefore, there is a vertical asymptote located at $s = \dfrac{T_1 - T_2}{2T_1T_2}$. For $K = 0$, the locus passes through the origin, and the locus on the real axis is between $-\dfrac{1}{T_1}$ and $-\dfrac{1}{T_2}$.

Fig. 2 shows the new locus. Since no part of the locus is in the right half-plane, the resulting system is stable for all values of K. In this example the introduction of a zero in the forward transfer function by an ideal derivative plus proportional compensator has improved the feedback control system.

Find an output feedback matrix which gives closed-loop eigenvalues at $\lambda_1 = -5$, $\lambda_2 = -6$ for an unstable but completely controllable system described by

$$A = \begin{bmatrix} -2 & 1 & 0 \\ 0 & -2 & 0 \\ 0 & 0 & 4 \end{bmatrix}, \quad B = \begin{bmatrix} 0 & 0 \\ 0 & 1 \\ 1 & 0 \end{bmatrix}$$

$$C = \begin{bmatrix} 0 & 0 & 1 \\ 1 & 0 & 0 \end{bmatrix}, \quad D = \begin{bmatrix} 1 & 0 \\ 0 & 0 \end{bmatrix}.$$

Solution: With output feedback, the state equation is

$$\dot{x} = \{A-BK'[I+DK']^{-1}C\}\, x + B[I+K'D]^{-1}u$$

The characteristic equation is

$$\Delta'(\lambda) = |\lambda I-A| \cdot | I + K'[I+DK']^{-1}C[\lambda I-A]^{-1}B|$$

Let

$$C[\lambda I-A]^{-1} B = \Psi'(\lambda)$$

and let $\Psi'_j(\lambda_i)$ be the jth column of $\Psi'(\lambda_i)$.

If n linearly independent $\Psi'_j(\lambda_i)$ can be found, where n is the number of outputs, we may form the matrices G' consisting of these $\Psi'_j(\lambda_i)$, and F' consisting of the corresponding jth columns of I.

Then

$$K' = -[I+F'(G')^{-1}D]^{-1}F'(G')^{-1} .$$

In our case we have

$$\Psi'(\lambda) = \begin{bmatrix} 0 & 0 & 1 \\ 1 & 0 & 0 \end{bmatrix} \begin{bmatrix} \lambda+2 & -1 & 0 \\ 0 & \lambda+2 & 0 \\ 0 & 0 & \lambda-4 \end{bmatrix}^{-1} \begin{bmatrix} 0 & 0 \\ 0 & 1 \\ 1 & 0 \end{bmatrix}$$

$$= \begin{bmatrix} \dfrac{1}{\lambda-4} & 0 \\ 0 & \dfrac{1}{(\lambda+2)^2} \end{bmatrix}$$

Since $\lambda_1 = -5$ and $\lambda_2 = -6$ we get

$$\Psi_1'(-5) = \begin{bmatrix} -\dfrac{1}{9} \\ 0 \end{bmatrix} \quad \text{and} \quad \Psi_2'\left(-6\right) = \begin{bmatrix} 0 \\ \dfrac{1}{16} \end{bmatrix}$$

and

$$G' = \begin{bmatrix} -\dfrac{1}{9} & 0 \\ 0 & \dfrac{1}{16} \end{bmatrix} \quad \text{and}$$

$$(G')^{-1} = \begin{bmatrix} -9 & 0 \\ 0 & 16 \end{bmatrix}$$

$$F' = \begin{bmatrix} 1 & 0 \\ 0 & 1 \end{bmatrix}.$$

For K' we obtain

$$K' = -\left\{ \begin{bmatrix} 1 & 0 \\ 0 & 1 \end{bmatrix} + \begin{bmatrix} 1 & 0 \\ 0 & 1 \end{bmatrix} \begin{bmatrix} -9 & 0 \\ 0 & 16 \end{bmatrix} \begin{bmatrix} 1 & 0 \\ 0 & 0 \end{bmatrix} \right\}^{-1} \begin{bmatrix} 1 & 0 \\ 0 & 1 \end{bmatrix} \begin{bmatrix} -9 & 0 \\ 0 & 16 \end{bmatrix}$$

$$= -\begin{bmatrix} -8 & 0 \\ 0 & 1 \end{bmatrix}^{-1} \begin{bmatrix} -9 & 0 \\ 0 & 16 \end{bmatrix} = \begin{bmatrix} \dfrac{1}{8} & 0 \\ 0 & -1 \end{bmatrix} \begin{bmatrix} -9 & 0 \\ 0 & 16 \end{bmatrix} = \begin{bmatrix} -\dfrac{9}{8} & 0 \\ 0 & -16 \end{bmatrix}$$

The characteristic equation is

$$\left| \lambda I - A + BK'[I + DK']^{-1}C \right| = \Delta'(\lambda)$$

Substituting we obtain

$$(\lambda + 5)(\lambda + 6)(\lambda - 2) = \Delta'(\lambda)$$

Thus, the system is still unstable.

A system is described by

$$\dot{x} = Ax + Bu$$

where

$$A = \begin{bmatrix} 0 & 2 \\ 0 & 3 \end{bmatrix} , \quad B = \begin{bmatrix} 1 & 0 \\ 0 & 1 \end{bmatrix}$$

It is desired that the closed-loop eigenvalues be $\lambda_1 = -3$, $\lambda_2 = -5$. To accomplish this, state feedback is used, so that $\dot{x}$ becomes

$$\dot{x} = (A-BK)x + Bu$$

Find the value of K.

Solution: The characteristic equation of the system is

$$|\lambda I - A - BK| = 0$$

Let us write it in the form

$$|\lambda I - A||I + (\lambda I - A)^{-1} BK| = 0$$

and denote

$$(\lambda I - A)^{-1}B = D(\lambda).$$

We shall determine K from the equation

$$|I + D(\lambda)K| = 0$$

Note that

$$|I + D(\lambda)K| = |I + KD(\lambda)|$$

Let us denote the jth column of I by e_j , and the jth column of $D(\lambda)$ by d_j. When a column of a matrix is zero, its determinant is zero. Thus, we obtain

$$K d_j(\lambda) = -e_j .$$

We form n linearly independent columns $d_{j(i)}(\lambda_i)$, then we have

$$K = = [e_{j_1} \ e_{j_2} \ \cdots \ e_{jn}][d_{j_1}(\lambda_1)d_{j_2}(\lambda_2) \ \cdots \ d_{jn}(\lambda_n)]^{-1}$$

In our problem,

$$D(\lambda) = \begin{bmatrix} \lambda & -2 \\ 0 & \lambda-3 \end{bmatrix}^{-1} \begin{bmatrix} 1 & 0 \\ 0 & 1 \end{bmatrix} = \frac{1}{\lambda(\lambda-3)} \begin{bmatrix} \lambda-3 & 2 \\ 0 & \lambda \end{bmatrix}$$

We have

$$d_1(\lambda_1) = \begin{bmatrix} -\dfrac{1}{3} \\ 0 \end{bmatrix}, \quad d_1(\lambda_2) = \begin{bmatrix} -\dfrac{1}{5} \\ 0 \end{bmatrix}$$

$$d_2(\lambda_1) = \begin{bmatrix} \dfrac{1}{9} \\ -\dfrac{1}{6} \end{bmatrix}, \quad d_2(\lambda_2) = \begin{bmatrix} \dfrac{1}{20} \\ -\dfrac{1}{8} \end{bmatrix}$$

We get for D,

$$D = - \begin{bmatrix} 1 & 0 \\ 0 & 1 \end{bmatrix} \begin{bmatrix} -3 & -\dfrac{6}{5} \\ 0 & -8 \end{bmatrix} = \begin{bmatrix} 3 & \dfrac{6}{5} \\ 0 & 8 \end{bmatrix}$$

Note that to determine D we used $d_1(\lambda_1)$ and $d_2(\lambda_2)$ as the independent columns, the other choice is $d_1(\lambda_2)$ $d_2(\lambda_1)$.

We obtain the desired eigenvalues $\lambda_1 = -3$, $\lambda_2 = -5$.

● **PROBLEM** 8-19

Fig. 1 shows the signal-flow graph of the control system.

x_1 and x_2 are state variables. Since an undamped response results from a step input or disturbance signal, the performance of the system is not satisfactory.

The equation of the system is

$$
\begin{bmatrix} \dot{x}_1 \\ \dot{x}_2 \end{bmatrix} = \begin{bmatrix} 0 & 1 \\ 0 & 0 \end{bmatrix} \begin{bmatrix} x_1 \\ x_2 \end{bmatrix} + \begin{bmatrix} 0 \\ 1 \end{bmatrix} u(t) \tag{1}
$$

$$
A = \begin{bmatrix} 0 & 1 \\ 0 & 0 \end{bmatrix} \tag{2}
$$

We choose a feedback control system so that

$$
u(t) = -k_1 x_1 - k_2 x_2 \tag{3}
$$

Rewrite eq. (1) as $\dot{x} = Dx$ and find a symmetric matrix P such that

$$
D^T P + PD = -I
$$

The performance index is given by $J = X^T(0)\ P\ x(0)$. Find the value of k_2 that will minimize J, take

$$
X^T(0) = [1,1] \qquad \text{and} \qquad k_1 = 1 .
$$

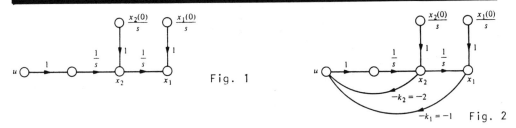

Fig. 1   Fig. 2

Solution: Substituting eq. (3) into eq. (1) we get

$$
\dot{x}_1 = x_2 \tag{4}
$$

$$
\dot{x}_2 = -k_1 x_1 - k_2 x_2 \tag{5}
$$

or

$$
\dot{x} = Dx = \begin{bmatrix} 0 & 1 \\ -k_1 & -k_2 \end{bmatrix} x \tag{6}
$$

We have

$$
D^T P + PD = -I
$$

or

396

$$\begin{bmatrix} 0 & -1 \\ 1 & -k_2 \end{bmatrix} \begin{bmatrix} p_{11} & p_{12} \\ p_{12} & p_{22} \end{bmatrix} + \begin{bmatrix} p_{11} & p_{12} \\ p_{12} & p_{22} \end{bmatrix} \begin{bmatrix} 0 & 1 \\ -1 & -k_2 \end{bmatrix} = \begin{bmatrix} -1 & 0 \\ 0 & -1 \end{bmatrix} \tag{7}$$

Multiplying we get the following equations:

$$-p_{12} - p_{12} = -1 \tag{8}$$

$$p_{11} - k_2 p_{12} - p_{22} = 0 \tag{9}$$

$$p_{12} - k_2 p_{22} + p_{12} - k_2 p_{22} = -1 \tag{10}$$

Thus,

$$p_{12} = \frac{1}{2} \ , \quad p_{22} = \frac{1}{k_2} \quad , \quad p_{11} = \frac{k_2^2 + 2}{2k_2} \tag{11}$$

The performance index is

$$J = x^T(0) \ P \ x(0) = [1,1] \begin{bmatrix} p_{11} & p_{12} \\ p_{12} & p_{22} \end{bmatrix} \begin{bmatrix} 1 \\ 1 \end{bmatrix} \tag{12}$$

$$= [1,1] \begin{bmatrix} (p_{11} + p_{12}) \\ (p_{12} + p_{22}) \end{bmatrix} = (p_{11} + p_{12}) + (p_{12} + p_{22}) = p_{11} + 2p_{12} + p_{22}$$

$$\tag{13}$$

Substituting the values of p_{ij} we get

$$J = \frac{k_2^2 + 2}{2k_2} + 1 + \frac{1}{k_2} = \frac{k_2^2 + 2k + 4}{2k_2} \tag{14}$$

To minimize J we take

$$\frac{dJ}{dk_2} = \frac{2k_2(2k_2 + 2) - 2(k_2^2 + 2k_2 + 4)}{(2k_2)^2} = 0 \tag{15}$$

Solving we find that J reaches a minimum for $k_2 = 2$.

From eq. (14) we have

$$J_{min} = 3.$$

The system matrix D is

$$D = \begin{bmatrix} 0 & 1 \\ -1 & -2 \end{bmatrix}$$ (17)

and the characteristic equation of the compensated system is

$$\det[\lambda I - D] = \det \begin{bmatrix} \lambda & -1 \\ 1 & \lambda+2 \end{bmatrix} = \lambda^2 + 2\lambda + 1$$ (18)

The characteristic equation of the second-order system is of the form $s^2 + 2\zeta\omega_n s + \omega_n^2$, thus the damping ratio of the compensated system is $\zeta = 1$. Fig. 2 shows the compensated system.

● **PROBLEM** 8-20

For the lag network shown below, derive the transfer function.

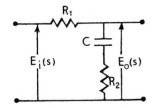

Solution: Using the Laplace version of the circuit elements, voltage division gives

$$\frac{E_0(s)}{E_i(s)} = \frac{R_2 + \dfrac{1}{sC}}{R_1 + R_2 + \dfrac{1}{sC}} = \frac{sCR_2 + 1}{s(R_1 + R_2)C + 1}$$

Let

$$(R_1 + R_2)C = T$$

and

$$\frac{R_2}{R_1 + R_2} = b$$

We have

$$\frac{E_0(s)}{E_i(s)} = \frac{1 + sbT}{1 + sT}$$

398

Since b < 1, the lag time constant in the denominator is larger than the lead time constant in the numerator.

The open-loop transfer function of the uncompensated system is

$$GH(s) = \frac{K}{s(s + 10)^2} \qquad (1)$$

The velocity constant of this system should be 20, while the damping ratio of the dominant roots is 0.707. The gain necessary for a K_V of 20 is

$$K_v = 20 = \frac{K}{(10)^2} \qquad (2)$$

We have K = 2000, and using the Routh's criterion, we find that for this value of K the roots of the characteristic equation lie on the $j\omega$-axis at $+ j10$. Thus the damping ratio is not satisfied. Satisfy the K_V and ζ requirements using a phase-lag network.

Solution: The uncompensated root locus of the system is shown in the sketch. The roots are shown when $\zeta = 0.707$ and $s = - 2.9 + j2.9$. Measuring the gain at these roots we find K = $23\overline{6}$. Next, we compute the necessary ratio of zero to pole of the conpensator.

$$\alpha = \left|\frac{z}{p}\right| = \frac{2000}{236} = 8.5 \qquad (3)$$

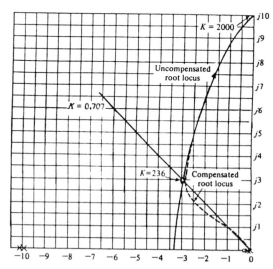

To allow a small margin of safety, we choose z = 0.1, $p = \frac{0.1}{9}$.

The difference between the angle from the pole and zero of $G_c(s)$ is negligible. For the compensated system we

399

have

$$G_c(s)GH(s) = \frac{236(s + 0.1)}{s(s + 10)^2(s + 0.0111)} \tag{4}$$

where $\frac{K}{\alpha} = 236$ and $\alpha = 9$.

● **PROBLEM** 8-22

Find the forward transfer function of the lead network shown below.

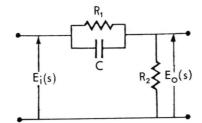

Solution: Let us find first the parallel impedance of R_1 and C.

$$Z = \frac{R_1 \dfrac{1}{sC}}{R_1 + \dfrac{1}{sC}} = \frac{R_1}{1 + sCR_1} = \frac{R_1}{1 + sT}$$

where $R_1C = T$

We obtain

$$\frac{E_o(s)}{E_i(s)} = \frac{R_2}{\dfrac{R_1}{1 + sT} + R_2} = \frac{R_2(1 + sT)}{R_1 + R_2 + R_2sT}$$

$$= \frac{R_2}{R_1 + R_2}\left(\frac{1 + sT}{1 + \dfrac{R_2}{R_1 + R_2}sT}\right)$$

Let

$$k = \frac{R_2}{R_1 + R_2}$$

The low-frequency attenuation of the network is

$$\frac{E_o(s)}{E_i(s)} = k\frac{1 + sT}{1 + ksT}$$

For $s = j\omega$ we obtain

400

$$\frac{E_o(j\omega)}{E_i(j\omega)} = k \cdot \frac{1 + j\omega T}{1 + jk\omega T}$$

The uncompensated open-loop transfer function is given by

$$GH(j\omega) = \frac{K_v}{j\omega (1 + j\frac{\omega}{4}) (1 + j\frac{\omega}{16})} \qquad (1)$$

Design a lag-lead network to provide an improved phase margin.

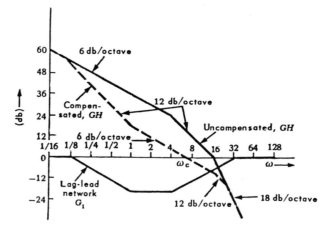

Solution: The uncompensated gain asymptote is shown in the figure.

In the vicinity of $\omega = 8 \frac{rad}{sec}$ additional phase margin is required for stability.

The design of the lag-lead network should provide required phase margin without additional amplifier gain. The transfer function of the lag-lead network is

$$G_1(j\omega) = \frac{(1 + j\omega) (1 + \frac{j\omega}{4})}{(1 + j8\omega) (1 + j\frac{\omega}{32})} \qquad (2)$$

The over-all, compensated open-loop transfer function is obtained through cascading GH and G_1.

$$G_1(j\omega) \, GH(j\omega) =$$

$$\frac{K_v (1 + j\omega)}{j\omega (1 + j8\omega) (1 + j\frac{\omega}{16}) (1 + j\frac{\omega}{32})} \qquad (3)$$

Note that the compensated curve has a slope of $6 \frac{db}{octave}$ at the cross-over point ω_c, which is shifted down-frequency from 16 to $6 \frac{rad}{sec}$. It is however permissible to cross the zero-db axis at a corner between a $6 \frac{db}{octave}$ and $12 \frac{db}{octave}$ region on the asymptotic attenuation diagram. When the asymptotes extend several octaves on either side of the corner, phase lag is approximately the average of $90°$ and $180°$, which is $135°$. This corresponds to a satisfactory phase margin of $45°$. We shall check the phase margin, checking the phase shift ϕ at $\omega_c = 6 \frac{rad}{sec}$

$$e^{j\phi} = e^{j[(\arctan 6 - \frac{\pi}{2} - \arctan 48 - \arctan \frac{3}{8} - \arctan \frac{3}{16}]} \tag{4}$$

or

$$\phi = -[(\frac{\pi}{2}) + \arctan 48 + \arctan 0.375 + \arctan 0.187$$
$$- \arctan 6] \tag{5}$$

We obtain

$$\phi = -129°$$

The phase margin is

$$\gamma = 180° - 129° = 51°.$$

● **PROBLEM** 8-24

Find the frequency at which the phase lead of a lead network is maximum. What is the maximum phase lead?

Solution: The transfer function for the lead network is

$$\frac{E_o}{E_i} (j\omega) = \frac{\alpha (1 + j\omega T)}{1 + j\omega\alpha T} \tag{1}$$

and the lead angle is

$$\phi = \arctan \omega T - \arctan \omega\alpha T = \phi_1 - \phi_2 \tag{2}$$

$$\tan \phi = \tan (\phi_1 - \phi_2) = \frac{\tan \phi_1 - \tan \phi_2}{1 + \tan \phi_1 \tan \phi_2} \tag{3}$$

$$= \frac{\omega T - \alpha\omega T}{1 + \alpha\omega^2 T^2} = \frac{\omega T(1 - \alpha)}{1 + \alpha\omega^2 T^2} = \frac{u(1 - \alpha)}{1 + \alpha u^2} \tag{4}$$

where $u = \omega T$ is dimensionless frequency. From the condition $\frac{d}{dx}(\frac{u}{v}) = 0$ we find the frequency where ϕ is a maximum, thus

$$\frac{d}{dx}(\frac{u}{v}) = \frac{vdu - udv}{v^2} = 0 \tag{5}$$

$$\frac{d}{du}(\tan \phi) = \frac{(1 + \alpha u^2)(1 - \alpha) - u(1 - \alpha)2\alpha u}{(1 + \alpha u^2)^2} \tag{6}$$

$$= 0$$

we get

$$1 + \alpha u^2 = 2\alpha u^2 \qquad \text{or } u = \sqrt{\frac{1}{\alpha}} \tag{7}$$

Since $u = \omega T$ we can write

$$\omega_{max} = \frac{u}{T} = \frac{1}{T\sqrt{\alpha}} \tag{8}$$

Substituting ω_{max} into eq. (1) and solving for the phase lead we obtain.

$$G(j\omega_{max}) = \frac{\alpha(1 + j\frac{1}{T\sqrt{\alpha}}T)}{1 + j\alpha\frac{1}{T\sqrt{\alpha}}T}$$

$$= \frac{\alpha(1 + j\frac{1}{\sqrt{\alpha}})}{1 + j\sqrt{\alpha}} \tag{9}$$

The maximum phase lead is

$$\phi_{max} = \text{arc tan} \frac{1}{\sqrt{\alpha}} - \text{arc tan} \sqrt{\alpha} \tag{10}$$

or

$$\tan \phi_{max} = \frac{\frac{1}{\sqrt{\alpha}} - \sqrt{\alpha}}{1 + \frac{1}{\sqrt{\alpha}}\sqrt{\alpha}} = \frac{1 - \alpha}{2\sqrt{\alpha}} \tag{11}$$

or

$$\phi_{max} = \text{arc tan} (\frac{1 - \alpha}{2\sqrt{\alpha}}) \tag{12}$$

This can be expressed as

$$\phi_{max} = \text{arc sin } (\frac{1 - \alpha}{1 + \alpha})$$ (13)

The frequency response plots for a lead network are shown.

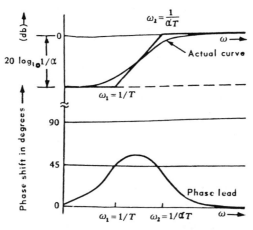

● PROBLEM 8-25

The open-loop transfer function GH of the Type 2 servomechanism is

$$G(j\omega) = \frac{K_a}{(j\omega)^2 (1 + j \frac{\omega}{4})}, \qquad K_a = \frac{1}{4}$$ (1)

$$H(j\omega) = 1.$$ (2)

Show the effect on stability of adding the lead network

$$G_1(j\omega) = \frac{1}{16}(\frac{1 + j 4\omega}{1 + j \frac{\omega}{4}})$$ (3)

and the preamplifier

$$K_{pa} = 16$$ (4)

to compensate for low-frequency attenuation of the lead network in series with the open-loop transfer function. Fig. 1 shows the modified system. Sketch Bode asymptotic plots and Nyquist plots, with and without G_1.

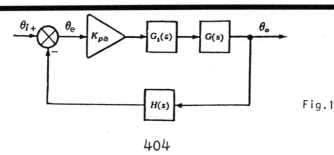

Fig.1

Solution: The loop gain, from eqs. (1), (2), (3) and (4) is

$$GH(j\omega) = G(j\omega)H(j\omega)G_1(j\omega)K_{pa}$$

$$= \frac{\frac{1}{4}(1 + j4\omega)}{(j\omega)^2(1 + j\frac{\omega}{4})^2} \qquad (5)$$

Fig. 2 shows the asymptotic attenuation and phase shift curves. The same curves sketched in polar coordinates are shown in Fig. 3. The uncompensated servo is unstable. Its initial (low-frequency) phase lag is 180° due to the fact that it is second order servo. Note that the lead network provides enough phase lead for a safe phase margin at the cross-over frequency of the compensated unit.

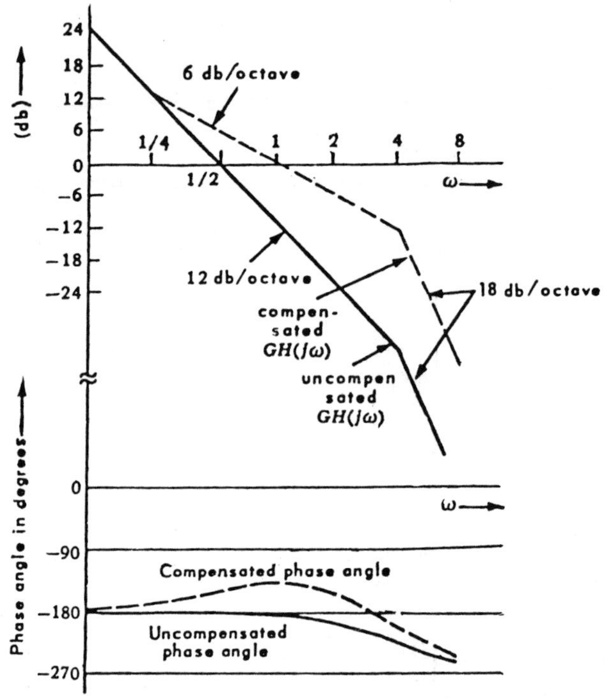

Fig.2

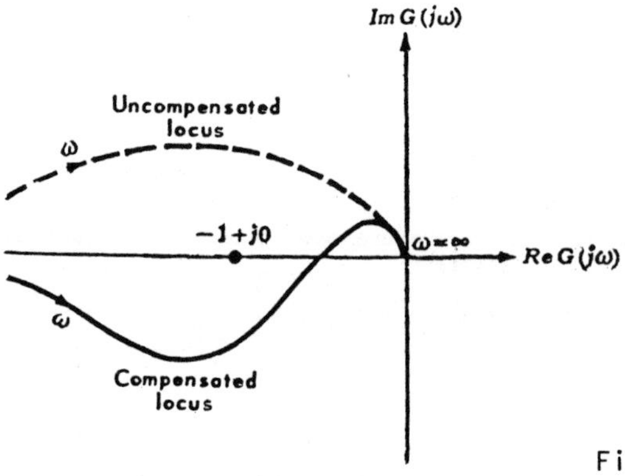

Fig.3

Determine the frequency at which the network lag angle is a maximum and then at this point compute the angle of lag.

Solution: The transfer function of a lag network is

$$\frac{E_o}{E_i}(j\omega) = \frac{1 + j\omega\alpha T}{1 + j\omega T} \tag{1}$$

We have

$$\phi \text{ (phase lag)} = \text{arc tan } \omega T - \text{arc tan } \alpha\omega T \tag{2}$$

Let $\phi_1 = \text{arc tan } \omega T$ and

$\phi_2 = \text{arc tan } \alpha\omega T$

Thus, we obtain

$$\phi = \phi_1 - \phi_2 \quad \text{and}$$

$$\tan \phi = \tan (\phi_1 - \phi_2) = \frac{\tan \phi_1 - \tan \phi_2}{1 + \tan \phi_1 \tan \phi_2}$$

$$= \frac{\omega T - \alpha\omega T}{1 + \alpha\omega^2 T^2} = \frac{\omega T(1 - \alpha)}{1 + \alpha\omega^2 T^2} \tag{3}$$

Eq. (3) gives the lag angle in terms of significant network parameters.

Phase lag ϕ is maximum where its tangent is maximum. Taking the derivative, we obtain

$$\frac{d(\tan \phi)}{d\omega}$$

$$= \frac{(1 + \alpha\omega^2 T^2) T(1 - \alpha) - \omega T(1 - \alpha) 2\alpha T^2 \omega}{(1 + \alpha\omega^2 T^2)^2} \tag{4}$$

From the condition $\dfrac{d(\tan \phi)}{d\omega} = 0$

we find

$$(1 + \alpha\omega^2 T^2) T(1 - \alpha)$$

$$= \alpha T(1 - \alpha)(2\alpha T^2 \omega) \tag{5}$$

Solving for ω,

$$1 + \alpha\omega^2 T^2 = 2\alpha T^2 \omega^2 \tag{6}$$

$$\omega_{max} = \sqrt{\frac{1}{\alpha T^2}} = \frac{1}{T}\sqrt{\frac{1}{\alpha}} \tag{7}$$

Note that the ratio of low- and high- corner frequencies is α and thus the value of ω for maxium phase lag is the geometric mean of the corner frequencies. On a logarithmic frequency scale, ω is located exactly midway between the corner frequencies. The Bode plot of the network is shown in Fig. 1.

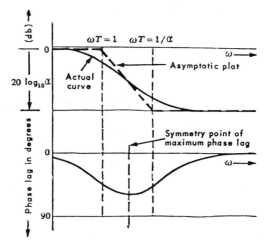

Fig.1

We determine the phase lag by substituting the derived value of ω in eq. (3).

$$\tan\phi_{max} = \frac{\omega T(1-\alpha)}{1+\alpha\omega^2 T^2} = \frac{\frac{1}{\sqrt{\alpha}}(1-\alpha)}{1+\alpha\left(\frac{1}{\alpha}\right)}$$

$$= \frac{1-\alpha}{2\sqrt{\alpha}} \tag{8}$$

$$\phi_{max} = \arctan\left(\frac{1-\alpha}{2\sqrt{\alpha}}\right) \tag{9}$$

or more conveniently

$$\phi_{max} = \arcsin\left(\frac{1-\alpha}{1+\alpha}\right) \tag{10}$$

● **PROBLEM** 8-27

Draw the polar plot and derive the transfer function of the lag-lead network shown in Fig. 1.

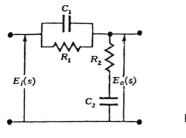

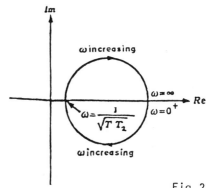

Fig.1

Fig.2

Solution: Using Laplace transform notation, the impedance of the parallel combination of R_1 and C_1 is

$$Z_p = \frac{R_1(\frac{1}{sC_1})}{R_1 + \frac{1}{sC_1}} \tag{1}$$

$$= \frac{R_1}{sC_1 R_1 + 1} = \frac{R_1}{1 + sT_1} \tag{2}$$

where $T_1 = C_1 R_1$.

The series impedance of R_2 and C_2 is

$$Z_{ser} = R_2 + \frac{1}{C_2 s} \tag{3}$$

Treating the network as a voltage divider

$$\frac{E_o}{E_i}(s) = \frac{Z_{ser}}{Z_{ser} + Z_p} \tag{4}$$

$$= \frac{R_2 + \frac{1}{sC_2}}{R_2 + \frac{1}{sC_2} + \frac{R_1}{1 + sT_1}}$$

$$= \frac{\frac{sC_2 R_2 + 1}{sC_2}}{\frac{sC_2 R_2 + 1}{sC_2} + \frac{R_1}{1 + sT_1}} \tag{5}$$

408

$$= \frac{(1 + sT_2)\ (1 + sT_1)}{(1 + sT_2)(1 + sT_1)+sT_{12}} \tag{6}$$

where $T_2 = R_2 C_2$, $\qquad T_{12} = R_1 C_2$

In the polynomial form, we have

$$\frac{E_o}{E_i}\ (s) = \frac{(1 + sT_1)\ (1 + sT_2)}{s^2 T_1 T_2 + S(T_1 + T_2 + T_{12}) + 1}$$

$$= \frac{s^2 T_1 T_2 + s(T_1 + T_2) + 1}{s^2 T_1 T_2 + s(T_1 + T_2 + T_{12}) + 1} \tag{7}$$

In terms of angular frequency ω, we obtain

$$\frac{E_o}{E_i}\ (j\omega) = \frac{(1 - \omega^2 T_1 T_2)\ + j\omega (T_1 + T_2)}{(1 - \omega^2 T_1 T_2)\ + j\omega (T_1 + T_2 + T_{12})} \tag{8}$$

At $\omega = 0$

$$\frac{E_o}{E_i}(j\omega) = 1 \qquad \text{with zero phase shift} \tag{9}$$

At $\omega = \infty$

$$\frac{E_o}{E_i}(j\omega) = 1 \qquad \text{with zero phase shift} \tag{10}$$

At $1 - \omega^2 T_1 T_2 = 0$ the phase shift is zero, but the amplitude is attenuated by the factor

$$\frac{T_1 + T_2}{T_1 + T_2 + T_{12}}$$

The term $\omega^2 T_1 T_2$ may be ignored for small ω. The transfer function for small ω becomes

$$\left. \frac{E_o}{E_i}(j\omega) \right|_{j\omega\ \text{small}}$$

$$= \frac{1 + j\omega (T_1 + T_2)}{1 + j\omega (T_1 + T_2 + T_{12})} \tag{11}$$

The polar plot of the transfer function is shown in Fig. 2. At low frequency the lag-lead network behaves like a lag network and at high frequency like a lead network.

For $\omega < \dfrac{1}{\sqrt{T_1 T_2}}$ lag network behavior occurs and for

$\omega > \dfrac{1}{\sqrt{T_1 T_2}}$ lead network behavior occurs.

The term $\dfrac{T_1 + T_2}{T_1 + T_2 + T_{12}}$ must be much less than unity if significant phase lag is to be accomplished.

In design, eq. (7) is more conveniently expressed as

$$\frac{E_o}{E_i}(s) = \frac{(1 + sT_1)\,(1 + sT_2)}{(1 + sT_A)\,(1 + sT_B)} \tag{12}$$

The constants T_A and T_B are found by equating coefficients, thus

$$T_A T_B = T_1 T_2 \tag{13}$$

$$T_A + T_B = T_1 + T_2 + T_{12} \tag{14}$$

The equations give a unique solution for T_A and T_B

● **PROBLEM** 8-28

Given the system shown in Fig. 1, the static velocity error coefficient 80 sec^{-1}, the transient response curves for the uncompensated system and the compensated system, Fig. 2 unit-step response curves, and Fig. 3 unit-ramp response curves, design an appropriate lag-lead compensator. The damping ratio and the undamped natural frequency of the dominant closed-loop poles are 0.5 and 5 respectively.

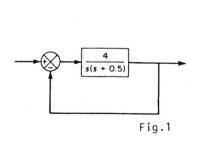

Fig.1

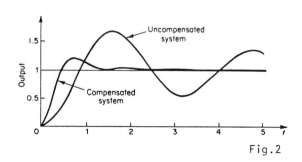

Fig.2

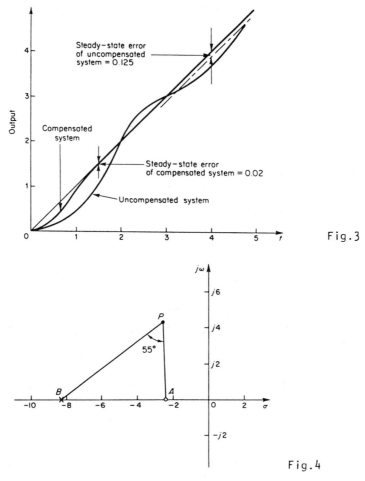

Fig.3

Fig.4

<u>Solution</u>: The requirement of the static velocity error
coefficient is that $K_V = 80 \text{ sec}^{-1}$.

From $\qquad K_V = \lim_{s \to 0} s K_c G(s)$

then $\qquad 80 = \lim_{s \to 0} K_c \dfrac{s^4}{s(s + 0.5)}$

$\qquad\qquad = \lim_{s \to 0} \dfrac{4K_c}{s + 0.5} = \dfrac{4K_c}{0.5}$

or $\qquad K_c = 10$

Then from $\zeta = \dfrac{1}{2}$, $\omega_n = 5$ and $s^2 + 2\zeta\omega_n s + \omega_n^2$, the location
of the closed loop dominant poles are at $s = -2.5$
$\pm j4.33$. The time constant T_1 and the value of β are
determined from the required phase and magnitude conditions.
Since

411

$$\angle \frac{4}{s(s + 0.5)} \Bigg|_{s = -2.5 + j4.33} = - 235°$$

so for stability the angle condition must be - 180° or less, the compensation must add 55° to the system at that point. Also at that point the magnitude must be 0 db or 1, resulting in the following:

$$\left| \frac{s + \dfrac{1}{T_1}}{s + \dfrac{\beta}{T_1}} \right| \left| \frac{40}{s(s + 0.5)} \right|_{s = - 2.5 + j \, 4.33}$$

$$= \left| \frac{s + \dfrac{1}{T_1}}{s + \dfrac{\beta}{T_1}} \right| \frac{5}{3} = 1$$

$$\angle \left| \frac{s + \dfrac{1}{T_1}}{s + \dfrac{\beta}{T_1}} \right|_{s = - 2.5 + j \, 4.33} = 55°$$

From the angle and magnitude conditions, referring to Fig. 4, we can easily locate points A and B so that

$$\angle APB = 55°, \quad \frac{\overline{PA}}{\overline{PB}} = \frac{3}{5}$$

The result is

$$\overline{AO} = 2.4, \quad \overline{BO} = 8.3$$

or

$$T_1 = \frac{1}{2.4} = 0.416, \qquad \beta = 8.3T_1 = 3.45$$

Thus, the phase lead portion of the lag-lead is

$$\frac{s + 2.4}{s + 8.3}$$

We may choose

$$T_2 = 10$$

for the phase lag portion

The lag-lead compensator becomes

$$G_c(s) = \left(\frac{s + 2.4}{s + 8.3} \right) \left(\frac{s + 0.1}{s + 0.029} \right) \tag{10}$$

The open-loop transfer function of the compensated system is

$$G_c(s)G(s) = \frac{(s + 2.4)(s + 0.1)40}{(s + 8.3)(s + 0.029)\,s(s + 0.5)}$$

The compensated system is of fourth order since no cancellation occurs.

The dominant closed-loop poles are located very near the desired location, because the angle contribution of the phase lag portion of the lag-lead network is small.

The two other closed-loop poles are

$$s = -0.09 \qquad \text{and } s = -3.74$$

Since the closed-loop pole at $s = -0.09$ and zero at $s = -0.1$ almost cancel each other, the effect of this closed-loop pole is very small. The other closed-loop pole, $s = -3.74$, is relatively close to the zero at $s = -2.4$; thus its effect on the transient response will be relatively small. Thus the dominant closed-loop poles are the poles at

$$s = -2.5 \pm j\,4.33$$

● **PROBLEM** 8-29

Fig. 1 shows the control system with the feedforward transfer function

$$G(s) = \frac{4}{s(s+0.5)}$$

The closed-loop poles of the system are

$$s = -0.25 \pm j\,1.98.$$

The static velocity error coefficient is 8 $\sec^{-1}$, the undamped natural frequency is 2 $\frac{rad}{sec}$, and the damping ratio is 0.125.

Make the damping ratio of the dominant closed-loop poles equal to 0.5 and increase the undamped natural frequency to 5 $\frac{rad}{sec}$ and the static velocity error coefficient to 50^{-1}.

Design an appropriate compensator to meet all the performance specifications.

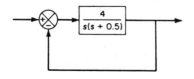

Fig.1

413

Solution: The dominant closed-loop poles are found by using the general form of the closed loop characteristic equation

$$s^2 + 2\zeta\omega_n s + \omega_n^{\;2}$$

where ζ is the damping ratio and ω_n is the undamped natural frequency. From the specifications $\zeta = \frac{1}{2}$ and $\omega_n = s$, we obtain

$$s^2 + 5s + 25.$$

Then the dominant closed-loop poles are at

$$s = -2.5 \pm j\,4.33.$$

Since

$$\left/ \underline{\frac{4}{s(s+0.5)}} \right|_{s = -2.5 + j\,4.33} = -235°$$

the phase lead portion of the lag-lead network must contribute 55°.

The lag-lead compensator has the transfer function

$$G_c(s) = \left(\frac{s + \dfrac{1}{T_1}}{s + \dfrac{\beta}{T_1}} \right) \left(\frac{s + \dfrac{1}{T_2}}{s + \dfrac{1}{\beta T_2}} \right) K_c$$

and the transfer function of the compensated system is

$$G_c(s)G(s) = \left(\frac{s + \dfrac{1}{T_1}}{s + \dfrac{\beta}{T_1}} \right) \left(\frac{s + \dfrac{1}{T_2}}{s + \dfrac{1}{\beta T_2}} \right) K_c G(s)$$

The static velocity error coefficient becomes

$$K_v = \lim_{s \to 0} s\, G_c(s)G(s) = \lim_{s \to 0} s\, K_c\, G(s)$$

Since the static velocity error coefficient is $K_v = 50 \text{ sec}^{-1}$, we get

$$K_v = \lim_{s \to 0} s\, K_c\, G(s) = \lim_{s \to 0} \frac{s4K_c}{s(s+0.5)} = 8K_c = 50$$

Thus

$$K_c = 6.25$$

and the compensated system's open-loop transfer function is

$$G_c(s)G(s) = \left(\frac{s + \dfrac{1}{T_1}}{s + \dfrac{\beta}{T_1}}\right)\left(\frac{s + \dfrac{1}{T_2}}{s + \dfrac{1}{\beta T_2}}\right)\frac{25}{s(s+0.5)}$$

T_2 has to be large enough that

$$\left|\frac{s + \dfrac{1}{T_2}}{s + \dfrac{1}{\beta T_2}}\right|_{s = -2.5 + j\,4.33} \doteq |1|$$

From the condition on the closed-loop poles

$$s = -2.5 \pm j\,4.33$$

we obtain the magnitude condition

$$\left|G_c(s)G(s)\right|_{s = -2.5 + j\,4.33} = \left|\frac{s + \dfrac{1}{T_1}}{s + \dfrac{\beta}{T_1}}\right|\left\{\left|\frac{25}{s(s+0.5)}\right|\right\}_{s = -2.5 + j4.33}$$

$$= \left|\frac{s + \dfrac{1}{T_1}}{s + \dfrac{\beta}{T_1}}\right|\frac{5}{4.77} = 1$$

and the angle condition

$$\left/\frac{s + \dfrac{1}{T_1}}{s + \dfrac{\beta}{T_1}}\right._{s = -2.5 + j\,4.33} = 55°.$$

We determine the values of T_1 and β that satisfy the magnitude and angle conditions, graphically.

From Fig. 2 we locate points A and B

$$\underline{/APB} = 55°, \quad \frac{\overline{PA}}{\overline{PB}} = \frac{4.77}{5}$$

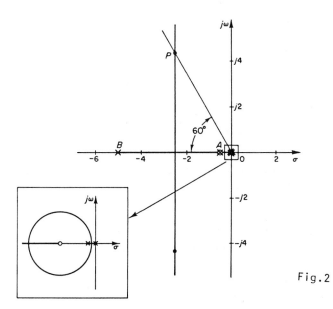

Fig.2

and

$$\overline{AO} = 0.5 , \quad \overline{BO} = 5$$

Hence

$$- \frac{1}{T_1} = -0.5 , \quad - \frac{\beta}{T_1} = -5$$

and

$$T_1 = 2 , \quad \beta = 10$$

The phase lead portion of the lag-lead network becomes

$$\frac{s + 0.5}{s + 5}$$

For the phase lag portion of the lag-lead network, it is required that

$$\left| \frac{s + \dfrac{1}{T_2}}{s + \dfrac{1}{10T_2}} \right| \doteqdot 1$$

$$s = -2.5 + j 4.33$$

$$0 < \left/ \frac{s + \dfrac{1}{T_2}}{s + \dfrac{1}{10T_2}} \right/ < 3°$$

$$S = -2.5 + j 4.33$$

In order to satisfy these relationships and simultaneously obtain the largest time constant ($10T_2$) of the lag-lead network which is not too large to be physically realized,

416

we choose

$$T_2 = 10.$$

The transfer function of the lag-lead compensator is

$$G_c(s) = \left(\frac{s + 0.5}{s + 5}\right)\left(\frac{s + 0.1}{s + 0.01}\right) \quad (6.25)$$

and the open-loop transfer function of the compensated system is

$$G_c(s)G(s) = \frac{(s+0.5)(s+0.1)25}{(s+5)(s+0.01)s(s+0.5)}$$

$$= \frac{25(s+0.1)}{s(s+5)(s+0.01)}$$

Note, that the compensated system is a third order system due to cancellation of the (s + 0.5) terms.

LAG COMPENSATION, ROOT LOCUS

● **PROBLEM** 8-30

Fig. 1 shows the system with the feedforward transfer function

$$G(s) = \frac{1.06}{s(s+1)(s+2)}$$

Fig. 2 shows the root-locus plot for the system.

The closed-loop transfer function is

$$\frac{C(s)}{R(s)} = \frac{1.06}{s(s+1)(s+2) + 1.06} = \frac{1.06}{(s+0.33-j0.58)(s+0.33+j0.58)(s+2.33)}$$

The dominant closed-loop poles are

$$s = -0.33 \pm j0.58,$$

the damping ratio of the dominant closed-loop poles is $\zeta = 0.5$, the undamped natural frequency of the dominant closed-loop poles is $0.67 \frac{rad}{sec}$, the static velocity error coefficient is 0.53 sec^{-1}. For the described system, increase the static velocity error coefficient K_v to about 5 sec^{-1} without significantly changing the location of the dominant closed-loop poles.

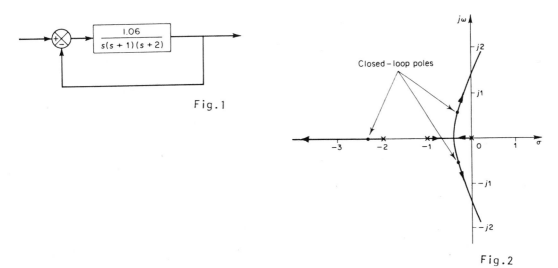

Fig.1

Closed–loop poles

Fig.2

Solution: In order to meet this specification, we shall insert a lag compensator consisting of a lag network and an amplifier in cascade with the given feedforward transfer function. To increase the static velocity error coefficient by a factor of about 10, we will place the pole and zero of the lag network at s = -0.01 and s = -0.1, respectively. The transfer function of the lag network is then

$$\frac{1}{10} \left(\frac{s + 0.1}{s + 0.01} \right) .$$

The angle contribution of this lag network near a dominant closed-loop pole is around seven degrees, therefore there is a small change in the new root locus near the desired dominant closed-loop poles. We cascade an amplifier of gain K_c, to account for the attenuation due to the lag network. The feedforward transfer function of the compensated system becomes

$$G_1(s) = \frac{1}{10} \left(\frac{s+0.1}{s+0.01} \right) (K_c) \frac{1.06}{s(s+1)(s+2)} = \frac{K(s+0.1)}{s(s+0.01)(s+1)(s+2)}$$

where

$$K = \frac{1.06 \ K_c}{10}$$

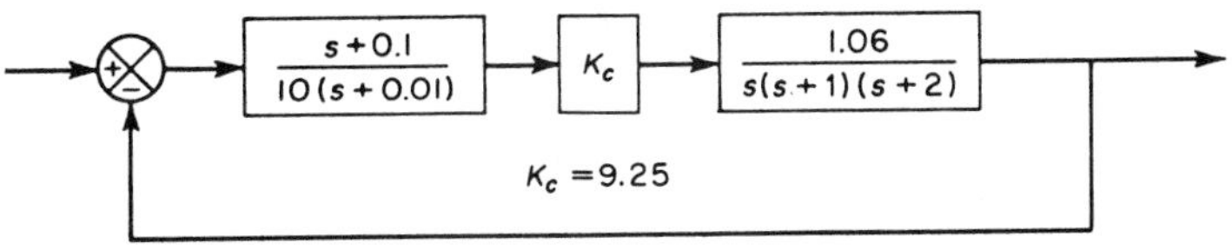

Fig.3

Fig. 3 shows the block diagram of the compensated system.

The root-locus for the compensated system near the dominant closed-loop poles is shown in Fig. 4.

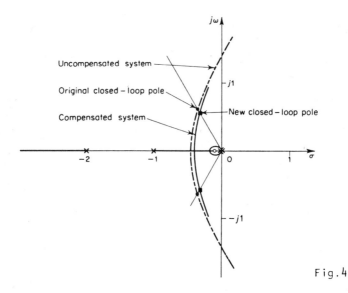

Fig.4

If the damping ratio of the new dominant closed-loop poles is kept the same, then the poles are obtained from the new root-locus plot as follows:

$$s_1 = -0.28 + j0.51, \qquad s_2 = -0.28 - j0.51$$

The open-loop gain K is

$$K = \left| \frac{s(s+0.01)(s+1)(s+2)}{s+0.1} \right|_{s = -0.28 + j0.51} = 0.98$$

and the amplifier gain K_c is

$$K_c = \frac{10 \, K}{1.06} = 9.25$$

Thus the compensated system has the following open-loop transfer function:

$$G_1(s) = \frac{0.98(s+0.1)}{s(s+0.01)(s+1)(s+2)} = \frac{4.9(10s+1)}{s(100s+1)(s+1)(0.5s+1)}$$

The static velocity error coefficient K_v is

$$K_v = \lim_{s \to 0} sG_1(s) = 4.9 \ \text{sec}^{-1}$$

● **PROBLEM** 8-31

The uncompensated transfer function of the system is

$$GH(j\omega) = \frac{K}{j\omega(j\omega + 2)} = \frac{K_v}{j\omega(0.5j\omega + 1)} \tag{1}$$

419

where

$$K_v = \frac{1}{2}K.$$

Design the phase-lag compensator, using Bode diagrams. It is desired that $K_v = 20$ while a phase margin of 45° is attained.

Solution: The phase margin of the uncompensated system is 20° and it should be increased. The solid line is the uncompensated Bode diagram of the system.

We take 5° for the phase-lag compensator and find the frequency ω where $\phi(\omega) = -130°$. We get $\omega_c' = 1.5$ which is the new crossover frequency. The attenuation causing ω_c' to be the new crossover frequency is 20db, the difference between the actual and asymptotic curves is 2 db. We have 20 db = 20 log α and $\alpha = 10$. Thus the zero is one decade below the crossover and $\omega_2 = \frac{\omega_c'}{10} = 0.15$. The pole is at $\omega_p = \frac{\omega_z}{10} = 0.015$. For the compensated system, we get

$$G_c(j\omega)\,GH(j\omega) = \frac{20(6.66j\omega + 1)}{j\omega(0.5j\omega + 1)(66.6j\omega + 1)} \tag{2}$$

The dotted lines show the frequency response of the compensated system. The attenuation, introduced by the phase lag, lowers the crossover frequency and increases the phase margin.

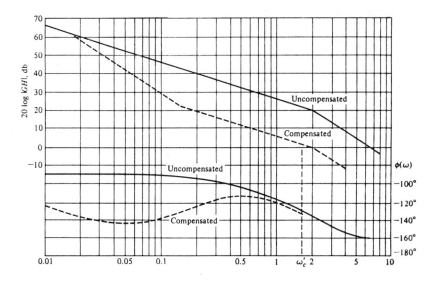

The open-loop uncompensated transfer function of the system is

$$GH(s) = \frac{K_1}{s^2} \tag{1}$$

and its characteristic equation is

$$1 + GH(s) = 1 + \frac{K_1}{s^2} = 0 \tag{2}$$

and the root locus is the $j\omega$-axis. We want, therefore to compensate this system with a network, $G_c(s)$, where

$$G_c(s) = \frac{s + z}{s + p} \tag{3}$$

$$|z| < |p|.$$

The system's specifications are

$T_s \leq 4$ sec, T_s - settling time

$p_0 \leq 30\%$, p_0 - percent overshoot for a step input.

Find z, p, K_1, and the error constants for the compensated system.

Solution: Percent overshoot is given by

$$p_0 = p_1 0_1 = 100e^{-\frac{\zeta \pi}{\sqrt{1 - \zeta^2}}} \tag{4}$$

Using the value of 30% we find the damping ratio to be $\zeta \geq 0.35$. The settling time is

$$T_s = \frac{4}{\zeta \omega_n} = 4 \tag{5}$$

thus

$$\zeta \omega_n = 1.$$

We choose a desired dominant root location as

$$r_1, \hat{r}_1 = -1 \pm j2 \tag{6}$$

and

$$\zeta = 0.45$$

We place the zero of the compensator directly below the desired location at $s = -z = -1$ as shown in the sketch. Measuring the angle at the desired root, we have

$$\phi = -2(116°) + 90° = -142° \tag{7}$$

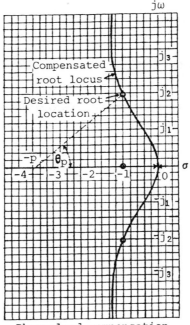

Phase-lead compensation

We evaluate the angle from the undetermined pole, θ_p as

$$-180° = -142° - \theta_p \tag{8}$$

$$\theta_p = 38°.$$

At the desired root the total is 180°. The line inter-secting the desired root location and the real axis is drawn at the angle $\theta_p = 38°$, the point of intersection with the real axis is $s = -p = -3.6$. Thus, the compensator is

$$G_c(s) = \frac{s + 1}{s + 3.6} \tag{9}$$

and the compensated transfer function for the system is

$$GH(s)G_c(s) = \frac{K_1(s + 1)}{s^2(s + 3.6)} \tag{10}$$

Measuring the vector lengths from the poles and zeros to the root location we evaluate the gain K_1.

$$K_1 = \frac{(2.23)^2(3.25)}{2} = 8.1 \tag{11}$$

Evaluating the error constants, we find that this system

422

will result in a zero steady-state error for a step and ramp input signal. The acceleration constant is

$$K_a = \frac{8.1}{3.6} = 2.25 \tag{12}$$

The compensation is complete, since the steady-state performance of the system is satisfactory.

CONTROLLER

● **PROBLEM** 8-33

Consider the linear process that has the transfer function

$$\frac{C(s)}{U(s)} = \frac{100}{s(s + 5)} \tag{1}$$

It will serve as an example of the design of a closed-loop system with observed state feedback.

The dynamic equations of the process are

$$\dot{x}(t) = Ax(t) + Bu(t) \tag{2}$$

$$c(t) = Dx(t) \tag{3}$$

where

$$A = \begin{bmatrix} 0 & 1 \\ 0 & -5 \end{bmatrix}, \quad B = \begin{bmatrix} 0 \\ 100 \end{bmatrix}$$

$$D = [1 \quad 0]$$

Design a state feedback control given by

$$u = r - Gx \tag{4}$$

$$G = [g_1 \quad g_2] \tag{5}$$

with the eigenvalues of the closed-loop system located at $\lambda = -7.07 \pm j7.07$.

The natural undamped frequency is 10 $\frac{rad}{sec}$ and the corresponding damping ratio is 0.707.

Design the system, assuming that the states x_1 and x_2 are inaccessible so that a state observer is to be designed.

Solution: Note that the pair [A,B] is completely controllable, so the eigenvalues of A - BG can be arbitrarily assigned. Since the pair [A,D] is completely observable, an observer may be constructed from c and u.

The matrices [B AB] and D' A'D'] are both of rank 2, that proves that [A B] is controllable and [A D] is observable.

The characteristic equation of the closed-loop system with state feedback is

$$|\lambda I - (A - BG)| = 0 \qquad (6)$$

or

$$\lambda^2 + (5 + 100g_2)\lambda + 100g_1 = 0 \qquad (7)$$

To obtain the desired eigenvalues, the coefficients of eq. (7) must satisfy the following conditions:

$$5 + 100g_2 = 14.14 \qquad (8)$$

$$100g_1 = 100 \qquad (9)$$

We have

$$g_1 = 1$$

$$g_2 = 0.0914$$

The figure shows the state diagram of the overall system.

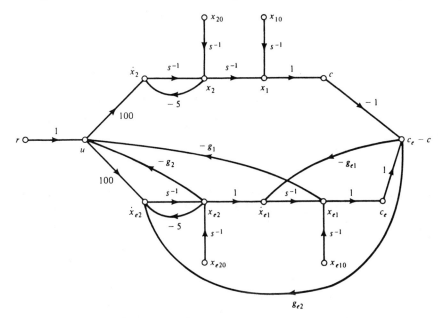

The characteristic equation of the observer is

$$|\lambda I - (A - G_e D)| = 0 \qquad (10)$$

where

$$G_e = \begin{bmatrix} g_{e1} \\ g_{e2} \end{bmatrix} \qquad (11)$$

We assume that the eigenvalues of the observer must be at $\lambda = -50, -50$. Since these eigenvalues have much

424

greater magnitude than the real parts of the eigenvalues of the system, the transient of the observer due to the difference between the initial states X_0 and X_{e0} should rapidly decay to zero.

With the chosen eigenvalues, multiplying out eq. (10) gives

$$\lambda^2 + (5 + g_{e1})\lambda + (5g_{e1} + g_2) = \lambda^2 + 100\lambda + 2500 \quad (12)$$

so

$$g_{e1} = 95$$

$$g_{e2} = 2025 \quad (13)$$

● **PROBLEM** 8-34

The controlled process of the digital control system shown below is described by

$$G(z) = \frac{z^{-2}}{1 - z^{-1} - z^{-2}} .$$

Design a digital controller so that a deadbeat response is obtained when the input is a unit-step function.

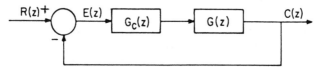

Solution: The transfer function $G(z)$ has two more poles than zeros. Thus we can not choose $M(z) = z^{-1}$.

Let $M(z) = z^{-2}$, then

$$G_c(z) = \frac{1}{G(z)} \frac{M(z)}{1 - M(z)} \quad \frac{1 - z^{-1} - z^{-2}}{z^{-2}} \quad \frac{z^{-2}}{1 - z^{-2}}$$

$$= \frac{1 - z^{-1} - z^{-2}}{1 - z^{-2}}$$

$G_c(z)$ is a physically realizable transfer function. The function $F(z)$ is

$$F(z) = \frac{1 - M(z)}{1 - z^{-1}} = 1 + z^{-1}$$

For the system

$$x[(k + 1)T] = Ax(kT) + Bu(kT)$$

$$c(kT) = Dx(kT)$$

where

$$A = \begin{bmatrix} 0 & 1 & 0 \\ 0 & 0 & 1 \\ -1 & 0 & 0 \end{bmatrix}, \quad B = \begin{bmatrix} 0 & 1 \\ 1 & 0 \\ 0 & 0 \end{bmatrix}$$

$$D = \begin{bmatrix} 1 & 0 & 0 \\ 1 & 0 & 0 \end{bmatrix}$$

find the feedback gain matrix G such that the output feedback

$$u(kT) = -GC(kT)$$

places the eigenvalues of A - BGD so that the system is stable.

Solution: Since D is of rank 1,

$$G*D = [g_1^* + g_2^* \ 0 \ 0]$$

There is only one independent parameter in $g_1^* + g_2^*$.

We have

$$G*D = [(MS')^{-1}(a - \phi)]'$$

and

$$\begin{bmatrix} g_1^* + g_2^* \\ 0 \\ 0 \end{bmatrix}' = \frac{1}{w_1^3 - w_2^3} \begin{bmatrix} -w_2^2 a_3 + w_1^2 a_2 - w_1 w_2 (a_1 - 1) \\ w_1^2 a_3 - a_2 w_1 w_2 + w_2^2 (a_1 - 1) \\ -w_1 w_2 a_3 + a_2 w_2^2 - w_1^2 (a_1 - 1) \end{bmatrix}$$

We can assign only one of w_1 and w_2 arbitrarily, since the last two rows are zero. Thus

$$a_3 w_1^2 - w_1 w_2 a_2 + w_2^2 (a_1 - 1) = 0$$

$$-a_3 w_1 w_2 + w_2^2 a_2 - w_1^2 (a_1 - 1) = 0$$

In order for the closed-loop eigenvalues for be at $z_1 = 0.1$ and $z_2 = 0.2$, the following equations must be satisfied.

$$a_2 + 0.3a_3 + 0.07 = 0$$

$$a_1 - 0.02a_3 - 0.06 = 0$$

We see that we have four equations and five unknowns a_1, a_2, a_3, w_1 and w_2. Thus only one of the unknowns can be arbitrarily assigned. Equations are nonlinear in w_1 and w_2.

In this case we can write

$$\left| zI - A + BGD \right| = z^3 + (g_{21} + g_{22})z^2 + (g_{11} + g_{12})z + 1 = 0$$

Only two of the three coefficients of this equation can be arbitrarily assigned. The system cannot be stabilized by output feedback with the matrix D given because the constant term is equal to one.

● **PROBLEM** 8-36

The digital control system is described by

$$x[(k + 1)T] = Ax(kT) + Bu(kT)$$

$$c(kT) = Dx(kT)$$

where

$$A = \begin{bmatrix} 0 & 1 & 0 \\ 0 & 0 & 1 \\ -1 & 0 & 0 \end{bmatrix}, \quad B = \begin{bmatrix} 0 & 1 \\ 1 & 0 \\ 0 & 0 \end{bmatrix}$$

$$D = \begin{bmatrix} 1 & 0 & 0 \\ 1 & 1 & 0 \end{bmatrix}.$$

Find the feedback gain matrix G such that the output feedback

$$u(kT) = -Gc(kT)$$

places the eigenvalues of A - BGD at $z_1 = 0.1$ and $z_2 = 0.2$. Since the rank of B and D is two, a maximum of two eigenvalues can be arbitrarily assigned.

Solution: The characterisitc equation of A is

$$\left| zI - A \right| = z^3 + 1 = z^3 + a_2z^2 + a_3z + a_4$$

We have for M,

$$M = \begin{bmatrix} a_4 & 0 & 0 \\ a_3 & a_4 & 0 \\ a_2 & a_3 & a_4 \end{bmatrix} = \begin{bmatrix} 1 & 0 & 0 \\ 0 & 1 & 0 \\ 0 & 0 & 1 \end{bmatrix}$$

427

From B* = Bw we have

$$B^* = \begin{bmatrix} 0 & 1 \\ 1 & 0 \\ 0 & 0 \end{bmatrix} \begin{bmatrix} w_1 \\ w_2 \end{bmatrix} = \begin{bmatrix} w_2 \\ w_1 \\ 0 \end{bmatrix}$$

B* has two independent parameters in w_1 and w_2. The controllability matrix for (A, B*) is

$$S = [B^* \quad AB^* \quad A^2B^*] = \begin{bmatrix} w_2 & w_1 & 0 \\ w_1 & 0 & -w_2 \\ 0 & -w_2 & -w_1 \end{bmatrix}$$

If $w_1^3 - w_2^3 \neq 0$, the matrix S is nonsingular.

If

$$G^* = [g_1^* \quad g_2^*]$$

then

$$G^*D = [g_1^* + g_2^* \quad g_2^* \quad 0]$$

G*D has two independent parameters in g_1^* and g_2^* since D is of rank two. Replacing G by G*, we have

$$G^*D = [(MS')^{-1}(\alpha - a)]'$$

or

$$\begin{bmatrix} g_1^* + g_2^* \\ g_2^* \\ 0 \end{bmatrix} = \frac{1}{w_1^3 - w_2^3} \begin{bmatrix} -w_2^2\alpha_3 + w_1^2\alpha_2 - w_1w_2(\alpha_1 - 1) \\ w_1^2\alpha_3 - w_1w_2\alpha_2 + w_2^2(\alpha_1 - 1) \\ -w_1w_2\alpha_3 + w_2^2\alpha_2 - w_1^2(\alpha_1 - 1) \end{bmatrix}$$

The last row is a constraint equation.

$$-w_1w_2\alpha_3 + w_2^2\alpha_2 - w_1^2(\alpha_1 - 1) = 0$$

Note that only two of the three coefficients of the closed-loop characteristic equation can be arbitrarily assigned.

For z = 0.1 and 0.2 to be roots of the characteristic equation

$$z^3 + \alpha_3 z^2 + a_2 z + \alpha_1 = 0$$

the following equations must be satisfied

$$\alpha_2 + 0.3a_3 + 0.07 = 0$$

$$\alpha_1 - 0.02\alpha_3 - 0.006 = 0$$

Solving the above equations and the contraint equation

428

$$-w_1 w_2 \alpha_3 + w_2^2 \alpha_2 - w_1^2 (\alpha_1 - 1) = 0$$

we obtain

$$\alpha_1 = \frac{0.02w_1^2 + 0.0004w_2^2 + 0.006w_1w_2}{0.3w_2^2 + w_1w_2 + 0.02w_1^2}$$

$$\alpha_2 = \frac{-0.2996w_1^2 - 0.07w_1w_2}{0.3w_2^2 + w_1w_2 + 0.02w_1^2}$$

$$\alpha_3 = \frac{0.994w_1^2 - 0.07w_2^2}{0.3w_2^2 + w_1w_2 + 0.02w_1^2}$$

Choosing the values of w_1, w_2 we determine α_1, α_2, α_3 and from $G = WG*$ we find G.

● **PROBLEM** 8-37

For the discrete control system

$$x[(k+1)T] = Ax(kT) + Bu(kT)$$

where

$$A = \begin{bmatrix} 0 & 1 \\ -1 & -2 \end{bmatrix}, \quad B = \begin{bmatrix} 1 & 0 \\ 0 & 1 \end{bmatrix}$$

(A,B) is controllable.

Obtain the feedback gain matrix G such that the state feedback

$$u(kT) = - Gx(kT)$$

places the closed-loop eigenvalues at

$$z_1 = 0.1 \quad \text{and} \quad z_2 = 0.2$$

Solution: Let us define

$$B* = Bw = \begin{bmatrix} 1 & 0 \\ 0 & 1 \end{bmatrix} \begin{bmatrix} w_1 \\ w_2 \end{bmatrix} = \begin{bmatrix} w_1 \\ w_2 \end{bmatrix}$$

The pair (A,B*) is controllable when the matrix [B* AB*] is nonsingular. Thus

$$|B* \ AB*| = \begin{vmatrix} w_1 & w_2 \\ w_2 & -w_1 - 2w_2 \end{vmatrix} = - (w_1 + w_2)^2$$

429

The matrix is nonsingular when

$$w_1 \neq w_2 \ .$$

For the single-input system the feedback matrix G^* is

$$G^* = -[\Delta_{01} \quad \Delta_{02}]K^{-1}$$

where

$$\Delta_{01} = \left| zI-A \right|_{z=0.1} = 1.21$$

$$\Delta_{02} = \left| zI-A \right|_{z=0.2} = 1.44$$

and

$$k(2) = adj(2I-A) \cdot B^*$$

$$= adj \begin{bmatrix} z & -1 \\ 1 & z+2 \end{bmatrix} \begin{bmatrix} w_1 \\ w_2 \end{bmatrix} = \begin{bmatrix} w_1(z+2) + w_2 \\ -w_1 + w_2 z \end{bmatrix}$$

We have

$$k_1 = k(z_1) = \begin{bmatrix} 2.1w_1 + w_2 \\ -w_1 + 0.1w_2 \end{bmatrix}$$

$$k_2 = k(z_2) = \begin{bmatrix} 2.2w_1 + w_2 \\ -w_1 + 0.2w_2 \end{bmatrix}$$

$$K = [k_1 \ k_2] = \begin{bmatrix} 2.1w_1 + w_2 & 2.2w_1 + w_2 \\ -w_1 + 0.1w_2 & -w_1 + 0.2w_2 \end{bmatrix}$$

Let us choose $w = [1 \ 1]'$, then

$$K = \begin{bmatrix} 3.1 & 3.2 \\ -0.9 & -0.8 \end{bmatrix}$$

Substituting we get

$$G^* = -[0.82 \quad 1.48]$$

The feedback matrix of the multiple-input system is

$$G = wG^* = - \begin{bmatrix} 0.82 & 1.48 \\ 0.82 & 1.48 \end{bmatrix}$$

and we get

$$|ZI-A + B^*G^*| = |zI-A + BG| = z^2 - 0.3z + 0.02$$

$$= (z-0.1)(z-0.2)$$

Thus we obtained the desired roots.

● **PROBLEM** 8-38

The digital control system is described by

$$x(k+1) = AX(k) + Bu(k) + v_2$$

where

$$A = \begin{bmatrix} 0 & 1 \\ -1 & 0 \end{bmatrix}, \quad B = \begin{bmatrix} 0 \\ 1 \end{bmatrix}$$

and v_2 is a constant disturbance.

Find the control scheme that satisfies the following conditions

1. The eigenvalues of the closed-loop system are at certain specified values.

2. x_1 reaches the reference input $r = v_1$ as $k \to \infty$.

Solution: Let us define the output variable

$$a(k) = v_1 - x_1(k)$$

From condition 2 we want $a(k)$ to be such that

$$\lim_{k \to \infty} a(k) = 0 .$$

Using the output equation model

$$a(k) = Dx(k) + Eu(k) + Hv$$

we have

$$D = [-1 \quad 0] \ , \qquad E = 0$$

$$H = [1 \quad 0] \ , \qquad F = \begin{bmatrix} 0 & 1 \\ 0 & 0 \end{bmatrix}$$

The pair (A,B) is completely controllable and

$$\begin{bmatrix} A-I_n & B \\ D & E \end{bmatrix} = \begin{bmatrix} -1 & 1 & 0 \\ -1 & -1 & 1 \\ -1 & 0 & 0 \end{bmatrix}$$

The above matrix has rank three.

The pair $(\hat{A},\hat{B})$ where

$$\hat{A} = \begin{bmatrix} 0 & 1 & 0 \\ -1 & 0 & 0 \\ -1 & 0 & 1 \end{bmatrix} \qquad \hat{B} = \begin{bmatrix} 0 \\ 1 \\ 0 \end{bmatrix}$$

is also completely controllable.

In the z-domain the control is given by

$$U(z) = -g_1 X_1(z) - g_2 X_2(z) - \frac{g_3}{z-1} A(z)$$

where g_1, g_2, g_3 are feedback gain constants.

The characteristic equation of the closed-loop system is

$$|zI-\hat{A} + \hat{B}G| = \begin{bmatrix} z & -1 & 0 \\ 1+g_1 & z+g_2 & g_3 \\ 1 & 0 & z-1 \end{bmatrix}$$

$$= z^3 + (g_2-1)z^2 + (1+g_1-g_2)z - (1+g_1+g_3) = 0$$

Let the closed-loop eigenvalues be at

$$z_1 = 0.3 + j0.3, \ z_2 = 0.3 - j0.3, \ z_3 = 0.5.$$

The characteristic equation becomes

$$(z-z_1)(z-z_2)(z-z_3) = z^3 - 1.1z^2 + 0.39z - 0.045 = 0$$

Comparing the coefficients we have

$$g_1 = -0.71, \qquad g_2 = -0.1, \qquad g_3 = -0.245.$$

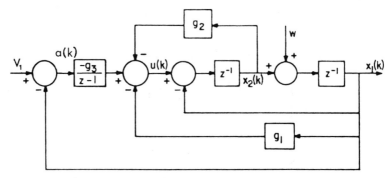

The block diagram of the closed-loop system is shown above.
The z-transforms of $x_1(k)$ and $x_2(k)$ are

$$
\begin{bmatrix} X_1(2) \\ \\ X_2(z) \end{bmatrix}
= \frac{1}{\Delta}
\begin{bmatrix}
\dfrac{-z^{-2}}{z-1} g_3 & z^{-1}\left(1 + g_2 z^{-1}\right) \\ \\
\dfrac{-z^{-1}}{z-1} g_3 & -z^{-2} - g_1 z^{-2} + \dfrac{g_3}{z-1} z^{-2}
\end{bmatrix}
\begin{bmatrix} v_1 \\ \\ v_2 \end{bmatrix}
\quad \frac{z}{z-1}
$$

where

$$\Delta = 1 + g_2 z^{-1} + z^{-2} + g_1 z^{-2} - \frac{g_3 z^{-2}}{z-1}$$

Applying the final value theorem of the z-transform
we obtain

$$
\lim_{k \to \infty}
\begin{bmatrix} x_1(k) \\ \\ x_2(k) \end{bmatrix}
=
\begin{bmatrix} 1 & 0 \\ \\ 0 & -1 \end{bmatrix}
\begin{bmatrix} v_1 \\ \\ v_2 \end{bmatrix}
=
\begin{bmatrix} v_1 \\ \\ v_1 - v_2 \end{bmatrix}
$$

The condition 1, $\lim_{k \to \infty} x_1(k) = v_1$, is satisfied.

COMPENSATOR, OBSERVER

● **PROBLEM** 8-39

Examine the response of the compensator described by

$$D(z) = \frac{E_2(z)}{E_1(z)} = \frac{a_0 + a_1 z^{-1}}{1 + b z^{-1}} .$$

Solution: The first-order compensator has the following
difference equation.

$$e_2(kT) = a_0e_1(kT) + a_1e_1[(k-1)T] - be_2[(k-1)T]$$

Below is shown the canonical discrete signal flow graph.

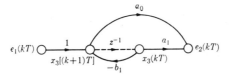

The first-order difference equation is

$$x_3[(k+1)T] = -bx_3(kT) + e_1(kT) .$$

Note that the discrete state variable $x_3(kT)$ is chosen
as the output of the ideal delay unit. The difference
equation for the output is

$$e_2(kT) = (a_1 - a_0b)x_3(kT) + a_0e_1(kT) .$$

From the above equations we can calculate the output,
when the computer input signal is given.

● **PROBLEM** 8-40

A system is described by

$$\dot{x} = Ax + Bu$$

$$y = Cx + Du$$

where

$$A = \begin{bmatrix} -2 & -2 & 0 \\ 0 & 0 & 1 \\ 0 & -3 & -4 \end{bmatrix} \qquad B = \begin{bmatrix} 1 & 0 \\ 0 & 0 \\ 0 & 1 \end{bmatrix}$$

$$C = \begin{bmatrix} 1 & 0 & 1 \\ 0 & 1 & 0 \end{bmatrix} , \qquad D = [0]$$

Find an observer whose output settles to the actual state
vector within one second.

Solution: To find the desired observer we have to
verify that the system is completely observable. Then
we can determine the eigenvalue locations. The response

434

of the error equation

$$\dot{e} = A_c e$$

consists of the terms involving $e^{\lambda_i t}$

The eigenvalues λ_i can be found from A_c . If we take e^{-4} as a close approximation of zero, then we can select $\lambda_1 = \lambda_2 = -5$ and $\lambda_3 = -6$. Note that all eigenvalues of A_c must have real parts smaller than -4. We get

$$\phi(\lambda) = (\lambda I - A)^{-1} = \begin{bmatrix} \dfrac{1}{\lambda+2} & \dfrac{-2(\lambda+4)}{(\lambda+1)(\lambda+2)(\lambda+3)} & \dfrac{-2}{(\lambda+1)(\lambda+2)(\lambda+3)} \\[3mm] & \dfrac{\lambda+4}{(\lambda+1)(\lambda+3)} & \dfrac{1}{(\lambda+1)(\lambda+3)} \\[3mm] 0 & \dfrac{-3}{(\lambda+1)(\lambda+3)} & \dfrac{\lambda}{(\lambda+1)(~+3)} \end{bmatrix}$$

and

$$C\phi(\lambda) = \begin{bmatrix} \dfrac{1}{(\lambda+2)} & \dfrac{-5\lambda-14}{(\lambda+1)(\lambda+2)(\lambda+3)} & \dfrac{\lambda+2\lambda-2}{(\lambda+1)(\lambda+2)(\lambda+3)} \\[3mm] 0 & \dfrac{\lambda+4}{(\lambda+1)(\lambda+3)} & \dfrac{1}{(\lambda+1)(\lambda+3)} \end{bmatrix}$$

For rows 1 and 2, we take $\lambda = -5$ and for row 2, we take $\lambda = -6$, thus

$$G_c = \begin{bmatrix} -\dfrac{1}{3} & -\dfrac{11}{24} & -\dfrac{13}{24} \\[3mm] 0 & -\dfrac{1}{8} & \dfrac{1}{8} \\[3mm] 0 & -\dfrac{2}{15} & \dfrac{1}{15} \end{bmatrix} \qquad J_c = \begin{bmatrix} 1 & 0 \\[2mm] 0 & 1 \\[2mm] 0 & 1 \end{bmatrix}$$

Then

$$B_c = -G_c^{-1} J_c$$

and

$$A_c = A - B_c C$$

so this gives

$$B_C = \begin{bmatrix} 3 & -8 \\ 0 & 7 \\ 0 & -1 \end{bmatrix}$$

and

$$A_C = \begin{bmatrix} -5 & 6 & -3 \\ 0 & -7 & 1 \\ 0 & -2 & -4 \end{bmatrix}.$$

We get for the observer

$$\dot{\hat{x}} = A_C \hat{x} + B_C y + Bu ,$$

where u and y are the input and the output of the system being observed.

● **PROBLEM** 8-41

The system is described by

$$A = \begin{bmatrix} 0 & 2 \\ 0 & 3 \end{bmatrix} \quad \text{and} \quad B = \begin{bmatrix} 0 \\ 1 \end{bmatrix}$$

Find an observer for the system such that its eigenvalues are $\lambda_1 = \lambda_2 = -8$. Let $C = [1 \quad 0]$.

<u>Solution:</u>

$$\Phi(\lambda) = (\lambda I - A)^{-1} = \begin{bmatrix} \dfrac{1}{\lambda} & \dfrac{-2}{\lambda(\lambda - 3)} \\ 0 & \dfrac{1}{\lambda - 3} \end{bmatrix}$$

and

$$C\Phi(\lambda) = \left\{ \dfrac{1}{\lambda} \quad \dfrac{2}{\lambda(\lambda - 3)} \right\}$$

one row of G_C is

$$C\Phi(-8) = \left[-\dfrac{1}{8} \quad \dfrac{1}{44} \right].$$

Differentiating $C\Phi(\lambda)$ we obtain the second independent row

$$\dfrac{d}{d\lambda}[C\Phi(\lambda)] \Big/_{\lambda=-8} = \left[-\dfrac{1}{\lambda^2} \quad \dfrac{-2(2\lambda - 3)}{\lambda^2(\lambda - 3)^2} \right]_{\lambda=-8}$$

436

We get

$$J_c = \begin{bmatrix} 1 \\ 0 \end{bmatrix}$$

and

$$B_c = G_c^{-1} J_c,$$

thus

$$B_c = -\begin{bmatrix} -\dfrac{1}{8} & \dfrac{1}{44} \\[2mm] -\dfrac{1}{64} & \dfrac{19}{3872} \end{bmatrix}^{-1} \begin{bmatrix} 1 \\ 0 \end{bmatrix} = \begin{bmatrix} 19 \\ 60.5 \end{bmatrix}$$

We can check, that

$$|\lambda I - A + B_c C| = (\lambda + 8)^2.$$

The observer has the desired eigenvalues $\lambda_1 = \lambda_2 = -8$.
We have

$$\dot{\hat{x}} = (A = B_c C)\hat{x} + B_c y + z$$

$$= \begin{bmatrix} 19 & 2 \\ -60.5 & 3 \end{bmatrix} \hat{x} + \begin{bmatrix} 19 \\ 60.5 \end{bmatrix} y + \begin{bmatrix} 0 \\ 1 \end{bmatrix} u$$

ROOT LOCUS

● **PROBLEM** 8-42

Design a feedback control system that satisfies the following specifications:

a)   Settling time of the system $\leq$ 3 sec.

b)   Damping ratio of dominant roots $\geq$ 0.707.

c)   Steady-state error for a ramp input $\leq$ 10% of input
     magnitude.

Fig. 1 shows the feedback control system:

Select the parameters K_1 and K_2.

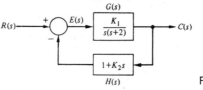

Fig. 1

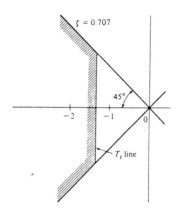

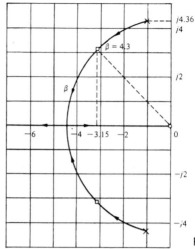

Fig. 2

Fig. 3

Solution: For a ramp, $R(s) = \frac{R}{s^2}$ so the steady-state error is

$$e_{ss} = \lim_{t \to \infty} e(t) = \lim_{s \to 0} sE(s) = \lim_{s \to 0} \frac{s \cdot \frac{|R|}{s^2}}{1 + GH(s)} \qquad (1)$$

with the condition

$$\frac{e_{ss}}{|R|} = \lim_{s \to 0} \frac{1}{sGH(s)} \leq 0.10 \qquad (2)$$

We have

$$GH(s) = \frac{K_1(1 + K_2 s)}{s(s + 2)} \qquad (3)$$

thus $K_1 \geq 20$.

For this system, we have

$$K_v = \frac{|R|}{e_{ss}} = \frac{K_1}{2} \qquad (4)$$

We rewrite the settling time condition in terms of the real part of the dominant roots.

$$T_5 = \frac{u}{\sigma} \leq 3 \text{ sec}; \ \sigma \geq \frac{4}{3} \qquad (5)$$

From the damping ratio condition we have the requirement that the roots of the closed-loop system are positioned below the line of 45° in the left-hand s-plane. Fig. 2 shows the area where $\sigma \geq \frac{4}{3}$ in the left-hand s-plane, together with the ζ-requirement. In order to satisfy the requirements, all the roots must lie within the shaded area of the left-hand plane.

We now select the parameters

$$\alpha = K_1$$

and

$$\beta = K_1 K_2.$$

The characteristic equation is

$$1 + GH(s) = s^2 + 2s + \beta s + \alpha = 0 \qquad (6)$$

From the equation

$$1 + \frac{\alpha}{s(s + 2)} = 0 \qquad (7)$$

we find the locus of roots as $\alpha = K_1$ varies, as shown in Fig. 3.

For $K_1 = 20$, the roots are shown on the locus. From the locus equation

$$1 + \frac{\beta s}{s^2 + 2s + \alpha} = 0 \qquad (8)$$

we determine the effect of varying β. The poles of this root locus are the roots of the locus of Fig. 3, for eq. (8). Roots with $\zeta = 0.707$ are obtained when

$$\beta = 4.3 = 20K_2, \quad K_2 = 0.215.$$

The settling time is 1.27 sec since the real part of the roots is $\sigma = 3.15$.

● **PROBLEM** 8-43

For the system shown in Fig. 1, determine the values of K and K_n so that the following requirements are satisfied:

1) Damping ratio of the closed-loop poles is 0.5.

2) Settling time is smaller or equal to 2 sec.

3) Velocity error coefficient $K_v \geq 50 \text{ sec}^{-1}$.

4) $0 < K_n < 1$.

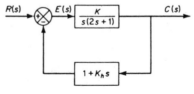

Fig.1

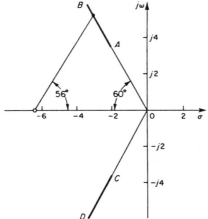

Fig. 2

Solution: From the equation $\zeta = \cos\theta$, where ζ is the damping ratio, we find that the closed-loop poles lie on lines at 60° in the left-half s plane. From the second requirement we have

$$t_s = \frac{4}{\sigma} \le 2 \text{ sec}$$

or

$$2 \le \sigma$$

Fig. 2 shows the possible locations for the closed-loop poles in the s plane. The closed-loop poles must lie on the heavy lines AB and CD in the left-half s plane.

From the definition of the velocity error coefficient K_v , where

$$K_v = \lim_{s \to 0} s\, G(s)H(s)$$

we have

$$K_v = \lim_{s \to 0} \frac{sK(1 + K_n s)}{s(2s+1)} = K$$

Thus

$$50 \le K$$

since from the third requirement we have

$$50 \le K_v$$

The open-loop poles are at s=0 and $s = -\frac{1}{2}$. The open-loop is at $s = -\frac{1}{K_n}$, where K_n is a constant to be

440

determined. Let points A and C be the closed-loop poles; they are located at $-2 \pm j\,3.4$. At the chosen closed-loop pole location the sum of the angles with the open-loop poles is $115° + 120° = 235°$.

To get the total sum of $-180°$ we need a $55°$ from the zero. To satisfy the angle condition, we choose the zero to be at $s = -4.4$. Then

$$K_n = \frac{1}{4.4} = 0.227.$$

From the magnitude condition, we get

$$\left| \frac{K(1 + 0.227s)}{s(2s+1)} \right|_{s = -2 + j\,3.4} = 1$$

or

$$K = 31.$$

Since we want K to be bigger than 50, the choice of the closed-loop poles at $-2 \pm j\,3.4$ is not acceptable.

Let us try $-3 \pm j\,5.1$. The sum of the angle contributions from the open-loop poles is $236°$. Thus, a $56°$ contribution is required from zero and zero must be at $s = -.6.4$. The magnitude condition gives $K = 70$ and $K_n = 0.156$. Thus all requirements are met.

CHAPTER 9

STATE SPACE REPRESENTATION

STATE SPACE REPRESENTATION OF TRANSFER FUNCTIONS

For the following transfer function obtain a state space equation:

$$G(S) = \frac{C(S)}{U(S)} = \frac{S + 2}{(S + 1)^2 (S + 3)}$$

Solution: The partial fractions expansion of the transfer function gives

$$G(S) = \frac{S + 2}{(S + 1)^2 (S + 3)} = \frac{\frac{1}{2}}{(S + 1)^2} + \frac{\frac{1}{4}}{S + 1} + \frac{-\frac{1}{4}}{S + 3}$$

Let us multiply $G(S)$ by S and then S^2 and expand into the partial fractions form.

$$SG(S) = \frac{S(S + 2)}{(S + 1)^2 (S + 3)} = \frac{-\frac{1}{2}}{(S + 1)^2} + \frac{\frac{1}{4}}{S + 1} + \frac{\frac{3}{4}}{S + 3}$$

$$S^2G(S) = \frac{S^2(S + 2)}{(S + 1)^2 (S + 3)} = 1 - \frac{3S^2 + 7S + 3}{(S + 1)^2(S + 3)}$$

$$= 1 + \frac{\frac{1}{2}}{(S + 1)^2} + \frac{-\frac{3}{4}}{S + 1} + \frac{-\frac{9}{4}}{S + 3}$$

Notice that

$$\frac{S^2C(S) - U(S)}{U(S)} = \frac{\frac{1}{2}}{(S + 1)^2} + \frac{-\frac{3}{4}}{S + 1} + \frac{-\frac{9}{4}}{S + 3}$$

442

Let us define

$$X_1(S) = C(S)$$

$$X_2(S) = SC(S)$$

$$X_3(S) = S^2C(S) - U(S)$$

We obtain

$$
\begin{bmatrix} \dfrac{X_1(S)}{U(S)} \\[2.5ex] \dfrac{X_2(S)}{U(S)} \\[2.5ex] \dfrac{X_3(S)}{U(S)} \end{bmatrix}
=
\begin{bmatrix} \dfrac{1}{2} & \dfrac{1}{4} & -\dfrac{1}{4} \\[2ex] -\dfrac{1}{2} & \dfrac{1}{4} & \dfrac{3}{4} \\[2ex] \dfrac{1}{2} & -\dfrac{3}{4} & -\dfrac{9}{4} \end{bmatrix}
\begin{bmatrix} \dfrac{1}{(S+1)^2} \\[2.5ex] \dfrac{1}{S+1} \\[2.5ex] \dfrac{1}{S+3} \end{bmatrix}
$$

Let us define the transformation

$$
\begin{bmatrix} x_1 \\[2ex] x_2 \\[2ex] x_3 \end{bmatrix}
=
\begin{bmatrix} \dfrac{1}{2} & \dfrac{1}{4} & -\dfrac{1}{4} \\[2ex] -\dfrac{1}{2} & \dfrac{1}{4} & \dfrac{3}{4} \\[2ex] \dfrac{1}{2} & -\dfrac{3}{4} & -\dfrac{9}{4} \end{bmatrix}
\begin{bmatrix} y_1 \\[2ex] y_2 \\[2ex] y_3 \end{bmatrix}
$$

We obtain

$$\frac{Y_1(S)}{U(S)} = \frac{1}{(S+1)^2}$$

$$\frac{Y_2(S)}{U(S)} = \frac{1}{S+1}$$

$$\frac{Y_3(S)}{U(S)} = \frac{1}{S+3}$$

or in the state space representation

$$\dot{y}_1 = -y_1 + y_2$$

$$\dot{y}_2 = -y_2 + u$$

$$\dot{y}_3 = -3y_3 + u$$

In the matrix form

$$
\begin{bmatrix} \dot{y}_1 \\ \dot{y}_2 \\ \dot{y}_3 \end{bmatrix} = \begin{bmatrix} -1 & 1 & 0 \\ 0 & -1 & 0 \\ 0 & 0 & -3 \end{bmatrix} \begin{bmatrix} y_1 \\ y_2 \\ y_3 \end{bmatrix} + \begin{bmatrix} 0 \\ 1 \\ 1 \end{bmatrix} [u]
$$

The initial conditions are

$$
\begin{bmatrix} y_1(0) \\ y_2(0) \\ y_3(0) \end{bmatrix} = \begin{bmatrix} \frac{1}{2} & \frac{1}{4} & -\frac{1}{4} \\ -\frac{1}{2} & \frac{1}{4} & \frac{3}{4} \\ \frac{1}{2} & -\frac{3}{4} & -\frac{9}{4} \end{bmatrix}^{-1} \begin{bmatrix} x_1(0) \\ x_2(0) \\ x_3(0) \end{bmatrix}
$$

where

$$
x_1(0) = C(0), \qquad x_2(0) = \dot{C}(0), \quad x_3(0) = \ddot{C}(0) - u(0)
$$

• **PROBLEM** 9-2

Given the transfer function

$$
G(s) = \frac{C(s)}{U(s)} = \frac{2}{s^3 + 6s^2 + 11s + 6}
$$

Obtain two differential state representations.

Solution: From the transfer function we obtain the following third-order differential equation relating the output $c(t)$ and the input $u(t)$.

$$
\dddot{c}(t) + 6\ddot{c}(t) + 11\dot{c}(t) + 6c(t) = 2u(t)
$$

For the state vector let us choose c and its variables $\dot{C}$

and $\ddot{C}$. Thus

$$
x = \begin{bmatrix} x_1 \\ x_2 \\ x_3 \end{bmatrix} = \begin{bmatrix} c \\ \dot{c} \\ \ddot{c} \end{bmatrix}
$$

Note that the state vector is not uniquely defined.

The third order differential equation is equivalent to the three first order differential equations

$$\dot{x}_1 = x_2$$

$$\dot{x}_2 = x_3$$

$$\dot{x}_3 = -6x_1 - 11x_2 - 6x_3 + 2u$$

Thus, the first model can be written

$$\dot{x} = Ax + Bu$$

$$c = Cx$$

where

$$A = \begin{bmatrix} 0 & 1 & 0 \\ 0 & 0 & 1 \\ -6 & -11 & -6 \end{bmatrix} \quad B = \begin{bmatrix} 0 \\ 0 \\ 2 \end{bmatrix} \quad C = [1 \quad 0 \quad 0]$$

The second state model can be obtained through the partial fractions expansion of the transfer function.

Thus

$$\frac{C(s)}{U(s)} = \frac{2}{s^3 + 6s^2 + 11s + 6} = \frac{1}{s+1} + \frac{-2}{s+2} + \frac{1}{s+3}$$

Multiplying by $U(s)$, we get

$$C(s) = \frac{1}{s+1} U(s) + \frac{-2}{s+2} U(s) + \frac{1}{s+3} U(s)$$

Let us, further define

$$X_1(s) = \frac{1}{s+1} U(s)$$

$$X_2(s) = -\frac{2}{s+2} U(s)$$

$$X_3(s) = \frac{1}{s+3} U(s)$$

Taking the inverse Laplace transform, we get

$$\dot{x}_1 = -x_1 + u$$

$$\dot{x}_2 = -2x_2 - 2u$$

$$\dot{x}_3 = -3x_3 + u$$

and

$$c(t) = x_1 + x_2 + x_3$$

In the matrix form we can write
$\dot{x} = Ax + Bu$

$c = Cx$

where

$$A = \begin{bmatrix} -1 & 0 & 0 \\ 0 & -2 & 0 \\ 0 & 0 & -3 \end{bmatrix} \qquad B = \begin{bmatrix} 1 \\ -2 \\ 1 \end{bmatrix} \qquad C = [1 \quad 1 \quad 1]$$

● **PROBLEM** 9-3

The transfer function of the system is

$$\frac{X(s)}{U(s)} = \frac{K(s - \alpha_1)(s - \alpha_2)}{(s - \lambda_1)(s - \lambda_2)(s - \lambda_3)}$$

where $\lambda_1 \neq \lambda_2 \neq \lambda_3$.

Obtain a state space equation of the system.

<u>Solution</u>: Let us define the matrix $B = [b_{ij}]$, whose elements are the coefficients of the partial fractions expansion.

$$\frac{X(s)}{U(s)} = \frac{b_{11}}{s - \lambda_1} + \frac{b_{12}}{s - \lambda_2} + \frac{b_{13}}{s - \lambda_3}$$

$$\frac{sX(s)}{U(s)} = K + \frac{b_{21}}{s - \lambda_1} + \frac{b_{22}}{s - \lambda_2} + \frac{b_{23}}{s - \lambda_3}$$

$$\frac{s^2X(s)}{U(s)} = Ks + K(\lambda_1+\lambda_2+\lambda_3-\alpha_1-\alpha_2) + \frac{b_{31}}{s - \lambda_1} + \frac{b_{32}}{s - \lambda_2} + \frac{b_{33}}{s - \lambda_3}$$

Let

$\quad X_1(s) = X(s)$

$\quad X_2(s) = sX(s) - K U(s)$ $\hspace{3cm}$ $(*)$

$\quad X_3(s) = s^2X(s) - KsU(s) - K(\lambda_1+\lambda_2+\lambda_3-\alpha_1-\alpha_2)U(s)$

We obtain

$$\begin{bmatrix} \dfrac{X_1(s)}{U(s)} \\[2ex] \dfrac{X_2(s)}{U(s)} \\[2ex] \dfrac{X_3(s)}{U(s)} \end{bmatrix} = \begin{bmatrix} b_{11} & b_{12} & b_{13} \\[1ex] b_{21} & b_{22} & b_{23} \\[1ex] b_{31} & b_{32} & b_{33} \end{bmatrix} \begin{bmatrix} \dfrac{1}{s - \lambda_1} \\[2ex] \dfrac{1}{s - \lambda_2} \\[2ex] \dfrac{1}{s - \lambda_3} \end{bmatrix}$$

One of the possible state-space representations of the system is

$$
\begin{bmatrix}
\dfrac{Y_1(s)}{U(s)} \\[2mm]
\dfrac{Y_2(s)}{U(s)} \\[2mm]
\dfrac{Y_3(s)}{U(s)}
\end{bmatrix}
=
\begin{bmatrix}
\dfrac{1}{s - \lambda_1} \\[2mm]
\dfrac{1}{s - \lambda_2} \\[2mm]
\dfrac{1}{s - \lambda_3}
\end{bmatrix}
$$

Taking the inverse Laplace transform, we get

$$\dot{y}_1 = \lambda_1 y_1 + u$$

$$\dot{y}_2 = \lambda_2 y_2 + u$$

$$\dot{y}_3 = \lambda_3 y_3 + u$$

The transformation matrix B is given by

$$x = By$$

where

$$x = x_1 = b_{11} y_1 + b_{12} y_2 + b_{13} y_3$$

with the initial conditions

$$
\begin{bmatrix}
y_1(0) \\
y_2(0) \\
y_3(0)
\end{bmatrix}
=
\begin{bmatrix}
b_{11} & b_{12} & b_{13} \\
b_{21} & b_{22} & b_{23} \\
b_{31} & b_{32} & b_{33}
\end{bmatrix}^{-1}
\begin{bmatrix}
x_1(0) \\
x_2(0) \\
x_3(0)
\end{bmatrix}
$$

Taking the inverse Laplace transform we express $x_1(0)$, $x_2(0)$, $x_3(0)$ in terms of $x(0)$ and $u(0)$

$$
\begin{bmatrix}
x_1(0) \\
x_2(0) \\
x_3(0)
\end{bmatrix}
=
\begin{bmatrix}
x(0) \\
\dot{x}(0) - K u(0) \\
\ddot{x}(0) - K \dot{u}(0) - K(\lambda_1 + \lambda_2 + \lambda_3 - \alpha_1 - \alpha_2) u(0)
\end{bmatrix}
$$

● **PROBLEM 9-4**

The transfer function of the system is

$$\frac{Y(s)}{U(s)} = T(s) = \frac{1}{(s + 1)(s + 2)(s + 3)}$$

Obtain a state space representation of this transfer function using the partial fractions expansion method.

Solution: We have

$$\frac{Y_1(s)}{U(s)} = T(s) = \frac{1}{(s+1)(s+2)(s+3)}$$

$$\frac{Y_2(s)}{U(s)} = sT(s) = \frac{s}{(s+1)(s+2)(s+3)}$$

$$\frac{Y_3(s)}{U(s)} = s^2 T(s) = \frac{s^2}{(s+1)(s+2)(s+3)}$$

Using the partial fractions expansion we get

$$\frac{Y_1(s)}{U(s)} = \frac{\frac{1}{2}}{s+1} + \frac{-1}{s+2} + \frac{\frac{1}{2}}{s+3}$$

$$\frac{Y_2(s)}{U(s)} = \frac{-\frac{1}{2}}{s+1} + \frac{2}{s+2} + \frac{-\frac{3}{2}}{s+3}$$

$$\frac{Y_3(s)}{U(s)} = \frac{\frac{1}{2}}{s+1} + \frac{-4}{s+2} + \frac{\frac{9}{2}}{s+3}$$

or in the matrix form

$$\begin{bmatrix} \dfrac{Y_1(s)}{U(s)} \\[2ex] \dfrac{Y_2(s)}{U(s)} \\[2ex] \dfrac{Y_3(s)}{U(s)} \end{bmatrix} = \begin{bmatrix} \dfrac{1}{2} & -1 & \dfrac{1}{2} \\[2ex] \dfrac{-1}{2} & 2 & -\dfrac{3}{2} \\[2ex] \dfrac{1}{2} & -4 & \dfrac{9}{2} \end{bmatrix} \begin{bmatrix} \dfrac{1}{s+1} \\[2ex] \dfrac{1}{s+2} \\[2ex] \dfrac{1}{s+3} \end{bmatrix}$$

Thus

$$A = \begin{bmatrix} \dfrac{1}{2} & -1 & \dfrac{1}{2} \\[2ex] -\dfrac{1}{2} & 2 & -\dfrac{3}{2} \\[2ex] \dfrac{1}{2} & -4 & \dfrac{9}{2} \end{bmatrix}$$

Let us make the following transformation

$$y = Ax$$

or

$$
\begin{bmatrix} y_1 \\ y_2 \\ y_3 \end{bmatrix} = \begin{bmatrix} \frac{1}{2} & -1 & \frac{1}{2} \\ -\frac{1}{2} & 2 & -\frac{3}{2} \\ \frac{1}{2} & -4 & \frac{9}{2} \end{bmatrix} \begin{bmatrix} x_1 \\ x_2 \\ x_3 \end{bmatrix}
$$

Then we obtain

$$
\begin{bmatrix} \dfrac{X_1(s)}{U(s)} \\[2ex] \dfrac{X_2(s)}{U(s)} \\[2ex] \dfrac{X_3(s)}{U(s)} \end{bmatrix} = \begin{bmatrix} \dfrac{1}{s+1} \\[2ex] \dfrac{1}{s+2} \\[2ex] \dfrac{1}{s+3} \end{bmatrix}
$$

The above transfer function can be transformed into

$$\dot{x}_1 = -x_1 + u$$

$$\dot{x}_2 = -2x_2 + u$$

$$\dot{x}_3 = -3x_3 + u$$

Note that x and y are related by

$$y = Ax$$

and

$$y_1 = y, \qquad y_2 = \dot{y}, \qquad y_3 = \ddot{y}.$$

The response is given by

$$Y(t) = \frac{1}{2} x_1(t) - x_2(t) + \frac{1}{2} x_3(t)$$

● **PROBLEM** 9-5

Obtain a state-space equation for the following transfer function

$$\frac{X(s)}{U(s)} = T(s) = \frac{5}{(s+1)^2 (s+2)}$$

449

Solution: We have

$$\frac{X_1(s)}{U(s)} = \frac{X(s)}{U(s)} = \frac{5}{(s+1)^2} + \frac{-5}{s+1} + \frac{5}{s+2}$$

$$\frac{X_2(s)}{U(s)} = s\frac{X(s)}{U(s)} = \frac{-5}{(s+1)^2} + \frac{10}{s+1} + \frac{-10}{s+2}$$

$$\frac{X_3(s)}{U(s)} = s^2\frac{X(s)}{U(s)} = \frac{5}{(s+1)^2} + \frac{-15}{s+1} + \frac{20}{s+2}$$

The general formula for the transformation matrix is

$$P = \begin{bmatrix} p_{11} & p_{12} & \cdots & p_{1n} \\ p_{21} & p_{22} & \cdots & p_{2n} \\ \cdot & \cdot & & \cdot \\ \cdot & \cdot & & \cdot \\ \cdot & \cdot & & \cdot \\ p_{n1} & p_{n2} & \cdots & p_{nn} \end{bmatrix}$$

where the elements p_{ij} of the matrix P are determined from

$$\frac{X_j(s)}{U(s)} = \frac{p_{j1}}{(s-\lambda_1)^2} + \frac{p_{j2}}{s-\lambda_1} + \frac{p_{j3}}{s-\lambda_3} + \cdots + \frac{p_{jn}}{s-\lambda_n}$$

where $j = 1, 2, \ldots n$.

We get

$$P = \begin{bmatrix} 5 & -5 & 5 \\ -5 & 10 & -10 \\ 5 & -15 & 20 \end{bmatrix}$$

Using the transformation $x = Py$ the transfer function can be transformed into

$$\begin{bmatrix} \dot{y}_1 \\ \dot{y}_2 \\ \dot{y}_3 \end{bmatrix} = \begin{bmatrix} -1 & 1 & 0 \\ 0 & -1 & 0 \\ 0 & 0 & -2 \end{bmatrix} \begin{bmatrix} y_1 \\ y_2 \\ y_3 \end{bmatrix} + \begin{bmatrix} 0 \\ 1 \\ 1 \end{bmatrix} [u]$$

For **the** differential equation we obtain the following state-space equation

$$(p + 1)^2 \, (p + 2)u = 5u \qquad p = \frac{d}{dt}$$

or

$$\dddot{x} + 4\ddot{x} + 5\dot{x} + 2x = 5\,u$$

The initial conditions are

$$y(0) = P^{-1} \begin{bmatrix} x(0) \\ \dot{x}(0) \\ \ddot{x}(0) \end{bmatrix}$$

x is given by

$$x = 5y_1 - 5y_2 + 5y_3$$

● **PROBLEM** 9-6

For a transfer function

$$G(s) = \frac{C(s)}{U(s)} = \frac{s + 4}{s^3 + 6s^2 + 11s + 6}$$

find a state model.

Solution: The corresponding differential equation is

$$\dddot{c} + 6\ddot{c} + 11\dot{c} + 6c = 4u + \dot{u} \tag{1}$$

Since the equation contains the derivative of the input $\dot{u}$ we can not use the standard definition of the state vector, namely

$$x = \begin{bmatrix} x_1 \\ x_2 \\ x_3 \end{bmatrix} = \begin{bmatrix} c \\ \dot{c} \\ \ddot{c} \end{bmatrix}$$

Instead, let us use

$$x = \begin{bmatrix} x_1 \\ x_2 \\ x_3 \end{bmatrix} = \begin{bmatrix} c \\ \dot{c} \\ \ddot{c} + au \end{bmatrix} \tag{2}$$

where a is a constant. The value of a can be determined by substitution of (2) into (1). We have

$$\dot{x}_3 - a\dot{u} + 6(x_3 - au) + 11x_2 + 6x_1 = 4u + \dot{u}$$

Taking a = -1 we eliminate $\dot{u}$, thus

$$\dot{x}_3 = - 6x_1 - 11x_2 - 6x_3 - 2u$$

and $\qquad \dot{x}_1 = x_2$

$$\dot{x}_2 = \ddot{c} = x_3 + u$$

In the matrix form

$$\begin{pmatrix} \dot{x}_1 \\ \dot{x}_2 \\ \dot{x}_3 \end{pmatrix} = \begin{bmatrix} 0 & 1 & 0 \\ 0 & 0 & 1 \\ -6 & -11 & -6 \end{bmatrix} \begin{pmatrix} x_1 \\ x_2 \\ x_3 \end{pmatrix} + \begin{pmatrix} 0 \\ 1 \\ -2 \end{pmatrix} u$$

● **PROBLEM** 9-7

Find three different state variable representations for a single-input, single-output system, whose transfer function is

$$\frac{y(s)}{u(s)} = T(s) = \frac{1}{s^3 + 10s^2 + 31s + 30}$$

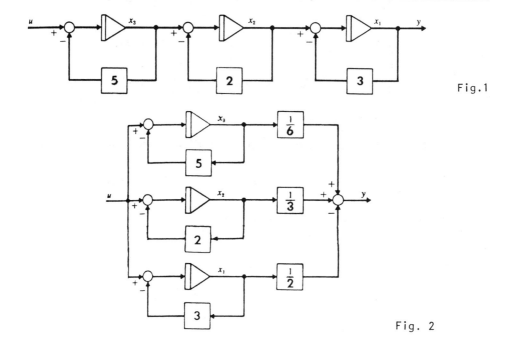

Fig.1

Fig. 2

<u>Solution</u>: 1)   The above transfer function represents the differential equation

$$\dddot{y} + 10\,\ddot{y} + 31\,\dot{y} + 30y = u$$

Setting

$$x_1 = y, \qquad x_2 = \dot{y}, \qquad x_3 = \ddot{y}$$

we obtain

$$\begin{bmatrix} \dot{x}_1 \\ \dot{x}_2 \\ \dot{x}_3 \end{bmatrix} = \begin{bmatrix} 0 & 1 & 0 \\ 0 & 0 & 1 \\ -30 & -31 & -10 \end{bmatrix} \begin{bmatrix} x_1 \\ x_2 \\ x_3 \end{bmatrix} + \begin{bmatrix} 0 \\ 0 \\ 1 \end{bmatrix} u$$

$$y = [1 \quad 0 \quad 0]\, x$$

2) In factored form

$$T(s) = \frac{1}{s^3 + 10s^2 + 31s + 30} = \frac{1}{(s+5)(s+2)(s+3)}$$

The simulation diagram is shown in Fig. 1.

Thus, we get

$$\begin{bmatrix} \dot{x}_1 \\ \dot{x}_2 \\ \dot{x}_3 \end{bmatrix} = \begin{bmatrix} -3 & 1 & 0 \\ 0 & -2 & 1 \\ 0 & 0 & -5 \end{bmatrix} \begin{bmatrix} x_1 \\ x_2 \\ x_3 \end{bmatrix} + \begin{bmatrix} 0 \\ 0 \\ 1 \end{bmatrix} u$$

$$y = [1 \quad 0 \quad 0]\, x$$

3)   Using partial fractions expansion, we have

$$T(s) = \frac{a}{s+5} + \frac{b}{s+2} + \frac{c}{s+3} = \frac{1}{(s+5)(s+2)(s+3)}$$

$$a = \frac{1}{(s+2)(s+3)} \bigg|_{s=-5} = \frac{1}{(-3)(-2)} = \frac{1}{6}$$

$$b = \frac{1}{(s+5)(s+3)} \bigg|_{s=-2} = \frac{1}{3 \cdot 1} = \frac{1}{3}$$

$$c = \frac{1}{(s+5)(s+2)} \bigg|_{s=-3} = \frac{1}{2 \cdot (-1)} = -\frac{1}{2}$$

Thus

$$T(s) = \frac{\frac{1}{6}}{s + 5} + \frac{\frac{1}{3}}{s + 2} - \frac{\frac{1}{2}}{s + 3}$$

The simulation diagram is shown in Fig. 2

The state equations are

$$\begin{bmatrix} \dot{x}_1 \\ \dot{x}_2 \\ \dot{x}_3 \end{bmatrix} = \begin{bmatrix} -3 & 0 & 0 \\ 0 & -2 & 0 \\ 0 & 0 & -5 \end{bmatrix} \begin{bmatrix} x_1 \\ x_2 \\ x_3 \end{bmatrix} + \begin{bmatrix} 1 \\ 1 \\ 1 \end{bmatrix} u \qquad y = [-\frac{1}{2} \quad \frac{1}{3} \quad \frac{1}{6}] x$$

TRANSFORMATION OF DIFFERENTIAL EQUATIONS INTO STATE-SPACE REPRESENTATION

● **PROBLEM** 9-8

Obtain an equivalent discrete-time representation of the system described by the following differential equation

$$\begin{bmatrix} \dot{x}_1 \\ \dot{x}_2 \end{bmatrix} = \begin{bmatrix} 0 & 1 \\ 0 & -2 \end{bmatrix} \begin{bmatrix} x_1 \\ x_2 \end{bmatrix} + \begin{bmatrix} 0 \\ 1 \end{bmatrix} [u]$$

Solution: The following equation gives an equivalent discrete-time system

$$x[(k + 1)T] = G(T)x(kT) + H(T)u(kT)$$

where

$$G(T) = e^{AT}$$

and

$$H(T) = [\int_0^T e^{A\tau}d\tau]B \qquad e^{AT} \text{ can be computed from}$$

$$e^{AT} = L^{-1}[(sI - A)^{-1}]$$

Thus

$$sI - A = \begin{bmatrix} s & -1 \\ 0 & s+2 \end{bmatrix}$$

454

det $(sI - A) = s(s+2)$

$$(sI - A)^{-1} = \frac{1}{s(s + 2)} \begin{bmatrix} s + 2 & 1 \\ 0 & s \end{bmatrix} = \begin{bmatrix} \dfrac{1}{s} & \dfrac{1}{s(s + 2)} \\ 0 & \dfrac{1}{s + 2} \end{bmatrix}$$

$$G(T) = e^{AT} = L^{-1}[(sI - A)^{-1}] = \begin{bmatrix} 1 & \dfrac{1}{2}(1-e^{-2T}) \\ 0 & e^{-2T} \end{bmatrix}$$

and

$$H(T) = \left(\int_0^T e^{A\tau} \, d\tau\right) B = \int_0^T \begin{bmatrix} 1 & \dfrac{1}{2}(1-e^{-2\tau}) \\ 0 & e^{-2\tau} \end{bmatrix} d\tau \cdot \begin{bmatrix} 0 \\ 1 \end{bmatrix}$$

$$= \begin{bmatrix} \dfrac{1}{2}(T + \dfrac{e^{-2T} -1}{2}) \\ \dfrac{1}{2}(1 - e^{-2T}) \end{bmatrix}$$

An equivalent discrete-time representation of the system can be written

$$\begin{bmatrix} x_1[(k + 1)T] \\ x_2[(k + 1)T] \end{bmatrix} = \begin{bmatrix} 1 & \dfrac{1}{2}(1-e^{-2T}) \\ 0 & e^{-2T} \end{bmatrix} \begin{bmatrix} x_1(kT) \\ x_2(kT) \end{bmatrix}$$

$$+ \frac{1}{2} \begin{bmatrix} T + \dfrac{e^{-2T} - 1}{2} \\ 1 - e^{-2T} \end{bmatrix} [u(kT)]$$

● PROBLEM 9-9

Transform the system

$$\dot{x} = Ax + Bu$$

into

$$\dot{z} = A' z + B'u$$

where

$$A = \begin{bmatrix} 0 & 1 & 0 \\ 3 & 0 & 2 \\ -12 & -7 & -6 \end{bmatrix} \quad \text{and} \quad A' = \begin{bmatrix} 0 & 1 & 0 \\ 0 & 0 & 1 \\ -6 & -11 & -6 \end{bmatrix}$$

$$B = \begin{bmatrix} -1 \\ 2 \\ 3 \end{bmatrix} \quad \text{and} \quad B' = \begin{bmatrix} 0 \\ 0 \\ 1 \end{bmatrix}$$

Solution: Let us define the transformation

$$x = T y$$

where

$$T = [B \,\vdots\, AB \,\vdots\, A^2 B]$$

We obtain

$$T = \begin{bmatrix} -1 & 2 & 3 \\ 2 & 3 & -34 \\ 3 & -20 & 75 \end{bmatrix}$$

and

$$T^{-1} = \frac{1}{196} \begin{bmatrix} 455 & 210 & 77 \\ 252 & 84 & 28 \\ 49 & 14 & 7 \end{bmatrix}$$

Since

$$A^3 B = \begin{bmatrix} -34 \\ 159 \\ -248 \end{bmatrix}$$

we get

$$\begin{bmatrix} -a_3 \\ -a_2 \\ -a_1 \end{bmatrix} = T^{-1}(A^3 B) = \begin{bmatrix} -6 \\ -11 \\ -6 \end{bmatrix}$$

Since

$$\dot{y} = T^{-1}ATy + T^{-1}Bu$$

Substituting the above results we can write

$$\begin{bmatrix} \dot{y}_1 \\ \dot{y}_2 \\ \dot{y}_3 \end{bmatrix} = \begin{bmatrix} 0 & 0 & -6 \\ 1 & 0 & -11 \\ 0 & 1 & -6 \end{bmatrix} \begin{bmatrix} y_1 \\ y_2 \\ y_3 \end{bmatrix} + \begin{bmatrix} 1 \\ 0 \\ 0 \end{bmatrix} [u]$$

Eliminating y_1 and y_2 we obtain

$$\dddot{y}_3 + 6\ddot{y}_3 + 11\dot{y}_3 + 6y_3 = u$$

Let

$$z_1 = y_3$$

$$z_2 = \dot{y}_3$$

$$z_3 = \ddot{y}_3$$

Then

$$\begin{bmatrix} \dot{z}_1 \\ \dot{z}_2 \\ \dot{z}_3 \end{bmatrix} = \begin{bmatrix} 0 & 1 & 0 \\ 0 & 0 & 1 \\ -6 & -11 & -6 \end{bmatrix} \begin{bmatrix} z_1 \\ z_2 \\ z_3 \end{bmatrix} + \begin{bmatrix} 0 \\ 0 \\ 1 \end{bmatrix} u$$

● **PROBLEM** 9-10

The system is described by

$$\dot{x} = Ax + Bu$$

where

$$A = \begin{bmatrix} -3 & -2 & -1 \\ 1 & 0 & -1 \\ 0 & 0 & -4 \end{bmatrix} \quad \text{and} \quad B = \begin{bmatrix} 1 \\ 0 \\ 0 \end{bmatrix}$$

Obtain a differential equation of the form

$$a_o \ddot{\ddot{y}} + a_1 \dddot{y} + a_2 \dot{y} + a_3 y = b_o \dddot{u} + b_1 \ddot{u} + b_2 \dot{u} + b_3 u$$

that describes the system.

Solution: For the matrix A we have

$$\det|A-\lambda I| = \begin{vmatrix} -3-\lambda & -2 & -1 \\ 1 & -\lambda & -1 \\ 0 & 0 & -4-\lambda \end{vmatrix} = -(\lambda^3 + 7\lambda^2 + 14\lambda + 8)$$

$$= -(\lambda + 1)(\lambda + 2)(\lambda + 4)$$

We shall define y as a linear function of x_1, x_2, x_3, thus

$$y = ax_1 + bx_2 + cx_3.$$

Furthermore, let

$$y = T^{-1}x = \begin{bmatrix} a & b & c \\ d & e & f \\ g & h & i \end{bmatrix} x$$

where T^{-1} is nonsingular. We choose T^{-1} such that

$$\begin{bmatrix} \dot{y}_1 \\ \dot{y}_2 \\ \dot{y}_3 \end{bmatrix} = \begin{bmatrix} 0 & 1 & 0 \\ 0 & 0 & 1 \\ -8 & -14 & -7 \end{bmatrix} \begin{bmatrix} y_1 \\ y_2 \\ y_3 \end{bmatrix} + \begin{bmatrix} c_1 \\ c_2 \\ c_3 \end{bmatrix} [u]$$

where c_1, c_2, c_3, are constants.

Parameters a, b, c ... f, g, h, i can be determined from the equation

$$\begin{bmatrix} a & b & c \\ d & e & f \\ g & h & i \end{bmatrix} \begin{bmatrix} -3 & -2 & -1 \\ 1 & 0 & -1 \\ 0 & 0 & -4 \end{bmatrix} = \begin{bmatrix} 0 & 1 & 0 \\ 0 & 0 & 1 \\ -8 & -14 & -7 \end{bmatrix} \begin{bmatrix} a & b & c \\ d & e & f \\ g & h & i \end{bmatrix}$$

One of the possible choices of T^{-1} is

$$T^{-1} = \begin{bmatrix} 0 & 1 & 1 \\ 1 & 0 & -5 \\ -3 & -2 & 19 \end{bmatrix}$$

The constants c_1, c_2, c_3 are

$$\begin{bmatrix} c_1 \\ c_2 \\ c_3 \end{bmatrix} = \begin{bmatrix} 0 & 1 & 1 \\ 1 & 0 & -5 \\ -3 & -2 & 19 \end{bmatrix} \begin{bmatrix} 1 \\ 0 \\ 0 \end{bmatrix} = \begin{bmatrix} 0 \\ 1 \\ -3 \end{bmatrix}$$

The parameters b_0, b_1, b_2, b_3 are

$$b_0 = c_0 = 0$$

$$b_1 = c_1 + a_1 c_0 = 0$$

$$b_2 = c_2 + a_1 c_1 + a_2 c_0 = 1$$

$$b_3 = c_3 + a_1 c_2 + a_2 c_1 + a_3 c_0 = 4$$

$$y = y_1 = x_2 + x_3$$

The differential equation that describes the system is

$$\dddot{y} + 7\ddot{y} + 14\dot{y} + 8y = \dot{u} + 4u$$

● **PROBLEM** 9-11

The continuous-time system is described by

$$\begin{bmatrix} \dot{x}_1 \\ \dot{x}_2 \end{bmatrix} = \begin{bmatrix} 0 & 0 \\ 1 & -1 \end{bmatrix} \begin{bmatrix} x_1 \\ x_2 \end{bmatrix} + \begin{bmatrix} 1 \\ 0 \end{bmatrix} [u]$$

Obtain a discrete-time state-space representation of the system.

Solution: The discrete-time state equation is given by

$$x[(k + 1)T] = G(T) x(kT) + H(T) u(kT)$$

where

$$G(T) = e^{AT}$$

We know that

$$e^{AT} = L^{-1}[(SI - A)^{-1}]$$

Thus

$$SI - A = \begin{bmatrix} S & 0 \\ -1 & S + 1 \end{bmatrix}$$

$$(SI - A)^{-1} = \frac{1}{S(S + 1)} \begin{bmatrix} S + 1 & 0 \\ 1 & S \end{bmatrix} = \begin{bmatrix} \frac{1}{S} & 0 \\ \frac{1}{S(S + 1)} & \frac{1}{S + 1} \end{bmatrix}$$

$$G(T) = e^{AT} = \begin{bmatrix} 1 & 0 \\ 1 - e^{-T} & e^{-T} \end{bmatrix}$$

$$H(T) = \left[\int_0^T e^{At} dt \right] B = \left[\int_0^T \begin{pmatrix} 1 & 0 \\ 1-e^{-t} & e^{-t} \end{pmatrix} dt \right] \cdot \begin{bmatrix} 1 \\ 0 \end{bmatrix}$$

$$= \begin{bmatrix} T & 0 \\ T + e^{-T} - 1 & -e^{-T} + 1 \end{bmatrix} \begin{bmatrix} 1 \\ 0 \end{bmatrix} = \begin{bmatrix} T \\ T + e^{-T} - 1 \end{bmatrix}$$

We can write now

$$\begin{bmatrix} x_1[(k + 1)T] \\ x_2[(k + 1)T] \end{bmatrix} = \begin{bmatrix} 1 & 0 \\ 1 - e^{-T} & e^{-T} \end{bmatrix} \begin{bmatrix} x_1(kT) \\ x_2(kT) \end{bmatrix} + \begin{bmatrix} T \\ T + e^{-T} - 1 \end{bmatrix} u(kT)$$

● **PROBLEM** 9-12

Form the vector-matrix state equation for the system whose normal-form state equations are

$$Dx_1 = 4x_1 + 2x_2 + 2u_1$$

$$Dx_2 = 3x_1 + 12x_2 + 3u_1 + 3u_2$$

$$Dx_3 = 2x_1 + u_1$$

Solution: The state vector and the input vector are, respectively

$$\bar{x} = \begin{bmatrix} x_1 \\ x_2 \\ x_3 \end{bmatrix} \qquad \bar{u} = \begin{bmatrix} u_1 \\ u_2 \end{bmatrix}$$

The equation

$$D\bar{x} = A\bar{x} + B\bar{u}$$

yields

$$D \begin{bmatrix} x_1 \\ x_2 \\ x_3 \end{bmatrix} = \begin{bmatrix} 4 & 2 & 0 \\ 3 & 12 & 0 \\ 2 & 0 & 0 \end{bmatrix} \begin{bmatrix} x_1 \\ x_2 \\ x_3 \end{bmatrix} + \begin{bmatrix} 2 & 0 \\ 3 & 3 \\ 1 & 0 \end{bmatrix} \begin{bmatrix} u_1 \\ u_2 \end{bmatrix}$$

Performing the multiplication it is easy to verify the above result.

Some phenomena like the spread of an epidemic disease or of an information through the population can be described by the following system of equations

$$\frac{dx_1}{dt} = -Ax_1 - Bx_2 + u_1(t)$$

$$\frac{dx_2}{dt} = Bx_1 - Cx_2 + u_2(t)$$

$$\frac{dx_3}{dt} = Ax_1 + Cx_2$$

For a closed system we have $u_1(t) = u_2(t) = 0$. Represent the system as a signal-flow graph. Assuming $u_1(t) = u_2(t) = 0$, what will be the steady state values of x_1 and x_2?

Determine constraints on A, B, and C so that the system is stable.

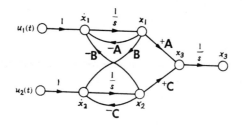

Solution: The state variables are x_1, x_2, and x_3. The signal-flow diagram representing the above set of differential equations is shown.

The system in the vector form is

$$\begin{bmatrix} \dot{x}_1 \\ \dot{x}_2 \\ \dot{x}_3 \end{bmatrix} = \begin{bmatrix} -A & -B & 0 \\ B & -C & 0 \\ A & C & 0 \end{bmatrix} \begin{bmatrix} x_1 \\ x_2 \\ x_3 \end{bmatrix} + \begin{bmatrix} 1 & 0 \\ 0 & 1 \\ 0 & 0 \end{bmatrix} \begin{bmatrix} u_1 \\ u_2 \end{bmatrix}$$

The state variable x_3 does not affect the variables x_1 and x_2 $\frac{\partial x_1}{\partial x_3} = \frac{\partial x_2}{\partial x_3} = 0$. From the equation $\frac{dx}{dt} = 0$ we obtain the equilibrium point.

It is the point to which the system settles in the equilibrium or rest condition. It is easy to see that the equilibrium point for this system is $x_1 = 0$ and $x_2 = 0$. From the signal-flow graph we obtain the flow graph determinant

$$\Delta(s) = 1 - (-As^{-1} - Cs^{-1} - B^2 s^{-2}) + ACs^{-2}$$

There are three loops, two of which are nontouching. The characteristic equation is

$$q(s) = s^2 \Delta(s) = s^2 + (A + C)s + (AC + B^2) = 0$$

Thus, we conclude that the system is stable when

     $A + C > 0$

and

     $AC + B^2 > 0$

● **PROBLEM** 9-14

A system is described by the second order differential equation

     $\ddot{c}(t) + 3\dot{c}(t) + 2c(t) = r(t)$

where $c(t)$ is the output and the $r(t)$ is a unit-step input. The initial conditions are given for $t = t_0$, $c(t_0)$ and $\dfrac{dc(t_0)}{dt}$. Find a state equation for the system.

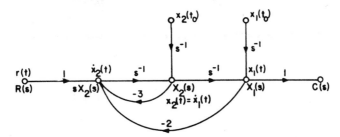

Solution: Let us rewrite the equation in the form

     $\ddot{c} = -3\dot{c} + 2c + r$

A state diagram of the system is shown.

The state variables of the system $x_1(t)$ and $x_2(t)$ are output variables of the integrators. Using Mason's gain formula, we obtain from the state diagram

$$\begin{bmatrix} X_1(s) \\ X_2(s) \end{bmatrix} = \frac{1}{(s+1)(s+2)} \begin{bmatrix} s+3 & 1 \\ -2 & s \end{bmatrix} \begin{bmatrix} x_1(t_0) \\ x_2(t_0) \end{bmatrix}$$

$$+ \begin{bmatrix} \dfrac{1}{(s+1)(s+2)} \\ \dfrac{s}{(s+1)(s+2)} \end{bmatrix} R(s) \qquad (1)$$

Note, that since the input is a unit-step function, we have

     $R(s) = \dfrac{1}{s}$

The inverse Laplace transform of (1) is

$$\begin{bmatrix} x_1(t) \\ x_2(t) \end{bmatrix} = \begin{bmatrix} 2e^{-(t-t_0)} - e^{-2(t-t_0)} & e^{-(t-t_0)} - e^{-2(t-t_0)} \\ -2e^{-(t-t_0)} + 2e^{-2(t-t_0)} & -e^{-(t-t_0)} + 2e^{-2(t-t_0)} \end{bmatrix}$$

$$\begin{bmatrix} x_1(t_0) \\ x_2(t_0) \end{bmatrix} + \begin{bmatrix} \frac{1}{2} - e^{-(t-t_0)} + \frac{1}{2}e^{-2(t-t_0)} \\ e^{-(t-t_0)} - e^{-2(t-t_0)} \end{bmatrix}$$

for $t \geq t_0$

Note that, since $t_0 \neq 0$, we have $L^{-1}[\frac{1}{s+a}] = e^{-a(t-t_0)}, t \geq t_0$.

We obtain the dynamic equations of the system from the state diagram by applying Mason's gain formula to the nodes $\dot{x}_1(t)$ and $\dot{x}_2(t)$. The inputs are $x_1(t)$, $x_2(t)$ and $r(t)$.

$$\begin{bmatrix} \dot{x}_1(t) \\ \dot{x}_2(t) \end{bmatrix} = \begin{bmatrix} 0 & 1 \\ -2 & -3 \end{bmatrix} \begin{bmatrix} x_1(t) \\ x_2(t) \end{bmatrix} + \begin{bmatrix} 0 \\ 1 \end{bmatrix} r(t)$$

$$c(t) = [1 \ \ 0] \begin{bmatrix} x_1(t) \\ x_2(t) \end{bmatrix}$$

● **PROBLEM** 9-15

Describe the system represented by the following differential equation

$$\frac{d^3c(t)}{dt^3} + 4\frac{d^2c(t)}{dt^2} + \frac{dc(t)}{dt} + 3c(t) = r(t)$$

in terms of state variables.

Solution: Rearranging the equation we get

$$\frac{d^3c(t)}{dt^3} = -4\frac{d^2c(t)}{dt^2} - \frac{dc(t)}{dt} - 3c(t) + r(t)$$

Let us define the state variables

$$x_1(t) = c(t)$$

$$x_2(t) = \frac{dc(t)}{dt}$$

$$x_3(t) = \frac{d^2c(t)}{dt^2}$$

and the state equations in the form

$$\dot{x}(t) = Ax + Br$$

where

$$A = \begin{bmatrix} 0 & 1 & 0 \\ 0 & 0 & 1 \\ -3 & -1 & -4 \end{bmatrix} \quad B = \begin{bmatrix} 0 \\ 0 \\ 1 \end{bmatrix}$$

The output equation is

$$c(t) = x_1(t)$$

463

Obtain a scalar differential equation of the form

$$\ddot{x} + a_1\dot{x} + a_2\dot{x} + a_3x = u$$

that describes the system

$$\begin{bmatrix} \dot{x}_1 \\ \dot{x}_2 \\ \dot{x}_3 \end{bmatrix} = \begin{bmatrix} -4 & 2 & 0 \\ 0 & 3 & 1 \\ -3 & 0 & 0 \end{bmatrix} \begin{bmatrix} x_1 \\ x_2 \\ x_3 \end{bmatrix} + \begin{bmatrix} 1 \\ 3 \\ 1 \end{bmatrix} [u] \tag{1}$$

Solution: From eq. (1) we have

$$A = \begin{bmatrix} -4 & 2 & 0 \\ 0 & 3 & 1 \\ -3 & 0 & 0 \end{bmatrix} \qquad B = \begin{bmatrix} 1 \\ 3 \\ 1 \end{bmatrix}$$

$$AB = \begin{bmatrix} 2 \\ 10 \\ -3 \end{bmatrix} \qquad A^2B = \begin{bmatrix} 12 \\ 27 \\ -6 \end{bmatrix}$$

Thus we can define matrix H

$$H = \begin{bmatrix} 1 & 2 & 12 \\ 3 & 10 & 27 \\ 1 & -3 & -6 \end{bmatrix}$$

Let

$$x = Hy$$

we obtain

$$\dot{y} = H^{-1}AHy + H^{-1}Bu$$

$$H^{-1} = \begin{bmatrix} -\dfrac{7}{39} & \dfrac{8}{39} & \dfrac{22}{39} \\ -\dfrac{5}{13} & \dfrac{2}{13} & -\dfrac{1}{13} \\ \dfrac{19}{117} & \dfrac{-5}{117} & -\dfrac{4}{117} \end{bmatrix} \qquad H^{-1}AH = \begin{bmatrix} 0 & 0 & -6 \\ 1 & 0 & 12 \\ 0 & 1 & -1 \end{bmatrix}$$

$$H^{-1}B = \begin{bmatrix} 1 \\ 0 \\ 0 \end{bmatrix}$$

Rewriting we obtain

$$\begin{bmatrix} \dot{y}_1 \\ \dot{y}_2 \\ \dot{y}_3 \end{bmatrix} = \begin{bmatrix} 0 & 0 & -6 \\ 1 & 0 & 12 \\ 0 & 1 & -1 \end{bmatrix} \begin{bmatrix} y_1 \\ y_2 \\ y_3 \end{bmatrix} + \begin{bmatrix} 1 \\ 0 \\ 0 \end{bmatrix} [u]$$

Eliminating y_1 and y_2 from the above system of equations we get

$$\dddot{y}_3 + \ddot{y}_3 - 12\dot{y}_3 + 6y_3 = u$$

The system is described by

$$\dddot{y} + 6\ddot{y} + 11\dot{y} + 6y = 6u$$

Obtain a state representation of the system using the partial fractions expansion method.

Solution: The transfer function of the system is given by

$$\frac{Y(s)}{U(s)} = \frac{6}{s^3 + 6s^2 + 11s + 6}$$

It is easy to notice that

$$s^3 + 6s^2 + 11s + 6 = (s + 1)(s + 2)(s + 3)$$

Thus, using the partial fractions expansion we obtain

$$\frac{Y(s)}{U(s)} = \frac{6}{(s + 1)(s + 2)(s + 3)} = \frac{3}{s + 1} + \frac{-6}{s + 2} + \frac{3}{s + 3}$$

or

$$Y(s) = \frac{3}{s + 1}U(s) + \frac{-6}{s + 2}U(s) + \frac{3}{s + 3}U(s)$$

$$= X_1(s) + X_2(s) + X_3(s)$$

We shall define $\begin{bmatrix} X_1(s) \\ X_2(s) \\ X_3(s) \end{bmatrix}$ as

$$\begin{bmatrix} X_1(s) \\ X_2(s) \\ X_3(s) \end{bmatrix} = \begin{bmatrix} \dfrac{3}{s + 1} \\ \dfrac{-6}{s + 2} \\ \dfrac{3}{s + 3} \end{bmatrix} U(s)$$

Taking the inverse Laplace transform we get

$$\begin{bmatrix} \dot{x}_1 \\ \dot{x}_2 \\ \dot{x}_3 \end{bmatrix} = \begin{bmatrix} -x_1 \\ -2x_2 \\ -3x_3 \end{bmatrix} + \begin{bmatrix} 3 \\ -6 \\ 3 \end{bmatrix} u$$

Since

$$Y(s) = X_1(s) + X_2(s) + X_3(s)$$

we can write

$$y(t) = x_1(t) + x_2(t) + x_3(t)$$

Using the matrix notation, we get

$$\begin{bmatrix} \dot{x}_1 \\ \dot{x}_2 \\ \dot{x}_3 \end{bmatrix} = \begin{bmatrix} -1 & 0 & 0 \\ 0 & -2 & 0 \\ 0 & 0 & -3 \end{bmatrix} \begin{bmatrix} x_1 \\ x_2 \\ x_3 \end{bmatrix} + \begin{bmatrix} 3 \\ -6 \\ 3 \end{bmatrix} [u]$$

$$y = \begin{bmatrix} 1 & 1 & 1 \end{bmatrix} \begin{bmatrix} x_1 \\ x_2 \\ x_3 \end{bmatrix}$$

● **PROBLEM** 9-18

The system is described by

$$\ddot{y} + 6\ddot{y} + 11\dot{y} + 6y = 6u$$

Obtain a state equation for the system in diagonal form.

Solution: Let

$$x_1 = y$$
$$x_2 = \dot{y}$$
$$x_3 = \ddot{y}$$

then

$$\dot{x}_1 = x_2$$
$$\dot{x}_2 = x_3$$
$$\dot{x}_3 = -6x_1 - 11x_2 - 6x_3 + 6u$$

or in the matrix form

$$\begin{bmatrix} \dot{x}_1 \\ \dot{x}_2 \\ \dot{x}_3 \end{bmatrix} = \begin{bmatrix} 0 & 1 & 0 \\ 0 & 0 & 1 \\ -6 & -11 & -6 \end{bmatrix} \begin{bmatrix} x_1 \\ x_2 \\ x_3 \end{bmatrix} + \begin{bmatrix} 0 \\ 0 \\ 6 \end{bmatrix} [u]$$

with
$$y = \begin{bmatrix} 1 & 0 & 0 \end{bmatrix} \begin{bmatrix} x_1 \\ x_2 \\ x_3 \end{bmatrix}$$

The transformation matrix T is given by

$$T = \begin{bmatrix} 1 & 1 & ..1 \\ \lambda_1 & \lambda_2 & \lambda_n \\ \lambda_1^2 & \lambda_2^2 & \lambda_n^2 \\ \cdot & \cdot & \cdot \\ \cdot & \cdot & \cdot \\ \cdot & \cdot & \cdot \\ \lambda_1^n & \lambda_2^n & \lambda_n^n \end{bmatrix}$$

where $\lambda_1 \neq \lambda_2 \neq \ldots \neq \lambda_n$ are the eigenvalues of the coefficient matrix.

In our case $\lambda_1 = -1$, $\lambda_2 = -2$, $\lambda_3 = -3$

and
$$x = Tz$$

where

$$T = \begin{bmatrix} 1 & 1 & 1 \\ -1 & -2 & -3 \\ 1 & 4 & 9 \end{bmatrix}$$

Substituting the above transformation into

$$\dot{x} = Ax + Bu$$

we obtain

$$T\dot{z} = ATz + Bu$$

or

$$\dot{z} = T^{-1}ATz + T^{-1}Bu$$

Thus, the system can be written as

$$\begin{bmatrix} \dot{z}_1 \\ \dot{z}_2 \\ \dot{z}_3 \end{bmatrix} = \begin{bmatrix} 3 & 2.5 & 0.5 \\ -3 & -4 & -1 \\ 1 & 1.5 & 0.5 \end{bmatrix} \begin{bmatrix} 0 & 1 & 0 \\ 0 & 0 & 1 \\ -6 & -11 & -6 \end{bmatrix} \begin{bmatrix} 1 & 1 & 1 \\ -1 & -2 & -3 \\ 1 & 4 & 9 \end{bmatrix} \begin{bmatrix} z_1 \\ z_2 \\ z_3 \end{bmatrix}$$

$$+ \begin{bmatrix} 3 & 2.5 & 0.5 \\ -3 & -4 & -1 \\ 1 & 1.5 & 0.5 \end{bmatrix} \begin{bmatrix} 0 \\ 0 \\ 6 \end{bmatrix} [u] = \begin{bmatrix} -1 & 0 & 0 \\ 0 & -2 & 0 \\ 0 & 0 & -3 \end{bmatrix} \begin{bmatrix} z_1 \\ z_2 \\ z_3 \end{bmatrix} + \begin{bmatrix} 3 \\ -6 \\ 3 \end{bmatrix} [u]$$

The output equation is

$$y = \begin{bmatrix} 1 & 0 & 0 \end{bmatrix} \begin{bmatrix} 1 & 1 & 1 \\ -1 & -2 & -3 \\ 1 & 4 & 9 \end{bmatrix} \begin{bmatrix} z_1 \\ z_2 \\ z_3 \end{bmatrix} = \begin{bmatrix} 1 & 1 & 1 \end{bmatrix} \begin{bmatrix} z_1 \\ z_2 \\ z_3 \end{bmatrix}$$

● **PROBLEM** 9-19

A system with two inputs and two outputs is described by

$$\ddot{y}_1 + 4\dot{y}_1 + 2y_2 = u_1 + 2u_2 + 2\dot{u}_2$$

$$\ddot{y}_2 + 5\dot{y}_1 + 3y_2 = \ddot{u}_2 + 3\dot{u}_2 + u_1$$

Define the state variables and find the state-space equations for the system.

Solution: Integrating each equation twice we obtain

$$y_1 = \int\int (-4\dot{y}_1 - 2y_2 + u_1 + 2u_2 + 2\dot{u}_2)dt'dt$$

$$= \int(-4y_1 + 2u_2 + \int[-2y_2 + u_1 + 2u_2]dt')dt$$

$$y_2 = \int\int (-5\dot{y}_1 - 3y_2 + \ddot{u}_2 + 3\dot{u}_2 + u_1)dt'dt$$

$$= u_2 + \int(-5y_1 + 3u_2 + \int[-3y_2 + u_1]dt')dt$$

The simulation diagram is shown in the Figure.

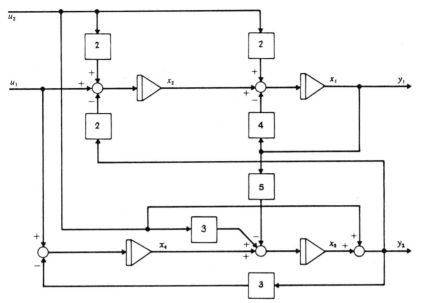

The state equations are

$$\dot{x}_1 = -4x_1 + x_2 + 2u_2$$

$$\dot{x}_2 = -2x_3 + u_1$$

$$\dot{x}_3 = -5x_1 + x_4 + 3u_2$$

$$\dot{x}_4 = -3x_3 + u_1 - 3u_2$$

or in the matrix form

$$\dot{x} = \begin{bmatrix} -4 & 1 & 0 & 0 \\ 0 & 0 & -2 & 0 \\ -5 & 0 & 0 & 1 \\ 0 & 0 & -3 & 0 \end{bmatrix} x + \begin{bmatrix} 0 & 2 \\ 1 & 0 \\ 0 & 3 \\ 1 & -3 \end{bmatrix} u$$

The output equation is

$$y = \begin{bmatrix} y_1 \\ y_2 \end{bmatrix} = \begin{bmatrix} 1 & 0 & 0 & 0 \\ 0 & 0 & 1 & 0 \end{bmatrix} x + \begin{bmatrix} 0 & 0 \\ 0 & 1 \end{bmatrix} u$$

● **PROBLEM** 9-20

Find the system of three state equations that represents the following differential equation

$$\frac{d^3c(t)}{dt^2} + 2\frac{d^2c(t)}{dt^2} + 3\frac{dc}{dt} + 4c(t) = \frac{dr(t)}{dt} + 5r(t).$$

Solution: The term $\frac{dr(t)}{dt}$ can not appear in the right side of the state equations. Let us rewrite the equation, in the form

468

$$\frac{d^3c(t)}{dt^3} - \frac{dr(t)}{dt} = -2\frac{d^2c(t)}{dt^2} - 3\frac{dc}{dt} - 4c(t) + 5r(t)$$

The state variables are

$$x_1(t) = c(t)$$

$$x_2(t) = \frac{dc(t)}{dt}$$

$$x_3(t) = \frac{d^2c(t)}{dt^2} - r(t)$$

Using the new state variables we obtain a system of equations that is equivalent to the initial equation.

$$\frac{dx_1(t)}{dt} = x_2(t)$$

$$\frac{dx_2(t)}{dt} = x_3(t) + r(t)$$

$$\frac{dx_3(t)}{dt} = -4x_1(t) - 3x_2(t) - 2x_3(t) + 3r(t).$$

In the matrix form

$$\begin{pmatrix} \dot{x}_1 \\ \dot{x}_2 \\ \dot{x}_3 \end{pmatrix} = \begin{bmatrix} 0 & 1 & 0 \\ 0 & 0 & 1 \\ -4 & -3 & -2 \end{bmatrix} \begin{pmatrix} x_1 \\ x_2 \\ x_3 \end{pmatrix} + \begin{pmatrix} 0 \\ 1 \\ 3 \end{pmatrix} r(t)$$

● **PROBLEM** 9-21

Consider a system with three inputs u_1, u_2 and u_3 and three outputs y_1, y_2, y_3. The input output equations are

$$\dddot{y}_1 + a_1\ddot{y}_1 + a_2\dot{y}_1 + a_3(y_1-y_2) = u_1(t)$$

$$\dot{y}_2 + a_4(y_2+y_3) + a_5(y_1-y_2) = u_2(t)$$

$$\ddot{y}_3 + a_6(y_3-y_2) = u_3(t)$$

Write state variable equations for the system.

Solution: We shall define state variables as the outputs and their derivatives up to the highest (n-1) order, thus

$$x_1 = y_1, \quad x_2 = \dot{y}_1, \quad x_3 = \ddot{y}_1, \quad x_4 = y_2, \quad x_5 = y_3, \quad x_6 = \dot{y}_3$$

Then

$$\dot{x}_1 = x_2, \quad \dot{x}_2 = x_3, \quad \dot{x}_5 = x_6$$

$$\dot{x}_3 = -a_1 x_3 - a_2 x_2 - a_3(x_1-x_4) + u_1$$

$$\dot{x}_4 = -a_4(x_4+x_5) - a_5(x_2-x_4) + u_2$$

$$\dot{x}_6 = -a_6(x_5-x_4) + u_3$$

In the matrix form

$$
\begin{bmatrix} \dot{x}_1 \\ \dot{x}_2 \\ \dot{x}_3 \\ \dot{x}_4 \\ \dot{x}_5 \\ \dot{x}_6 \end{bmatrix}
=
\begin{bmatrix}
0 & 1 & 0 & 0 & 0 & 0 \\
0 & 0 & 1 & 0 & 0 & 0 \\
-a_3 & -a_2 & -a_1 & a_3 & 0 & 0 \\
0 & -a_5 & 0 & (a_5-a_4) & -a_4 & 0 \\
0 & 0 & 0 & 0 & 0 & 1 \\
0 & 0 & 0 & a_6 & -a_6 & 0
\end{bmatrix}
\begin{bmatrix} x_1 \\ x_2 \\ x_3 \\ x_4 \\ x_5 \\ x_6 \end{bmatrix}
+
\begin{bmatrix}
0 & 0 & 0 \\
0 & 0 & 0 \\
1 & 0 & 0 \\
0 & 1 & 0 \\
0 & 0 & 0 \\
0 & 0 & 1
\end{bmatrix}
\begin{bmatrix} u_1 \\ u_2 \\ u_3 \end{bmatrix}
$$

The output equation is

$$
\begin{bmatrix} y_1 \\ y_2 \\ y_3 \end{bmatrix}
=
\begin{bmatrix}
1 & 0 & 0 & 0 & 0 & 0 \\
0 & 0 & 0 & 1 & 0 & 0 \\
0 & 0 & 0 & 0 & 1 & 0
\end{bmatrix} x
$$

● **PROBLEM** 9-22

Obtain the state-space representation for the system, whose system equation is

$$\dddot{x} + 6\ddot{x} + 11\dot{x} + 6x = \dddot{u} + 8\ddot{u} + 17\dot{u} + 8u$$

Diagonalize the system and then transform it so that the driving functions become unity.

Solution: An equation

$$\dddot{x} + a_1\ddot{x} + a_2\dot{x} + a_3 x = b_0\dddot{u} + b_1\ddot{u} + b_2\dot{u} + b_3 u$$

can be written in the form

$$
\begin{bmatrix} \dot{x}_1 \\ \dot{x}_2 \\ \dot{x}_3 \end{bmatrix}
=
\begin{bmatrix}
-a_1 & 1 & 0 \\
-a_2 & 0 & 1 \\
-a_3 & 0 & 0
\end{bmatrix}
\begin{bmatrix} x_1 \\ x_2 \\ x_3 \end{bmatrix}
+
\begin{bmatrix} c_1 \\ c_2 \\ c_3 \end{bmatrix} u
$$

470

where

$$c_1 = b_1 - a_1 b_0$$

$$c_2 = b_2 - a_2 b_0$$

$$c_3 = b_3 - a_3 b_0$$

If we choose the state variables as

$$x_1 = x - b_0 u$$

$$\dot{x}_1 = -a_1 x_1 + x_2 + (b_1 - a_1 b_0) u$$

$$\dot{x}_2 = -a_2 x_1 + x_3 + (b_2 - a_2 b_0) u$$

$$\dot{x}_3 = -a_3 x_1 + (b_3 - a_3 b_0) u$$

we get

$$\begin{bmatrix} \dot{x}_1 \\ \dot{x}_2 \\ \dot{x}_3 \end{bmatrix} = \begin{bmatrix} -6 & 1 & 0 \\ -11 & 0 & 1 \\ -6 & 0 & 0 \end{bmatrix} \begin{bmatrix} x_1 \\ x_2 \\ x_3 \end{bmatrix} + \begin{bmatrix} 2 \\ 6 \\ 2 \end{bmatrix} [u]$$

The roots of the characteristic equation are

$$\lambda_1 = -1, \qquad \lambda_2 = -2, \qquad \lambda_3 = -3$$

Let us define the following transformation

$$x = Ty$$

where T is the matrix of eigenvectors of A

$$T = [t_1\ t_2\ t_3]$$

We have

$$(\lambda_i I - A) t_i = 0$$

for $\qquad \lambda_1 = -1, \qquad \lambda_2 = -2, \qquad \lambda_3 = -3$

If we choose $t_{i1} = 1$, then

multiplying out gives

$$t_{i2} = \lambda_i + 6, \qquad t_{i3} = \frac{6}{\lambda_i}.$$

Thus

$$T = \begin{bmatrix} 1 & 1 & 1 \\ 5 & 4 & 3 \\ 6 & 3 & 2 \end{bmatrix}$$

and

$$T^{-1} = -\frac{1}{2} \begin{bmatrix} -1 & 1 & -1 \\ 8 & -4 & 2 \\ -9 & 3 & -1 \end{bmatrix}$$

Hence

$$\begin{bmatrix} \dot{y}_1 \\ \dot{y}_2 \\ \dot{y}_3 \end{bmatrix} = \begin{bmatrix} -1 & 0 & 0 \\ 0 & -2 & 0 \\ 0 & 0 & -3 \end{bmatrix} \begin{bmatrix} y_1 \\ y_2 \\ y_3 \end{bmatrix} + \begin{bmatrix} \frac{1}{2} & -\frac{1}{2} & \frac{1}{2} \\ -4 & 2 & -1 \\ \frac{9}{2} & -\frac{3}{2} & \frac{1}{2} \end{bmatrix} \begin{bmatrix} 2 \\ 6 \\ 2 \end{bmatrix} [u]$$

$$= \begin{bmatrix} -1 & 0 & 0 \\ 0 & -2 & 0 \\ 0 & 0 & -3 \end{bmatrix} \begin{bmatrix} y_1 \\ y_2 \\ y_3 \end{bmatrix} + \begin{bmatrix} -1 \\ 2 \\ 1 \end{bmatrix} [u]$$

Let us define the transformation

$$y = T'z$$

where

$$T' = \begin{bmatrix} -1 & 0 & 0 \\ 0 & 2 & 0 \\ 0 & 0 & 1 \end{bmatrix}$$

We obtain

$$\begin{bmatrix} \dot{z}_1 \\ \dot{z}_2 \\ \dot{z}_3 \end{bmatrix} = \begin{bmatrix} -1 & 0 & 0 \\ 0 & -2 & 0 \\ 0 & 0 & -3 \end{bmatrix} \begin{bmatrix} z_1 \\ z_2 \\ z_3 \end{bmatrix} + \begin{bmatrix} 1 \\ 1 \\ 1 \end{bmatrix} [u]$$

The initial condition is given by

$$z(0) = (T')^{-1} T^{-1} x(0)$$

● **PROBLEM** 9-23

Consider the following n-th order differential equation

$$a_n \frac{d^n y}{dt^n} + a_{n-1} \frac{d^{n-1} y}{dt^{n-1}} + \ldots + a_1 \frac{dy}{dt} + a_0 y =$$

$$\beta_n \frac{d^n u}{dt^n} + \beta_{n-1} \frac{d^{n-1} u}{dt^{n-1}} + \ldots + \beta_1 \frac{du}{dt} + \beta_0 u$$

We assume that all coefficients are constant. Find the
state variable representation of the above equation by sim-
ulating the system with an analog computer diagram. Then
find the same answer by taking integrator outputs as state
variables.

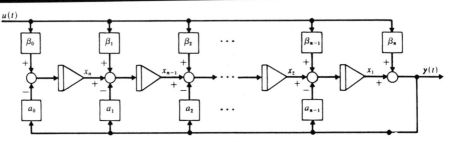

Solution: Integrating the equation n times we obtain

$$a_n y(t) - \beta_n u(t) = \underbrace{\int\!\!\int}_{n \text{ times}} \cdots \int \left[\left(\beta_{n-1} \frac{d^{n-1} u}{dt^{n-1}} - \alpha_{n-1} \frac{d^{n-1} y}{dt^{n-1}} \right) + \right.$$

$$\left. \cdots + \left(\beta_1 \frac{du}{dt} - \alpha_1 \frac{du}{dt} \right) + (\beta_0 u - \alpha_0 u) \right] dt' \ldots dt^n$$

The k th derivative can be integrated k times, so it appears
to the right of n-k integral signs.

$$a_n y(t) = \beta_n u + \int \{ (\beta_{n-1} u - a_{n-1} y) + \int [\beta_{n-2} u - a_{n-2} y]$$

$$+ \int (\ldots + \int \{ \beta_1 u - a_1 y + \int (\beta_0 u - a_0 y) dt \} \ldots) dt'] dt'' dt'''$$

From the above equation we draw the analog computer, assuming $a_n = 1$ we obtain the figure shown.

Let us number the outputs of integrators from the right,
thus we get

$$\dot{x}_1 = -a_{n-1}[x_1 + \beta_n u] + \beta_{n-1} u + x_2 = -a_{n-1} x_1 + x_2 + (\beta_{n-1} - a_{n-1}\beta_n) u$$

$$\dot{x}_2 = -a_{n-2} x_1 + x_3 + (\beta_{n-2} - a_{n-2} \beta_n) u$$

$$\dot{x}_{n-1} = -a_1 x_1 + x_n + (\beta_1 - a_1 \beta_n) u$$

$$\dot{x}_n = -a_0 x_1 + (\beta_0 - a_0 \beta_n) u$$

where the output equation is given by

$$y = x_1 + \beta_n u = [1 \; 0 \; 0 \; \ldots \; 0] x + \beta_n u$$

● **PROBLEM** 9-24

Obtain a state space equation of the following system

$$(D+1)(D+2)(D+3) x = (D+1) u,$$

$$D = \frac{d}{dt} \qquad (1)$$

Solution: Let us define

$$x = x_1$$

$$\dot{x}_1 = x_2 + a_1 u$$

$$\dot{x}_2 = x_3 + a_2 u$$

$$\dot{x}_3 = -6x_1 - 11x_2 - 6x_3 + a_3 u$$

In the matrix form

$$\begin{bmatrix} \dot{x}_1 \\ \dot{x}_2 \\ \dot{x}_3 \end{bmatrix} = \begin{bmatrix} 0 & 1 & 0 \\ 0 & 0 & 1 \\ -6 & -11 & -6 \end{bmatrix} \begin{bmatrix} x_1 \\ x_2 \\ x_3 \end{bmatrix} + \begin{bmatrix} a_1 \\ a_2 \\ a_3 \end{bmatrix} [u] \qquad (2)$$

To find constants a_1, a_2, a_3 let us note that

$$x = x_1$$

$$\dot{x} = \dot{x}_1 = x_2 + a_1 u$$

$$\ddot{x} = \ddot{x}_1 = \dot{x}_2 + a_1 \dot{u} = x_3 + a_2 u + a_1 \dot{u}$$

$$\dddot{x} = \dddot{x}_1 = \dot{x}_3 + a_2 \dot{u} + a_1 \ddot{u} = -6x_1 - 11x_2 - 6x_3 + a_3 u + a_2 \dot{u} + a_1 \ddot{u}$$

Substituting into (1) and equating coefficients we get

$$a_3 + 6a_2 + 11a_1 = 1$$

$$a_2 + 6a_1 = 1$$

$$a_1 = 0$$

Thus $a_1 = 0$, $a_2 = 1$, $a_3 = -5$ and (2) becomes

$$\begin{bmatrix} \dot{x}_1 \\ \dot{x}_2 \\ \dot{x}_3 \end{bmatrix} = \begin{bmatrix} 0 & 1 & 0 \\ 0 & 0 & 1 \\ -6 & -11 & -6 \end{bmatrix} \begin{bmatrix} x_1 \\ x_2 \\ x_3 \end{bmatrix} + \begin{bmatrix} 0 \\ 1 \\ -5 \end{bmatrix} [u]$$

The above equation can be simplified into

$$\begin{bmatrix} \dot{y}_1 \\ \dot{y}_2 \\ \dot{y}_3 \end{bmatrix} = \begin{bmatrix} -1 & 0 & 0 \\ 0 & -2 & 0 \\ 0 & 0 & -3 \end{bmatrix} \begin{bmatrix} y_1 \\ y_2 \\ y_3 \end{bmatrix} + \begin{bmatrix} 0 \\ 1 \\ 1 \end{bmatrix} [u]$$

where x and y are related by

$$\begin{bmatrix} x_1 \\ x_2 \\ x_3 \end{bmatrix} = \begin{bmatrix} 1 & 1 & 1 \\ -1 & -2 & -3 \\ 1 & 4 & 9 \end{bmatrix} \begin{bmatrix} 1 & 0 & 0 \\ 0 & 1 & 0 \\ 0 & 0 & -1 \end{bmatrix} \begin{bmatrix} y_1 \\ y_2 \\ y_3 \end{bmatrix}$$

● **PROBLEM** 9-25

The system is described by the following equation

$$\dot{x} = Ax + Bu$$

where

$$A = \begin{bmatrix} 0 & 1 & 0 \\ 0 & 0 & 1 \\ -6 & -11 & -6 \end{bmatrix}, \quad B = \begin{bmatrix} 0 \\ 0 \\ 1 \end{bmatrix}$$

Obtain the explicit forms of $G(\tau)$ and $H(\tau)$, where

$$x(kT+\tau) = G(\tau) x(kT) + H(\tau) u(kT)$$

Solution: The matrix $G(\tau)$ is defined by

$$G(\tau) = e^{A\tau}$$

and

$$e^{A\tau} = L^{-1}[(sI-A)^{-1}]$$

We have

$$sI - A = \begin{bmatrix} s & -1 & 0 \\ 0 & s & -1 \\ 6 & 11 & s+6 \end{bmatrix}$$

and

$$(sI-A)^{-1} = \frac{1}{(s+1)(s+2)(s+3)} \begin{bmatrix} s(s+6)+11 & s+6 & 1 \\ -6 & s(s+6) & s \\ -6s & -11s-6 & s^2 \end{bmatrix}$$

Taking the inverse Laplace transform of each element of the above matrix, we find

$G(\tau) =$

$$\begin{bmatrix} 3e^{-\tau}-3e^{-2\tau}+e^{-3\tau} & , & \frac{5}{2}e^{-\tau}-4e^{-2\tau}+\frac{3}{2}e^{-3\tau} \\ -3e^{-\tau}+6e^{-2\tau}-3e^{-3\tau} & , & -\frac{5}{2}e^{-\tau}+8e^{-2\tau}-\frac{9}{2}e^{-3\tau} \\ 3e^{-\tau}-12e^{-2\tau}+9e^{-3\tau} & , & \frac{5}{2}e^{-\tau}-16e^{-2\tau}+\frac{27}{2}e^{-3\tau} \end{bmatrix}$$

$$\begin{bmatrix} , & \frac{1}{2}e^{-\tau}-e^{-2\tau}+\frac{1}{2}e^{-3\tau} \\ , & -\frac{1}{2}e^{-\tau}+2e^{-2\tau}-\frac{3}{2}e^{-3\tau} \\ , & \frac{1}{2}e^{-\tau}-4e^{-2\tau}+\frac{9}{2}e^{-3\tau} \end{bmatrix}$$

$H(\tau)$ is given by

$$H(\tau) = \int_0^\tau G(t)\,dt \cdot B \qquad \text{where} \qquad B = \begin{bmatrix} 0 \\ 0 \\ 1 \end{bmatrix}$$

Integrating the elements of $G(t)$ and multiplying by B we find

$H(\tau) =$

$$\begin{bmatrix} -\frac{1}{2}e^{-\tau}+\frac{1}{2}e^{-2\tau}-\frac{1}{6}e^{-3\tau}+\frac{1}{6} \\ \frac{1}{2}e^{-\tau}-e^{-2\tau}+\frac{1}{2}e^{-3\tau} \\ -\frac{1}{2}e^{-\tau}+2e^{-2\tau}-\frac{3}{2}e^{-3\tau} \end{bmatrix}$$

STATE SPACE-REPRESENTATION FROM BLOCK DIAGRAMS AND DIFFERENCE EQUATIONS

Obtain a state representation of a digital system described by

$$c(k+2) + 2c(k+1) + 3c(k) = r(k)$$

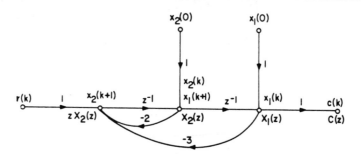

Solution: We can rewrite the above equation in the form

$$c(k+2) = -2c(k+1) - 3c(k) + r(k)$$

The state diagram of the system is shown.

To write the state equations of the system we shall designate the state variables as the output variables of the time delay units. Thus

$$\begin{bmatrix} x_1(k+1) \\ x_2(k+1) \end{bmatrix} = \begin{bmatrix} 0 & 1 \\ -3 & -2 \end{bmatrix} \begin{bmatrix} x_1(k) \\ x_2(k) \end{bmatrix} + \begin{bmatrix} 0 \\ 1 \end{bmatrix} r(k)$$

and the output equation is

$$c(k) = \begin{bmatrix} 1 & 0 \end{bmatrix} \begin{bmatrix} x_1(k) \\ x_2(k) \end{bmatrix}$$

Note that we have neglected the initial states and the branches with the gain z^{-1}.

Using Mason's gain formula and $x_1(z)$ and $x_2(z)$ as output nodes and $R(z)$, $x_1(0)$ and $x_2(0)$ as input nodes, we obtain

$$\begin{bmatrix} x_1(z) \\ x_2(z) \end{bmatrix} = \frac{1}{\Delta} \begin{bmatrix} 1+2z^{-1} & z^{-1} \\ -3z^{-1} & 1 \end{bmatrix} \begin{bmatrix} x_1(0) \\ x_2(0) \end{bmatrix} + \frac{1}{\Delta} \begin{bmatrix} z^{-2} \\ z^{-1} \end{bmatrix} R(z)$$

where

$$\Delta = 1 + 2z^{-1} + 3z^{-2}$$

The above equation is a particular case of

$$x(z) = (zI-A)^{-1} z x(0) + (zI - A)^{-1} BR(z)$$

● PROBLEM 9-27

Rewrite the following scalar difference equation

$$x(k+2) + 4x(k+1) + 7x(k) = 5u(k+1) + 2u(k)$$

in the form of a state space equation.

Solution: The equation in the general form

$$x(k+2) + a_1(k)x(k+1) + a_2(k)x(k) = b_1(k)u(k+1) + b_2(k)u(k)$$

can be written in the form of a state space equation as

$$x_1(k+1) = x_2(k) + b_1(k-1)u(k)$$

$$x_2(k+1) = -a_2(k)x_1(k) - a_1(k)x_2(k) + [b_2(k) - a_1(k)b_1(k-1)]u(k)$$

or in the matrix form

$$\begin{bmatrix} x_1(k+1) \\ x_2(k+1) \end{bmatrix} = \begin{bmatrix} 0 & 1 \\ -a_2(k) & -a_1(k) \end{bmatrix} \begin{bmatrix} x_1(k) \\ x_2(k) \end{bmatrix} + \begin{bmatrix} b_1(k-1) \\ b_2(k) - a_1(k)b_1(k-1) \end{bmatrix} [u(k)]$$

In our case

$$a_1(k) = 4 \qquad\qquad a_2(k) = 7$$

$$b_1(k) = b_1(k-1) = 5 \qquad b_2(k) = 2$$

Thus we have

$$\begin{bmatrix} x_1(k+1) \\ x_2(k+1) \end{bmatrix} = \begin{bmatrix} 0 & 1 \\ -7 & -4 \end{bmatrix} \begin{bmatrix} x_1(k) \\ x_2(k) \end{bmatrix} + \begin{bmatrix} 5 \\ -18 \end{bmatrix} [u(k)]$$

● PROBLEM 9-28

For the following difference equation

$$x(k+3) + 5x(k+2) + 7x(k+1) + 3x(k) = u(k+1) + 2u(k)$$

obtain a state space equation.

Solution: In the general form the above equation can be written

$a_0(k) x(k+3) + a_1(k) x(k+2) + a_2(k) x(k+1) + a_3(k) x(k)$

$= b_0(k) u(k+3) + b_1(k) u(k+2) + b_2(k) u(k+1) + b_3(k) u(k)$

Let us define

$$x_1(k) = x(k) - c_0 u(k)$$

$$x_1(k+1) = x_2(k) + c_1 u(k)$$

$$x_2(k+1) = x_3(k) + c_2 u(k)$$

$$x_3(k+1) = -3x_1(k) - 7x_2(k) - 5x_3(k) + c_3 u(k)$$

where the parameters c are given by

$$c_0(k) = b_0(k-3)$$

$$c_1(k) = b_1(k-2) - a_1(k-2) b_0(k-3)$$

$$c_2(k) = b_2(k-1) - a_2(k-1) b_0(k-3) - a_1(k-1) [b_1(k-2) - a_1(k-2) b_0(k-3)]$$

$$c_3(k) = b_3(k) - a_3(k) b_0(k-3) - [a_2(k) - a_1(k) a_1(k-1)]$$

$$[b_1(k-2) - a_1(k-2) b_0(k-3)]$$

$$-a_1(k) [b_2(k-1) - a_2(k-1) b_0(k-3)]$$

Thus
$$c_0 = 0$$
$$c_1 = 0$$
$$c_2 = 1$$
$$c_3 = -3$$

In the matrix form

$$X(k+1) = Ax(k) + Cu(k) \qquad (1)$$

where

$$x(k+1) = \begin{bmatrix} x_1(k+1) \\ x_2(k+1) \\ x_3(k+1) \end{bmatrix}, A = \begin{bmatrix} 0 & 1 & 0 \\ 0 & 0 & 1 \\ -3 & -7 & -5 \end{bmatrix}, C = \begin{bmatrix} c_1(k) \\ c_2(k) \\ c_3(k) \end{bmatrix}$$

The characteristic equation is

$$\det(A - \lambda I) = \begin{vmatrix} -\lambda & 1 & 0 \\ 0 & -\lambda & 1 \\ -3 & -7 & (-5-\lambda) \end{vmatrix} = 0$$

479

The roots are

$$\lambda_1 = -1, \quad \lambda_2 = -1, \quad \lambda_3 = -3$$

To simplify the state space equation let

$$x(k) = Qy(k)$$

where

$$Q = \begin{bmatrix} 1 & 0 & 1 \\ \lambda_1 & 1 & \lambda_3 \\ \lambda_1^2 & 2\lambda_1 & \lambda_3^2 \end{bmatrix} = \begin{bmatrix} 1 & 0 & 1 \\ -1 & 1 & -3 \\ 1 & -2 & 9 \end{bmatrix}$$

Equation (1) becomes

$$y(k+1) = Q^{-1} AQy(k) + Q^{-1}Cu(k)$$

We compute matrix Q^{-1}

$$Q^{-1} = \begin{bmatrix} \dfrac{3}{4} & -\dfrac{1}{2} & -\dfrac{1}{4} \\ \dfrac{3}{2} & 2 & \dfrac{1}{2} \\ \dfrac{1}{4} & \dfrac{1}{2} & \dfrac{1}{4} \end{bmatrix}$$

and

$$Q^{-1}AQ = \begin{bmatrix} -1 & 1 & 0 \\ 0 & -1 & 0 \\ 0 & 0 & -3 \end{bmatrix}, \quad Q^{-1}C = \begin{bmatrix} \dfrac{1}{4} \\ \dfrac{1}{2} \\ -\dfrac{1}{4} \end{bmatrix}$$

where

$$C = \begin{bmatrix} 0 \\ 1 \\ -3 \end{bmatrix}$$

Let us define transformation U

$$U : y(k) \rightarrow z(k)$$

where

$$U = \begin{bmatrix} \dfrac{1}{2} & \dfrac{1}{4} & 0 \\ 0 & \dfrac{1}{2} & 0 \\ 0 & 0 & -\dfrac{1}{4} \end{bmatrix}$$

$$y(k) = Uz(k)$$

We get

$$U^{-1}Q^{-1}AQU = \begin{bmatrix} -1 & 1 & 0 \\ 0 & -1 & 0 \\ 0 & 0 & -3 \end{bmatrix}$$

and

$$U^{-1}Q^{-1}C = \begin{bmatrix} 2 & -1 & 0 \\ 0 & 2 & 0 \\ 0 & 0 & -4 \end{bmatrix} \begin{bmatrix} \frac{1}{4} \\ \frac{1}{2} \\ -\frac{1}{4} \end{bmatrix} = \begin{bmatrix} 0 \\ 1 \\ 1 \end{bmatrix}$$

We obtained the simplified state space equation

$$\begin{bmatrix} z_1(k+1) \\ z_2(k+1) \\ z_3(k+1) \end{bmatrix} = \begin{bmatrix} -1 & 1 & 0 \\ 0 & -1 & 0 \\ 0 & 0 & -3 \end{bmatrix} \begin{bmatrix} z_1(k) \\ z_2(k) \\ z_3(k) \end{bmatrix} + \begin{bmatrix} 0 \\ 1 \\ 1 \end{bmatrix} [u(k)]$$

The initial conditions are

$$z(0) = U^{-1}y(0) = U^{-1}Q^{-1}x(0)$$

where

$$x(0) = \begin{bmatrix} x_1(0) \\ x_2(0) \\ x_3(0) \end{bmatrix} = \begin{bmatrix} x(0) \\ x(1) \\ x(2) - u(0) \end{bmatrix}$$

● **PROBLEM** 9-29

Obtain a state-space representation of the system described by

$$y(k+2) + y(k+1) + 0.7y(k) = u(k+1) + 3u(k)$$

<u>Solution:</u> We define the state variables as follows

$$x_1(k) = y(k)$$

$$x_2(k) = x_1(k+1) - u(k)$$

The difference equation can be written

$$x_1(k+1) = x_2(k) + u(k)$$

481

$$x_2(k+1) = -0.7x_1(k) - x_2(k) + 2u(k)$$

$$y(k) = x_1(k)$$

In the matrix form

$$\begin{bmatrix} x_1(k+1) \\ x_2(k+1) \end{bmatrix} = \begin{bmatrix} 0 & 1 \\ -0.7 & -1 \end{bmatrix} \begin{bmatrix} x_1(k) \\ x_2(k) \end{bmatrix} + \begin{bmatrix} 1 \\ 2 \end{bmatrix} [u(k)]$$

$$y(k) = [1 \; 0] \begin{bmatrix} x_1(k) \\ x_2(k) \end{bmatrix}$$

The initial conditions are

$$\begin{bmatrix} x_1(0) \\ x_2(0) \end{bmatrix} = \begin{bmatrix} y(0) \\ y(1) - u(0) \end{bmatrix}$$

● **PROBLEM** 9-30

The block diagram of a control system is shown.

x_1, x_2, x_3 constitute the state vector. Write the state-space equation of the system, transform it in such a way that the coefficient matrix is in the diagonal form and the coefficient matrix of u is

$$\begin{bmatrix} 1 \\ 1 \\ 1 \end{bmatrix}.$$

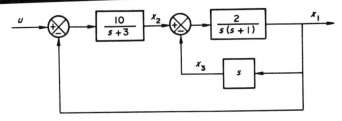

Solution: We shall write the input output relationship for each transfer function

$$sX_1(s) = X_3(s)$$

$$(s+3)X_2(s) = 10[U(s) - X_1(s)]$$

$$s(s+1)X_1(s) = 2[X_2(s) - X_3(s)]$$

482

Taking the inverse Laplace transform, we get

$$\dot{x}_1 = x_3$$

$$\dot{x}_2 = -10x_1 - 3x_2 + 10u$$

$$\dot{x}_3 = 2x_2 - 3x_3$$

In the matrix form

$$\begin{bmatrix} \dot{x}_1 \\ \dot{x}_2 \\ \dot{x}_3 \end{bmatrix} = \begin{bmatrix} 0 & 0 & 1 \\ -10 & -3 & 0 \\ 0 & 2 & -3 \end{bmatrix} \begin{bmatrix} x_1 \\ x_2 \\ x_3 \end{bmatrix} + \begin{bmatrix} 0 \\ 10 \\ 0 \end{bmatrix} [u]$$

or $\qquad \dot{x} = Ax + Bu$

From the characteristic equation

$$\det \quad |\lambda I - A| = 0$$

we obtain the eigenvalues of A

$$\lambda_1 = -5, \quad \lambda_2 = -\frac{1}{2} - j\frac{\sqrt{15}}{2}, \quad \lambda_3 = -\frac{1}{2} + j\frac{\sqrt{15}}{2}$$

A transformation matrix P is obtained as the matrix of eigenvectors of A with first elements 1.

Thus

$$\begin{bmatrix} \lambda_i & 0 & -1 \\ 10 & \lambda_i + 3 & 0 \\ 0 & -2 & \lambda_i + 3 \end{bmatrix} \begin{bmatrix} 1 \\ p_2 \\ p_3 \end{bmatrix} = 0$$

and, we get

$$p_3 = \lambda_i, \qquad p_2 = \frac{1}{2}\lambda_i(\lambda_i + 3)$$

We can write the transformation matrix P

$$P = \begin{bmatrix} 1 & 1 & 1 \\ 5 & -\frac{5}{2} - j\frac{\sqrt{15}}{2} & -\frac{5}{2} + j\frac{\sqrt{15}}{2} \\ -5 & -\frac{1}{2} - j\frac{\sqrt{15}}{2} & -\frac{1}{2} + j\frac{\sqrt{15}}{2} \end{bmatrix}$$

Let us define the transformation

$$x = Py$$

Thus

$$\dot{y} = P^{-1}APy + P^{-1}Bu$$

483

where

$$P^{-1} = \begin{bmatrix} \dfrac{1}{6} & \dfrac{1}{12} & -\dfrac{1}{12} \\[2mm] \dfrac{5}{12} + j\,\dfrac{\sqrt{15}}{12} & -\dfrac{1}{24} + j\,\dfrac{\sqrt{15}}{40} & \dfrac{1}{24} + j\,\dfrac{\sqrt{15}}{24} \\[2mm] \dfrac{5}{12} - j\,\dfrac{\sqrt{15}}{12} & -\dfrac{1}{24} - j\,\dfrac{\sqrt{15}}{40} & \dfrac{1}{24} - j\,\dfrac{\sqrt{15}}{24} \end{bmatrix}$$

Matrices $P^{-1}\,AP$ and $P^{-1}\,B$ are

$$P^{-1}\,AP = \begin{bmatrix} -5 & 0 & 0 \\[2mm] 0 & -\dfrac{1}{2} - j\,\dfrac{\sqrt{15}}{2} & 0 \\[2mm] 0 & 0 & -\dfrac{1}{2} + j\,\dfrac{\sqrt{15}}{2} \end{bmatrix}$$

$$P^{-1}\,B = \begin{bmatrix} \dfrac{5}{6} \\[2mm] -\dfrac{5}{12} + j\,\dfrac{\sqrt{15}}{4} \\[2mm] -\dfrac{5}{12} - j\,\dfrac{\sqrt{15}}{4} \end{bmatrix}$$

Let us define the second transformation

$$y = Vz$$

where

$$V = \begin{bmatrix} \dfrac{5}{6} & 0 & 0 \\[2mm] 0 & -\dfrac{5}{12} + j\,\dfrac{\sqrt{15}}{4} & 0 \\[2mm] 0 & 0 & -\dfrac{5}{12} - j\,\dfrac{\sqrt{15}}{4} \end{bmatrix}$$

The state-space equation of the system can be written in the following form

$$\begin{bmatrix} \dot{z}_1 \\[2mm] \dot{z}_2 \\[2mm] \dot{z}_3 \end{bmatrix} = \begin{bmatrix} -5 & 0 & 0 \\[2mm] 0 & -\dfrac{1}{2} - j\,\dfrac{\sqrt{15}}{2} & 0 \\[2mm] 0 & 0 & -\dfrac{1}{2} + j\,\dfrac{\sqrt{15}}{2} \end{bmatrix} \begin{bmatrix} z_1 \\[2mm] z_2 \\[2mm] z_3 \end{bmatrix} + \begin{bmatrix} 1 \\[2mm] 1 \\[2mm] 1 \end{bmatrix} [u]$$

Obtain a state space representation of a control system shown.

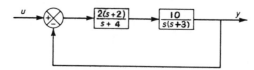

Solution: The closed-loop transfer function of the system is

$$\frac{Y(s)}{U(s)} = \frac{G(s)}{1 + G(s)}$$

where

$$G(s) = \frac{20(s+2)}{s(s+3)(s+4)}$$

Thus

$$\frac{Y(s)}{U(s)} = \frac{\dfrac{20(s+2)}{s(s+3)(s+4)}}{1 + \dfrac{20(s+2)}{s(s+3)(s+4)}} = \frac{20(s+2)}{s(s+3)(s+4)} \cdot$$

$$\frac{s(s+3)(s+4)}{[s(s+3)(s+4) + 20(s+2)]}$$

$$= \frac{20(s+2)}{s(s+3)(s+4) + 20(s+2)} = \frac{20(s+2)}{s^3 + 7s^2 + 32s + 40}$$

The corresponding differential equation is

$$\dddot{y} + 7\ddot{y} + 32\dot{y} + 40y = 20\dot{u} + 40u$$

Let us define the state variables

$$x_1 = y - \alpha_0 u$$

$$x_2 = \dot{y} - \alpha_0\dot{u} - \alpha_1 u = \dot{x}_1 - \alpha_1 u$$

$$x_3 = \ddot{y} - \alpha_0\ddot{u} - \alpha_1\dot{u} - \alpha_2 u = \dot{x}_2 - \alpha_2 u$$

where the coefficients are given by

$$y^{(n)} + a_1 y^{(n-1)} + a_2 y^{(n-2)} + \ldots + a_{n-1}\dot{y} + a_n y = b_0 u^{(n)} +$$

$$b_1 u^{(n-1)} + \ldots + b_{n-1}\dot{u} + b_n u$$

$\alpha_0 = b_0$ in our case $b_0 = 0$

$\alpha_1 = b_1 - a_1\alpha_0$ -''- $\alpha_1 = 0$

$\alpha_2 = b_2 - a_1\alpha_1 - a_2\alpha_0 = 20 - 7;0 - 32;0 = 20$

$\alpha_3 = b_3 - a_1\alpha_2 - a_2\alpha_1 - a_3\alpha_0 = 40 - 7;20 = -100$

The state equation is of the form

$$\begin{bmatrix} \dot{x}_1 \\ \dot{x}_2 \\ . \\ . \\ . \\ \dot{x}_n \end{bmatrix} = \begin{bmatrix} 0 & 1 & 0 & . & . & . & 0 \\ 0 & 0 & 1 & & & & 0 \\ . & . & . & . & & & . \\ . & . & . & . & & & . \\ . & . & . & . & & & . \\ & & & & & & 1 \\ -a_n & -a_{n-1} & -a_{n-2} & . & . & . & -a_1 \end{bmatrix} \begin{bmatrix} x_1 \\ x_2 \\ . \\ . \\ . \\ x_n \end{bmatrix} + \begin{bmatrix} \alpha_1 \\ \alpha_2 \\ . \\ . \\ . \\ \alpha_n \end{bmatrix} u$$

$$y = [1 \quad 0 \quad . \quad . \quad . \quad 0] \begin{bmatrix} x_1 \\ x_2 \\ . \\ . \\ . \\ x_n \end{bmatrix} + \alpha_0 u$$

In our case

$$\begin{bmatrix} \dot{x}_1 \\ \dot{x}_2 \\ \dot{x}_3 \end{bmatrix} = \begin{bmatrix} 0 & 1 & 0 \\ 0 & 0 & 1 \\ -40 & -32 & -7 \end{bmatrix} \begin{bmatrix} x_1 \\ x_2 \\ x_3 \end{bmatrix} + \begin{bmatrix} 0 \\ 20 \\ -100 \end{bmatrix} [u]$$

and

$$y = [1 \quad 0 \quad 0] \begin{bmatrix} x_1 \\ x_2 \\ x_3 \end{bmatrix}$$

● **PROBLEM 9-32**

For the system shown in Fig. 1 obtain the state-space representation.

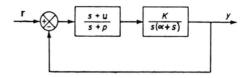

Fig. 1

Solution: A state-space representation will be obtained from the block diagram.

Let us expand $\dfrac{s+u}{s+p}$ into partial fractions

$$\frac{s+u}{s+p} = 1 + \frac{u-p}{s+p} \tag{1}$$

and

$$\frac{K}{s(s+\alpha)} = \frac{K}{s} \cdot \frac{K}{s+\alpha} \tag{2}$$

From eq.(1) we see that the first element of the block diagram can be represented in an equivalent form as shown in Fig. 2.

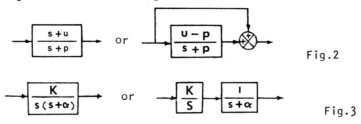

Fig.2

Fig.3

Whereas the second element's equivalent form is shown in Fig. 3.

We can now redraw the block diagram of the system, indicating the state variables as shown in Fig. 4.

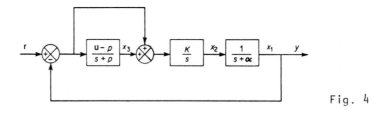

Fig. 4

We obtain the following set of equations

$$\dot{x}_1 = -\alpha x_1 + x_2$$

$$\dot{x}_2 = -Kx_1 + Kx_3 + Kr$$

$$\dot{x}_3 = -(u-p)x_1 - px_3 + (u-p)r$$

$$y = x_1$$

or in the matrix form

$$
\begin{bmatrix} \dot{x}_1 \\ \dot{x}_2 \\ \dot{x}_3 \end{bmatrix} = \begin{bmatrix} -\alpha & 1 & 0 \\ -K & 0 & K \\ -(u-p) & 0 & -p \end{bmatrix} \begin{bmatrix} x_1 \\ x_2 \\ x_3 \end{bmatrix} + \begin{bmatrix} 0 \\ K \\ u-p \end{bmatrix} [r]
$$

$$
y = [1 \ 0 \ 0] \begin{bmatrix} x_1 \\ x_2 \\ x_3 \end{bmatrix}
$$

● **PROBLEM** 9-33

For the system shown obtain a state space equation.

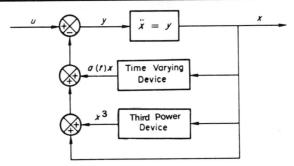

Solution: From the diagram we get the following relationship between x and u

$$\ddot{x} + x + x^3 + a(t)x = u$$

Let us define

$$x_1 = x$$

$$x_2 = \dot{x}$$

The state space representation of the system is

$$\dot{x}_1 = x_2$$

$$\dot{x}_2 = -x_1 - a(t)x_1 - x_1^3 + u.$$

● **PROBLEM** 9-34

The block diagram of a discrete-time system is shown in Fig. 1.

Obtain a discrete-time state-space representation of the system.

Solution: From the diagram we get the open-loop transfer function of the system.

$$G(s) = \frac{1 - e^{-Ts}}{s} \cdot \frac{1}{s(s+a)}$$

Taking the z-transform, we obtain

$$G(z) = \frac{T}{a(z-1)} - \frac{1 - e^{-aT}}{a^2(z-e^{-aT})}$$

Thus, the block diagram can be drawn in the modified form as shown in Fig. 2.

Using the equality $zZ[x(k)] = Z[x(k+1)]$

we obtain the discrete-time state equations

$$\begin{bmatrix} x_1(k+1) \\ \\ x_2(k+1) \end{bmatrix} = \begin{bmatrix} 1 - \dfrac{T}{a} & -\dfrac{T}{a} \\ \\ \dfrac{1 - e^{-aT}}{a^2} & \dfrac{1 + (a^2-1)e^{-aT}}{a^2} \end{bmatrix} \begin{bmatrix} x_1(k) \\ \\ x_2(k) \end{bmatrix} \begin{bmatrix} \dfrac{T}{a} \\ \\ \dfrac{-1+e^{-aT}}{a^2} \end{bmatrix} [u(k)]$$

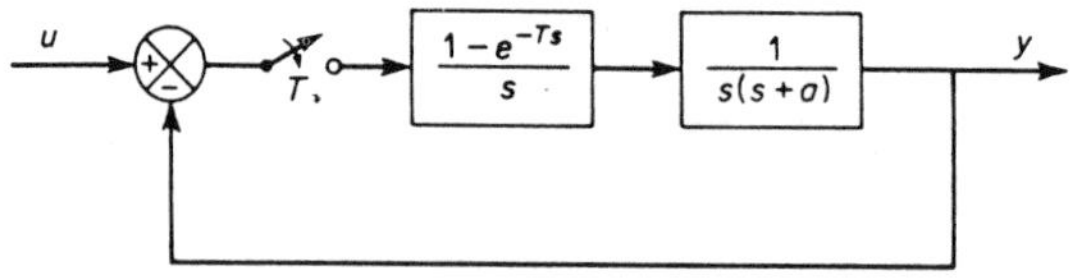

Fig.1

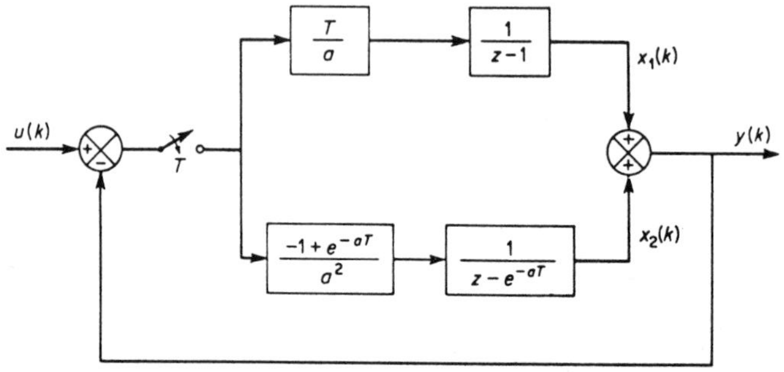

Fig.2

● **PROBLEM** 9-35

A digital control system with nonsynchronous sampling is shown in Fig. 1.

Write the state transition equations for the system assuming that the sampling instants of the sampler S_2 are behind that of S_1 by T_1 seconds, as shown in the sampling schemes in Fig. 2.

Solution: Let us obtain the state diagrams shown in Fig. 3 of the system. Note that S_1 and S_2 close at different

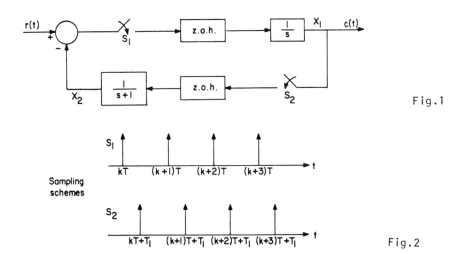

Fig.1

Sampling schemes

Fig.2

sampling instants. The forward path has a transition interval from $t = kT$ to $t = (k+1)T$ and the feedback path has a transition interval from $t = kT + T_1$ to $t = (k+1)T + T_1$.

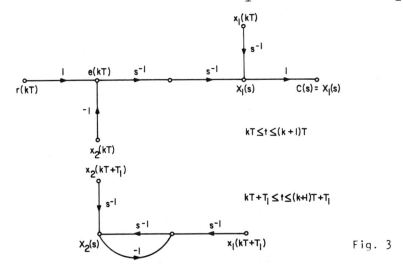

Fig. 3

From the diagram we obtain the state transition equations in the Laplace domain.

$$x_1(s) = \frac{1}{s}x_1(kT) - \frac{1}{s^2}x_2(kT) + \frac{1}{s^2}r(kT) ; \quad kT \leq t \leq (k+1)T$$

$$x_2(s) = \frac{1}{s(s+1)}x_1(kT+T_1) + \frac{1}{s+1}x_2(kT+T_1)$$

$$kT + T_1 \leq t \leq kT + T + T_1$$

Taking the inverse Laplace transform, we obtain

$$x_1(t) = x_1(kT) - (t-kT)x_2(kT) + (t-kT)r(kT)$$

$$x_2(t) = [1 - e^{-(t-kT-T_1)}]x_1(kT+T_1) + e^{-(t-kT-T_1)}x_2(kT+T_1)$$

490

The initial states are

$$x_1(kT) \quad, \quad x_1(kT+T_1)$$
$$x_2(kT) \quad, \quad x_2(kT+T_1)$$

To compute the initial states let us set

$t = kT + T_1$ and $t = (k+1)T$, thus

$$x_1(kT+T_1) = x_1(kT) - T_1 x_2(kT) + T_1 r(kT)$$

$$x_1(k+1)T = x_1(kT) - T x_2(kT) + T r(kT)$$

To find x_2 we set $t = (1+k)T + T_1$ and $t = (k+1)T$

$$x_2(k+1)T = [1 - e^{-(T-T_1)}]x_1(kT+T_1) + e^{-(T-T_1)}x_2(kT+T_1)$$

$$x_2[(k+1)T + T_1] = (1-e^{-T})x_1(kT+T_1) + e^{-T}x_2(kT+T_1)$$

From the above system of equations we can compute the states and the output of the system by setting $k = 0,1,2,3,...$

● **PROBLEM** 9-36

Obtain a state representation for the sampled data discrete system shown in Fig. 1.

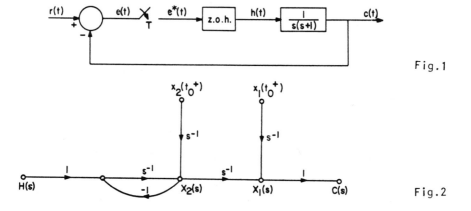

Fig.1

Fig.2

Solution: The controlled process is

$$G(s) = \frac{1}{s(s+1)}$$

whose state diagram is shown in Fig. 2.

We have

$$e(kT) = r(kT) - c(kT) = r(kT) - x_1(kT)$$

let $t_0 = kT$

$$h(kT^+) = h(t) = e(kT) \qquad kT \le t \le (k+1)T.$$

We add now the z.o.h. (zero-order-hold) to the system. The complete state diagram is shown in Fig. 3.

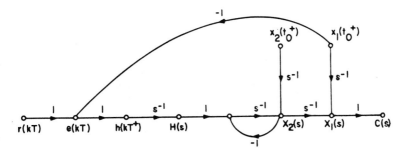

Fig.3

The Mason's gain formula gives

$$\begin{bmatrix} x_1(s) \\ x_2(s) \end{bmatrix} = \begin{bmatrix} \dfrac{1}{s} - \dfrac{1}{s^2(s+1)} & \dfrac{1}{s(s+1)} \\ -\dfrac{1}{s(s+1)} & \dfrac{1}{s+1} \end{bmatrix} \begin{bmatrix} x_1(kT) \\ x_2(kT) \end{bmatrix} + \begin{bmatrix} \dfrac{1}{s^2(s+1)} \\ \dfrac{1}{s(s+1)} \end{bmatrix} r(kT)$$

for $kT \le t \le (k+1)T$

The inverse Laplace transform of the above equation is

$$\begin{bmatrix} x_1(t) \\ x_2(t) \end{bmatrix} = \begin{bmatrix} 2 - (t-kT) - e^{-(t-kT)} & 1 - e^{-(t-kT)} \\ -1 + e^{-(t-kT)} & e^{-(t-kT)} \end{bmatrix} \begin{bmatrix} x_1(kT) \\ x_2(kT) \end{bmatrix}$$
$$+ \begin{bmatrix} (t-kT) - 1 + e^{-(t-kT)} \\ 1 - e^{-(t-kT)} \end{bmatrix} r(kT)$$

Substituting $t = (k+\Delta)T$ in the above equation, we obtain

$$\begin{bmatrix} x_1(kT+\Delta T) \\ x_2(kT+\Delta T) \end{bmatrix} = \begin{bmatrix} 2 - \Delta T - e^{-\Delta T} & 1 - e^{-\Delta T} \\ -1 + e^{-\Delta T} & e^{-\Delta T} \end{bmatrix} \begin{bmatrix} x_1(kT) \\ x_2(kT) \end{bmatrix} +$$
$$\begin{bmatrix} \Delta T - 1 + e^{-\Delta T} \\ 1 - e^{-\Delta T} \end{bmatrix} r(kT)$$

The iteration procedure is completed by letting $t = (k+1)T$ and $t_o = (k+\Delta)T$. We obtain

$$x[(k+1)T] = \phi(T-\Delta T)x[(k+\Delta)T] + \theta(T-\Delta T)r[(k+\Delta)T].$$

The state transition equation is

$$
\begin{bmatrix} x_1[(k+1)T] \\ x_2[(k+1)T] \end{bmatrix} = \begin{bmatrix} 2 - (1-\Delta)T - e^{-(1-\Delta)T} & 1 - e^{-(1-\Delta)T} \\ -1 + e^{-(1-\Delta)T} & e^{-(1-\Delta)T} \end{bmatrix} \begin{bmatrix} x_1[(k+\Delta)T] \\ x_2[(k+\Delta)T] \end{bmatrix}
$$

$$
+ \begin{bmatrix} (1-\Delta)T - 1 + e^{-(1-\Delta)T} \\ 1 - e^{-(1-\Delta)T} \end{bmatrix} r[(k+\Delta)T]
$$

$$(k+\Delta)T = t_0 \le t \le (k+1)T$$

● **PROBLEM** 9-37

Obtain a state representation of the system shown.

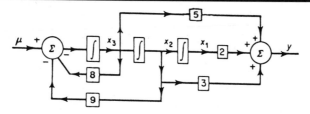

Solution: From the diagram we obtain

$$x_1 = \int x_2 \quad \text{or} \quad \dot{x}_1 = x_2$$

$$x_2 = \int x_3 \quad \text{or} \quad \dot{x}_2 = x_3$$

$$x_3 = \int(-8x_3 - 9x_2 + u) \quad \text{or} \quad \dot{x}_3 = -8x_3 - 9x_2 + u$$

$$y = 2x_1 + 3x_2 + 5x_3$$

In the matrix form

$$\dot{x} = Ax + Bu$$

$$y = Cx + Du$$

where

$$A = \begin{bmatrix} 0 & 1 & 0 \\ 0 & 0 & 1 \\ 0 & -9 & -8 \end{bmatrix} \qquad B = \begin{bmatrix} 0 \\ 0 \\ 1 \end{bmatrix}$$

$$C = [2 \; 3 \; 5] \qquad D = [0]$$

Obtain a state representation for the multiple input multiple output system shown.

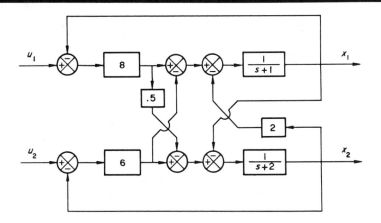

Solution: From the diagram we have

$$x_1(s) = (\frac{1}{s+1})[8(u_1 - x_1) - 6(u_2 - x_2) - 2x_2]$$

or

$$(s+1)x_1 = 8u_1 - 8x_1 - 6u_2 + 6x_2 - 2x_2$$

$$sx_1(s) = -9x_1(s) + 8u_1(s) - 6u_2(s) + 4x_2(s)$$

Taking the inverse Laplace transfrom we obtain

$$\dot{x}_1(t) = -9x_1(t) + 4x_2(t) + 8u_1(t) - 6u_2(t)$$

For x_2 we have

$$x_2(s) = \frac{1}{s+2} \cdot [6(u_2 - x_2) - 4(u_1 - x_1) - x_1]$$

or

$$x_2 \cdot (s+2) = 6u_2 - 6x_2 - 4u_1 + 4x_1 - x_1$$

$$sx_2(s) = 6u_2 - 8x_2 - 4u_1 + 3x_1$$

Taking the inverse Laplace transform we get

$$\dot{x}_2(t) = 3x_1(t) - 8x_2(t) - 4u_1(t) + 6u_2(t)$$

In matrix form

$$\begin{bmatrix} \dot{x}_1(t) \\ \dot{x}_2(t) \end{bmatrix} = \begin{bmatrix} -9 & 4 \\ 3 & -8 \end{bmatrix} \begin{bmatrix} x_1(t) \\ x_2(t) \end{bmatrix} + \begin{bmatrix} 8 & -6 \\ -4 & 6 \end{bmatrix} \begin{bmatrix} u_1(t) \\ u_2(t) \end{bmatrix}$$

● **PROBLEM** 9-39

The system, whose block diagram is shown consists of integrators, adders and scalers.

$u(t)$ is the input, $y(t)$ output and $x_1(t)$, $x_2(t)$, $x_3(t)$ are the state variables. The initial conditions are

$$x_1(0) = x_2(0) = x_3(0) = 0.$$

Write state equations for the system and the differential equation which relates the input $u(t)$ and the output $y(t)$.

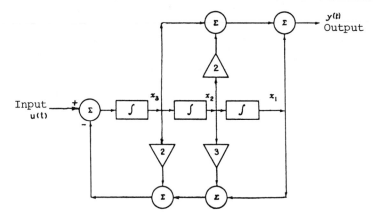

Solution: From the diagram we have

$$\dot{x}_1(t) = x_2(t) \tag{1}$$

$$\dot{x}_2(t) = x_3(t)$$

$$\dot{x}_3(t) = -x_1(t) - 3x_2(t) - 2x_3(t) + u(t)$$

and

$$y(t) = x_1(t) + 2x_2(t) + x_3(t) \tag{2}$$

To obtain an equation relating $u(t)$ and $y(t)$ let us differentiate the last equation (2) and substitute values for $\dot{x}_1, \dot{x}_2, \dot{x}_3$ from (1)

$$\frac{dy}{dt} = \dot{x}_1(t) + 2\dot{x}_2(t) + \dot{x}_3(t) = -x_1(t) - 2x_2(t) + u(t)$$

Differentiating again we get

$$\frac{d^2y}{dt^2} = -\dot{x}_1(t) - 2\dot{x}_2(t) + \frac{du}{dt} = -x_2(t) - 2x_3(t) + \frac{du}{dt}$$

and

$$\frac{d^3y}{dt^3} = -\dot{x}_2(t) - 2\dot{x}_3(t) + \frac{d^2u}{dt^2} = 2x_1(t) + 6x_2(t) + 3x_3(t)$$

$$- 2u(t) + \frac{d^2u}{dt^2}$$

Eliminating x_1, x_2, x_3 we obtain the equation relating $u(t)$ and $y(t)$, thus

$$\frac{d^3y}{dt^3} + 2\frac{d^2y}{dt^2} + 3\frac{dy}{dt} + y = \frac{d^2u}{dt^2} + 2\frac{du}{dt} + u$$

● **PROBLEM** 9-40

Determine the state equations of the state diagram shown in Fig. 1.

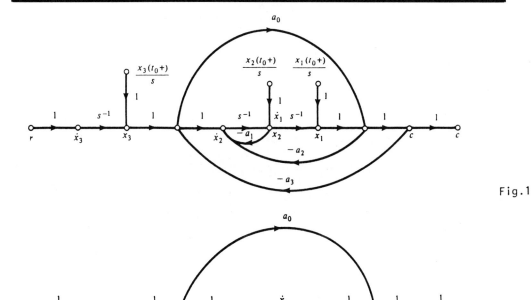

Fig.1

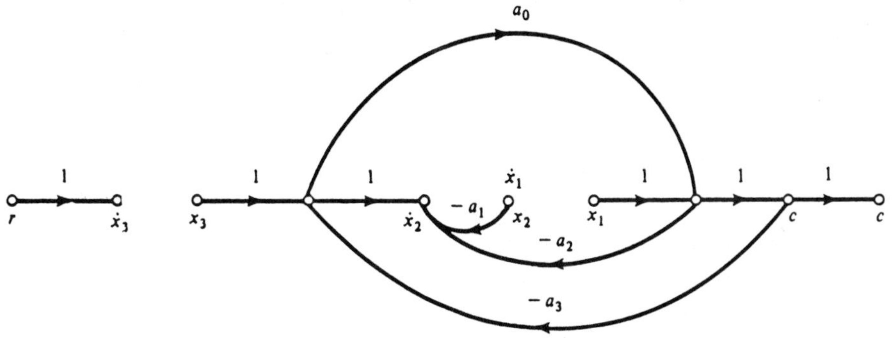

Fig.2

Solution: Let us remove from the state diagram all initial conditions and integrators. We obtain the state diagram shown in Fig. 2.

The coefficients of the state equations are the gains between $\dot{x}_1$, $\dot{x}_2$ and $\dot{x}_3$ and x_1, x_2, x_3 and r. The new state diagram still contains a loop. We shall determine the state equations from application of the gain formula to the diagram. Where, $\dot{x}_1$, $\dot{x}_2$, and $\dot{x}_3$ are the output node variables and x_1, x_2, and x_3 and r are the input nodes. We have

$$\begin{bmatrix} \dot{x}_1 \\ \dot{x}_2 \\ \dot{x}_3 \end{bmatrix} = \begin{bmatrix} 0 & 1 & 0 \\ \dfrac{-(a_2 + a_3)}{1 + a_0 a_3} & -a_1 & \dfrac{1 - a_0 a_2}{1 + a_0 a_3} \\ 0 & 0 & 0 \end{bmatrix} \begin{bmatrix} x_1 \\ x_2 \\ x_3 \end{bmatrix} + \begin{bmatrix} 0 \\ 0 \\ 1 \end{bmatrix} r$$

● **PROBLEM** 9-41

Obtain a state representation for the system shown.

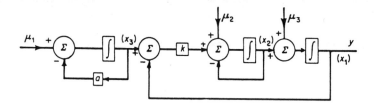

Solution:

From the diagram we have

$$x_1 = \int (x_2 + \mu_3) \quad \text{or} \quad \dot{x}_1 = x_2 + \mu_3$$

$$x_2 = \int [-x_2 - kx_1 + kx_3 + \mu_2] \quad \text{or} \quad \dot{x}_2 = -x_2 - kx_1 + kx_3 + \mu_2$$

$$x_3 = \int [-ax_3 + \mu_1] \quad \text{or} \quad \dot{x}_3 = -ax_3 + \mu_1$$

$$y = x_1$$

Thus in standard form we obtain

$$\begin{bmatrix} \dot{x}_1 \\ \dot{x}_2 \\ \dot{x}_3 \end{bmatrix} = \begin{bmatrix} 0 & 1 & 0 \\ -k & -1 & k \\ 0 & 0 & -a \end{bmatrix} \begin{bmatrix} x_1 \\ x_2 \\ x_3 \end{bmatrix} + \begin{bmatrix} 0 & 0 & 1 \\ 0 & 1 & 0 \\ 1 & 0 & 0 \end{bmatrix} \begin{bmatrix} u_1 \\ u_2 \\ u_3 \end{bmatrix}$$

$$y = \begin{bmatrix} 1 & 0 & 0 \end{bmatrix} \begin{bmatrix} x_1 \\ x_2 \\ x_3 \end{bmatrix} + \begin{bmatrix} 0 & 0 & 0 \end{bmatrix} \begin{bmatrix} u_1 \\ u_2 \\ u_3 \end{bmatrix}$$

STATE-SPACE REPRESENTATION OF ELECTRICAL AND MECHANICAL SYSTEMS

● PROBLEM 9-42

For a RLC network shown obtain a state-space representation.

Solution: Let us take i(t) and e_c(t) as the state variables for the system. This is one of many possible choices.

The following equations describe the system

$$e(t) = L\frac{di}{dt} + Ri + e_c$$

$$i = C\frac{de_c}{dt}$$

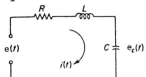

Thus we can write a state-space representation for the given network in the matrix notation

$$\begin{bmatrix} \dot{i} \\ \dot{e}_c \end{bmatrix} = \begin{bmatrix} -\frac{R}{L} & -\frac{1}{L} \\ \frac{1}{C} & 0 \end{bmatrix} \begin{bmatrix} i \\ e_c \end{bmatrix} + \begin{bmatrix} \frac{1}{L} \\ 0 \end{bmatrix} [e]$$

● PROBLEM 9-43

The diagram of the RLC network is shown.

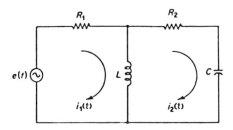

Find a state equation for the network.

Solution: Let us define the state variables as

$$x_1 = i_1 - i_2$$

x_1 is the current through the inductance L and

$$x_2 = v_c$$

x_2 is the voltage across the capacitor. e(t) is the input of the system.

We have the following equations for the system

$$L\left(\frac{di_1}{dt} - \frac{di_2}{dt}\right) + R_1 i_1 = e(t)$$

$$\frac{1}{C}\int i_2 dt + R_2 i_2 + L\left(\frac{di_2}{dt} - \frac{di_1}{dt}\right) = 0$$

Using the definitions of x_1 and x_2 we get

$$x_2 = \frac{1}{C}\int i_2 dt$$

$$L\frac{dx_1}{dt} + R_1(x_1 + i_2) = e(t)$$

$$x_2 + R_2 i_2 + R_1(x_1 + i_2) = e(t)$$

To eliminate $i_2(t)$ let us take the Laplace transforms of the equations:

$$L[sX_1(s) - x_1(0)] + R_1[X_1(s) + I_2(s)] = E(s)$$

$$X_2(s) + R_2 I_2(s) + R_1[X_1(s) + I_2(s)] = E(s)$$

$$sX_2(s) - x_2(0) = \frac{1}{C}I_2(s)$$

The last equation we got from

$$\dot{x}_2 = \frac{1}{C}i_2$$

Eliminating $I_2(s)$ we obtain

$$L[sX_1(s) - x_1(0)] = \frac{-R_1 R_2}{R_1 + R_2}X_1(s) + \frac{R_1}{R_1 + R_2}X_2(s) + \frac{R_2}{R_1 + R_2}E(s)$$

$$sX_2(s) - x_2(0) = \frac{-R_1}{C(R_1 + R_2)}X_1(s) + \frac{-1}{C(R_1 + R_2)}X_2(s) +$$

$$\frac{1}{C(R_1 + R_2)}E(s)$$

Dividing the first equation by L and taking the inverse Laplace transforms, we get

$$\dot{x}_1(t) = \frac{-R_1 R_2}{L(R_1 + R_2)}x_1(t) + \frac{R_1}{L(R_1 + R_2)}x_2(t) + \frac{R_2}{L(R_1 + R_2)}e(t)$$

$$\dot{x}_2(t) = \frac{-R_1}{C(R_1 + R_2)}x_1(t) + \frac{-1}{C(R_1 + R_2)}x_2(t) + \frac{1}{C(R_1 + R_2)}e(t)$$

The state equation for the system in the matrix notation is

$$\begin{bmatrix} \dot{x}_1(t) \\ \dot{x}_2(t) \end{bmatrix} = \begin{bmatrix} \dfrac{-R_1 R_2}{L(R_1 + R_2)} & \dfrac{R_1}{L(R_1 + R_2)} \\ \dfrac{-R_1}{C(R_1 + R_2)} & \dfrac{-1}{C(R_1 + R_2)} \end{bmatrix}\begin{bmatrix} x_1(t) \\ x_2(t) \end{bmatrix} + \begin{bmatrix} \dfrac{R_2}{L(R_1 + R_2)} \\ \dfrac{1}{C(R_1 + R_2)} \end{bmatrix}[e(t)]$$

The RLC network is shown.

Its loop equation is

$$e(t) = Ri(t) + L\frac{di(t)}{dt} + \frac{1}{C}\int i(t)dt$$

Write the state equations for the circuit.

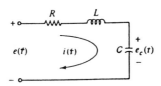

Solution: Since the last term of the network equation is a time integral this equation is not in the form of a state equation. Let us define the state variables as follows

$$x_1(t) = i(t)$$

$$x_2(t) = \int i(t)dt$$

Substituting the new variables we obtain

$$e(t) = Rx_1(t) + L\frac{dx_1(t)}{dt} + \frac{1}{C}x_2(t)$$

Differentiating the equation

$$x_2(t) = \int i(t)dt$$

we have

$$\dot{x}_2(t) = i(t) = x_1(t)$$

We have two state equations of the network

$$\frac{dx_1(t)}{dt} = -\frac{R}{L}x_1(t) - \frac{1}{LC}x_2(t) + \frac{1}{L}e(t)$$

$$\frac{dx_2(t)}{dt} = x_1(t)$$

The temperature-controlled oven shown is a double capacitance heat system.

u is the heat input that can be controlled. The system parameters are

T_o, T_j, T_i - the outside, jacket, inside temperatures

S_i, S_o - inside and outside jacket surfaces

C_i, C_j - heat capacities of inside space and jacket.

h_i, h_o - film coefficients for inside and outside surfaces.

We have two equations for heat balance in the jacket and in the inside space.

$$C_j \dot{T}_j = S_o h_o (T_o - T_j) + S_i h_i (T_i - T_j) + u$$

$$C_i \dot{T}_i = S_i h_i (T_j - T_i)$$

For the system described above obtain a state representation.

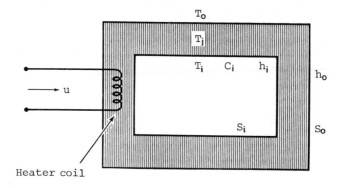

Heater coil

<u>Solution</u>: Let us define the state variables as the temperature differences $T_i - T_o$ and $T_j - T_o$

$$x = \begin{bmatrix} x_1 \\ x_2 \end{bmatrix} = \begin{bmatrix} T_j - T_o \\ T_i - T_o \end{bmatrix}$$

Let us further assume that the temperature of the environment is constant, thus

$$\dot{T}_o = 0$$

and

$$C_j \dot{x}_1 = -S_o h_o x_1 + S_i h_i (x_2 - x_1) + u$$

$$C_i \dot{x}_2 = S_i h_i (x_1 - x_2)$$

Dividing the first equation by C_j and the second by C_i, we can write the system in the form

$$\dot{x} = Ax + Bu$$

where

$$A = \begin{bmatrix} -\dfrac{1}{C_j}(S_o h_o + S_i h_i) & \dfrac{S_i h_i}{C_j} \\ \dfrac{S_i h_i}{C_i} & -\dfrac{S_i h_i}{C_i} \end{bmatrix}$$

$$B = \begin{bmatrix} \dfrac{1}{C_j} \\ 0 \end{bmatrix}$$

● **PROBLEM** 9-46

Given the electrical network shown, obtain a state representation.

We assume that the two voltage sources e_1 and e_2 are adjustable, thus the system can be controlled.

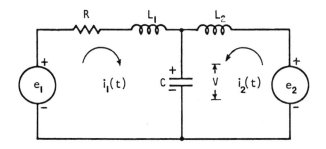

Solution: Let

$$u = \begin{bmatrix} u_1 \\ u_2 \end{bmatrix} = \begin{bmatrix} e_1 \\ e_2 \end{bmatrix}$$

The electrical network shown in the figure is a third-order system. We choose the following three-dimensional state vector

$$x = \begin{bmatrix} x_1 \\ x_2 \\ x_3 \end{bmatrix} = \begin{bmatrix} i_1 \\ i_2 \\ v \end{bmatrix}$$

The system is described by three differential equations. Two are obtained by requiring voltage equilibrium in the two loops, thus

$$e_1 = R i_1 + L_1 \frac{di_1}{dt} + v$$

$$e_2 = L_2 \frac{di_2}{dt} + v$$

For the capacitor C we have

$$\frac{dv(t)}{dt} = \frac{1}{C}(i_1 + i_2)$$

Rearranging the terms and substituting the state variables we obtain

$$\dot{x}_1 = -\frac{R}{L_1}x_1 - \frac{1}{L_1}x_3 + \frac{1}{L_1}u_1$$

$$\dot{x}_2 = -\frac{1}{L_2}x_3 + \frac{1}{L_2}u_2$$

$$\dot{x}_3 = \frac{1}{C}x_1 + \frac{1}{C}x_2$$

In the vector form, we have

$$\dot{x} = Ax + Bu$$

where

$$A = \begin{bmatrix} -\dfrac{R}{L_1} & 0 & -\dfrac{1}{L_1} \\ 0 & 0 & -\dfrac{1}{L_2} \\ \dfrac{1}{C} & \dfrac{1}{C} & 0 \end{bmatrix} \qquad B = \begin{bmatrix} \dfrac{1}{L_1} & 0 \\ 0 & \dfrac{1}{L_2} \\ 0 & 0 \end{bmatrix}$$

● **PROBLEM 9-47**

Obtain the state representation for the LRC network shown in Fig. 1.

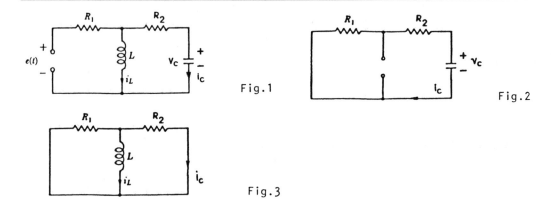

Fig.1

Fig.2

Fig.3

Solution: Let us choose the state variables as the inductor current and the capacitor voltage. Thus

$$x = \begin{bmatrix} x_1 \\ x_2 \end{bmatrix} = \begin{bmatrix} V_c \\ i_L \end{bmatrix}$$

We shall write the vector differential equation

$$\dot{x} = \begin{bmatrix} a_{11} & a_{12} \\ a_{21} & a_{22} \end{bmatrix} x + \begin{bmatrix} b_1 \\ b_2 \end{bmatrix} e(t)$$

To determine a_{11} let us set

$$x_2 = i_L = 0 \quad \text{and} \quad e(t) = 0$$

We obtain the reduced network shown in Fig. 2.

The Kirchhoff's current law for this network is

$$C\frac{dv_c}{dt} = C\frac{dx_1}{dt} = \frac{-x_1}{R_1 + R_2}$$

thus

$$a_{11} = -\frac{1}{C(R_1 + R_2)}$$

To find a_{12} let $x_1 = V_c = 0$ and $e(t) = 0$.

The reduced network is shown in Fig. 3.

and the resulting equation is

$$i_c = \frac{R_1}{R_1 + R_2} i_L = C\frac{dv}{dt}$$

or

$$C\frac{dx_1}{dt} = \frac{R_1}{R_1 + R_2} x_2$$

and

$$a_{12} = \frac{R_1}{(R_1 + R_2)C}$$

In a similar way we find a_{21} and a_{22}. Thus

$$A = \begin{bmatrix} \dfrac{-1}{(R_1 + R_2)C} & \dfrac{R_1}{(R_1 + R_2)C} \\[3mm] \dfrac{R_1}{L(R_1 + R_2)} & \dfrac{-R_1 R_2}{L(R_1 + R_2)} \end{bmatrix}$$

To find b_1 we set $x_1 = x_2 = 0$ and obtain the following relation

$$i_c = \frac{1}{R_1 + R_2} e(t)$$

or

$$\frac{dx_1}{dt} = \frac{1}{C(R_1 + R_2)} e(t)$$

$$b_1 = \frac{1}{C(R_1 + R_2)}$$

We find b_2 to be

$$b_2 = \frac{R_2}{L(R_1 + R_2)}$$

504

The vector differential equation of the system is

$$\frac{dx}{dt} = \begin{bmatrix} \dfrac{-1}{(R_1 + R_2)C} & \dfrac{R_1}{(R_1 + R_2)C} \\ \\ \dfrac{R_1}{L(R_1 + R_2)} & \dfrac{-R_1R_2}{L(R_1 + R_2)} \end{bmatrix} \begin{bmatrix} x_1 \\ \\ x_2 \end{bmatrix} + \begin{bmatrix} \dfrac{1}{C(R_1 + R_2)} \\ \\ \dfrac{R_2}{L(R_1 + R_2)} \end{bmatrix} e(t)$$

● **PROBLEM** 9-48

The inverted pendulum is mounted on a cart as shown. We assume that the rod between the pendulum and the cart is stiff and does not bend or expand.

The cart must be moved so that mass m is always in an upright position. The following equations describe the motion of the system

$$M\ddot{x} + ml\ddot{\theta} - f(t) = 0$$

$$ml\ddot{x} + ml^2\ddot{\theta} - mlg\theta = 0$$

We assume that $M \gg m$ and the angle of rotation θ is small, f(t) is the force on the cart. Obtain the state variable representation for the system and show that the system is unstable.

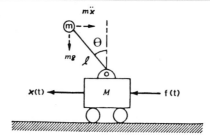

Solution: The state variables are

$$(x_1, x_2, x_3, x_4) = (x, \dot{x}, \theta, \dot{\theta})$$

Then, we have

$$M\dot{x}_2 + ml\dot{x}_4 - f(t) = 0$$

$$\dot{x}_2 + l\dot{x}_4 - gx_3 = 0$$

From the second equation we have

$$l\dot{x}_4 = gx_3 - \dot{x}_2$$

Substituting into the first

$$M\dot{x}_2 + mgx_3 - m\dot{x}_2 - f(t) = 0$$

Since $M \gg m$

we can write

505

$$M\dot{x}_2 + mgx_3 = f(t)$$

Substituting $\dot{x}_2$ we get

$$Ml\dot{x}_4 - Mgx_3 + f(t) = 0$$

Thus we can write the following system of equations

$$\dot{x}_1 = x_2$$

$$\dot{x}_2 = -\frac{mg}{M}x_3 + \frac{1}{M}f(t)$$

$$\dot{x}_3 = x_4$$

$$\dot{x}_4 = \frac{g}{l}x_3 - \frac{1}{Ml}f(t)$$

The system matrix is

$$A = \begin{bmatrix} 0 & 1 & 0 & 0 \\ 0 & 0 & -\frac{mg}{M} & 0 \\ 0 & 0 & 0 & 1 \\ 0 & 0 & \frac{g}{l} & 0 \end{bmatrix}$$

The characteristic equation is

$$\det(\lambda I - A) = \det \begin{bmatrix} \lambda & -1 & 0 & 0 \\ 0 & \lambda & \frac{mg}{M} & 0 \\ 0 & 0 & \lambda & -1 \\ 0 & 0 & -\frac{g}{l} & \lambda \end{bmatrix} = \lambda \cdot \lambda \cdot (\lambda^2 - \frac{g}{l}) = 0$$

The roots are

$$\lambda_1 = 0 \quad \text{and} \quad \lambda_2 = \sqrt{\frac{g}{l}}, \quad \lambda_3 = -\sqrt{\frac{g}{l}}$$

The system is unstable since there is a root in the right-hand plane, that is at $\lambda = +\sqrt{\frac{g}{l}}$.

● **PROBLEM** 9-49

The position ϕ and the velocity $\dot{\phi}$ of a radar antenna is controlled by the servomechanism shown.

The position coordinate ϕ of the inertia load is controlled by a d-c motor via a gear. The torque of the motor can be varied by means of the control voltage from the amplifier. With the switch in position B we control the position ϕ of the inertia load; the reference voltage r is compared with the potentiometer voltage. With the switch in position A we compare the voltage from the tachometer TM, thus controlling the velocity $\dot{\phi}$.

Obtain a state representation of the velocity servo.

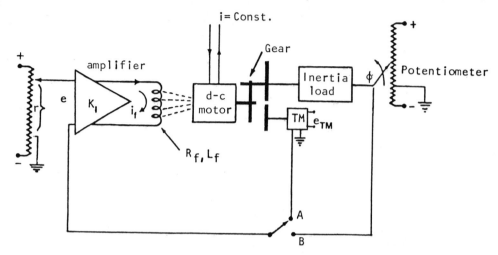

Solution: The torque of the motor is proportional to the field current i_f, thus

$$T = \alpha i_f, \quad \alpha \text{ is a proportional constant.} \quad \text{Furthermore}$$

the torque is proportional to the acceleration $\ddot{\phi}$ of the inertia load.

$$T = \beta \ddot{\phi} \quad \beta \text{ is a proportional constant}$$

We can write

$$\ddot{\phi} = \frac{T}{\beta} = \frac{\alpha}{\beta} i_f = k_2 i_f$$

For the field current we have

$$u = R_f i_f + L_f \frac{di_f}{dt}$$

We choose the state vector

$$x = \begin{bmatrix} x_1 \\ x_2 \\ x_3 \end{bmatrix} = \begin{bmatrix} i_f \\ \phi \\ \dot{\phi} \end{bmatrix}$$

Rewriting in the matrix form

$$\begin{bmatrix} \dot{x}_1 \\ \dot{x}_2 \\ \dot{x}_3 \end{bmatrix} = \begin{bmatrix} -\dfrac{R_f}{L_f} & 0 & 0 \\ 0 & 0 & 1 \\ k_2 & 0 & 0 \end{bmatrix} \begin{bmatrix} x_1 \\ x_2 \\ x_3 \end{bmatrix} + \begin{bmatrix} \dfrac{1}{L_f} \\ 0 \\ 0 \end{bmatrix} u$$

In the above example the output is the antenna velocity $\dot{\phi}$. But we are actually comparing the tachometer voltage with the reference input r. Let us use the tachometer voltage c as the output

$$c = k_3 \dot{\phi} = [0 \ 0 \ k_3]x$$

or

$$C = [0 \ 0 \ k_3]$$

where k_3 is a constant.

To obtain the closed-loop model, we compute

$$B k_1 C = \begin{bmatrix} \dfrac{1}{L_f} \\ 0 \\ 0 \end{bmatrix} k_1 [0 \ 0 \ k_3] = \begin{bmatrix} 0 & 0 & \dfrac{k_1 k_3}{L_f} \\ 0 & 0 & 0 \\ 0 & 0 & 0 \end{bmatrix}$$

Thus

$$A_{cl} = A - Bk_1 C = \begin{bmatrix} -\dfrac{R_f}{L_f} & 0 & -\dfrac{k_1 k_3}{L_f} \\ 0 & 0 & 1 \\ k_2 & 0 & 0 \end{bmatrix}$$

and

$$B_{cl} = Bk_1 = \begin{bmatrix} \dfrac{k_1}{L_f} \\ 0 \\ 0 \end{bmatrix}$$

For the closed-loop model we have the following system of equations

$$\begin{bmatrix} \dot{x}_1 \\ \dot{x}_2 \\ \dot{x}_3 \end{bmatrix} \begin{bmatrix} -\dfrac{R_f}{L_f} & 0 & -\dfrac{k_1 k_3}{L_f} \\ 0 & 0 & 1 \\ k_2 & 0 & 0 \end{bmatrix} \begin{bmatrix} x_1 \\ x_2 \\ x_3 \end{bmatrix} \begin{bmatrix} \dfrac{k_1}{L_f} \\ 0 \\ 0 \end{bmatrix} r$$

● **PROBLEM** 9-50

Consider a gyro with a single degree of freedom. Gyros sense the angular motion of a system and are used in automatic pilot systems. Shown in the figure are the schematic and functional diagrams of a gyro.

Axis OA is perpendicular to OB and OH, and OB is perpendicular to OH. The gimbal moves about the output axis OB. The input is measured around the input axis OA. The equation of motion

508

about the output axis is obtained by equating the rate of change of angular momentum to the sum of the external torques.

We have:

The change in angular momentum = $I\ddot{\theta} - H\omega\cos\theta$

The external torques = $-f\dot{\theta} - k\theta$

Obtain a state-space representation of the gyro system.

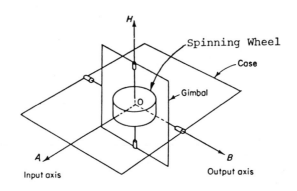

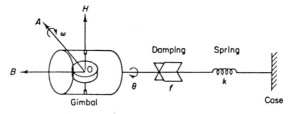

Solution: Let us choose the state variables as

$$x = \begin{bmatrix} x_1 \\ x_2 \end{bmatrix} = \begin{bmatrix} \theta \\ \dot{\theta} \end{bmatrix}$$

ω is the input variable

$$u = \omega$$

The equation describing the system is

$$I\ddot{\theta} - H\omega\cos\theta = -f\dot{\theta} - k\theta$$

It can be written using the state variables in the form

$$\dot{x}_1 = x_2$$

$$\dot{x}_2 = -\frac{k}{I}x_1 - \frac{f}{I}x_2 + \frac{H}{I}u \cos x_1$$

To avoid nonlinearity of the second equation let us expand $\cos x_1$

$$\cos x_1 = 1 - \frac{1}{2} x_1^2 + \dots$$

Since θ is a very small angle we can approximate

$$\cos x_1 \approx 1$$

We obtain the following state equation

$$\begin{bmatrix} \dot{x}_1 \\ \dot{x}_2 \end{bmatrix} = \begin{bmatrix} 0 & 1 \\ -\dfrac{k}{I} & -\dfrac{f}{I} \end{bmatrix} \begin{bmatrix} x_1 \\ x_2 \end{bmatrix} + \begin{bmatrix} 0 \\ \dfrac{H}{I} \end{bmatrix} [u]$$

Since θ is the output, we have

$$y = \theta = [1 \ 0] \begin{bmatrix} x_1 \\ x_2 \end{bmatrix}$$

510

CHAPTER 10

STATE TRANSITION MATRIX

METHODS OF DETERMINING THE STATE TRANSITION MATRIX

• **PROBLEM** 10-1

Using the Cayley-Hamilton theorem obtain e^{At}, where

$$A = \begin{bmatrix} 0 & 1 \\ -2 & -3 \end{bmatrix}$$

Solution: The characteristic equation is

$$\det(\lambda I - A) = \begin{vmatrix} \lambda & -1 \\ 2 & \lambda+3 \end{vmatrix} = (\lambda + 1)(\lambda + 2)$$

Thus $\lambda_1 = -1$, $\lambda_2 = -2$ and

$$e^{At} = \alpha_0(t)I + \alpha_1(t)A$$

or in the scalar form

$$e^{\lambda t} = \alpha_0(t) + \alpha_1(t)\lambda$$

Substituting $\lambda_1 = -1$, $\lambda_2 = -2$ we get

$$e^{-t} = \alpha_0(t) - \alpha_1(t)$$

$$e^{-2t} = \alpha_0(t) - 2\alpha_1(t)$$

Solving the above system of equations, we obtain

$$\alpha_0(t) = 2e^{-t} - e^{-2t}$$

511

$$\alpha_1(t) = e^{-t} - e^{-2t}$$

and

$$e^{At} = \alpha_0(t)I + \alpha_1(t)A$$

$$= (2e^{-t} - e^{-2t}) \begin{bmatrix} 1 & 0 \\ 0 & 1 \end{bmatrix} + (e^{-t} - e^{-2t}) \begin{bmatrix} 0 & 1 \\ -2 & -3 \end{bmatrix}$$

$$= \begin{bmatrix} 2e^{-t} - e^{-2t} & e^{-t} - e^{-2t} \\ -2e^{-t} + 2e^{-2t} & -e^{-t} + 2e^{-2t} \end{bmatrix}$$

● **PROBLEM** 10-2

Compute matrix e^{At}, where

$$A = \begin{bmatrix} \lambda & 1 & 0 \\ 0 & \lambda & 1 \\ 0 & 0 & \lambda \end{bmatrix}$$

<u>Solution:</u> Matrix A can be written as a sum

$$\begin{bmatrix} \lambda & 1 & 0 \\ 0 & \lambda & 1 \\ 0 & 0 & \lambda \end{bmatrix} = \lambda \begin{bmatrix} 1 & 0 & 0 \\ 0 & 1 & 0 \\ 0 & 0 & 1 \end{bmatrix} + \begin{bmatrix} 0 & 1 & 0 \\ 0 & 0 & 1 \\ 0 & 0 & 0 \end{bmatrix}$$

Since both matrices on the right-hand side of the above equation commute, we can write

$$\exp \left\{ \begin{bmatrix} \lambda & 1 & 0 \\ 0 & \lambda & 1 \\ 0 & 0 & \lambda \end{bmatrix} t \right\} = \left\{ \exp \begin{bmatrix} \lambda t & 0 & 0 \\ 0 & \lambda t & 0 \\ 0 & 0 & \lambda t \end{bmatrix} \right\} \left\{ \exp \begin{bmatrix} 0 & t & 0 \\ 0 & 0 & t \\ 0 & 0 & 0 \end{bmatrix} \right\}$$

Since

$$\exp \begin{bmatrix} 0 & t & 0 \\ 0 & 0 & t \\ 0 & 0 & 0 \end{bmatrix} = \begin{bmatrix} 1 & t & \frac{1}{2}t^2 \\ 0 & 1 & t \\ 0 & 0 & 1 \end{bmatrix}$$

and

$$\exp \begin{bmatrix} \lambda t & 0 & 0 \\ 0 & \lambda t & 0 \\ 0 & 0 & \lambda t \end{bmatrix} = e^{\lambda t}$$

We have

$$e^{At} = e^{\lambda t} \begin{bmatrix} 1 & t & \frac{1}{2}t^2 \\ 0 & 1 & t \\ 0 & 0 & 1 \end{bmatrix}$$

where

$$A = \begin{bmatrix} \lambda & 1 & 0 \\ 0 & \lambda & 1 \\ 0 & 0 & \lambda \end{bmatrix}$$

● **PROBLEM** 10-3

Compute e^{At} where

$$A = \begin{bmatrix} 0 & 1 \\ 0 & -2 \end{bmatrix}$$

using the following methods

1. eigenvalue method

2. series method

3. Laplace transform method

4. Sylvester's interpolation method

Solution: 1) The characteristic equation of A is

$$|\lambda I - A| = \begin{vmatrix} \lambda & -1 \\ 0 & \lambda + 2 \end{vmatrix} = 0$$

The eigenvalues are

$$\lambda_1 = 0 \qquad \lambda_2 = -2$$

The transformation matrix P is

$$P = \begin{bmatrix} 1 & 1 \\ 0 & -2 \end{bmatrix}$$

From

$$e^{At} = Pe^{Dt}P^{-1} = P \begin{bmatrix} e^{\lambda_1 t} & & & & & 0 \\ & e^{\lambda_2 t} & & & & \\ & & & & \ddots & \\ & & & & & \\ 0 & & & & & e^{\lambda_n t} \end{bmatrix} P^{-1}$$

$$= \begin{bmatrix} 1 & 1 \\ 0 & -2 \end{bmatrix} \begin{bmatrix} e^0 & 0 \\ 0 & e^{-2t} \end{bmatrix} \begin{bmatrix} 1 & \frac{1}{2} \\ 0 & -\frac{1}{2} \end{bmatrix} = \begin{bmatrix} 1 & \frac{1}{2}(1 - e^{-2t}) \\ 0 & e^{-2t} \end{bmatrix}$$

2) Using the series method we find

$$e^{At} = I + \sum_{k=1}^{\infty} \frac{A^k t^k}{k!} = \begin{bmatrix} 1 & 0 \\ 0 & 1 \end{bmatrix} + \begin{bmatrix} 0 & 1 \\ 0 & -2 \end{bmatrix} t + \begin{bmatrix} 0 & 1 \\ 0 & -2 \end{bmatrix}^2 \frac{t^2}{2!} + \cdots$$

$$= \begin{bmatrix} 1 & \frac{1}{2} - \frac{1}{2}(1 - 2t + \frac{(2t)^2}{2!} - \frac{(2t)^3}{3!} + \cdots) \\ 0 & 1 - 2t + \frac{(2t)^2}{2!} - \frac{(2t)^3}{3!} + \frac{(2t)^4}{4!} - \cdots \end{bmatrix} = \begin{bmatrix} 1 & \frac{1}{2}(1 - e^{-2t}) \\ 0 & e^{-2t} \end{bmatrix}$$

3) We have

$$sI - A = \begin{bmatrix} s & -1 \\ 0 & s+2 \end{bmatrix}$$

and

$$(sI-A)^{-1} = \begin{bmatrix} \frac{1}{s} & \frac{1}{s(s+2)} \\ 0 & \frac{1}{s+2} \end{bmatrix}$$

$$e^{At} = \mathcal{L}^{-1}[(sI-A)^{-1}] = \begin{bmatrix} 1 & \frac{1}{2}(1 - e^{-2t}) \\ 0 & e^{-2t} \end{bmatrix}$$

4) Sylvester's formula

$$\begin{vmatrix} 1 & \lambda_1 & \lambda_1^2 & \cdots & \lambda_1^{m-1} & e^{\lambda_1 t} \\ 1 & \lambda_2 & \lambda_2^2 & & \lambda_2^{m-1} & e^{\lambda_2 t} \\ \cdot & \cdot & \cdot & & \cdot & \cdot \\ \cdot & \cdot & \cdot & \cdots & \cdot & \cdot \\ \cdot & \cdot & \cdot & & \cdot & \cdot \\ 1 & \lambda_m & \lambda_m^2 & \cdots & \lambda_m^{m-1} & e^{\lambda_m t} \\ I & A & A^2 & \cdots & A^{m-1} & e^{At} \end{vmatrix} = 0$$

In our case, we get

$$
\begin{vmatrix}
1 & \lambda_1 & e^{\lambda_1 t} \\
1 & \lambda_2 & e^{\lambda_2 t} \\
I & A & e^{At}
\end{vmatrix} = 0
$$

The eigenvalues are $\lambda_1 = 0$, $\lambda_2 = -2$, thus

$$
\begin{vmatrix}
1 & 0 & 1 \\
1 & -2 & e^{-2t} \\
I & A & e^{At}
\end{vmatrix} = 0
$$

or

$$
-2e^{At} + A + 2I - Ae^{-2t} = 0
$$

Solving for e^{At} we get

$$
e^{At} = \frac{1}{2}(A + 2I - Ae^{-2t}) = \begin{bmatrix} 1 & \frac{1}{2}(1 - e^{-2t}) \\ 0 & e^{-2t} \end{bmatrix}
$$

● **PROBLEM** 10-4

Using the Cayley-Hamilton theorem compute e^{At} where

$$
A = \begin{bmatrix} 2 & 1 & 4 \\ 0 & 2 & 0 \\ 0 & 3 & 1 \end{bmatrix}
$$

Solution: From the characteristic equation

$$
\det|\lambda I - A| = 0
$$

we find the eigenvalues of A

$$
\lambda_1 = 2, \ \lambda_2 = 2, \ \lambda_3 = 1
$$

Thus, the minimal polynomial is given by

$$
\phi(\lambda) = (2 - \lambda)^2 (1 - \lambda)
$$

The Cayley-Hamilton theorem states that every matrix satisfies its own characteristic equation.

That is, if A is an n x n matrix with characteristic equation

$$\Delta(\lambda) = \lambda^n + a_{n-1}\lambda^{n-1} + a_{n-2}\lambda^{n-2} + \ldots + a_1\lambda + a_0 = 0$$

then

$$\Delta(A) = A^n + a_{n-1}A^{n-1} + a_{n-2}A^{n-2} + \ldots + a_1A + a_0I = 0$$

From the above equation it follows that any A^N, $N \geq n$ may be written as a linear combination of I, A, A^2 ... A^{n-1} and any power series in A may therefore also be written as a linear combination of the same matrices. Thus, if

$$f(A) = \sum_{k=0}^{\infty} a_k A^k$$

we have

$$f(A) = \sum_{k=0}^{n-1} \alpha_k A^k$$

and

$$f(\lambda_i) = \sum_{k=0}^{n-1} \alpha_k \lambda_i^k$$

If $\lambda_i \neq \lambda_j$ for all eigenvalues of A, the last equation gives n equations in n unknowns and all coefficients α_k can be determined. If multiple eigenvalues occur the method must be modified. Let λ_i have multiplicity m. Then

$$\Delta(\lambda) = (\lambda - \lambda_i)^m \Delta_1(\lambda)$$

Differentiating the last equation we get

$$\Delta'(\lambda) = m(\lambda - \lambda_i)^{m-1} \Delta_1(\lambda) + (\lambda - \lambda_i)^m \Delta_1'(\lambda)$$

$$= (\lambda - \lambda_i)^{m-1} \Delta_2(\lambda)$$

Thus, we see that

$$\Delta(\lambda_i) = \Delta'(\lambda_i) = \Delta''(\lambda_i) = \ldots = \Delta^{(m-1)}(\lambda_i) = 0$$

We see that the equation

$$f(\lambda_i) = \sum_{k=0}^{n-1} \alpha_k \lambda_i^k$$

can be differentiated to give (m - 1) independent equations from which we find the coefficients α_k.

In our case we have

$$e^{At} = \alpha_o I + \alpha_1 A + \alpha_2 A^2$$

We obtain the following system of equations.

$$te^{\lambda_1 t} = \alpha_1 + 2\alpha_2 \lambda_1$$

$$e^{\lambda_1 t} = \alpha_o + \alpha_1 \lambda_1 + \alpha_2 \lambda_1^2$$

$$e^{\lambda_3 t} = \alpha_o + \alpha_1 \lambda_3 + \alpha_2 \lambda_3^2$$

Substituting $\lambda_1 = 2$ and $\lambda_3 = 1$ and solving the equations we obtain

$$\alpha_o = 4e^t - 3e^{2t} + 2te^{2t}$$

$$\alpha_1 = -4e^t + 4e^{2t} - 3te^{2t}$$

$$\alpha_2 = e^t - e^{2t} + te^{2t}$$

Thus, for e^{At} we obtain

$$e^{At} = (4e^t - 3e^{2t} + 2te^{2t}) \begin{bmatrix} 1 & 0 & 0 \\ 0 & 1 & 0 \\ 0 & 0 & 1 \end{bmatrix}$$

$$+ (-4e^t + 4e^{2t} - 3te^{2t}) \begin{bmatrix} 2 & 1 & 4 \\ 0 & 2 & 0 \\ 0 & 3 & 1 \end{bmatrix}$$

$$+ (e^t - e^{2t} + te^{2t}) \begin{bmatrix} 4 & 16 & 12 \\ 0 & 4 & 0 \\ 0 & 9 & 1 \end{bmatrix}$$

$$= \begin{bmatrix} e^{2t} & 12e^t - 12e^{2t} + 13te^{2t} & -4e^t + 4e^{2t} \\ 0 & e^{2t} & 0 \\ 0 & -3e^t + 3e^{2t} & e^t \end{bmatrix}$$

Compute e^{At} where

$$A = \begin{bmatrix} \lambda & 1 & 0 & \ldots & 0 \\ 0 & \lambda & 1 & \ldots & 0 \\ \vdots & \vdots & \vdots & & \vdots \\ 0 & 0 & 0 & \ldots & \lambda 1 \\ 0 & 0 & 0 & \ldots & 0\lambda \end{bmatrix}$$

Solution: We can represent matrix At as a sum of two matrices

$$At = Bt + Ct$$

where

$$Bt = \begin{bmatrix} \lambda t & & 0 \\ & \lambda t & \\ & & \ddots \\ 0 & & \lambda t \end{bmatrix}$$

and

$$Ct = \begin{bmatrix} 0 & t & 0 & \ldots & 0 \\ 0 & 0 & t & \ldots & 0 \\ \vdots & & & & \vdots \\ & & & & t \\ 0 & \ldots & \ldots & \ldots & 0 \end{bmatrix}$$

Note that B and C commute, thus

$$e^{At} = e^{Bt} e^{Ct}$$

$$= e^{Bt} [I + Ct + \frac{1}{2!} (Ct)^2 + \ldots + \frac{1}{(n-1)!}(Ct)^{n-1}]$$

where n is the dimension of the matrix A. Since

$$e^{Bt} = e^{\lambda t}$$

we have

518

$$e^{At} = e^{\lambda t} \begin{bmatrix} 1 & t & \dfrac{t^2}{2!} & \cdots & \dfrac{t^{n-1}}{(n-1)!} \\ 0 & 1 & t & \cdots & \dfrac{t^{n-2}}{(n-2)!} \\ \vdots & \vdots & & & \vdots \\ 0 & 0 & \cdots & \cdots & t \\ 0 & 0 & 0 & \cdots & \cdots 1 \end{bmatrix}$$

● **PROBLEM** 10-6

Using Sylvester's interpolation method compute e^{At}, where

$$A = \begin{bmatrix} 0 & 0 & 1 & 0 \\ 1 & 0 & 0 & 1 \\ 0 & 0 & 0 & -1 \\ 0 & 1 & 0 & 0 \end{bmatrix}$$

<u>Solution</u>: The characteristic equation of A is

$$|A - \lambda I| = \begin{vmatrix} -\lambda & 0 & 1 & 0 \\ 1 & -\lambda & 0 & 1 \\ 0 & 0 & -\lambda & -1 \\ 0 & 1 & 0 & -\lambda \end{vmatrix} = \lambda^4 - \lambda^2 + 1 = 0$$

The roots are

$$\lambda_1 = \frac{-\sqrt{3} + j}{2} \qquad \lambda_2 = \frac{\sqrt{3} - j}{2}$$

$$\lambda_3 = \frac{-\sqrt{3} - j}{2} \qquad \lambda_4 = \frac{\sqrt{3} + j}{2}$$

From Sylvester's interpolation formula we have the following equation.

$$
\begin{vmatrix}
1 & \lambda_1 & \lambda_1^2 & \lambda_1^3 & e^{\lambda_1 t} \\
1 & \lambda_2 & \lambda_2^2 & \lambda_2^3 & e^{\lambda_2 t} \\
1 & \lambda_3 & \lambda_3^2 & \lambda_3^3 & e^{\lambda_3 t} \\
1 & \lambda_4 & \lambda_4^2 & \lambda_4^3 & e^{\lambda_4 t} \\
I & A & A^2 & A^3 & e^{At}
\end{vmatrix} = 0
$$

Let us expand the determinant about the last column. We have

$$
\begin{vmatrix}
1 & \lambda_1 & \lambda_1^2 & \lambda_1^3 \\
1 & \lambda_2 & \lambda_2^2 & \lambda_2^3 \\
1 & \lambda_3 & \lambda_3^2 & \lambda_3^3 \\
1 & \lambda_4 & \lambda_4^2 & \lambda_4^3
\end{vmatrix} e^{At} = -
\begin{vmatrix}
1 & \lambda_2 & \lambda_2^2 & \lambda_2^3 \\
1 & \lambda_3 & \lambda_3^2 & \lambda_3^3 \\
1 & \lambda_4 & \lambda_4^2 & \lambda_4^3 \\
I & A & A^2 & A^3
\end{vmatrix} e^{\lambda_1 t}
$$

$$
+
\begin{vmatrix}
1 & \lambda_1 & \lambda_1^2 & \lambda_1^3 \\
1 & \lambda_3 & \lambda_3^2 & \lambda_3^3 \\
1 & \lambda_4 & \lambda_4^2 & \lambda_4^3 \\
I & A & A^2 & A^3
\end{vmatrix} e^{\lambda_2 t} -
\begin{vmatrix}
1 & \lambda_1 & \lambda_1^2 & \lambda_1^3 \\
1 & \lambda_2 & \lambda_2^2 & \lambda_2^3 \\
1 & \lambda_4 & \lambda_4^2 & \lambda_4^3 \\
I & A & A^2 & A^3
\end{vmatrix} e^{\lambda_3 t}
$$

$$
+
\begin{vmatrix}
1 & \lambda_1 & \lambda_1^2 & \lambda_1^3 \\
1 & \lambda_2 & \lambda_2^2 & \lambda_2^3 \\
1 & \lambda_3 & \lambda_3^2 & \lambda_3^3 \\
I & A & A^2 & A^3
\end{vmatrix} e^{\lambda_4 t}
$$

Evaluating 4 x 4 determinants and substituting into the last equation we get

$$
e^{At} = -\frac{1}{12} \{ [-(3 - j\sqrt{3})I + 2\sqrt{3}A - j2\sqrt{3}A^2 - (\sqrt{3} - j3)A^3]e^{\lambda_1 t}
$$

$$
+ [-(3 - j\sqrt{3})I - 2\sqrt{3}A - j2\sqrt{3}A^2 + (\sqrt{3} - j3)A^3]e^{\lambda_2 t}
$$

$$
+ [-(3 + j\sqrt{3})I + 2\sqrt{3}A + j2\sqrt{3}A^2 - (\sqrt{3} + j3)A^3]e^{\lambda_3 t}
$$

$$
+ [-(3 + j\sqrt{3})I - 2\sqrt{3}A + j2\sqrt{3}A^2 + (\sqrt{3} + j3)A^3]e^{\lambda_4 t}\}
$$

Using the identity

$$e^{j\theta} = \cos \theta + j \sin \theta$$

we shall group the real and imaginary terms. Thus

$$e^{At} = \frac{\sqrt{3}}{6}\left\{ e^{\frac{\sqrt{3}}{2}t} \left[(\sqrt{3}I + 2A - A^3)\cos \frac{t}{2} + (-I + 2A^2 + \sqrt{3}A^3)\sin \frac{t}{2}\right] \right.$$

$$\left. + e^{-\frac{\sqrt{3}}{2}t} \left[(\sqrt{3}I - 2A + A^3)\cos \frac{t}{2} + (I - 2A^2 + \sqrt{3}A^3)\sin \frac{t}{2}\right] \right\}$$

Let us denote

$$\alpha = e^{\frac{\sqrt{3}}{2}t} + e^{-\frac{\sqrt{3}}{2}t}$$

$$\beta = e^{\frac{\sqrt{3}}{2}t} - e^{\frac{\sqrt{3}}{2}t}$$

Thus, e^{At} can be written as

$$e^{At} = \frac{\sqrt{3}}{6}\left[I\left(\sqrt{3}\,\alpha \cos \frac{t}{2} - \beta \sin \frac{t}{2}\right) + A\, 2\beta \cos \frac{t}{2}\right.$$

$$\left. + A^2 2\beta \sin \frac{t}{2} + A^3\left(-\beta \cos \frac{t}{2} + \sqrt{3}\,\alpha \sin \frac{t}{2}\right)\right]$$

Multiplying we find A^2 and A^3

$$A^2 = \begin{bmatrix} 0 & 0 & 0 & -1 \\ 0 & 1 & 1 & 0 \\ 0 & -1 & 0 & 0 \\ 1 & 0 & 0 & 1 \end{bmatrix}, \quad A^3 = \begin{bmatrix} 0 & -1 & 0 & 0 \\ 1 & 0 & 0 & 0 \\ -1 & 0 & 0 & -1 \\ 0 & 1 & 1 & 0 \end{bmatrix}$$

Substituting we obtain for e^{At}

$$e^{At} = \begin{bmatrix} \frac{\alpha}{2}\cos \frac{t}{2} - \frac{\beta}{2\sqrt{3}}\sin \frac{t}{2} & \frac{\beta}{2\sqrt{3}}\cos \frac{t}{2} - \frac{\alpha}{2}\sin \frac{t}{2} \\[4mm] \frac{\beta}{2\sqrt{3}}\cos \frac{t}{2} + \frac{\alpha}{2}\sin -\frac{t}{2} & \frac{\alpha}{2}\cos \frac{t}{2} + \frac{\beta}{2\sqrt{3}}\sin \frac{t}{2} \\[4mm] \frac{\beta}{2\sqrt{3}}\cos \frac{t}{2} - \frac{\alpha}{2}\sin \frac{t}{2} & -\frac{\beta}{\sqrt{3}}\sin \frac{t}{2} \\[4mm] \frac{\beta}{\sqrt{3}}\sin \frac{t}{2} & \frac{\beta}{2\sqrt{3}}\cos \frac{t}{2} + \frac{\alpha}{2}\sin \frac{t}{2} \end{bmatrix}$$

$$\frac{\beta}{\sqrt{3}} \cos \frac{t}{2} \qquad\qquad -\frac{\beta}{\sqrt{3}} \sin \frac{t}{2}$$

$$\frac{\beta}{\sqrt{3}} \sin \frac{t}{2} \qquad\qquad \frac{\beta}{\sqrt{3}} \cos \frac{t}{2}$$

$$\frac{\alpha}{2} \cos \frac{t}{2} - \frac{\beta}{2\sqrt{3}} \sin \frac{t}{2} \qquad -\frac{\beta}{2\sqrt{3}} \cos \frac{t}{2} - \frac{\alpha}{2} \sin \frac{t}{2}$$

$$-\frac{\beta}{2\sqrt{3}} \cos \frac{t}{2} + \frac{\alpha}{2} \sin \frac{t}{2} \qquad \frac{\alpha}{2} \cos \frac{t}{2} + \frac{\beta}{2\sqrt{3}} \sin \frac{t}{2}$$

STATE TRANSITION MATRIX OF SYSTEMS

● PROBLEM 10-7

The system is described by

$$\dot{x} = Ax(t), \quad x(t_o) = x_o$$

Assuming that the eigenvalues of A are distinct, show that:

1) a fundamental matrix can be given by

$$X(t) = \begin{bmatrix} e^{\lambda_1(t-t_o)}k_{11} & e^{\lambda_2(t-t_o)}k_{12} & \cdots & e^{\lambda_n(t-t_o)}k_{1n} \\ e^{\lambda_1(t-t_o)}k_{21} & e^{\lambda_2(t-t_o)}k_{22} & \cdots & e^{\lambda_n(t-t_o)}k_{2n} \\ \vdots & & & \vdots \\ e^{\lambda_1(t-t_o)}k_{n_1} & e^{\lambda_2(t-t_o)}k_{n_2} & \cdots & e^{\lambda_n(t-t_o)}k_{nn} \end{bmatrix}$$

2) the unique fundamental matrix $\phi(t)$ which satisfies

$$\dot{\phi}(t) = A\phi(t), \quad \phi(t_o) = I$$

can be written as

$$\phi(t) = \begin{bmatrix} k_{11} & k_{12} & \cdots & k_{1n} \\ \vdots & & & \\ k_{n1} & k_{n2} & \cdots & k_{nn} \end{bmatrix} \begin{bmatrix} e^{\lambda_1(t-t_o)} & & & o \\ & e^{\lambda_2(t-t_o)} & & \\ & & \ddots & \\ o & & & e^{\lambda_n(t-t_o)} \end{bmatrix}$$

$$\begin{bmatrix} k_{11} & k_{12} & \cdots & k_{1n} \\ \vdots & & & \\ k_{n1} & k_{n2} & \cdots & k_{nn} \end{bmatrix}^{-1}$$

<u>Solution:</u> Let us denote the solution of the equation

$$\dot{x}(t) = Ax(t); \qquad x(t_o) = x_o$$

by

$$x(t) = e^{\lambda(t-t_o)}k$$

Differentiating

$$\dot{x}(t) = \lambda e^{\lambda(t-t_o)}k$$

Combining the above equations we get

$$(A - \lambda I)k = 0$$

or

$$\lambda e^{\lambda(t-t_o)}k = Ae^{\lambda(t-t_o)}k$$

Equation $(A - \lambda I)k = 0$ has a nontrivial solution $k \neq 0$ if and only if

$$\det(A - \lambda I) = 0$$

For each eigenvalue λ_i there corresponds an eigenvector k_i. Setting $\lambda = \lambda_i$ and solving the n simultaneous equations, we find the coordinates of k_i.

$$(a_{11} - \lambda_i)k_{1i} + a_{12}k_{2i} + a_{13}k_{3i} + \cdots + a_{1n}k_{ni} = 0$$

$$a_{22}k_{1i} + (a_{22} - \lambda_i)k_{2i} + a_{23}k_{3i} + \cdots + a_{2n}k_{ni} = 0$$

$$\cdots \qquad \qquad \cdots$$

$$a_{n1}k_{1i} + a_{n2}k_{2i} + \cdots + (a_{nn} - \lambda_i)k_{ni} = 0$$

for i = 1, 2, ... n.

Vectors $X_i(t) = e^{\lambda i(t-t_o)}k_i$ are n solutions of the initial equation and are linearly independent over any interval. Thus the fundamental matrix is given by

$$X(t) = \begin{bmatrix} e^{\lambda_1(t-t_o)}k_{11} & e^{\lambda_2(t-t_o)}k_{12} & \cdots & e^{\lambda n(t-t_o)}k_{1n} \\ \vdots & \vdots & & \\ e^{\lambda_1(t-t_o)}k_{n1} & e^{\lambda_2(t-t_o)}k_{n2} & \cdots & e^{\lambda n(t-t_o)}k_{nn} \end{bmatrix}$$

2) The unique fundamental matrix $\phi(t)$ which satisfies

$$\dot{\phi}(t) = A\phi(t) \quad ; \quad \phi(t_o) = I$$

where A is a constant matrix n x n, can be written as X(t) times a constant nonsingular matrix.

Thus, we obtain

$$\phi(t) = X(t)X^{-1}(t_o)$$

$$= \begin{bmatrix} k_{11} & k_{12} & \cdots & k_{1n} \\ \vdots & & \vdots & \\ k_{n1} & k_{n2} & \cdots & k_{nn} \end{bmatrix} \begin{bmatrix} e^{\lambda_1(t-t_o)} & & & 0 \\ & \ddots & & \\ & & \ddots & \\ 0 & & & e^{\lambda n(t-t_o)} \end{bmatrix} \begin{bmatrix} k_{11} & k_{12} & \cdots & k_{1n} \\ k_{21} & & & \\ \vdots & & & \\ k_{n1} & & \cdots & k_{nn} \end{bmatrix}^{-1}$$

• PROBLEM 10-8

Find the transition matrix of the system

$$\dot{x} = \begin{bmatrix} 0 & 1 & 0 \\ 0 & 0 & 1 \\ -6 & -11 & -6 \end{bmatrix} x \qquad (1)$$

Solution: Taking the Laplace transformation of eq. (1) and rearranging the elements of the equation we obtain

$$\begin{bmatrix} x_1(s) \\ x_2(s) \\ x_3(s) \end{bmatrix} = \begin{bmatrix} s & -1 & 0 \\ 0 & s & -1 \\ 6 & 11 & s+6 \end{bmatrix}^{-1} \begin{bmatrix} x_1(0) \\ x_2(0) \\ x_3(0) \end{bmatrix}$$

524

Let us denote by $[K_1]$, $[K_2]$ and $[K_3]$, 3 x 3 matrices such that

$$\begin{bmatrix} s & -1 & 0 \\ 0 & s & -1 \\ 6 & 11 & s+6 \end{bmatrix}^{-1} = \frac{[K_1]}{s+1} + \frac{[K_2]}{s+2} + \frac{[K_3]}{s+3}$$

To determine $[K_1]$ we multiply each side by $(s + 1)$ and set $s = -1$

$$(s + 1) \quad \begin{bmatrix} s & -1 & 0 \\ 0 & s & -1 \\ 6 & 11 & s+6 \end{bmatrix}^{-1} \Bigg|_{s \to -1} = [K_1]$$

We shall evaluate $[K_1]$ in terms of the adjoint matrix and determinant

$$[K_1] = \frac{[L(s)]|_{s \to -1}}{(-1 + 2)(-1 + 3)}$$

To find $[L(s)]|_{s \to -1}$ we use Faddeev's method.

$$[L_1] = [L(s)]|_{s \to -1} = \begin{bmatrix} -1 & -1 & 0 \\ 0 & -1 & -1 \\ 6 & 11 & 5 \end{bmatrix}$$

$\text{tr}[L_1] = 3$

where $\quad \text{tr } A = a_{11} + a_{22} + \ldots + a_{nn}$

$$[M_1] = [L_1] - \text{tr}[L_1] \cdot [1] = \begin{bmatrix} -4 & -1 & 0 \\ 0 & -4 & -1 \\ 6 & 11 & 2 \end{bmatrix}$$

where $[I]$ is the identity matrix.

$$[L_2] = [L_1][M_1] = \begin{bmatrix} 4 & 5 & 1 \\ -6 & -7 & -1 \\ 6 & 5 & -1 \end{bmatrix}$$

$\text{tr}[L_2] = -4$

$$[M_2] = [L_2] - \frac{1}{2}\text{tr}[L_2][I] = \begin{bmatrix} 6 & 5 & 1 \\ -6 & -5 & -1 \\ 6 & 5 & 1 \end{bmatrix}$$

525

For $[K_1]$ we obtain

$$[K_1] = \frac{M_2}{(-1 + 2)(-1 + 3)} = \begin{bmatrix} 3 & \frac{5}{2} & \frac{1}{2} \\ -3 & -\frac{5}{2} & -\frac{1}{2} \\ 3 & \frac{5}{2} & \frac{1}{2} \end{bmatrix}$$

Next, we find $[K_2]$ and $[K_3]$.

$$[K_2] = \begin{bmatrix} 3 & -4 & -1 \\ 6 & 8 & 2 \\ -12 & -16 & -4 \end{bmatrix} \qquad [K_3] = \begin{bmatrix} 1 & \frac{3}{2} & \frac{1}{2} \\ -3 & -\frac{9}{2} & -\frac{3}{2} \\ 9 & \frac{27}{2} & \frac{9}{2} \end{bmatrix}$$

The transition matrix is given by

$$[\phi(t)] = [K_1]e^{s_1 t} + [K_2]e^{s_2 t} + [K_3]e^{s_3 t}$$

$$= e^{-t} \begin{bmatrix} 3+3e^{-t}+e^{-2t} & \frac{5}{2}-4e^{-t}+\frac{3}{2}e^{-2t} & \frac{1}{2}e^{-t}+\frac{1}{2}e^{-2t} \\ -3+6e^{-t}-3e^{-2t} & -\frac{5}{2}+8e^{-t}-\frac{9}{2}e^{-2t} & -\frac{1}{2}+2e^{-t}-\frac{3}{2}e^{-2t} \\ 3-12e^{-t}+9e^{-2t} & \frac{5}{2}-16e^{-t}+\frac{27}{2}e^{-2t} & \frac{1}{2}-4e^{-t}+\frac{9}{2}e^{-2t} \end{bmatrix}$$

● **PROBLEM** 10-9

Find the closed-loop transfer matrix for the system shown below.

where

$$G(s) = \begin{bmatrix} \dfrac{2}{s+1} & 0 \\ \dfrac{1}{s} & \dfrac{1}{s+3} \end{bmatrix}$$

and

$$H(s) = \begin{bmatrix} 1 & 0 \\ 0 & 1 \end{bmatrix}$$

<u>Solution:</u> We shall evaluate the closed-loop transfer matrix from the following formula

$$M(s) = [I + G(s)H(s)]^{-1}G(s)$$

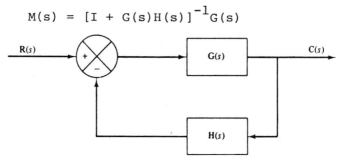

Let us compute $I + G(s)H(s)$.

$$I+G(s)H(s) = \begin{bmatrix} 1 & 0 \\ 0 & 1 \end{bmatrix} + \begin{bmatrix} \dfrac{2}{s+1} & 0 \\ \dfrac{1}{s} & \dfrac{1}{s+3} \end{bmatrix} \begin{bmatrix} 1 & 0 \\ 0 & 1 \end{bmatrix}$$

$$= \begin{bmatrix} 1 & 0 \\ 0 & 1 \end{bmatrix} + \begin{bmatrix} \dfrac{2}{s+1} & 0 \\ \dfrac{1}{s} & \dfrac{1}{s+3} \end{bmatrix} = \begin{bmatrix} 1 + \dfrac{2}{s+1} & 0 \\ \dfrac{1}{s} & 1 + \dfrac{1}{s+3} \end{bmatrix}$$

$$= \begin{bmatrix} \dfrac{s+3}{s+1} & 0 \\ \dfrac{1}{s} & \dfrac{s+4}{s+3} \end{bmatrix}$$

$$\det [I + G(s)H(s)] = \frac{s+3}{s+1} \cdot \frac{s+4}{s+3} = \frac{s+4}{s+1}$$

We can compute now the closed-loop transfer matrix.

$$M(s) = [I + G(s)H(s)]^{-1}G(s)$$

$$= \frac{s+1}{s+4} \begin{bmatrix} \dfrac{s+4}{s+3} & 0 \\ -\dfrac{1}{s} & \dfrac{s+3}{s+1} \end{bmatrix} \begin{bmatrix} \dfrac{2}{s+1} & 0 \\ \dfrac{1}{s} & \dfrac{1}{s+3} \end{bmatrix}$$

$$= \begin{bmatrix} \dfrac{s+1}{s+3} & 0 \\ -\dfrac{s+1}{s(s+4)} & \dfrac{s+3}{s+4} \end{bmatrix} \begin{bmatrix} \dfrac{2}{s+1} & 0 \\ \dfrac{1}{s} & \dfrac{1}{s+3} \end{bmatrix}$$

$$= \begin{bmatrix} \dfrac{2}{s+3} & 0 \\ \dfrac{s+1}{s(s+4)} & \dfrac{1}{s+4} \end{bmatrix}$$

For the time-varying system obtain $\phi(t,0)$

$$\dot{x} = \begin{bmatrix} 0 & 1 \\ 0 & t \end{bmatrix} x$$

Solution: To compute $\phi(t, 0)$ we shall use the formula

$$\phi(t, t_o) = I + \int_{t_o}^{t} A(\tau)d\tau + \int_{t_o}^{t} A(\tau_1) \left[\int_{0}^{\tau_1} A(\tau_2)d\tau_2 \right] d\tau_1 + \ldots$$

For $t_o = 0$, the terms are

$$\int_0^t A(\tau)d\tau = \int_0^t \begin{bmatrix} 0 & 1 \\ 0 & \tau \end{bmatrix} d\tau = \begin{bmatrix} 0 & t \\ 0 & \dfrac{t^2}{2} \end{bmatrix}$$

$$\int_0^t \begin{bmatrix} 0 & 1 \\ 0 & \tau_1 \end{bmatrix} \left\{ \int_0^{\tau_1} \begin{bmatrix} 0 & 1 \\ 0 & \tau_2 \end{bmatrix} d\tau_2 \right\} d\tau_1 = \int_0^t \begin{bmatrix} 0 & 1 \\ 0 & \tau_1 \end{bmatrix} \begin{bmatrix} 0 & \tau_1 \\ 0 & \dfrac{\tau_1^2}{2} \end{bmatrix} d\tau_1$$

$$= \begin{bmatrix} 0 & \dfrac{t^3}{6} \\ 0 & \dfrac{t^4}{8} \end{bmatrix}$$

Thus, for $\phi(t, 0)$ we have

$$\phi(t,0) = \begin{bmatrix} 1 & 0 \\ 0 & 1 \end{bmatrix} + \begin{bmatrix} 0 & t \\ 0 & \dfrac{t^2}{2} \end{bmatrix} + \begin{bmatrix} 0 & \dfrac{t^3}{6} \\ 0 & \dfrac{t^4}{8} \end{bmatrix} + \begin{bmatrix} 0 & \dfrac{t^5}{10} \\ 0 & \dfrac{t^6}{12} \end{bmatrix} + \ldots$$

$$= \begin{bmatrix} 1 & t + \dfrac{t^3}{6} + \dfrac{t^5}{10} + \ldots \\ 0 & 1 + \dfrac{t^2}{2} + \dfrac{t^4}{8} + \dfrac{t^6}{12} + \ldots \end{bmatrix}$$

Obtain the fundamental matrix $\phi(t,0)$ with $\phi(0,0) = I$ for the system described by

$$\dot{x} = \begin{bmatrix} 0 & 1 \\ -1 & t \end{bmatrix} x$$

Solution: We have the following system of equations

$$\dot{x}_1 = x_2$$

$$\dot{x}_2 = -x_1 + tx_2$$

Differentiating the last equation with respect to t

$$\ddot{x}_2 = -\dot{x}_1 + t\dot{x}_2 + x_2 = t\dot{x}_2$$

Let

$$y = \dot{x}_2$$

Then

$$\dot{y} - ty = 0$$

Solving the last equation we get

$$y = c_1 e^{\int_0^t \tau d\tau} = c_1 e^{\frac{t^2}{2}}$$

For $x_2(t)$ we have

$$x_2(t) = \int y dt = c_1 \int_0^t e^{\frac{\tau^2}{2}} d\tau + x_2(0)$$

$$x_1(t) = \int x_2 dt = c_1 \int_0^t \int_0^s e^{\frac{\tau^2}{2}} d\tau ds + tx_2(0) + x_1(0)$$

At t = 0 we have

$$\ddot{x}_1(0) + x_1(0) = 0$$

and from the above condition we get

$$c_1 = -x_1(0)$$

529

Substituting into the equations for

$x_1(t)$ and $x_2(t)$ we obtain

$$x_1(t) = x_1(0)\left[1 - \int_0^t\int_0^s e^{\frac{\tau^2}{2}} \, d\tau ds\right] + t\, x_2(0)$$

$$x_2(t) = -x_1(0)\int_0^t e^{\frac{\tau^2}{2}} \, d\tau + x_2(0)$$

Thus, for $\phi(t,0)$ we have

$$\phi(t,0) = \begin{bmatrix} 1 - \int_0^t\int_0^s e^{\frac{\tau^2}{2}} \, d\tau ds & t \\ & \\ -\int_0^t e^{\frac{\tau^2}{2}} \, d\tau & 1 \end{bmatrix}$$

● **PROBLEM** 10-12

Obtain the transformation matrix for the system shown below.

Solution: In the matrix form the differential equation of the system can be written

$$\dot{x} = Ax$$

where

$$A = \begin{bmatrix} 0 & 1 \\ 0 & 0 \end{bmatrix}$$

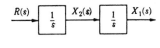

The transition matrix is given by

$$\phi(t) = e^{At}$$

Using the expansion formula, we get

$$\phi(t) = e^{At} = I + At + \frac{A^2 t^2}{2!} + \ldots + \frac{A^n t^n}{n!} + \ldots$$

Note that

$$A = \begin{bmatrix} 0 & 1 \\ 0 & 0 \end{bmatrix}$$

$$A^2 = \begin{bmatrix} 0 & 0 \\ 0 & 0 \end{bmatrix} \qquad A^3 = \begin{bmatrix} 0 & 0 \\ 0 & 0 \end{bmatrix} \qquad \cdots$$

Thus we have

$$\phi(t) = \begin{bmatrix} 1 & 0 \\ 0 & 1 \end{bmatrix} + \begin{bmatrix} 0 & 1 \\ 0 & 0 \end{bmatrix} t = \begin{bmatrix} 1 & t \\ 0 & 1 \end{bmatrix}$$

The solution of the equation is

$$x(t) = \begin{bmatrix} 1 & t \\ 0 & 1 \end{bmatrix} x(t_0)$$

● **PROBLEM 10-13**

The system is described by

$$\begin{bmatrix} \dot{x}_1 \\ \dot{x}_2 \end{bmatrix} = \begin{bmatrix} 0 & 1 \\ -2 & -3 \end{bmatrix} \begin{bmatrix} x_1 \\ x_2 \end{bmatrix} + \begin{bmatrix} 0 \\ 1 \end{bmatrix} r(t)$$

where $r(t) = 1$ $\quad (r(t) = u_s(t))$ for $t \geq 0$.

Determine the state vector $x(t)$ for $t \geq 0$.

Solution: The coefficient matrices A and B are

$$A = \begin{bmatrix} 0 & 1 \\ -2 & -3 \end{bmatrix} \qquad B = \begin{bmatrix} 0 \\ 1 \end{bmatrix}$$

Thus

$$sI - A = \begin{bmatrix} s & 0 \\ 0 & s \end{bmatrix} - \begin{bmatrix} 0 & 1 \\ -2 & -3 \end{bmatrix} = \begin{bmatrix} s & -1 \\ 2 & s+3 \end{bmatrix}$$

The inverse matrix is

$$(sI - A)^{-1} = \frac{1}{(s+1)(s+2)} \begin{bmatrix} s+3 & 1 \\ -2 & s \end{bmatrix}$$

$$= \begin{bmatrix} \dfrac{s+3}{(s+1)(s+2)} & \dfrac{1}{(s+1)(s+2)} \\[4mm] \dfrac{-2}{(s+1)(s+2)} & \dfrac{s}{(s+1)(s+2)} \end{bmatrix}$$

531

Taking the partial fractions expansion of each element of the matrix and then the inverse Laplace transform we get

$$\phi(t) = \pounds^{-1}[(sI - A)]^{-1} = \begin{bmatrix} 2e^{-t} - e^{-2t} & e^{-t} - e^{-2t} \\ \\ -2e^{-t} + 2e^{-2t} & -e^{-t} + 2e^{-2t} \end{bmatrix}$$

From the equation

$$x(t) = \phi(t)x(0+) + \int_0^t \phi(t-\tau)Br(\tau)d\tau$$

substituting B, $\phi(t)$ and r(t) we get

$$x(t) = \begin{bmatrix} 2e^{-t} - e^{-2t} & e^{-t} - e^{-2t} \\ \\ -2e^{-t} + 2e^{-2t} & -e^{-t} + 2e^{-2t} \end{bmatrix} x(0+)$$

$$+ \int_0^t \begin{bmatrix} 2e^{-(t-\tau)} - e^{-2(t-\tau)} & e^{-(t-\tau)} - e^{-2(t-\tau)} \\ \\ -2e^{-(t-\tau)} + 2e^{-2(t-\tau)} & -e^{-(t-\tau)} + 2e^{-2(t-\tau)} \end{bmatrix} \begin{bmatrix} 0 \\ 1 \end{bmatrix} d\tau$$

$$= \begin{bmatrix} 2e^{-t} - e^{-2t} & e^{-t} - e^{-2t} \\ \\ -2e^{-t} + 2e^{-2t} & -e^{-t} + 2e^{-2t} \end{bmatrix} x(0+) + \begin{bmatrix} \frac{1}{2} - e^{-t} + \frac{1}{2}e^{-2t} \\ \\ e^{-t} - e^{-2t} \end{bmatrix}$$

for $t \geq 0$.

● **PROBLEM** 10-14

Determine the transition matrix for the following system

$$\frac{d^2x}{dt^2} - \alpha^2 x = 0$$

Solution: This system can be represented by

$$\dot{x} = Ax + Bu$$

$$y = Gx + Hu$$

532

where

$$B = H = 0$$

$$A = \begin{bmatrix} 0 & 1 \\ \alpha^2 & 0 \end{bmatrix} \qquad G = (1,0)$$

Using

$$\frac{d}{dt} \phi = A\phi$$

we get

$$\frac{d\phi^{(1)}}{dt} = \begin{pmatrix} 0 & 1 \\ \alpha^2 & 0 \end{pmatrix} \phi^{(1)}$$

$$\phi^{(1)}(0) = \begin{pmatrix} 1 \\ 0 \end{pmatrix}$$

which is equivalent to

$$\frac{d\phi_{11}}{dt} = \phi_{21}$$

$$\frac{d\phi_{21}}{dt} = \alpha^2 \phi_{11}$$

or

$$\frac{d^2 \phi_{11}}{dt^2} = \alpha^2 \phi_{11}$$

The general solution of the above equation is

$$\phi_{11} = C_1 \sinh\alpha (t-t_0) + C_2 \cosh\alpha (t-t_0)$$

Differentiating, we get

$$\phi_{21} = C_1 \alpha \cosh\alpha (t-t_0) + C_2 \alpha \sinh\alpha (t-t_0)$$

We find coefficients C_1 and C_2 from the initial conditions, thus

$$C_1 = 0 , \qquad C_2 = 1$$

Hence

$$\phi^{(1)} = \begin{pmatrix} \phi_{11} \\ \\ \phi_{21} \end{pmatrix} = \begin{pmatrix} \cosh\alpha (t-t_0) \\ \\ \alpha \sinh\alpha (t-t_0) \end{pmatrix}$$

For the second column

$$\frac{d}{dt} \phi_{12} = \phi_{22}$$

$$\frac{d\phi_{22}}{dt} = \phi_{12}$$

and

$$\frac{d^2\phi_{12}}{dt^2} = \phi_{12}$$

The solution is

$$\phi_{12} = \frac{C_1 \sinh\alpha(t-t_0) + C_2 \cosh\alpha(t-t_0)}{\alpha}$$

$$\phi_{22} = C_1 \cosh\alpha(t-t_0) + C_2 \sinh\alpha(t-t_0)$$

The coefficients are

$$C_1 = 1, \quad C_2 = 0$$

Thus

$$\phi^{(2)} = \begin{pmatrix} \phi_{12} \\ \phi_{22} \end{pmatrix} = \begin{pmatrix} \dfrac{\sinh\alpha(t-t_0)}{\alpha} \\ \cosh\alpha(t-t_0) \end{pmatrix}$$

and

$$\phi = \left(\phi^{(1)}, \phi^{(2)}\right) = \begin{bmatrix} \cosh\alpha(t-t_0) & \dfrac{\sinh\alpha(t-t_0)}{\alpha} \\ \alpha\sinh\alpha(t-t_0) & \cosh\alpha(t-t_0) \end{bmatrix}$$

● **PROBLEM** 10-15

Obtain the state-transition matrix $\phi(t)$ of the following system:

$$\frac{d\dot{x}_1}{dt} = 2x_2$$

$$\frac{d\dot{x}_2}{dt} = -2x_1 + x_2$$

Then, obtain the inverse of the state-transition matrix, that is $\phi^{-1}(t)$.

Solution: Let us write the equations in the matrix form

$$\begin{bmatrix} \dot{x}_1 \\ \dot{x}_2 \end{bmatrix} = \begin{bmatrix} 0 & 2 \\ -2 & 1 \end{bmatrix} \begin{bmatrix} x_1 \\ x_2 \end{bmatrix}$$

534

or

$$\dot{x} = Ax$$

where

$$A = \begin{bmatrix} 0 & 3 \\ -1 & -4 \end{bmatrix}$$

The state-transition matrix is given by

$$\phi(t) = e^{At} = \mathcal{L}^{-1}[(sI-A)^{-1}]$$

Let us first find the matrix $(sI-A)^{-1}$.

Since

$$sI - A = \begin{bmatrix} s & -3 \\ 1 & s+4 \end{bmatrix}$$

and $\det(sI-A) = s(s+4) + 3 = (s+1)(s+3)$

we have

$$(sI - A)^{-1} = \frac{1}{(s+1)(s+3)} \begin{bmatrix} s+4 & 3 \\ -1 & s \end{bmatrix}$$

$$= \begin{bmatrix} \dfrac{s+4}{(s+1)(s+3)} & \dfrac{3}{(s+1)(s+3)} \\ \dfrac{-1}{(s+1)(s+3)} & \dfrac{s}{(s+1)(s+3)} \end{bmatrix}$$

Taking the inverse Laplace transformation we obtain:

$$\phi(t) = e^{At} = \mathcal{L}^{-1} \begin{bmatrix} \dfrac{\frac{3}{2}}{s+1} - \dfrac{\frac{1}{2}}{s+3} & \dfrac{\frac{3}{2}}{s+1} - \dfrac{\frac{3}{2}}{s+3} \\ -\dfrac{\frac{1}{2}}{s+1} + \dfrac{\frac{1}{2}}{s+3} & -\dfrac{\frac{1}{2}}{s+1} & \dfrac{\frac{3}{2}}{s+3} \end{bmatrix}$$

$$= \begin{bmatrix} \frac{3}{2}e^{-t} - \frac{1}{2}e^{-3t} & \frac{3}{2}e^{-t} - \frac{3}{2}e^{-3t} \\ -\frac{1}{2}e^{-t} + \frac{1}{2}e^{-3t} & -\frac{1}{2}e^{-t} + \frac{3}{2}e^{-3t} \end{bmatrix}$$

535

To obtain $\phi^{-1}(t)$ let us note that $\phi^{-1}(t) = \phi(-t)$, thus

$$\phi^{-1}(t) = e^{-At} = \begin{bmatrix} \frac{3}{2} e^t - \frac{1}{2} e^{3t} & \frac{3}{2} e^t - \frac{3}{2} e^{3t} \\ \\ -\frac{1}{2} e^t + \frac{1}{2} e^{3t} & -\frac{1}{2} e^t + \frac{3}{2} e^{3t} \end{bmatrix}$$

● **PROBLEM** 10-16

Obtain the fundamental matrix $\Phi(t)$ of the system

$$\frac{d^3 x}{dt^3} = ax$$

where a is a constant. Assume $\Phi(0) = I$.

Solution: Let $x_1 = x$

$$x_2 = \dot{x}$$
$$x_3 = \ddot{x}$$

We have $\dot{x}_1 = x_2$

$$\dot{x}_2 = x_3$$
$$\dot{x}_3 = ax_1$$

In the matrix form

$$\begin{bmatrix} \dot{x}_1 \\ \dot{x}_2 \\ \dot{x}_3 \end{bmatrix} = \begin{bmatrix} 0 & 1 & 0 \\ 0 & 0 & 1 \\ a & 0 & 0 \end{bmatrix} \begin{bmatrix} x_1 \\ x_2 \\ x_3 \end{bmatrix}$$

Let us compute the powers of A where

$$A = \begin{bmatrix} 0 & 1 & 0 \\ 0 & 0 & 1 \\ a & 0 & 0 \end{bmatrix}$$

$$A^2 = \begin{bmatrix} 0 & 0 & 1 \\ a & 0 & 0 \\ 0 & a & 0 \end{bmatrix}, \quad A^3 = \begin{bmatrix} a & 0 & 0 \\ 0 & a & 0 \\ 0 & 0 & a \end{bmatrix}, \quad A^4 = \begin{bmatrix} 0 & a & 0 \\ 0 & 0 & a \\ a^2 & 0 & 0 \end{bmatrix}$$

$$A^5 = \begin{bmatrix} 0 & 0 & a \\ a^2 & 0 & 0 \\ 0 & a^2 & 0 \end{bmatrix} \qquad \text{etc.}$$

536

From the definition

$$\Phi(t) = e^{At} = I + At + \frac{A^2 t^2}{2!} + \frac{A^3 t^3}{3!} + \frac{A^4 t^4}{4!} + \ldots$$

Substituting the powers of A we get

$$\Phi(t) = \begin{bmatrix} 1 + \frac{a}{3!}t^3 + \frac{a^2}{6!}t^6 + \ldots, & t + \frac{a}{4!}t^4 + \frac{a^2}{7!}t^7 + \ldots, \\[2ex] \frac{a}{2!}t^2 + \frac{a^2}{5!}t^5 + \frac{a^3}{8!}t^8 + \ldots, & 1 + \frac{a}{3!}t^3 + \frac{a^2}{6!}t^6 + \ldots, \\[2ex] at + \frac{a^2}{4!}t^4 + \frac{a^3}{7!}t^7 + \ldots, & \frac{a}{2!}t^2 + \frac{a^2}{5!}t^5 + \frac{a^3}{8!}t^8 + \ldots, \end{bmatrix}$$

$$\begin{matrix} \frac{1}{2!}t^2 + \frac{a}{5!}t^5 + \frac{a^2}{8!}t^8 + \ldots \\[2ex] t + \frac{a}{4!}t^4 + \frac{a^2}{7!}t^7 + \ldots \\[2ex] 1 + \frac{a}{3!}t^3 + \frac{a^2}{6!}t^6 + \ldots \end{matrix}$$

● **PROBLEM** 10-17

The discrete-time system is given by

$$x(k + 1) = Ax(k) + Bu(k)$$

where

$$A = \begin{bmatrix} 0 & 1 \\ -0.16 & -1 \end{bmatrix}, \quad B = \begin{bmatrix} 1 \\ 1 \end{bmatrix}$$

The initial conditions are

$$\begin{bmatrix} x_1(0) \\ x_2(0) \end{bmatrix} = \begin{bmatrix} 1 \\ -1 \end{bmatrix}$$

Obtain the state-transition matrix of the system and for
u(k) = 1 obtain x(k) for k = 0, 1, 2, ...

Solution: For the general discrete-time system

$$x(k+1) = Ax(k) + Bu(k)$$

$$y(k) = Cx(k) + Du(k)$$

The state-transition matrix is

$$\phi(k) = A^k$$

where

$$A^k = Z^{-1}[(zI - A)^{-1}z]$$

Let us find first $(zI - A)^{-1}$.

We have

$$zI - A = \begin{bmatrix} z & -1 \\ 0.16 & z + 1 \end{bmatrix}$$

and

$$(zI-A)^{-1} = \begin{bmatrix} \dfrac{z + 1}{(z + 0.2)(z + 0.8)} & \dfrac{1}{(z + 0.2)(z + 0.8)} \\[3mm] \dfrac{-0.16}{(z + 0.2)(z + 0.8)} & \dfrac{z}{(z + 0.2)(z + 0.8)} \end{bmatrix}$$

Taking the partial fractions expansion of the elements of the above matrix and using the formula for $\phi(k)$

we obtain

$$\phi(k) = A^k = Z^{-1}[(zI - A)^{-1}z]$$

$$= Z^{-1} \begin{bmatrix} \dfrac{4}{3}(\dfrac{z}{z+0.2}) - \dfrac{1}{3}(\dfrac{z}{z+0.8}) & \dfrac{5}{3}(\dfrac{z}{z+0.2}) - \dfrac{5}{3}(\dfrac{z}{z+0.8}) \\[3mm] -\dfrac{0.8}{3}(\dfrac{z}{z+0.2}) + \dfrac{0.8}{3}(\dfrac{z}{z+0.8}) & -\dfrac{1}{3}(\dfrac{z}{z+0.2}) + \dfrac{4}{3}(\dfrac{z}{z+0.8}) \end{bmatrix}$$

$$= \begin{bmatrix} \dfrac{4}{3}(-0.2)^k - \dfrac{1}{3}(-0.8)^k & \dfrac{5}{3}(-0.2)^k - \dfrac{5}{3}(-0.8)^k \\[3mm] -\dfrac{0.8}{3}(-0.2)^k + \dfrac{0.8}{3}(-0.8)^k & -\dfrac{1}{3}(-0.2)^k + \dfrac{4}{3}(-0.8)^k \end{bmatrix}$$

To compute $x(k)$ let us find the z transform of $x(k)$.

$$Z[x(k)] = X(z)$$

$$= (zI - A)^{-1}zx(0) + (zI - A)^{-1}BU(z)$$

$$= (zI - A)^{-1}[zx(0) + BU(z)].$$

Since

$$U(z) = \frac{z}{z - 1}$$

we have

538

$$zx(0) + BU(z) = \begin{bmatrix} z \\ -z \end{bmatrix} + \begin{bmatrix} \dfrac{z}{z-1} \\ \dfrac{z}{z-1} \end{bmatrix} = \begin{bmatrix} \dfrac{z^2}{z-1} \\ \dfrac{-z^2+2z}{z-1} \end{bmatrix}$$

Thus

$$X(z) = (zI - A)^{-1}[zx(0) + BU(z)]$$

$$= \begin{bmatrix} \dfrac{-\dfrac{17}{6}\,z}{z+0.2} + \dfrac{\dfrac{22}{9}\,z}{z+0.8} + \dfrac{\dfrac{25}{18}\,z}{z-1} \\[4ex] \dfrac{\dfrac{3.4}{6}\,z}{z+0.2} + \dfrac{-\dfrac{17.6}{9}\,z}{z+0.8} + \dfrac{\dfrac{7}{18}\,z}{z-1} \end{bmatrix}$$

Hence

$$x(k) = Z^{-1}[X(z)] = \begin{bmatrix} -\dfrac{17}{6}(-0.2)^k + \dfrac{22}{9}(-0.8)^k + \dfrac{25}{18} \\[3ex] \dfrac{3.4}{6}(-0.2)^k - \dfrac{17.6}{9}(-0.8)^k + \dfrac{7}{18} \end{bmatrix}$$

CHAPTER 11

SOLUTIONS TO STATE EQUATIONS

Find the solution of the system

$$\begin{bmatrix} \dot{x}_1 \\ \dot{x}_2 \end{bmatrix} = \begin{bmatrix} 1 & 0 \\ 1 & t \end{bmatrix} \begin{bmatrix} x_1 \\ x_2 \end{bmatrix}$$

The initial conditions are $x_1(0) = 1$, $x_2(0) = 0$.

<u>Solution:</u> To solve the above problem we shall use Neumann's series. The first three terms of the series are

$$A = \begin{bmatrix} 1 & 0 \\ 1 & t \end{bmatrix}$$

$$\int_0^t \begin{bmatrix} 1 & 0 \\ 1 & \tau \end{bmatrix} d\tau = \begin{bmatrix} t & 0 \\ t & \dfrac{t^2}{2} \end{bmatrix}$$

$$\int_0^t \begin{bmatrix} 1 & 0 \\ 1 & \tau_1 \end{bmatrix} \left\{ \int_0^{\tau_1} \begin{bmatrix} 1 & 0 \\ 1 & \tau_2 \end{bmatrix} d\tau_2 \right\} d\tau_1 = \int_0^t \begin{bmatrix} 1 & 0 \\ 1 & \tau_1 \end{bmatrix} \begin{bmatrix} \tau_1 & 0 \\ \tau_1 & \dfrac{\tau_1^2}{2} \end{bmatrix} d\tau_1$$

$$= \int_0^t \begin{bmatrix} \tau_1 & 0 \\ \tau_1 + \tau_1^2 & \dfrac{\tau_1^3}{2} \end{bmatrix} d\tau_1 = \begin{bmatrix} \dfrac{t^2}{2} & 0 \\ \dfrac{t^2}{2} + \dfrac{t^3}{3} & \dfrac{t^4}{8} \end{bmatrix}$$

The solution is given by

$$x(t) = \left\{ \begin{bmatrix} 1 & 0 \\ 0 & 1 \end{bmatrix} + \begin{bmatrix} t & 0 \\ t & \frac{t^2}{2} \end{bmatrix} + \begin{bmatrix} \frac{t^2}{2} & 0 \\ \frac{t^2}{2} + \frac{t^3}{3} & \frac{t^4}{8} \end{bmatrix} + \cdots \right\} \begin{bmatrix} 1 \\ 0 \end{bmatrix}$$

$$= \begin{bmatrix} 1 + t + \frac{1}{2} t^2 + \cdots \\ t + \frac{1}{2} t^2 + \frac{1}{3} t^3 + \cdots \end{bmatrix}$$

● **PROBLEM** 11-2

Obtain the solution of the system

$$\begin{bmatrix} \dot{x}_1 \\ \dot{x}_2 \end{bmatrix} = \begin{bmatrix} 1 & 2 \\ 4 & 3 \end{bmatrix} \begin{bmatrix} x_1 \\ x_2 \end{bmatrix}$$

by use of the eigenvectors. The initial conditions are
$x_1(0) = 1$, $x_2(0) = 2$.

<u>Solution</u>:   The characteristic equation is

$$\det | A - \lambda I | = \det \begin{bmatrix} 1 - \lambda & 2 \\ 4 & 3 - \lambda \end{bmatrix}$$

$$= (1 - \lambda)(3 - \lambda) - 8 = 0$$

The eigenvalues are $\lambda_1 = -1$, $\lambda_2 = 5$

From the definition of the eigenvector
$$Ae_i = \lambda_i e_i \qquad i = 1,2$$

we have
$$Ae_1 = -e_1$$

$$Ae_2 = 5e_2 \tag{1}$$

Solving equations (1) we obtain

$$e_1 = \begin{bmatrix} e_{11} \\ -e_{11} \end{bmatrix} \qquad\qquad e_2 = \begin{bmatrix} e_{12} \\ 2e_{12} \end{bmatrix}$$

or in the normalized form

$$|e_1| = |e_{11}^2 + e_{11}^2| = |2e_{11}^2| = 2|e_{11}^2| = 1$$

$$e_{11} = \frac{1}{\sqrt{2}}$$

and

$$|e_2| = |e_{12}^2 + 4e_{12}^2| = 5|e_{12}^2| = 1$$

$$e_{12} = \frac{1}{\sqrt{5}}$$

$$\hat{e}_1 = \begin{bmatrix} \dfrac{1}{\sqrt{2}} \\ -\dfrac{1}{\sqrt{2}} \end{bmatrix} \qquad \hat{e}_2 = \begin{bmatrix} \dfrac{1}{\sqrt{5}} \\ \dfrac{2}{\sqrt{5}} \end{bmatrix}$$

We can write the solution

$$x(t) = \alpha_1 \hat{e}_1 e^{-t} + \alpha_2 \hat{e}_2 e^{5t}$$

Using the initial conditions we have

$$\begin{bmatrix} 1 \\ 2 \end{bmatrix} = \alpha_1 \begin{bmatrix} \dfrac{1}{\sqrt{2}} \\ -\dfrac{1}{\sqrt{2}} \end{bmatrix} + \alpha_2 \begin{bmatrix} \dfrac{1}{\sqrt{5}} \\ \dfrac{2}{\sqrt{5}} \end{bmatrix}$$

Hence

$$\alpha_1 = 0 \qquad \alpha_2 = \sqrt{5}$$

Thus the solution is

$$\begin{bmatrix} x_1(t) \\ x_2(t) \end{bmatrix} = \begin{bmatrix} 1 \\ 2 \end{bmatrix} e^{5t}$$

Obtain the response to the following system

$$\begin{bmatrix} \dot{x}_1 \\ \dot{x}_2 \end{bmatrix} = \begin{bmatrix} 1 & 0 \\ 1 & 1 \end{bmatrix} \begin{bmatrix} x_1 \\ x_2 \end{bmatrix}$$

with $x_1(0) = 1$, $x_2(0) = 1$

Solution: For a scalar system

$$\dot{x} = ax \quad ; \quad x(0) = x_0$$

the solution is

$$x = x_0 e^{at}$$

Analogously for the vector equation

$$\dot{x} = Ax$$

$$x = e^{At} x(0)$$

where

$$e^{At} = I + At + \frac{1}{2!} A^2 t^2 + \frac{1}{3!} A^3 t^3 + \ldots$$

Including only the first three terms of the expansion, we obtain

$$e^{At} = I + \begin{bmatrix} 1 & 0 \\ 1 & 1 \end{bmatrix} t + \frac{1}{2} \begin{bmatrix} 1 & 0 \\ 2 & 1 \end{bmatrix} t^2 = \begin{bmatrix} 1+t+\frac{1}{2}t^2 + \ldots, & 0 \\ t+t^2 + \ldots, & 1+t+\frac{1}{2}t^2 + \ldots \end{bmatrix}$$

Thus, the solution is

$$\begin{bmatrix} x_1(t) \\ x_2(t) \end{bmatrix} = \begin{bmatrix} 1 + t + 0.5t^2 + \ldots \\ 1 + 2t + 1.5t^2 + \ldots \end{bmatrix}$$

Given the system

$$\dot{x} = Ax$$

where

$$A = \begin{bmatrix} 1 & 0 \\ 1 & 1 \end{bmatrix}$$

Determine the response of the system.

Solution: Taking the Laplace transform of the equation we get

$$sIX(s) - x(0) = AX(s)$$

Solving for X(s)

$$X(s) = [sI - A]^{-1} x(0)$$

In our case

$$sI-A = \begin{bmatrix} s-1 & 0 \\ -1 & s-1 \end{bmatrix}$$

$$\det(sI-A) = (s-1)^2$$

$$(sI-A)^{-1} = \frac{1}{(s-1)^2} \begin{bmatrix} s-1 & 0 \\ 1 & s-1 \end{bmatrix} = \begin{bmatrix} \dfrac{1}{s-1} & 0 \\ \dfrac{1}{(s-1)^2} & \dfrac{1}{s-1} \end{bmatrix}$$

We have

$$L^{-1}\left[(sI-A)^{-1}\right] = e^{At}$$

and

$$x(t) = e^{At} x(0)$$

Thus

$$e^{At} = \begin{bmatrix} e^t & 0 \\ te^t & e^t \end{bmatrix}$$

Let us assume the initial state to be

$$x_1(0) = 1 \qquad x_2(0) = -1$$

Thus the explicit solution for $x(t)$ is

$$x(t) = e^{At} x(0) = \begin{bmatrix} e^t & 0 \\ te^t & e^t \end{bmatrix} \begin{bmatrix} 1 \\ -1 \end{bmatrix} = \begin{bmatrix} e^t \\ te^t - e^t \end{bmatrix}$$

● PROBLEM 11-5

Consider the following system

$$\dot{x}(t) = \begin{bmatrix} 0 & 1 \\ -2 & -3 \end{bmatrix} x + \begin{bmatrix} 2 \\ 0 \end{bmatrix} u$$

where

$$u(t) = \begin{cases} 0 & t < 0 \\ e^{-t} & t \geq 0 \end{cases}$$

The initial conditions are $x_1(0) = x_2(0) = 0$. Obtain the response of the following system.

Solution: For the matrix

$$A = \begin{bmatrix} 0 & 1 \\ -2 & -3 \end{bmatrix}$$

let us find the van der Monde matrix.

$$V = \begin{bmatrix} 1 & 1 & \cdots & 1 \\ \lambda_1 & \lambda_2 & & \lambda_n \\ \lambda_1^2 & \lambda_2^2 & & \lambda_n^2 \\ \vdots & \vdots & & \vdots \\ \lambda_1^n & \lambda_2^n & \cdots & \lambda_n^n \end{bmatrix}$$

The above definition of the Van der Monde matrix is valid

545

when all eigenvalues λ_1, λ_2, ... λ_n are distinct.

In our case

$$|\lambda I-A| = \lambda(\lambda+3) + 2 = 0$$

The eigenvalues are $\lambda_1 = -1$, $\lambda_2 = -2$.

Thus

$$V = \begin{bmatrix} 1 & 1 \\ -1 & -2 \end{bmatrix}$$

Let us define the transformation

$$x = Vy$$

We have

$$\begin{bmatrix} \dot{y}_1 \\ \dot{y}_2 \end{bmatrix} = \begin{bmatrix} -1 & 0 \\ 0 & -2 \end{bmatrix} \begin{bmatrix} y_1 \\ y_2 \end{bmatrix} + \begin{bmatrix} 2 & 1 \\ -1 & -1 \end{bmatrix} \begin{bmatrix} 2 \\ 0 \end{bmatrix} [u]$$

$$= \begin{bmatrix} -1 & 0 \\ 0 & -2 \end{bmatrix} \begin{bmatrix} y_1 \\ y_2 \end{bmatrix} + \begin{bmatrix} 4 \\ -2 \end{bmatrix} [u]$$

In the matrix notation

$$\dot{y} = Cy + Du$$

The solution of the above equation is given by

$$y(t) = e^{Ct} [y(0) + \int_0^t e^{-C\tau} Du(\tau)d\tau]$$

where

$$e^{Ct} = \begin{bmatrix} e^{-t} & 0 \\ 0 & e^{-2t} \end{bmatrix}$$

We have, in the explicit form

$$\begin{bmatrix} y_1(t) \\ y_2(t) \end{bmatrix} = \begin{bmatrix} e^{-t} & 0 \\ 0 & e^{-2t} \end{bmatrix} \left\{ \begin{bmatrix} y_1(0) \\ y_2(0) \end{bmatrix} + \int_0^t \begin{bmatrix} e^{\tau} & 0 \\ 0 & e^{2\tau} \end{bmatrix} \begin{bmatrix} 4 \\ -2 \end{bmatrix} e^{-\tau}d\tau \right\}$$

$$= \begin{bmatrix} e^{-t} & 0 \\ & \\ 0 & e^{-2t} \end{bmatrix} \begin{bmatrix} 4t \\ \\ -2e^{t} + 2 \end{bmatrix} = \begin{bmatrix} 4te^{-t} \\ \\ -2e^{-t} + 2e^{-2t} \end{bmatrix}$$

To find $x(t)$, we use

$$x(t) = Vy(t) = \begin{bmatrix} 1 & 1 \\ & \\ -1 & -2 \end{bmatrix} \begin{bmatrix} 4te^{-t} \\ \\ -2e^{-t} + 2e^{-2t} \end{bmatrix} = \begin{bmatrix} 4te^{-t} - 2e^{-t} + 2e^{-2t} \\ \\ -4te^{-t} + 4e^{-t} - 4e^{-2t} \end{bmatrix}$$

● **PROBLEM** 11-6

Using the Laplace transform method obtain the solution of

$$\begin{bmatrix} \dot{x}_1 \\ \\ \dot{x}_2 \end{bmatrix} = \begin{bmatrix} -1 & 1 \\ & \\ 1 & 1 \end{bmatrix} \begin{bmatrix} x_1 \\ \\ x_2 \end{bmatrix}$$

Solution: The coefficient matrix is

$$A = \begin{bmatrix} -1 & 1 \\ & \\ 1 & 1 \end{bmatrix}$$

The solution is given by

$$x(t) = e^{At} x(0)$$

where

$$e^{At} = L^{-1}[(sI-A)^{-1}]$$

We have

$$(sI-A)^{-1} = \begin{bmatrix} s+1 & -1 \\ & \\ -1 & s-1 \end{bmatrix}^{-1} = \frac{1}{s^2-2} \begin{bmatrix} s-1 & 1 \\ & \\ 1 & s+1 \end{bmatrix}$$

547

$$= \begin{bmatrix} \dfrac{s-1}{(s+\sqrt{2})(s-\sqrt{2})} & \dfrac{1}{(s+\sqrt{2})(s-\sqrt{2})} \\[4mm] \dfrac{1}{(s+\sqrt{2})(s-\sqrt{2})} & \dfrac{s+1}{(s+\sqrt{2})(s-\sqrt{2})} \end{bmatrix}$$

Thus

$x(t)$ is given by

$$x(t) = e^{At} x(0) = L^{-1}[(sI-A)^{-1}] x(0)$$

$$= \frac{1}{2\sqrt{2}} \begin{bmatrix} (\sqrt{2}+1)e^{-\sqrt{2}t} + (\sqrt{2}-1)e^{\sqrt{2}t} & -e^{-\sqrt{2}t} + e^{\sqrt{2}t} \\[4mm] -e^{-\sqrt{2}t} + e^{\sqrt{2}t} & (\sqrt{2}-1)e^{-\sqrt{2}t} + (\sqrt{2}+1)e^{\sqrt{2}t} \end{bmatrix} \begin{bmatrix} x_1(0) \\[4mm] x_2(0) \end{bmatrix}$$

● **PROBLEM** 11-7

Obtain the solution of the following state space equation

$$\begin{bmatrix} \dot{x}_1 \\ \dot{x}_2 \\ \dot{x}_3 \end{bmatrix} = \begin{bmatrix} \alpha & 0 & 0 \\ 0 & \sigma-j\omega & 0 \\ 0 & 0 & \sigma+j\omega \end{bmatrix} \begin{bmatrix} x_1 \\ x_2 \\ x_3 \end{bmatrix}$$

Solution: The above system is homogeneous and is in diagonal form. The solution is given by

$$x(t) = \begin{bmatrix} e^{\lambda_1 t} \\ e^{\lambda_2 t} \\ \vdots \\ e^{\lambda_n t} \end{bmatrix} x(0) = [e^{\lambda_1 t} x_1(0), \ e^{\lambda_2 t} x_2(0), \ \ldots \ e^{\lambda_n t} x_n(0)]$$

where $\lambda_1, \ \ldots \ \lambda_n$ are the elements of the diagonal matrix.

We have

548

$$\begin{bmatrix} x_1(t) \\ x_2(t) \\ x_3(t) \end{bmatrix} = \begin{bmatrix} e^{\alpha t} x_1(0) \\ e^{(\sigma-j\omega)t} x_2(0) \\ e^{(\sigma+j\omega)t} x_3(0) \end{bmatrix} = \begin{bmatrix} e^{\alpha t} x_1(0) \\ e^{\sigma t}(\cos\omega t - j\sin\omega t) x_2(0) \\ e^{\sigma t}(\cos\omega t + j\sin\omega t) x_3(0) \end{bmatrix}$$

● **PROBLEM** 11-8

The system is given by

$$\begin{bmatrix} \dot{x}_1 \\ \dot{x}_2 \end{bmatrix} = \begin{bmatrix} 1 & 0 \\ 1 & 1 \end{bmatrix} \begin{bmatrix} x_1 \\ x_2 \end{bmatrix} + \begin{bmatrix} 1 \\ 1 \end{bmatrix} u(t)$$

where

u(t) is a unit step function

$$u(t) = \begin{cases} 0 & \text{for } t < 0 \\ 1 & \text{for } t \geq 0 \end{cases}$$

The initial conditions are $x_1(0) = 1$, $x_2(0) = 1$.

Find the response of the system.

<u>Solution</u>: Let us find the transition matrix, we have

$$A = \begin{bmatrix} 1 & 0 \\ 1 & 1 \end{bmatrix} \qquad A^2 = \begin{bmatrix} 1 & 0 \\ 2 & 1 \end{bmatrix} \qquad A^3 = \begin{bmatrix} 1 & 0 \\ 3 & 1 \end{bmatrix} \quad \cdots$$

Using the identity

$$e^{\pm At} = I \pm At + \frac{1}{2!} A^2 t^2 \pm \frac{1}{3!} A^3 t^3 + \cdots$$

we obtain

$$e^{\pm At} = \begin{bmatrix} 1 \pm t + \frac{1}{2} t^2 \pm \cdots & 0 \\ \pm t + t^2 \pm \cdots & 1 \pm t + \frac{1}{2} t^2 \pm \cdots \end{bmatrix}$$

Since

$$e^{-A\tau} Bu(\tau) = \begin{bmatrix} 1 - \tau + \frac{1}{2} \tau^2 - \cdots \\ \\ 1 - 2\tau + \frac{3}{2} \tau^2 - \cdots \end{bmatrix}$$

We get

$$\int_0^t e^{-A\tau} Bu(\tau) d\tau = \begin{bmatrix} t - \frac{1}{2} t^2 + \frac{1}{6} t^3 - \cdots \\ \\ t - t^2 + \frac{1}{2} t^3 - \cdots \end{bmatrix}$$

Thus the solution is

$$x(t) = e^{At}[x(0) + \int_0^t e^{-A\tau} Bu(\tau) d\tau]$$

To verify the above result, let us take

$$\frac{dx}{dt} = Ae^{At} x(0) + Ae^{At} \int_0^t e^{-A\tau} Bu(\tau) d\tau + e^{At} e^{-At} Bu(t)$$

$$= A[e^{At} \{x(0) + \int_0^t e^{-A\tau} Bu(\tau) d\tau\}] + Bu(t)$$

$$= Ax(t) + Bu(t)$$

We can write $x(t)$ in the form

$$x(t) = e^{At} \begin{bmatrix} 1 + t - \frac{1}{2} t^2 + \frac{1}{6} t^3 - \cdots \\ \\ 1 + t - t^2 + \frac{1}{2} t^3 - \cdots \end{bmatrix} = \begin{bmatrix} 1 + 2t + t^2 + \cdots \\ \\ 1 + 3t + \frac{5}{2} t^2 + \cdots \end{bmatrix}$$

550

Obtain the solution of the following state-space equation

$$\begin{bmatrix} \dot{x}_1 \\ \dot{x}_2 \\ \dot{x}_3 \end{bmatrix} = \begin{bmatrix} -2 & 0 & 0 \\ 0 & -3 & 0 \\ 0 & 0 & -4 \end{bmatrix} \begin{bmatrix} x_1 \\ x_2 \\ x_3 \end{bmatrix}$$

Solution: Since the matrix A is in diagonal form the elements along the diagonal are the eigenvalues of the system. Also since the system is homogeneous, i.e., B=0, the solution to the system is given by

$$\begin{bmatrix} x_1(t) \\ x_2(t) \\ x_3(t) \end{bmatrix} = \begin{bmatrix} e^{-2t} x_1(0) \\ e^{-3t} x_2(0) \\ e^{-4t} x_3(0) \end{bmatrix}$$

where $x_1(0)$, $x_2(0)$, and $x_3(0)$ are the initial conditions.

In the matrix notation we get

$$x(t) = Q(t) x(0)$$

where

$$Q(t) = \begin{bmatrix} e^{-2t} & 0 & 0 \\ 0 & e^{-3t} & 0 \\ 0 & 0 & e^{-4t} \end{bmatrix}$$

Notice that the above system is decoupled, that is, the matrix differential equation represents three independent scalar differential equations.

Using Laplace transforms solve and verify the solution of the equation

$$\dot{x}(t) = Ax(t) + B(t)u(t)$$

Solution: Let us denote

$$B(t) u(t) = f(t)$$

The original equation becomes

$$\dot{x} = Ax + f(t)$$

Taking the Laplace transform we obtain

$$sX(s) - x(0) = AX(s) + F(s)$$

or

$$[sI-A]X = x(0) + F(s)$$

Solving for X we get

$$X = [sI-A]^{-1} x(0) + [sI-A]^{-1} F(s)$$

Using the following identities

$$L^{-1}\{[sI-A]^{-1}\} = e^{At}$$

and

$$L^{-1}\{f(s)g(s)\} = f(t)*g(t) = \int_0^t f(t-\tau)g(\tau)d\tau$$

and taking the inverse Laplace transform we get

$$x(t) = e^{At} x(0) + \int_0^t e^{A(t-\tau)} f(\tau)d\tau$$

or

$$x(t) = e^{At} x(0) + \int_0^t e^{A(t-\tau)} B(\tau)u(\tau)d\tau.$$

If the initial conditions are given at t_0 we have

$$x(t) = e^{A(t-t_0)} x(t_0) + \int_{t_0}^t e^{A(t-\tau)} B(\tau)u(\tau)d\tau$$

To verify the solution let us rewrite the above equation

$$x(t) = e^{A(t-t_0)} x(t_0) + e^{At} \int_{t_0}^{t} e^{-A\tau} B(\tau) u(\tau) d\tau$$

Differentiating, we get

$$\dot{x}(t) = Ae^{A(t-t_0)} x(t_0) + Ae^{At} \int_{t_0}^{t} e^{-A\tau} B(\tau) u(\tau) d\tau$$

$$+ e^{At} e^{-At} B(t) u(t) = A \left\{ e^{A(t-t_0)} x(t_0) + \int_{t_0}^{t} e^{A(t-\tau)} B(\tau) u(\tau) d\tau \right\}$$

$$+ B(t) u(t) = Ax(t) + B(t) u(t).$$

• **PROBLEM** 11-11

Find the response of the system

$$\dot{x}(t) = \begin{bmatrix} 0 & 1 \\ 0 & 0 \end{bmatrix} x(t) + \begin{bmatrix} 0 \\ 1 \end{bmatrix} u(t)$$

where $u(t) = t$ for $t \geq 0$. The initial conditions are
$x_1(0) = 2$, $x_2(0) = 3$.

Solution: The general solution is given by

$$x(t) = e^{At} x(0) + e^{At} \int_{0}^{t} e^{-A\tau} Bu \, d\tau$$

where

553

$$A = \begin{bmatrix} 0 & 1 \\ 0 & 0 \end{bmatrix} \quad \text{and} \quad B = \begin{bmatrix} 0 \\ 1 \end{bmatrix}$$

To find the matrix e^{At} let us notice that

$$A^2 = \begin{bmatrix} 0 & 1 \\ 0 & 0 \end{bmatrix}\begin{bmatrix} 0 & 1 \\ 0 & 0 \end{bmatrix} = \begin{bmatrix} 0 & 0 \\ 0 & 0 \end{bmatrix}$$

Thus $A^k = 0$ for $k > 1$. We have

$$e^{At} = I + At = \begin{bmatrix} 1 & t \\ 0 & 1 \end{bmatrix}$$

$$e^{-A\tau} = \begin{bmatrix} 1 & -\tau \\ 0 & 1 \end{bmatrix}$$

To evaluate the integral we have to find the value of $e^{-A\tau} Bu$

$$e^{-A\tau} Bu = \begin{bmatrix} 1 & -\tau \\ 0 & 1 \end{bmatrix}\begin{bmatrix} 0 \\ \tau \end{bmatrix} = \begin{bmatrix} -\tau^2 \\ \tau \end{bmatrix}$$

Thus

$$\int_0^t e^{-A\tau} Bu \, d\tau = \int_0^t \begin{bmatrix} -\tau^2 \\ \tau \end{bmatrix} d\tau = \begin{bmatrix} -\frac{1}{3} t^3 \\ \frac{1}{2} t^2 \end{bmatrix}$$

Substituting the above results into the general formula we obtain

$$x(t) = \begin{bmatrix} 1 & t \\ 0 & 1 \end{bmatrix}\left\{ \begin{bmatrix} 2 \\ 3 \end{bmatrix} + \begin{bmatrix} -\frac{1}{3} t^3 \\ \frac{1}{2} t^2 \end{bmatrix} \right\} = \begin{bmatrix} 2 + 3t - \frac{1}{3} t^3 + \frac{1}{2} t^3 \\ 3 + \frac{1}{2} t^2 \end{bmatrix}$$

or

$$x_1(t) = 2 + 3t + \frac{1}{6} t^3$$

$$x_2(t) = 3 + \frac{1}{2} t^2$$

● **PROBLEM** 11-12

Obtain the time response of the system given by

$$\dot{x} = \begin{bmatrix} 0 & 1 \\ -2 & -3 \end{bmatrix} x + \begin{bmatrix} 0 \\ 1 \end{bmatrix} u$$

where u(t) is the unit step function occurring at t = 0.

<u>Solution</u>: Let us find the state transition matrix $\phi(t)$

where

$$\phi(t) = e^{At} = L^{-1}[(sI-A)^{-1}]$$

$$sI-A = \begin{bmatrix} s & -1 \\ 2 & s+3 \end{bmatrix}$$

$$\det(sI-A) = s^2 + 3s + 2$$

and

$$(sI-A)^{-1} = \frac{1}{s^2+3s+2} \begin{bmatrix} s+3 & 1 \\ -2 & s \end{bmatrix}$$

$$= \begin{bmatrix} \dfrac{2}{s+1} - \dfrac{1}{s+2} & \dfrac{1}{s+1} - \dfrac{1}{s+2} \\ \dfrac{-2}{s+1} + \dfrac{2}{s+2} & \dfrac{-1}{s+1} + \dfrac{2}{s+2} \end{bmatrix}$$

Taking the inverse Laplace transform, we find

555

$$\phi(t) = e^{At} = \begin{bmatrix} 2e^{-t} - e^{-2t} & e^{-t} - e^{-2t} \\ \\ -2e^{-t} + 2e^{-2t} & -e^{-t} + 2e^{-2t} \end{bmatrix}$$

The response to the unit-step input is

$$x(t) = e^{At} x(0) + \int_0^t e^{A(t-\tau)} Bu(\tau) d\tau$$

$$= e^{At} x(0) + \int_0^t \begin{bmatrix} 2e^{-(t-\tau)} - e^{-2(t-\tau)} & e^{-(t-\tau)} - e^{-2(t-\tau)} \\ \\ -2e^{-(t-\tau)} + 2e^{-2(t-\tau)} & -e^{-(t-\tau)} + 2e^{-2(t-\tau)} \end{bmatrix}$$

$$\begin{bmatrix} 0 \\ \\ 1 \end{bmatrix} [1] d\tau$$

$$= \begin{bmatrix} 2e^{-t} - e^{-2t} & e^{-t} - e^{-2t} \\ \\ -2e^{-t} + 2e^{-2t} & -e^{-t} + 2e^{-2t} \end{bmatrix} \begin{bmatrix} x_1(0) \\ \\ x_2(0) \end{bmatrix} + \begin{bmatrix} \frac{1}{2} - e^{-t} + \frac{1}{2} e^{-2t} \\ \\ e^{-t} - e^{-2t} \end{bmatrix}$$

● **PROBLEM 11-13**

The system is described by

$$x = Ax + Bu$$

Assuming that each input is given at t = 0, obtain the response to each of the following inputs:

1. u is a ramp function

2. u is a step function

3. u is an impulse function.

Solution: 1) The ramp input u(t) is

$$u(t) = tv$$

556

where v is a vector whose components are magnitudes of ramp functions applied at t = 0. The solution is given by

$$x(t) = e^{At} x(0) + \int_0^t e^{A(t-\tau)} B\tau \, v \, d\tau$$

$$= e^{At} x(0) + e^{At} \int_0^t e^{-A\tau} \tau \, d\tau \, Bv$$

$$= e^{At} x(0) + e^{At} \left(\frac{I}{2} t^2 - \frac{2A}{3!} t^3 + \frac{3A^2}{4!} t^4 - \frac{4A^3}{5!} t^5 + \ldots \right) Bv$$

$$= e^{At} x(0) + [A^{-2}(e^{At} - I) - A^{-1}t] Bv \, .$$

We assumed that A is nonsingular.

2) For the step function we have

$$u(t) = k$$

The solution is

$$x(t) = e^{At} x(0) + \int_0^t e^{A(t-\tau)} Bk \, d\tau$$

$$= e^{At} x(0) + e^{At} \left[\int_0^t (I - A\tau + \frac{A^2\tau^2}{2!} - \ldots) d\tau \right] Bk$$

$$= e^{At} x(0) + e^{At} (It - \frac{At^2}{2!} + \frac{A^2t^3}{3!} - \ldots) Bk.$$

If A is nonsingular, then we can simplify the above equation to

$$x(t) = e^{At} x(0) + A^{-1} (e^{At} - I) Bk$$

3) For the impulse response we have

$$u(t) = \delta(t) m$$

The solution x(t) is given by

$$x(t) = e^{At} x(0_-) + \int_{0_-}^t e^{A(t-\tau)} B\delta(\tau) m \, d\tau$$

$$= e^{At} x(0_-) + e^{At} Bm.$$

Find the solution of the system, using two methods

$$\dot{x}(t) = \begin{bmatrix} 7 & -2 \\ 15 & -4 \end{bmatrix} x(t)$$

The initial conditions are $x_1(0) = x_2(0) = 1$.

Solution: 1) The characteristic equation of the matrix A is

$$\det |\lambda I - A| = \det \begin{bmatrix} \lambda-7 & 2 \\ -15 & \lambda+4 \end{bmatrix} = (\lambda-7)(\lambda+4) + 30 = 0$$

The eigenvalues of A are $\lambda_1 = 1$, $\lambda_1 = 2$.

The diagonal matrix of the eigenvalues is

$$\Lambda = \begin{bmatrix} \lambda_1 & 0 \\ 0 & \lambda_2 \end{bmatrix} = \begin{bmatrix} 1 & 0 \\ 0 & 2 \end{bmatrix}$$

Let us define matrix V, such that

$$AV = V\Lambda$$

$$\begin{bmatrix} 7 & -2 \\ 15 & -4 \end{bmatrix} \begin{bmatrix} v_{11} & v_{12} \\ v_{21} & v_{22} \end{bmatrix} = \begin{bmatrix} v_{11} & v_{12} \\ v_{21} & v_{22} \end{bmatrix} \begin{bmatrix} 1 & 0 \\ 0 & 2 \end{bmatrix}$$

Multiplying we get the system of equations

$$7v_{11} - 2v_{21} = v_{11}$$

$$7v_{12} - 2v_{22} = 2v_{12}$$

$$15v_{11} - 4v_{21} = v_{21}$$

$$15v_{12} - 4v_{22} = 2v_{22}$$

Solving the above system we obtain

$$V = \begin{bmatrix} 1 & 2 \\ 3 & 5 \end{bmatrix}$$

and

$$V^{-1} = \begin{bmatrix} -5 & 2 \\ 3 & -1 \end{bmatrix}$$

The matrix $e^{\Lambda t}$ is given by

$$e^{\Lambda t} = \begin{bmatrix} e^{\lambda_1 t} & 0 \\ 0 & e^{\lambda_2 t} \end{bmatrix} = \begin{bmatrix} e^t & 0 \\ 0 & e^{2t} \end{bmatrix}$$

The matrix e^{At} is found from

$$e^{At} = V e^{\Lambda t} V^{-1}$$

Thus

$$e^{At} = \begin{bmatrix} 1 & 2 \\ 3 & 5 \end{bmatrix} \begin{bmatrix} e^t & 0 \\ 0 & e^{2t} \end{bmatrix} \begin{bmatrix} -5 & 2 \\ 3 & -1 \end{bmatrix} = \begin{bmatrix} -5e^t+6e^{2t} & 2e^t-2e^{2t} \\ -15e^t+15e^{2t} & 6e^t-5e^{2t} \end{bmatrix}$$

The solution $x(t)$ is given in the form

$$x(t) = e^{At} x(0)$$

or

$$\begin{bmatrix} x_1(t) \\ x_2(t) \end{bmatrix} = \begin{bmatrix} -3e^t + 4e^{2t} \\ -9e^t + 10e^{2t} \end{bmatrix}$$

2) The eigenvectors of the matrix A are

$$V_1 = \begin{bmatrix} 1 \\ 3 \end{bmatrix} \quad \text{and} \quad V_2 = \begin{bmatrix} 2 \\ 5 \end{bmatrix}$$

Since both vectors are linearly independent we can represent the initial state $x(0)$ as a linear combination of V_1 and V_2

$$x(0) = \alpha_1 V_1 + \alpha_2 V_2$$

Thus

$$\begin{bmatrix} 1 \\ 1 \end{bmatrix} = \alpha_1 \begin{bmatrix} 1 \\ 3 \end{bmatrix} + \alpha_2 \begin{bmatrix} 2 \\ 5 \end{bmatrix}$$

Solving the above system we get

$$\alpha_1 = -3 \qquad \alpha_2 = 2$$

Thus, the solution $x(t)$ is

$$x(t) = \alpha_1 e^{\lambda_1 t} V_1 + \alpha_2 e^{\lambda_2 t} V_2$$

$$= -3e^t \begin{bmatrix} 1 \\ 3 \end{bmatrix} + 2e^{2t} \begin{bmatrix} 2 \\ 5 \end{bmatrix}$$

or

$$\begin{bmatrix} x_1(t) \\ x_2(t) \end{bmatrix} = \begin{bmatrix} -3e^t + 4e^{2t} \\ -9e^t + 10e^{2t} \end{bmatrix}$$

● **PROBLEM** 11-15

Using

a) the partial fractions expansion method

b) the Laplace transform

obtain the response of the system

$$\frac{X(s)}{U(s)} = \frac{2(s+3)}{(s+1)(s+2)}$$

where

$$u(t) = \begin{cases} 0 & t < 0 \\ e^{-t} & t \geq 0 \end{cases}$$

The initial conditions are

$$x(0) = \dot{x}(0) = 0.$$

560

Solution:

a)

$$\frac{X(s)}{U(s)} = \frac{X_1(s)}{U(s)} = \frac{4}{s+1} + \frac{-2}{s+2}$$

$$\frac{sX(s) - 2U(s)}{U(s)} = \frac{X_2(s)}{U(s)} = \frac{-4}{s+1} + \frac{4}{s+2}$$

Thus

$$\begin{bmatrix} \dfrac{X_1(s)}{U(s)} \\[3mm] \dfrac{X_2(s)}{U(s)} \end{bmatrix} = \begin{bmatrix} 4 & -2 \\[2mm] -4 & 4 \end{bmatrix} \begin{bmatrix} \dfrac{1}{s+1} \\[3mm] \dfrac{1}{s+2} \end{bmatrix}$$

Let

$$\begin{bmatrix} X_1(s) \\[2mm] X_2(s) \end{bmatrix} = \begin{bmatrix} 4 & -2 \\[2mm] -4 & 4 \end{bmatrix} \begin{bmatrix} Y_1(s) \\[2mm] Y_2(s) \end{bmatrix}$$

We obtain

$$\begin{bmatrix} \dfrac{Y_1(s)}{U(s)} \\[3mm] \dfrac{Y_2(s)}{U(s)} \end{bmatrix} = \begin{bmatrix} \dfrac{1}{s+1} \\[3mm] \dfrac{1}{s+2} \end{bmatrix}$$

In the time domain

$$y_1(t) = L^{-1}[Y_1(s)] = L^{-1}\left[\frac{1}{(s+1)^2}\right] = te^{-t}$$

$$y_2(t) = L^{-1}[Y_2(s)] = L^{-1}\left[\frac{1}{(s+1)(s+2)}\right] = e^{-t} - e^{-2t}$$

Thus $x_1(t)$ and $x_2(t)$ are

$$x_1(t) = 4y_1(t) - 2y_2(t) = 4te^{-t} - 2e^{-t} + 2e^{-2t}$$

$$x_2(t) = -4te^{-t} + 4e^{-t} - 4e^{-2t}$$

b)

$$\frac{X(s)}{U(s)} = \frac{2(s+3)}{(s+1)(s+2)}$$

Notice that

$$U(s) = L[u(t)] = L[e^{-t}] = \frac{1}{s+1}$$

$$X(s) = \frac{2(s+3)}{(s+1)^2(s+2)} = \frac{-2}{s+1} + \frac{4}{(s+1)^2} + \frac{2}{s+2}$$

The inverse Laplace transform of $X(s)$ is

$$x(t) = L^{-1}[X(s)] = -2e^{-t} + 4te^{-t} + 2e^{-2t}$$

● **PROBLEM** 11-16

Obtain the response of the following time-varying system

$$\frac{dx}{dt} + tx = u$$

where u is an arbitrary input.

Solution: The solution of the system is given in the form

$$x(t) = \phi(t,t_0)x(t_0) + \phi(t,t_0)\int_{t_0}^{t} \phi^{-1}(\tau,t_0)u(\tau)\ d\tau$$

or

$$x(t) = \phi(t,t_0)x(t_0) + \int_{t_0}^{t} g(t,\tau)u(\tau)d\tau$$

where

$$g(t,\tau) = \phi(t,t_0)\phi^{-1}(\tau,t_0)$$

In this system t and $\int_{t_0}^{t} \tau\ d\tau$

commute for all t. Thus

$$\phi(t,t_0) = e^{-\int_{t_0}^{t}\tau d\tau} = e^{-\left[\frac{t^2}{2} - \frac{t_0^2}{2}\right]}$$

562

and

$$\phi^{-1}(\tau, t_0) = e^{\left(\frac{\tau^2}{2} - \frac{t_0^2}{2}\right)}$$

For $g(t, \tau)$ we obtain

$$g(t, \tau) = \begin{cases} e^{\left(-\frac{t^2}{2} + \frac{\tau^2}{2}\right)} & \text{for} \quad t \geq \tau \\ \\ 0 & t < \tau \end{cases}$$

The output $x(t)$ is

$$x(t) = e^{\left(-\frac{t^2}{2} + \frac{t_0^2}{2}\right)} x(t_0) + e^{-\frac{t^2}{2}} \int_{t_0}^{t} e^{\frac{\tau^2}{2}} u(\tau) \, d\tau \, .$$

● **PROBLEM** 11-17

The state equations of the system are

$$\begin{bmatrix} \dot{x}_1 \\ \dot{x}_2 \end{bmatrix} = \begin{bmatrix} 0 & 1 \\ -2 & -3 \end{bmatrix} \begin{bmatrix} x_1 \\ x_2 \end{bmatrix} + \begin{bmatrix} 0 \\ 1 \end{bmatrix} u(t)$$

$$c(t) = x_1(t)$$

$c(t)$ is the scalar output, $u(t)$ is the scalar input and $x_1(t)$ and $x_2(t)$ are the state variables. $u(t)$ is the output of the zero-order-hold

$$u(t) = u(kT) = r(kT) \quad \text{for} \quad kT \leq t < (k+1)T$$

Determine the system response.

Solution: Let us write the state equations in the matrix notation

$$\dot{x} = Ax + B[u]$$

where

$$A = \begin{bmatrix} 0 & 1 \\ -2 & -3 \end{bmatrix} \quad \text{and} \quad B = \begin{bmatrix} 0 \\ 1 \end{bmatrix}$$

and

$$(sI - A) = \begin{bmatrix} s & -1 \\ 2 & s + 3 \end{bmatrix}$$

$$\det(sI - A) = s^2 + 3s + 2$$

The inverse matrix is

$$(sI - A)^{-1} = \frac{1}{s^2 + 3s + 2} \begin{bmatrix} s + 3 & 1 \\ -2 & s \end{bmatrix}$$

The state transition matrix is obtained by taking the inverse Laplace transform of $(sI - A)^{-1}$. Thus

$$\phi(t) = L^{-1}[(sI - A)^{-1}] = \begin{bmatrix} 2e^{-t} - e^{-2t} & e^{-t} - e^{-2t} \\ -2e^{-t} + 2e^{-2t} & -e^{-t} + 2e^{-2t} \end{bmatrix}$$

Substituting B and $\phi(t)$ into

$$\theta(T) = \int_0^T \phi(T - \tau) B d\tau$$

gives

$$\theta(T) = \int_0^T \begin{bmatrix} e^{-(T-\tau)} - e^{-2(T-\tau)} \\ -e^{-(T-\tau)} + 2e^{-2(T-\tau)} \end{bmatrix} d\tau = \begin{bmatrix} \frac{1}{2} - e^{-T} + \frac{1}{2}e^{-2T} \\ e^{-T} - e^{-2T} \end{bmatrix}$$

Now, since

$$x[(k + 1)T] = \phi(T) x(kT) + \theta(T) u(kT)$$

the discrete state equation of the system is

$$\begin{bmatrix} x_1[(k + 1)T] \\ x_2[(k + 1)T] \end{bmatrix} = \begin{bmatrix} 2e^{-T} - e^{-2T} & e^{-T} - e^{-2T} \\ -2e^{-T} + 2e^{-2T} & -e^{-T} + 2e^{-2T} \end{bmatrix} \begin{bmatrix} x_1(kT) \\ x_2(kT) \end{bmatrix}$$

$$+ \begin{bmatrix} \frac{1}{2} - e^{-T} + \frac{1}{2}e^{-2T} \\ e^{-T} - e^{-2T} \end{bmatrix} u(kT)$$

Let us take the sampling period $T = 1$ sec. We obtain

$$\begin{bmatrix} x_1(k + 1) \\ x_2(k + 1) \end{bmatrix} = \begin{bmatrix} 0.6 & 0.233 \\ -0.466 & -0.097 \end{bmatrix} \begin{bmatrix} x_1(k) \\ x_2(k) \end{bmatrix} + \begin{bmatrix} 0.2 \\ 0.233 \end{bmatrix} u(k)$$

Using the formula

$$x(NT) = \phi(NT) x(0) + \sum_{k=0}^{N-1} \phi[(N - k - 1)T] \theta(T) u(kT)$$

where

$$\phi (NT) = \underbrace{\phi (T) \cdot \phi (T) \cdot \ldots \cdot \phi (T)}_{N} = \phi^N (T)$$

we obtain

$$\begin{bmatrix} x_1 (N) \\ x_2 (N) \end{bmatrix} = \begin{bmatrix} 2e^{-N} - e^{-2N} & e^{-N} - e^{-2N} \\ -2e^{-N} + 2e^{-2N} & -e^{-N} + 2e^{-2N} \end{bmatrix} \begin{bmatrix} x_1 (0) \\ x_2 (0) \end{bmatrix}$$

$$+ \sum_{k=0}^{N-1} \begin{bmatrix} 0.633e^{-(N-k-1)} - 0.433e^{-2(N-k-1)} \\ -0.633e^{-(N-k-1)} + 0.866e^{-2(N-k-1)} \end{bmatrix} u(k)$$

where N and k are positive integers.

Using two methods obtain the impulse response function of the following system

$$\frac{dx}{dt} + a(t)x = \delta (t - \tau)$$

$$x(t) = 0 \quad \text{for} \quad t < \tau.$$

Solution:

1.) Let us denote the impulse response function as $y(t, \tau)$. We have

$$\frac{\partial y(t,\tau)}{\partial t} + a(t)y(t,\tau) = \delta (t - \tau)$$

$$0 \leq \tau \leq t$$

The homogeneous equation of the form

$$\frac{\partial y(t,\tau)}{\partial t} + a(t)y(t,\tau) = 0$$

has a solution

$$\ln[y(t,\tau)] = -\int a(t)dt + f(\tau) \qquad t \geq \tau$$

Thus

$$y(t,\tau) = g(\tau)e^{-\int a(t)dt} \qquad t \geq \tau$$

where

$$g(\tau) = e^{f(\tau)}$$

Note that

$$x(t) = 0 \quad \text{for} \quad t < \tau$$

565

thus

$$y(t,\tau) = 0 \quad \text{for} \quad t < \tau$$

We write

$$y(t,\tau) = \begin{cases} g(\tau)e^{-\int a(t)dt} & \text{for} \quad \begin{array}{l} t \geq \tau \\ t \leq \tau \end{array} \\ 0 \end{cases}$$

Thus

$$\frac{\partial y(t,\tau)}{\partial t} = \begin{cases} \delta(t-\tau)g(\tau)e^{-\int a(t)dt} - a(t)g(\tau)e^{-\int a(t)dt} & t \geq \tau \\ 0 & t < \tau \end{cases}$$

Substituting the above result into the initial equation we obtain

$$\delta(t - \tau)g(\tau)e^{-\int a(t)dt} = \delta(t - \tau) \tag{1}$$

At $t = \tau$ we have

$$g(\tau)e^{-\int a(\tau)d\tau} = 1$$

and

$$g(\tau) = e^{\int a(\tau)d\tau}$$

For $t \neq \tau$ equation (1) is satisfied, since $\delta(t - \tau) = 0$.

Thus the impulse response function is

$$y(t,\tau) = e^{\int a(\tau)d\tau - \int a(t)dt} = e^{\int_{t_0}^{\tau} a(s)ds - \int_{t_0}^{t} a(s)ds}$$

$$\begin{cases} = e^{-\int_{\tau}^{t} a(s)ds} & \text{for } t \geq \tau \\ = 0 & \text{for } t < \tau \end{cases}$$

2.) Let us rewrite the system equation

$$\frac{dx}{dt} = -a(t)x + \delta(t - \tau); \quad x(t) = 0 \text{ for } t < \tau$$

Let us express the impulse response function

$$y(t,\tau) = \alpha(t,t_0)\beta(\tau,t_0)$$

where $\alpha(t,t_0)$ is the state transition function

and $\beta(\tau,t_0) = \alpha^{-1}(\tau,t_0)$.

$a(t)$ is a scalar and, $a(t)$ and $\int_{t_0}^{t} a(s)ds$ commute for

566

all t.

Thus

$$\alpha(t, t_0) = e^{-\int_{t_0}^{t} a(s)\,ds}$$

and

$$\beta(\tau, t_0) = \alpha^{-1}(\tau, t_0) = e^{\int_{t_0}^{\tau} a(s)\,ds}$$

The impulse response function is

$$y(t, \tau) = e^{-\int_{t_0}^{t} a(s)\,ds + \int_{t_0}^{\tau} a(s)\,ds} = e^{-\int_{\tau}^{t} a(s)\,ds} \quad ; \quad t \geq \tau$$

• **PROBLEM** 11-19

A system is described by

$$\begin{bmatrix} \dot{x}_1 \\ \dot{x}_2 \\ \dot{x}_3 \end{bmatrix} = \begin{bmatrix} -2 & -2 & 0 \\ 0 & 0 & 1 \\ 0 & -3 & -4 \end{bmatrix} \begin{bmatrix} x_1 \\ x_2 \\ x_3 \end{bmatrix} + \begin{bmatrix} 1 & 0 \\ 0 & 1 \\ 1 & 1 \end{bmatrix} \begin{bmatrix} u_1 \\ u_2 \end{bmatrix}$$

a) find the transformation

$$x = Ky$$

which uncouples this system.

b) find $x(t)$ for $x(0) = [0 \quad 1 \quad 2]^T$, $u(t) = [t \quad 1]^T$

Solution: Let us first find the modal matrix K. We have

$$\det|A - \lambda I| = \det \begin{bmatrix} -2-\lambda & -2 & 0 \\ 0 & -\lambda & 1 \\ 0 & -3 & -4-\lambda \end{bmatrix}$$

$$= (\lambda + 1)(\lambda + 2)(\lambda + 3)$$

The eigenvalues are $\lambda_1 = -1$, $\lambda_2 = -2$, $\lambda_3 = -3$.

$$\text{adj}[A - \lambda I] = \begin{bmatrix} \lambda^2 + 4\lambda + 3 & -2(\lambda + 4) & -2 \\ 0 & (\lambda + 2)(\lambda + 4) & 2 + \lambda \\ 0 & -3(2 + \lambda) & \lambda(2 + \lambda) \end{bmatrix} \quad (1)$$

Substituting the eigenvalues into (1) we obtain the eigenvectors

567

$$\eta_1 = \begin{bmatrix} -2 \\ 1 \\ -1 \end{bmatrix} \qquad \eta_2 = \begin{bmatrix} 1 \\ 0 \\ 0 \end{bmatrix} \qquad \eta_3 = \begin{bmatrix} -2 \\ -1 \\ 3 \end{bmatrix}$$

and

$$K = \begin{bmatrix} -2 & 1 & -2 \\ 1 & 0 & -1 \\ -1 & 0 & 3 \end{bmatrix}$$

The transformation

$$x = \begin{bmatrix} -2 & 1 & -2 \\ 1 & 0 & -1 \\ -1 & 0 & 3 \end{bmatrix} y$$

is the decoupling transformation.

Note that

$$K^{-1} = \begin{bmatrix} 0 & \dfrac{3}{2} & \dfrac{1}{2} \\ 1 & 4 & 2 \\ 0 & \dfrac{1}{2} & \dfrac{1}{2} \end{bmatrix}$$

We have

$$\dot{x} = Ax + Bu$$

Substituting $x = Ky$

$$K\dot{y} = AKy + Bu$$

Solving for $\dot{y}$

$$\dot{y} = K^{-1}AKy + K^{-1}Bu$$

Thus

$$\dot{y} = \begin{bmatrix} -1 & 0 & 0 \\ 0 & -2 & 0 \\ 0 & 0 & -3 \end{bmatrix} y + \begin{bmatrix} \dfrac{1}{2} & 2 \\ 3 & 6 \\ \dfrac{1}{2} & 1 \end{bmatrix} \begin{bmatrix} u_1 \\ u_2 \end{bmatrix}$$

b)   For $u(t) = \begin{bmatrix} t \\ 1 \end{bmatrix}$ we have

$$\dot{y}_1 = -y_1 + \tfrac{1}{2}t + 2$$

$$\dot{y}_2 = -2y_2 + 3t + 6$$

$$\dot{y}_3 = -3y_2 + \tfrac{1}{2}t + 1$$

Solving the above equations we obtain

$$y_1(t) = e^{-t}y_1(0) + \frac{1}{2}t + \frac{3}{2}(1 - e^{-t})$$

$$y_2(t) = e^{-2t}y_2(0) + \frac{3}{2}t + \frac{9}{4}(1 - e^{-2t})$$

$$y_3(t) = e^{-3t}y_3(0) + \frac{1}{6}t + \frac{5}{18}(1 - e^{-3t})$$

Since

$$y(0) = K^{-1}x(0) = \begin{bmatrix} 0 & \frac{3}{2} & \frac{1}{2} \\ 1 & 4 & 2 \\ 0 & \frac{1}{2} & \frac{1}{2} \end{bmatrix} \begin{bmatrix} 0 \\ 1 \\ 2 \end{bmatrix} = \begin{bmatrix} \frac{5}{2} \\ 8 \\ \frac{3}{2} \end{bmatrix}$$

and since x = Ky we obtain the solution of the system

$$x(t) = \begin{bmatrix} -2 & 1 & -2 \\ 1 & 0 & -1 \\ -1 & 0 & 3 \end{bmatrix} \begin{bmatrix} \frac{5}{2}e^{-t} + \frac{1}{2}t + \frac{3}{2}(1 - e^{-t}) \\ 8e^{-2t} + \frac{3}{2}t + \frac{9}{4}(1 - e^{-2t}) \\ \frac{3}{2}e^{-3t} + \frac{1}{6}t + \frac{5}{18}(1 - e^{-3t}) \end{bmatrix}$$

$$= \begin{bmatrix} -2e^{-t} + \frac{23}{4}e^{-2t} - \frac{22}{9}e^{-3t} + \frac{1}{6}t - \frac{47}{36} \\ e^{-t} - \frac{11}{9}e^{-3t} + \frac{1}{3}t + \frac{11}{9} \\ -e^{-t} + \frac{11}{3}e^{-3t} - \frac{2}{3} \end{bmatrix}$$

• **PROBLEM** 11-20

The continuous-time system is described by

$$\ddot{x} = u$$

Discretize the system and derive a vector matrix difference equation for the system.

Solution: Let us define

$$x_1 = x$$

$$x_2 = \dot{x}$$

The scalar equation can be written

$$\begin{bmatrix} \dot{x}_1 \\ \dot{x}_2 \end{bmatrix} = \begin{bmatrix} 0 & 1 \\ 0 & 0 \end{bmatrix} \begin{bmatrix} x_1 \\ x_2 \end{bmatrix} + \begin{bmatrix} 0 \\ 1 \end{bmatrix} [u]$$

or

$$\dot{x} = Ax + Bu$$

The general equation for the discrete-time function is

$$x(kT + \tau) = G(\tau)x(kT) + H(\tau) u(kT)$$

where

$$0 < \tau \leq T$$

In the above expression T is the sampling period. We have

$$G(\tau) = L^{-1}[(sI - A)^{-1}] = L^{-1}\left[\begin{bmatrix} s & -1 \\ 0 & s \end{bmatrix}^{-1}\right]$$

$$= L^{-1}\begin{bmatrix} \frac{1}{s} & \frac{1}{s^2} \\ 0 & \frac{1}{s} \end{bmatrix} = \begin{bmatrix} 1 & \tau \\ 0 & 1 \end{bmatrix}$$

$H(\tau)$ is found from

$$H(\tau) = \int_0^\tau G(\sigma)B d\sigma$$

$$H(\tau) = \int_0^\tau \begin{bmatrix} 1 & \sigma \\ 0 & 1 \end{bmatrix}\begin{bmatrix} 0 \\ 1 \end{bmatrix} d\sigma = \begin{bmatrix} \frac{1}{2}\tau^2 \\ \tau \end{bmatrix}$$

The discrete-time equation is

$$\begin{bmatrix} x_1(kT + \tau) \\ x_2(kT + \tau) \end{bmatrix} = \begin{bmatrix} 1 & \tau \\ 0 & 1 \end{bmatrix}\begin{bmatrix} x_1(kT) \\ x_2(kT) \end{bmatrix} + \begin{bmatrix} \frac{1}{2}\tau^2 \\ \tau \end{bmatrix}[u(kT)]$$

● **PROBLEM** 11-21

The scalar equation is

$$\frac{d^2x(t)}{dt^2} + a(t)\frac{dx(t)}{dt} + b(t)x = u(t)$$

with the initial conditions

$$x(t_0) = \dot{x}(t_0) = 0$$

Show that the solution of the equation is

$$x(t) = x_1(t) = -x_{11}(t)\int_{t_0}^t \frac{x_{12}(\tau)u(\tau)}{W(x_{11},x_{12})}d\tau$$

$$+ x_{12}(t)\int_{t_0}^t \frac{x_{11}(\tau)u(\tau)}{W(x_{11},x_{12})}d\tau$$

where $x_{11}(t)$ and $x_{12}(t)$ are linearly independent solutions of

570

$$\frac{d^2x}{dt^2} + a(t)\frac{dx}{dt} + b(t)x = 0$$

and W is the Wronskian

$$W(x_{11},x_{12}) = \begin{vmatrix} x_{11}(t) & x_{12}(t) \\ \dot{x}_{11}(t) & \dot{x}_{12}(t) \end{vmatrix} \neq 0$$

__Solution:__ Let $x_1 = x$, $x_2 = \dot{x}$

In the matrix form

$$\begin{bmatrix} \dot{x}_1 \\ \dot{x}_2 \end{bmatrix} = \begin{bmatrix} 0 & 1 \\ -b(t) & -a(t) \end{bmatrix} \begin{bmatrix} x_1 \\ x_2 \end{bmatrix} + \begin{bmatrix} 0 \\ 1 \end{bmatrix}[u]$$

The solution of the above equation is

$$x(t) = \phi(t,t_0)x(t_0) + \Phi(t,t_0)\int_{t_0}^{t} \Phi^{-1}(\tau,t_0)Bud\tau$$

where $\Phi(t,t_0)$ is the fundamental matrix such that $\Phi(t_0,t_0)$ $= I$ and $B = \begin{bmatrix} 0 \\ 1 \end{bmatrix}$. Since $x(t_0) = \dot{x}(t_0) = 0$ we have $x(t_0) = 0$ and

$$x(t) = \Phi(t,t_0)\int_{t_0}^{t} \Phi^{-1}(\tau,t_0)Bud\tau$$

Any fundamental matrix can be written as a product of Φ and a constant matrix C

$$X = \phi C$$

$$\begin{bmatrix} x_{11} & x_{12} \\ \dot{x}_{11} & \dot{x}_{12} \end{bmatrix} = \begin{bmatrix} \phi_{11} & \phi_{12} \\ \phi_{21} & \phi_{22} \end{bmatrix} \begin{bmatrix} c_{11} & c_{12} \\ c_{21} & c_{22} \end{bmatrix}$$

or

$$\Phi = XC^{-1}$$

We obtain for $x(t)$

$$x(t) = X(t)C^{-1}\int_{t_0}^{t} CX^{-1}(\tau)Bud\tau = X(t)\int_{t_0}^{t} X^{-1}(\tau)Bud\tau$$

X^{-1} is given by

$$X^{-1} = \frac{1}{W(x_{11},x_{12})}\begin{bmatrix} \dot{x}_{12} & -x_{12} \\ -\dot{x}_{11} & x_{11} \end{bmatrix}$$

Thus

$$\begin{bmatrix} x_1(t) \\ x_2(t) \end{bmatrix} = \begin{bmatrix} x_{11} & x_{12} \\ \dot{x}_{11} & \dot{x}_{12} \end{bmatrix} \int_{t_0}^{t} \frac{1}{W(x_{11}, x_{12})} \begin{bmatrix} -x_{12}(\tau) \\ x_{11}(\tau) \end{bmatrix} u(\tau) d\tau$$

Multiplying we obtain the solution $x(t)$

$$x(t) = x_1(t) = -x_{11}(t) \int_{t_0}^{t} \frac{x_{12}(\tau) u(\tau)}{W(x_{11}, x_{12})} d\tau + x_{12}(t) \int_{t_0}^{t} \frac{x_{11}(\tau) u(\tau)}{W(x_{11}, x_{12})} d\tau$$

● **PROBLEM 11-22**

For the system

$$\frac{d^2 x_1}{dt^2} + \frac{dx_2}{dt} + 4x_1 = y_1(t)$$

$$\frac{d^2 x_2}{dt^2} + 3\frac{dx_1}{dt} + 7\frac{dx_2}{dt} + x_2 = y_2(t)$$

obtain the expression for the impulsive response of the output x_2 with respect to the input function $y_1(t)$.

Solution: Let us assign the state variables as follows

$$u_1 = x_2, \qquad u_2 = x_1, \qquad u_3 = \dot{x}_2, \qquad u_4 = \dot{x}_1$$

Thus the state variable presentation of the performance equation is

$$\dot{u}_1 = u_3$$

$$\dot{u}_2 = u_4$$

$$\dot{u}_3 = -3u_4 - 7u_3 - u_1 + y_2(t)$$

$$\dot{u}_4 = -4u_2 - u_3 + y_1(t)$$

In the matrix notation

$$\dot{u} = \begin{bmatrix} 0 & 0 & 1 & 0 \\ 0 & 0 & 0 & 1 \\ -1 & 0 & -7 & -3 \\ 0 & -4 & -1 & 0 \end{bmatrix} u + \begin{bmatrix} 0 & 0 \\ 0 & 0 \\ 0 & 1 \\ 1 & 0 \end{bmatrix} y$$

or

$$\dot{u} = Au + By \qquad (1)$$

The characteristic polynomial is

$$\Delta = |sI - A| = \begin{vmatrix} s & 0 & -1 & 0 \\ 0 & s & 0 & -1 \\ 1 & 0 & s+7 & 3 \\ 0 & 4 & 1 & s \end{vmatrix}$$

$$= s \begin{vmatrix} s & 0 & -1 \\ 0 & s+7 & 3 \\ 4 & 1 & s \end{vmatrix} + \begin{vmatrix} 0 & -1 & 0 \\ s & 0 & -1 \\ 4 & 1 & s \end{vmatrix}$$

$$= s^4 + 7s^3 + 2s^2 + 28s + 4$$

In the Laplace domain eq. (1) becomes

$$sIU = AU + By$$

thus

$$U = [sI - A]^{-1}By$$

Since we want the impulse response to y_1, we have $y = \begin{bmatrix} 1 \\ 0 \end{bmatrix}$ and

By is $\begin{bmatrix} 0 \\ 0 \\ 0 \\ 1 \end{bmatrix}$. Postmultiplying $[sI - A]^{-1}$ by $\begin{bmatrix} 0 \\ 0 \\ 0 \\ 1 \end{bmatrix}$ will give the

fourth column of $[sI - A]^{-1}$ as the result. And since we are only interested in $u_1 = x_2$ we want only the first element. Thus, we have to find element a_{14} of $[sI - A]^{-1}$.

Since $[sI - A]^{-1} = \frac{1}{\Delta} \text{Adj} [sI - A]$, the required element is

$$LI_{14} = \frac{1}{\Delta} \begin{vmatrix} 0 & 0 & -1 & 0 \\ 0 & s & 0 & -1 \\ 0 & 0 & s+7 & 3 \\ 1 & 4 & 1 & s \end{vmatrix} = \frac{-3}{s^4 + 7s^3 + 2s^2 + 28s + 4}$$

$$LI_{14} = \frac{-3s}{(s + 0.28)(s + 0.88)(s^2 + 7.96s + 16.5)}$$

The above expression can be inverted to

$$I_{14} = 0.074e^{-0.28t} - 0.18e^{-0.88t} + 0.73e^{-3.98t} \cos(0.78t + \theta)$$

where

$$\theta = 1.72 \text{rd}.$$

● **PROBLEM** 11-23

The difference equation is

$$x(k + 2) + \alpha x(k + 1) + \beta x(k) = 0 \qquad (1)$$

where α, β are constants and $k \geq 1$. The initial data are $x(0) = a$, $x(1) = b$. Obtain the solution of the equation.

Solution: Let E be a unit delay operator, we have

573

$$Ex(k) = x(k + 1)$$

$$E^2 x(k) = x(k + 2)$$

The initial equation can be written

$$(E^2 + \alpha E + \beta) x(k) = 0$$

Let us write $x(k)$ in the form

$$x(k) = Cq^k$$

where C is an arbitrary constant and q has to be determined in such a way that $Cq^k = x(k)$ is a solution of (1). We have

$$E^2 x(k) = E^2 Cq^k = q^2 Cq^k = q^2 x(k)$$

$$Ex(k) = ECq^k = qCq^k = qx(k)$$

Thus

$$(q^2 + \alpha q + \beta) x(k) = 0$$

or

$$q^2 + \alpha q + \beta = 0$$

$$q_1 = \frac{-\alpha + \sqrt{\alpha^2 - 4\beta}}{2} \qquad q_2 = \frac{-\alpha - \sqrt{\alpha^2 - 4\beta}}{2}$$

Thus

$$x(k) = Cq^k = C_1 q_1{}^k + C_2 q_2{}^k$$

From the initial data

$$x(0) = a = C_1 + C_2$$

$$x(1) = b = C_1 q_1 + C_2 q_2$$

Hence

$$C_1 = \frac{aq_2 - b}{q_2 - q_1}$$

$$C_2 = \frac{aq_1 - b}{q_1 - q_2}$$

The solution of the system is

$$x(k) = \frac{aq_2 - b}{q_2 - q_1} \cdot q_1{}^k + \frac{aq_1 - b}{q_1 - q_2} q_2{}^k$$

where

$$q_1 = \frac{-\alpha + \sqrt{\alpha^2 - 4\beta}}{2}, \ q_2 = \frac{-\alpha - \sqrt{\alpha^2 - 4\beta}}{2}$$

assuming $q_1 \neq q_2$ or $\alpha^2 - 4\beta \neq 0$

574

Given a system described by the following differential equation

$$\frac{d^3x}{dt^3} + 6\frac{d^2x}{dt^2} + 11\frac{dx}{dt} + 6x = 0$$

obtain the state representation and the response of the system.

Solution: Let us choose the state variables as

$$\dot{x}_1 = x_2$$

$$\dot{x}_2 = x_3$$

$$\dot{x}_3 = -6x_1 - 11x_2 - 6x_3$$

$$x_1 = x$$

In the matrix form we can write

$$\begin{bmatrix} \dot{x}_1 \\ \dot{x}_2 \\ \dot{x}_3 \end{bmatrix} = \begin{bmatrix} 0 & 1 & 0 \\ 0 & 0 & 1 \\ -6 & -11 & -6 \end{bmatrix} \begin{bmatrix} x_1 \\ x_2 \\ x_3 \end{bmatrix}$$

Taking the Laplace transforms, we obtain

$$\begin{bmatrix} sX_1(s) \\ sX_2(s) \\ sX_3(s) \end{bmatrix} - \begin{bmatrix} x_1(0) \\ x_2(0) \\ x_3(0) \end{bmatrix} = \begin{bmatrix} 0 & 1 & 0 \\ 0 & 0 & 1 \\ -6 & -11 & -6 \end{bmatrix} \begin{bmatrix} X_1(s) \\ X_2(s) \\ X_3(s) \end{bmatrix}$$

or

$$\left\{ s\begin{bmatrix} 1 & 0 & 0 \\ 0 & 1 & 0 \\ 0 & 0 & 1 \end{bmatrix} - \begin{bmatrix} 0 & 1 & 0 \\ 0 & 0 & 1 \\ -6 & -11 & -6 \end{bmatrix} \right\} \begin{bmatrix} X_1(s) \\ X_2(s) \\ X_3(s) \end{bmatrix} = \begin{bmatrix} x_1(0) \\ x_2(0) \\ x_3(0) \end{bmatrix}$$

Solving for $\begin{bmatrix} X_1(s) \\ X_2(s) \\ X_3(s) \end{bmatrix}$ we obtain

$$\begin{bmatrix} X_1(s) \\ X_2(s) \\ X_3(s) \end{bmatrix} = \begin{bmatrix} s & -1 & 0 \\ 0 & s & -1 \\ 6 & 11 & s+6 \end{bmatrix}^{-1} \begin{bmatrix} x_1(0) \\ x_2(0) \\ x_3(0) \end{bmatrix}$$

$$
\begin{bmatrix} s & -1 & 0 \\ 0 & s & -1 \\ 6 & 11 & s+6 \end{bmatrix}^{-1} = \frac{1}{s^3 + 6s^2 + 11s + 6} \begin{bmatrix} s^2+6s+11 & s+6 & 1 \\ -6 & s^2+6s & s \\ -6s & -11s-6 & s^2 \end{bmatrix}
$$

Taking the inverse Laplace transformation, we get

$$
\begin{bmatrix} x_1(t) \\ x_2(t) \\ x_3(t) \end{bmatrix} = \begin{bmatrix} 3e^{-t} - 3e^{-2t} + e^{-3t} \\ -3e^{-t} + 6e^{-2t} - 3e^{-3t} \\ 3e^{-t} - 12e^{-2t} + 9e^{-3t} \end{bmatrix}
$$

$$
\begin{matrix}
\frac{5}{2}e^{-t} - 4e^{-2t} + \frac{3}{2}e^{-3t} \\
-\frac{5}{2}e^{-t} + 8e^{-2t} - \frac{9}{2}e^{-3t} \\
\frac{5}{2}e^{-t} - 16e^{-2t} + \frac{27}{2}e^{-3t}
\end{matrix}
$$

$$
\begin{bmatrix}
\frac{1}{2}e^{-t} - e^{-2t} + \frac{1}{2}e^{-3t} \\
-\frac{1}{2}e^{-t} + 2e^{-2t} - \frac{3}{2}e^{-3t} \\
\frac{1}{2}e^{-t} - 4e^{-2t} + \frac{9}{2}e^{-3t}
\end{bmatrix}
$$

$$
\begin{bmatrix} x_1(0) \\ x_2(0) \\ x_3(0) \end{bmatrix}
$$

In the matrix form

$$
x(t) = \phi(t)x(0)
$$

● **PROBLEM 11-25**

The system is given by

$$
\frac{d^2x}{dt^2} + (\alpha + \beta) \frac{dx}{dt} + \alpha\beta x = u
$$

where α and β are constants and $u(t) = \delta(t)$ is a unit impulse. The initial conditions are

$$
x(0_-) = \frac{dx(0_-)}{dt} = 0
$$

576

Obtain the response of the system using

1.   Laplace transform method

2.   State-space method.

Solution:

1. The transfer function of the system is

$$\frac{X(s)}{U(s)} = \frac{1}{s^2 + (\alpha + \beta)s + \alpha\beta}$$

Since u(t) is a unit impulse we have

$$X(s) = \frac{1}{s^2 + (\alpha + \beta)s + \alpha\beta} = \frac{1}{(s+\alpha)(s+\beta)} = \frac{1}{\beta - \alpha}\left[\frac{1}{s+\alpha} - \frac{1}{s+\beta}\right]$$

Taking the inverse Laplace transform we get

$$x(t) = \frac{1}{\alpha - \beta}\left(-e^{-\alpha t} + e^{-\beta t}\right)$$

2. Let us rewrite the initial equation

$$\begin{bmatrix} \dot{x}_1 \\ \dot{x}_2 \end{bmatrix} = \begin{bmatrix} 0 & 1 \\ -\alpha\beta & -(\alpha+\beta) \end{bmatrix}\begin{bmatrix} x_1 \\ x_2 \end{bmatrix} + \begin{bmatrix} 0 \\ 1 \end{bmatrix} u$$

where

$$x_1 = x$$

$$x_2 = \dot{x}$$

Let us define the transformation

$$\begin{bmatrix} x_1 \\ x_2 \end{bmatrix} = \begin{bmatrix} 1 & 1 \\ -a & -b \end{bmatrix}\begin{bmatrix} y_1 \\ y_2 \end{bmatrix} \tag{1}$$

We get

577

$$
\begin{bmatrix} \dot{y}_1 \\ \dot{y}_2 \end{bmatrix} = \begin{bmatrix} -\alpha & 0 \\ 0 & -\beta \end{bmatrix} \begin{bmatrix} y_1 \\ y_2 \end{bmatrix} + \begin{bmatrix} \dfrac{-\beta}{\alpha-\beta} & \dfrac{-1}{\alpha-\beta} \\ \dfrac{\alpha}{\alpha-\beta} & \dfrac{1}{\alpha-\beta} \end{bmatrix} \begin{bmatrix} 0 \\ 1 \end{bmatrix} u
$$

$$
= \begin{bmatrix} -\alpha & 0 \\ 0 & -\beta \end{bmatrix} \begin{bmatrix} y_1 \\ y_2 \end{bmatrix} + \begin{bmatrix} \dfrac{-1}{\alpha-\beta} \\ \dfrac{1}{\alpha-\beta} \end{bmatrix} u
$$

The general solution formula is

$$
x(t) = e^{At} x(0_-) + \int_0^t e^{A(t-\tau)} B u(\tau) d\tau
$$

$$
= e^{At} B
$$

Since $x(0_-) = 0$ and $u(t)$ is a unit impulse

we have

$$
e^{At} = \begin{bmatrix} e^{-\alpha t} & 0 \\ 0 & e^{-\beta t} \end{bmatrix} \qquad B = \begin{bmatrix} \dfrac{-1}{\alpha-\beta} \\ \dfrac{1}{\alpha-\beta} \end{bmatrix}
$$

Thus

$$
\begin{bmatrix} y_1(t) \\ y_2(t) \end{bmatrix} = \begin{bmatrix} \dfrac{-e^{-\alpha t}}{\alpha-\beta} \\ \dfrac{e^{-\beta t}}{\alpha-\beta} \end{bmatrix}
$$

Using the transformation (1) we find

$$
x(t) = \begin{bmatrix} x_1(t) \\ x_2(t) \end{bmatrix} = \begin{bmatrix} 1 & 1 \\ -\alpha & -\beta \end{bmatrix} \begin{bmatrix} \dfrac{-e^{-\alpha t}}{\alpha-\beta} \\ \dfrac{e^{-\beta t}}{\alpha-\beta} \end{bmatrix} = \begin{bmatrix} -\dfrac{e^{-\alpha t}}{\alpha-\beta} + \dfrac{e^{-\beta t}}{\alpha-\beta} \\ \dfrac{\alpha e^{-\alpha t}}{\alpha-\beta} - \dfrac{\beta e^{-\beta t}}{\alpha-\beta} \end{bmatrix}
$$

578

Since $x = x_1$, the response of the system is

$$x(t) = \frac{1}{\alpha - \beta} \left(e^{-\beta t} - e^{-\alpha t} \right)$$

● **PROBLEM** 11-26

The state diagram of a system is shown below.

Find $x_1(t)$ and $x_2(t)$, when $r(t)$ is a unit-step function applied at $t = t_0$.

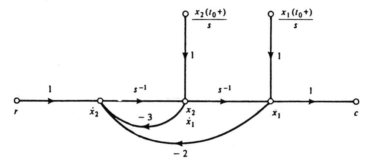

Solution: From Mason's gain formula we have

$$C(s) = \frac{\sum_i P_i \Delta_i}{\Delta} R(s)$$

where P_i are the forward path gains, Δ is the loop determinant.

$\Delta = 1 - \Sigma$ (all loop gains) $+ \Sigma$ (product of gains of all possible combinations of two non-touching loops) $- \Sigma$ (product of gains of all possible combinations of 3 non-touching loops) $+ \Sigma \ldots$

Δ_i is Δ less any term containing a loop touched by P_i .

In our case $X_1(s)$ and $X_2(s)$ are output nodes and $R(s)$, $x_1(t_0+)$, $x_2(t_0+)$ are input nodes, thus

$$X_1(s) = \frac{s^{-1}(1+3s^{-1})}{\Delta} x_1(t_0+) + \frac{s^{-2}}{\Delta} x_2(t_0+) + \frac{s^{-2}}{\Delta} R(s)$$

$$X_2(s) = \frac{-2s^{-2}}{\Delta} x_1(t_0+) + \frac{s^{-1}}{\Delta} x_2(t_0+) + \frac{s^{-1}}{\Delta} R(s)$$

where

$$\Delta = 1 + 3s^{-1} + 2s^{-2}$$

We shall rewrite the above equations in the matrix form

$$
\begin{bmatrix} X_1(s) \\ \\ X_2(s) \end{bmatrix} = \frac{1}{(s+1)(s+2)} \begin{bmatrix} s+3 & 1 \\ \\ -2 & s \end{bmatrix} \begin{bmatrix} x_1(t_0+) \\ \\ x_2(t_0+) \end{bmatrix} + \begin{bmatrix} \dfrac{1}{(s+1)(s+2)} \\ \\ \dfrac{s}{(s+1)(s+2)} \end{bmatrix} R(s)
$$

The input of the system $r(t)$ is a unit step function applied at $t = t_0$. We have

$$L^{-1}\left(\frac{1}{s+a}\right) = e^{-(t-t_0)} u_s(t-t_0) \qquad t \geq t_0$$

$$L^{-1}\left(\frac{1}{s}\right) = u_s(t-t_0) \qquad t \geq t_0$$

$$R(s) = \frac{1}{s}$$

Applying the partial fractions expansion we obtain

$$
\begin{bmatrix} X_1(s) \\ \\ X_2(s) \end{bmatrix} = \begin{bmatrix} \dfrac{s+3}{(s+1)(s+2)} & \dfrac{1}{(s+1)(s+2)} \\ \\ \dfrac{-2}{(s+1)(s+2)} & \dfrac{s}{(s+1)(s+2)} \end{bmatrix} \begin{bmatrix} x_1(t_0+) \\ \\ x_2(t_0+) \end{bmatrix} + \begin{bmatrix} \dfrac{1}{s(s+1)(s+2)} \\ \\ \dfrac{1}{(s+1)(s+2)} \end{bmatrix}
$$

$$
= \begin{bmatrix} \dfrac{2}{s+1} - \dfrac{1}{s+2} & \dfrac{1}{s+1} - \dfrac{1}{s+2} \\ \\ \dfrac{-2}{s+1} + \dfrac{2}{s+2} & \dfrac{-1}{s+1} + \dfrac{2}{s+2} \end{bmatrix} \begin{bmatrix} x_1(t_0+) \\ \\ x_2(t_0+) \end{bmatrix} + \begin{bmatrix} \dfrac{\frac{1}{2}}{s} - \dfrac{1}{s+1} + \dfrac{\frac{1}{2}}{s+2} \\ \\ \dfrac{1}{s+1} - \dfrac{1}{s+2} \end{bmatrix}
$$

Taking the inverse Laplace transform, we have

$$
\begin{bmatrix} x_1(t) \\ \\ x_2(t) \end{bmatrix} = \begin{bmatrix} 2e^{-(t-t_0)} - e^{-2(t-t_0)} & e^{-(t-t_0)} - e^{-2(t-t_0)} \\ \\ -2e^{-(t-t_0)} + 2e^{-2(t-t_0)} & -e^{-(t-t_0)} + 2e^{-2(t-t_0)} \end{bmatrix}
$$

$$\begin{bmatrix} x_1(t_0+) \\ \\ x_2(t_0+) \end{bmatrix}$$

$$+ \begin{bmatrix} \frac{1}{2} u_s(t-t_0) - e^{-(t-t_0)} + \frac{1}{2} e^{-2(t-t_0)} \\ \\ e^{-(t-t_0)} - e^{-2(t-t_0)} \end{bmatrix} \qquad t \geq t_0$$

● **PROBLEM** 11-27

Obtain the system response to a unit step input for the system shown below.

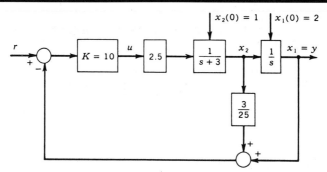

Solution: From the block diagram we get

$$\dot{x}_1 = x_2$$

$$\dot{x}_2 = -3x_2 + 25(r - \frac{3}{25} x_2 - x_1) = -25x_1 - 6x_2 + 25r$$

In the matrix form we can write

$$\dot{x} = Ax + Bu$$

where

$$A = \begin{bmatrix} 0 & 1 \\ -25 & -6 \end{bmatrix} \qquad \text{and} \qquad B = \begin{bmatrix} 0 \\ 25 \end{bmatrix}$$

The transition matrix is given by

$$\phi(s) = (sI-A)^{-1}$$

581

where

$$(sI-A) = \begin{bmatrix} s & -1 \\ 25 & s+6 \end{bmatrix}$$

and

$$(sI-A)^{-1} = \frac{1}{(s+3)^2 + 4^2} \begin{bmatrix} s+6 & 1 \\ -25 & s \end{bmatrix}$$

From

$$X(s) = \phi(s)x(0) + \phi(s)Br(s)$$

we have

$$X(s) = \frac{1}{(s+3)^2 + 4^2} \begin{bmatrix} s+6 & 1 \\ -25 & s \end{bmatrix} \begin{bmatrix} 2 \\ 1 \end{bmatrix} + \frac{1}{(s+3)^2 + 4^2} \begin{bmatrix} s+6 & 1 \\ -25 & s \end{bmatrix}$$

$$\begin{bmatrix} 0 \\ 25 \end{bmatrix} r(s)$$

$$= \frac{1}{(s+3)^2 + 4^2} \begin{bmatrix} 2(s+6) + 1 + 25r(s) \\ s - 50 + 25sr(s) \end{bmatrix}$$

Separating the variables we get

$$X(s) = \frac{1}{(s+3)^2 + 4^2} \begin{bmatrix} 2s + 13 \\ s - 50 \end{bmatrix} + \frac{1}{(s+3)^2 + 4^2} \begin{bmatrix} 25 \\ 25s \end{bmatrix} r(s)$$

Let us assume that r(t) is a unit step input, then

$$r(s) = \frac{1}{s}$$

and

582

$$X(s) = \frac{1}{s[(s+3)^2+4^2]} \begin{bmatrix} 2s^2 + 13s + 25 \\ \\ s^2 - 25s \end{bmatrix} \qquad (1)$$

Let us expand $X(s)$ in the following form

$$X(s) = \frac{K_1}{s} + \frac{K_2 s + K_3}{(s+3)^2+4^2} \qquad (2)$$

We have

$$K_1 = \frac{1}{(s+3)^2+4^2} \begin{bmatrix} 2s^2 + 13s + 25 \\ \\ s^2 - 25s \end{bmatrix}\Bigg|_{s=0} = \begin{bmatrix} 1 \\ \\ 0 \end{bmatrix}$$

Substituting K_1 into (2) and comparing the coefficients of equal powers of s in (1) and (2) we find

$$K_2 = \begin{bmatrix} 1 \\ \\ 1 \end{bmatrix} \qquad \text{and} \qquad K_3 = \begin{bmatrix} 7 \\ \\ -25 \end{bmatrix}$$

Thus

$$X(s) = \frac{1}{s} \begin{bmatrix} 1 \\ \\ 0 \end{bmatrix} + \frac{1}{(s+3)^2+4^2} \left(s \begin{bmatrix} 1 \\ \\ 1 \end{bmatrix} + \begin{bmatrix} 7 \\ \\ -25 \end{bmatrix} \right)$$

We shall find the inverse transform of

$$f(t) = L^{-1} \left(\frac{\alpha s + \beta}{(s+a)^2 + b^2} \right)$$

Let us rewrite

$$f(t) = \alpha L^{-1} \left(\frac{s + a}{(s+a)^2 + b^2} \right) + \frac{\beta - a\alpha}{b} L^{-1} \left(\frac{b}{(s+a)^2 + b^2} \right)$$

$$= e^{-at} [\alpha \cos bt + \frac{\beta - \alpha a}{b} \sin bt]$$

$$= e^{-at} \frac{1}{b} \sqrt{\alpha^2 b^2 + (\beta - a\alpha)^2} \ \sin(bt + \tan^{-1} \frac{b\alpha}{\beta - a\alpha})$$

Finally substituting the parameters we obtain

$$\begin{bmatrix} x_1(t) \\ \\ x_2(t) \end{bmatrix} = \begin{bmatrix} 1 \\ \\ 0 \end{bmatrix} + \begin{bmatrix} 1.414 e^{-3t} \sin(4t + 45°) \\ \\ 7.08 e^{-3t} \sin(4t + 171.9°) \end{bmatrix}$$

CHAPTER 12

CONTROLLABILITY AND OBSERVABILITY

For the system described by

$$x_1(k+1) = ax_1(k) + b_1u$$

$$x_2(k+1) = ax_2(k) + b_2u$$

where $b_1 \neq 0$

and $b_2 \neq 0$,

examine the controllability of the system.

Solution: By inspection, this system is uncontrollable. The two equations are decoupled, hence you cannot change one by changing the other. Note that Λ is diagonal and $\Gamma = [b_1 \; b_2]'$ contains no zero rows. We form the matrix $[\Gamma \; \Lambda\Gamma]$ and show that it is singular, the system is thus uncontrollable.

We shall verify that the system is not state controllable.

$$Q = [B \; AB] = \begin{bmatrix} b_1 & ab_1 \\ \\ b_2 & ab_2 \end{bmatrix}$$

Q is of rank 1, thus the pair [AB] is uncontrollable.

We may use

$$W = \sum_{i=0}^{N-1} A^{N-i-1} \; BB' \left(A^{N-i-1}\right)'$$

For N = 2 we have

$$W = A \, BB'A' + BB' = \begin{bmatrix} b_1^2 & b_1 b_2 \\ \\ b_1 b_2 & b_2^2 \end{bmatrix} (1 + a^2)$$

Since W is singular, the pair [AB] is uncontrollable.

Using the z-transforms,

$$(zI-A)^{-1} B = \begin{bmatrix} z-a & 0 \\ \\ 0 & z-a \end{bmatrix}^{-1} \begin{bmatrix} b_1 \\ \\ b_2 \end{bmatrix} = \begin{bmatrix} \dfrac{b_1}{z-a} \\ \\ \dfrac{b_2}{z-a} \end{bmatrix}$$

Since the rows of $(ZI-A)^{-1}B$ are dependent, the pair [AB] is uncontrollable.

● **PROBLEM 12-2**

The following differential equation describes the input-output relationship of a linear system.

$$\frac{d^2 c(t)}{dt^2} + 2 \frac{dc(t)}{dt} + C(t) = \frac{du(t)}{dt} + u(t)$$

Make two sets of state assignments, one of which is state controllable, the other not. Determine if the output is controllable.

Solution: A system is described by state equations as

$$\dot{x}(t) = Ax(t) + Bu(t)$$

$$C(t) = Dx(t) + Eu(t)$$

The system has n states, r inputs and p outputs.

If the matrix S, where

$$S = [B \; AB \; A^2 B \; . \; . \; A^{n-1}B] \quad \text{has rank n.}$$

It is output controllable if the matrix

$$T = [DB \ DAB \ DA^2B \ . \ . \ DA^{n-1}BE]$$

has rank p.

In our case we have $n = 2$, $r = p = 1$.

Let us define the state variables as

$$x_1 = c$$

$$x_2 = \dot{c} - u$$

We shall write the equation describing the system in the form

$$\frac{d^2c}{dt^2} - \frac{du}{dt} = -2\frac{dc}{dt} - c + u$$

Then the state equations become

$$\dot{x}_1 = x_2 + u$$

$$\dot{x}_2 = -x_1 - 2x_2 - u$$

In the matrix form we have

$$\begin{bmatrix} \dot{x}_1 \\ \dot{x}_2 \end{bmatrix} = \begin{bmatrix} 0 & 1 \\ -1 & -2 \end{bmatrix} \begin{bmatrix} x_1 \\ x_2 \end{bmatrix} + \begin{bmatrix} 1 \\ -1 \end{bmatrix} u$$

and the output equation is

$$c = x_1$$

Matrix S is

$$S = [B \ AB] = \begin{bmatrix} 1 & -1 \\ -1 & 1 \end{bmatrix}$$

The rank of the matrix S is óne, thus the system is not state controllable.

We have from the output equation

$$D = [1 \ 0] \ , \ E = 0$$

Matrix T is given by

$$T = [DB \quad DAB \quad E] = [1 \quad -1 \quad 0]$$

T is of rank 1, which is the same as the number of output and the system is output controllable.

We shall use the method of direct decomposition to define the state variables in a different way.

We have

$$\frac{C(s)}{U(s)} = \frac{s + 1}{s^2 + 2s + 1} = \frac{(s^{-1} + s^{-2})x(s)}{(1 + 2s^{-1} + s^{-2})x(s)}$$

so

$$c(s) = s^{-1} x(s) + s^{-2} x(s)$$

$$U(s) = x(s) + 2s^{-1} x(s) + s^{-2} x(s)$$

From the last equation

$$x(s) = -2s^{-1} x(s) - s^{-2} x(s) + U(s)$$

We choose state variables

$$x_1 = s^{-1} x \qquad \text{and} \qquad x_2 = s^{-2} x$$

Then

$$\dot{x} = x_1$$

$$\dot{x_1} = x_2$$

and the state equations are

$$\begin{bmatrix} \dot{x_1} \\ \dot{x_2} \end{bmatrix} = \begin{bmatrix} 0 & 1 \\ -1 & -2 \end{bmatrix} \begin{bmatrix} x_1 \\ x_2 \end{bmatrix} + \begin{bmatrix} 0 \\ 1 \end{bmatrix} u$$

The output equation is

$$c = x_1 + x_2$$

587

Matrix S is

$$S = [B \ AB] = \begin{bmatrix} 0 & 1 \\ 1 & -2 \end{bmatrix}$$

and nonsingular, thus the system is completely state controllable.

It's easy to check that the system is output controllable since

$$T = [DB \ DAB \ E] = [1 \ -1 \ 0]$$

which is of rank 1.

From the above example we see that for a given linear system state controllability depends on how the state variables are defined.

● **PROBLEM** 12-3

The input-output equations of a system are given by

$$\dot{y}_1 + 2(y_1 - y_2) = 4u_1 - u_2$$

$$\dot{y}_2 + 3(y_2 - y_1) = 4u_1 - u_2$$

Draw the simulation diagram for the system. Write the state equations. Derive the transfer matrix. Using the transfer matrix realize the system in several ways. What is the minimum number of integrators needed?

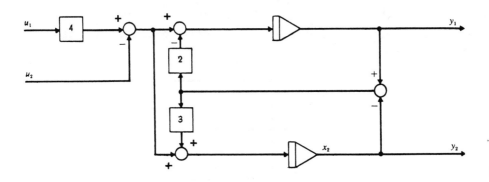

Fig.1

Solution: The simulation diagram is shown in Fig. 1.

From the diagram we obtain the state equations

$$\begin{bmatrix} \dot{x}_1 \\ \\ \dot{x}_2 \end{bmatrix} = \begin{bmatrix} -2 & 2 \\ \\ 3 & -3 \end{bmatrix} \begin{bmatrix} x_1 \\ \\ x_2 \end{bmatrix} + \begin{bmatrix} 4 & -1 \\ \\ 4 & -1 \end{bmatrix} \begin{bmatrix} u_1 \\ \\ u_2 \end{bmatrix}$$

$$\begin{bmatrix} y_1 \\ \\ y_2 \end{bmatrix} = \begin{bmatrix} 1 & 0 \\ \\ 0 & 1 \end{bmatrix} \begin{bmatrix} x_1 \\ \\ x_2 \end{bmatrix}$$

The state vector has dimension 2. Using Laplace transforming and a matrix inversion we derive the transfer matrix.

$$(s+2)y_1 - 2y_2 = 4u_1 - u_2$$

$$- 3y_1 + (s+3)y_2 = 4u_1 - u_2$$

so $\quad y = \begin{bmatrix} s+2 & -2 \\ \\ -3 & s+3 \end{bmatrix}^{-1} \begin{bmatrix} 4 & -1 \\ \\ 4 & -1 \end{bmatrix} u = H(s)u$

Then

$$H(s) = \frac{1}{s(s+5)} \begin{bmatrix} s+3 & 2 \\ \\ 3 & s+2 \end{bmatrix} \begin{bmatrix} 4 & -1 \\ \\ 4 & -1 \end{bmatrix}$$

$$= \frac{1}{s(s+5)} \begin{bmatrix} 4(s+5) & -(s+5) \\ \\ 4(s+5) & -(s+5) \end{bmatrix} = \begin{bmatrix} \dfrac{4}{s} & -\dfrac{1}{s} \\ \\ \dfrac{4}{s} & -\dfrac{1}{s} \end{bmatrix}$$

We can obtain the fourth-order system realization as shown in Fig. 2.

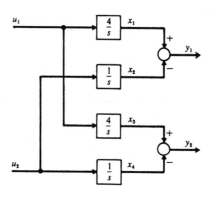

Fig.2

In the above example, the cancellation of pole and zero allows the reduction of the system. The above realization does not resemble the original system.

● **PROBLEM** 12-4

Construct an observer for the system shown below when only the state variable x_1 is available for an output.

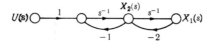

Solution: The following differential equation represents the system

$$\dot{x} = \begin{bmatrix} -2 & 1 \\ 0 & -1 \end{bmatrix} x + \begin{bmatrix} 0 \\ 1 \end{bmatrix} u = Ax + Bu$$

The output is

$$V = [1 \quad 0]x = Cx$$

since only x_1 is measurable, and we may choose an observer of $n - m = 1$, first order.

z satisfies the following differential equation

$$\dot{z} = \lambda z + Cx + Gu$$

We shall choose the characteristic root λ to be -3; it has to be larger than the roots of S_1.

For the transformation matrix T we have

$$TA - DT = C$$

and

$$\begin{bmatrix} t_{11} & t_{12} \end{bmatrix} \begin{bmatrix} -2 & 1 \\ 0 & -1 \end{bmatrix} + 3 \begin{bmatrix} t_{11} & t_{12} \end{bmatrix} = [1,0]$$

Thus

$$T = \begin{bmatrix} 1 & -\frac{1}{2} \end{bmatrix}$$

For G we have

$$G = TB,$$

590

thus

$$G = g_{11} = \begin{bmatrix} 1 & -\frac{1}{2} \end{bmatrix} \begin{bmatrix} 0 \\ 1 \end{bmatrix} = -\frac{1}{2}$$

From z, using the T matrix, we obtain x_2.

$$x_2 = -2z + 2x_1$$

The total system is shown below.

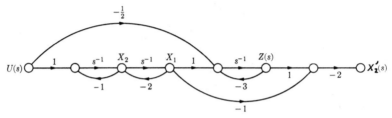

$x_2{'}$ represents the observer's estimate of x_2.

● **PROBLEM 12-5**

The system is given by

$$\dot{x} = Ax + Bu$$

$$y = Cx$$

where

$$A = \begin{bmatrix} 0 & 1 & 0 \\ 0 & 0 & 1 \\ -6 & -11 & -6 \end{bmatrix} \qquad B = \begin{bmatrix} 0 \\ 0 \\ 1 \end{bmatrix}$$

$$C = \begin{bmatrix} 4 & 5 & 1 \end{bmatrix}$$

Show that the system is not completely observable.

Solution: We may set $u = 0$, since the control function u does not affect the complete observability of the system. The observability matrix is given by

$$[C' \ A'C' \ (A')^2 C'] = \begin{bmatrix} 4 & -6 & 6 \\ 5 & -7 & 5 \\ 1 & -1 & -1 \end{bmatrix}$$

591

The rank of the above matrix is less than 3 since

$$
\begin{bmatrix}
4 & -6 & 6 \\
5 & -7 & 5 \\
1 & -1 & -1
\end{bmatrix} = 0
$$

Thus the system is not completely observable. The cancellation occurs in the transfer function of the system. The transfer function between $X_1(s)$ and $U(s)$ is

$$
\frac{X_1(s)}{U(s)} + \frac{1}{(s+1)(s+2)(s+3)}
$$

and between $Y(s)$ and $X_1(s)$ is

$$
\frac{Y(s)}{X_1(s)} = (s+1)(s+4)
$$

The transfer function between $Y(s)$ and $U(s)$ is

$$
\frac{Y(s)}{U(s)} = \frac{(s+1)(s+4)}{(s+1)(s+2)(s+3)}
$$

The factors $(s+1)$ cancel each other, thus there are nonzero initial states $x(0) \neq 0$ which cannot be determined from the measurement of $y(t)$.

● **PROBLEM 12-6**

The following vector-matrix state equation represents the digital model of the dynamics of a space vehicle

$$
x(k+1) = Ax(k) + Bu(k)
$$

where A is an 11 × 11 matrix and B is an 11 × 1 matrix, (see following page).

Examine the controllability of this system.

Solution: The above system has one input, to examine its controllability, we must evaluate the determinant of the 11 × 11 matrix

$$
S = [B \ AB \ . \ . \ . \ A^{10}B]^1
$$

We find that

$$
|S| = 4.46 \times 10^{18}
$$

thus the system is completely state controllable.

$$
A = \begin{bmatrix}
0 & 0 & 0 & 0 & 0 & 1 & 0 & 0 & 0 & 0 \\
0 & 0 & 0 & 0 & 0 & 0 & 1 & 0 & 0 & 0 \\
0 & 0 & 0 & 0 & 0 & 0 & 0 & 1 & 0 & 0 \\
0 & 0 & 0 & 0 & 0 & 0 & 0 & 0 & 1 & 0 \\
0 & 0 & 0 & 0 & 0 & 0 & 0 & 0 & 0 & 1 \\
0.176 & 0 & 0 & -85.26 & -2.56 & 0 & -6.1 & -0.188 & 0 & -0.077 \\
0 & -134.6 & -1.06 & 0 & 0 & -24.0 & 0 & 0 & 0.02 & 0 \\
0 & -0.69 & 10.17 & 0 & 0 & 0.69 & 0 & 0 & 0.304 & 0 \\
0 & 19.9 & -293.0 & 0 & 0 & -19.9 & 0 & 0 & -8.76 & 0 \\
-0.366 & 0 & 0 & -52.86 & -0.168 & -3.66 & 2.36 & -11.67 & -7.0 & -0.005 \\
6.02 & 0 & 0 & -85.4 & -18.84 & 0 & -6.1 & -0.188 & 0 & -0.388
\end{bmatrix}
$$

$$
B = \begin{bmatrix} 0 & 0 & 0 & -7.28 & 0 & 0 & 0 & 0 & -0.478 & -7.28 \end{bmatrix}
$$

In the case of the systems of greater complexity, the numerical value of the controllability matrix S is sometimes not sufficient. Another way of testing the controllability of such systems is to transform A into a diagonal matrix by a similarity transformation. First we find the eigenvalues of A. We have

$$\lambda_{1,2} = -0.603 \pm j\ 30.15$$

$$\lambda_{3,4} = -0.0328 \pm j\ 1.76$$

$$\lambda_{5,6} = -0.0276 \pm j\ 1.36$$

$$\lambda_{7,8} = \pm j$$

$$\lambda_{9,11} = -0.00056 \pm j\ 0.29$$

$$\lambda_{12} = 0$$

The Γ matrix of the decoupled system is given by

$$
\Gamma =
\begin{bmatrix}
-8.16 \times 10^{-2} \\
1.67 \times 10^{-3} \\
-9.32 \times 10^{-1} \\
2.99 \times 10^{-2} \\
-1.32 \times 10^{-3} \\
9.00 \times 10^{-2} \\
-1.76 \times 10^{-1} \\
1.54 \times 10^{-10} \\
1.97 \times 10^{-3} \\
-3.19 \\
-2.79 \times 10^{-16}
\end{bmatrix}
$$

Since the matrix S has a rank of eleven and all the elements of Γ are different from zero, the system should be completely state controllable. But the last element of Γ is -2.79×10^{-16}; that is nearly equal to zero.

For practical purposes, we can assume that the system is uncontrollable, since the magnitudes of elements of Γ indicate the controllability of the states.

Examine the controllability of the following system

$$\begin{bmatrix} \dot{x}_1 \\ \dot{x}_2 \end{bmatrix} = \begin{bmatrix} -3 & 1 \\ -2 & 1.5 \end{bmatrix} \begin{bmatrix} x_1 \\ x_2 \end{bmatrix} + \begin{bmatrix} 0 \\ 1 \end{bmatrix} [u]$$

Solution: To find matrix S, let us compute

$$AB = \begin{bmatrix} -3 & 1 \\ -2 & 1.5 \end{bmatrix} \begin{bmatrix} 0 \\ 1 \end{bmatrix} = \begin{bmatrix} 1 \\ 1.5 \end{bmatrix}$$

Thus

$$S = \begin{bmatrix} 0 & 1 \\ 1 & 1.5 \end{bmatrix}$$

The rank of the matrix S is two and the system is completely stable controllable.

Determine the controllability of the following system

$$\begin{bmatrix} \dot{x}_1 \\ \dot{x}_2 \end{bmatrix} = \begin{bmatrix} 1 & 1 \\ 0 & -1 \end{bmatrix} \begin{bmatrix} x_1 \\ x_2 \end{bmatrix} + \begin{bmatrix} 1 \\ 0 \end{bmatrix} [u]$$

Solution: Since

$$AB = \begin{bmatrix} 1 & 1 \\ 0 & -1 \end{bmatrix} \begin{bmatrix} 1 \\ 0 \end{bmatrix} = \begin{bmatrix} 1 \\ 0 \end{bmatrix}$$

Matrix S is given by

$$S = [B \quad AB] = \begin{bmatrix} 1 & 1 \\ 0 & 0 \end{bmatrix} = \text{singular}$$

Thus the system is not completely state controllable.

The system is described by

$$\dot{x} = Ax + Bu$$

where

$$A = \begin{bmatrix} 1 & 1 \\ 2 & -1 \end{bmatrix} \quad \text{and B} \quad = \begin{bmatrix} 0 \\ 1 \end{bmatrix}$$

Determine the controllability of the system.

Solution: For the above system to be completely state controllable, it is necessary and sufficient that matrix S has a rank of 2 where $S = [B \quad AB]$.

We have

$$[B \quad AB] = \begin{bmatrix} 0 & 1 \\ 0 & -1 \end{bmatrix} = \text{nonsingular}$$

The system is thus completely state controllable.

The system is given in the Jordan canonical form, after the transformation

$$\begin{bmatrix} \dot{q}_1 \\ \dot{q}_2 \\ \dot{q}_3 \end{bmatrix} = \begin{bmatrix} -5 & 0 & 0 \\ 0 & 1 & 0 \\ 0 & 0 & -3 \end{bmatrix} \begin{bmatrix} q_1 \\ q_2 \\ q_3 \end{bmatrix} + \begin{bmatrix} 1 & 0 \\ 0 & 1 \\ 0 & 0 \end{bmatrix} \begin{bmatrix} u_1 \\ u_2 \end{bmatrix}$$

$$\begin{bmatrix} y_1 \\ y_2 \end{bmatrix} = \begin{bmatrix} 1 & 0 & 0 \\ 0 & 0 & 1 \end{bmatrix} \begin{bmatrix} q_1 \\ q_2 \\ q_3 \end{bmatrix} + \begin{bmatrix} 0 & 0 \\ 0 & 1 \end{bmatrix} \begin{bmatrix} u_1 \\ u_2 \end{bmatrix}$$

Reduce the system based on controllability and observability. Show that the transfer matrix of the reduced system is the same as that of the original.

Solution: The Jordan form system has distinct eigenvalues, thus controllability and observability can be easily determined. The third row of B_n is zero, thus mode q_3 is uncontrollable. The second column of C_n is zero, thus q_2 is an unobservable mode. Modes q_2 and q_3 can be dropped since they have no effect on the input-output behavior. An irreducible realization is

$$\dot{q}_1 = -5q_1 + [1 \quad 0]\, u$$

$$y = \begin{bmatrix} 1 \\ 0 \end{bmatrix} q_1 + \begin{bmatrix} 0 & 0 \\ 0 & 1 \end{bmatrix} u$$

$H = C(sI - A)^{-1}B + D$, so for the original system

$$H(s) = \begin{bmatrix} 1 & 0 & 0 \\ 0 & 0 & 1 \end{bmatrix} \begin{bmatrix} s+5 & 0 & 0 \\ 0 & s-1 & 0 \\ 0 & 0 & s+3 \end{bmatrix}^{-1} \begin{bmatrix} 1 & 0 \\ 0 & 1 \\ 0 & 0 \end{bmatrix} + \begin{bmatrix} 0 & 0 \\ 0 & 1 \end{bmatrix}$$

$$= \begin{bmatrix} 1 & 0 & 0 \\ 0 & 0 & 1 \end{bmatrix} \begin{bmatrix} \frac{1}{s+5} & 0 & 0 \\ 0 & \frac{1}{s-1} & 0 \\ 0 & 0 & \frac{1}{s+3} \end{bmatrix} \begin{bmatrix} 1 & 0 \\ 0 & 1 \\ 0 & 0 \end{bmatrix} + \begin{bmatrix} 0 & 0 \\ 0 & 1 \end{bmatrix}$$

$$= \begin{bmatrix} \frac{1}{s+5} & 0 & 0 \\ 0 & 0 & \frac{1}{s+3} \end{bmatrix} \begin{bmatrix} 1 & 0 \\ 0 & 1 \\ 0 & 0 \end{bmatrix} + \begin{bmatrix} 0 & 0 \\ 0 & 1 \end{bmatrix} = \begin{bmatrix} \frac{1}{s+5} & 0 \\ 0 & 1 \end{bmatrix}$$

For the reduced system we have

$$H(s) = \begin{bmatrix} 1 \\ 0 \end{bmatrix} [s+5]^{-1} [1 \quad 0] + \begin{bmatrix} 0 & 0 \\ 0 & 1 \end{bmatrix}$$

$$= \begin{bmatrix} 1 \\ 0 \end{bmatrix} \begin{bmatrix} \frac{1}{s+5} \end{bmatrix} [1 \quad 0] + \begin{bmatrix} 0 & 0 \\ 0 & 1 \end{bmatrix} = \begin{bmatrix} \frac{1}{s+5} & 0 \\ 0 & 1 \end{bmatrix}$$

Thus the transfer matrix is the same for both forms of the state equations.

● PROBLEM 12-11

The matrices A and B are

$$A = \begin{bmatrix} 0 & 1 & -1 \\ 1 & 0 & 1 \\ 0 & 1 & -1 \end{bmatrix}, \quad B = \begin{bmatrix} 1 & 0 \\ 0 & 1 \\ 0 & 0 \end{bmatrix}$$

Determine if [A, B] is a controllable pair.

<u>Solution</u>: Since A is 3 × 3, B is 3 × 2 matrix S has to be 3 × 6.

$$S = [B \quad AB \quad A^2B]$$

We find

$$AB = \begin{bmatrix} 0 & 1 \\ 1 & 0 \\ 0 & 1 \end{bmatrix} \quad \text{and} \quad A^2B = \begin{bmatrix} 1 & -1 \\ 0 & 2 \\ 1 & -1 \end{bmatrix}$$

S can be written

$$S = \begin{bmatrix} 1 & 0 & 0 & 1 & 1 & -1 \\ 0 & 1 & 1 & 0 & 0 & 2 \\ 0 & 0 & 0 & 1 & 1 & -1 \end{bmatrix}.$$

It can be easily checked that rank S = 3 and the system is controllable.

● **PROBLEM** 12-12

Examine the controllability of the system

$$\frac{X(s)}{V(s)} = \frac{K(s + a)\ (s + b)}{(s + a)\ (s + b)\ (s + c)\ (s + d)}$$

<u>Solution</u>: The system is not completely state controllable becuase the transfer function has cancellations. The state space representation of the above system can be written as follows.

$$\begin{bmatrix} \dot{z}_1 \\ \dot{z}_2 \\ \dot{z}_3 \\ \dot{z}_4 \end{bmatrix} = \begin{bmatrix} -a & 0 & 0 & 0 \\ 0 & -b & 0 & 0 \\ 0 & 0 & -c & 0 \\ 0 & 0 & 0 & -d \end{bmatrix} \begin{bmatrix} z_1 \\ z_2 \\ z_3 \\ z_4 \end{bmatrix} + \begin{bmatrix} 0 \\ 0 \\ 1 \\ 1 \end{bmatrix} [u]$$

It can be easily seen that the motions of z_1 and z_2 do not depend on u, therefore it can not be controlled.

The system is described by

$$x(k + 1) = Ax(k) + Bu(k)$$

$$c(k) = Dx(k)$$

where

$$A = \begin{bmatrix} 0 & 1 \\ -2 & -3 \end{bmatrix} \quad \text{and} \quad B = \begin{bmatrix} 1 \\ 1 \end{bmatrix} \quad D = \begin{bmatrix} 1 & 2 \end{bmatrix}$$

Determine the controllability and observability of the open-loop system, that is, u(k) is not a function of x(k) and of the closed-loop system where

$$u(k) = r(k) - Gx(k)$$

$$G = [g_1 \quad g_2].$$

Solution: We have, for the open-loop system

$$B = \begin{bmatrix} 1 \\ 1 \end{bmatrix}, \quad AB = \begin{bmatrix} 1 \\ -5 \end{bmatrix}$$

$$D' = \begin{bmatrix} 1 \\ 2 \end{bmatrix}, \quad A'D' = \begin{bmatrix} -4 \\ -5 \end{bmatrix}$$

Thus $|B \quad AB| \neq 0$ and the system is controllable and $|D' \quad A'D'| \neq 0$, so the system is observable.

For the closed-loop system described by

$$x(k + 1) = Ax(k) + B[r(k) = Gx(k)].$$

or

$$x(k + 1) = (A - BG) x(k) + Br(k)$$

$$A - BG = \begin{bmatrix} -g_1 & 1-g_2 \\ -2-g_1 & -3-g_2 \end{bmatrix}$$

The observability matrix is given by

$$V = [D' \quad (A - BG)' \quad D'] = \begin{bmatrix} 1 & -3g_1 & -4 \\ 2 & -3g_2 & -5 \end{bmatrix}$$

and

det $V = 6g_1 - 3g_2 + 3$.

We see that the closed-loop system is unobservable if g_1, g_2 are such that $|V| = 0$.

$$(A - BG)B = \begin{bmatrix} 1 - g_1 - g_2 \\ -5 - g_1 - g_2 \end{bmatrix}$$

The controllability matrix is

$$H = [B \quad (A - BG)B] = \begin{bmatrix} 1 & 1 - g_1 - g_2 \\ 1 & -5 - g_1 - g_2 \end{bmatrix}$$

and det $H = -6$, thus the closed-loop system is controllable for all g_1 and g_2.

● **PROBLEM** 12-14

Compute an irreducible realization of

$$H(s) = \begin{bmatrix} \dfrac{1}{(s + 2)^3(s + 5)} & \dfrac{1}{s + 5} \\ \dfrac{1}{s + 2} & 0 \end{bmatrix}$$

Solution: We shall use the partial fraction expansion.

$$H(s) = \frac{\begin{bmatrix} \frac{1}{3} & 0 \\ 0 & 0 \end{bmatrix}}{(s + 2)^3} + \frac{\begin{bmatrix} -\frac{1}{9} & 0 \\ 0 & 0 \end{bmatrix}}{(s + 2)^2} + \frac{\begin{bmatrix} \frac{1}{27} & 0 \\ 1 & 0 \end{bmatrix}}{s + 2} + \frac{\begin{bmatrix} -\frac{1}{27} & 1 \\ 0 & 0 \end{bmatrix}}{s + 5}$$

$$= \frac{\begin{bmatrix} \frac{1}{3} \\ 0 \end{bmatrix}[1 \quad 0]}{(s + 2)^3} + \frac{\begin{bmatrix} -\frac{1}{9} \\ 0 \end{bmatrix}[1 \quad 0]}{(s + 2)^2} + \frac{\begin{bmatrix} \frac{1}{27} \\ 1 \end{bmatrix}[1 \quad 0]}{s + 2} + \frac{\begin{bmatrix} 1 \\ 0 \end{bmatrix}\begin{bmatrix} -\frac{1}{27} & 1 \end{bmatrix}}{s + 5}$$

The vector $b_1^T = [1 \quad 0]$ is associated with the three terms involving $s + 2$. There is only one such vector. The three terms can be simulated from a series connection of $\dfrac{1}{s + 2}$ terms with a single input $b_1^T u$. These terms form a single Jordan block.

The state equations can be written from the simulation
diagram shown below.

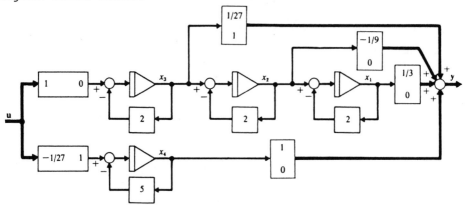

We have:

$$x_1(s) = \frac{1}{(s + 2)^3} [1 \quad 0]u(s)$$

$$x_2(s) = \frac{1}{(s + 2;^2} [1 \quad 0]u(s)$$

$$x_3(s) = \frac{1}{(s + s)} [1 \quad 0]u(s)$$

From the diagram we have

$$
\begin{bmatrix} \dot{x}_1 \\ \dot{x}_2 \\ \dot{x}_3 \\ \dot{x}_4 \end{bmatrix}
=
\begin{bmatrix} -2 & 1 & 0 & 0 \\ 0 & -2 & 1 & 0 \\ 0 & 0 & -2 & 0 \\ 0 & 0 & 0 & -5 \end{bmatrix}
\begin{bmatrix} x_1 \\ x_2 \\ x_3 \\ x_4 \end{bmatrix}
+
\begin{bmatrix} 0 & 0 \\ 0 & 0 \\ 1 & 0 \\ -\frac{1}{27} & 1 \end{bmatrix} u
$$

and

$$
y = \begin{bmatrix} \frac{1}{3} \\ 0 \end{bmatrix} x_1 + \begin{bmatrix} -\frac{1}{9} \\ 0 \end{bmatrix} x_2 + \begin{bmatrix} \frac{1}{27} \\ 1 \end{bmatrix} x_3 + \begin{bmatrix} 1 \\ 0 \end{bmatrix} x_4
$$

$$
= \begin{bmatrix} \frac{1}{3} & -\frac{1}{9} & \frac{1}{27} & 1 \\ 0 & 0 & 1 & 0 \end{bmatrix} x
$$

This realization is completely observable and completely
controllable. It is thus irreducible.

Consider the system described by

$$\frac{d^2 c(t)}{dt^2} + 2 \frac{dc(t)}{dt} + c(t) = \frac{du(t)}{dt} + u(t)$$

Assign the state variables two ways and show that observability depends on state assignment.

Solution:

Assigning

$$x_1 = c$$

$$x_2 = \dot{c} - u$$

leads to the state equation

$$\begin{bmatrix} \dot{x}_1 \\ \dot{x}_2 \end{bmatrix} = \begin{bmatrix} 0 & 1 \\ -1 & -2 \end{bmatrix} \begin{bmatrix} x_1 \\ x_2 \end{bmatrix} + \begin{bmatrix} 1 \\ -1 \end{bmatrix} u$$

Thus

$$A = \begin{bmatrix} 0 & 1 \\ -1 & -2 \end{bmatrix} \quad \text{and} \quad D = [1 \quad 0]$$

the output is $C = Dx$

Thus

$$[D' \quad A'D'] = \begin{bmatrix} 1 & 0 \\ 0 & 1 \end{bmatrix}$$

and the system is completely observable.

By direct decomposition we get the state equation

$$\begin{bmatrix} \dot{x}_1 \\ \dot{x}_2 \end{bmatrix} = \begin{bmatrix} 0 & 1 \\ -1 & -2 \end{bmatrix} \begin{bmatrix} x_1 \\ x_2 \end{bmatrix} + \begin{bmatrix} 0 \\ 1 \end{bmatrix} u$$

and the output is

$$c = x_1 + x_2 \text{ , so}$$

$$A = \begin{bmatrix} 0 & 1 \\ -1 & -2 \end{bmatrix} \quad\quad D = [1 \quad 1]$$

Then

$$[D' \quad A' D'] = \begin{bmatrix} 1 & -1 \\ 1 & -1 \end{bmatrix}$$

This matrix is singular and the system is observable.

We conclude thus that the ovservability of the system depends on how the state variables are defined.

● **PROBLEM** 12-16

Determine if the system is controllable and observable.

$$\begin{bmatrix} \dot{x}_1 \\ \dot{x}_2 \end{bmatrix} = \begin{bmatrix} 1 & 1 \\ -2 & -1 \end{bmatrix} \begin{bmatrix} x_1 \\ x_2 \end{bmatrix} + \begin{bmatrix} 0 \\ 1 \end{bmatrix} [u]$$

$$y = [1 \quad 0] \begin{bmatrix} x_1 \\ x_2 \end{bmatrix}$$

Solution: The controllability matrix is

$$[B \quad AB] = \begin{bmatrix} 0 & 1 \\ 1 & -1 \end{bmatrix}$$

and its rank is two, thus the system is completely state controllable.

To determine the output controllability we shall find the rank of the matrix

$$[CB \quad CAB] = [0 \quad 1]$$

The rank of this matrix is one, therefore the system is completely output controllable. To determine the observability, we compute

$$[C' \quad A' C'] = \begin{bmatrix} 1 & 1 \\ 0 & 1 \end{bmatrix}$$

The rank of this matrix is two, thus the system is completely observable.

● **PROBLEM** 12-17

Determine the observability of the following system

$$\dot{x} = Ax + Bu$$

$$y = Cx$$

where

$$A = \begin{bmatrix} 0 & 1 & 0 \\ 0 & 0 & 1 \\ -6 & -11 & -6 \end{bmatrix},$$

$$B = \begin{bmatrix} 0 \\ 0 \\ 1 \end{bmatrix}, \qquad C = \begin{bmatrix} 4 & 5 & 1 \end{bmatrix}$$

<u>Solution</u>: Let us compute the matrix S given by

$$S = [C', \quad A'C', \quad (A')^2 C']$$

We obtain

$$C' = \begin{bmatrix} 4 \\ 5 \\ 1 \end{bmatrix}$$

$$A'C' = \begin{bmatrix} -6 \\ -7 \\ -1 \end{bmatrix}$$

$$(A')^2 C' = \begin{bmatrix} 6 \\ 5 \\ -1 \end{bmatrix}$$

$$S = \begin{bmatrix} 4 & -6 & 6 \\ 5 & -7 & 5 \\ 1 & -1 & -1 \end{bmatrix}$$

The rank of the matrix S is two, thus the system is not completely observable. From the block diagram of the system

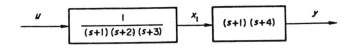

604

we see that the two factors (s + 1) cancel each other.

Thus all the measurements of y(t) do not give estimation of all the initial values of x(t).

Determine the observability of the following system

$$\dot{x} = Ax + Bu$$

$$y = Cx$$

where

$$A = \begin{bmatrix} 0 & 1 & 0 \\ 0 & 0 & 1 \\ -6 & -11 & -6 \end{bmatrix}, \quad B = \begin{bmatrix} 0 \\ 0 \\ 1 \end{bmatrix},$$

$$C = \begin{bmatrix} 20 & 9 & 1 \end{bmatrix}$$

Solution: Let us compute the matrices

$$C', \qquad A'C', \qquad (A')^2 C'$$

$$C' = \begin{bmatrix} 20 \\ 9 \\ 1 \end{bmatrix}$$

$$A'C' = \begin{bmatrix} 0 & 0 & -6 \\ 1 & 0 & -11 \\ 0 & 1 & -6 \end{bmatrix} \begin{bmatrix} 20 \\ 9 \\ 1 \end{bmatrix} = \begin{bmatrix} -6 \\ 9 \\ 3 \end{bmatrix}$$

$$(A')^2 C' = \begin{bmatrix} 0 & -6 & 36 \\ 0 & -11 & 60 \\ 1 & -6 & 25 \end{bmatrix} \begin{bmatrix} 20 \\ 9 \\ 1 \end{bmatrix} = \begin{bmatrix} -18 \\ -39 \\ -9 \end{bmatrix}$$

Since

$$\begin{vmatrix} C' & A'C' & (A')^2 C' \end{vmatrix} = \begin{vmatrix} 20 & -6 & -18 \\ 9 & 9 & -39 \\ 1 & 3 & -9 \end{vmatrix} \neq 0$$

the rank of the matrix is 3 and the system is completely observable.

605

Determine if the linear system described below is observable.

$$\begin{bmatrix} \dot{x}_1 \\ \dot{x}_2 \end{bmatrix} = \begin{bmatrix} 1 & -1 \\ 1 & 1 \end{bmatrix} \begin{bmatrix} x_1 \\ x_2 \end{bmatrix} + \begin{bmatrix} 2 & -1 \\ 1 & 0 \end{bmatrix} \begin{bmatrix} u_1 \\ u_2 \end{bmatrix}$$

$$\begin{bmatrix} c_1 \\ c_2 \end{bmatrix} = \begin{bmatrix} 1 & 0 \\ -1 & 1 \end{bmatrix} \begin{bmatrix} x_1 \\ x_2 \end{bmatrix}$$

Solution: The observability matrix is given by

$$[D' \quad A'D'] = \begin{bmatrix} 1 & -1 & 1 & 0 \\ 0 & 1 & -1 & 2 \end{bmatrix}$$

The system is completely observable since the observability matrix has a rank of 2 and the system has two inputs.

The system is given by

$$\begin{bmatrix} \dot{x}_1 \\ \dot{x}_2 \\ \dot{x}_3 \end{bmatrix} = \begin{bmatrix} 2 & 0 & 0 \\ 0 & 2 & 0 \\ 0 & 3 & 1 \end{bmatrix} \begin{bmatrix} x_1 \\ x_2 \\ x_3 \end{bmatrix}$$

The output is $y = \begin{bmatrix} 1 & 1 & 1 \end{bmatrix} \begin{bmatrix} x_1 \\ x_2 \\ x_3 \end{bmatrix}$.

a) Show that the system is not completely observable.

b) Show that the system is completely observable if the output is given by

$$\begin{bmatrix} y_1 \\ y_2 \end{bmatrix} = \begin{bmatrix} 1 & 1 & 1 \\ 1 & 2 & 3 \end{bmatrix} \begin{bmatrix} x_1 \\ x_2 \\ x_3 \end{bmatrix}$$

<u>Solution</u>: Let us find the observability matrix

$$A = \begin{bmatrix} 2 & 0 & 0 \\ 0 & 2 & 0 \\ 0 & 3 & 1 \end{bmatrix} , \qquad C = [1 \quad 1 \quad 1]$$

$$[C' \quad A'C' \quad (A')^2 C'] = \begin{bmatrix} 1 & 2 & 4 \\ 1 & 5 & 13 \\ 1 & 1 & 1 \end{bmatrix}$$

The rank of this matrix is two, thus the system is not completely observable.

b)   For $C = \begin{bmatrix} 1 & 1 & 1 \\ 1 & 2 & 3 \end{bmatrix}$, the observability matrix is

$$[C' \quad A'C' \quad (A')^2 C'] = \begin{bmatrix} 1 & 1 & 2 & 2 & 4 & 4 \\ 1 & 2 & 5 & 13 & 13 & 35 \\ 1 & 3 & 1 & 3 & 1 & 3 \end{bmatrix}$$

Since $\begin{bmatrix} 1 & 1 & 2 \\ 1 & 2 & 5 \\ 1 & 3 & 1 \end{bmatrix} \neq 0$

the rank of the matrix is 3 and the system is completely observable.

● **PROBLEM** 12-21

Show that the following system is completely state controllable and completely observable.

$$\dot{x} = \begin{bmatrix} -1 & -2 & -2 \\ 0 & -1 & 1 \\ 1 & 0 & -1 \end{bmatrix} x + \begin{bmatrix} 2 \\ 0 \\ 1 \end{bmatrix} [u]$$

$$y = [1 \quad 1 \quad 0] \begin{bmatrix} x_1 \\ x_2 \\ x_3 \end{bmatrix}$$

<u>Solution</u>: Let us find the vectors B, AB and $A^2 B$.

607

$$B = \begin{bmatrix} 2 \\ 0 \\ 1 \end{bmatrix}, \qquad AB = \begin{bmatrix} -4 \\ 1 \\ 1 \end{bmatrix},$$

$$A^2B = \begin{bmatrix} 0 \\ 0 \\ -5 \end{bmatrix}$$

Since

$$\det \begin{bmatrix} 2 & -4 & 0 \\ 0 & 1 & 0 \\ 1 & 1 & -5 \end{bmatrix} = -10 \neq 0$$

the vectors are linearly independent thus the system is completely state controllable.

The system is completely observable when the vectors C*, A*C* (A*)^{2}C* are linearly independent.

$$C^* = \begin{bmatrix} 1 \\ 1 \\ 0 \end{bmatrix}, \qquad A^*C^* = \begin{bmatrix} -1 \\ -3 \\ -1 \end{bmatrix},$$

$$(A^*)^2C^* = \begin{bmatrix} 0 \\ 5 \\ 0 \end{bmatrix}$$

and

$$\det \begin{bmatrix} 1 & -1 & 0 \\ 1 & -3 & 5 \\ 1 & -1 & 0 \end{bmatrix} = 5 \neq 0$$

Thus the vectors are linearly independent and the system is completely observable.

● **PROBLEM 12-22**

The continuous-time system is described by

$$\dot{x} = Ax + Bu$$

$$y = \begin{bmatrix} 1 & 0 \end{bmatrix} \begin{bmatrix} x_1 \\ x_2 \end{bmatrix}$$

where

$$A = \begin{bmatrix} 0 & 1 \\ -1 & 0 \end{bmatrix} \quad \text{and} \quad B = \begin{bmatrix} 0 \\ 1 \end{bmatrix}$$

Show that the system is completely state controllable and completely observable.

Solution: We shall find the rank of the matrix S

$$S = [B \quad AB] = \begin{bmatrix} 0 & 1 \\ 1 & 0 \end{bmatrix}.$$

The rank of [B   AB] is two, thus the system is completely state controllable.

To find if the system is completely observable we shall find the rank of

$$[C' \quad A' C'] = \begin{bmatrix} 1 & 0 \\ 0 & 1 \end{bmatrix}$$

The rank of the above matrix is two and the system is completely observable.

● **PROBLEM** 12-23

The third-order system has the coefficient matrices

$$A = \begin{bmatrix} 1 & 2 & -1 \\ 0 & 1 & 0 \\ 1 & -4 & 3 \end{bmatrix} \quad \text{and} \quad B = \begin{bmatrix} 0 \\ 0 \\ 1 \end{bmatrix}$$

Show that the system is not controllable

1) by forming the matrix $S = [B \quad AB \quad A^2 B]$,

2) by transforming the state equations into Jordan canonical form.

Solution: We have

$$AB = \begin{bmatrix} 1 & 2 & -1 \\ 0 & 1 & 0 \\ 1 & -4 & 3 \end{bmatrix} \begin{bmatrix} 0 \\ 0 \\ 1 \end{bmatrix} = \begin{bmatrix} -1 \\ 0 \\ 3 \end{bmatrix}$$

$$A^2 B = \begin{bmatrix} 1 & 2 & -1 \\ 0 & 1 & 0 \\ 1 & -4 & 3 \end{bmatrix} \begin{bmatrix} -1 \\ 0 \\ 3 \end{bmatrix} = \begin{bmatrix} -4 \\ 0 \\ 8 \end{bmatrix}$$

Then

$$
S \quad = \quad [B \quad AB \quad A^2B] \quad = \quad \begin{bmatrix} 0 & -1 & -4 \\ 0 & 0 & 0 \\ 1 & 3 & 8 \end{bmatrix}
$$

Matrix S is singular, therefore the system is not state controllable.

2)   We shall find the eigenvalues of A.

$$
|\lambda I - A| \quad = \quad \begin{bmatrix} \lambda-1 & -2 & 1 \\ 0 & \lambda-1 & 0 \\ -1 & 4 & \lambda-3 \end{bmatrix}
$$

$$
= (\lambda - 1) [(\lambda - 1) (\lambda - 3) + 1] \quad = \quad (\lambda - 1) (\lambda - 2)^2
$$

The eigenvalues are $\lambda_1 = 2$, $\lambda_2 = 2$, $\lambda_3 = 1$.

The corresponding eigenvectors for λ_1 are

$$
[\lambda_1 I - A] \ p_1 = 0
$$

$$
\begin{bmatrix} 1 & -2 & 1 \\ 0 & 1 & 0 \\ -1 & 4 & -1 \end{bmatrix} \begin{bmatrix} p_{11} \\ p_{21} \\ p_{31} \end{bmatrix} = \quad 0
$$

Let us choose $p_{11} = 1$, then $p_{31} = -1$ and $p_{21} = 0$.
For λ_2 we have

$$
(\lambda_2 I - A) \ p_2 = p_1 \qquad \text{since } \lambda_2 = \lambda_1
$$

or

$$
\begin{bmatrix} 1 & -2 & 1 \\ 0 & 1 & 0 \\ -1 & 4 & -1 \end{bmatrix} \begin{bmatrix} p_{12} \\ p_{22} \\ p_{32} \end{bmatrix} = \begin{bmatrix} 1 \\ 0 \\ -1 \end{bmatrix}
$$

and $p_{22} = 0$. Choosing $p_{32} = 1$ we get $p_{12} = 0$.

For λ_3 we have

$$
(\lambda_3 I - A) \ p_3 = 0
$$

or

$$\begin{bmatrix} 0 & -2 & 1 \\ 0 & 0 & 0 \\ -1 & 4 & -2 \end{bmatrix} \begin{bmatrix} p_{13} \\ p_{23} \\ p_{33} \end{bmatrix} = 0$$

Solving the above equation we get $p_{13} = 0$

Choosing $p_{23} = 1$ gives $p_{33} = 2$.

Thus

$$P = [p_1 \quad p_2 \quad p_3] = \begin{bmatrix} 1 & 0 & 0 \\ 0 & 0 & 1 \\ -1 & 1 & 2 \end{bmatrix}$$

Then

$$\Lambda = P^{-1}AP = \begin{bmatrix} 2 & 1 & 0 \\ 0 & 2 & 0 \\ 0 & 0 & 1 \end{bmatrix}$$

and

$$\Gamma = P^{-1}B = \begin{bmatrix} 0 \\ -1 \\ 0 \end{bmatrix}$$

Since the last row of Γ is zero, the state variable y_3 is uncontrollable. x_2 is uncontrollable since $x_2 = y_3$.

● **PROBLEM** 12-24

The system is described by

$$\begin{bmatrix} \dot{x}_1(t) \\ \dot{x}_2(t) \end{bmatrix} = \begin{bmatrix} 0 & 1 \\ -1 & 0 \end{bmatrix} \begin{bmatrix} x_1(t) \\ x_2(t) \end{bmatrix} + \begin{bmatrix} 0 \\ 1 \end{bmatrix} u(t)$$

Determine the state controllability.

<u>Solution</u>: We have

$$A = \begin{bmatrix} 0 & 1 \\ -1 & 0 \end{bmatrix}, \quad B = \begin{bmatrix} 0 \\ 1 \end{bmatrix}$$

We shall apply the following criterion for state controllability:

The matrix S is given by

$$S = [B \quad AB \quad A^2B \ldots A^{N-1}B]$$

where N is the number of states.

The system is completely state controllable if S has rank N. In our case, N = 2. Then

$$AB = \begin{bmatrix} 1 \\ 0 \end{bmatrix}$$

and

$$s = [B \quad AB] = \begin{bmatrix} 0 & 1 \\ 1 & 0 \end{bmatrix}$$

The rank of S is 2, thus the system is completely state controllable.

● **PROBLEM** 12-25

Determine if the following time-variable system is completely controllable.

$$\begin{bmatrix} \dot{x}_1 \\ \dot{x}_2 \end{bmatrix} = \frac{1}{12} \begin{bmatrix} 5 & 1 \\ 1 & 5 \end{bmatrix} \begin{bmatrix} x_1 \\ x_2 \end{bmatrix} + e^{\frac{t}{2}} \begin{bmatrix} 1 \\ 1 \end{bmatrix} u(t)$$

<u>Solution</u>: Let us write the controllability matrix

$$G(t_1, t_o) = \Phi(t_1, 0) \int_{t_o}^{t_1} \Phi^{-1}(\tau, 0)B(\tau)B^T(\tau)[\Phi^{-1}(\tau, 0)^T d\tau \Phi^T(t, 0)$$

For the system to be controllable $G(t_1, t_o)$ must be nonsingular. The eigenvalues of

$$A = \begin{bmatrix} \frac{5}{12} & \frac{1}{12} \\ \frac{1}{12} & \frac{5}{12} \end{bmatrix} \text{ are } \lambda_1 = \frac{1}{2}, \ \lambda_2 = \frac{1}{3} .$$

The transition matrix is $\Phi(t, 0) = e^{At}$.

Using the Cayley-Hamilton theorem we get

$$e^{At} = \alpha_o I + \alpha_1 A = \begin{bmatrix} \alpha_o + \frac{5}{12}\alpha_1 & \frac{1}{12}\alpha_1 \\ \frac{1}{12}\alpha_1 & \alpha_o + \frac{5}{12}\alpha_1 \end{bmatrix}$$

$$e^{\frac{1}{2}t} = \alpha_o + \frac{1}{2}\alpha_1$$

$$e^{\frac{1}{3}t} = \alpha_o + \frac{1}{3}\alpha_1$$

We have

$$\alpha_1 = 6(e^{\frac{1}{2}t} - e^{\frac{1}{3}t})$$

$$\alpha_o = -2e^{\frac{1}{2}t} + 3e^{\frac{1}{3}t}$$

Thus

$$\Phi(t, 0) = \begin{bmatrix} \frac{1}{2}e^{\frac{1}{2}t} + \frac{1}{2}e^{\frac{1}{3}t} & \frac{1}{2}e^{\frac{1}{2}t} - \frac{1}{2}e^{\frac{1}{3}t} \\ \\ \frac{1}{2}e^{\frac{1}{2}t} - \frac{1}{2}e^{\frac{1}{3}t} & \frac{1}{2}e^{\frac{1}{2}t} + \frac{1}{2}e^{\frac{1}{3}t} \end{bmatrix}$$

Inverting, we obtain

$$\Phi^{-1}(\tau, 0) = \Phi(-\tau, 0) = \frac{1}{2}\begin{bmatrix} e^{-\frac{\tau}{2}} + e^{-\frac{\tau}{3}} & e^{-\frac{\tau}{2}} - e^{-\frac{\tau}{3}} \\ \\ e^{-\frac{\tau}{2}} - e^{-\frac{\tau}{3}} & e^{-\frac{\tau}{2}} + e^{-\frac{\tau}{3}} \end{bmatrix}$$

Thus

$$\Phi^{-1}(\tau, 0)B(\tau) = \begin{bmatrix} 1 \\ 1 \end{bmatrix}$$

Thus

$$G(t_1, t_o) = \left| \Phi(t_1, 0) \right| \left| \begin{bmatrix} t_1 - t_o & t_1 - t_o \\ t_1 - t_o & t_1 - t_o \end{bmatrix} \right| \left| \Phi^T(t_1, 0) \right| = 0$$

Since the above equation is true for all t_o, t_1 the system is not completely controllable.

The system is given by

$$\dot{x} = Ax + Bu$$

where

$$A = \begin{bmatrix} -3 & 1 \\ -2 & 1.5 \end{bmatrix} \quad \text{and} \quad B = \begin{bmatrix} 1 \\ 4 \end{bmatrix}$$

Examine state controllability of the system.

Solution: To find matrix S let us note that

$$AB = \begin{bmatrix} -3 & 1 \\ -2 & 1.5 \end{bmatrix} \begin{bmatrix} 1 \\ 4 \end{bmatrix} = \begin{bmatrix} 1 \\ 4 \end{bmatrix}$$

Thus the matrix S

$$S = \begin{bmatrix} 1 & 1 \\ 4 & 4 \end{bmatrix}$$

has rank one and the system is not completely state controllable. To verify the above results, let us solve the equations

$$\dot{x}_1 = -3x_1 + x_2 + u$$

$$\dot{x}_2 = -2x_1 + 1.5 x_2 + 4u$$

Eliminating x_2 we obtain

$$\ddot{x}_1 + 1.5 \dot{x}_1 - 2.5x_1 = \dot{u} + 2.5 u$$

In the form of the transfer function we get

$$\frac{X_1(s)}{V(s)} = \frac{s + 2.5}{(s + 2.5)(s - 1)}$$

We lose one degree of freedom since the factor $(s + 2.5)$ occurs in the numerator and denominator of the transfer function. This is the reason why the system is not completely state controllable.

The following system is completely controllable and completely observable over the interval $[t_0, t_1]$.

$$\dot{x} = A(t)x(t) + B(t)u(t)$$

$$y(t) = C(t)x(t) + D(t)u(t)$$

a)   Derive an explicit expression for an input which transfers the state from $x(t_0)$ to $x(t_1)$.

b)   If the input is zero, derive an explicit expression for $x(t_0)$ in terms of the output function $y(t)$, $t_0 \leq t \leq t_1$.

Solution:

a)   We shall write the solution for the state at t_1 in terms of the transformation

$$A_c: U \to \Sigma$$

The transformation is

$$A_c(v) = \int_{t_0}^{t_1} \Phi(t_1, \tau)B(\tau)u(\tau)d\tau$$

The adjoint transformation A_c^* maps back from Σ to U.

It is given by

$$A_c^* = \int_{t_0}^{t_1} \bar{B}^T(\tau) \bar{\Phi}^T(t_1, \tau) \omega(\tau)d\tau$$

$A_c A_c^*$ is a matrix given by

$$A_c A_c^* = \int_{t_0}^{t_1} \Phi(t_1, \tau)B(\tau)\bar{B}^T(\tau)\bar{\Phi}^T(t_1, \tau)d\tau$$

It maps from U to U.

The solution of the equation

$$\dot{x} = A(t)x(t) + B(t)u(t)$$

is

$$x(t_1) = \Phi(t_1, t_o)x(t_o) + \int_{t_o}^{t} \Phi(t_1, \tau)B(\tau)u(\tau)d\tau$$

so

$$x(t_1) - \Phi(t_1, t_o)x(t_o) \quad A_c(u)$$

Let w be an unknown vector in Σ such that $u(t) = A_c^*(w)$.

We have

$$x(t_1) - \Phi(t_1, t_o)x(t_o) = A_c A_c^*(w).$$

The system is completely controllable when the n x n matrix $A_c A_c^*$ has an inverse.

We get

$$u(t) = A_c^*(A_c A_c^*)^{-1}[x(t_1) - \Phi(t_1, t_o)x(t_o)]$$

$$= \bar{B}^T(t)\bar{\Phi}^T(t_1, t)[\int_{t_o}^{t_1} \Phi(t_1, \tau)B(\tau)\bar{B}^T(\tau)\bar{\Phi}^T(t_1, \tau)d\tau]^{-1}$$

$$\cdot [x(t_1) - \Phi(t_1, t_o)x(t_o)].$$

b)   The output of the unforced system is

$$y(t) = A_o(x(t_o)).$$

Taking the adjoint transformation A_o^* we get

$$A_o^*(y(t) = A_o^* A_o(x(t_o))$$

The system is completely observable when the matrix $A_o^* A_o$ has an inverse, thus

$$x(t_o) = (A_o^* A_o)^{-1} A_o^*(y(t))$$

$$= \left[\int_{t_o}^{t_1} \bar{\Phi}^{-T}(\tau, t_o)\bar{C}^T(\tau)C(\tau)\Phi(\tau, t_o)d\tau\right]^{-1} \int_{t_o}^{t_1} \bar{\Phi}^T(\tau, t_o)\bar{C}^T(\tau)y(\tau)d\tau$$

● **PROBLEM** 12-28

Examine the controllability and observability of a linear digital control system whose input-output relation is described by the difference equation

$$c(k + 2) + 2c(k + 1) + c(k) = u(k + 1) + u(k)$$

Solution: We shall decompose the above difference equation into the following dynamic equations

$$x(k + 1) = Ax(k) + Bu(k)$$

$$c(k) = Dx(k)$$

where

$$A = \begin{bmatrix} 0 & 1 \\ -1 & -2 \end{bmatrix}, \quad B = \begin{bmatrix} 0 \\ 1 \end{bmatrix},$$

$$D = [1 \quad 1]$$

Matrix S is given by

$$S = [B \quad AB] = \begin{bmatrix} 0 & 1 \\ 1 & -2 \end{bmatrix}$$

and is nonsingular, thus the system is completely state controllable for the state variables $x_1(k)$ and $x_2(k)$.

To determine the observability of the system let us form the matrix

$$[D' \quad A'D'] = \begin{bmatrix} 1 & -1 \\ 1 & -1 \end{bmatrix}$$

This matrix is singular therefore the system is unobservable.

● **PROBLEM** 12-29

Find an irreducible realization of H(s)

$$H(s) = \begin{bmatrix} \dfrac{s^2 + 2s + 3}{(s + 1)^2} & \dfrac{s^2 + 6s + 3}{(s + 1)} \\[3mm] \dfrac{1}{s + 1} & \dfrac{-1}{s + 1} \\[3mm] \dfrac{-s + 2}{(s + 1)^2} & \dfrac{7s + 4}{(s + 1)^2} \end{bmatrix}$$

Solution: Let us expand each element of H(s), we obtain

$$H(s) = \begin{bmatrix} 1 + \dfrac{2}{(s + 1)^2} & 1 - \dfrac{2}{(s + 1)^2} + \dfrac{4}{s + 1} \\[3mm] \dfrac{1}{s + 1} & -\dfrac{1}{s + 1} \\[3mm] \dfrac{3}{(s + 1)^2} - \dfrac{1}{s + 1} & \dfrac{-3}{(s + 1)^2} + \dfrac{7}{s + 1} \end{bmatrix}$$

617

$$= \frac{\begin{bmatrix} 2 & -2 \\ 0 & 0 \\ 3 & -3 \end{bmatrix}}{(s+1)^2} + \frac{\begin{bmatrix} 0 & 4 \\ 1 & -1 \\ -1 & 7 \end{bmatrix}}{s+1} + \begin{bmatrix} 1 & 1 \\ 0 & 0 \\ 0 & 0 \end{bmatrix}$$

Thus

$$D = \begin{bmatrix} 1 & 1 \\ 0 & 0 \\ 0 & 0 \end{bmatrix}$$

The other two matrices are of rank 1 and 2. We shall write them as outer products.

$$H(S) = \frac{\begin{bmatrix} 2 \\ 0 \\ 3 \end{bmatrix} [1 \ -1]}{(s+1)^2} + \frac{\begin{bmatrix} 0 \\ 1 \\ -1 \end{bmatrix} [1 \ -1]}{s+1} + \frac{\begin{bmatrix} 4 \\ 0 \\ 6 \end{bmatrix} [0 \quad 1]}{s+1} + D$$

We shall use the linearly independent vectors $b_1^T = [1 \quad -1]$, $b_3^T = [0 \quad 1]$ to form the last rows associated with two Jordan blocks $B = \begin{bmatrix} 0 & 0 \\ 1 & -1 \\ 0 & 1 \end{bmatrix}$. The first and third columns of the corresponding output matrix $C = \begin{bmatrix} 2 & 0 & 4 \\ 0 & 1 & 0 \\ 3 & -1 & 6 \end{bmatrix}$ are linearly dependent.

This leads to a realization which is completely controllable but not observable and thus reducible.

Since $c_3 = 2 c_1$, the first and third terms can be combined.

$$H(s) = \frac{c_1 \rangle \langle [b_1 + 2b_3(s + 1)]}{(s + 1)^2} + \frac{c_2 \rangle \langle b_1}{s + 1} + D$$

The irreducible realization is

$$\begin{bmatrix} \dot{x}_1 \\ \dot{x}_2 \end{bmatrix} = \begin{bmatrix} -1 & 1 \\ 0 & -1 \end{bmatrix} \begin{bmatrix} x_1 \\ x_2 \end{bmatrix} + \begin{bmatrix} 0 & 2 \\ 1 & -1 \end{bmatrix} \begin{bmatrix} u_1 \\ u_2 \end{bmatrix}$$

$$\begin{bmatrix} y_1 \\ y_2 \\ y_3 \end{bmatrix} = \begin{bmatrix} 2 & 0 \\ 0 & 1 \\ 3 & -1 \end{bmatrix} \begin{bmatrix} x_1 \\ x_2 \end{bmatrix} + \begin{bmatrix} 1 & 1 \\ 0 & 0 \\ 0 & 0 \end{bmatrix} \begin{bmatrix} u_1 \\ u_2 \end{bmatrix}$$

Determine the controllability and observability of the systems (a) and (b) and both systems connected in series (c).

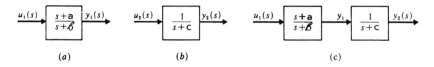

$$(a) \qquad\qquad (b) \qquad\qquad (c)$$

Solution: For the system shown in (a) we have

$$\dot{y}_1 + by_1 = \dot{u}_1 + au_1$$

For $x_1 = y_1 - u_1$, we obtain the state equation

$$\dot{x}_1 = - bx_1 + (a - b)u_1$$

Then the controllability matrix is

$$P = [a - b]$$

and the observability matrix is

$$Q = [1]$$

so this system is completely controllable if $a \neq b_1$ and is always completely observable.

For the system (b) we have

$$\dot{y}_2 + cy_2 = u_2$$

Let $x_2 = y_2$, the state equation is

$$\dot{x}_2 = - cx_2 + u_2$$

Here $P = [1]$ and $Q = [1]$, therefore the system is completely controllable and observable.

For the system shown in (c) we set

$$x_1 = y_1 - u_1 \qquad \text{and } x_2 = y_2. \quad \text{The state}$$

equations are

$$\begin{bmatrix} \dot{x}_1 \\ \dot{x}_2 \end{bmatrix} = \begin{bmatrix} -b & 0 \\ 1 & -c \end{bmatrix} \begin{bmatrix} x_1 \\ x_2 \end{bmatrix} + \begin{bmatrix} a - b \\ 1 \end{bmatrix} u_1$$

The controllability matrix is

$$P = \begin{bmatrix} a - b & -b(a - b) \\ 1 & a - b - c \end{bmatrix} \text{ and}$$

$$|P| = (a - b)(a - c)$$

For $a \neq b$ and $a \neq c$ the rank of P is 2 and the system is completely controllable.

The observability matrix is $Q = \begin{bmatrix} 0 & 1 \\ 1 & -c \end{bmatrix}$ its rank is 2 and the system is completely observable.

● **PROBLEM** 12-31

Consider the control system shown in Fig. 1. The controller is a nonlinear controller. Assuming that there is no input signal and the system is subjected to only the initial condition, determine the region of controllability for the following two cases:

Case 1: The control signal can assume only two values, $u \pm 1$.

Case 2: The control signal can assume any value between 1 and -1, or $- 1 \leq u \leq 1$.

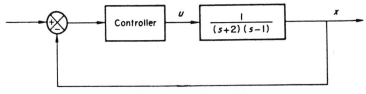

Control System   Fig.1

Solution: The system equation is given by

$$\ddot{x} + \dot{x} - 2x = u \qquad\qquad 1$$

Case 1: For $u = 1$ this equation becomes

$$\ddot{x} + \dot{x} - 2x = 1$$

from which the equilibrium point is

$$x = - \frac{1}{2}, \qquad \dot{x} = 0$$

Since the characteristic equation

$$\lambda^2 + \lambda - 2 = 0$$

has roots $\lambda = 1$ and $\lambda = - 2$, the equilibrium point $(- \frac{1}{2}, 0)$

is a saddle point. The separatrices which pass through
the saddle point have slopes of -2 and 1. Figure 2 is a
state plane diagram showing the saddle point, separatrices,
and several trajectories.

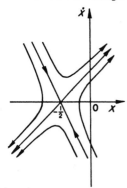

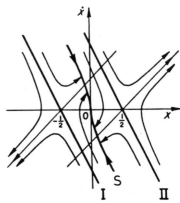

Diagram showing saddle
point, separatrices, and
  several trajectories   Fig.2

Diagram showing domain
of controllability.   Fig.3

        For u = - 1, Eq. 1 becomes

$$\ddot{x} + \dot{x} - 2x = - 1$$

The equilibrium point is now

$$x = \frac{1}{2}, \qquad \dot{x} = 0$$

It is also a saddle point. The state plane diagram is
identical to the foregoing, except that it is shifted to
the right by 1. Figure 3 is the superposition of the two
sets of diagrams corresponding to u = 1 and u = - 1.

        Graphically we see that the origin can be reached
from any point between separatrices I and II shown in Fig.
3. To reach the origin with u=± 1 it is necessary that
the trajectory starting from a given initial point
intersects the curve S. (The curve S is the switching
curve consisting of the trajectory with u = - 1 in the
upper half plane and that with u = 1 in the lower half
plane both passing through the origin). To reach the
curve S from points lying between separatrices I and II
but to the left of S it is necessary to use u = 1 until
the curve S is reached. Similarly, for the points lying
between separatrices I and II but to the right of the
curve S, it is necessary to use u = - 1 until the curve
S is reached.

        Graphically, it is easily seen that it is not possible
to enter the region bounded by separatrices I and II from
the exterior. Hence the region bounded by two separatrices
I and II is the domain of controllability for u=± 1.

        Case 2: The domain of controllability in this case
can be determined similarly to Case 1. For u = constant
and bounded by ± 1 the saddle points are

$$x = -\frac{1}{2} u, \qquad \dot{x} = 0 \qquad (-1 \leq u \leq 1)$$

For $-1 \leq u \leq 1$ (u is bounded but not necessarily constant), the maximum width of the domain of controllability is obtained when $|u|$ is maximum. Hence for Case 2 the domain of controllability is the same as Case 1 and is bounded by separatrices I and II shown in Fig. 3.

● **PROBLEM 12-32**

Examine the controllability of the system shown in Fig. 1.

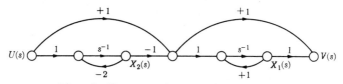

Signal flow graph of cascade system   Fig.1

Solution: The equations of the separate systems may be written as

$$S_a = 1 - \frac{1}{s + 2}$$

and

$$S_b = 1 + \frac{1}{s - 1}$$

The state variables are defined as the outputs of the integrators.

Therefore the state differential equation is

$$\dot{x} = \begin{bmatrix} +1 & -1 \\ 0 & -2 \end{bmatrix} x + \begin{bmatrix} 1 \\ 1 \end{bmatrix} u = Ax + Bu$$

The equation for the output is

$$v = [1, -1]x = Cx$$

The roots of the characteristic equation are 1 and -2. Therefore we have

$$\Lambda = \begin{bmatrix} 1 & 0 \\ 0 & -2 \end{bmatrix}$$

Also the transformation matrix and its inverse are

$$T = \begin{bmatrix} \frac{2}{3} & \frac{1}{3} \\ 0 & 1 \end{bmatrix}$$

and

$$T^{-1} = \begin{bmatrix} \dfrac{3}{2} & -\dfrac{1}{2} \\ 0 & 1 \end{bmatrix}$$

Where T was formed from the eigenvectors of A. Completing the necessary multiplication of matrices, we obtain

$$\dot{y} = \Lambda y + T^{-1}Bu = \begin{bmatrix} 1 & 0 \\ 0 & -2 \end{bmatrix} y + \begin{bmatrix} 1 \\ 1 \end{bmatrix} u$$

and

$$v = CTy = [\tfrac{2}{3}, -\tfrac{2}{3}]y.$$

Since $T^{-1}B$ contains all nonzero elements, the system is controllable. Furthermore, since CT contains all nonzero elements, the system is observable. This situation is also clearly illustrated by the flow graph shown in Fig. 2.

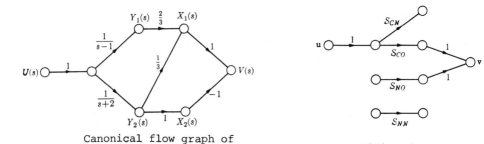

Canonical flow graph of
cascaded system

Fig.2

Partitioned system

Fig.3

Examining Fig. 2, and the differential equation in terms of the canonical state variables, we note that each state is uncoupled and may be classified as either controllable or not controllable, and either observable or not observable. In fact, Gilbert has shown that a system S may be always partitioned into four possible subsystems as shown in Fig. 3. The four possible subsystems are (1) a system S_{CN} which is controllable and not observable, (2) a system S_{CO} which is controllable and observable, (3) a system S_{NO} which is not controllable and is observable, and (4) a system S_{NN} which is not controllable and is not observable.

Determine if the system shown in Fig. 1 is controllable.

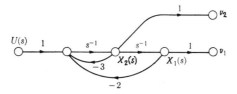

Open-loop second-order system.   Fig.1

Solution: The output vector is simply the state vector for this physical system. Then the equations describing the behavior of the system are

$$\dot{x} = \begin{bmatrix} 0 & 1 \\ -2 & -3 \end{bmatrix} x + \begin{bmatrix} 0 \\ 1 \end{bmatrix} u,$$

$$v = \begin{bmatrix} 1 & 0 \\ 0 & 1 \end{bmatrix} x = Ix.$$

Use the transformation

$$x = \begin{bmatrix} 1 & -1 \\ -1 & 2 \end{bmatrix} \quad y = Ty.$$

The transformed equations describing the uncoupled state variables are

$$\dot{y} = \Lambda y + T^{-1}Bu$$

$$= \begin{bmatrix} -1 & 0 \\ 0 & -2 \end{bmatrix} y + \begin{bmatrix} 1 \\ 1 \end{bmatrix} u(t)$$

Therefore, the system is controllable since all the rows of $T^{-1}B$ have nonzero elements. The equation for the output is

$$v = CTy$$

$$= ITy = Ty.$$

Hence the system is observable, since all the columns of CT possess at least one nonzero element. The canonical variable flow graph is shown in Fig. 2. The input is connected to each canonical state variable, and therefore each variable is controllable. Furthermore, the canonical state variables are connected to the output variables, and therefore the system is observable.

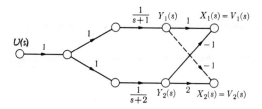

Flow graph of canonical state variables.   Fig. 2

It is in the interconnection of multivariable systems that the analyst must be particularily concerned with testing for controllability and observability.

CHAPTER 13

AUTOMATIC CONTROL STABILITY

ROUTH AND HURWITZ CRITERIA

Determine the characteristic equation and evaluate the stability of the system, whose transfer function is

$$\frac{\theta_0}{\theta_i} = \frac{K}{D^3 + 2D^2 - 5D - 6}$$

Solution: The characteristic equation of the system is

$$D^3 + 2D^2 - 5D - 6 = 0$$

Factoring the equation yields

$$D^3 + 2D^2 - 5D - 6 = (D+1)(D^2+D-6) = (D+1)(D+3)(D-2) = 0$$

The roots of the equation are

$$D_1 = -1, \ D_2 = -3, \ D_3 = 2$$

The characteristic equation has a positive root $D_3=2$ thus the system is unstable. That is, one term of the complementary solution component of the system response to a forcing function is of the form

$$Ae^{2t}$$

which increases continually in time.

Determine the stability conditions of the system whose transfer function is

$$\frac{\theta_0}{\theta_i} = \frac{K}{D^2 + 4D + 5}$$

Solution: The characteristic equation is

$$D^2 + 4D + 5 = 0$$

We calculate the roots

$$D_1 = \frac{-4 + (16-20)^{\frac{1}{2}}}{2} = -2 + j \cdot 2$$

$$D_2 = \frac{-4 - (16-20)^{\frac{1}{2}}}{2} = -2 - j \cdot 2$$

The real parts of both roots are negative hence the system is stable.

Determine the stability of the system whose transfer function is

$$\frac{\theta_0}{\theta_i} = \frac{K}{4D^2 + 13D + 3}$$

Solution: The characteristic equation of the system is

$$4D^2 + 13D + 3 = 0$$

One of the roots of the equation is $D_1 = -3$. Dividing $(4D^2 + 13D + 3)$ by $(D+3)$ we obtain $(4D+1)$. Thus

$$(4D+1)(D+3) = 0$$

The roots $D_1 = -3$, $D_2 = -\frac{1}{4}$ are both negative and hence the system is stable. We have

$$\theta_{02} = A_1 e^{-3t} + A_2 e^{-\frac{1}{4}t}$$

and it is easy to notice that

$$\lim_{t \to \infty} [A_1 e^{-3t} + A_2 e^{-\frac{1}{4}t}] = 0$$

In a closed-loop system the open-loop and closed-loop zeros are not identical. Prove the above statement.

Solution: In a closed-loop system, let the feedforward transfer function be

$$G(s) = p(s)/q(s)$$

and feedback transfer function be

$$H(s) = r(s)/n(s)$$

where $p(s)$, $q(s)$, $r(s)$, $n(s)$ are polynomials in s. We have

$$\frac{C(s)}{R(s)} = \frac{G(s)}{1+G(s)H(s)} = \frac{p(s)n(s)}{q(s)n(s) + p(s) r(s)}$$

The zeros of the closed-loop transfer function are the solutions s of $p(s)n(s) = 0$ and of the open-loop transfer function the solutions of $p(s)r(s) = 0$. That proves above. Some of the zeros of the closed-loop and of the open-loop transfer functions are the same.

Determine the stability of the system whose characteristic equation is

$$D(s) = (s+2)(s-1)(s-5)$$

Solution: We see that two roots of the polynomial are located in the right-half plane, thus the system is not stable. As an exercise we write the Routh tabulation

$$D(s) = (s+2)(s-1)(s-5) = s^3-4s^2-7s+10$$

1	-7
-4	10
-4.5	0
10	

and reach the same conclusion, that the system is not stable.

Examine the stability condition of the system whose characteristic equation is

$$D^4 + 3D^3 + 2D^2 + 10D + 2 = 0$$

Solution: The Routh array is

$$
\begin{array}{ccc}
1 & 2 & 2 \\
3 & 10 & 0 \\
-\dfrac{4}{3} & 2 & \\
\dfrac{29}{2} & 0 & \\
2 & & \\
0 & &
\end{array}
$$

The system is unstable, since there are two sign changes in the first column.

Check the stability of the following system

$$F(s) = s^4 + 2s^3 + 3s^2 + s + 5 = 0$$

Solution: The necessary condition for stability, that all coefficients be of the same sign and none of them zero, is satisfied. To determine the stability we write the Routh array

$$
\begin{array}{cccc}
s^4 & 1 & 3 & 5 \\
s^3 & 2 & 1 & 0 \\
s^2 & 2.5 & 5 & \\
s^1 & -3 & 0 & \\
s^0 & 5 & &
\end{array}
$$

There are two sign changes in the first column, thus two roots of the equation are in the right-hand plane; the system, therefore is not stable.

Using the Routh array determine the stability of the system described by the following equation

$$F(s) = s^5 + 2s^4 + 3s^3 + 9s^2 + 2s + 10$$

Solution: The Routh array is

s^5	1	3	2
s^4	2	9	10
s^3	-1.5	-3	0
s^2	5	10	
s^1	0	0	

We form the auxiliary equation

$$P(s) = 5s^2 + 10 = 5(s^2 + 2) = 5(s + j\sqrt{2})(s - j\sqrt{2})$$

Both roots of the equation are on the imaginary axis. In order to examine the remaining roots we can take the derivative of P(s) and continue with the Routh array or divide

$$\frac{F(s)}{P(s)}$$

and examine the result.

$$\frac{F(s)}{s^2 + 2} = s^3 + 2s^2 + s + 5$$

The new Routh array is

s^3	1	1
s^2	2	5
s^1	-1.5	0
s^0	5	

There are two changes in sign in the first column, thus the polynomial has two roots in the right-hand plane.

The system is unstable.

Investigate the stability of systems with the following characteristic equations. Use the Routh criterion.

a) $2D^3 + 4D^2 + D + 1 = 0$

b) $D^5 + 3D^4 + 2D^3 + 12D^2 + 10D + 100 = 0$

c) $3D^2 + 4D + 5 = 0$

d) $D^5 + 3D^4 + 2D^3 + 4D^2 + 5D + 65 = 0$

Solution:

a) D^3 2   1

         D^2 4   1

         D^1 0.5   0

         D^0 1

There are no sign changes in the first column therefore the system is stable.

b)   The Routh tabulation of

$$D^5 + 3D^4 + 2D^3 + 12D^2 + 10D + 100 = 0$$

is

D^5	1	2	10
D^4	3	12	100
D^3	-2	$-\dfrac{70}{3}$	0
D^2	-23	100	0
D^1	-32	0	
D^0	100		

There are two sign changes, thus the characteristic equation has two positive roots and the system is unstable.

c) $3D^2 + 4D + 5 = 0$

D^2	3	5
D^1	4	0
D^0	5	

No sign changes, the system is stable.

d) $D^5 + 3D^4 + 2D^3 + 4D^2 + 5D + 65 = 0$

D^5	1	2	5
D^4	3	4	65
D^3	$\dfrac{2}{3}$	$-\dfrac{50}{3}$	0
D^2	79	65	0
D^1	-17.2	0	
D^0	65		

631

The first column has two sign changes thus the system is unstable.

Determine the stability of the system described by the equation

$$F(s) = s^3 + 2s^2 + s + 2 = 0$$

Solution: We shall use the Routh array.

s^3	1	1
s^2	2	2
s^1	0	
s^0	∞	

Because of the zero in the first column, we have to transform either the Routh array or the initial equation.

One method is to change the Routh array by replacing the zero element with ε - a small positive number. We have

s^3	1	1
s^2	2	2
s^1	ε	0
s^0	2	

Since $\varepsilon > 0$ all elements of the first column are positive and the system is stable.

The other method is to multiply the equation $F(s) = 0$ by $(s+a)$ where $a > 0$. Since $a > 0$ we add one root in the left half-plane and that does not change the condition of stability.

The characteristic equation of the system is

$$D^3 + 16D^2 + 650D + 800K = 0$$

What is the range of K for the system to be stable?

Solution: The Routh array is

1	650
16	800K
$\dfrac{16 \cdot 650 - 800K}{16}$	0

800K   0

0

For the system to be stable all the elements of the first
column must be positive.

$$800K > 0 \qquad \rightarrow \qquad K > 0$$

$$16 \cdot 650 - 800K > 0 \qquad 13 > K$$

Hence the system is stable when

$$0 < K < 13.$$

● **PROBLEM** 13-12

Let

$$\frac{C(s)}{R(s)} = \frac{K}{(s^4+4s^3+3s^2+s+1) + K}$$

be the closed loop transfer function of a single-input
single-output system and K be the gain of the system.
Using the Routh criterion determine the range of K for
which the system is asymptotically stable.

Solution: The characteristic equation is

$$s^4+4s^3+3s^2+s+(K+1) = 0$$

The Routh array is

s^4	1	3	K+1
s^3	4	1	0
s^2	$\frac{11}{4}$	K+1	
s^1	$1 - \frac{16(K+1)}{11}$	0	
s^0	K+1		

For the system to be asymptotically stable all the coefficients in the first column must be positive. We
obtain $K + 1 > 0 \rightarrow K > - 1$

$$1 - \frac{16(K+1)}{11} > 0 \rightarrow - \frac{5}{16} > K$$

$$- 1 < K < - \frac{5}{16}$$

● **PROBLEM** 13-13

The characteristic equation of the system is

$$D^4+KD^3+2D^2+D+3 = 0$$

For what values of K is the equilibrium state of the system asymptotically stable.

Solution: The Routh array is

$$D^4 \quad 1 \quad\quad 2 \quad 3$$

$$D^3 \quad K \quad\quad 1 \quad 0$$

$$D^2 \quad \frac{2K-1}{K} \quad 3 \quad 0$$

$$D^1 \quad 1 - \frac{3K^2}{2K-1} \quad 0$$

$$D^0 \quad 3$$

For the system to be asymptotically stable all the elements of the first column must be positive. We have

$$K > 0$$

$$\frac{2K-1}{K} > 0$$

$$1 - \frac{3K^2}{2K-1} > 0$$

The first two conditions give us $K > \frac{1}{2}$, the third

$-3K^2 + 2K - 1 > 0$. The quadratic equation $-3K^2 + 2K - 1 = 0$

has two imaginary roots, the value of the polynominal on the left hand side is always negative for any real K. Thus the three conditions for K cannot be fulfilled simultaneously. Therefore there is no value of K for which the above system is stable.

● **PROBLEM 13-14**

Examine the stability of the system with the following characteristic equation

$$2D^4 + 3D^2 + D + 1 = 0$$

Use the Routh criterion.

Solution: The following theorem is useful in determining instability of the system without performing any calculations:

Theorem: In order that there be no roots with positive real parts, it is necessary (but not sufficient!) that

a) All the coefficients of the polynomial have the same sign,

b) None of the coefficients vanishes.

In our example coefficient of D^3 is zero, thus the system is unstable. As an exercise we write the Routh array

```
2       3       1
0       1       0
∞       ∞
```

Element zero in the first column can be removed by multi-plying the characteristic equation by a factor $D+\alpha$ where $\alpha > 0$ and rewriting the Routh array. This adds a new root $-\alpha$ to the equation, but since that root is negative it does not affect the stability condition. Let $\alpha = 3$

$$(2D^4+3D^2+D+1)(D+3) = 2D^5+6D^4+3D^3+10D^2+4D+3$$

D^5	2	3	4
D^4	6	10	3
D^3	$-\dfrac{1}{3}$	3	0
D^2	64	3	
D^1	$3\dfrac{1}{64}$	0	
D^0	3		

There are two sign changes, thus the system is unstable.

● **PROBLEM** 13-15

Determine the values of K for stability of the system whose characteristic equation is:

$$D^3+3KD^2+(K+3)D+5 = 0$$

Solution: The Routh array is

1	K+3
3K	5
$\dfrac{3K(K+3)-5}{3K}$	0
5	

Since $1>0$ all the elements of the first column have to be positive for the system to be stable. We obtain

$$3K>0 \qquad K>0$$

$$3K(K+3)-5>0 \equiv (K-0.48)(K+3.48)>0$$

Thus, the system is stable when $K>0.48$.

The characteristic equation of a closed-loop control system is

$$s^3 + 2Ks^2 + (3K+5)s + 9K = 0$$

For what values of K is the system stable?

Solution: We use the Routh array to find the range of K.

s^3	1	3K+5
s^2	2K	9K
s^1	$\dfrac{2K(3K+5)-9K}{2K}$, 0	
s^0	9K	

For the system to be stable the elements of the first column must be of the same sign, in our case positive

s^2 gives 2K>0, k>0

s^1 gives $\dfrac{2K(3K+5)-9K}{2K} > 0$

Since K>0 we can write

$$2K(3K+5)-9K = 6K^2+K = K(6K+1) > 0$$

$$K>0 \text{ and } K>-\frac{1}{6}$$

The first condition (K>0) is stronger and we conclude that for K>0 the system is stable.

The characteristic equation of the system is given in the following form

$$F(s) = s^4 + s^3 + s^2 + s + K$$

Determine the gain K which results in borderline stability.

Solution: The Routh array is

s^4	1	1	K
s^3	1	1	
s^2	0	K	

We replace the zero element with ε and obtain

$$
\begin{array}{ccccc}
s^4 & 1 & 1 & K \\
s^3 & 1 & 1 & \\
s^2 & \varepsilon & K & \\
s^1 & \dfrac{\varepsilon - K}{\varepsilon} & \cong & \dfrac{-K}{\varepsilon} \\
s^0 & K & & \\
\end{array}
$$

We note that for $\varepsilon > 0$

$$s^1 = \frac{-K}{\varepsilon} \text{ is positive only when } K < 0$$

For $K > 0$ and $\varepsilon > 0$ $\quad s^1 = \frac{-K}{\varepsilon} < 0$

We conclude that the system is unstable for all values of K.

● **PROBLEM** 13-18

Examine the stability of the system described by the characteristic equation

$$D(s) = 26s^4 + 104s^3 + 26s^2 + 208s + 104$$

Solution: The Routh array is

$$
\begin{array}{cccc}
s^4 & 26 & 26 & 104 \\
s^3 & 104 & 208 & \\
s^2 & A_{21} & & \\
s^1 & A_{31} & & \\
s^0 & A_{41} & & \\
\end{array}
$$

We have $A_{21} = \dfrac{104 \cdot 26 - 26 \cdot 208}{104} = -26$

The system is unstable because there is a sign change in the first column. There is no need to continue the tabulation.

The following property of the Routh array is very useful:

Any row of the Routh array can be multiplied or divided by any positive number without changing the sign of any element in the array. In our case we can divide the first row by 26 and the second by 104. We obtain

$$
\begin{array}{cccc}
s^4 & 1 & 1 & 4 \\
s^3 & 1 & 2 & \\
s^2 & -1 & & \\
\end{array}
$$

The numerical values of the array are different but the signs are the same.

The above property makes the calculations easier.

637

Write the Routh tabulation and determine the stability of the system

$$\frac{\theta_0}{\theta_i} = \frac{K(2+3D)}{D^5+10D^4+38D^3+68D^2+57D+18}$$

Solution: The characteristic equation is

$$D^5+10D^4+38D^3+68D^2+57D+18 = 0$$

The Routh array

s^5	1	38	57
s^4	10	68	18
s^3	31.2=A	55.2=B	0
s^2	50.3=C	18=D	
s^1	44=E	0	
s^0	18=F		

Where $A = \dfrac{10 \cdot 38 - 1 \cdot 68}{10}$, $B = \dfrac{10 \cdot 57 - 18 \cdot 1}{10}$

$C = \dfrac{A \cdot 68 - 10 \cdot B}{A}$, $D = \dfrac{A \cdot 18 - 10 \cdot 0}{A}$

etc.

All the elements of the first column are positive, thus the system is stable.

Using Routh criterion examine the stability of the following system

$$\lambda^3 + 3\lambda^2 + \lambda + 3 = 0$$

Solution: The Routh array is

λ^3	1	1
λ^2	3	3
λ^1	0	
λ^0		

In the first column we replace 0 by any small positive number ε and obtain

λ^3	1	1
λ^2	3	3
λ^1	$\varepsilon > 0$	0
λ^0	3	

Since ε is positive there is no sign change in the first column and the system is stable.

The signs above and below the zero element (ε) are the same, it indicates that there is a pair of imaginary roots. If the sign of the coefficient above the zero (ε) is opposite that below it, it indicates one sign change. Let us consider the following example

$$\lambda^3 - 2\lambda^2 - \lambda + 2 = 0$$

the Routh array

one sign change	1	-1
	-2	2
one sign change	$\varepsilon > 0$	0
	2	

Thus there are two sign changes in the first column, which agrees with the result obtained from the factored form of the characteristic equation

$$\lambda^3 - 2\lambda^2 - \lambda + 2 = (\lambda+1)(\lambda-1)(\lambda-2) = 0$$

● **PROBLEM** 13-21

Examine the stability of the system whose characteristic equation is

$$\lambda^5 + 2\lambda^4 + 15\lambda^3 + 30\lambda^2 - 20\lambda - 40 = 0$$

Solution: The Routh array is

λ^5	1	15	-20
λ^4	2	30	-40
λ^3	0	0	

All the elements of the λ^3 row of the Routh tabulation are zeros, it indicates that one or more of the following conditions may exist:

1) Pairs of imaginary roots

639

2) Pairs of real roots with opposite signs

3) Pairs of complex conjugate roots.

Using row λ^4 we form the auxiliary equation

$$P(\lambda) = 2\lambda^4 + 30\lambda^2 - 40$$

and take the derivative of $P(\lambda)$ with respect to λ

$$\frac{d\ P(\lambda)}{d\lambda} = 8\lambda^3 + 60\lambda.$$

We replace the row of zeros with the obtained coefficients of

$$\frac{d\ P(\lambda)}{d\lambda}\ .$$

λ^5	1	15	-20
λ^4	2	30	-40
λ^3	8	60	0
λ^2	15	-40	0
λ^1	81.3	0	
λ^0	-40		

There is one change in sign in the first column, thus the characteristic equation has one root with a positive real part and the system is not stable.

Note that the roots of the auxiliary equation are also roots of the original equation, which is under the Routh test. In our example

$$P(\lambda) = 2\lambda^4 + 30\lambda^2 - 40 = 0$$

the roots are

$$\lambda_1 = \sqrt{\frac{\sqrt{1220} - 30}{4}} \qquad \lambda_2 = -\sqrt{\frac{\sqrt{1220} - 30}{4}}$$

$$\lambda_3 = j\sqrt{\frac{30 + \sqrt{1220}}{4}} \qquad \lambda_4 = -j\sqrt{\frac{30 + \sqrt{1220}}{4}}$$

We can write

$$\lambda^5 + 2\lambda^4 + 15\lambda^3 + 30\lambda^2 - 20\lambda - 40$$

$$= (2\lambda^4 + 30\lambda^2 - 40)(0.5\lambda + 1)$$

and see that in agreement with the results obtained by the Routh test; the characteristic equation of the system has one root with the positive real part (the root λ_1 of $P(\lambda) = 0$).

The block diagram of the system is shown in Fig. 1, using the Routh criterion determine the stability of the system.

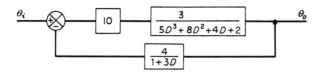

Fig. 1

Solution: First we find the characteristic equation of the system

$$\frac{\theta_o}{\theta_i} = \frac{10 \cdot \dfrac{3}{5D^3+8D^2+4D+2}}{1 + \dfrac{4 \cdot 10 \cdot 3}{(1+3D)(5D^3+8D^2+4D+2)}}$$

$$= \frac{30(1+3D)}{(1+3D)(5D^3+8D^2+4D+2) + 4 \cdot 10 \cdot 3}$$

The characteristic equation is

$$(1+3D)(5D^3+8D^2+4D+2) + 4 \cdot 10 \cdot 3 =$$

$$15D^4+29D^3+20D^2+10D+122 = 0$$

The Routh array is

D^4	15	20	122
D^3	29	10	0
D^2	A_1	A_2	
D^1	B_1	B_2	
D^0	C_1		

where

$$A_1 = \frac{29 \cdot 20 - 15 \cdot 10}{29} = 14.8$$

$$A_2 = 122$$

$$B_1 = \frac{A_1 \cdot 10 - 29 \cdot A_2}{A_1} = \frac{148 - 3538}{14.8} < 0$$

There is a sign change in the first column and the system is not stable.

Using the Routh criterion, find the range of values K for which the system is stable

$$\frac{\theta_0}{\theta_i} = \frac{K}{D(1 + \frac{D}{3})(1 + \frac{D}{4}) + K}$$

Solution: The characteristic equation is

$$\frac{1}{12} D^3 + \frac{7}{12} D^2 + D + K = 0$$

The Routh array is

D^3	$\frac{1}{12}$	1	0
D^2	$\frac{7}{12}$	K	0
D^1	$\frac{7-K}{7}$	0	
D^0	K	0	

and we obtain

K > 0 and 7 - K > 0, thus the system is stable for 7 > K > 0 .

Using the Routh's criterion determine the stability limits of the servomechanism. The open-loop transfer function is

$$G(s) = \frac{K_a (2 + sT_1)}{s^2 (1 + sT_2)}$$

Solution: We find the characteristic equation of the system

$$P(s) = \frac{G(s)}{1 + G(s)} = \frac{K_a (2 + sT_1)}{s^2 (1 + sT_2) + K_a (2 + sT_1)}$$

$$= \frac{2 + sT_1}{s^3 \frac{T_2}{K_a} + \frac{s^2}{K_a} + sT_1 + 2}$$

The Routh array is

s^3	$\dfrac{T_2}{K_a}$	T_1	

s^2	$\dfrac{1}{K_a}$	2	

s^1	$T_1 - 2T_2$	0	

s^0	2	

All the elements of the first column are of the same sign when

$$T_1 > 2T_2 .$$

The marginal stability occurs when $T_1 = 2T_2$ and the open-loop transfer function is

$$G(s) = \frac{2K_a}{s^2}$$

The system is stable when $T_1 > 2T_2$.

● **PROBLEM 13-25**

Determine the stability of the system whose characteristic equation is

$$D(s) = s^4 + 3s^3 + 4s^2 + 3s + 3$$

Solution: The Routh array of this polynomial is

s^4	1	4	3
s^3	3	3	0
s^2	3	3	0
s^1	0	0	

We obtained the row of zeros; using row s^2 we form the auxiliary equation

$$A(s) = 3s^2 + 3$$

and its derivative

$$\frac{d\,A(s)}{ds} = 6s$$

The new Routh array is

s^4	1	4	3
s^3	3	3	0
s^2	3	3	0
s^1	6	0	
s^0	3		

We see that the system is stable.

● **PROBLEM** 13-26

Determine the stability of the system with the characteristic equation

$$D^6 + D^5 - 4D^4 - 5D^3 - 9D^2 - 4D - 4 = 0$$

Use Routh criterion.

Solution: It is obvious that this system is not stable, because the characteristic equation has both positive and negative coefficients. We will reach the same conclusion using the Routh criterion.

$$\begin{vmatrix} 1 & -4 & -9 & -4 \\ 1 & -5 & -4 & 0 \\ 1 & -5 & -4 & 0 \\ 0 & 0 & 0 & \\ \infty & \infty & \infty & \end{vmatrix}$$

The fourth row of the Routh array consists of zeros thus in the fifth row we obtain infinities.

The standard method of solving this problem is to write the subsidiary equation represented by the row preceding the row of zeros, then differentiating it and substituting for the row of zeros.

The subsidiary equation is

$$D^4 - 5D^2 - 4$$

The first derivative

$$4D^3 - 10D$$

We rewrite the Routh tabulation

1	-4	-9	-4
1	-5	-4	0

644

1	-5	-4	0
4	-10	0	
-2.5	-4		
-16.4	0		
-4			
0			

In the first column the sign changes (from 4 to -2.5) thus the system is not stable.

● **PROBLEM** 13-27

Consider the system shown in the figure.

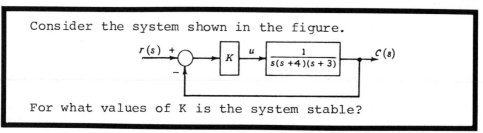

For what values of K is the system stable?

Solution: The closed-loop transfer function is

$$\frac{c(s)}{r(s)} = \frac{K}{s(s+4)(s+3) + K}$$

The characteristic equation is

$$D(s) = s(s+4)(s+3) + K = s^3 + 7s^2 + 12s + K$$

and the Routh array is

s^3	1	12
s^2	7	K
s^1	$\frac{84-K}{7}$	0
s^0	K	

For the system to be stable all the elements of the first column must be positive.   We have

$$84 - K > 0$$

$$K > 0$$

or $0 < K < 84$

If the above condition on K is satisfied the system is stable.

In this simple case it is possible to define a stability

margin for parameter K.

$$\text{Stability margin} = \frac{\text{maximum stable value}}{\text{actual value}}$$

Stability margin of the gain K, called the gain margin (GM), indicates how many times gain may be increased before stability occurs.

In our case if K = 4

$$GM = \frac{84}{4} = 21$$

● **PROBLEM** 13-28

Examine the stability of the system shown in Fig. 1.

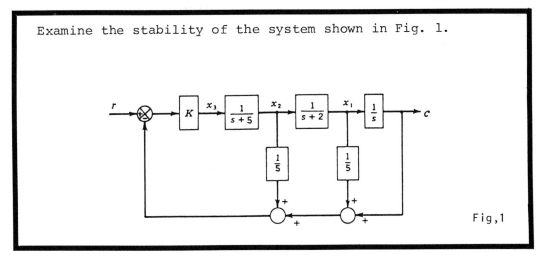

Fig,1

Solution: First we find the transfer function and the characteristic equation of the system.

$$\frac{K}{(s+5)(s+2)s}\left[r - \left(c + \frac{1}{5}x_1 + \frac{1}{5}x_2\right)\right] = c$$

where

$$x_1 = cs, \qquad \frac{x_2}{s+2} = x_1$$

$$x_2 = x_1(s+2) = cs(s+2)$$

We obtain

$$\frac{c}{r} = \frac{K}{s(s+5)(s\cdot\,\,)+\frac{1}{5}K(s^2+3s+5)}$$

The characteristic equation of the system is

$$D(s) = s^3 + s^2\left(7+\frac{1}{5}K\right) + s\left(10+\frac{3}{5}K\right) + K$$

Since the coefficient at s^3 is positive it is necessary for

646

the system to be stable that all the coefficients of the characteristic equation are positive.

We get

$$7 + \frac{1}{5}K > 0$$

$$10 + \frac{3}{5}K > 0 \quad \Bigg\} \quad \longrightarrow \quad K > 0$$

$$K > 0$$

To get the exact results we write the Routh array.

s^3	1	$10 + \frac{3}{5}K$
s^2	$7 + \frac{1}{5}K$	K
s^1	A	0
s^0	K	

where

$$A = \frac{(7 + \frac{1}{5}K)(10 + \frac{3}{5}K) - K}{7 + \frac{1}{5}K}$$

$$= \frac{70 + \frac{26}{5}K + \frac{3}{25}K^2}{7 + \frac{1}{5}K}$$

For the system to be stable all the elements of the first column must be positive.

We see that for $K > 0$ all the elements are positive.

● **PROBLEM** 13-29

In modern computers data are stored on disks. During the process of storing or retrieving the information the data head has to move to different positions on the spinning disk with high accuracy.

The disk storage data head positioning system is shown on the block diagram.

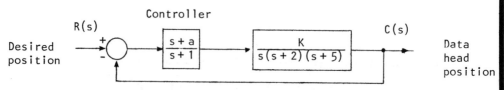

K and a are parameters of the system. Determine the range of K and a for which the system is stable.

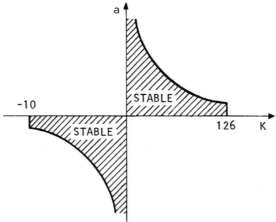

Fig. 1

Solution: First we find the characteristic equation of the system.

$$1 + G(s) = 1 + \frac{K(s+a)}{s(s+1)(s+2)(s+5)} = 0$$

The characteristic equation is

$$s(s+1)(s+2)(s+5) + K(s+a) = 0$$

or

$$s^4 + 8s^3 + 17s^2 + (K+10)s + Ka = 0$$

The Routh array is

s^4	1	17	Ka
s^3	8	$K+10$	0
s^2	$\frac{126-K}{8}$	Ka	
s^1	A	0	
s^0	Ka		

For the system to be stable all elements of the first column must be of the same sign. We have

$$\frac{126-K}{8} > 0$$

$$A = \frac{\frac{126-K}{8} \cdot (K+10) - 8Ka}{\frac{126-K}{8}} > 0$$

$$Ka > 0$$

We have

648

$$\begin{cases} 126 > K \\ Ka > 0 \\ (K+10)(126-K) - 64Ka > 0 \end{cases} \quad (1)$$

We will show the solution in Fig. (1).

The shaded area indicates where all three inequalities are satisfied and the system is stable.

● **PROBLEM** 13-30

Find the value of K for which the system shown is stable.

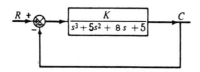

Solution: The characteristic equation is

$$s^3 + 5s^2 + 8s + 5 + K = 0$$

We write the Hurwitz determinant

$$\begin{vmatrix} 5 & 5+K \\ \\ 1 & 8 \end{vmatrix} > 0$$

and obtain

$$40 - 5 - K > 0$$

$$35 > K$$

For K < 35 the system is stable.

● **PROBLEM** 13-31

Determine the stability of the system whose characteristic equation is

$$P(s) = s^3 + s^2 + 1000s + 100$$

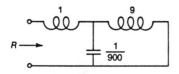

Solution:

$$R(s) = \frac{s^3 + 1000s}{s^2 + 100}$$

649

Dividing two polynomials

$$s^2 + 100 \overline{\smash{\big)}\ s^3 + 1000s} \quad \genfrac{}{}{0pt}{}{s}{}$$

$$\underline{s^3 + 100s} \quad \genfrac{}{}{0pt}{}{s}{900}$$

$$900s \overline{\smash{\big)}\ s^2 + 100}$$

$$\underline{s^2} \quad \genfrac{}{}{0pt}{}{9s}{}$$

$$100 \overline{\smash{\big)}\ 900s}$$

we obtain

$$R(s) = s + \cfrac{1}{\dfrac{s}{900} + \dfrac{1}{9s}}$$

Because all three divisions have positive coefficients the system is stable.

The L-C network for the impedance $R(s)$ is shown in the figure

● **PROBLEM** 13-32

Show that the Hurwitz stability criterion and the Routh stability criterion are equivalent.

Solution: The characteristic equation of the system is

$$F(s) = a_0 s^n + a_1 s^{n-1} + a_2 s^{n-2} + \ldots + a_{n-1} s + a_n = 0$$

The Hurwitz determinants are given by

$$D_1 = a_1 \qquad D_2 = \begin{vmatrix} a_1 & a_3 \\ a_0 & a_2 \end{vmatrix} \qquad D_3 = \begin{vmatrix} a_1 & a_3 & a_5 \\ a_0 & a_2 & a_4 \\ 0 & a_1 & a_3 \end{vmatrix}$$

$$\ldots \qquad D_n = \begin{vmatrix} a_1 & a_3 & a_5 & \cdots & a_{2n-1} \\ a_0 & a_2 & a_4 & \cdots & a_{2n-2} \\ 0 & a_1 & a_3 & \cdots & a_{2n-3} \\ & & & & \\ 0 & 0 & 0 & \cdots & a_n \end{vmatrix}$$

The Hurwitz criterion states that the system is stable if and only if all the Hurwitz determinants are positive.

We show that the above criterion is equivalent to the Routh criterion.

Let us write the following array of the coefficients of the characteristic equation called the Routh array.

s^n $\qquad a_0 \qquad\qquad a_2 \qquad\qquad a_4 \qquad\qquad a_6 \ldots$

s^{n-1} $\qquad a_1 \qquad\qquad a_3 \qquad\qquad a_5 \qquad\qquad a_7 \ldots$

$s^{n-2} \qquad \dfrac{a_1 a_2 - a_0 a_3}{a_1} = A; \qquad \dfrac{a_1 a_4 - a_0 a_5}{a_1} = B; \qquad \dfrac{a_1 a_6 - a_0 a_7}{a_1} = C \ldots$

$s^{n-3} \qquad \dfrac{A a_3 - a_1 B}{A} = D \qquad \dfrac{A a_5 - a_1 C}{A} = E$

$s^{n-4} \qquad \dfrac{DB - AE}{D} = F$

$\quad \vdots$

$s^0 \qquad a_n \qquad 0 \ldots \ldots \ldots \ldots \ldots \ldots 0$

The relations between the elements of the first column of the Routh array and the Hurwitz determinants are:

$s^n \qquad\qquad a_0 = a_0$

$s^{n-1} \qquad\quad a_1 = D_1$

$s^{n-2} \qquad\quad A = \dfrac{D_2}{D_1}$

$s^{n-3} \qquad\quad D = \dfrac{D_3}{D_2}$

$s^{n-4} \qquad\quad F = \dfrac{D_4}{D_3}$

We see that the following two conditions are equivalent

1. All the Hurwitz determinants are positive.

2. All the elements of the first column of the Routh array are positive.

That proves that the Routh and the Hurwitz criteria are equivalent.

Note: We can multiply the characteristic equation by -1 and replace "positive" by "negative ".

651

Examine the stability of the following general third order equation.

$$D(s) = s^3 + d_1 s^2 + d_2 s + d_3$$

Solution: We use both the Hurwitz criterion and the Routh criterion and compare the results. The Hurwitz determinants are

$$D_1 = d_1 \qquad D_2 = \begin{vmatrix} d_1 & d_3 \\ 1 & d_2 \end{vmatrix} \qquad D_3 = \begin{vmatrix} d_1 & d_3 & 0 \\ 1 & d_2 & 0 \\ 0 & d_1 & d_3 \end{vmatrix}$$

For the system to be stable all the Hurwitz determinants must be positive. Thus

$$d_1 > 0$$

$$d_1 d_2 > d_3$$

$$d_1 d_2 d_3 > d_3^2$$

The above system of inequalities is equivalent to:

$$d_1 > 0$$

$$d_3 > 0 \qquad\qquad (1)$$

$$d_1 d_2 > d_3$$

The Routh tabulation gives

s^3	1	d_2	0
s^2	d_1	d_3	0
s^1	$\dfrac{d_1 d_2 - d_3}{d_1}$;	0	
s^0	d_3 ;	0	

For the system to be stable all the elements of the first column must have the same sign, in our case be positive. We obtain

$$d_1 > 0$$

$$d_1 d_2 > d_3 \qquad\qquad (2)$$

$$d_3 > 0$$

We see that both methods give the same results.

For the characteristic equation

$$s^n + a_1 s^{n-1} + a_2 s^{n-1} + \ldots + a_{n-1} s + a_n = 0 \qquad (1)$$

show that the first column of the Routh tabulation is given by

$$1, \ \Delta_1, \ \frac{\Delta_2}{\Delta_1}, \ \frac{\Delta_3}{\Delta_2}, \ \ldots \ \frac{\Delta_n}{\Delta_{n-1}}$$

where

$$\Delta_r = \begin{vmatrix} a_1 & 1 & 0 & 0 \ldots 0 \\ a_3 & a_2 & a_1 & 1 \ldots 0 \\ a_5 & a_4 & a_3 & a_2 \ldots 0 \\ \cdot & \cdot & \cdot & \cdot & \cdot \\ \cdot & \cdot & \cdot & \cdot & \cdot \\ \cdot & \cdot & \cdot & \cdot & \cdot \\ a_{2r-1} & & & \ldots a_r \end{vmatrix}$$

$$a_k = 0 \text{ if } k > n \qquad (2)$$

Δ_r are called Hurwitz determinants.

Solution: The Routh tabulation for Eq. (1) has the form

1	a_2	a_4	a_6	. . .	a_n
a_1	a_3	a_5 . . .			
b_1	b_2	b_3 . . .			
c_1	c_2 . . .				
d_1	. . .				

From Eq. (2) we have $\Delta_1 = a_1$, a_1 is the second element of the first column.

$$b_1 = \frac{a_1 a_2 - a_3}{a_1} = \frac{\Delta_2}{\Delta_1}$$

We compute the next element c_1

$$c_1 = \frac{b_1 a_3 - a_1 b_2}{b_1} = \frac{[a_1 a_2 - a_3]\,\dfrac{a_3}{a_1} - a_1\left(\dfrac{a_1 a_4 - a_5}{a_1}\right)}{\left(\dfrac{a_1 a_2 - a_3}{a_1}\right)}$$

$$= \frac{\begin{vmatrix} a_1 & 1 & 0 \\ a_3 & a_2 & a_1 \\ a_5 & a_4 & a_3 \end{vmatrix}}{\begin{vmatrix} a_1 & 1 \\ a_3 & a_2 \end{vmatrix}} = \frac{\Delta_3}{\Delta_2}$$

The last nonzero terms of the columns are the same, if the array is given by

a_0	a_2	a_4	a_6
a_1	a_3	a_5	a_7
b_1	b_2	b_3	
c_1	c_2	c_3	
d_1	d_2		
e_1	e_2		
f_1			
g_1			

We conclude that $g_1 = a_7$. In general

$$g_1 = e_2 = c_3 = a_7 \neq 0$$

The last term of the first column is equal to a_n. We have

$$a_n = \frac{\Delta_{n-1}\, a_n}{\Delta_{n-1}} = \frac{\Delta_n}{\Delta_{n-1}}$$

Thus we proved that the first column of the Routh tabulation is given by

$$1, \ \Delta_1, \ \frac{\Delta_2}{\Delta_1}, \ \frac{\Delta_3}{\Delta_2}, \ \cdots \ \frac{\Delta_n}{\Delta_{n-1}} \ .$$

Note:

If any of the determinants Δ_r is equal to 0, we can write $\Delta_r = \varepsilon$ where ε is a small positive number, and continue calculations.

Determine conditions for asymptotic stability of the equilibrium state x = 0 for the following system:

$$\dot{x}_1 = f_1(x_1) + f_2(x_2) \tag{1}$$

$$\dot{x}_2 = x_1 + ax_2$$

where

$$f_1(0) = f_2(0) = 0$$

and both f_1, f_2 are real differentiable functions.

We have

$$[f_1(x_1) + f_2(x_2)]^2 + [x_1 + ax_2]^2 \rightarrow \infty \qquad \text{as} \qquad \| x \| \rightarrow \infty$$

Solution: We compute

$$f(x) = \begin{bmatrix} f_1(x_1) + f_2(x_2) \\ \\ x_1 + ax_2 \end{bmatrix}$$

and

$$F(x) = \begin{bmatrix} \dfrac{\partial f_1}{\partial x_1} & \dfrac{\partial f_2}{\partial x_2} \\ \\ 1 & a \end{bmatrix}$$

$$\hat{F}(x) = F'(x) + F(x) = \begin{bmatrix} 2\dfrac{\partial f_1}{\partial x_1} & 1 + \dfrac{\partial f_2}{\partial x_2} \\ \\ 1 + \dfrac{\partial f_2}{\partial x_2} & 2a \end{bmatrix}$$

The Krasovskii theorem states that, the equilibrium state x = 0 of the system (1) is asymptotically stable in the large if $\hat{F}(x)$ is negative definite.

Hence, if

1) $\dfrac{\partial f_1}{\partial x_1} < 0$ for all $x_1 \neq 0$

2) $4a \dfrac{\partial f_1}{\partial x_2} - \left[1 + \dfrac{\partial f_2}{\partial x_2}\right]^2 > 0$ for all $x_1 \neq 0$ and $x_2 \neq 0$

then the equilibrium state x = 0 is asymptotically stable in the large.

KRASOVSKII THEOREM

Examine the stability of the equilibrium state $x = 0$ of the following system

$$\dot{x}_1 = -2x_1$$

$$\dot{x}_2 = x_1 - 2x_2 - x_2^3$$

Use the Krasovskii theorem.

Solution: Let

$$\dot{x}_1 = f_1(x)$$

$$\dot{x}_2 = f_2(x)$$

$$f'(x) = [f_1(x) \quad f_2(x)]$$

$F(x)$ is the Jacobian of $f(x)$

$$F(x) = \frac{\partial f(x)}{\partial x} = \begin{bmatrix} \dfrac{\partial f_1}{\partial x_1} & \dfrac{\partial f_1}{\partial x_2} \\[2mm] \dfrac{\partial f_2}{\partial x_1} & \dfrac{\partial f_2}{\partial x_2} \end{bmatrix}$$

If $\hat{F}(x) = F(x) + F'(x)$ is negative definite for $x \neq 0$, then $x = 0$ is asymptotically stable. If, in addition

$$\lim_{\|x\| \to \infty} f'(x) \ f(x) \to \infty$$

then the equilibrium state $x = 0$ is asymptotically stable in the large.

In our case we have

$$f(x) = \begin{bmatrix} -2x_1 \\[3mm] x_1 - 2x_2 - x_2^3 \end{bmatrix}$$

and

$$F(x) = \begin{bmatrix} -2 & 0 \\[3mm] 1 & -2-3x_2^2 \end{bmatrix}$$

656

$$\hat{F} = \begin{bmatrix} -4 & 1 \\ \\ 1 & -4-6x_2^2 \end{bmatrix}$$

Since $\hat{F}(x)$ is negative definite for all $x \neq 0$ the equilibrium state $x = 0$ is asymptotically stable. We obtain

$$f'(x)f(x) = 4x_1^2 + (x_1 - 2x_2 - x_2^3)^2 \to \infty$$

$$\text{for} \quad \|x\| \to \infty$$

therefore the equilibrium state is asymptotically stable in the large.

● **PROBLEM** 13-37

The system is described by the following equations

$$\dot{x}_1 = -5x_1 + 2x_2$$

$$\dot{x}_2 = 2x_1 - x_2 - x_2^3$$

Prove that the equilibrium state $x = 0$ of the system is asymptotically stable in the large.

Use the Krasovskii theorem.

<u>Solution</u>: The function $f(x)$ is given by

$$f(x) = \begin{bmatrix} -5x_1 + 2x_2 \\ \\ 2x_1 - x_2 - x_2^3 \end{bmatrix}$$

and its Jacobian

$$F(x) = \frac{\partial f}{\partial x} = \begin{bmatrix} -5 & 2 \\ \\ 2 & -1-3x_2^2 \end{bmatrix}$$

$$\hat{F}(x) = F'(x) + F(x)$$

$$= \begin{bmatrix} -10 & 4 \\ \\ 4 & -2-6x_2^2 \end{bmatrix}$$

We show that the above matrix is negative definite.

By Sylvester's theorem, a matrix is positive definite if its leading determinants are positive. For

657

$$-\hat{F}(x)$$

$$|10| = 10 > 0$$

$$10(2 + 6x_2^2) - 16 > 0$$

Thus $-\hat{F}(x)$ is positive definite, so $F(x)$ is negative definite, and the equilibrium state $x = 0$ is asymptotically stable.

Since

$$f'(x)\ f(x) = (-5x_1 + 2x_2)^2 + (2x_1 - x_2 - x_2^3)^2 \rightarrow \infty$$

for $\quad \|x\| \rightarrow \infty$

the equilibrium state is asymptotically stable in the large.

● **PROBLEM** 13-38

Determine the stability of the nonlinear system (1) using Krasovskii's method.

$$\dot{x} = \begin{bmatrix} (x_2) \\ (-Kg(x_1) - x_2) \end{bmatrix} = \begin{bmatrix} f_1(x) \\ f_2(x) \end{bmatrix}, \qquad (1)$$

where $\quad Kg(x_1) = x_1^3$.

Solution: We select an arbitrary δ of the form

$$\delta = \begin{bmatrix} \delta_1 \\ \delta_2 \end{bmatrix} = \begin{bmatrix} (\alpha_{11}x_1 + \alpha_{12}x_2) \\ (\alpha_{21}x_1 + 2x_2) \end{bmatrix}.$$

and we obtain

$$\dot{V} = \delta^T \dot{x} = \begin{bmatrix} (\alpha_{11}x_1 + \alpha_{12}x_2) \\ (\alpha_{21}x_1 + 2x_2) \end{bmatrix}^T \begin{bmatrix} x_2 \\ (-x_1^3 - x_2) \end{bmatrix} \qquad (2)$$

$$= x_1 x_2 (\alpha_{11} - \alpha_{21} - 2x_1^2) + x_2^2 (\alpha_{12} - 2) - \alpha_{21} x_1^4.$$

When the system is asymptotically stable we can use a large number of Lyapunov functions. We can require that $\dot{V}$ be at least negative semi-definite. To achieve this we choose coefficients

$$\alpha_{11} - \alpha_{21} - 2x_1^2 = 0$$

$$\alpha_{21} > 0$$

$$0 < \alpha_{12} < 2$$

From the first we have

658

$$\alpha_{11} = \alpha_{21} + 2x_1^2$$

In this case we will choose $\alpha_{12} = 1$ and determine the required value of α_{21} from the curl equation. Thus we obtain

$$\delta = \begin{bmatrix} (\alpha_{21}x_1 + 2x_1^3 + x_2) \\ \\ (\alpha_{21}x_1 + 2x_2) \end{bmatrix}.$$

Using the curl relation for the second-order system,

$$\frac{\partial \delta_1}{\partial x_2} = \frac{\partial \delta_2}{\partial x_1}$$

or

$$1 = \alpha_{21}$$

Therefore we acquire the equation for the gradient as follows:

$$(\nabla V) = \delta = \begin{bmatrix} (2x_1^3 + x_1 + x_2) \\ \\ (x_1 + 2x_2) \end{bmatrix}.$$

The choice of coefficients in (2) makes $\dot{V}$ negative definite. To prove stability we have to show that V is positive definite. Integration in the state space from the origin to the arbitrary point x gives

$$V = \int_0^x \delta \cdot dx$$

$$= \int_0^{x_1 \, (x_2 = 0)} (2x_1^3 + x_1 + x_2) dx_1 + \int_0^{x_2} (x_1 + 2x_2) dx_2$$

$$= \frac{x_1^4}{2} + \frac{x_1^2}{2} + x_1 x_2 + x_2^2 .$$

For all points x in the state space V is positive definite, therefore the system is asymptotically stable.

LIAPUNOV FUNCTION

The system is described by the following equation

$$\dot{x} = \begin{pmatrix} -4 & 3 \\ -1 & -1 \end{pmatrix} x$$

Determine the stability of the system.

Solution: We use the following equation to find the symmetric matrix P.

$$A^T P + PA = -Q$$

We choose Q = I and obtain

$$\begin{pmatrix} -4 & -1 \\ 3 & -1 \end{pmatrix} \begin{pmatrix} p_{11} & p_{12} \\ p_{12} & p_{22} \end{pmatrix} + \begin{pmatrix} p_{11} & p_{12} \\ p_{12} & p_{22} \end{pmatrix} \begin{pmatrix} -4 & 3 \\ -1 & -1 \end{pmatrix} = \begin{pmatrix} -1 & 0 \\ 0 & -1 \end{pmatrix}$$

Solving the system of equations

$$-4p_{11} - p_{12} - 4p_{11} - p_{12} = -1$$

$$-4p_{12} - p_{22} + 3p_{11} - p_{12} = 0$$

$$3p_{12} - p_{22} + 3p_{12} - p_{22} = -1$$

We get the following matrix P.

$$P = \begin{pmatrix} \dfrac{9}{70} & -\dfrac{1}{70} \\ -\dfrac{1}{70} & \dfrac{32}{70} \end{pmatrix}$$

which is positive definite since

$$\Delta_1 = \frac{9}{70} > 0$$

$$\Delta_2 = \frac{9}{70} \cdot \frac{32}{70} - \frac{1}{70} \cdot \frac{1}{70} > 0$$

(i.e., principal minors are positive). Thus the system is asymptotically stable.

The linear system is described by

$$\dot{x} = Ax$$

where A is a constant matrix. Using the Lyapunov direct method determine the stability of the system.

Solution: Let us assume that the quadratic Lyapunov function is

$$V(x) = x'Px$$

where P is a symmetric, positive definite matrix. The time derivative of V is

$$\dot{V}(x) = \dot{x}'Px + x'P\dot{x}$$

Since $\dot{x} = Ax$ and $(Ax)' = x'A'$ we have

$$\dot{V}(x) = (Ax)'Px + x'PAx$$

$$= x'A'Px + x'PAx = x'[A'P + PA]x$$

If $A'P + PA = -Q$ for some positive definite matrix Q, then the system is asymptotically stable.

If the matrix Q is defined such that Q is positive definite, then the equation

$$A'P + PA = -Q$$

defines matrix P.

The system is asymptotically stable when matrix P is positive definite.

Determine the stability of the system described by the following equation

$$\dot{x} = \begin{bmatrix} -2 & 0 \\ 1 & -1 \end{bmatrix} x$$

Solution: From the equation

$$A^T P + PA = -Q$$

we find the symmetric matrix P.

Let us choose Q = I, where I by definition is positive definite.

$$\begin{bmatrix} -2 & 1 \\ 0 & -1 \end{bmatrix} \begin{bmatrix} p_{11} & p_{12} \\ p_{12} & p_{22} \end{bmatrix} + \begin{bmatrix} p_{11} & p_{12} \\ p_{12} & p_{22} \end{bmatrix} \begin{bmatrix} -2 & 0 \\ 1 & -1 \end{bmatrix} = \begin{bmatrix} -1 & 0 \\ 0 & -1 \end{bmatrix}$$

Solving the system of equations

$$-2p_{11} + p_{12} - 2p_{11} + p_{12} = -1$$

$$-2p_{12} + p_{22} - p_{12} = 0$$

$$- p_{22} - p_{22} = -1$$

We obtain

$$P = \begin{bmatrix} \dfrac{1}{3} & \dfrac{1}{6} \\ \dfrac{1}{6} & \dfrac{1}{2} \end{bmatrix}$$

The principal minors are

$$\frac{1}{3} > 0$$

$$\frac{1}{3} \cdot \frac{1}{2} - \frac{1}{6} \cdot \frac{1}{6} > 0$$

positive, thus the system is asymptotically stable.

● **PROBLEM 13-42**

The system is described by the equation

$$\dot{x} = Ax$$

where

$$A = \begin{bmatrix} a & 0 \\ 1 & -1 \end{bmatrix}$$

Determine the range of the parameter a for which the system is asymptotically stable.

Solution: From the equation

$$A^T P + PA = -Q$$

we find the symmetric matrix P. Choosing Q = I we obtain

662

$$\begin{bmatrix} a & 1 \\ 0 & -1 \end{bmatrix} \begin{bmatrix} p_{11} & p_{12} \\ p_{12} & p_{22} \end{bmatrix} + \begin{bmatrix} p_{11} & p_{12} \\ p_{12} & p_{22} \end{bmatrix} \begin{bmatrix} a & 0 \\ 1 & -1 \end{bmatrix} = \begin{bmatrix} -1 & 0 \\ 0 & -1 \end{bmatrix}$$

or

$$2ap_{11} + 2p_{12} = -1$$

$$ap_{12} + p_{22} - p_{12} = 0$$

$$-p_{22} - p_{22} = -1$$

Solving the above system we have

$$P = \begin{bmatrix} \dfrac{2-a}{2a(a-1)} & \dfrac{1}{2(1-a)} \\ \dfrac{1}{2(1-a)} & \dfrac{1}{2} \end{bmatrix}$$

The following inequalities must be satisfied for the matrix P to be positive definite

$$\frac{2-a}{a(a-1)} > 0 \quad \text{or} \quad a(2-a)(a-1) > 0$$

$$\frac{2-a}{a(a-1)} - \frac{1}{(1-a)^2} > 0 \qquad \text{or} \quad a(2-a)(a-1) > a$$

From the first equation we get

$$a < 0$$

or $\quad 1 < a < 2$

The second equation is satisfied for $a < 0$.

For $\quad 1 < a < 2$ we have

$$a(2-a)(a-1) > a$$

dividing by a

$$(2-a)(a-1) > 1$$

$$a^2 - 3a + 3 < 0$$

Thus, both inequalities can not be satisfied.

We conclude that the system is asymptotically stable for $a < 0$.

A linear RC network is described by

$$\dot{x} = Ax \tag{1}$$

where

$$A = \begin{bmatrix} -3k & 3k \\ 2k & -5k \end{bmatrix} \tag{2}$$

Determine the restrictions on the parameter k in order to guarantee stability.

Solution: To solve the above problem we shall use the method of Lyapunov. Let P be a positive-definite symmetric matrix and let

$$V = x^T P x \tag{3}$$

For the system to be stable it is necessary that $\dot{V} < 0$. We have

$$\dot{V} = \dot{x}^T P x + x^T P \dot{x} \tag{4}$$

Substituting Eq. (1)

$$\dot{V} = (Ax)^T Px + x^T P Ax =$$
$$x^T A^T Px + x^T PAx = x^T (A^T P + PA)x \tag{5}$$

If we find P such that

$$A^T P + PA = -Q \tag{6}$$

where Q is some positive definite, symmetric matrix, then $\dot{V}$ is negative definite and the system is stable.

Let

$$P = \begin{bmatrix} P_{11} & P_{12} \\ P_{12} & P_{22} \end{bmatrix} \tag{7}$$

Since Q is an arbitrary symmetric positive definite matrix, let us try Q = I. Where I is the identity matrix.

Eq. (6) becomes

$$\begin{bmatrix} P_{11} & P_{12} \\ P_{12} & P_{22} \end{bmatrix} \begin{bmatrix} -3k & 3k \\ 2k & -5k \end{bmatrix} + \begin{bmatrix} -3k & 3k \\ 2k & -5k \end{bmatrix} \begin{bmatrix} P_{11} & P_{12} \\ P_{12} & P_{22} \end{bmatrix} = \begin{bmatrix} -1 & 0 \\ 0 & -1 \end{bmatrix} \tag{8}$$

We obtain the set of following equations

$$-6kp_{11} + 4kp_{12} = -1$$

664

$$3kp_{11} - 8kp_{12} + 2kp_{22} = 0$$

$$6kp_{12} - 10kp_{22} = -1$$

Solving the system we get

$$p_{11} = \frac{19}{12k} \quad , \quad p_{12} = \frac{7}{48k} \quad , \quad p_{22} = \frac{15}{80k}$$

$$P = \frac{1}{4k} \begin{bmatrix} \frac{19}{3} & \frac{7}{12} \\ \frac{7}{12} & \frac{15}{20} \end{bmatrix} \qquad (9)$$

P is positive definite when $k > 0$, since

$$\frac{19}{3} > 0$$

$$\frac{19}{3} \cdot \frac{15}{20} - \frac{7}{12} \cdot \frac{7}{12} > 0$$

Therefore the system is asymptotically stable when $k > 0$. The Lyapunov function is

$$V(x) = x^T P x$$

$$= \frac{1}{4k} [x_1 \ x_2] \begin{bmatrix} \frac{19}{3} & \frac{7}{12} \\ \frac{7}{12} & \frac{15}{20} \end{bmatrix} \begin{bmatrix} x_1 \\ x_2 \end{bmatrix}$$

$$= \frac{1}{4k} \cdot \frac{7}{12} \left[\frac{19 \cdot 4}{7} x_1^2 + 2x_1 x_2 + \frac{9}{7} x_2^2 \right]$$

$$= \frac{1}{4k} \cdot \frac{7}{12} \left[(x_1 + x_2)^2 + \frac{69}{7} x_1^2 + \frac{2}{7} x_2^2 \right]$$

The Lyapunov function $V(x)$ is positive-definite.

● **PROBLEM 13-44**

Examine the stability of the equilibrium state of the system

$$\dot{x}_1 = -x_1 - 2x_2$$

$$\dot{x}_2 = x_1 - 4x_2$$

(1)

Solution: We see that the equilibrium state is the origin or $x = 0$.

We solve the general matrix equation

665

$$A'P + PA = -I$$

where $P = p'$

or

$$\begin{bmatrix} -1 & 1 \\ -2 & -4 \end{bmatrix} \begin{bmatrix} p_{11} & p_{12} \\ p_{12} & p_{22} \end{bmatrix} + \begin{bmatrix} p_{11} & p_{12} \\ p_{12} & p_{22} \end{bmatrix} \begin{bmatrix} -1 & -2 \\ 1 & -4 \end{bmatrix} = \begin{bmatrix} -1 & 0 \\ 0 & -1 \end{bmatrix} \qquad (2)$$

The term $x'Px$ is a Lyapunov function and the origin is asymptotically stable when the matrix P is positive definite. We obtain the following system of equations from (2):

$$-2p_{11} + 2p_{12} = -1$$

$$-2p_{11} - 5p_{12} + p_{22} = 0$$

$$-4p_{12} - 8p_{22} = -1$$

Solving the system we obtain

$$p_{11} = \frac{23}{60}, \qquad p_{12} = -\frac{7}{60}, \qquad p_{22} = \frac{11}{60}$$

$$P = \begin{bmatrix} \dfrac{23}{60} & -\dfrac{7}{60} \\ -\dfrac{7}{60} & \dfrac{11}{60} \end{bmatrix}$$

Using Sylvester's criterion we check that P is positive definite. Thus the origin of the system is asymptotically stable and the Lyapunov function is given by

$$V(x) = x'Px = [x_1 \ x_2] \begin{bmatrix} \dfrac{23}{60} & -\dfrac{7}{60} \\ -\dfrac{7}{60} & \dfrac{11}{60} \end{bmatrix} \begin{bmatrix} x_1 \\ x_2 \end{bmatrix}$$

$$= \frac{1}{60} [23x_1^2 - 14x_1x_2 + 11x_2^2]$$

and

$$\dot{V} = -x_1^2 - x_2^2$$

● **PROBLEM** 13-45

The equilibrium state of a system is described by the following equation

$$\begin{bmatrix} \dot{x}_1 \\ \dot{x}_2 \end{bmatrix} = \begin{bmatrix} -4 & -1-j \\ -1+j & -3 \end{bmatrix} \begin{bmatrix} x_1 \\ x_2 \end{bmatrix}$$

determine the stability of the system.

Solution: Since the matrix A is non-singular (det A ≠ 0), the only equilibrium state is the origin x = 0. The following matrix equation

$$\begin{bmatrix} -4 & -1-j \\ -1+j & -3 \end{bmatrix} \begin{bmatrix} p_{11} & p_{12} \\ \overline{p}_{12} & p_{22} \end{bmatrix} + \begin{bmatrix} p_{11} & p_{12} \\ \overline{p}_{12} & p_{22} \end{bmatrix} \begin{bmatrix} -4 & -1-j \\ -1+j & -3 \end{bmatrix} = \begin{bmatrix} -1 & 0 \\ 0 & -1 \end{bmatrix}$$

gives

$$4p_{11} + (1-j)p_{12} + (1+j)\overline{p}_{12} = 1$$
$$(1-j)p_{11} + 7\overline{p}_{12} + (1-j)p_{22} = 0$$
$$(1+j)p_{11} + 7p_{12} + (1+j)p_{22} = 0 \qquad (1)$$
$$(1-j)p_{12} + (1+j)\overline{p}_{12} + 6p_{22} = 1$$

p_{11} and p_{22} are real ($p_{11} = \overline{p}_{11}$ and $p_{22} = \overline{p}_{22}$)

Thus the third equation is conjugate to the second.

Solving equation (1) we obtain

$$p_{11} = \frac{21}{64}, \quad p_{12} = -\frac{5}{64} - j\frac{5}{64}, \quad p_{22} = \frac{63}{128}$$

or

$$P = \begin{bmatrix} \dfrac{21}{64} & -\dfrac{5}{64} - j\dfrac{5}{64} \\ -\dfrac{5}{64} + j\dfrac{5}{64} & \dfrac{63}{128} \end{bmatrix}$$

P is positive definite. Hence we conclude that the origin of the system is asymptotically stable.

● **PROBLEM** 13-46

Find the conditions on the elements a_{ij} which are required for stability of the following system

$$\dot{x}(t) = Ax(t)$$

where

$$A = \begin{bmatrix} a_{11} & a_{12} \\ a_{21} & a_{22} \end{bmatrix}$$

Solution: If for any symmetric positive definite matrix Q there exists a symmetric positive definite matrix P which is the unique solution of

$$A^T P + PA = -Q$$

then the system $\dot{x}(t) = Ax(t)$ is asymptotically stable at the origin.

Let $Q = I$; we obtain the following equation

$$\begin{bmatrix} a_{11} & a_{21} \\ a_{12} & a_{22} \end{bmatrix} \begin{bmatrix} p_{11} & p_{12} \\ p_{12} & p_{22} \end{bmatrix} + \begin{bmatrix} p_{11} & p_{12} \\ p_{12} & p_{22} \end{bmatrix} \begin{bmatrix} a_{11} & a_{12} \\ a_{21} & a_{22} \end{bmatrix} = \begin{bmatrix} -1 & 0 \\ 0 & -1 \end{bmatrix}$$

We used the fact that $P = P^T$.

The above equation can be reduced to

$$\begin{bmatrix} 2a_{11} & 2a_{21} & 0 \\ a_{12} & a_{11} + a_{22} & a_{21} \\ 0 & 2a_{12} & 2a_{22} \end{bmatrix} \begin{bmatrix} p_{11} \\ p_{12} \\ p_{22} \end{bmatrix} = \begin{bmatrix} -1 \\ 0 \\ -1 \end{bmatrix}$$

We find the solution

$$\begin{bmatrix} p_{11} \\ p_{12} \\ p_{22} \end{bmatrix} = \frac{-1}{2(a_{11}+a_{22})(a_{11}a_{22}-a_{12}a_{21})} \begin{bmatrix} a_{11}a_{22}-a_{12}a_{21}+a_{21}^2+a_{22}^2 \\ -a_{12}a_{22}-a_{11}a_{21} \\ a_{11}a_{22}-a_{12}a_{21}+a_{11}^2+a_{12}^2 \end{bmatrix}$$

Thus for P to be positive-definite it is necessary that

$$a_{11}a_{22} - a_{12}a_{21} > 0$$

$$a_{11} + a_{22} < 0 .$$

For the system to be stable the above inequalities for the elements of matrix A must hold.

To find the eigenvalues we write

$$|A - \lambda I| = \begin{vmatrix} a_{11}-\lambda & a_{12} \\ a_{21} & a_{22}-\lambda \end{vmatrix} = 0$$

We obtain the equation

$$\lambda^2 - (a_{11} + a_{22})\lambda + (a_{11}a_{22} - a_{12}a_{21}) = 0$$

For the system to be stable the roots of the above equation must have negative real parts. The necessary condition is that all the coefficient of the characteristic equation must be of the same sign, that yields

$$a_{11} + a_{22} < 0$$

$$a_{11}a_{22} - a_{12}a_{21} > 0$$

For the system shown in the diagram determine the stability range for the gain K.

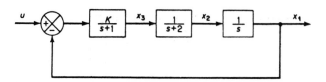

Solution: From the diagram we have the following state equation of the system

$$\dot{x} = Ax + Bu$$

where

$$A = \begin{bmatrix} 0 & 1 & 0 \\ 0 & -2 & 1 \\ -K & 0 & -1 \end{bmatrix} \quad \text{and} \quad B = \begin{bmatrix} 0 \\ 0 \\ K \end{bmatrix}$$

We assume the input u to be zero and choose the real positive-semidefinite symmetric matrix Q

$$Q = \begin{bmatrix} 0 & 0 & 0 \\ 0 & 0 & 0 \\ 0 & 0 & 1 \end{bmatrix}$$

For $\dot{V}(x)$ we have

$$\dot{V}(x) = -x'Qx = -x_3^2$$

The choice of Q is right since $\dot{V}(x)$ cannot be identically equal to zero except at the origin.

If $\dot{V} = 0$ then $x_3 = 0$

and $0 = -Kx_1 - 0$

thus, $x_1 = 0$.

From $\dot{x}_1 = x_2$ we conclude that if $x_1 = 0$ then $x_2 = 0$.
Thus

$$\dot{V}(x) = 0$$

only at the origin.

To find the range of K let us solve the following equation

$$A'P + PA = -Q$$

or

$$\begin{bmatrix} 0 & 0 & -K \\ 1 & -2 & 0 \\ 0 & 1 & -1 \end{bmatrix} \begin{bmatrix} P_{11} & P_{12} & P_{13} \\ P_{12} & P_{22} & P_{23} \\ P_{13} & P_{23} & P_{33} \end{bmatrix} + \begin{bmatrix} P_{11} & P_{12} & P_{13} \\ P_{12} & P_{22} & P_{23} \\ P_{13} & P_{23} & P_{33} \end{bmatrix} \begin{bmatrix} 0 & 1 & 0 \\ 0 & -2 & 1 \\ -K & 0 & -1 \end{bmatrix} = \begin{bmatrix} 0 & 0 & 0 \\ 0 & 0 & 0 \\ 0 & 0 & -1 \end{bmatrix}$$

Solving the above equation we get

$$P = \begin{bmatrix} \dfrac{K^2 + 12K}{12 - 2K} & \dfrac{6K}{12 - 2K} & 0 \\[2ex] \dfrac{6K}{12 - 2K} & \dfrac{3K}{12 - 2K} & \dfrac{K}{12 - 2K} \\[2ex] 0 & \dfrac{K}{12 - 2K} & \dfrac{6}{12 - 2K} \end{bmatrix}$$

P is positive definite when

$$0 < K < 6$$

We conclude that for $0 < K < 6$ the origin of the system is asymptotically stable in the large.

● **PROBLEM 13-48**

Find the Lyapunov function for the following system

$$\begin{bmatrix} \dot{x}_1 \\ \dot{x}_2 \end{bmatrix} = \begin{bmatrix} 0 & 1 \\ -1 & -1 \end{bmatrix} \begin{bmatrix} x_1 \\ x_2 \end{bmatrix}$$

Obtain an estimate of the minimum value of μ where

$$\mu = -\frac{\dot{V}}{V}$$

Solution: $A'P + PA = -Q$

where $P' = P.$

We assume that $Q = I$ and compute P.

$$\begin{bmatrix} 0 & -1 \\ 1 & -1 \end{bmatrix} \begin{bmatrix} P_{11} & P_{12} \\ P_{12} & P_{22} \end{bmatrix} + \begin{bmatrix} P_{11} & P_{12} \\ P_{12} & P_{22} \end{bmatrix} \begin{bmatrix} 0 & 1 \\ -1 & -1 \end{bmatrix} = \begin{bmatrix} -1 & 0 \\ 0 & -1 \end{bmatrix}$$

$$P = \begin{bmatrix} \dfrac{3}{2} & \dfrac{1}{2} \\[2ex] \dfrac{1}{2} & 1 \end{bmatrix}$$

The Lyapunov function is given in the form

$$V = x'Px = \frac{1}{2} (3x_1^2 + 2x_1x_2 + 2x_2^2)$$

$V(x)$ is positive definite.

$$\dot{V}(x) = -x'x = -(x_1^2 + x_2^2)$$

We have the following equation for μ

$$\mu = \frac{-\dot{V}}{V} = \frac{2(x_1^2 + x_2^2)}{3x_1^2 + 2x_1x_2 + 2x_2^2}$$

Note that the minimum value of μ is the minimum eigenvalue of QP^{-1}. We assumed that $Q = I$, and obtain

$$|QP^{-1} - \lambda I| = |P^{-1} - \lambda I| = 0$$

The eigenvalues determined by the above equation are equal to those of

$$|I - P\lambda| = 0$$

We have

$$\begin{vmatrix} 1 - \frac{3}{2}\lambda & -\frac{1}{2} \\ -\frac{1}{2} & 1 - \lambda \end{vmatrix} = 0$$

$$\frac{5}{4}\lambda^2 - \frac{5}{2}\lambda + 1 = 0$$

$$\lambda_1 = 1 + \sqrt{\frac{1}{5}} \qquad \lambda_2 = 1 - \sqrt{\frac{1}{5}}$$

Thus

$$\mu_{min} = 1 - \sqrt{\frac{1}{5}} = 0.553$$

Note that the same value for μ_{min} can be obtained from the equation

$$\frac{\partial \mu}{\partial x_1} = 0 .$$

● **PROBLEM** 13-49

For the system described by

$$x_1(k+1) = -0.7 \, x_1(k)$$

$$x_2(k+1) = -0.7 \, x_2(k)$$

(1)

Determine the stability of the equilibrium state.

671

Solution: We can write Eq. (1) in the matrix form

$$x(k+1) = A \, x(k)$$

where

$$A = \begin{bmatrix} -0.7 & 0 \\ 0 & -0.7 \end{bmatrix}$$

The general equation for P is

$$A'PA - P = -Q \qquad\qquad (2)$$

We set   Q = I

P is a symmetric matrix in the form

$$P = \begin{bmatrix} p_{11} & p_{12} \\ p_{12} & p_{22} \end{bmatrix}$$

Substituting A, P and Q into (2) we get

$$\begin{bmatrix} -0.7 & 0 \\ 0 & -0.7 \end{bmatrix} \begin{bmatrix} p_{11} & p_{12} \\ p_{12} & p_{22} \end{bmatrix} \begin{bmatrix} -0.7 & 0 \\ 0 & -0.7 \end{bmatrix} - \begin{bmatrix} p_{11} & p_{12} \\ p_{12} & p_{22} \end{bmatrix} = \begin{bmatrix} -1 & 0 \\ 0 & -1 \end{bmatrix}$$

Multiplying we obtain

$$(0.7)^2 \, p_{11} - p_{11} = -1$$
$$(0.7)^2 \, p_{22} - p_{22} = -1$$
$$(0.7)^2 \, p_{12} - p_{12} = 0$$

Solving the above equations gives matrix P

$$P = \begin{bmatrix} 1.96 & 0 \\ 0 & 1.96 \end{bmatrix}$$

which is positive definite.

Therefore the function V(k)

$$V(x) = x'(k) \, P \, x(k)$$

is a Lyapunov function and is positive definite.

The function $\Delta V(x) = -x'(k) \, Q \, x(k)$ is negative definite thus the equilibrium state is asymptotically stable.

Determine the stability of the second-order system.

$$\begin{bmatrix} \dot{x}_1 \\ \dot{x}_2 \end{bmatrix} = \begin{bmatrix} 0 & -1 \\ 1 & -1 \end{bmatrix} \begin{bmatrix} x_1 \\ x_2 \end{bmatrix} \tag{1}$$

Use the Lyapunov method.

Solution: We assume the Lyapunov function to be

$$V(x) = x'Px$$

where

P is defined by the equation

$$A'P + PA = -I$$

$$P' = P$$

Substituting matrix A from (1) we have

$$\begin{bmatrix} 0 & 1 \\ -1 & -1 \end{bmatrix} \begin{bmatrix} P_{11} & P_{12} \\ P_{12} & P_{22} \end{bmatrix} + \begin{bmatrix} P_{11} & P_{12} \\ P_{12} & P_{22} \end{bmatrix} \begin{bmatrix} 0 & -1 \\ 1 & -1 \end{bmatrix} = \begin{bmatrix} -1 & 0 \\ 0 & -1 \end{bmatrix}$$

and multiplying the matrices and comparing the elements we obtain

$$2p_{12} = -1$$

$$p_{22} - p_{11} - p_{12} = 0$$

$$-2p_{22} - 2p_{12} = -1$$

$$P = \begin{bmatrix} \dfrac{3}{2} & -\dfrac{1}{2} \\ \dfrac{1}{2} & 1 \end{bmatrix}$$

Both determinants of the principal minors are positive thus P is positive definite.

$$p_{11} = \frac{3}{2} > 0$$

and

$$\begin{vmatrix} \dfrac{3}{2} & -\dfrac{1}{2} \\ -\dfrac{1}{2} & 1 \end{vmatrix} = \frac{3}{2} - \frac{1}{4} > 0$$

The equilibrium state at the origin is asymptotically stable in the large and the Lyapunov function is given by

$$V = x'Px = \frac{3}{2} x_1^2 - \frac{1}{2} x_1 x_2 - \frac{1}{2} x_1^2 + x_2^2$$

$$= x_1^2 - \frac{1}{2} x_1 x_2 + x_2^2 = \left(\frac{x_1}{2} - \frac{x_2}{2} \right)^2 + \frac{3}{4} x_1^2 + \frac{3}{4} x_2^2 > 0$$

and its time derivative

$$\dot{V}(x) = - \frac{1}{2}\left(x_1 - \frac{x_2}{2}\right)^2 - \frac{11}{8} x_2^2 < 0$$

● **PROBLEM** 13-51

The system is described by the equation

$$\dot{x} = Ax$$

where A is the matrix with the distinct eigenvalues having negative real parts.

Given the transformation

$$x = Py$$

such that

$$\dot{y} = P^{-1}APy = Dy$$

where D is diagonal, obtain the Lyapunov function for the system.

Solution: Let $V(y) = y*y$

V is positive definite.

Its time derivative

$$\dot{V}(y) = \dot{y}*y + y*\dot{y}$$

$$= y*D*y + y*Dy = 2y*Dy$$

The eigenvalues of A have negative real parts and the eigenvalues of D and A are identical, thus

$$\dot{V}(y)$$

is negative definite.   That proves that the function $V(y)$ is the Lyapunov function for the system.

● **PROBLEM** 13-52

The system is described by the following equation

$$\dot{x} = Ax + Bu \qquad (1)$$

where A and B are real matrices, x and the control vector u are real vectors.

Show that the origin of the system is asymptotically stable in the large if

$$u = -B'Px \tag{2}$$

where P is a real positive definite symmetric matrix such that

$$A'P + PA = -I \tag{3}$$

<u>Solution</u>: Let us choose the following function

$$V(x) = x'Px$$

which is positive definite since P is positive definite.

For $\dot{V}$ we get

$$\dot{V} = \dot{x}'Px + x'P\dot{x}$$

$$= x'(A'P + PA)x + u'B'Px + x'PBu$$

$$= -x'Ix - 2x'PBB'Px$$

$\dot{V}$ is negative definite since PBB'P is always positive semidefinite and I is positive definite.

We conclude that the origin of the system is asymptotically stable in the large since

$$V > 0, \ \dot{V} < 0$$

and

$$\lim_{\|x\| \to \infty} V = \infty \ .$$

● **PROBLEM** 13-53

Consider the second-order system

$$\ddot{x} + \dot{x} + x = 0$$

Using V function obtain an estimate of the largest time constant of the system.

<u>Solution</u>: Let us substitute $x = x_1$, $\dot{x}_1 = x_2$

and get the following associated system

$$\dot{x}_1 = x_2$$
$$\dot{x}_2 = -x_1 - x_2$$

$$A = \begin{bmatrix} 0 & 1 \\ -1 & -1 \end{bmatrix}$$

We assume that

$$V = x'Px$$

675

where

$$A'P + PA = -I$$

$$P' = P$$

Substituting for A

$$\begin{bmatrix} 0 & -1 \\ 1 & -1 \end{bmatrix} \begin{bmatrix} P_{11} & P_{12} \\ P_{12} & P_{22} \end{bmatrix} + \begin{bmatrix} P_{11} & P_{12} \\ P_{12} & P_{22} \end{bmatrix} \begin{bmatrix} 0 & 1 \\ -1 & -1 \end{bmatrix} = \begin{bmatrix} -1 & 0 \\ 0 & -1 \end{bmatrix}$$

Computing P we have

$$P = \begin{bmatrix} \dfrac{3}{2} & \dfrac{1}{2} \\ \dfrac{1}{2} & 1 \end{bmatrix}$$

and for V

$$V = x'Px = \frac{1}{2}\left((x_1 + x_2)^2 + 2x_1^2 + x_2^2 \right)$$

V is positive definite.

$$\dot{V} = -x'x = -(x_1^2 + x_2^2)$$

Thus

$$\mu = -\frac{\dot{V}}{V} = \frac{2(x_1^2 + x_2^2)}{3x_1^2 + 2x_1x_2 + 2x_2^2}$$

Differentiating μ with respect to x_1

$$\frac{\partial \mu}{\partial x_1} = \frac{4x_1^2 x_2 - 4x_1 x_2^2 - 4x_2^3}{(3x_1^2 + 2x_1x_2 + 2x_2^2)^2}$$

and letting

$$\frac{\partial \mu}{\partial x_1} = 0$$

gives

$$x_1 = 1.618x_2$$

or

$$x_1 = -0.618x_2$$

That gives

$$\mu_{min} = 0.553$$

$$\mu_{max} = 1.447$$

676

Thus the upper limit of the convergence time is

$$\frac{1}{\mu_{min}} = 1.81 \text{ sec.}$$

Note that 1.81 sec. is an estimate of the convergence time for $V(x)$, since

$$V(x) = F(x_i^2(t)),$$

thus the upper limit on the time constant for $x_1(t)$ is

$$\tau = 2 \cdot 1.81 = 3.62 \text{ sec.}$$

The system time constant is $T = 2$; note that

$$T < \tau$$

● **PROBLEM** 13-54

Prove that for $A < 0$ and $B < 0$ the system

$$\dot{x}_1 = x_2$$

$$\dot{x}_2 = Ax_1 + Bx_2$$

is asymptotically stable at the origin.

Solution: As one of the possible Lyapunov functions, we shall use

$$V(x_1, x_2) = (-Bx_1 + x_2)^2 - A(1 - A)x_1^2 - Ax_2^2$$

Since

$$A < 0 \quad \text{and} \quad B < 0$$

$$V(x_1, x_2) \geq 0$$

$$\dot{V}(x_1, x_2) = -2AB(x_1^2 + x_2^2)$$

$\dot{V}$ is negative definite, therefore the system is asymptotically stable at the origin.

We shall find the Lyapunov function using another approach. Let us try the following function

$$V(x_1, x_2) = \alpha x_1^2 + x_2^2$$

For $\alpha > 0$ the above function is positive definite. We shall calculate the time derivative of V and choose α to make

$$\frac{dV}{dt}$$

negative definite.

$$\frac{dV}{dt} = 2\alpha \, x_1 x_2 + 2x_2 \, (Ax_1 + Bx_2)$$

thus

$$2\alpha \, x_1 x_2 + 2Ax_1 x_2 = 0$$

$$\alpha = -A$$

We have

$$\frac{dV}{dt} = 2Bx_2^2 \leq 0$$

which is negative semidefinite

(it is zero for $(x_1, 0)$).

● **PROBLEM** 13-55

Determine the stability of the origin of the system

$$\dot{x}_1 = x_2$$

$$\dot{x}_2 = -2x_1 - 3x_2 - 2x_1^3$$

Use the Lyapunov function.

Solution: We choose the following V function

$$V(x_1, x_2) = \frac{1}{2} \, (x_1^4 + 2x_1^2 + x_2^2)$$

V is positive definite.

$$\dot{V} = x_2 (-2x_1 - 3x_2 - 2x_1^3) + x_2 (2x_1 + 2x_1^3) = -3x_2^2$$

$\dot{V}$ is negative definite and the origin of the system is asymptotically stable.

● **PROBLEM** 13-56

Determine the stability of the following system

$$\dot{x} = Ax$$

where

$$A = \begin{bmatrix} 0 & 1 \\ -1 & -1 \end{bmatrix}$$

Solution: As a possible Lyapunov function let us take

$$V(x) = x_1^2 + x_2^2$$

Clearly $V(x)$ is positive definite. The only equilibrium

state is the origin, $x = 0$.

For the time derivative we have

$$\dot{V}(x) = 2x_1\dot{x}_1 + 2x_2\dot{x}_2 = 2x_1x_2 - 2x_2x_1 - 2x_2^2$$

$$\dot{V}(x) = -2x_2^2$$

is negative semidefinite. Observe that $\dot{V}(x)$ vanishes identically only at the origin. Indeed, if $\dot{V}(x)$ is to vanish identically for $t_1 \geq t_0$ then x_2 must be zero for $t_1 \geq t_0$; x_1 must be also zero since

$$\dot{x}_2 = -x_1 - x_2.$$

The Lyapunov function is not unique and for the above system the function

$$V(x) = \frac{1}{2}[(x_1 + x_2)^2 + 2x_1^2 + x_2^2]$$

is positive definite.

Its time derivative

$$\dot{V}(x) = (x_1 + x_2)(\dot{x}_1 + \dot{x}_2) + 2x_1\dot{x}_1 + x_2\dot{x}_2$$

$$= -(x_1^2 + x_2^2)$$

is negative definite. The equilibrium state at the origin is asymptotically stable in the large, since

$$\lim_{\|x\| \to \infty} V(x) = \infty$$

● **PROBLEM** 13-57

In the system shown, $U(s)$ consists of an external input plus state-variable feedback. Determine the conditions for the feedback of the system to be stable.

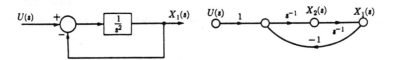

Solution: We shall use Lyapunov's direct method to determine a control signal $u(t)$ which will result in a stable system response. The matrix differential equation is

$$\dot{x} = \begin{bmatrix} 0 & 1 \\ -1 & 0 \end{bmatrix} x + \begin{bmatrix} 0 \\ 1 \end{bmatrix} u.$$

The Lyapunov function to be used is

$$V(x) = x^TIx = x_1^2 + x_2^2 .$$

The time derivative of the Lyapunov function is then

$$\dot{V} = 2(x_1\dot{x}_1 + x_2\dot{x}_2)$$

$$= 2(x_1 x_2 + x_2(-x_1 + u))$$

$$= 2x_2 u .$$

Therefore in order to assure stability we require that

$$u(t) = -K_1 x_2 ,$$

where K_1 is constant. The value of K_1 is chosen on the basis of the desired response. If an external input $R(s)$ affects the system then we use

$$u(t) = r(t) - K_1 x_2(t) .$$

● **PROBLEM** 13-58

Determine the stability of a harmonic oscillator with frequency w. Use both the Lyapunov function and the exact solution and compare the results.

Solution: The equation of motion of a harmonic oscillator with frequency w is

$$\frac{d^2 x}{dt^2} + w^2 x = 0$$

Let us write

$$x = x_1 \quad , \quad \frac{dx_1}{dt} = x_2$$

and get an equivalent system

$$\frac{dx_1}{dt} = x_2$$

$$\frac{dx_2}{dt} = -w^2 x_1$$

In the matrix notation we have

$$\frac{dx}{dt} = Ax ,$$

where

$$A = \begin{pmatrix} 0 & 1 \\ -w^2 & 0 \end{pmatrix}$$

We shall use the following positive definite function

$$V(x_1 , x_2) = w^2 x_1^2 + x_2^2$$

and see that it is the Lyapunov function. We have

$$\frac{\partial V}{\partial x} = \begin{bmatrix} 2w^2 x_1 \\ 2x_2 \end{bmatrix}$$

and

$$\frac{dV}{dt} = \left(\frac{\partial V}{\partial x}\right)' \cdot \frac{dx}{dt} = (2w^2 x_1 \ , \ 2x_2) \begin{pmatrix} 0 & 1 \\ -w^2 & 0 \end{pmatrix} \begin{pmatrix} x_1 \\ x_2 \end{pmatrix}$$

$$= 2w^2 x_1 x_2 - 2w^2 x_1 x_2 = 0$$

We conclude that the system is stable. We shall use the exact solution of the equations and the definition of stability to prove that the system is stable.

The initial state of the system is

$$x(t_0) = (x_1^0 \ , \ x_2^0)$$

The solution is

$$\begin{pmatrix} x_1(t) \\ x_2(t) \end{pmatrix} = \begin{pmatrix} x_1^0 \cos(wt + \alpha) + \dfrac{x_2^0}{w} \sin(wt + \alpha) \\ -wx_1^0 \sin(wt + \alpha) + x_2^0 \cos(wt + \alpha) \end{pmatrix}$$

where $\alpha = n \cdot 2\Pi$.

We shall use the following norm of x

$$\|x\| = \underset{i \ j}{\Sigma \ \Sigma} \ |x_{ij}|$$

and get

$$\|x(t)\| = \left| x_1^0 \cos(wt + \alpha) + \frac{x_2^0}{w} \sin(wt + \alpha) \right|$$

$$+ \left| -wx_1^0 \sin(wt + \alpha) + x_2^0 \cos(wt + \alpha) \right| \le |x_1^0| + |w||x_1^0| + \frac{|x_2^0|}{|w|} + |x_2^0|$$

$$\le |x_1^0| + |w||x_1^0| + \frac{|x_1^0|}{|w|} + |x_2^0| + |w| \cdot |x_2^0| + \frac{|x_2^0|}{|w|}$$

$$= (1 + \frac{1}{|w|} + |w|)(|x_1^0| + |x_2^0|) = (1 + \frac{1}{|w|} + |w|) \cdot \|x(t_0)\|$$

For a given $\varepsilon > 0$ we choose the following δ

$$\delta = \frac{\varepsilon}{1 + \dfrac{1}{|w|} + |w|}$$

and see that

If

$$\|x(t_0)\| < \delta \text{ then } \|x(t)\| < \varepsilon .$$

That proves that the system is stable.

Determine the stability of the origin of the system

$$\dot{x}_1 = -x_1 + x_2 + x_1(x_1^2 + x_2^2)$$

$$\dot{x}_2 = -x_1 - x_2 + x_2(x_1^2 + x_2^2)$$

Use

$$V(x_1 x_2) = x_1^2 + x_2^2$$

Solution: $V(x_1 x_2)$ is positive and its time derivative

$$\dot{V}(x_1, x_2) = -2x_1^2 - 2x_2^2 + 2(x_1^2 + x_2^2)^2 < 0$$

$$2(x_1^2 + x_2^2)^2 < 2(x_1^2 + x_2^2) \qquad x_1^2 + x_2^2 \neq 0$$

$$x_1^2 + x_2^2 < 1$$

The region where $\dot{V}(x_1, x_2)$ is negative is the inside of the circle of radius 1. We see that the origin of the system is asymptotically stable.

• **PROBLEM** 13-60

Determine the stability of the origin of the following system

$$\dot{x}_1 = -\frac{7}{4} x_1 + \frac{1}{4} x_2$$

$$\dot{x}_2 = \frac{3}{4} x_1 - \frac{5}{4} x_2$$

Solution: The origin is the only equilibrium point. The function

$$V(x_1, x_2) = 12x_1^2 + 12x_1 x_2 + 20x_2^2$$

is the Lyapunov function. Indeed

$$V(x_1, x_2) = 12x_1^2 + 12x_1 x_2 + 20x_2^2$$

$$= 3x_1^2 + (3x_1 + 2x_2)^2 + 16x_2^2 > 0$$

is positive definite.

Its time derivative is

$$\dot{V}(x_1, x_2) = -(33x_1^2 + 47x_2^2)$$

and negative definite.

The system's origin is asymptotically stable.

We should note that an equally good choice (among many others) for V function would have been

$$V(x_1, x_2) = 3x_1^2 + 6x_1x_2 + 11x_2^2$$

Show that the following system

$$\dot{x}_1 = a\varepsilon x_1 - 2x_2 - 3ax_1^3$$

$$\dot{x}_2 = 2x_1 + a\varepsilon x_2 - 3ax_2^3$$

where $a > 0$ and ε is a small positive constant, is unstable in the origin.

Solution: The above system near the origin becomes linear

$$\dot{x}_1 = a\varepsilon x_1 - 2x_2$$

$$\dot{x}_2 = 2x_1 + a\varepsilon x_2$$

$$\begin{vmatrix} a\varepsilon - \lambda & -2 \\ 2 & a\varepsilon - \lambda \end{vmatrix} = 0$$

$$(a\varepsilon - \lambda)^2 + 4 = 0$$

Solving the equation we obtain

$$\lambda_1 = a\varepsilon + 2j$$

$$\lambda_2 = a\varepsilon - 2j$$

Since $a\varepsilon > 0$ the characteristic roots have positive real parts and the origin is unstable.

It's easy to see that if we disregard small oscillations within a small circle containing the origin, the system becomes "asymptotically stable" in the large. We shall use the following Lyapunov function

$$V(x_1, x_2) = x_1^2 + x_2^2$$

and compute its time derivative

$$\dot{V} = 2x_1(a\varepsilon x_1 - 2x_2 - 3ax_1^3) + 2x_2(2x_1 + a\varepsilon x_2 - 3x_2^3 a)$$

$$= -6a(x_1^4 + x_2^4) + 2a\varepsilon(x_1^2 + x_2^2)$$

Note that for (x_1, x_2) outside the circle K with its center at the origin and radius

$$\sqrt{\frac{2\varepsilon}{3}}$$

we have

$$\sqrt{\frac{2\varepsilon}{3}} < \sqrt{x_1^2 + x_2^2}$$

$$\dot{V} < -6a[(x_1^4 + x_2^4) - \frac{1}{2}(x_1^2 + x_2^2)^2] = -3a(x_1^2 - x_2^2)^2 \leq 0$$

Hence outside the circle K we have $V > 0$ and $\dot{V} < 0$. If we neglect small oscillations of amplitude less than

$$\sqrt{\frac{2\varepsilon}{3}}$$

around the origin, the origin may be regarded as "asymptotically stable" in the large. That means that any solution starting from the outside of the circle K approaches a limit cycle located inside the circle as $t \to \infty$.

● **PROBLEM** 13-62

Using the Lyapunov function determine the stability of the origin of the system.

$$\dot{x}_1 = \frac{-2x_1}{(1+x_1^2)^2} + \frac{4x_2}{(1+x_1^2)^2}$$

$$\dot{x}_2 = \frac{-4x_1}{(1+x_1^2)^4} - 36.7x_2$$

Solution: Let us try the following $V(x_1, x_2)$ function

$$V(x_1, x_2) = \frac{x_1^2}{1+x_1^2} + x_2^2$$

This function is positive definite for the whole space x_1, x_2 except the origin. The time derivative is

$$\frac{dV(x_1, x_2)}{dt} = \frac{\partial V}{\partial x_1} \frac{dx_1}{dt} + \frac{\partial V}{\partial x_2} \frac{dx_2}{dt} =$$

$$\frac{2x_1}{(1+x_1^2)^2} \left[\frac{-2x_1}{(1+x_1^2)^2} + \frac{4x_2}{(1+x_1^2)^2} \right] + 2x_2 \left[\frac{-4x_1}{(1+x_1^2)^4} - 36.7x_2 \right]$$

We have

$$\dot{V}(x_1, x_2) = \frac{-4x_1^2}{(1+x_1^2)^4} - 73.4x_2^2$$

Thus the time derivative is negative definite and V is the Lyapunov function. The origin of the system is asymptotically stable.

Determine the stability of the system described by

$$\dot{x}_1 = Ax_2 - Bx_1(x_1^2 + x_2^2)$$

$$\dot{x}_2 = -Ax_1 - Bx_2(x_1^2 + x_2^2)$$

where A and B are real parameters.

Solution: We choose the following scalar function

$$V(x) = x_1^2 + x_2^2$$

Its time derivative is

$$\dot{V}(x) = 2x_1[Ax_2 - Bx_1(x_1^2 + x_2^2)] + 2x_2[-Ax_1 - Bx_2(x_1^2 + x_2^2)]$$

$$= -2B(x_1^2 + x_2^2)^2$$

which is negative definite for B > 0. Thus V(x) is a Lyapunov function and the equilibrium state at the origin of the system is asymptotically stable in the large when B > 0.

Determine the stability of the system

$$\dot{x}_1 = x_2$$

$$\dot{x}_2 = -x_1^3 - x_2$$

(1)

Solution: The origin is the equilibrium state. We choose the following Lyapunov function

$$V(x_1, x_2) = x_1^4 + 2x_2^2$$

and obtain

$$\frac{dV(x_1, x_2)}{dt} = \frac{\partial V}{\partial x_1} \frac{dx_1}{dt} + \frac{\partial V}{\partial x_2} \frac{dx_2}{dt}$$

$$= 4x_1^3 \dot{x}_1 + 4x_2 \dot{x}_2 = 4x_1^3 x_2 + 4x_2(-x_2 - x_1^3)$$

$$= -4x_2^2 \leq 0$$

Let A be the set of points (x_1, x_2) for which

$$\dot{V}(x_1, x_2) = 0$$

$$A = \{(x_1, x_2); \quad \dot{V}(x_1, x_2) = 0\}$$

$$= \{(x_1, 0); \quad -\infty < x_1 < \infty\}$$

and B the maximum invariant subset of A.

For $x_2 = 0$ we have

$$\dot{x}_2 = -x_1^3 = 0 \qquad x_1 = 0$$

and $B = \{(0,0)\}$.

Note that

$$\lim_{\|x\| \to \infty} V(x) = \infty$$

thus each solution is bounded for $t \geq 0$ and all solutions approach B (the origin $x = 0$) when $t \to \infty$. Hence the origin of the system is asymptotically stable in the large.

● **PROBLEM** 13-65

The motion of a rotating object about the principal axes of inertia is described by the Euler equations. Let us assume that the object is a space vehicle with the control torques T_x, T_y, T_z equal to

$$T_x = k_1 A w_x$$

$$T_y = k_2 B w_y$$

$$T_z = k_3 C w_z$$

The equations of motion are

$$A\dot{w}_x - (B-C)w_y w_z = T_x$$

$$B\dot{w}_y - (C-A)w_z w_x = T_y$$

$$C\dot{w}_z - (A-B)w_x w_y = T_z$$

A, B, C and w_x, w_y, w_z are moments of inertia and the angular velocities about the principal axes respectively. Application of the control torques can stop possible rolling about of the vehicle.

Determine the sufficient conditions (inequalities on k_1, k_2, k_3) for the system to be asymptotically stable.

Solution: Let us rewrite the equations, assuming $A \neq 0$, $B \neq 0$, $C \neq 0$

$$\dot{w}_x - \left(\frac{B}{A} - \frac{C}{A}\right)w_y w_z = k_1 w_x$$

$$\dot{w}_y - \left(\frac{C}{B} - \frac{A}{B}\right)w_z w_x = k_2 w_y$$

$$\dot{w}_z - \left(\frac{A}{C} - \frac{B}{C}\right)w_x w_y = k_3 w_z$$

686

or
$$\dot{w} = Aw$$

where

$$A = \begin{bmatrix} k_1 & \dfrac{B}{A}\,w_z & -\dfrac{C}{A}\,w_y \\[2ex] -\dfrac{A}{B}\,w_z & k_2 & \dfrac{C}{B}\,w_x \\[2ex] \dfrac{A}{C}\,w_y & -\dfrac{B}{C}\,w_x & k_3 \end{bmatrix}$$

Clearly the equilibrium state is the origin $w = 0$.

If we choose

$$V(w) = w'Pw = w' \begin{bmatrix} A^2 & 0 & 0 \\[1ex] 0 & B^2 & 0 \\[1ex] 0 & 0 & C^2 \end{bmatrix} w$$

$$= A^2 w_x^2 + B^2 w_y^2 + C^2 w_z^2$$

$V(w)$ is positive definite, its time derivative is

$$\dot{V}(w) = \dot{w}'Pw + w'P\dot{w}$$

$$= w' \begin{bmatrix} k_1 & -\dfrac{A}{B}\,w_z & \dfrac{A}{C}\,w_y \\[2ex] \dfrac{B}{A}\,w_z & k_2 & -\dfrac{B}{C}\,w_x \\[2ex] -\dfrac{C}{A}\,w_y & \dfrac{C}{B}\,w_x & k_3 \end{bmatrix} \begin{bmatrix} A^2 & 0 & 0 \\[1ex] 0 & B^2 & 0 \\[1ex] 0 & 0 & C^2 \end{bmatrix} w +$$

$$w' \begin{bmatrix} A^2 & 0 & 0 \\[1ex] 0 & B^2 & 0 \\[1ex] 0 & 0 & C^2 \end{bmatrix} \begin{bmatrix} k_1 & \dfrac{B}{A}\,w_z & -\dfrac{C}{A}\,w_y \\[2ex] -\dfrac{A}{B}\,w_z & k_2 & \dfrac{C}{B}\,w_x \\[2ex] \dfrac{A}{C}\,w_y & -\dfrac{B}{C}\,w_x & k_3 \end{bmatrix} w$$

$$= w' \begin{bmatrix} 2k_1 A^2 & 0 & 0 \\[1ex] 0 & 2k_2 B^2 & 0 \\[1ex] 0 & 0 & 2k_3 C^2 \end{bmatrix} w = -w'Qw$$

The sufficient condition for asymptotic stability
is that Q be positive definite, hence we have

$$k_1 < 0 , \qquad k_2 < 0 , \qquad k_3 < 0$$

The equilibrium state is asymptotically stable in the
large since for negative k_i

$$V(w) \to \infty \qquad \text{as} \qquad ||w|| \to \infty .$$

● **PROBLEM** 13-66

Investigate the stability of the system

$$\frac{dx_1}{dt} = - x_1$$

$$\frac{dx_2}{dt} = -x_2 - x_1^2$$

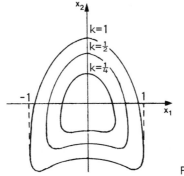

Fig. 1

Solution: It is easy to find the general solution of the
system. We assume the initial conditions to be

$$x_1(t_0) = x_{10}$$

$$x_2(t_0) = x_{20}$$

The first equation is independent of x_2 and its solution is

$$x_1(t) = x_{10} e^{-(t-t_0)}$$

Integration of the second equation gives

$$x_2(t) = (x_{20} - x_{10}^2) e^{-(t-t_0)} + x_{10}^2 e^{-2(t-t_0)}$$

We see that the system is asymptotically stable in-the-
large.

In the same example we shall apply the Lyapunov function
and see the limitations. To prove the stability we use
the following theorem:

Given a scalar function $V(x)$, with continuous partial derivatives, which satisfies the following conditions in the region P

1. $V(x) < k$

2. $V(x) > 0$ for $x \neq 0$

3. $V(0) = 0$

4. $\dfrac{dV}{dt} < 0$

Then the origin is asymptotically stable, and every solution, where $x(t_0) \, \varepsilon \, P$, approaches the origin asymptotically as $t \to \infty$.

Let us choose

$$V(x_1, x_2) = (x_1^2 + x_2)^2 + x_1^2$$

For $\dfrac{dV}{dt}$ we get

$$\frac{dV}{dt} = -2[(x_2 + 2x_1^2)^2 + x_1^2(1 - x_1^2)]$$

$$\frac{dV}{dt} < 0 \qquad \text{for} \qquad -1 < x_1 < 1$$

We see that in order to apply the quoted theorem we have to find the region P for which

$$\left. \begin{array}{c} V(x) < k \\[20pt] -1 < x_1 < 1 \end{array} \right\} x \, \varepsilon \, P$$

The origin is asymptotically stable and every solution with the initial condition in this region approaches the origin as $t \to \infty$.

There are many possible choices of such a region.

Let

$$(x_1^2 + x_2)^2 + x_1^2 = k$$

Setting different values of $k = \dfrac{1}{4}, \dfrac{1}{2}$, 1 we get different contours, see Fig. (1), the largest contour we obtain for $k = 1$. For this contour the condition $-1 < x_1 < 1$ is satisfied.

From the above example we clearly see that the Lyapunov conditions are sufficient but not necessary. Using the Lyapunov function we could not establish the fact that the system is asymptotically stable in-the-large.

Investigate the stability of the system associated with the van der Pol equation

$$\ddot{x} + \varepsilon(x^2 - 1)\dot{x} + x = 0 \tag{1}$$

where $\quad \varepsilon = \text{const} < 0$

Solution: Let us find the associated system of equations

$$x = \begin{bmatrix} x_1 \\ x_2 \end{bmatrix} \triangleq \begin{bmatrix} x \\ \dot{x} \end{bmatrix} \tag{2}$$

Substituting (2) into (1) we get

$$\dot{x}_1 = x_2$$

$$\dot{x}_2 = -x_1 - \varepsilon(x_1^2 - 1)x_2 \tag{3}$$

We choose the following V function

$$V(x) = x_1^2 + x_2^2$$

Then

$$\dot{V}(x) = 2x_1\dot{x}_1 + 2x_2\dot{x}_2 = 2x_1x_2 + 2x_2[-x_1 - \varepsilon(x_1^2 - 1)x_2]$$

$$= -2\varepsilon(x_1^2 - 1)x_2^2$$

Since $\quad \varepsilon < 0$, for $\dot{V}(x)$ to be negative definite we must have

$$x_1^2 - 1 < 0 \qquad \text{or} \qquad |x_1| < 1$$

Let us try the following choice of the state vector

$$x = \begin{bmatrix} x_1 \\ x_2 \end{bmatrix} \triangleq \begin{bmatrix} x \\ \int_0^t x\, dt \end{bmatrix}$$

Equation (3) becomes

$$\dot{x}_1 = -x_2 - \varepsilon\left(\frac{x_1^2}{3} - 1\right)x_1 \tag{4}$$

$$\dot{x}_2 = x_1$$

We choose the same V function

$$V(x_1\ x_2) = x_1^2 + x_2^2$$

and get

$$\dot{V}(x) = 2x_1\dot{x}_1 + 2x_2\dot{x}_2 = 2x_1\left[-x_2 - \varepsilon\left(\frac{x_1^2}{3} - 1\right)x_1\right] + 2x_1x_2$$

$$\dot{V}(x) = -2 \, \varepsilon \, x_1^2 \left(\frac{x_1^2}{3} - 1 \right)$$

Since $\varepsilon < 0$, for $\dot{V}(x)$ to be negative definite we must have

$$x_1^2 < 3$$

or

$$|x_1| < \sqrt{3}$$

The second choice of the state variables is better since

$$1 < \sqrt{3}$$

A simplified model of an inverted pendulum system is described by the state equations

$$\dot{x}_1 = x_2 \tag{1}$$

$$\dot{x}_2 = g/\ell \; x_1 - \frac{1}{\ell} \; u(t)$$

The control function is

$$u(t) = hx = h_1 x_1 + h_2 x_2 \tag{2}$$

For what values of h_1 and h_2 is the system stable?

Solution: Equation (1) written in the matrix form is

$$\frac{d}{dt} \begin{bmatrix} x_1 \\ x_2 \end{bmatrix} = \begin{bmatrix} 0 & 1 \\ g/\ell & 0 \end{bmatrix} \begin{bmatrix} x_1 \\ x_2 \end{bmatrix} + \begin{bmatrix} 0 \\ -\frac{1}{\ell} \end{bmatrix} u(t) \tag{3}$$

The characteristic equation of the matrix A is

$$\det(\lambda I - A) = \begin{vmatrix} \lambda & -1 \\ -g/\ell & \lambda \end{vmatrix} = \lambda^2 - \frac{g}{\ell} = 0$$

The equation has one root in the right-hand s-plane. The system is unstable, to stabilize it we generate a control signal

$$u(t) = hx \tag{4}$$

Substituting (4) into Eq. (3) we get

$$\begin{bmatrix} \dot{x}_1 \\ \dot{x}_2 \end{bmatrix} = \begin{bmatrix} 0 & 1 \\ g/\ell & 0 \end{bmatrix} \begin{bmatrix} x_1 \\ x_2 \end{bmatrix} + \begin{bmatrix} 0 \\ -\frac{1}{\ell} \, (h_1 x_1 + h_2 x_2) \end{bmatrix}$$

or

$$\begin{pmatrix} \dot{x}_1 \\ \dot{x}_2 \end{pmatrix} = \begin{bmatrix} 0 & 1 \\ \frac{1}{\ell}(g-h_1) & -\frac{h_2}{\ell} \end{bmatrix} \begin{pmatrix} x_1 \\ x_2 \end{pmatrix}$$

The characteristic equation is

$$\det \begin{bmatrix} \lambda & -1 \\ -\frac{1}{\ell}(g-h_1) & \lambda + \frac{h_2}{\ell} \end{bmatrix} = \lambda\left(\lambda + \frac{h_2}{\ell}\right) - \frac{1}{\ell}(g-h_1) = 0$$

or

$$\lambda^2 + \lambda\left(\frac{h_2}{\ell}\right) + \frac{1}{\ell}(h_1-g) = 0$$

We conclude that the system is stable when $\frac{h_2}{\ell} > 0$ and $h_1 > g$.

The control function $u = h_1 x_1 + h_2 x_2$

where

$$\frac{h_2}{\ell} > 0$$

and
$$h_1 > g$$

stabilizes an unstable system.

● **PROBLEM** 13-69

Determine the stability of the nonlinear system shown in Fig. 1, where the nonlinear function is single valued.

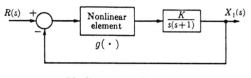

Nonlinear second-order system.   Fig. 1

Solution: The differential equation for the system is

$$\dot{x} = F(x) = \begin{bmatrix} f_1(x) \\ f_2(x) \end{bmatrix} = \begin{bmatrix} x_2 \\ -Kg(x_1) - x_2 \end{bmatrix}.$$

The Liapunov function is

$$V = F^T P F = \begin{bmatrix} f_1 \\ f_2 \end{bmatrix}^T \begin{bmatrix} P_{11} & P_{12} \\ P_{12} & P_{22} \end{bmatrix} \begin{bmatrix} f_1 \\ f_2 \end{bmatrix}$$

$$=p_{11}f_1^2 + 2p_{12}f_1f_2 + p_{22}f_2^2,$$

where P is a positive definite symmetric matrix.

$$\dot{V} = \frac{\partial V}{\partial f_1}\frac{\partial f_1}{\partial t} + \frac{\partial V}{\partial f_2}\frac{\partial f_2}{\partial t}$$

$$= \frac{\partial V}{\partial f_1}\left(\frac{\partial f_1}{\partial x_1}\frac{dx_1}{dt} + \frac{\partial f_1}{\partial x_2}\frac{dx_2}{dt}\right) + \frac{\partial V}{\partial f_2}\left(\frac{\partial f_2}{\partial x_1}\frac{dx_1}{dt} + \frac{\partial f_2}{\partial x_2}\frac{dx_2}{dt}\right).$$

Since

$$\frac{\partial V}{\partial f_1} = 2p_{11}f_1 + 2p_{12}f_2,$$

$$\frac{\partial V}{\partial f_2} = 2p_{12}f_1 + 2p_{22}f_2,$$

$$\frac{\partial f_2}{\partial x_2} = -1, \quad \frac{\partial f_2}{\partial x_1} = \frac{-K\partial g(x_1)}{\partial x_1}, \frac{\partial f_1}{\partial x_2} = 1, \quad \frac{\partial f_1}{\partial x_1} = 0,$$

we obtain

$$\dot{V} = \left(-2p_{12}K\frac{\partial g(x_1)}{\partial x_1}\right)f_1^2 + \left(2p_{11} - 2p_{22}K\frac{\partial g(x_1)}{\partial x_1} - 2p_{12}\right)f_1f_2$$

$$+ (2p_{12} - 2p_{22})f_2^2 . \tag{1}$$

In order to determine the requirements for the coefficients, we will constrain $-\dot{V}$ to be positive definite. We note that Eq. (1) is of the form

$$\dot{V} = af_1^2 + 2bf_1f_2 + cf_2^2 = FQF ,$$

where

$$Q = \begin{bmatrix} a & b \\ b & c \end{bmatrix}.$$

-Q is positive definite, thus

$$a < 0 \tag{2}$$

and

$$ac - b^2 > 0 . \tag{3}$$

Equation (2) requires that

$$2p_{12}K\frac{\partial g(x_1)}{\partial x_1} > 0$$

or

$$p_{12} > 0 \quad \text{and} \quad \frac{\partial g(x_1)}{\partial x_1} > 0$$

when $K > 0$. Equation (3) requires that

$$\left(-2p_{12} K \frac{\partial g(x_1)}{\partial x_1}\right)(2p_{12} - 2p_{22}) - \left(p_{11} - p_{22} K \frac{\partial g(x_1)}{\partial x_1} - p_{12}\right)^2 > 0$$

$$> 0$$

If $g(x_1) = x_1^3$, then it is necessary that

$$12p_{12} Kx_1^2 (p_{22} - p_{12}) > (p_{11} - 3p_{22} Kx_1^2 - p_{12})^2 . \qquad (4)$$

If $p_{22} = \beta p_{12}$, where $\beta > 1$, and $p_{11} = p_{12}$, then we may write Eq. (4) as follows:

$$4(\beta - 1) > 3\beta^2 Kx_1^2 ,$$

where x_1 equals the magnitude of x_1 at the operating point.

● **PROBLEM** 13-70

Determine the stability of the nonlinear system described by

$$\dot{x}_1 = -2x_1 + x_1 x_2 = f_1(x) \qquad (1)$$

$$\dot{x}_2 = -x_2 + x_1 x_2 = f_2(x) \qquad (2)$$

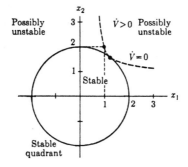

Fig. 1

Boundaries of assuredly stable solutions.

Solution: There are two equilibrium points, $x = 0$ and $x^T = \{1,2\}$, for both of them $\dot{x} = 0$.

Let us evaluate the simplest Lyapunov function--a quadratic function which is always positive definite:

$$V = x^T x = x_1^2 + x_2^2$$

$$\dot{V} = 2x_1 \dot{x}_1 + 2x_2 \dot{x}_2$$

$$\dot{V} = 2x_1(-2x_1 + x_1 x_2) + 2x_2(-x_2 + x_1 x_2)$$

$$= 2x_1^2(x_2 - 2) + 2x_2^2(x_1 - 1)$$

For asymptotic stability, we require that

$$\dot{V} = 2x_1^2(x_2 - 2) + 2x_2^2(x_1 - 1) < 0 . \qquad (4)$$

The limiting condition for stability is

$$x_1^2 (x_2 - 2) + x_2^2 (x_1 - 1) = 0 .$$ (5)

Investigation of Eq. (5) shows that there are curves dividing the stable region from the region where $\dot{V} > 0$.

(3)

Substituting (1) and (2) into (3), we obtain

In the third quadrant $(x_1 < 0 , x_2 < 0)$ we set

$$x_1 = -|x_1| \qquad x_2 = - |x_2|$$

and obtain

$$\dot{V} = 2|x_1|^2 (-|x_2| - 2) + 2|x_2|^2 (-|x_1| - 1)$$

for all values of x_1 and x_2 , $\dot{V}$ is less than zero, hence the system is stable in the third quadrant. Using La Salle results that the region within the curve V = const. tangent to $\dot{V} = 0$ curve in quadrant one contains stable solutions. In our case

$$V = x_1^2 + x_2^2 = \text{const.}$$

is a circle. We can find the radius of the circle from the equations

$$x_1^2 + x_2^2 = r^2$$

$$x_1^2 (x_2 - 2) + x_2^2 (x_1 - 1) = 0$$

and the requirement that there be only one solution of the above system. Fig. 1 shows the results for our choice of Eq. (3) of the Lyapunov function.

We shall use the gradient method to generate a Lyapunov function in order to obtain more information about stability of the system.

Let

$$\delta = \begin{bmatrix} \alpha_{11}x_1 + \alpha_{12}x_2 \\ \alpha_{12}x_1 + 2x_2 \end{bmatrix}$$

to satisfy the curl equation we set $\alpha_{12} = \alpha_{21}$.

We obtain

$$\dot{V} = \delta_1 f_1 + \delta_2 f_2$$

$$= (\alpha_{11}x_1 + \alpha_{12}x_2)(-2x_1 + x_1x_2) + (\alpha_{12}x_1 + 2x_2)(-x_2 + x_1x_2)$$

$$= x_1^2(-2\alpha_{11}) + x_1x_2(-3\alpha_{12}) - 2x_2^2 + (2 + \alpha_{12})x_1x_2^2$$

$$+ (\alpha_{11} + \alpha_{12})x_1^2x_2 \quad .$$

Choosing $\alpha_{11} = 1$ and $\alpha_{12} = -2$, we have

$$\dot{V} = -2x_1^2 + 6x_1x_2 - 2x_2^2 - x_1^2x_2 \quad , \tag{6}$$

which is not always negative. Solving for V, we get

$$V = \int_0^X \delta \cdot dx = \int_0^{x_1} \delta_1 dx_1 + \int_0^{x_2} \delta_2 dx_2 = \frac{x_1^2}{2} - 2x_1x_2 + x_2^2 \quad . \tag{7}$$

Unfortunately, V is not always positive. However, information about stability can definitely be obtained from Eqs. (6) and (7). For example, for the second quadrant we set

$x_1 = -|x_1|$ and $x_2 = +|x_2|$. Then Eq. (6) becomes

$$\dot{V} = -2|x_1|^2 - 6|x_1||x_2| - 2|x_2|^2 - |x_1|^2|x_2| \quad ,$$

which is always negative. Equation (7) is then

$$V = \frac{|x_1|^2}{2} + 2|x_1||x_2| + |x_2|^2 \quad ,$$

which is always positive. Therefore the system is always stable in the second quadrant. In the fourth quadrant, we let $x_1 = |x_1|$ and $x_2 = -|x_2|$ and obtain from Eq. (7)

$$V = \frac{|x_1|^2}{2} + 2|x_1||x_2| + |x_2|^2 \quad ,$$

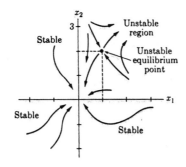

Fig. 2

Phase plane trajectories.

which is always positive. Then Eq. 6 becomes

$$\dot{V} = -2|x_1|^2 - 6|x_1||x_2| - 2|x_2|^2 \quad ,$$

and $\dot{V}$ is always negative. We conclude that the second, third and fourth quadrants are stable, part of the first quadrant is assuredly stable.

To reach the above conclusion we calculated for two Lyapunov functions

$$V = x_1^2 + x_2^2$$

and

$$V = \frac{x_1^2}{2} - 2x_1x_2 + x_2^2$$

Consider the following third-order equation

$$\dddot{x} + f(\ddot{x}) + h(\dot{x}) + g(x) = 0$$

for simplicity we put

$$f(\ddot{x}) = \alpha\ddot{x}$$

The above equation is equivalent to the following system.

$$\frac{dx_1}{dt} = x_2$$

$$\frac{dx_2}{dt} = x_3 \tag{1}$$

$$\frac{dx_3}{dt} = -g(x_1) - h(x_2) - \alpha x_3$$

where g and h are differentiable functions and

$$g(0) = h(0) = 0 .$$

Prove the following theorem.

Theorem 1:

System (1) is asymptotically stable in-the-large if:

1. $\alpha > 0$

2. $\frac{g(x_1)}{x_1} \geq \varepsilon_1 > 0$ for $x_1 \neq 0$

3. $\frac{\alpha h(x_2)}{x_2} - g'(x_1) \geq \varepsilon_2 > 0$ for $x_2 \neq 0$

ε_1 , ε_2 are constants and

$$g'(x_1) \triangleq \frac{dg}{dx_1}$$

Solution: We shall use the following theorem to prove theorem 1.

Theorem 2.

The system is asymptotically stable in the large if the

697

Lyapunov function has the following properties

1. $V(x) > 0$ for $x \neq 0$

2. $V(0) = 0$

3. $V(x)$ has continuous partial derivatives

4. $\dfrac{dV}{dt} \leq 0$

5. There is no initial state $x(t_0) \neq 0$ such that

$$\frac{dV}{dt} = 0$$

for all $t \geq t_0$

6. $V(x) \to \infty$ as $\|x\| \to \infty$.

Let us define the function $V(x)$

$$V(x) = \alpha G(x_1) + x_2 g(x_1) + H(x_2) + \frac{(\alpha x_2 + x_3)^2}{2} \qquad (2)$$

where

$$G(x_1) \triangleq \int_0^{x_1} g(x)\,dx$$

$$H(x_2) \triangleq \int_0^{x_2} h(x)\,dx$$

We show that this function satisfies all the properties 1-6 of the theorem 2. Obviously $V(0) = 0$.

Equation (2) for $V(x)$ can be written in an equivalent form, when $x_2 \neq 0$

$V(x) =$

$$\frac{[2H(x_2) + x_2 g(x_1)]^2 + 4\int_0^{x_1} g(x_1)\left[\int_0^{x_2}\left[\frac{\alpha h(x_2)}{x_2} - g'(x_1)\right]x_2\,dx_2\right]dx_1}{4H(x_2)}$$

$$+ \frac{(\alpha x_2 + x_3)^2}{2} \qquad (3)$$

If $x_2 = 0$, $x_1 \neq 0$ and $x_3 \neq 0$ from (2) we conclude that $V(x) > 0$.

Properties 1., 2., and 3. of Theorem 1 show that

$$g'(0) = \left(\frac{dg}{dx_1}\right)_{x_1 = 0} > 0$$

698

and

$$\frac{h(x_2)}{x_2} > 0$$

thus

$$G(x_1) > 0 \quad \text{and } H(x_2) > 0 \quad \text{for } x_1, x_2 \neq 0 .$$

We conclude that

$$V(x) > 0 \quad \text{for } x \neq 0$$

and

$$V(0) = 0 \quad \text{for } x = 0$$

$$\frac{\partial V}{\partial x} \quad \text{is continuous.}$$

The time derivative of V is

$$\frac{dV}{dt} = - x_2^2 \left[\frac{\alpha h(x_2)}{x_2} - g'(x_1) \right]$$

For $x_2 \neq 0$, $\quad \dfrac{dV}{dt} < 0$

and for

$$x_2 = 0 \qquad \frac{dV}{dt} = 0$$

thus 4. of theorem 2 is satisfied.
Let us investigate property 5. If the initial state exists such that

$$x(t_0) = 0 ,$$

then

$$x_2(t) = 0 \quad \text{for} \quad t \geq t_0 ,$$

but this requires

$$\frac{dx_2}{dt} = 0 \quad \text{for } t \geq t_0$$

and we must have $\qquad x_3(t) = 0 \quad \text{for } t \geq t_0$

and

$$\frac{dx_3}{dt} = 0 \quad \text{for } t \geq t_0 .$$

From (1) we conclude that $x_1(t) = 0$ for $t \geq t_0$ and see that only $x(t_0) = 0$ can give

$$\frac{dV}{dt} = 0 \text{ for } t \geq t_0 .$$

699

Property 6 is satisfied because of the lower bounds ϵ_1 and ϵ_2 . That completes the proof of theorem 1.

● **PROBLEM** 13-72

Examine the stability of the general system described by the equation

$$\dot{x} = Ax + Bu(t)$$

Solution: If we choose the Lyapùnov function

$$V = x^T Px ,$$

then

$$\dot{V} = x^T (PA + A^T P) x + 2x^T PBu(t) .$$

In order to improve stability we can see that we will require a $u(t)$ of the form

$$u^T = -k_2 (x^T PB) . \tag{1}$$

Let us examine the same problem in terms of the canonical state variables y. The differential equation in terms of y, where $x = Ty$ and T is a diagonalizing matrix for A, is

$$\dot{y} = \Lambda y + T^{-1} Bu ,$$

where u is a vector representing j control signals. Then the time derivative of $V(y)$ is

$$\dot{V} = y^T (\Lambda^T + \Lambda) y + 2y^T T^{-1} Bu .$$

In order to improve the stability of the system, the control vector should be

$$u = -Dy \tag{2}$$

within the physical constraints of the signals, where D is a positive definite matrix. If we can select the matrix B so that $T^{-1} B$ is diagonal, then each control variable is a function of only one state, that is,

$$u_j = -k_j y_j \quad .$$

The solution is a control signal which is a function of the state variables. If any of the equations (1) or (2) is going to be a practical solution the state vector x or y must be given. In most cases it is impossible to measure all the state variables. Usually we measure some and compute the remaining ones.

Determine the stability of a discrete system

$$x_1(k+1) = -1.8x_1(k)$$

$$x_2(k+1) = -0.4x_2(k) \, .$$

<u>Solution</u>: Assume that the Lyapunov function is given in the form

$$V(x) = x_1^2(k) + x_2^2(k)$$

We have

$$\Delta V(x) = V[x(k+1)] - V[x(k)]$$

$$= 3.24x_1^2(k) + 0.16x_2^2(k) - x_1^2(k) - x_2^2(k)$$

$$= 2.24x_1^2(k) - 0.84x_2^2(k)$$

Function $V(x)$ is indefinite in sign and can not be used as a Lyapunov function for the above system. Let us temporarily assume that the system is unstable and apply the instability theorem of Lyapunov.

Let

$$V(x) = a_1x_1^2(k) + 2a_2x_1(k)x_2(k) + a_3x_2^2(k)$$

and $\qquad \Delta V(x)$ be

$$\Delta V(x) = -x_1^2(k) - x_2^2(k)$$

which is negative for all $x_1(k) \neq 0$, $x_2(k) \neq 0$.

Calculating the difference and comparing the coefficients we obtain

$$\Delta V(x) = V[x(k+1)] - V[x(k)]$$

$$= 2.24a_1x_1^2(k) - 0.56a_2x_1x_2 - 0.84a_3x_3^2$$

$$a_1 = -0.446, \quad a_2 = 0 , \quad a_3 = 1.19$$

We get the following function V

$$V(x) = -0.446x_1^2(k) + 1.19x_3^2(k) \, .$$

$V(x)$ is indefinite and cannot be used to determine stability. Note that the A matrix where

$$A = \begin{bmatrix} -1.8 & 0 \\ 0 & -0.4 \end{bmatrix}$$

has eigenvalues $\lambda_1 = -1.8$ and $\lambda_2 = -0.4$

thus the system is unstable.

Determine the stability of the following discrete-data system

$$x_1(k+1) = -0.7x_1(k)$$

$$x_2(k+1) = -0.65x_2(k)$$

Solution: We choose the following Lyapunov function

$$V(x) = x_1^2(k) + x_2^2(k)$$

$V(x)$ is positive for all non-zero values of $x_1(k)$ and $x_2(k)$.

$$\Delta V(x) = V[x(k+1)] - V[x(k)]$$

$$= x_1^2(k+1) + x_2^2(k+1) - x_1^2(k) - x_2^2(k)$$

$$= x_1^2(k) \cdot 0.49 + 0.4225 \, x_2^2(k) - x_1^2(k) - x_2^2(k)$$

$$= -0.51 \, x_1^2(k) - 0.58 \, x_2^2(k)$$

For all $x \neq 0$, $\Delta V(x)$ is negative and the system is asymptotically stable.

Examine the stability of the equilibrium state $x = 0$ of the system.

$$\dot{x}_1 = \frac{-5x_1}{1 + x_1^2} + 2x_2$$

$$\dot{x}_2 = -2x_1 - \frac{3x_2}{1 + x_1^2} \tag{1}$$

Solution: We shall use the variable gradient method to find a Lyapunov function.

Let

$$\nabla V = \begin{bmatrix} a_{11}x_1 + a_{12} x_2 \\ \\ a_{21}x_1 + a_{22} x_2 \end{bmatrix} \tag{2}$$

Using (1) and (2) we obtain

$$\dot{V} = (\nabla V)'\dot{x} = [a_{11}x_1 + a_{12}x_2, \; a_{21}x_1 + a_{22}x_2] \begin{bmatrix} \dot{x}_1 \\ \\ \dot{x}_2 \end{bmatrix}$$

$$= \frac{-5a_{11}}{1 + x_1^2} x_1^2 + 2a_{12}x_2^2 + x_1x_2 \left(- \frac{5a_{12}}{1 + x_1^2} + 2a_{11} - 2a_{22} - \frac{3a_{21}}{1 + x_1^2} \right)$$

$$- 2a_{21}x_1^2 - 3 \frac{a_{22}}{1 + x_1^2} x_2^2$$

$\dot{V}$ must be at least negative semidefinite, thus we want the coefficient of x_1x_2 to be zero; that gives

$$a_{12} = a_{21} = 0 \qquad , \qquad a_{11} = a_{22}$$

We also choose $a_{11} = 2$ and obtain

$$\dot{V} = \frac{-10}{1 + x_1^2} x_1^2 - \frac{6}{1 + x_1^2} x_2^2$$

To obtain V we integrate

$$V = \int_0^{x_1, (x_2=0)} 2x_1 dx_1 + \int_0^{x_2} 2x_2 dx_2 = x_1^2 + x_2^2$$

V is positive definite and is the Lyapunov function.

Since
$$\lim_{\|x\| \to \infty} v(x) = \infty$$

the equilibrium state $x = 0$ is asymptotically stable in the large.

● **PROBLEM** 13-76

Using the variable-gradient method construct a Lyapunov function for the system

$$\dot{x}_1 = -x_1 + 2x_1^2 x_2$$

$$\dot{x}_2 = -x_2$$

and determine its stability.

Solution: Let us assume that the gradient of V is given in the form

$$\nabla V = \begin{bmatrix} a_{11} x_1 + a_{12} x_2 \\ a_{21} x_1 + 2x_2 \end{bmatrix}$$

Since V does not explicitly depend on time, that is

$$\frac{\partial V}{\partial t} = 0 ,$$

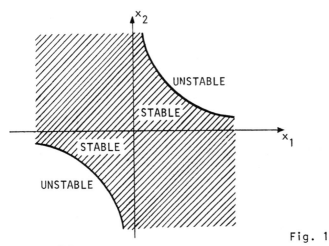

Fig. 1

the value of $\frac{dV}{dt}$ is in the form

$$\dot{V} = (\nabla V)'\dot{x} = (a_{11}x_1 + a_{12}x_2)\,\dot{x}_1 + (a_{21}x_1 + 2x_2)\,\dot{x}_2$$

$$= -a_{11}x_1^2 + 2a_{11}x_1^3x_2 - a_{12}x_1x_2 + 2a_{12}x_1^2x_2^2 - a_{21}x_1x_2 - 2x_2^2$$

We shall choose

$$a_{11} = 1 \quad\text{and}\quad a_{12} = a_{21} = 0$$

and get

$$\dot{V} = -x_1^2(1 - 2x_1x_2) - 2x_2^2$$

Since $\dot{V}$ has to be negative definite

$$1 - 2x_1x_2 > 0$$

The gradient of V for such choice of parameters a_{11} and a_{12} becomes

$$\nabla V = \begin{bmatrix} x_1 \\ 2x_2 \end{bmatrix}$$

in this case the curl equation is satisfied since

$$\frac{\partial}{\partial x_2}\left(\frac{\partial V}{\partial x_1}\right) = \frac{\partial}{\partial x_1}\left(\frac{\partial V}{\partial x_2}\right) = 0$$

Integrating ∇V we get

$$V = \int_0^{x_1(x_2=0)} x_1 dx_1 + \int_0^{x_2} 2x_2 dx_2 = \frac{x_1^2}{2} + x_2^2$$

Fig. (1) shows the region of stability for the Lyapunov function

$$V(x) = \frac{-x_1^2}{2} + x_2^2$$

To make the region of instability smaller we can try another choice of a_{ij} coefficients.

Let

$$a_{11} = \frac{2}{(1 - x_1 x_2)^2} \qquad a_{12} = \frac{-x_1^2}{(1 - x_1 x_2)^2} \quad , \qquad a_{21} = -a_{12}$$

We get

$$\dot{V} = -2x_1^2 - 2x_2^2$$

which is negative definite.

The gradient of V becomes

$$\nabla V = \begin{bmatrix} \dfrac{2x_1}{(1 - x_1 x_2)^2} - \dfrac{x_1^2 x_2}{(1 - x_1 x_2)^2} \\[4mm] \dfrac{x_1^3}{(1 - x_1 x_2)^2} + 2x_2 \end{bmatrix}$$

The curl identity is satisfied since

$$\frac{\partial}{\partial x_1} \left(\frac{\partial V}{\partial x_2} \right) = \frac{\partial}{\partial x_2} \left(\frac{\partial V}{\partial x_1} \right)$$

Linear integration gives

$$V = \int_0^{x_1, (x_2=0)} \left[\frac{2x_1}{(1 - x_1 x_2)^2} - \frac{x_1^2 x_2}{(1 - x_1 x_2)^2} \right] dx_1$$

$$+ \int_0^{x_2 (x_1=x_1)} \left[\frac{x_1^3}{(1 - x_1 x_2)^2} + 2x_2 \right] dx_2$$

$$= x_1^2 + x_2^2 + \frac{x_1^2}{1 - x_1 x_2}$$

which is positive definite for $1 - x_1 x_2 > 0$.

We see that the second choice of the Lyapunov function gives the smaller region of instability.

Determine the stability of the following system

$$\dot{x}_1 = -2x_1 + 4x_1^2 \, x_2$$

$$\dot{x}_2 = -2x_2$$

Use the variable gradient method to construct a Lyapunov function.

Solution: Let

$$\nabla V = \begin{bmatrix} a_{11} \, x_1 + a_{12} \, x_2 \\ a_{21} \, x_1 + a_{22} \, x_2 \end{bmatrix}$$

the derivative of V is then given by

$$\dot{V} = (\nabla V)' \dot{x} = (a_{11}x_1 + a_{12}x_2) \, \dot{x}_1 + (a_{21}x_1 + a_{22}x_2) \, \dot{x}_2$$

$$= (a_{11}x_1 + a_{12}x_2)(-2x_1 + 4x_1^2 x_2) + (a_{21}x_1 + a_{22}x_2)(-2x_2)$$

As one of many possible choices let us take

$$a_{11} = 1 \qquad \text{and} \qquad a_{12} = a_{21} = 0$$

$$a_{22} = 2$$

$\dot{V}$ becomes

$$\dot{V} = -2x_1^2 + 4x_1^3 x_2 - 4x_2^2 = -2x_1^2(1 - 2x_1x_2) - 4x_2^2$$

We see that $\dot{V}$ is negative definite when

$$1 - 2x_1 x_2 > 0$$

The gradient ∇V becomes

$$\nabla V = \begin{bmatrix} x_1 \\ 2x_2 \end{bmatrix}$$

Obviously the curl equation is satisfied.

Integrating we obtain V

$$V = \int_0^{x_1(x_2=0)} x_1 dx_1 + \int_0^{x_2} 2x_2 dx_2 = \frac{x_1^2}{2} + x_2^2$$

V is positive definite.

We conclude that the origin of the system is asymptotically stable.

The linear time-varying system is described by

$$\dot{x} = A(t)x$$

where

$$x = \begin{bmatrix} x_1 \\ x_2 \end{bmatrix}, \quad A(t) = \begin{bmatrix} 0 & 1 \\ -\dfrac{1}{t+3} & -15 \end{bmatrix} \quad (t \geq 0)$$

Using the variable-gradient method determine the stability of the origin of the system.

<u>Solution</u>: Let ∇V be defined by

$$\nabla V = \begin{bmatrix} a_{11}\,x_1 + a_{12}\,x_2 \\ a_{21}\,x_1 + a_{22}\,x_2 \end{bmatrix}$$

The time derivative of V is

$$(\nabla V)'\dot{x} = (a_{11}x_1 + a_{12}x_2)\,x_2 + (a_{21}x_1 + a_{22}x_2)\left(-\frac{x_1}{t+3} - 15x_2\right)$$

let us choose

$$a_{12} = a_{21} = 0$$

then we get

$$\dot{V} = a_{11}x_1x_2 + a_{22}x_2\left(-\frac{1}{t+3}\,x_1 - 15x_2\right)$$

For $\quad a_{11} = 1$

$$a_{22} = t+3$$

we get

$$\nabla V = \begin{bmatrix} x_1 \\ (t+3)\,x_2 \end{bmatrix}$$

The linear integration gives

$$\int_0^{x_1, (x_2=0)} x_1 dx_1 + \int_0^{x_2} (t+3)\,x_2 dx_2 = \frac{1}{2}\left[x_1^2 + (t+3)\,x_2^2 \right]$$

Thus

$$V = \frac{1}{2}\left[x_1^2 + (t+3)\,x_2^2 \right]$$

The time derivative of V is given by

$$\frac{dV}{dt} = \frac{\partial V}{\partial t} + (\nabla V)'\dot{x} = x_1\dot{x}_1 + (t+3)\ x_2 \cdot \dot{x}_2 + \frac{x_2^2}{2}$$

$$= \frac{x_2^2}{2} - 15(t+3)x_2^2 = -\frac{x_2^2}{2}(30t + 89)$$

Thus $\dot{V}$ is negative semidefinite.

$\dot{V}$ is identically zero only at the origin $x_1 = x_2 = 0$,

therefore the equilibrium state at the origin $x = 0$ is asymptotically stable in the large.

● **PROBLEM** 13-79

For a general class of nonlinear, autonomous systems

$$\dot{x} = f(x)$$

describe a method of finding a Lyapunov function, starting with an assumed form for the gradient of the function ∇V.

Solution: The system is described by

$$\dot{x} = f(x)$$

where $\quad f(0) = 0$

The gradient ∇V is

$$\nabla V = \begin{bmatrix} \dfrac{\partial V}{\partial x_1} \\[2mm] \dfrac{\partial V}{\partial x_2} \\[2mm] \vdots \\[2mm] \dfrac{\partial V}{\partial x_n} \end{bmatrix}$$

and we have

$$\frac{dV(x)}{dt} = (\nabla V)'\dot{x} = (\nabla V)'f(x)$$

We assume that the gradient ∇V is given in the linear form

$$\nabla V = \begin{bmatrix} a_{11}x_1 + a_{12}x_2 + \ldots + a_{1n}\,x_n \\[1ex] a_{21}x_1 + a_{22}x_2 + \ldots + a_{2n}\,x_n \\[1ex] \vdots \\[1ex] a_{n1}x_1 + a_{n2}x_2 + \ldots + a_{nn}\,x_n \end{bmatrix}$$

In the most general case the coefficients a_{ij} are functions of x . They satisfy the following equation

$$\text{curl } \nabla V = 0 \quad \text{or} \quad \frac{\partial^2 V}{\partial x_i \partial x_j} = \frac{\partial^2 V}{\partial x_j \partial x_i}$$

From

$$V(x) = \int_0^x (\nabla V)'dx$$

we determine $V(x)$.

Since the matrix is symmetric

$$\left[\frac{\partial^2 V}{\partial x_i \partial x_j} \right] i,j=1 \ldots n$$

the above integral can be written

$$V(x) = \int_0^{x_1} \frac{\partial V}{\partial x_1} dx_1 \Big|_{x_2 = \ldots = x_n = 0}$$

$$+ \int_0^{x_2} \frac{\partial V}{\partial x_2} dx_2 \Big|_{\substack{x_1 = x_1 \\ x_3 = x_4 = \ldots = x_n = 0}} + \ldots$$

$$+ \int_0^{x_n} \frac{\partial V}{\partial x_n} dx_n \Big|_{\substack{x_1 = x_1 \\ x_2 = x_2 \\ x_{n-1} = x_{n-1}}}$$

We have to choose the coefficients a_{ij} in such a way that

1. $\dot{V}$ is at least semidefinite

2. V is positive definite.

When the above conditions are met V(x) is the Lyapunov function of the system and the system is stable.

● **PROBLEM** 13-80

Determine the stability of the equilibrium state x = 0 of the system described by the following equation

$$\dddot{x} + a\ddot{x} + (b\dot{x} + x)^3 = 0 \qquad (1)$$

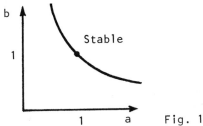

Fig. 1

Solution: Note that equation (1) is equivalent to the associated system

$$\dot{x}_1 = x_2$$

$$\dot{x}_2 = x_3$$

$$\dot{x}_3 = -(x_1 + bx_2)^3 - ax_3$$

where $x = x_1$

Now we will try to construct the Lyapunov function for the above system, using the variable gradient method.

Let

$$\nabla V = \begin{bmatrix} a_{11}x_1 + a_{12}x_2 + a_{13}x_3 \\ a_{21}x_1 + a_{22}x_2 + a_{23}x_3 \\ a_{31}x_1 + a_{32}x_2 + a_{33}x_3 \end{bmatrix}$$

the time derivative of V is

$$\dot{V} = (\nabla V)' \cdot \dot{x} = (a_{11}x_1 + a_{12}x_2 + a_{13}x_3)x_2 \qquad (2)$$

$$+ (a_{21}x_1 + a_{22}x_2 + a_{23}x_3)x_3 - (a_{31}x_1 + a_{32}x_2 + a_{33}x_3)\, ax_3$$

$$- (a_{31}x_1 + a_{32}x_2 + a_{33}x_3)(x_1 + bx_2)^3$$

Remembering that $\dot{V}$ must be negative definite we shall impose some restrictions upon the parameters a_{ij}.

Let

$$a_{22} = a^2, \quad a_{13} = a_{31} = 0$$

$$a_{23} = a_{32} = a \quad , \quad a_{33} = 1$$

Thus ∇V becomes

$$\nabla V = \begin{bmatrix} a_{11} \, x_1 + a_{12} \, x_2 \\ a_{21} \, x_1 + a^2 x_2 + a x_3 \\ a x_2 + x_3 \end{bmatrix} \tag{3}$$

We shall use the identity

$$\text{curl grad } f = 0$$

for f continuous with continuous derivatives, to further restrain parameters a_{ij}

$$\text{curl } \nabla V = 0$$

Substituting eq (3) we get

$$\frac{\partial (a_{11} x_1)}{\partial x_2} + \frac{\partial (a_{12} x_2)}{\partial x_2} = \frac{\partial (a_{21} x_1)}{\partial x_1}$$

$$\frac{\partial (a_{11} x_1)}{\partial x_3} + \frac{\partial (a_{12} x_2)}{\partial x_3} = 0$$

$$\frac{\partial (a_{21} x_1)}{\partial x_3} + a = a$$

We see that $\qquad \dfrac{\partial a_{21}}{\partial x_3} = 0$

Let us choose

$$a_{21} = \frac{1}{x_1} (x_1 + b x_2)^3$$

and

$$a_{12} = \frac{1}{b x_2} (x_1 + b x_2)^3$$

and we have

$$\frac{\partial (a_{21} x_1)}{\partial x_1} = 3 (x_1 + b x_2)^2$$

$$\frac{\partial (a_{12} x_2)}{\partial x_2} = 3 (x_1 + b x_2)^2$$

Let us further assume that

$$a_{11} = a_{11}(x_1)$$

711

We have the following equation for ∇V

$$\nabla V = \begin{bmatrix} a_{11}(x_1) \, x_1 + \frac{1}{b} \, (x_1 + bx_2)^3 \\ (x_1 + bx_2)^3 + a^2 x_2 + ax_3 \\ ax_2 + x_3 \end{bmatrix}$$

Equation (2) becomes

$$\dot{V} = (b^2 - ab^3) \, x_2^4 + 3(1 - ab) \, x_1^2 x_2^2 + 3(b - b^2 a) \, x_1 x_2^3$$

$$+ \left(\frac{1}{b} - a \right) x_2 x_1^3 + a_{11}(x_1) \, x_1 x_2$$

Let us choose

$$a_{11}(x_1) = (a - \frac{1}{b}) \, x_1^2$$

and get

$$\dot{V} = x_2^2 \, (1 - ab) \, (b^2 x_2^2 + 3x_1^2 + 3bx_1 x_2)$$

$3x_1^2 + 3bx_1 x_2 + b^2 x_2^2$ is positive definite since

$$3 > 0 \quad \text{and} \quad \begin{vmatrix} 3 & 1.5b \\ 1.5b & b^2 \end{vmatrix} > 0$$

For $\dot{V} < 0$ we obtain

$$1 - ab < 0 \text{ or } 1 < ab$$

The line integration gives us the value of V

$$V = \int_0^x (\nabla V)' dx = \int_0^{x_1 (x_2 = x_3 = 0)} ax_1^3 dx_1 + \int_0^{x_2 (x_3 = 0)} [(x_1 + bx_2)^3 + a^2 x_2] dx_2$$

$$+ \int_0^{x_3} (ax_2 + x_3) dx_3 = \frac{a}{4} \, x_1^4 + \frac{1}{4b} \, (x_1 + bx_2)^4 + \frac{1}{2} \, (ax_2 + x_3)^2$$

The requirement that V be positive definite gives $a > 0$ and $b > 0$.

We see that $V(x) \to \infty$ as $\|x\| \to \infty$.

V is a Lyapunov function of the system when

$$a > 0 , \quad b > 0 , \quad ab > 1$$

The equilibrium state $x = 0$ is asymptotically stable when

$$a > 0 , \quad b > 0 , \quad ab > 1 .$$

For the following system

$$\dot{x}_1 = x_2$$

$$\dot{x}_2 = a_1(t)x_1 + a_2(t)x_2$$

find conditions of stability, that is the range of functions $a_1(t)$ and $a_2(t)$ for which the system is stable.

Solution: In this case we apply the variable gradient method to find the Lyapunov function of the system.

The equilibrium state is the origin.

Let

$$\nabla V = \begin{bmatrix} a_{11} x_1 + a_{12} x_2 \\ a_{21} x_1 + a_{22} x_2 \end{bmatrix}$$

then

$$(\nabla V)'\dot{x} = a_{11}x_1x_2 + a_{12}x_2^2 + a_{21}a_1(t)x_1^2 + a_{22}a_2(t)x_2^2$$

$$+ a_{21}a_2(t)x_1x_2 + a_{22}a_1(t)x_1x_2$$

We choose the following restrictions on a_{ij}

$$a_{12} = a_{21} = 0$$

and obtain

$$(\nabla V)'\dot{x} = a_{11}x_1x_2 + a_{22}a_2(t)x_2^2 + a_{22}a_1(t)x_1x_2$$

Now let

$$a_{11} + a_{22} a_1(t) = 0$$

assuming that

$$a_1(t) \neq 0$$

we choose $a_{11} = 2$

and get

$$a_{22} = -\frac{2}{a_1(t)}$$

Finally we have

$$\nabla V = \begin{bmatrix} 2x_1 \\ -\dfrac{2}{a_1(t)} x_2 \end{bmatrix}$$

Integrating we obtain

$$\int_0^{x_1, (x_2=0)} 2x_1 dx_1 + \int_0^{x_2} - \frac{2}{a_1(t)} x_2 dx_2 = x_1^2 - \frac{1}{a_1(t)} x_2^2$$

$$V = x_1^2 - \frac{1}{a_1(t)} x_2^2$$

Then

$$\dot{V} = 2x_1\dot{x}_1 + \frac{\dot{a}_1(t)}{[a_1(t)]^2} x_2^2 - \frac{1}{a_1(t)} 2x_2\dot{x}_2$$

$$= - \left(\frac{2a_2(t)}{a_1(t)} - \frac{\dot{a}_1(t)}{a_1^2(t)} \right) x_2^2$$

Note that since V depends on t in the explicit way
V = V(t) we had

$$\dot{V} = \frac{\partial V}{\partial t} + (\nabla V)' \cdot \dot{x}$$

We see that if $a_1(t) < 0$ for $t \geq t_0$ then V is positive
definite, and if

$$- \frac{2a_2(t)}{a_1(t)} - \frac{\dot{a}_1(t)}{a_1^2(t)} < 0 \qquad \text{for } t \geq t_0$$

$\dot{V}$ is negative semidefinite. In such a case V fulfills
criteria for Lyapunov function. We conclude that the system
is stable when

$$a_1(t) < 0$$

$$- \frac{2a_2(t)}{a_1(t)} + \frac{\dot{a}_1(t)}{a_1^2(t)} < 0 \qquad \text{for } t \geq t_0$$

or

$$\frac{\dot{a}_1(t)}{a_1(t)} > 2a_2(t)$$

and

$$a_1(t) < 0 \quad \text{for} \quad t \geq t_0$$

● **PROBLEM** 13-82

The equations describing a homogeneous atomic reactor with
constant power extraction are

$$\frac{d}{dt} \log W = - \frac{K}{\tau} T$$

$$\frac{d}{dt} T = (W-1) \frac{1}{\varepsilon}$$

where W(t) is the instantaneous reactor power, T(t) is the temperature, K the temperature coefficient, ε the heat capacity of the reactor and τ the average life of a neutron. Determine the conditions for stability of the reactor.

Solution: Let us substitute

$$x_1 = \log W$$

$$x_2 = T$$

We obtain the following system of equations

$$\dot{x}_1 = - \frac{K}{\tau} x_2$$

$$\dot{x}_2 = \frac{1}{\varepsilon} (e^{x_1} - 1) = f(x_1)$$

The origin $x_1 = x_2 = 0$ is the equilibrium state. We shall use the variable gradient method to find the Lyapunov function for the system.

$$\nabla V = \begin{bmatrix} a_{11} x_1 + a_{12} x_2 \\ a_{21} x_1 + x_2 \end{bmatrix}$$

The following identity must be satisfied

$$\text{curl } \nabla V = 0$$

We have

$$\frac{\partial}{\partial x_2} \left(\frac{\partial V}{\partial x_1} \right) = \frac{\partial}{\partial x_1} \left(\frac{\partial V}{\partial x_2} \right)$$

$$x_1 \frac{\partial a_{11}}{\partial x_2} = \frac{\partial (a_{12} x_2)}{\partial x_2} = \frac{\partial (a_{21} x_1)}{\partial x_1}$$

and choose

$$a_{12} = a_{21} = 0 \quad \text{and} \quad a_{11} = a_{11}(x_1)$$

Then

$$\dot{V} = (\nabla V)'\dot{x} = a_{11}(x_1) \cdot x_1 \cdot \left(- \frac{K}{\tau} x_2 \right) + x_2 \frac{1}{\varepsilon} (e^{x_1} - 1)$$

We now choose

$$a_{11}(x_1) = \frac{1}{x_1} \left(\frac{\tau}{K\varepsilon} \right) (e^{x_1} - 1)$$

715

then $\dot{V}$ becomes

$$\dot{V} = - \frac{x_2}{\varepsilon} (e^{x_1} - 1) + \frac{x_2}{\varepsilon} (e^{x_1} - 1) = 0$$

Substituting the value of a_{11} we write

$$\nabla V = \begin{bmatrix} \frac{\tau}{K\varepsilon} (e^{x_1} - 1) \\ \\ x_2 \end{bmatrix}$$

To find V we integrate ∇V as follows

$$V = \int_0^{x_1(x_2=0)} \left(\frac{\tau}{K\varepsilon}\right) (e^{x_1} - 1) dx_1 + \int_0^{x_2} x_2 dx_2$$

$$= \frac{\tau}{K\varepsilon} (e^{x_1} - x_1 - 1) + \frac{x_2^2}{2}$$

Since τ and ε are positive by definition, for

$K > 0$ we get

$V > 0$ for x_1 , $x_2 \neq 0$

$V = 0$ for $x_1 = x_2 = 0$

V is a positive definite scalar function and the equilibrium state, the origin is stable.

Since $x_1 = \log W$, $x_2 = T$

we conclude that for $W = 1$, $T = 0$ the system is stable for $K > 0$.

● **PROBLEM** 13-83

The following system

$$\dot{x}_1 = -4x_1 x_2 x_3$$

$$\dot{x}_2 = \frac{x_2}{t+3} + x_1^2 x_3$$

$$\dot{x}_3 = \frac{x_3}{t+3} + x_1^2 x_2$$

has the unstable origin. The parameter $t \geq 0$ can be interpreted as time. Prove that the origin of the system is unstable.

Solution: We shall use the following theorem.

Theorem:

If there exists a scalar function $V(x,t)$, with continuous first partial derivatives and the function satisfies the following conditions:

I $V(x,t) > 0$ for all $x \neq 0$ in the region Γ and all $t \geq 0$

II $V(0,t) = 0$ for all $t \geq 0$

III $\dot{V}(x,t) > 0$ for all $x \neq 0$ in the region Γ and all $t \geq 0$

IV $\dot{V}(0,t) = 0$ for all $t \geq 0$

then the origin of the system $\dot{x} = f(x,t)$ is unstable.

Note that we can replace I and III conditions with $V(x,t) < 0$ and $\dot{V}(x,t) < 0$.

The equilibrium state $\dot{x} = 0$ is $(a,0,0)$ where a is any constant. We choose (as one of many) the following gradient of V

$$\nabla V = \begin{bmatrix} 4x_1 \\ 2x_2 + 2\left(\dfrac{t+2}{t+3}\right)x_3 \\ 2\left(\dfrac{t+2}{t+3}\right)x_2 + 14x_3 \end{bmatrix}$$

Integrating ∇V we obtain

$$\int_0^x (\nabla V)'dx = \int_0^{x_1 (x_2=x_3=0)} 4x_1 dx_1 + \int_0^{x_2, (x_3=0,x_1=x_1)} 2x_2 dx_2 +$$

$$\int_0^{x_3} (2\,\frac{t+2}{t+3}\,x_2 + 14x_3)dx_3 = 2x_1^2 + x_2^2 + 7x_3^2 + 2\,\frac{t+2}{t+3}\,x_2 x_3$$

$$= 2x_1^2 + \frac{1}{t+3}\,(x_2^2 + x_3^2) + \frac{t+2}{t+3}\,(x_2 + x_3)^2 + 6x_3^2$$

$$V = 2x_1^2 + \frac{1}{t+3}\,(x_2^2 + x_3^2) + \frac{t+2}{t+3}\,(x_2 + x_3)^2 + 6x_3^2$$

We see that

    $V(x_1, x_2, x_3, t) > 0$ for $x \neq 0$, for all $t \geq 0$

    $V(0,t) = 0$ for all $t \geq 0$

717

We have to show that $\dot{V}(x,t) > 0$.

$$\frac{dV}{dt} = \frac{\partial V}{\partial t} + (\nabla v)'\dot{x}$$

$$= \frac{2}{t+3}\left[\frac{1}{2(t+3)}(x_2 + x_3)^2 + (x_2 + \frac{t+2}{t+3}x_3)^2 + x_3^2(6t^2 + 38t + 59)\right.$$

$$\left. + (t+2)(x_2^2 + x_3^2)x_1^2\right]$$

We see that since $t \geq 0$

$$\dot{V}(x,t) > 0 \quad \text{and} \quad \dot{V}(0,t) = 0 \quad \text{for all } t \geq 0$$

that completes the proof.

● **PROBLEM** 13-84

The problem is to find the controls u_1 and u_2 as functions of time so that the system reaches the origin (that is, x = 0) as fast as possible. We shall consider the system to be sufficiently close to the origin if

$$|x_i| < 0.001 \text{ for } i = 1,2,\ldots,6 .$$

We state the problem as follows. The normalized gas absorber equations for a six-plate absorber are

$$\dot{x} = Ax + Bu , \tag{1}$$

where the values of A and B are

$$A = \begin{bmatrix} -1.173113 & .634115 & 0 & \cdots & & 0 \\ .538998 & -1.173113 & .634115 & 0 & & \vdots \\ 0 & .538998 & -1.173113 & .634115 & & \\ 0 & 0 & .538998 & -1.173113 & .634115 & 0 \\ 0 & 0 & & .538998 & -1.173113 & .634115 \\ 0 & 0 & \cdots & 0 & 0.538998 & -1.173113 \end{bmatrix} \tag{2}$$

and

$$B = \begin{bmatrix} .538998 & 0 \\ 0 & 0 \\ 0 & 0 \\ 0 & 0 \\ 0 & 0 \\ 0 & .634115 \end{bmatrix} \tag{3}$$

The initial value of x is

$$x(0) = \begin{bmatrix} -.0306632 \\ -.0567271 \\ -.0788812 \\ -.0977124 \\ -.1137188 \\ -.1273242 \end{bmatrix} \qquad (4)$$

and the control vector u is bounded from below so that

$$0 \le u_1, \qquad -0.4167 \le u_2. \qquad (5)$$

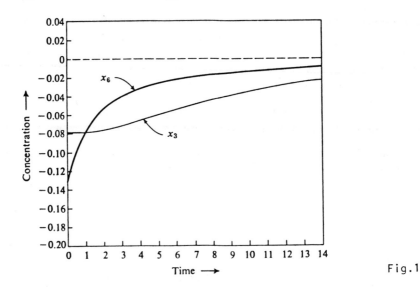

Fig.1

Solution: First, integrate Equation 1 over a time interval T to give

$$x(k + 1) = \phi x(k) + \Delta, \qquad (6)$$

where $\phi = \exp(AT)$

and

$$\Delta = \left[\int_0^T \exp(A\lambda)\,d\lambda \right] Bu(k). \qquad (7)$$

It should be noted that the coefficient matrix, A, is constant, and therefore ϕ is a constant in this problem.

We shall choose a Lyapunov function of the positive-definite form

$$V[x(k)] = x(k)Qx(k), \qquad (8)$$

719

so that

$$\Delta V[x(k)] = [Px(k) + s]'x(k) + R , \qquad (9)$$

where

$$P = \phi'Q\phi - Q , \qquad s = 2\phi'Q\Delta , \qquad R = \Delta'Q\Delta , \qquad (10)$$

and in this example we shall use two forms for Q to show the effect of the choice of Lyapunov function on the resulting trajectory. In selecting the controls, we choose u_1 and u_2 from a discretized set to satisfy the constraints and to minimize $\Delta V[x(k)]$ at each time interval k. To show the improvement of using control, the uncontrolled response ($u_1 = u_2 = 0$) is plotted in Figure 1. By choosing the third and the sixth trays an idea is obtained about the behavior of the entire system. The approach to the origin is rather slow and the criterion that all $|x_i| < .001$ is met after 39 minutes have passed when no control action is used.

Now let us consider the response of the absorber to the applied controls u_1 and u_2 chosen in such a way as to minimize $\Delta V[x(k)]$ at each interval of time.

(a) $Q = I .$ \qquad\qquad\qquad (11)

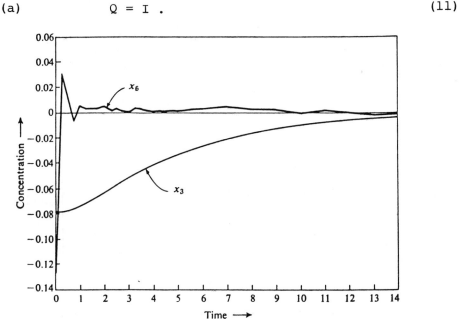

Fig.2

In this part of the problem, we shall weigh each absorption tray evenly. The results are tabulated in part in Table 1 and shown graphically in Figure 2 where again the concentrations from the third and sixth trays have been plotted as a function of time. It is noticed at once that the approach to steady-state values is considerably faster. As a matter of fact, x_6 remains very close to the desired value. However, since x_1 and x_6 are kept very close to the desired values, the freedom of using u_1 and u_2 is lost to some extent since the response of the trays 2 to 5 is the same as for a four-tray absorber without control. Greater benefit is therefore expected if the inner trays are more heavily weighted. We therefore turn to the second part of the problem.

Table 1

Q = I

Time	u_1	u_2	x_1	x_2	x_3	x_4	x_5	x_6
0	—	—	−.03066	−.05673	−.07888	−.09771	−.11372	−.12732
.25	.5	1.0	.02792	−.05297	−.07869	−.09710	−.10188	.02967
.50	0	0	.01395	−.04790	−.07769	−.09421	−.08450	.01137
.75	0	0	.00403	−.04529	−.07601	−.08998	−.07313	−.00065
1.00	.05	.10	.00274	−.04369	−.07390	−.08527	−.06442	.00526
1.25	.05	.05	.00198	−.04230	−.07140	−.08038	−.05711	.00373
1.50	.05	.05	.00160	−.04098	−.06888	−.07555	−.05121	.00336
1.75	.05	.05	.00150	−.03965	−.06612	−.07089	−.04625	.00370
2.00	.05	.05	.00161	−.03827	−.06328	−.06646	−.04194	.00450
2.25	.05	.02	.00188	−.03683	−.06040	−.06228	−.03840	.00143
2.50	.05	.05	.00229	−.03531	−.05753	−.05838	−.03531	.00364
2.75	.05	.02	.00281	−.03374	−.05469	−.05473	−.03259	.00151
3.00	.05	.02	.00342	−.03211	−.05190	−.05135	−.03038	.00021
3.25	.02	.05	.00058	−.03067	−.04920	−.04821	−.02818	.00361
3.50	.05	.02	.00216	−.02928	−.04660	−.04526	−.02607	.00228
3.75	.05	.02	.00354	−.02772	−.04410	−.04249	−.02431	.00152
4.00	.02	.02	.00127	−.02629	−.04169	−.03992	−.02275	.00114
4.25	.05	.02	.00327	−.02489	−.03939	−.03752	−.02134	.00103
4.50	.02	.02	.00146	−.02356	−.03719	−.03528	−.02001	.00110
4.75	.02	.02	.00027	−.02243	−.03511	−.03319	−.01875	.00131
5.00	.05	.02	.00304	−.02121	−.03315	−.03122	−.01754	.00161
6.00	.02	.02	.00035	−.01704	−.02624	−.02437	−.01305	.00326
7.00	.02	.02	.00086	−.01363	−.02069	−.01876	−.00905	.00510
8.00	.02	.01	.00225	−.01036	−.01607	−.01425	−.00655	.00296
9.00	.01	.01	.00055	−.00803	−.01232	−.01083	−.00482	.00290

$$
\text{(b)} \quad Q = \begin{bmatrix}
1.0 & 0 & & \cdots & & 0 \\
0 & 7.39 & 0 & & & \\
\vdots & 0 & 230.0 & 0 & & \vdots \\
\vdots & & 0 & 230.0 & 0 & \\
& & & 0 & 7.39 & 0 \\
0 & \cdots & & & 0 & 1.0
\end{bmatrix} \quad (12)
$$

Since the effect of one tray on the next is related approximately logarithmically, we have chosen Q so that the second tray is weighted as $e^2 (=7.39)$ and the third tray as $(e^e)^2 = 230$. The effect is squared, since Q weights the square of x.

Table 2 shows that the response is considerably better than by using the identity matrix for Q. Figure 3 shows the concentrations on the third and sixth trays as functions of time. The criterion that all $|x_i| < 0.001$ is met in 9 min which, compared to 39 min for the uncontrolled case, is a considerable improvement.

Table 2

Weighted Q

Time	u_1	u_2	x_1	x_2	x_3	x_4	x_5	x_6
0	—	—	−.03066	−.05673	−.07888	−.09771	−.11372	−.12732
.25	1.0	2.0	.08651	−.04922	−.07851	−.09656	−.09149	.16753
.50	.05	0	.06442	−.03870	−.07652	−.09120	−.06010	.11634
.75	.20	.10	.06662	−.03149	−.07313	−.08352	−.04069	.09476
1.00	.10	.10	.05746	−.02604	−.06874	−.07517	−.02767	.08050
1.25	.10	.10	.05124	−.02221	−.06379	−.06690	−.01856	.07114
1.50	.10	.10	.04707	−.01924	−.05860	−.05908	−.01187	.06505
1.75	.10	.05	.04432	−.01671	−.05339	−.05185	−.00718	.05431
2.00	.05	.05	.03675	−.01474	−.04829	−.04535	−.00412	.04674
2.25	.10	.05	.03719	−.01296	−.04343	−.03956	−.00199	.04138
2.50	.05	0	.03192	−.01130	−.03886	−.03448	−.00091	.03071
2.75	.05	.05	.02819	−.00997	−.03463	−.03007	−.00029	.02971
3.00	.05	0	.02558	−.00879	−.03077	−.02624	.00003	.02215
3.25	.05	.05	.02379	−.00765	−.02727	−.02293	.00030	.02340
3.50	0	0	.01676	−.00690	−.02411	−.02003	.00050	.01751
4.00	.10	.10	.02836	−.00408	−.01869	−.01504	.00331	.03406
4.50	.10	.05	.03578	.00057	−.01370	−.01028	.00718	.03207
5.00	0	0	.02045	.00307	−.00905	−.00605	.00827	.01948
5.50	0	−.05	.01215	.00320	−.00540	−.00307	.00567	.00030
6.00	0	.01	.00747	.00275	−.00291	−.00152	.00325	.00340
6.50	0	0	.00475	.00224	−.00135	−.00061	.00229	.00244
7.00	0	0	.00312	.00183	−.00041	−.00004	.00171	.00175
7.50	0	0	.00213	.00152	.00015	.00032	.00134	.00128
8.00	0	0	.00152	.00129	.00047	.00054	.00110	.00096
8.50	0	0	.00113	.00112	.00065	.00066	.00093	.00074
9.00*	0	0	.00088	.00099	.00074	.00072	.00082	.00059

* After this time all $|x_i| < 0.001$ simply by keeping the control variables zero.

While we are still considering this example, a very important point should be noted about the nature of ΔV. It is sometimes misunderstood that if ΔV and V are both positive, the system is unstable. Both ΔV and V may be positive in a finite interval of time and the system can still be asymptotically stable. This is illustrated by looking at the changes in Lyapunov's functions for the time interval (7.50,8.00). The changes in Lyapunov function are tabulated as functions of the control variables in Table 3. As usual, the minimum value for ΔV determines the control to be used for this interval. Here the minimum value for ΔV is 0.000077 as given by $u_1 = 0$, $u_2 = 0$, so with these values for control, the vector x(8.00) is evaluated and the calculation continued until all $|x_i|$ are less than 0.001 and also remain less than 0.001.

This condition is met at t = 9.0. Hence, the system is asymptotically stable in spite of the fact that ΔV and V are both positive for some intervals of time. The ΔV, of course, becomes negative again later.

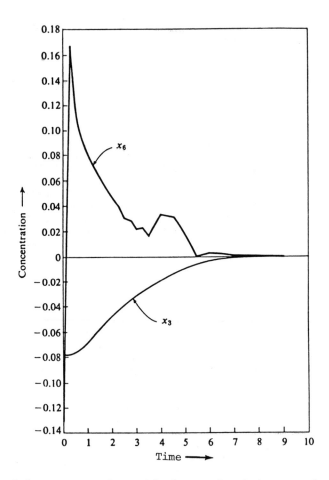

Fig.3

With Lyapunov's method constraints can be placed on both the control and the state variables. Let us consider the constraints on the state variables and use the example for illustration. Suppose that we use the weighted Q as was used to construct Table 2, but demand that no x_i should exceed some value, let us say, 0.10. Then, even though $u_1 = 1.0$ and $u_2 = 2.0$ give the minimum value for ΔV, the resulting value of x_6, namely 0.16753, is outside the state constraint. Thus, the minimum value for ΔV is sought in the range where the resulting state variables lie inside to constraint bounds. For example, here the controls $u_1 = 1.0$, $u_2 = 1.0$ give a value of 0.02967 for x_6 and a ΔV of -7.41×10^{-2}. Again, the constraints on the state variables simplify the problem since a smaller range of controls has to be considered. Since $u_2 = 2.0$ produces a response which lies outside bounds, it is not necessary to try values of u_2 which are greater than 2.0 for this interval of time, or the other interval for that matter.

These two examples illustrate the virtues of Lyapunov's second method as used to establish the control of some linear process. First of all, constraints can be placed on both the control and the state variables, and actually aid the calculations by reducing the range of variables to be scanned by the computer. This is equivalent to dynamic programming

723

Table 3

Changes in Lyapunov Function at
Time Interval from 7.50 to 8.00

u_1	u_2	ΔV
.10	.10	.001628
.05	.10	.001183
.02	.10	.001031
.01	.10	.001000
.00	.10	.000979
.10	.05	.000998
.05	.05	.000554
.02	.05	.000403
.01	.05	.000372
.00	.05	.000351
.10	.02	.000789
.05	.02	.000346
.02	.02	.000196
.01	.02	.000165
.00	.02	.000144
.10	.01	.000748
.05	.01	.000345
.02	.01	.000155
.01	.01	.000124
.00	.01	.000103
.10	.00	.000721
.05	.00	.000278
.02	.00	.000128
.01	.00	.000098
.00	.00	.000077
.10	-.05	.000798
.05	-.05	.000356
.02	-.05	.000207
.01	-.05	.000177
.00	-.05	.000156
.10	-.10	.001228
.05	-.10	.000788
.02	-.10	.000640
.01	-.10	.000610
.00	-.10	.000589

numerically. Second, the dimension of the state does not
matter; the problem of 10 stages instead of 6 can be handled
just as easily. This is very important and quite different
from the effect in dynamic programming, where the state dimension is usually more important than the number of control
variables. Here, as in dynamic programming, the addition of
an extra control variable complicates the problem considerably by increasing the permutations of control variables to
be scanned. However, the problem can still be handled.
Third, the size of time interval that is used does not really
matter so long as enough terms are used in the series expansion of the exponential. Finally, the programming on a digital computer is straightforward since only matrix multiplica-

tion and addition is involved, that is, no inverse matrices need be calculated. Also, the computation time is small.

We have not mentioned the effect of sampling time on the rate of approach to the origin. This can be rather important, especially when a nonidentity weighting matrix is used for the Lyapunov function.

● **PROBLEM** 13-85

Use the linear approximation theorem to examine the stability of the system described by

$$\dot{x}_1 = -x_1 - x_2^2$$

$$\dot{x}_2 = -x_2$$

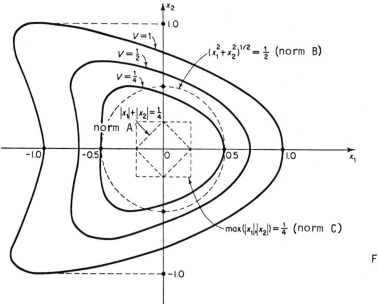

Fig. 1

Solution: The linear approximation method states that given a system described by

$$\frac{d}{dt} x = F(x)$$

we define a constant matrix A and a function g(x) by

$$F(x) = Ax + g(x)$$

If $\frac{d}{dt} x = Ax$ is asymptotically stable in the large and

$$\lim_{\|x\| \to \infty} \frac{\|g(x)\|}{\|x\|} = 0$$

then the system described by

725

$$\frac{d}{dt} x = F(x)$$

is asymptotically stable at the origin.

$\|x\|$ denotes the norm of x -- it is the real positive function (i.e., $\|x\| \in R'_+$)

with the following properties:

for every x and y

1) $\qquad \|x\| \geq 0$

2) $\qquad \|x\| = 0$ if and only if $x = 0$

3) $\qquad \|x + y\| \leq \|x\| + \|y\|$

4) $\qquad \|Ax\| \leq M \|A\| \cdot \|x\|$ for some constant

$\qquad M > 0$

We use

$$A = \left(\frac{\partial F}{\partial x}\right)_{x=0}$$

where

$$\frac{\partial F}{\partial x} = \begin{bmatrix} \frac{\partial F_1}{\partial x_1} & \frac{\partial F_1}{\partial x_2} & \cdots & \frac{\partial F_1}{\partial x_n} \\ \cdot \\ \cdot \\ \cdot \\ \frac{\partial F_n}{\partial x_1} & & & \frac{\partial F_n}{\partial x_n} \end{bmatrix}$$

and for the given system of equations we have

$$A = \begin{pmatrix} -1 & -2x_2 \\ 0 & -1 \end{pmatrix}_{x=0} = -I$$

From $F(x) = Ax + g(x)$ we obtain

$$g(x) = F(x) - Ax = \begin{pmatrix} -x_1 - x_2^2 \\ -x_2 \end{pmatrix} - Ax = \begin{pmatrix} -x_2^2 \\ 0 \end{pmatrix}$$

We compute the transition matrix exp A $(t-t_0)$ corresponding

726

to A

$$\Phi(t-t_0) = \begin{pmatrix} e^{-(t-t_0)} & 0 \\ 0 & e^{-(t-t_0)} \end{pmatrix}$$

We shall examine the stability of the system using different norms $\| \cdot \|$.

A)   Let $\| A \| = \sum_{i=1}^{n} \sum_{j=1}^{m} |a_{ij}|$; we have

$$\| \Phi(t-t_0) \| = 2e^{-(t-t_0)}$$

From $\| \Phi(t-t_0) \| \leq \alpha e^{-\beta(t-t_0)}$, where $t \geq t_0$ we choose $\alpha = 2$ and $\beta = 1$. For this norm $\beta = 1$ is the largest value to guarantee the inequality.

The stability region

$$\| x(t_0) \| < \frac{\gamma}{\alpha M}$$

should be as large as possible, so we have to take the largest value of β. In such case the inequality

$$\| g(x) \| \leq \beta k \| x \| /_{\alpha M}$$

for all $\| x \| \leq \gamma$ can be met with the largest possible γ. We should choose $k = 1 - \varepsilon$ where $\varepsilon > 0$ is very small.

$\| g \| = x_2^2$ and we are looking for the largest value of γ such that for all x_1 x_2

$$|x_1| + |x_2| \leq \gamma$$

the following inequality is true

$$x_2^2 \leq \frac{(1-\varepsilon)}{2} (|x_1| + |x_2|) < \frac{|x_1| + |x_2|}{2}$$

We find that $\gamma = \frac{1}{2}$. For the above norm $M = 1$ and the stability region $|x_1| + |x_2| < \frac{\gamma}{\alpha M} = \frac{1}{4}$. This norm is sometimes called the diamond norm, see Fig. 1.

B)   Let us define the norm

$$\| A \| = \left[\sum_{i=1}^{n} \sum_{j=1}^{m} |a_{ij}|^2 \right]^{\frac{1}{2}}$$

727

we have

$$\| \Phi(t-t_0) \| = \sqrt{2} \quad e^{-(t-t_0)}$$

$$\| g(x) \| = x_2^2$$

We find $\alpha = \sqrt{2}$, $\beta = 1$ and choose the largest γ guaranteeing

$$x_2^2 < \frac{(x_1^2 + x_2^2)^{\frac{1}{2}}}{\sqrt{2}}$$

for all x such that

$$(x_1^2 + x_2^2)^{\frac{1}{2}} \leq \gamma$$

The critical values are $x_1 = 0$, $x_2 = \pm \gamma$ and

$$\gamma^2 = \frac{\gamma}{\sqrt{2}} \quad \text{or} \quad \gamma = \frac{1}{\sqrt{2}}$$

The stability region is

$$(x_1^2 + x_2^2)^{\frac{1}{2}} < \frac{1}{2} \quad , \text{ see Fig. 1.}$$

C) Finally we use the norm

$$\| A \| = \max_{i,j} |a_{ij}|$$

We have

$$\| \Phi(t-t_0) \| = e^{-(t-t_0)}$$

$$\| g(x) \| = x_2^2$$

$$\alpha = \beta = 1, \text{ for the above norm } M=2$$

We choose the largest γ guaranteeing

$$x_2^2 < \frac{\max(|x_1|, |x_2|)}{2}$$

for all x such that

$$\max(|x_1|, |x_2|) < \frac{\gamma}{M}$$

We find $\gamma = \frac{1}{2}$ and the stability region $\max(|x_1|, |x_2|) < \frac{1}{4}$ shown in the Fig. 1.

$$* \qquad\qquad * \qquad\qquad *$$

728

Using Lyapunov method we find that the system is stable in the entire plane. Fig. 1 shows contours of constant V (the value of the Lyapunov function).

For the system shown in the block diagram

where $\qquad G_c(s) = K$

$$G_p(s) = \frac{1}{s(s+1)}$$

$$U(s) = R(s) = 0$$

choose K so that the following performance index is minimized

$$V(x) = \int_{t_0}^{\infty} P(y(\tau))d\tau$$

where $\qquad \dfrac{dy}{d\tau} = F(y) \quad ; \quad y(t_0) = x$

$p(x) > 0$

$\qquad\qquad$ for $x \neq 0$

$p(0) = 0$

p(x) is a continuous scalar function, and the system is described by

$$\frac{dx}{dt} = F(x)$$

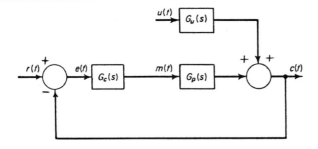

Solution: We assume that the initial value of the loop output is nonzero, but the initial derivative of the loop output is zero.

We can describe the loop by the equations

$$\frac{d^2c}{dt^2} + \frac{dc}{dt} = m$$

$$m = Ke$$

729

$$e = -c$$

substituting e and m

$$\frac{d^2c}{dt^2} + \frac{dc}{dt} + Kc = 0$$

The state variables are $x_1 = c$, $x_2 = \frac{dc}{dt}$

and we obtain

$$\frac{dx}{dt} = Ax$$

where

$$A = \begin{pmatrix} 0 & 1 \\ -K & -1 \end{pmatrix}$$

we choose

$$\rho(x) = x_1^2 + \alpha^2 x_2^2 \quad , \qquad \alpha > 0$$

From

$$\rho(x) = x^T Q x$$

we obtain

$$Q = \begin{pmatrix} 1 & 0 \\ 0 & \alpha^2 \end{pmatrix}$$

If the system is stable there exists matrix P such that

$$A^T P + PA = -Q$$

Substituting A and Q we obtain

$$P = \frac{1}{2} \begin{pmatrix} \alpha^2 K + 1 + \frac{1}{K} & \frac{1}{K} \\ \frac{1}{K} & \alpha^2 + \frac{1}{K} \end{pmatrix}$$

We have

$$\rho(x) = x^T Q x = -x^T(A^T P + PA)x$$
$$= -x^T A^T P x - x^T P A x$$

since

$$\dot{x} = Ax$$

$$\rho(x) = -\dot{x}^T P x - x^T P \dot{x} = -\frac{d}{dt}(x^T P x)$$

Then

730

$$\int_{t_0}^{\infty} \rho(x) d\,\tau = -x^T P x \Big|_{t_0} = x^T P x \Big|_{t_0} - x^T P x \Big|_{\infty}$$

The system is asymptotically stable, so

$$x^T P x \Big|_{\infty} = 0 .$$

Thus we must minimize

$$x^T P x \Big|_{t_0} .$$

We have

$$V(x) = x^T P x = \frac{1}{2} (\alpha^2 K + 1 + \frac{1}{K}) x_1^2 + \frac{x_1 x_2}{K} + \frac{1}{2} (\alpha^2 + \frac{1}{K}) x_2^2$$

We assumed that the initial

$$\frac{dc}{dt} = 0 , \quad x_2 = 0 ,$$

hence we minimize

$$V(x) = \frac{1}{2} (\alpha^2 K + 1 + \frac{1}{K}) x_1^2$$

differentiating we obtain

$$K_{min} = \frac{1}{\alpha}$$

● **PROBLEM** 13-87

Examine the stability of the VAN DER POL equation subject to a sinusoidal forcing function

$$\ddot{x} + \varepsilon(x^2 - 1)\dot{x} + x = A \sin wt \qquad (1)$$

where A is the amplitude of the forcing function and w is the frequency.

Solution: When A=0 the system has a stable limit cycle of radius 2, and if $\varepsilon < 1$, $w \cong 1$. When w is equal to the frequency of the unforced case (A≠0) the forcing function has maximum effect. To analyze such a situation we will use averaging technique.

We write the system of two first-order differential equations which is equivalent to Eq. (1)

$$\dot{x} = y, \quad \dot{y} = -x + \varepsilon(1-x^2)y + A \sin wt$$

Transfering to polar coordinates we have

$$\frac{dr}{dt} = \frac{1}{r} (\varepsilon y^2 - \varepsilon x^2 y^2 + Ay \sin wt)$$

where
$$r = \sqrt{x^2 + y^2} .$$

Let w = frequency of the unforced system, using averaging technique we have

$$\frac{dr}{dt} = \frac{\varepsilon r}{2} - \frac{\varepsilon r^3}{8} + \frac{A}{2} \qquad (2)$$

and the radius is given by

$$r_0^3 - 4r_0 - \frac{4A}{\varepsilon} = 0 \qquad (3)$$

Using Cardan formula

$$r_0 = \sqrt[3]{\frac{2A}{\varepsilon} + \sqrt{\frac{4A^2}{\varepsilon^2} - \frac{64}{27}}} + \sqrt[3]{\frac{2A}{\varepsilon} - \sqrt{\frac{4A^2}{\varepsilon^2} - \frac{64}{27}}} \qquad (4)$$

Since r_0 must be positive and real we can drop the other two values for r_0.

ε is small in most cases and

$$\frac{4A^2}{\varepsilon^2} - \frac{64}{27} \geq 0$$

is true.   If $\frac{4A^2}{\varepsilon^2} - \frac{64}{27} < 0$ we use (3) to compute r_0.   If A = 0 Eq (3) becomes

$$r_0^3 - 4r_0 = 0$$

and r_0 = 2 — the radius for the unforced system.   From (2) we conclude that the oscillation is stable, since when $r>r_0$ then

$\frac{dr}{dt} < 0$ and $r<r_0$, $\frac{dr}{dt} > 0$, the radius approaches a constant value.

DIFFERENT KINDS OF STABILITY-JURY TEST

● PROBLEM 13-88

Which of the following transfer functions define a system which is i) output-Lyapunov stable; ii) asymptotically stable; iii) bibo stable.

a) $H(s) = \dfrac{s + 3}{s^4 + 4s^3 + 5s^2 + 4s + 4}$

b) $H(s) = \dfrac{K}{2s^5 + 5s^4 + 4s^3 + 2s + 1}$

c) $H(s) = \dfrac{(s - 1)^2}{8s^4 + 5s^3 + 9s^2 + 7s + 2}$

<u>Solution:</u>   If the system has Hurwitz property then it has all three stability properties. If the system does not have the Hurwitz property, then it cannot be bibo stable or asymptotically stable. If only simple zeros occur on the jω axis, the system can be output-Lyapunov stable. We will start by checking the Hurwitz property of the denominator polynomials.

a) $P(s) = s^4 + 4s^3 + 5s^2 + 4s + 4$

$$R(s) = \frac{4s^3 + 4s}{s^4 + 5s^2 + 4} = \frac{1}{\frac{s}{4} + \frac{1}{s}} = \frac{4s}{s^2 + 4}$$

R(s) has degree-2 and P(s) has degree-4, thus P(s) has an even polynomial factor and cannot be Hurwitz.

Let $R = \frac{A}{B}$ and $P = C \cdot D$ then $D = A + B = s^2 + 4s + 4$. We find

$$\frac{s^4 + 4s^3 + 5s^2 + 4s + 4}{s^2 + 4s + 4} = s^2 + 1$$

Thus, $P(s) = (s^2 + 1)(s^2 + 4s + 4)$ and we see that P(s) is a Hurwitz polynomial times $s^2 + 1$, a polynomial with simple zeros on the jω axis, $s = \pm j$. This system is output-Lyapunov stable but not bibo nor asymptotically stable.

b) $P(s) = 2s^5 + 5s^4 + 4s^3 + 2s + 1$

$$R(s) = \frac{2s^5 + 4s^3 + 2s}{5s^4 + 1} = \frac{2}{5}s + \cfrac{1}{\frac{5}{4}s + \cfrac{1}{-2s + \cfrac{1}{-\frac{5}{9}s + \cfrac{1}{\frac{18}{5}s}}}}$$

There are two negative signs in the fraction expansion (indicating two right half-plane zeros) thus the system is not stable in any sense. That could also be seen from the missing s^2 term in P(s).

c) $P(s) = 8s^4 + 5s^3 + 9s^2 + 7s + 2$

$$R(s) = \frac{5s^3 + 7s}{8s^4 + 9s^2 + 2} = \cfrac{1}{\frac{8}{5}s + \cfrac{1}{-\frac{25}{11}s + \ldots}}$$

Thus P(s) is not Hurwitz. We have to check for the factor s-1 in P(s), because of the numerator, but P(+1)>0 and the system is not stable in any sense.

Determine the stability of the following systems

a) $H(s) = (s-5)/(s+4)$

b) $H(s) = \dfrac{s+k}{(s+m)^2}$ $k \geq m$

Solution: a) $\dfrac{H(s)}{s} = s$ has a simple pole on the $j\omega$ axis

at $s = 0$ and no poles at infinity nor in the right half-plane thus the system is bibo stable. This system is also output-Lyapunov stable since all the poles of $H(s)$, $s = -4$, are in the left half-plane.

b)   For $m<0$ the system is unstable in all senses since it has a pole in the right half-plane. If $k \neq 0$ and $m = 0$, there is a double pole on the imaginary axis and the system is unstable.

For $k = m = 0$ $H(s) = \dfrac{1}{s}$; one simple pole on the imaginary

axis, and the system is output-Lyapunov stable but not bibo or asymptotically stable.

For $m>0$ the system is stable in all senses.

Discuss the stability of the systems:

a) $H(s) = \dfrac{1}{(s + k)^3}$

b) $H(s) = s^2$

Solution: a)   For $k<0$, $H(s) = \dfrac{1}{(s + k)^3}$ has a pole in the

right half-plane and the system is unstable in all senses.

For $k>0$ there are no poles in the right half-plane and the system is bibo stable. The poles of $H(s)$ determine the form of the zero input-output

$$y_0(t)1(t) = [a_1 e^{-kt} + a_2 t e^{-kt} + a_3 t^2 e^{-kt}]1(t)$$

thus the system is output-Lyapunov stable. If $k = 0$ the system is neither bibo stable nor output-Lyapunov stable.

b)   Since $h * 1 = \delta$ is not a function the system is not bibo stable. The system is output-Lyapunov stable since $y = \ddot{u}$ and the zero-input response is $y = 0$.

By determining the form of the transient response evaluate the stability of the closed-loop system described by the differential equation

$$\frac{d^2c}{dt^2} + 4\frac{dc}{dt} + 13c = r(t)$$

Solution: We transform the differential equation describing the system and obtain

$$\frac{C(s)}{R(s)} = \frac{1}{s^2 + 4s + 13}$$

This equation can be written as follows

$$\frac{C(s)}{R(s)} = \frac{1}{3} \cdot \frac{3}{(s + 2)^2 + 9}$$

Since

$$\alpha^{-1}\{F(s + a)\} = e^{-at}\,\alpha^{-1}\{F(s)\}$$

we have

$$\alpha^{-1}\left[\frac{C(s)}{R(s)}\right] = \frac{1}{3}\alpha^{-1}\left[\frac{3}{(s + 2)^2 + 9}\right] = \frac{1}{3}e^{-2t}\alpha^{-1}\left[\frac{3}{s^2 + 3^2}\right]$$

Since $\alpha^{-1}\left\{\dfrac{w}{s^2 + w^2}\right\} = \sin wt$ we get

$$\alpha^{-1}\left[\frac{C(s)}{R(s)}\right] = \frac{1}{3}e^{-2t}\sin 3t.$$

The transient response of the system is

$$y = \frac{1}{3}e^{-2t}\sin 3t$$

The amplitude of this oscillatory response decreases with increasing time. Hence the system is stable.

Determine the stability of a digital control system. The characteristic equation of the system is

$$F(z) = z^3 + (68.2T^2 + 15.6T - 4)z^2$$
$$+ (6.271 \times 10^{-3}KT^3 - 31.2T + 4)z$$
$$+ (6.271 \times 10^{-3}KT^3 - 68.2T^2 + 15.6T - 1) = 0$$

K is a constant parameter and T is the sampling period.

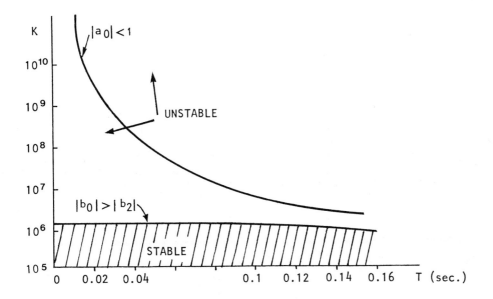

Solution: We will find the range of parameters K and T for which the system is asymptotically stable. Using the Jury's test we have

(1) $F(1) = 2 \cdot 6.271 \cdot 10^{-3} KT^3 > 0$ thus $KT^3 > 0$

Since T is time, we get

$T > 0$ and $K \geqslant 0$

(2) $F(-1) = -10 + 62.4T < 0$

$T < 0.16$ sec

The coefficients of the equation are

$a_0 = 6.271 \times 10^{-3} KT^3 - 68.2T^2 + 15.6T - 1$

$a_1 = 6.271 \times 10^{-3} KT^3 - 31.2T + 4$

$a_2 = 68.2T^2 + 15.6T - 4$

$a_3 = 1$

$b_0 = a_0^2 - a_3^2$

$b_2 = a_0 a_2 - a_1 a_3$

The conditions for stability are

(3) $|a_0| < a_3$ or

$|6.271 \times 10^{-3} KT^3 - 68.2T^2 + 15.6T - 1| < 1$

(4) $|b_0| > |b_2|$

We have the following system of inequalities

736

K > 0

0 < T < 0.16

$| 6.271 \times 10^{-3} KT^3 - 68.2T^2 + 15.6T - 1 | < 1$

$| a_0^2 - 1 | > | a_0 a_2 - a_1 |$

The exact general solution of the above system does not exist, term a_0^2 gives terms T^6 and the sixth order equation does not have a general solution. We shall use the graphic method, T and K being x and y axis, furthermore for K we use logarithmic scale (because of the term $6.271 \times 10^{-3} KT^3$)

● **PROBLEM** 13-93

Using the Raible's method determine the stability of the system described by the following equation

$$F(z) = z^3 + 2.7z^2 + 2.26z + 0.6 = 0$$

Solution: We find the values of the function F(z) for Z = 1 and Z = -1 and apply the Jury test.

F(1) = 6.56 > 0

F(-1) = 0.04 > 0

Thus, since n is odd, F(z) has at least one root outside the unit circle. To find the exact number of roots that are outside the unit circle we apply the Raible tabulation and obtain

	1	2.7	2.26	0.6	$k_a = 0.6$
-)	0.36	1.356	1.62		
	0.64	0.644	0.64		$k_b = 1$
-)	0.64	0.644			
	0	0			

We got the row of zeros, this is a singular case. To solve this problem let us apply the following transformation

$$z \rightarrow (1 + \varepsilon) z$$

where ε is a small number.

For z^2 we have

$$z^2 \rightarrow (1 + \varepsilon)^2 z^2 = (1 + 2\varepsilon + \varepsilon^2) z^2$$

Since ε is small we can neglect ε^2 term

Thus $\qquad z^2 \rightarrow \qquad (1 + 2\varepsilon)z^2$

In general

$$z^n \rightarrow \qquad (1 + \varepsilon)^n z^n = (1 + n\varepsilon + \dots)z^n$$

We can neglect all powers of ε higher than ε.

i.e. $\varepsilon^2,\ \varepsilon^3,\ \dots \qquad$ etc.

We have $\qquad z^n \rightarrow \qquad (1 + n\varepsilon)z^n$

Applying this transformation to our initial equation we have

$$F[(1+\varepsilon)Z] \cong (1+3\varepsilon)z^3 + 2.7(1+2\varepsilon)z^2 + 2.26(1+\varepsilon)z + 0.6$$

The Raible tabulation gives

$1+3\varepsilon$	$2.7(1+2\varepsilon)$	$2.26(1+\varepsilon)$	0.6
$-)\quad \dfrac{0.36}{1+3\varepsilon}$	$\dfrac{1.356(1+\varepsilon)}{1+3\varepsilon}$	$\dfrac{1.62(1+2\varepsilon)}{1+3\varepsilon}$	$k_a = \dfrac{0.6}{1+3\varepsilon}$

$$b_0 = \frac{0.64 + 6\varepsilon}{1+3\varepsilon} \qquad \frac{1.344 + 12.144\varepsilon}{1+3\varepsilon} \qquad \frac{0.64 + 5.8\varepsilon}{1+3\varepsilon} \qquad k_b = \frac{0.64 + 5.8\varepsilon}{0.64 + 6\varepsilon}$$

$$-)\quad \frac{(0.64 + 5.8\varepsilon)^2}{(1+3\varepsilon)(0.64 + 6\varepsilon)} \qquad \frac{(1.344 + 12.144\varepsilon)(0.64 + 5.8\varepsilon)}{(1+3\varepsilon)(0.64 + 6\varepsilon)}$$

$$c_0 = \frac{0.256\varepsilon}{(1+3\varepsilon)(0.64 + 6\varepsilon)} \qquad \frac{0.27\varepsilon}{(1+3\varepsilon)(0.64 + 6\varepsilon)} \qquad k_c = 1.05$$

$$-)\quad \frac{0.283\varepsilon}{(1+3\varepsilon)(0.64 + 6\varepsilon)}$$

$$d_0 = \frac{-0.027\varepsilon}{(1+3\varepsilon)(0.64 + 6\varepsilon)}$$

We see that

$$\text{for}\quad \varepsilon > 0 : \quad b_0 > 0,\quad c_0 \gtrless 0,\quad d_0 < 0$$

$$\text{for}\quad \varepsilon < 0 : \quad b_0 > 0,\quad c_0 < 0,\quad d_0 > 0$$

There is one negative element in both cases when $\varepsilon > 0$ and when $\varepsilon < 0$, we conclude that there is one root outside the unit circle and none on it.

Given the equation

$$F(z) = z^3 + z^2 + z + 1 = 0 \qquad\qquad (1)$$

examine the stability.

Solution: Since all the coefficients of the equation are identical this is a singular case. We use the following transformation

$$z \rightarrow (1 + \varepsilon)z$$

where ε is small, so we can neglect all the terms with ε^2, ε^3 etc.

We have

$$F(z) \rightarrow F[(1 + \varepsilon)z] \cong (1 + 3\varepsilon)z^3 + (1 + 2\varepsilon)z^2 + (1 + \varepsilon)z + 1$$

Raible's tabulation gives

$1 + 3\varepsilon$	$1 + 2\varepsilon$	$1 + \varepsilon$	1

$-)\quad \dfrac{1}{1 + 3\varepsilon} \qquad \dfrac{1 + \varepsilon}{1 + 3\varepsilon} \qquad \dfrac{1 + 2\varepsilon}{1 + 3\varepsilon}\qquad\qquad k_a = \dfrac{1}{1 + 3\varepsilon}$

$b_0 = \dfrac{6\varepsilon}{1 + 3\varepsilon} \qquad \dfrac{4\varepsilon}{1 + 3\varepsilon} \qquad \dfrac{2\varepsilon}{1 + 3\varepsilon} \qquad\qquad k_b = \dfrac{1}{3}$

$-)\qquad \dfrac{2\varepsilon}{3(1 + 3\varepsilon)} \qquad \dfrac{4\varepsilon}{3(1 + 3\varepsilon)}$

$c_0 = \dfrac{16\varepsilon}{3(1 + 3\varepsilon)} \qquad \dfrac{8\varepsilon}{3(1 + 3\varepsilon)}$

$\qquad\qquad\qquad\qquad\qquad\qquad\qquad\qquad k_c = \dfrac{1}{2}$

$-)\qquad \dfrac{4\varepsilon}{3(1 + 3\varepsilon)}$

$d_0 = \dfrac{4\varepsilon}{1 + 3\varepsilon}$

We see from the first column of the tabulation that the elements b_0, c_0, d_0 have the same sign as ε. When ε changes sign from negative to positive all three roots move from the outside to the inside of the unit circle.

Thus we conclude that all three roots of the equation (1) are on the unit circle. It is easy to check that equation (1) has the roots at $z = -1$ and $z = \pm j$.

Examine the stability of the system whose characteristic equation is

$$F(z) = z^2 + z + 0.15 = 0$$

Solution: We shall use the Jury test. The order of the equation is $n = 2$ and for an even order equation the requirements for stability are $F(1) > 0$ and $F(-1) > 0$

We have

$$F(1) = 2.15 \quad \text{and} \quad F(-1) = 0.15 > 0$$

The Jury tabulation will consist of one row, since $2n-3 = 1$. We have

z^0	z^1	z^2
0.15	1	1

and $\quad |a_0| = 0.15 < a_2 = 1.$

We conclude that the system is asymptotically stable and all the roots are inside the unit circle.

As an exercise we shall use the Raible tabulation for the above example. We obtain

	1	1	0.15	$k_a = 0.15$
-)	0.0225	0.15		
	$b_0 = 0.9775$	0.85		$k_b = 0.87$
-)	0.739			
	$c_0 = 0.2385$			

Since the coefficients in the first column are positive, two roots of the equation are inside the unit circle.

DISCRETE SYSTEMS

● PROBLEM 13-96

Determine the stability of both discrete systems

$$\begin{bmatrix} x_1(k+1) \\ x_2(k+1) \end{bmatrix} = \begin{bmatrix} 1 & 3 \\ 2 & 1 \end{bmatrix} \begin{bmatrix} x_1(k) \\ x_2(k) \end{bmatrix} \tag{1}$$

and

$$\begin{bmatrix} x_1(k+1) \\ x_2(k+1) \end{bmatrix} = \begin{bmatrix} 0.4 & 1 \\ 0 & 0.3 \end{bmatrix} \begin{bmatrix} x_1(k) \\ x_2(k) \end{bmatrix} \tag{2}$$

Solution: From $\det|\lambda I - A| = 0$ we compute the eigenvalues of the matrix A for the first system.

$$(\lambda - 1)^2 - 6 = 0$$

$$\lambda = 1 + \sqrt{6}$$

The system is unstable since eigenvalues are greater than unity.

For the second system we have

$$\det \begin{pmatrix} \lambda-0.4 & -1 \\ 0 & \lambda-0.3 \end{pmatrix} = 0$$

$$(\lambda - 0.4)(\lambda - 0.3) = 0$$

$$\lambda_1 = 0.4$$

$$\lambda_2 = 0.3$$

The system (2) is stable since

$$|\lambda_1| < 1 \quad \text{and} \quad |\lambda_2| < 1.$$

● PROBLEM 13-97

Show that the system described by the equation

$$\begin{bmatrix} x_1(k+1) \\ x_2(k+1) \end{bmatrix} = \begin{bmatrix} 0 & 1 \\ -4 & -5 \end{bmatrix} \begin{bmatrix} x_1(k) \\ x_2(k) \end{bmatrix} + \begin{bmatrix} 0 \\ 1 \end{bmatrix} u(k)$$

$$y(k) = \begin{bmatrix} 1 & 0 \end{bmatrix} \begin{bmatrix} x_1(k) \\ x_2(k) \end{bmatrix}$$

where k is a discrete parameter, is unstable. Find a scalar function H such that

$$u(k) = -Hy(k)$$

will stabilize the system.

<u>Solution</u>: The eigenvalues of the matrix A

$$A = \begin{bmatrix} 0 & 1 \\ -4 & -5 \end{bmatrix}$$

are $\lambda_1 = -1$, $\lambda_2 = -4$ so the open-loop system is unstable.

In order to stabilize the system we seek a scalar function H, such that taking into account the output feedback

$$u(k) = -Hy(k)$$

the eigenvalues of $\overline{A}$, where $\overline{A}$ is a closed-loop matrix are less than unity in magnitude.

$$\overline{A} = A - BHC = \begin{bmatrix} 0 & 1 \\ -4 & -5 \end{bmatrix} - \begin{bmatrix} 0 \\ 1 \end{bmatrix} H[1 \ 0]$$

$$= \begin{bmatrix} 0 & 1 \\ -4-H & -5 \end{bmatrix}$$

The characteristic polynomial of $\overline{A}$ is

$$\det |\lambda I - \overline{A}| = \det \begin{vmatrix} \lambda & -1 \\ 4 + H & \lambda + 5 \end{vmatrix} = 0$$

$$\lambda^2 + 5\lambda + 4 + H = 0$$

$$\lambda_1 = -\frac{5}{2} + \sqrt{25-4(4+H)}$$

$$\lambda_2 = -\frac{5}{2} - \sqrt{25-4(4+H)}$$

We notice that there is no value of H that can make both roots of the polynomial less than unity in magnitude, hence the system is not stable and cannot be stabilized using output feedback.

Check the stability of the following system

$$x_1(k + 1) = x_2(k)$$

$$x_2(k + 1) = 2.5x_1(k) + x_2(k) + u(k)$$

$$y(k) = x_1(k)$$

If the system is unstable, use the output feedback, $u(k) = -Hy(k)$ to stabilize the system, determine the range of H.

Solution: We rewrite the equations in the vector notation

$$\begin{bmatrix} x_1(k + 1) \\ x_2(k + 1) \end{bmatrix} = \begin{bmatrix} 0 & 1 \\ 2.5 & 1 \end{bmatrix} \begin{bmatrix} x_1(k) \\ x_2(k) \end{bmatrix} + \begin{bmatrix} 0 \\ 1 \end{bmatrix} u(k)$$

$$y(k) = \begin{bmatrix} 1 & 0 \end{bmatrix} \begin{bmatrix} x_1(k) \\ x_2(k) \end{bmatrix}$$

$$x(k + 1) = Ax(k) + Bu(k)$$

$$y(k) = Cx(k)$$

From $\det|\lambda I - A| = 0$ we calculate the eigenvalues of the matrix A.

$$\lambda^2 - \lambda - 2.5 = 0$$

$$\lambda_1 = \frac{1}{2} + \sqrt{\frac{11}{4}} = 2.16$$

$$\lambda_2 = \frac{1}{2} - \sqrt{\frac{11}{4}} = -1.16$$

The open-loop system is unstable. We shall try to stabilize the system using output feedback

$$u(k) = -Hy(k)$$

and determining the range of H. The closed-loop matrix $\bar{A}$ is

$$\bar{A} = A - BHC = \begin{bmatrix} 0 & 1 \\ 2.5 & 1 \end{bmatrix} - \begin{bmatrix} 0 \\ 1 \end{bmatrix} H \begin{bmatrix} 1 & 0 \end{bmatrix} = \begin{bmatrix} 0 & 1 \\ 2.5 - H & 1 \end{bmatrix}$$

The eigenvalues of $\bar{A}$ are

$$\det|\lambda I - \bar{A}| = 0$$

$$\det \begin{vmatrix} \lambda & -1 \\ H - 2.5 & \lambda - 1 \end{vmatrix} = 0$$

$$\lambda^2 - \lambda + (H - 2.5) = 0$$

$$\lambda_1 = \frac{1}{2} + \frac{1}{2}\sqrt{1 - 4(H - 2.5)} = \frac{1}{2} + \frac{1}{2}\sqrt{11 - 4H}$$

$$\lambda_2 = \frac{1}{2} - \frac{1}{2}\sqrt{1 - 4(H - 2.5)} = \frac{1}{2} - \frac{1}{2}\sqrt{11 - 4H}$$

If we assume that λ_1, λ_2 are real we get the following range of H

$$2.5 \leq H \leq 2.75$$

for which the system is stable. If the roots are complex we have

$$2.75 \leq H \leq 3.5$$

We conclude that the system is stable when $2.5 \leq H \leq 3.5$

● **PROBLEM** 13-99

The discrete-time system is shown in Fig. (1).

The open-loop transfer function of the system is

$$G(s) = \frac{5}{s \cdot (s + 1)}$$

Examine the stability of the system.

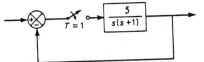

Fig. 1

Solution: We find the z - transformation of G(s).

$$G(s) = \frac{5}{s(s + 1)} = 5\left(\frac{1}{s} - \frac{1}{s + 1}\right)$$

and

$$g(t) = 5\left(u(t) - e^{-t}\right).$$

Substituting $t = kT$, where $T = 1$ we obtain

$$g(kT) = g(k) = 5(u(k) - e^{-k}).$$

Then

$$G(z) = \sum_{k=0}^{\infty} g(k) z^{-k} = \sum_{k=0}^{\infty} 5(u(k) - e^{-k}) z^{-k} = 5\left(\frac{z}{z - 1} - \frac{z}{z - e^{-1}}\right)$$

Thus

$$G(z) = \frac{5(1 - e^{-1})z}{(z - 1)(z - e^{-1})}$$

The characteristic equation of the system is

$$1 + G(z) = 0$$

or

$$5(1 - e^{-1})z + (z - 1)(z - e^{-1}) = 0$$

$$e^{-1} = 0.368$$

$$z^2 - 0.368z - z + 0.368 + 5z - 1.84z = 0$$

$$z^2 + 1.792z + 0.368 = 0$$

$$z_1 = -0.236$$

$$z_2 = -1.555$$

We see that root z_2 of the characteristic equation has a magnitude greater than unity, thus the system is unstable. It can be easily shown that for the system shown in Fig.(1) with the transfer function

$$G(s) = \frac{K}{s(s + 1)}$$

the condition of stability is

$$0 < K < 4.32.$$

In case of a characteristic equation of higher degree, where the exact solution is difficult or impossible to find, we use the following transformation of z

$$z = \frac{r + 1}{r - 1}.$$

To check the stability of resulting equation one applies the Routh criterion. In the above example we obtain

$$\left(\frac{r + 1}{r - 1}\right)^2 + 1.792\left(\frac{r + 1}{r - 1}\right) + 0.368 = 0$$

$$3.16r^2 + 1.264r - 1.16 = 0$$

All the coefficients of the equation must be non-zero and have the same sign for the system to be stable, this is the necessary but not sufficient condition. We conclude that the system is not stable. This corresponds to the result obtained previously.

Determine   the stability of a closed-loop discrete-time system
shown below

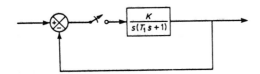

Solution: The gain of the system is

$$G(s) = \frac{K}{s(T_1 s + 1)}$$

To find the z - transform of G(s) note that

$$G(s) = \frac{K}{s(s + \frac{1}{T_1})T_1} = K\frac{\frac{1}{T_1}}{s(s + \frac{1}{T_1})}$$

The z transform of

$$\frac{a}{s(s + a)} \text{ is } \frac{(1 - e^{-aT})z}{(z - 1)(z - e^{-aT})}$$

thus

$$G(z) = \frac{K\left(1 - e^{-\frac{T}{T_1}}\right)z}{(z-1)(z-e^{\frac{-T}{T_1}})}$$

The characteristic equation is

$$1 + G(z) = 0$$

or

$$1 + \frac{K\left(1 - e^{-\frac{T}{T_1}}\right)z}{(z - 1)(z - e^{-\frac{T}{T_1}})} = 0$$

After some calculations we get

$$z^2 + \left[K\left(1 - e^{-\frac{T}{T_1}}\right) - \left(1 + e^{-\frac{T}{T_1}}\right)\right]z + e^{-\frac{T}{T_1}} = 0$$

In order to use the Routh criterion we transform z,

$$z = \frac{r + 1}{r - 1}$$

The characteristic equation becomes

$$r^2 \left[K \left(1 - e^{-\frac{T}{T_1}} \right) \right] + r \cdot 2 \left(1 - e^{-\frac{T}{T_1}} \right) + 2 \left(1 + e^{-\frac{T}{T_1}} \right)$$

$$- K \left(1 - e^{-\frac{T}{T_1}} \right) = 0$$

The Routh tabulation is

$$r^2 \quad K \left(1 - e^{-\frac{T}{T_1}} \right) \qquad\qquad 2 \left(1 + e^{-\frac{T}{T_1}} \right) - K \left(1 - e^{-\frac{T}{T_1}} \right)$$

$$r^1 \quad 2 \left(1 - e^{-\frac{T}{T_1}} \right)$$

$$\qquad\qquad\qquad\qquad\qquad\qquad 0$$

$$r^0 \quad 2 \left(1 + e^{-\frac{T}{T_1}} \right) - K \left(1 - e^{-\frac{T}{T_1}} \right)$$

Note that $1 - e^{-\frac{T}{T_1}} > 0$ and the system is stable if and only if $K > 0$

$$\frac{1 + e^{-\frac{T}{T_1}}}{1 - e^{-\frac{T}{T_1}}} > \frac{K}{2}$$

Since $\dfrac{1 + e^{-\frac{T}{T_1}}}{1 - e^{-\frac{T}{T_1}}} = \coth \left(\dfrac{T}{2T_1} \right)$

We conclude that the system is stable when

$$0 < K < 2 \coth \left(\frac{T}{2T_1} \right)$$

PHASE PLANE

● PROBLEM 13-101

Determine the stability for the system described by the equation

$$x(k+1) = Ax(k) \tag{1}$$

where

$$A = \begin{bmatrix} 0 & -1 \\ 1 & 0 \end{bmatrix}$$

Solution: Using the z-transform of both sides of Eq. 1 and solving for X(z) we obtain

$$X(z) = (zI-A)^{-1} zx(0)$$

or

$$\begin{bmatrix} X_1(z) \\ \\ X_2(z) \end{bmatrix} = \frac{1}{z^2 + 1} \begin{bmatrix} z^2 & -z \\ \\ z & z^2 \end{bmatrix} \begin{bmatrix} x_1(0) \\ \\ x_2(0) \end{bmatrix} \qquad (2)$$

The functions $x_1(k)$ and $x_2(k)$ can be found by taking the inverse z-transform of Eq. 2. We obtain

$$\begin{bmatrix} x_1(k) \\ \\ x_2(k) \end{bmatrix} = \begin{bmatrix} \cos \frac{k\pi}{2} & -\sin \frac{k\pi}{2} \\ \\ \sin \frac{k\pi}{2} & \cos \frac{k\pi}{2} \end{bmatrix} \begin{bmatrix} x_1(0) \\ \\ x_2(0) \end{bmatrix}$$

We will find the state trajectories of $x_1(k)$ and $x_2(k)$ for two arbitrary initial states $x(0)$.

$$x_1(k) = \cos \frac{k\pi}{2} x_1(0) - \sin \frac{k\pi}{2} x_2(0)$$

$$x_2(k) = \sin \frac{k\pi}{2} x_1(0) + \cos \frac{k\pi}{2} x_2(0) \qquad (3)$$

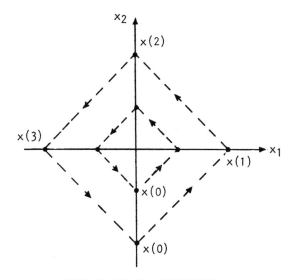

STATE-PLANE TRAJECTORIES

From Eq. (3) we conclude that trajectories form limit cycle; $x_1(k)$ and $x_2(k)$ are periodic functions which neither grow nor decay in amplitude. Therefore the system is stable but not asymptotically stable.

We can determine the stability of the system without finding the trajectories.

748

The characteristic equation of the system is

$$|zI-A| = z^2 + 1 = 0$$

The characteristic equation has two roots $z_1 = j$ $z_2 = -j$ which are inside the unit circle $|z| = 1$ in the z-plane.

Therefore the system is stable.

Given the system

$$x(k+1) = Ax(k)$$

where k is a discrete independent variable and matrix A is given

$$A = \begin{bmatrix} -0.5 & 0 \\ 0 & 0.5 \end{bmatrix}$$

determine the stability of the system.

Solution: Using the z-transform we obtain

$$X(z) = \begin{bmatrix} \dfrac{z}{z+0.5} & 0 \\ 0 & \dfrac{z}{z-0.5} \end{bmatrix} x(0)$$

where

$$X(z) = (zI-A)^{-1} zx(0)$$

The inverse z-transform of X(z) gives

$$x(k) = \begin{bmatrix} (-0.5)^k & 0 \\ 0 & (0.5)^k \end{bmatrix} x(0)$$

For any initial state x(0) the state trajectories of x(k) will converge to the equilibrium state x = 0 as k tends to infinity.

Therefore the system is asymptotically stable in the large.

Let x(0) = [1 1]' and [-1 -1]' be initial states.   The state trajectories of x(k) are shown below.

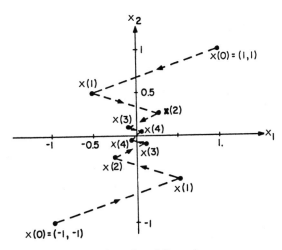

State-plane trajectories of the system

The characteristic equation of the system is

$$|zI-A| = (z-0.5)(z+0.5) = 0$$

The equation has two real roots $z_1 = -0.5$ $z_2 = 0.5$
They both are inside the unit circle $|z| = 1$.

● **PROBLEM** 13-103

The system is described by the equation

$$x(k+1) = Ax(k)$$

where k is a discrete independent variable and A is a
diagonal matrix

$$A = \begin{bmatrix} -2.5 & 0 \\ 0 & 0.5 \end{bmatrix}$$

Examine the stability of the system.

<u>Solution:</u> From $X(z) = (zI-A)^{-1} zx(0)$

which is the solution in the z domain we obtain

$$X(z) = \begin{bmatrix} \dfrac{z}{z+2.5} & 0 \\ 0 & \dfrac{z}{z-0.5} \end{bmatrix} x(0)$$

We note that the characteristic equation of the system has a
root at z = -2.5 which is outside the unit circle.

Therefore the system is unstable.

The inverse z-transform of X(z) gives

$$x(k) = \begin{bmatrix} (-2.5)^k & 0 \\ 0 & (0.5)^k \end{bmatrix} x(0)$$

It is easy to show that $|x_1(k)| \to \infty$ as k approaches infinity.

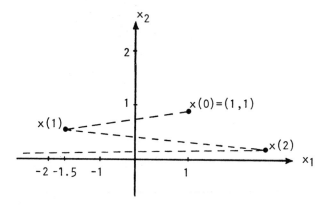

State-plane trajectories of the system

Let the initial state (k=0) be

$x(0) = [1,1]$. We have

$$\begin{bmatrix} 1 \\ 1 \end{bmatrix} \to \begin{bmatrix} -2.5 \\ 0.5 \end{bmatrix} \to \begin{bmatrix} 6.25 \\ 0.25 \end{bmatrix} \to \begin{bmatrix} -15.6 \\ 0.125 \end{bmatrix}$$

$$\qquad k=0 \qquad\quad k=1 \qquad\quad k=2 \qquad\qquad k=3$$

● **PROBLEM** 13-104

Describe the motion of a pendulum and determine its stability characteristics. Assume that the whole mass is concentrated at the end of the pendulum.

Solution: The motion of a pendulum is described by the following differential state equation

$$\ddot{\phi} = -\frac{g}{L} \sin \phi$$

We can choose the system of physical units where $\frac{g}{L} = 1$.

We have

$$\ddot{\phi} = -\sin \phi$$

Let us set

$$x_1 = \phi$$

$$x_2 = \dot{\phi}$$

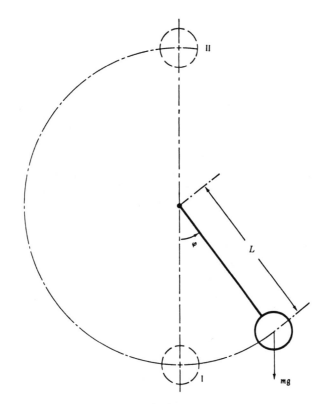

Fig. 1

Simple pendulum.

and obtain

$$\dot{x}_1 = x_2$$
$$\dot{x}_2 = -\sin x_1 \qquad\qquad (1)$$

The equilibrium states are

$$x_1^0 = \begin{pmatrix} 0 \\ \\ \\ 0 \end{pmatrix} \qquad \text{and} \qquad x_{11}^0 = \begin{pmatrix} \pm\pi \\ \\ 0 \end{pmatrix}$$

as shown in Fig(1).
The Jacobian matrix of (1) is

$$A = \begin{bmatrix} 0 & 1 \\ -\cos x_1 & 0 \end{bmatrix}$$

and for the equilibrium states is

$$A_1 = \begin{bmatrix} 0 & 1 \\ -1 & 0 \end{bmatrix} \qquad A_{11} = \begin{bmatrix} 0 & 1 \\ 1 & 0 \end{bmatrix}$$

From the equation $|A - \lambda I| = 0$ we calculate the eigenvalues

752

$$\det \begin{vmatrix} -\lambda & 1 \\ -1 & -\lambda \end{vmatrix} = 0 \qquad \text{which gives for}$$

x_1 , $\lambda = \pm j$ λ a vortex,

and

$$\det \begin{vmatrix} -\lambda & 1 \\ 1 & -\lambda \end{vmatrix} = 0 \qquad \text{which gives for}$$

x_2 , $\lambda = \pm 1$ a saddle.

To determine the stability in the finite phase plane we have to find the phase trajectories, that is the relationship $x_2 = x_2(x_1)$.

From (1) we get

$$x_2 dx_2 = - \sin x_1 \, dx_1$$

and integrating

$$x_2^2 = 2 \cos x_1 + C$$

where C is a constant.

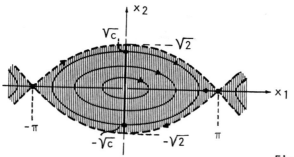

phase trajectories $x_2^2 = 2 \cos x_1 + c$

Fig. 2

Different values of C give different trajectories. To find the trajectory that passes through the saddle points at $x_1 = \pm \pi$

substitute $\qquad x = \begin{pmatrix} \pi \\ 0 \end{pmatrix}$. We have

$0 = 2 \cos \pi + C$ thus

$x_2^2 = 2 \cos x_1 + 2$

The trajectories that pass through $x = \pm \pi$ are called separatrices, they separate the stable regions from the unstable ones. The shaded area represents the stable region.

Examine the stability of the second order regulator with nonlinear damping shown in Fig. (1).

The transfer function of the nonlinear plant is

$$G_p = \frac{1}{s^2}$$

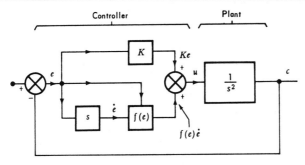

Proportional-plus-nonlinear derivative control.

Fig. 1

Solution: From the diagram we have

$$u = Ke + f(e)\dot{e}$$

where u is the control force. The amount of the derivative control depends upon f(e) - the magnitude of the error.

The following equation describes the dynamic of the system

$$u = \ddot{e}$$

To obtain the set of state equations let $e = x_1$ and $\dot{e} = x_2$

$$\dot{x}_1 = x_2$$

$$\dot{x}_2 = -Kx_1 - f(x_1) \ x_2$$

Let us assume for simplicity that K = 1. The equilibrium points are determined from the equation

$$0 = x_2^0$$

$$0 = -x_1^0 - f(x_1^0)x_2^0$$

Thus $x_1^0 = x_2^0 = 0$ is the only point.

The Jacobian matrix is

$$A = \begin{bmatrix} 0 & 1 \\ -1 & -f(0) \end{bmatrix}$$

The linear behavior of the system in the close neighborhood of the origin is described by

754

$$\dot{x}_1 = x_2$$

$$\dot{x}_2 = -x_1 - f(0)x_2$$

and the characteristic equation is

$$s[s + f(0)] + 1 = 0$$

with the eigenvalues

$$\lambda_1 = -\frac{1}{2} f(0) + \sqrt{\frac{1}{4} f^2(0) - 1}$$

$$\lambda_2 = -\frac{1}{2} f(0) - \sqrt{\frac{1}{4} f^2(0) - 1}$$

From the table below

Stable		Unstable	
Trajectory type	Eigenvalues	Trajectory type	Eigenvalues
Stable focus		Unstable focus	
Stable node		Unstable node	
Vortex		Saddle	

VARIOUS TYPES OF SINGULARITIES

we obtain the following possible responses:

1. $f(0) > 2$ stable node

2. $0 < f(0) < 2$ stable focus

3. $f(0) = 0$ vortex

4. $-2 < f(0) < 0$ unstable focus

5. $f(0) < -2$ unstable node

The first two cases ensure asymptotic stability, and the third case local stability around the origin.

Next we shall investigate the response of the system subjected to large input steps, i.e., the problem of the finite and global stability of the system.

Using the phase plane we shall demonstrate that it is possible to control and shape the behavior of a nonlinear system. For this purpose let us assume the following types of derivative control:

755

I. $f(e) = 2$

II. $f(e) = 1$

III. $f(e) = 0$

IV. $f(e) = \dfrac{0.1}{|e|}$

The first three cases are linear, thus

$\dot{x}_1 = x_2$

$\dot{x}_2 = -x_1 - f(0)x_2$

holds throughout the entire phase plane. We get the following slope equations using the isocline method:

I. $\quad S = -\dfrac{x_1 + 2x_2}{x_2} = -2 - \dfrac{x_1}{x_2}$

II. $\quad S = -1 - \dfrac{x_1}{x_2}$

III. $\quad S = -\dfrac{x_1}{x_2}$

In all three cases the slope is constant for constant ratio

$$\dfrac{x_1}{x_2},$$

therefore the isoclines are straight lines through origin.

Fig. 2 shows the phase trajectories and Fig. 3 the corresponding time responses.

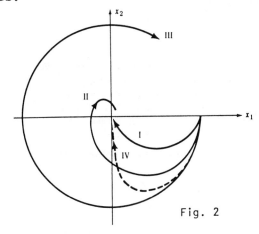

Fig. 2

Note that in Fig. 3 the time responses range from undamped to overdamped. By increasing the damping we sacrifice the response of the system.

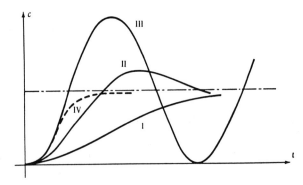

Fig. 3

To find the right combination of time response and damping let us investigate case IV:

$$f(e) = \frac{0.1}{|e|}$$

We obtain the isocline equation

$$\frac{dx_2}{dx_1} = S = -\frac{x_1 + 0.1 \frac{x_2}{|x_1|}}{x_2} = -\frac{x_1}{x_2} - 0.1 \frac{1}{|x_1|}$$

The phase trajectory and the corresponding time response are shown in Fig(2) and (3).

ROOT LOCUS

● **PROBLEM** 13-106

For the system of Fig. 1, where $K \geq 0$

a) Sketch the root locus for the system, giving starting and ending points, asymptote intersects and angles.
b) Using Hurwitz test, find the values of $K \geq 0$ for which the system becomes unstable in the Lyapunov sense.
c) Calculate the overall transfer function $H(s)$. Using it, analyze results of a) and b).

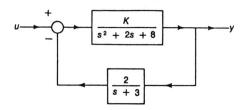

Fig. 1

Solution: We have

$$H_1 H_f = \frac{1}{(s + 3)(s^2 + 2s + 8)}$$

757

The poles are -3, $-1 + j\sqrt{7}$, $-1 - j\sqrt{7}$. The locus starts at the poles and ends at the zeros, all of which are at infinity. The asymptote angles are

$$-\frac{1}{3}(2k + 1)\pi \quad \text{or} \quad -\frac{\pi}{3}, \ -\pi, \ -\frac{5}{3}\pi \ .$$

The asymptote intersection on the real axis is

$$\sigma_a = \frac{-3 - 1 + j\sqrt{7} - 1 - j\sqrt{7}}{3} = -\frac{5}{3}$$

The results are shown in Fig. 2

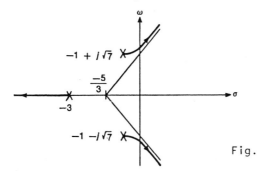

Fig. 2

b) There is a K for which the system becomes unstable. We test the polynomial

$$1 + 2KH_1H_f = 1 + \frac{2K}{(s + 3)(s^2 + 2s + 8)}$$

or

$$P(s) = s^3 + 5s^2 + 14s + 24 + 2K$$

The test function

$$R(s) = \frac{s^3 + 14s}{5s^2 + 24 + 2K} = \frac{s}{5} \frac{(s + 14)}{\left(s^2 + \frac{24 + 2K}{5}\right)}$$

Using the fact that the poles and zeros must alternate on the imaginary axis for a reactance function, we have

$$14 > \frac{24 + 2K}{7} \quad \text{or}$$

$$K < 37$$

as a condition for stability.

c) The overall transfer function is

$$H(s) = \frac{2K/(s + 3)(s^2 + 2s + 8)}{1 + 2K/(s + 3)(s^2 + 2s + 8)}$$

$$= \frac{2K}{s^3 + 5s^2 + 14s + 24 + 2K}$$

758

The system is bibo and output-Lyapunov asymptotically stable if $K < 37$. If $K = 37$ there are simple poles on the $j\omega$ axis and the system is output-Lyapunov but not bibo or asymptotically stable.

● **PROBLEM** 13-107

Fig. 1 shows a block diagram of a simplified minesweeper, K_m is a variable non-negative constant.

(a) Plot a root locus for the system; label important points.
(b) Can the system ever be stable?
(c) Give an analogue computer realization for the complete (compensated minesweeper) system; how many state variables are there? Discuss implementation of your analogue realization where K_m is variable.

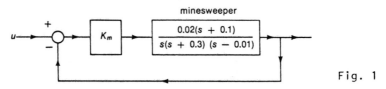

Fig. 1

Solution: We have,

$$K = 0.02K_m \quad \text{and} \quad H_1 H_f = \frac{s + 0.1}{s(s + 0.3)(s - 0.01)} .$$

The asymptotes are at angles

$$\frac{(2k + 1)\pi}{1 - 3} = \frac{-\pi}{2} , \frac{-3\pi}{2} , \ldots \text{ for } k = 0, 1, \ldots$$

and intersect at the real axis point

$$\sigma_0 = \frac{[0 + (-0.3) + 0.01] - (-0.1)}{3 - 1} = \frac{-0.19}{2} = -0.095.$$

We then have for (a) the root locus of Fig. 2

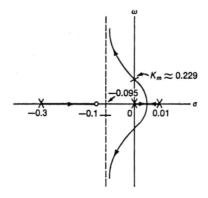

Fig. 2

(b) From the curve, it is clear that there are $K_m > 0$ for which the system can be stable. The "characteristic polynomial" is

759

$$P(s) = s(s + 0.3)(s - 0.01) + K(s + 0.1)$$

$$= s^3 + 0.29s^2 + (K - 0.003)s + 0.1K,$$

from which the "stable" K can be determined by the Hurwitz test applied to

$$R(s) = \frac{s^3 + (K - 0.003)s}{0.29s^2 + 0.1K}$$

$$
\begin{array}{r}
\frac{1}{0.29}s \\
0.29s^2 + 0.1K \overline{\smash{\big)}\ s^3 + (K - 0.003)s} \\
s^3 + \frac{10}{29}Ks \\
\hline
\left(\frac{19}{29}K - 0.003\right)s
\end{array}
$$

$$
\begin{array}{r}
\frac{0.29}{\left(\frac{19}{29}K - 0.003\right)}s \\
\left(\frac{19}{29}K - 0.003\right)s \overline{\smash{\big)}\ 0.29s^2 + 0.1K} \\
0.29s^2 \qquad\qquad \left(\frac{19}{29}K-0.003\right)s/0.1K \\
\hline
0.1K \left(\frac{19}{29}K-0.003\right)s
\end{array}
$$

The factors have to be positive, therefore

$$\frac{19}{29}K - 0.003 > 0 \quad \text{where} \quad K = 0.02K_m$$

is required for stability, or

$$K_m > \frac{29 \times 3}{19 \times 20} = \frac{87}{380} \approx 0.229.$$

(c) We desire the transfer function

$$H(s) = \frac{K_m \dfrac{0.02(s+0.1)}{s(s+0.3)(s-0.01)}}{1+K_m \dfrac{(0.02)(s+0.1)}{s(s+0.3)(s-0.01)}} = \frac{Ks+0.1K}{s^3+0.29s^2+(K-0.003)s+0.1K}$$

Setting

$$H(s) = \frac{Y(s)}{U(s)}$$

we obtain

$$\frac{Y(s)}{U(s)} = \frac{Ks + 0.1K}{s^3 + 0.29s^2 + (K-0.003)s + 0.1K}$$

cross multiplying and dividing by s^3

$$\left[1 + \frac{0.29}{s} + \frac{K-0.003}{s^2} + \frac{0.1K}{s^3}\right]Y(s) = \left[\frac{K}{s^2} + \frac{0.1K}{s^3}\right]U(s)$$

$$Y(s) = \left[\frac{-0.29}{s} - \frac{K-0.003}{s^2} - \frac{0.1K}{s^3}\right]Y(s) + \left[\frac{K}{s^2} + \frac{0.1K}{s^3}\right]U(s)$$

The above formula can be arranged

$$Y(s) = \frac{1}{s} \left\{ -0.29Y + \frac{1}{s} \left[KU - (K-0.003)Y + \frac{1}{s} (0.1KU - 0.1KY) \right] \right\} ,$$

which gives the analogue computer realization as shown in Fig. 3.

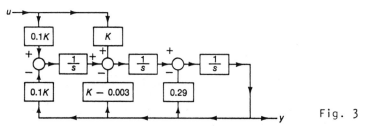

Fig. 3

Note that there are three state variables and that if K_m varies, then so must four of the gain blocks, making this realization somewhat unwieldy as far as variation in $K_m = K/0.02$ is concerned.

● **PROBLEM** 13-108

The characteristic equation of a feedback control system is

$$1 + F(s) = 1 + \frac{K(s+1)}{s(s+2)(s+4)^2}$$

Determine the effect of the gain K from the root locus.

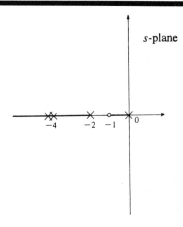

Solution: First we plot the poles and zeros of the characteristic equation.

0,-2 are poles of the first order; -4 is a pole of the second order.

The root loci on the real axis are shown as heavy lines; they must be located to the left of an odd number of poles and zeros.

From

$$\sigma_A = \frac{1}{n_p - n_z} (\Sigma p_i - \Sigma z_j)$$

we find the intersection of the asymptotes

$$\sigma_A = \frac{-2 + 2(-4) - (-1)}{4 - 1} = -3$$

and the angles of the asymptotes are

$$\phi_A = +60° \qquad q = 0$$

$$\phi_A = +180° \qquad q = 1$$

$$\phi_A = 300° \qquad q = 2$$

Since $n_p - n_z = 3$, we have three asymptotes. The root loci
must begin at the poles and therefore two loci leave the second
order pole at s = -4. Having the asymptotes and the breakaway
point we draw the root locus.

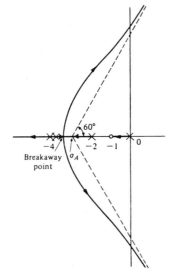

From the root locus we conclude that if K is sufficiently
increased the system becomes unstable.

● **PROBLEM** 13-109

We desire to plot the root locus for the characteristic
equation of a system when

$$1 + \frac{K}{s(s + 4)(s + 4 + j4)(s + 4 - j4)} = 0 \qquad (1)$$

as K varies from zero to infinity.

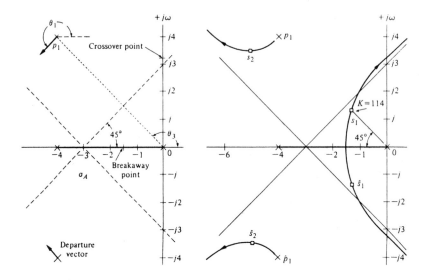

Fig. 1   Fig. 2

<u>Solution</u>: The poles are located on the s-plane as shown in Fig. 1. Let s be a point on the complex plane. Let θ_{p_i} be the angle that pole p_i makes with s, and let θ_{z_j} be the angle that zero z_j makes with s. Then, if s is a point on the root locus this relationship must hold, for K > 0 :

$$\sum_i \theta_{p_i} - \sum_j \theta_{z_j} = (2n + 1)180°) \tag{2}$$

On the real axis, this condition can be met only if the total number of poles and zeros to the right is odd. (Poles and zeros on the left contribute 0°, while those on the right contribute ±180°.) In our case, this condition exists only between s = 0 and s = -4. Thus, a segment of the root locus exists on the real axis between s = 0 and s = -4. Since the number of poles n_p is equal to four and there are no zeros, we have $n_p - n_z$ = 4 separate loci. The angles of the asymptotes are

$$\phi_A = \frac{(2q + 1)}{n_p - n_z} 180° , \quad q = 0,1,2,3, \tag{3}$$

$$\phi_A = +45°, \ 135°, \ 225°, \ 315°$$

The center of the asymptotes is

$$\sigma_A = \frac{1}{n_p - n_z} \left(\sum_i Re\{p_i\} - \sum_j Re\{z_j\} \right) \tag{4}$$

$$\sigma_A = \frac{-4 - 4 - 4}{4} = -3 \ .$$

763

Then the asymptotes are drawn as shown in Fig. 1.

The breakaway point is estimated by evaluating

$$K = p(s) = -s(s + 4)(s + 4 + j4)(s + 4 - j4) \qquad (5)$$

between $s = -4$ and $s = 0$.

Breakaway points exist between two adjacent poles or two adjacent zeros (including zeros at infinity) on the real axis, and sometimes at other points not on the real axis. At a breakaway point, the value of K in terms of s is an extreme point. We expect the breakaway point to lie between $s = -3$ and $s = -1$ and, therefore, we search for a maximum value of $p(s)$ in that region. The resulting values of $p(s)$ for several values of s are given in the table. The maximum of $p(s)$ is found to lie at approximately $s = -1.5$ as indicated in the table. A more accurate estimate of the breakaway point is

$p(s)$	0	51	68.5	80	85	75	0
s	-4.0	-3.0	-2.5	-2.0	-1.5	-1.0	0

normally not necessary or worthwhile. The breakaway point is then indicated on Fig. 1.

The characteristic equation is rewritten as

$$s(s + 4)(s^2 + 8s + 32) + K = s^4 + 12s^3 + 64s^2 + 128s + K = 0 \quad (6)$$

Therefore, the Routh-Hurwitz array is

$$
\begin{array}{c|ccc}
s^4 & 1 & 64 & K \\
s^3 & 12 & 128 & \\
s^2 & b_1 & K & \\
s^1 & c_1 & & \\
s^0 & K & &
\end{array}
$$

where

$$b_1 = \frac{12(64) - 128}{12} = 53.33 \text{ and } c_1 = \frac{53.33(128) - 12K}{53.33} \qquad (7)$$

For a certain value of K, the system becomes marginally stable. At that point, a row in the Routh-Hurwitz array becomes all zero. The auxiliary equation of the row above then gives the points at which the locus crosses the imaginary axis. Here, we can only set $c_1 = 0$.

Hence the limiting value of gain for stability is $K = 570$ and the roots of the auxiliary equation are

$$53.33s^2 + 570 = 53.33(s^2 + 10.6) = 53.33(s + j3.25)(s - j3.25).$$

$$(8)$$

The points where the locus crosses the imaginary axis is shown in Fig. 1.

The angle of departure at the complex pole p_1 can be estimated by utilizing the angle criterion as follows:

$$\theta_1 + 90° + 90° + \theta_3 = 180° ,\tag{9}$$

where θ_3 is the angle subtended by the vector from pole p_3. The angles from the pole at $s = -4$ and $s = -4 - j4$ are each equal to $90°$. Since $\theta_3 = 135°$, we find that

$$\theta_1 = -135° = +225°$$

as shown in Fig. 1.

Utilizing all the information obtained from the steps of the root locus method, the complete root locus is plotted by using a protractor or Spirule to locate points that satisfy the angle criterion. The root locus for this system is shown in Fig. 2. When complex roots near the origin have a damping ratio of $\zeta = 0.707$, the gain K can be determined graphically as shown in Fig. 2. The vector lengths to the root location s_1 from the open-loop poles are evaluated and result in a gain at s_1 of

$$K = |s_1||s_1 + 4||s_1 - p_1||s_1 - \hat{p}_1|\tag{10}$$

$$= (1.3)(3.2)(4.4)(6.2) = 114 .$$

The remaining pair of complex roots occurs at s_2 and $\hat{s}_2$ when $K = 114$. The effect of the complex roots at s_2 and $\hat{s}_2$ on the transient response will be negligible compared to the roots s_1 and $\hat{s}_1$. This fact can be ascertained by considering the damping of the response due to each pair of roots. The damping due to s_1 and $\hat{s}_1$ is

$$e^{-\zeta_1 \omega n_1 t} = e^{-\sigma_1 t} ,\tag{11}$$

and the damping factor due to s_2 and $\hat{s}_2$ is

$$e^{-\zeta_2 \omega n_2 t} = e^{-\sigma_2 t} ,\tag{12}$$

where σ_2 is approximately five times as large as σ_1. Therefore, the transient response term due to s_2 will decay much more rapidly than the transient response term due to s_1. Thus the response to a unit step input may be written as

$$c(t) = 1 + c_1 e^{-\sigma_1 t} \sin(\omega_1 t + \theta_1) + c_2 e^{-\sigma_2 t} \sin(\omega_2 t + \theta_2)$$

$$\cong 1 + c_1 e^{-\sigma_1 t} \sin(\omega_1 t + \theta_1) .\tag{13}$$

The complex conjugate roots near the origin of the s-plane relative to the other roots of the closed-loop system are labeled the dominant roots of the system since they represent or dominate the transient response. The relative dominance of the roots is determined by the ratio of the real parts of the complex roots and will result in reasonable dominance for ratios exceeding five.

Of course, the dominance of the second term of Eq. 13 also depends upon the relative magnitudes of the coefficients c_1 and c_2. These coefficients, which are the residues evaluated at the complex roots, in turn depend upon the location of the zeros in the s-plane. Therefore, the concept of dominant roots is useful for estimating the response of a system but must be used with caution and with a comprehension of the underlying assumptions.

● **PROBLEM** 13-110

In some control systems a positive feedback inner loop may appear. This loop is usually stabilized by the outer loop. For the system shown in Fig. 1, with the positive feedback, sketch the root locus plot.

Assume that $H(s) = 1$

$$G(s) = \frac{K(s + 3)}{(s + 4)(s^2 + 2s + 2)} \qquad K > 0$$

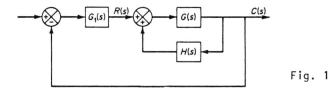

Fig. 1

Solution: The procedure for plotting positive feedback control system is similar to that for the negative feedback control system with slight modifications.

We write the transfer function of the inner loop

$$\frac{C(s)}{R(s)} = \frac{G(s)}{1 - G(s)H(s)}$$

and the characteristic equation

$$1 - G(s)H(s) = 0$$
or $G(s)H(s) = 1$.

This is equivalent to two equations:

$$\underline{/G(s)H(s)} = 0° \pm k360° \quad (k = 0, 1, 2, \ldots)$$
$$|G(s)H(s)| = 1$$

The total sum of all angles from the open-loop poles and zeros is equal to $0° \pm k360°$. We use the following rules for constructing root loci of the positive feedback system.

1. If the total number of real zeros and real poles to the right of a test point on the real axis is even, then this test point lies on the root locus.

2. Angles of asymptotes $= \dfrac{\pm 360°k}{n - m}$

   where

   n = number of finite poles of $G(s)$ $H(s)$
   m = number of finite zeros of $G(s)$ $H(s)$

3. To calculate the angle of departure (or angle of arrival) from a complex open-loop pole (or at a complex zero) we subtract from $0°$ the sum of all the angles of the complex quantities from all the other poles and zeros to the complex pole (or complex zero) in question, including the appropriate signs.

For the positive feedback system the closed-loop transfer function is

$$\frac{C(s)}{R(s)} = \frac{G(s)}{1 - G(s)\,H(s)} = \frac{k(s + 3)}{(s + 4)(s^2 + 2s + 2) - k(s + 3)}$$

1. We find the open-loop poles

   $s = -1 + j$, $s = -1 - j$, $s = -4$ and zero $s = -3$.
   We note that when k increases from 0 to ∞, the closed-loop poles start at the open-loop poles and terminate at the open-loop zeros.

2. The root loci on the real axis exist between -3 and $+\infty$ and between -4 and $-\infty$.

3. The asymptotes of the root loci.
   We have
   $$\text{Angle of asymptote} = \frac{\pm k360°}{3-1} = \pm 180°$$
   That means that root loci branches are located on the real axis.

4. Breakaway points and break-in points.
   The characteristic equation is

   $$(s + 4)(s^2 + 2s + 2) - k(s + 3) = 0$$

   we calculate k and $\dfrac{dk}{ds}$

   $$k = \frac{(s + 4)(s^2 + 2s + 2)}{(s + 3)}$$

   $$\frac{dk}{ds} = \frac{2s^3 + 15s^2 + 36s + 22}{(s + 3)^2}$$

   Solving the equation we obtain

   $$2s^3 + 15s^2 + 36s + 22 = 2(s + 0.9)(s^2 + 6.6s + 12.2) =$$

   $$2(s + 0.9)(s + 3.3 - j1.15)(s + 3.3 + j1.15)$$

points s = -3.3 + jl.15 and s = -3.3 - jl.15
do not satisfy the angle condition. At s = -0.9 the
value of k is positive. Thus the break-in point is
s = -0.9.

5. The angle of departure of the root locus from a complex
    pole. For the complex pole s = -1 + j, the angle of de-
    parture φ is

$$45° - 29° - 90° - φ = 0°$$

$$φ = - 74°$$

and for s = - 1 - j, φ = 74°.

6. The test point should be in the broad neighborhood of the
    imaginary axis and the origin; we apply the angle condition.

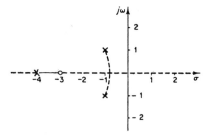

Root-locus plot for the positive
feedback system with
$$G(s) = \frac{K (s + 3)}{(s + 4)(s^2 + 2s + 2)} \quad K > 0$$
$$H(s) = 1$$

For the positive feedback system we obtain the following
root-locus plot using results 1 - 6.

NYQUIST-BODE

● **PROBLEM** 13-111

The closed-loop system is shown in Fig. (1).

Using the Nyquist criterion determine the critical value of
k for stability of the system.

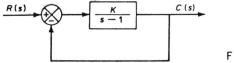

Fig. 1

Solution: The gain function of the system is

$$G(s) = \frac{k}{s - 1}$$

Let us draw the polar plot of G(jw)

$$G(jw) = \frac{k}{jw - 1} = \frac{-k - jwk}{w^2 + 1}$$

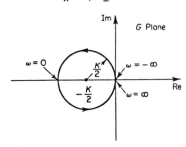

Fig. 2

The direction is counterclockwise. There is one pole of G(s) in the right-half s plane, thus P = 1. The condition for stability of a closed-loop system is Z = 0. Therefore N must be equal to -1, since N = Z - P or there must be one counter-clockwise encirclement of the -1 + j0 point for stability.

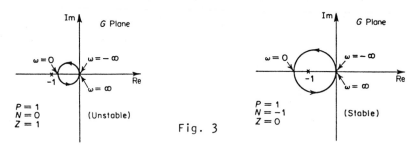

Fig. 3

Fig. 4

For stability of the system we have k>1

k = 1 is the stability limit.

The two figures 3, 4, illustrate the results.

● **PROBLEM** 13-112

For the system whose open-loop transfer function is

$$\frac{\Theta_{on}}{\Theta} = \frac{10}{D(1 + \frac{D}{4})(1 + \frac{D}{16})}$$

determine the stability.

Solution: Using the Nyquist Criterion we have:

D = jw

then

$$\frac{\Theta_{on}}{\Theta} = \frac{10}{jw(1 + \frac{jw}{4})(1 + \frac{jw}{16})} = \frac{640}{jw(jw + 4)(jw + 16)}$$

$$= \frac{640}{-20w^2 + j(64w - w^2)}$$

Accurate plot in the region of (-1, 0)

We can draw the Nyquist plot as shown in the figure.

We see that the point (-1,0) is not encircled and the system is stable.

● **PROBLEM** 13-113

Determine for what values of k the system with the following open-loop transfer function is stable

$$G(s) \ H(s) = \frac{k}{s(s + 1)(2s + 1)}$$

Solution: Let s = jw we have

$$G(jw) \ H(jw) = \frac{k}{jw(jw + 1)(2jw + 1)} =$$

$$\frac{k}{-3w^2 + jw(1 - 2w^2)}$$

We see that the open-loop transfer function has no poles in the right half s-plane. For the system to be stable is thus enough that the Nyquist plot does not encircle the -1 + 0j point. To find the point where the Nyquist plot crosses the negative real axis let $I_m \ G(jw) \ H(jw) = 0$ thus

$$1 - 2w^2 = 0$$

and $w = \pm \frac{1}{\sqrt{2}}$

$$G(j\frac{1}{\sqrt{2}}) \ H(j\frac{1}{\sqrt{2}}) \ = \ \frac{-2k}{3}$$

To get the critical value of k let

$$- \ \frac{2k}{3} = - \ 1$$

$$k = \frac{3}{2}$$

We have

$$0 < k < \frac{3}{2}$$

for stability of the system.

● **PROBLEM** 13-114

A closed-loop system has the following open-loop transfer function

$$G(s) \ H(s) \ = \ \frac{k(s + 3)}{s(s - 1)}$$

Investigate the stability of the system.

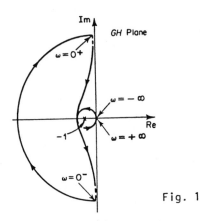

Fig. 1

Polar plot of the system

Solution: It is clear that the open-loop transfer function has one pole for s = 1 in the right-half s plane, therefore the open-loop system is unstable. From the Nyquist plot see Fig. (1), we see that the point - 1 + j0 is encircled by the G(s) H(s) locus once in the counterclockwise direction. Thus, N = -1. Since P = 1 from Z = N + P we find Z to be zero, that indicates that there is no zero of 1 + G(s) H(s) in the right-half s plane and the closed-loop system is stable.

It is worthwhile to note that in the above example an unstable open-loop system becomes stable when the loop is closed.

The loop transfer function of a single-loop feedback control system is

$$G(s) \ H(s) \ = \ \frac{k}{s(s \ + \ \alpha)}$$

where k and α are positive constants. Investigate the stability of the system.

Solution: We start with construction of the Nyquist path for the system. Since $\alpha > 0$, G(s) H(s) does not have any poles in the right-half s-plane, thus $P_0 = P_{-1} = 0$.

G(s) H(s) has a pole at the origin, the Nyquist path must not pass through any singularity of G(s) H(s) thus we draw a small semicircle around s = 0.

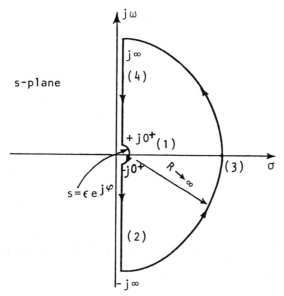

THE NYQUIST PATH FOR $G(s)H(s) = \frac{K}{s(s+\alpha)}$

We divided the Nyquist path into four sections.

The points of section (1) may be represented by the phasor

$$S = \varepsilon e^{j\phi}$$

where $\varepsilon \to 0$ and ε, ϕ denote the magnitude and phase.

We see that as the Nyquist path is traversed from $+j0^+$ to $-j0^+$ along section (1) ϕ changes from $+90°$ to $-90°$.

Since $e^{j\phi} = \cos\phi + j \sin\phi$

for $\phi = +90°$ $s = j\varepsilon$.

The Nyquist plot for section (1) is

$$G(s)H(s) = \left. \frac{K}{s(s+\alpha)} \right|_{s=\varepsilon e^{j\phi}} = \frac{K}{\varepsilon e^{j\phi}(\varepsilon e^{j\phi}+\alpha)} \xrightarrow{\varepsilon \to 0} \frac{K}{\alpha \varepsilon e^{j\phi}} = \infty e^{-j\phi}$$

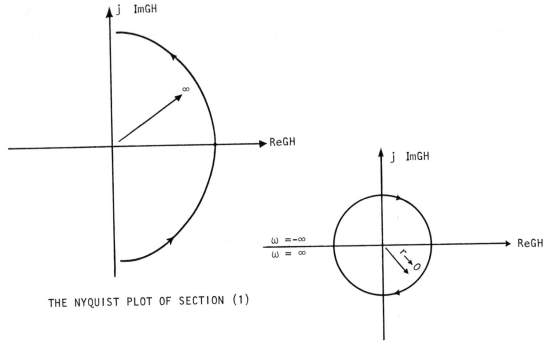

THE NYQUIST PLOT OF SECTION (1)

THE NYQUIST PLOT OF SECTION (3)

The points corresponding to section (1) have an infinite magnitude and the phase is opposite to that of the s-plane locus.

For section (3) we have

$$s = Re^{j\phi}$$

where $R \to \infty$, and ϕ changes from $-90°$ to $+90°$.

$$G(s)H(s) = \left. \frac{k}{s(s+\alpha)} \right|_{s=Re^{j\phi}} = \frac{k}{Re^{j\phi}(Re^{j\phi}+\alpha)} \xrightarrow{R \to \infty} \frac{k}{R^2 e^{2j\phi}} = 0 e^{-2j\phi}$$

We see that the magnitude is infinitesimally small, and the phasor rotates 2 x 180° = 360° in the clockwise direction.

We are left with sections (2) and (4). For section (4) we substitute s = jw

$$G(s)H(s) = G(jw)H(jw) = \frac{k}{jw(jw + \alpha)} = \frac{k(-w^2 - j\alpha w)}{w^4 + \alpha^2 w^2}$$

To find the intersect of $G(jw)H(jw)$ on the real axis we equate

$$\text{Im}G(jw)H(jw) = 0$$

$$- \frac{k\alpha w}{w^4 + w^2\alpha^2} = \frac{-k\alpha}{w(w^2 + \alpha^2)} = 0$$

which gives $w = \infty$.

We can draw the complete Nyquist plot of $G(s)H(s)$.

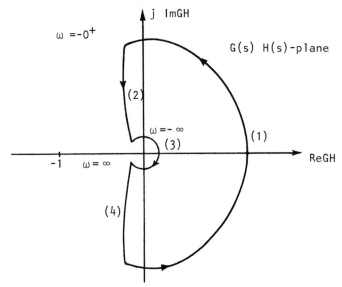

THE NYQUIST PLOT OF G(s) H(s)

We see that

$N_0 = N_{-1} = 0$ where N_0 is the number of encirclements of the origin made by $G(s)H(s)$.

Since $G(s)H(s) = \frac{k}{s(s + \alpha)}$, $k>0$, $\alpha>0$
we have $Z_0 = 0$ and $P_0 = 0$.

Since $P_{-1} = P_0 = 0$
we have $Z_{-1} = N_{-1} + P_{-1} = 0$

We conclude that the closed-loop system is stable.

● **PROBLEM** 13-116

The open loop transfer function of a system is

$$\frac{\Theta_{on}}{\Theta} = \frac{80}{D(1 + \frac{D}{5})(1 + \frac{D}{25})}$$

Determine the stability of the system.

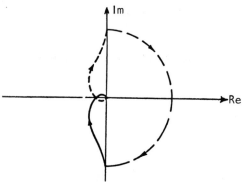

THE GENERAL SHAPE OF THE NYQUIST PLOT

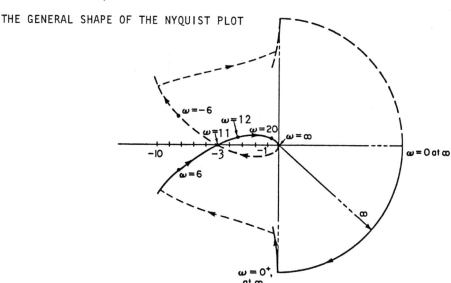

ACCURATE PLOT

Solution: Let us substitute jw = D. We have

$$\left(\frac{\Theta_{on}}{\Theta}\right)_{D=jw} = \frac{80}{jw(1 + \frac{jw}{5})(1 + \frac{jw}{25})} = \frac{80}{-\frac{6}{25}w^2 + j(w - \frac{w^3}{125})}$$

For amplitude and phase we get:

$$G_0 = \frac{80}{w(1 + \frac{w^2}{25})^{\frac{1}{2}}(1 + \frac{w^2}{625})^{\frac{1}{2}}}$$

$$\phi_0 = -90° - \tan^{-1}\frac{w}{5} - \tan^{-1}\frac{w}{25}$$

For some values of w we shall calculate G_0 and ϕ_0.

w =	0+	1	2	4	6	10	20	50	∞
G_0 =	∞	78	37	15	8	3.32	0.75	0.07	0
ϕ_0 =	-90°	-103°	-116°	-137°	-153°	-175°	-204°	-237°	-270°

775

For the negative values of w, we plot the mirror images of the positive values. The point $(-1, 0_j)$ is encircled, hence the system with the open loop transfer function

$$\frac{80}{D(1 + \frac{D}{5})(1 + \frac{D}{25})} \quad \text{is unstable.}$$

● **PROBLEM** 13-117

The open-loop transfer function of a closed-loop system is given by

$$G(s)H(s) = \frac{k}{(T_1 s + 1)(T_2 s + 1)}$$

Determine the stability of the system.

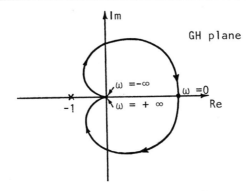

POLAR PLOT OF $G(j\omega)\ H(j\omega) = \dfrac{K}{(T_1 j\omega + 1)\ (T_2 j\omega + 1)}$

Solution: We shall plot $G(jw)H(jw)$.

Substituting $s = jw$ one gets

$$G(jw)H(jw) = \frac{k}{(T_1 jw + 1)(T_2 jw + 1)}$$

We transform the above equation to get the real and imaginary parts

$$G(jw)H(jw) = \frac{k(1 - T_1 jw)(1 - T_2 jw)}{(T_1 jw + 1)(T_1 jw - 1)(T_2 jw + 1)(T_2 jw - 1)}$$

$$= \frac{k(1 - T_1 jw)(1 - T_2 jw)}{(T_1^2 w^2 + 1)(T_2^2 w^2 + 1)} = \frac{k(1 - T_1 T_2 w^2) - jw(T_1 + T_2)k}{(T_1^2 w^2 + 1)(T_2^2 w^2 + 1)}$$

The angle is given by

$$\tan^{-1}\left[\frac{\text{imaginary part}}{\text{real part}}\right] = \tan^{-1}\left[\frac{-w(T_1 + T_2)k}{k(1 - T_1 T_2 w^2)}\right]$$

Then

$$w \to 0+ \Rightarrow +1 \; -j0 \quad \text{angle} \approx -\frac{1}{w}$$

$$\text{so} \quad \underline{/-90°}$$

$$w \to \infty \Rightarrow -0 \; -j0 \quad \text{angle} \approx -\frac{1}{w^2}$$

$$\text{so} \quad \underline{/-180°}$$

$$w \to -\infty \Rightarrow \quad -0 \; +j0 \quad \text{angle} \approx \frac{1}{w^2}$$

$$\underline{/180°}$$

$$w \to 0^- \Rightarrow +1 \; +j0 \quad \text{angle} \approx \frac{1}{w}$$

$$\underline{/90°}$$

We have all the information to plot $G(jw)H(jw)$.

Since $G(s)H(s)$ does not have any poles in the right-half s plane and the point $(-1, 0j)$ is not encircled by the $G(jw)H(jw)$ locus, the system is stable for

$$K > 0$$
$$T_1 > 0$$
$$T_2 > 0.$$

● **PROBLEM** 13-118

The open-loop transfer function of the system is

$$G(s)H(s) = \frac{k}{S(T_1s + 1)(T_2s + 1)}$$

Investigate the stability of the system in two cases

a) the value of the gain k is small
b) the value of the gain k is large.

Solution:

We shall draw the Nyquist plot for the small and large values of k.

The number of poles of $G(s)H(s)$ in the right-half s plane is zero. Thus for stability of the system it is necessary that

$$Z = N = 0$$

or that the $G(s)H(s)$ plot does not encircle $(-1, 0j)$. From the plot we conclude that for small values of K there is no encirclement of the $(-1, j0)$ point and the system is stable.

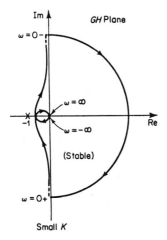

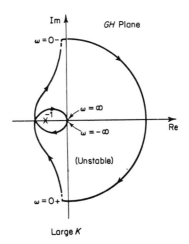

For small k we have

$$P = 0, \quad N = 0, \quad Z = 0.$$

For large K there are two encirclements of the point $(-1, 0j)$ in the clockwise direction indicating two closed-loop poles in the right-half s plane and the system is unstable.

For large K we have

$$P = 0, \quad N = Z = 2.$$

In the case of the above system, large K increases accuracy but decreases stability.

● **PROBLEM** 13-119

The loop transfer function of a control system with a single feedback loop is

$$G(s)H(s) = \frac{K(s - 1)}{s(s + 1)}$$

Using the Nyquist criterion determine for which values of K the system is stable.

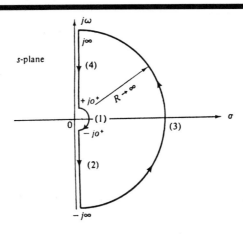

778

Solution:

We shall start with drawing the Nyquist path of the system.

To obtain the Nyquist plot let us investigate the separate sections of the path.

Section (1).

Let $s = \varepsilon\, e^{j\phi}$ where $\varepsilon \to 0$

we have

$$G(s)H(s)\ \Big|_{s\, =\, \varepsilon\, e^{j\phi}} = \frac{k}{s} = \infty\ e^{-j(\phi + \pi)}$$

For this section the magnitude is infinite and the angle changes from $+ 90°$ to $-90°$ counterclockwise.

Section (3).

$$s = R\, e^{j\phi}$$

$$\lim_{s \to \infty} G(s)H(s) = \lim_{s \to \infty} \frac{k}{s} = 0\ e^{-j\phi}$$

The magnitude is zero and the angle changes $180°$ in the clockwise direction.

Section (4).

Let $s = jw$

$$G(jw)H(jw) = \frac{K(jw - 1)}{jw(jw + 1)} = K\,\frac{2w + j(1 - w^2)}{w(w^2 + 1)}$$

From the equation

$$\text{Im } G(jw)H(jw) = 0$$

We obtain

$$W = \pm 1 \text{ rad/sec.}$$

Then $G(j1)H(j1) = K.$

Gathering the results we can draw the Nyquist plot of

$$G(s)H(s) = \frac{K(s - 1)}{s(s + 1)}$$

We see that

$$Z_0 = 1, \quad P_0 = P_{-1} = 0$$

$$N_0 = N_{-1} = 1$$

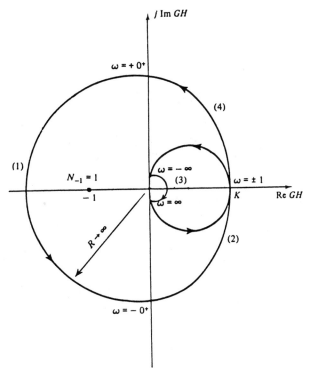

THE NYQUIST PLOT

then $Z_{-1} = N_{-1} + P_{-1} = 1$

Thus the closed-loop system is unstable. From the Nyquist plot we conclude that the system can not be stabilized by changing the value of parameter K.

● **PROBLEM** 13-120

Below is shown the Nyquist diagram for a system whose open-loop transfer function may be one of the following

a) no poles and zeros in the right-half s plane

b) no poles and one zero in the right-half s plane

c) one pole and no zeros in the right-half s plane

d) two poles and no zeros in the right-half s plane

e) two poles and two zeros in the right-half s plane.

Decide on stability for each case if the point $-1 + j0$ is located first in region I and then in II. Assume that the feedback loop is closed on each one of the above transfer functions.

780

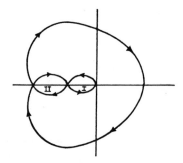

Solution:

First of all let us note that for all the transfer functions a
point in region I is not encircled and a point in region II is
encircled twice. The general formula is

$$N_{-1} = P_{-1} - Z_{-1}$$

where P_{-1} is the number of poles of $1 + G$, and $P_{-1} = P_0$.

For stability the number Z_{-1} of zeros of $1 + G$ must be zero.
Thus, for stability $N_{-1} = P_{-1}$ or P_0

a) $P_0 = 0$ so N_{-1} must be 0, therefore
      I - stable, II unstable

b) $P_0 = 0$, $N_{-1} = 0$ so I - stable, II - unstable

c) $P_0 = 1$ $N_{-1} = 1$ thus I and II unstable

d) $P_0 = 2$ $N_{-1} = 2$ I unstable II stable

e) $P_0 = 2$ $N_{-1} = 2$ I unstable II stable

● **PROBLEM** 13-121

For the function shown in Fig. (1) determine the
stability margin.

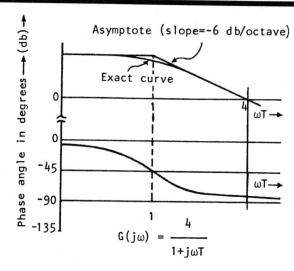

Fig. 1

At wT = 4 the magnitude plot crosses the OdB line. On the phase
curve the point corresponding to wT = 4 is   (=tan^{-1}4) about 75°.
The phase margin is then 180° - 75° = 105° and there is no gain
margin since the phase never reaches 180°.

● **PROBLEM** 13-122

Determine the range of parameter K for which the system shown
on the block diagram is stable.

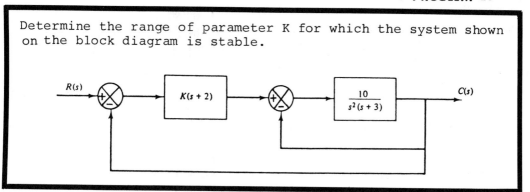

Solution:

The transfer function $G(s) = \dfrac{10K(s + 2)}{s^2(s + 3)}$

and the open-loop transfer function is

$$K(s + 2) \dfrac{10/s^2(s + 3)}{1 + 10/s^2(s + 3)}$$

Since this function does not have any zeros or poles on jω
axis the Nyquist path is

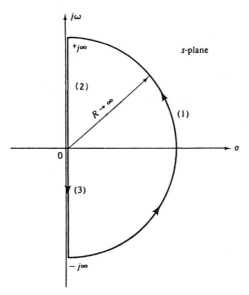

THE NYQUIST PATH

782

We shall construct the Nyquist plot for the (1), (2) and (3) sections.

Section (1):

Let $s = R\,e^{j\theta}$ where $R \to \infty$

$$\lim_{s\to\infty} G(s) = \frac{10K}{s^2} = 0e^{-j2\theta}$$

We see that the magnitude is zero and the angle changes from $+180°$ to $-180°$ clockwise, since θ changes from $-90°$ to $+90°$ counterclockwise.

Section (2): $s = j\omega$

$$G(j\omega) = \frac{10K(j\omega+2)}{(10-3\omega^2) - j\omega^3}$$

To rationalize the above fraction we multiply by

$$(10-3\omega^2) + j\omega^2 ,$$

thus

$$G(j\omega) = \frac{10K[2(10-3\omega^2) - \omega^4 + j\omega(10-3\omega^2) + j2\omega^3]}{(10-3\omega^2)^2 + \omega^6}$$

From $\text{Im}\{G(j\omega)\} = 0$ we get

$\omega = 0$ or $\omega = \pm\sqrt{10}$

which are the values of the intersects of the real axis of the G(s) plane.

To determine the intersection of the G(s) plot on the imaginary axis we set

$$\text{Re}\{G(s)\} = 0$$

and obtain
$$\omega^4 + 6\omega^2 - 20 = 0$$

$$\omega = \pm\sqrt{2}$$

Thus the intersects of the real axis are

$$G(j0) = 2K$$

$$G(j\sqrt{10}) = -K$$

and the imaginary axis

$$G(j\sqrt{2}) = j10\sqrt{2}\,\frac{K}{3}$$

We have all the information to draw the Nyquist plot.

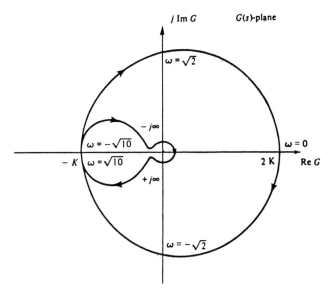

From the plot we see that $N_0 = -2$ and since $Z_0 = 0$ we get $P_0 = 2$. Thus $P_{-1} = 2$.

From the Nyquist criterion we have

$$N_{-1} = Z_{-1} - P_{-1} = Z_{-1} - 2$$

For the closed-loop system to be stable $Z_{-1} = 0$ and we get $N_{-1} = -2$.

Thus the point $(-1, j0)$ must be encircled twice in the clock-wise direction.

We have this when point $(-1, j0)$ is inside the circle, thus the condition for stability is

$$K > 1.$$

● **PROBLEM** 13-123

The block diagram of a system is

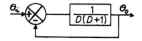

Find the stability condition of the system.

Solution: The open loop transfer function of the system is

$$\frac{\theta_0}{\theta} = \frac{1}{D(D+1)}$$

Let us substitute $D = j\omega$

$$\left(\frac{\theta_0}{\theta}\right)_{D=j\omega} = \frac{1}{j\omega(j\omega+1)} = \frac{1}{j\omega-\omega^2}$$

We shall compute the magnitude ratio and phase of the open loop transfer function

$$G_0 = \frac{1}{\omega(1+\omega^2)^{\frac{1}{2}}}$$

and

$$\phi_0 = -\tan^{-1}\frac{\omega}{-\omega^2} = -\tan^{-1}\frac{1}{-\omega}$$

$$= -90 - \tan^{-1}\omega .$$

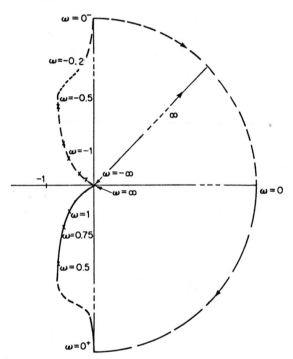

THE NYQUIST DIAGRAM FOR THE SYSTEM WITH OPEN LOOP TRANSFER FUNCTION 1/[D(D + 1)].

The plot for negative ω is a mirror image of the positive ω plot. Point (-1,0) is not encircled, thus the closed loop system is stable.

785

CHAPTER 14

PHASE PLANE ANALYSIS

INITIAL CONDITIONS

Consider the first-order systems described by

$$\dot{x} = -x \qquad\qquad (1)$$

and

$$\dot{x} = -x + x^3 \qquad\qquad (2)$$

Draw the phase trajectories and show where the systems are stable and unstable.

Solution: In the phase plane, or $x - \dot{x}$ plane, the phase-plane plot of eq. (1) is a straight line. For any initial condition, $x(0)$, the system returns to its singular point, the origin, after an infinite time.

The starting point of the trajectory is determined by the initial condition $x(0)$. For the system $\dot{x} = -x + x^3$, the trajectory is shown below.

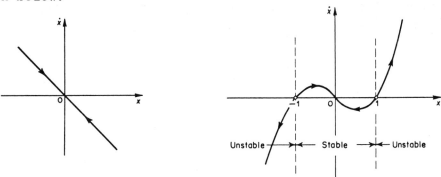

The trajectory is divided into three parts, two unstable and one stable part.

If $x(0) > 1$, then $x(\infty) \to \infty$,

if $-1 < x(0) < 1$, then $x(\infty) \to 0$,

if $x(0) < -1$, then $x(\infty) \to -\infty$.

The arrows are determined as follows. If $\dot{x}$ is positive, x is increasing, and if $\dot{x}$ is negative x is decreasing. Thus, above the x-axis, the arrows point right and below point left. Arrows pointing to the origin indicate a stable system. Arrows pointing to infinity indicate an unstable system.

● **PROBLEM** 14-2

The system is represented by the following equation

$$\ddot{x} + \dot{x} + x = 0.$$

Draw a phase portrait for the system. Locate singular points.

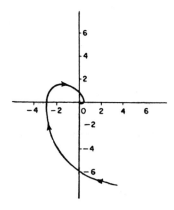

Solution: A phase portrait is a plot of x vs. $\dot{x}$. The slope at any point is given by

$$\frac{\ddot{x}}{\dot{x}} = \frac{\frac{d\dot{x}}{dt}}{\frac{dx}{dt}} = \frac{d\dot{x}}{dx} = \frac{-\dot{x} - x}{\dot{x}}$$

At the origin,

$$\frac{d\dot{x}}{dx} = \frac{0}{0}$$

Thus the origin is a singular point. That is, the system is in equilibrium at the origin.

For the initial conditions $x(0) = 2$ and $\dot{x}(0) = -7$, we get the trajectory shown.

Obtain a phase plane portrait for the equation

$$\ddot{x} = -K \quad (K = \text{constant}) \tag{1}$$

with the initial conditions

$$x(0) = x_0, \quad \dot{x}(0) = 0 \tag{2}$$

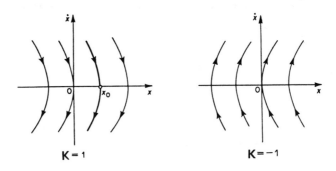

$$K = 1 \qquad\qquad K = -1$$

Solution: Let us rewrite the equation

$$\dot{x} \frac{d\dot{x}}{dx} = -K \tag{3}$$

Separating the variables and integrating we obtain

$$\dot{x}^2 = 2(x_0 - x)K \tag{4}$$

The initial conditions determined the integration constant.

Equation (4) represents a trajectory passing through $(x_0, 0)$ of (x,y) plane. The other way of obtaining eq. (4) is by finding $x = \psi_1(t)$, $\dot{x} = \psi_2(t)$ and eliminating t from the equations. In our case we have

$$\dot{x}(t) = -Kt$$

$$x(t) = -\frac{1}{2}Kt^2 + x_0.$$

Eliminating t we have

$$\dot{x}^2 = 2(x_0 - x)K.$$

The phase-plane portraits of the system $\ddot{x} = -K$ for $K = 1$ and $K = -1$ are shown.

The system is described by

$$\ddot{x} + \dot{x} + x^3 = 0 \tag{1}$$

with the initial conditions

$$x(0) = 1, \dot{x}(0) = 0.$$

Construct the phase plane trajectory from the initial point. Use the delta method.

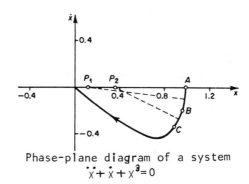

Phase-plane diagram of a system
$$\ddot{x} + \dot{x} + x^3 = 0$$

Solution: Since eq. (1) does not contain a term in x with a positive coefficient we have to add $\omega^2 x$ to both sides of the equation. We shall choose $\omega = 1$. Thus

$$\ddot{x} + x = -\dot{x} - x^3 + x$$

$$\delta = -\dot{x} - x^3 + x$$

The value of δ depends on $\dot{x}$ and x. The starting point of the trajectory is A(x = 1, $\dot{x}$ = 0). We have

$$\delta = -0 - 1 + 1 = 0$$

in the neighborhood of A.

The initial arc is thus centered at (0,0) with radius equal to unity. We draw a short arc. To obtain a more accurate value of δ we use the mean value of x and of $\dot{x}$. We get successive arcs. The first arc AB is centered at point P_1 (x = 0.12, $\dot{x}$ = 0). The second arc BC is centered at point P_2, (x = 0.37, $\dot{x}$ = 0).

Using this method we construct the trajectory as far as desired. Below is shown a phase-plane diagram starting from point x(0) = 1, $\dot{x}$(0) = 0.

METHOD OF ISOCLINES

The unexcited linear second-order system is described by

$$\ddot{x} + 2\zeta\omega_n\dot{x} + \omega_n^2 x = 0 \qquad (1)$$

Sketch the phase portrait of the system using the method of isoclines.

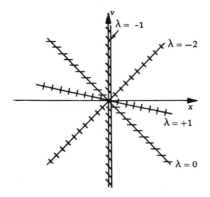

Fig. 1

Solution: The equation for the slope is

$$\lambda = \frac{\ddot{x}}{\dot{x}} = \frac{\dot{v}}{\dot{x}} \qquad (2)$$

Combining equations (1) and (2), we get

$$\ddot{x} = -2\zeta\omega_n\dot{x} - \omega_n^2 x$$

$$\lambda = \frac{\ddot{x}}{\dot{x}} = \frac{\dot{v}}{\dot{x}} = -2\zeta\omega_n - \omega_n^2\frac{x}{\dot{x}}$$

Let $\zeta = 0.5$ and $\omega_n = \frac{1\ \text{rad}}{\text{sec}}$, we have $\lambda = -1 - \frac{x}{\dot{x}}$

or

$$v(\lambda + 1) + x = 0$$

Fig. 1 shows some of the constant slope lines for different λ.

Fig. 2 shows phase portrait of the system.

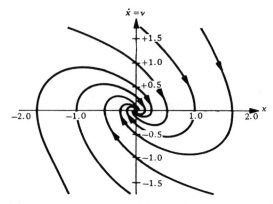

Fig. 2

● **PROBLEM 14-6**

Using the isocline method obtain a phase-plane portrait of the following equation

$$\ddot{x} + \alpha|\dot{x}| + x = 0 \quad \text{where} \quad \alpha > 0 \tag{1}$$

Solution:Since we have the absolute value of $|\dot{x}|$ we shall replace the nonlinear differential equation by two linear differential equations as follows:

$$\ddot{x} + \alpha\dot{x} + x = 0 \qquad \text{for } \dot{x} > 0$$
$$\ddot{x} - \alpha\dot{x} + x = 0 \qquad \text{for } \dot{x} < 0 \tag{2}$$

We find the slope at any point on the $x\dot{x}$ plane

$$\lambda = \frac{d\dot{x}}{dx} = \frac{\dfrac{d\dot{x}}{dt}}{\dfrac{dx}{dt}} = \frac{\ddot{x}}{\dot{x}} \tag{3}$$

From eq. (1),

$$\lambda = \frac{\ddot{x}}{\dot{x}} = -\alpha\frac{|\dot{x}|}{\dot{x}} - \frac{x}{\dot{x}} = -\alpha \; \text{sgn}(\dot{x}) - \frac{x}{\dot{x}} \tag{4}$$

Solving for the isoclines,

$$\frac{\dot{x}}{x} = \frac{-1}{\alpha \; \text{sgn}(\dot{x}) + \lambda}$$

In the upper half of the $x,\dot{x}$ plane the equation of isoclines is

791

$$\frac{\dot{x}}{x} = \frac{-1}{\alpha + \lambda}$$

in the lower half

$$\frac{\dot{x}}{x} = \frac{-1}{-\alpha + \lambda}$$

Below is shown a phase-plane portrait with isoclines for the system $\ddot{x} + |\dot{x}| + x = 0$.

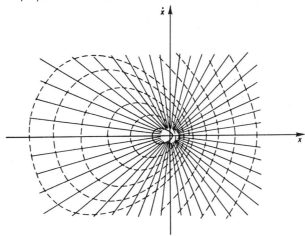

For this system, the trajectories are concentric ovals and the motion is periodic for any given initial condition, except at the origin. The straight lines are isoclines and short line segments on each line indicate the field of directions of tangents to the trajectories.

● **PROBLEM 14-7**

The system is given by

$$\ddot{x} + 0.5\dot{x} + 2x + x^2 = 0 \qquad (1)$$

Obtain a phase-plane portrait of the system.

Solution: From eq. (1) we get the singular points for the system.

$$\left.\begin{array}{c} x = 0 \\ \dot{x} = 0 \end{array}\right\}, \qquad \left.\begin{array}{c} x = -2 \\ \dot{x} = 0 \end{array}\right\} \qquad (2)$$

The characteristic equation is

$$\lambda^2 + 0.5\lambda + 2 = 0$$

with the roots $\qquad\qquad\qquad\qquad\qquad\qquad\qquad$ (3)

$$\lambda_1 = -0.25 + j1.39$$

$$\lambda_2 = -0.25 - j1.39$$

We see that this singular point is a stable focus. Eq. (1) in the neighborhood of the singular point (-2, 0) becomes

$$y = x + 2$$

we get

$$\ddot{y} + 0.5\dot{y} - 2y + y^2 = 0 \tag{4}$$

Eq. (4) near $y = 0$, $\dot{y} = 0$ becomes

$$\ddot{y} + 0.5\dot{y} - 2y = 0 \tag{5}$$

The characteristic equation is

$$\alpha^2 + 0.5\alpha - 2 = 0$$

and its roots are

$$\alpha_1 = 1.19, \quad \alpha_2 = -1.69 \tag{6}$$

The singular point (-2,0) is a saddle point.

We shall use the isocline method to obtain the phase-plane portrait of a system, see below.

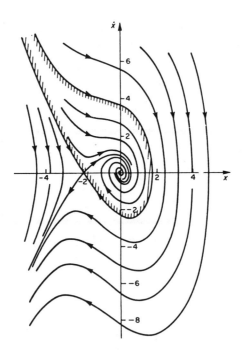

The two trajectories which enter the saddle point (-2, 0) are separatrices. In this case, the type of trajectory depends on the initial condition. If the initial point is located within the region bounded by one of the two separatrices the trajectories converge to the origin.

Draw the phase portrait corresponding to the differential
equation

$$\ddot{\phi} + \frac{g}{\ell} \sin \phi = 0 \tag{1}$$

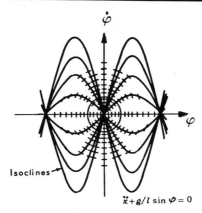

Fig. 1

Solution: Equation (1) describes the motion of a pendulum
of length ℓ, where g is the acceleration of gravity and ϕ
the angular displacement from the vertical. A phase plane
plot is a plot of ϕ vs. $\dot{\phi}$. We shall use the method of
isoclines. The slope at any point is given by

$$\lambda = \frac{d\dot{\phi}}{d\phi} = \frac{\dfrac{d\dot{\phi}}{dt}}{\dfrac{d\phi}{dt}} = \frac{\ddot{\phi}}{\dot{\phi}} \tag{2}$$

From eq. (1) we have

$$\ddot{\phi} = -\frac{g}{\ell} \sin \phi \tag{3}$$

and

$$\lambda = -\frac{g}{\ell} \frac{\sin \phi}{\dot{\phi}} \tag{4}$$

We computed the slopes of the phase trajectories at any
point in the phase plane. We draw short line segments of
slope λ at many points (see Fig. 1) and join them in smooth
curves (Fig. 2).

λ is constant along the curves

$$\dot{\phi} = -\frac{g}{\ell\lambda} \sin \phi \tag{5}$$

which are sinusoids.

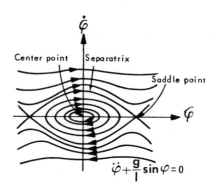

Fig. 2

Fig. 2 can be interpreted as follows. For small initial conditions (low initial velocity) the pendulum swings back and forth, so that ϕ and $\dot{\phi}$ both go between positive and negative values, which is shown by the closed curves around the center point.

If initial velocity is high enough the pendulum will continuously rotate, with small variations in $\dot{\phi}$ and constantly increasing ϕ. This is shown by the trajectories above the separatrix in Fig. 2. The separatrix represents the condition of unstable equilibrium, where the inital velocity is just enough to make the pendulum poise upside-down at the height of its rotation.

The initial conditions determine which trajectory is taken.

● **PROBLEM 14-9**

A simple positional control system with a nonlinear gain is shown in the block diagram.

Draw a phase plane portrait for the system with no input.

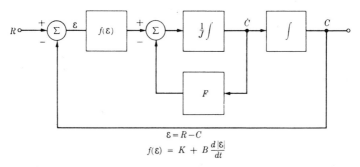

$$\varepsilon = R - C$$
$$f(\varepsilon) = K + B\frac{d|\varepsilon|}{dt}$$

Solution: For the zero input we get the following equations:

$$\ddot{C} + \frac{F}{J}\dot{C} + \frac{1}{J}f(C)C = 0 \tag{1}$$

$$F(C) = K + B\frac{d|C|}{dt} \tag{2}$$

Let us assume for simplicity that

$$\frac{F}{J} = \frac{K}{J} = \frac{B}{J} = 1$$

We choose

$$x_1 = C, \quad x_2 = \dot{C} \tag{3}$$

and obtain the following system of equations:

$$\dot{x}_1 = x_2$$

$$\dot{x}_2 = \begin{cases} -x_2 - (1 + x_2)x_1 & x_1 > 0 \\ -x_2 - (1 - x_2)x_1 & x_1 \leq 0 \end{cases} \tag{4}$$

The isocline equation is

$$M = \frac{\dot{x}_2}{\dot{x}_1} = \begin{cases} -1 - \dfrac{(1 + x_2)x_1}{x_2} & x_1 > 0 \\ -1 - \dfrac{(1 - x_2)x_1}{x_2} & x_1 \leq 0 \end{cases} \tag{5}$$

or

$$x_2 = \begin{cases} \dfrac{-x_1}{M + 1 + x_1} & x_1 > 0 \\ \dfrac{-x_1}{M + 1 - x_1} & x_1 \leq 0 \end{cases} \tag{6}$$

The eqs. (5) and (6) indicate that isoclines consist of families of hyperbolas whose asymptotes are functions of M.

asymptotes are for $\begin{cases} x_1 > 0; \quad x_1 = -(1 + M); \quad x_2 = -1 \\ x_1 < 0; \quad x_1 = 1 + M; \quad x_2 = 1 \end{cases}$

Phase portrait of the system is shown below.

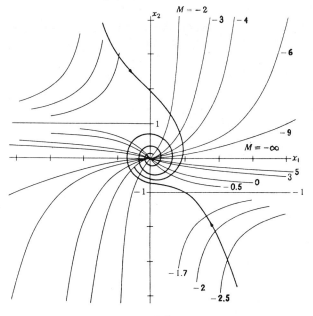

796

From the phase portrait we see that the nonlinear gain
causes the trajectories to get into a circle of radius 1.
That limits the overshoot.

The network is shown in Fig. 1.

The piecewise linear branch characteristic of the resistive
element is shown in Fig. 2.

Draw a phase portrait of the system.

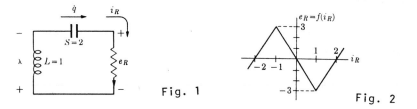

Fig. 1   Fig. 2

Solution: Let $q = x_1$ and $\lambda = x_2$, we obtain the following
equations for the network.

$$\dot{x}_1 = x_2$$
$$\dot{x}_2 = -2x_1 - f(x_2)$$

(1)

where

$$f(x_2) = \begin{cases} -3x_2 & |x_2| \leq 1 \\ 3(x_2 - 2) & x_2 > 1 \\ 3(x_2 + 2) & x_2 < 1 \end{cases}$$

We may break up the problem into three linear parts.
We choose the following regions:

Region I $|x_2| \leq 1$

II $x_2 > 1$ (2)

III $x_2 < -1$

and consider them separately.

Region I:

Equations (1) become

$$\dot{x}_1 = x_2$$
$$\dot{x}_2 = -2x_1 + 3x_2$$

(3)

The normal modes and the eigenvalues are determined from

797

$$\left| MI - \begin{bmatrix} 0 & 1 \\ -2 & 3 \end{bmatrix} \right| = M^2 - 3M + 2$$

$$= (M - 2)(M - 1) = 0 \qquad (4)$$

There are two real unstable normal modes with slopes
1 and 2, thus the origin is an unstable node.

From the isocline equation

$$M = \frac{\dot{x}_2}{\dot{x}_1} = \frac{-2x_1 + 3x_2}{x_2} \qquad (5)$$

we get

$$\frac{x_2}{x_1} = \frac{-2}{M - 3} \qquad (6)$$

We can now determine a family of isoclines that is valid
in region I.
Region II:

Equations (1) become

$$\dot{x}_1 = x_2$$

$$\dot{x}_2 = -2x_1 - 3x_2 + 6$$

We get a singular point at $x_1 = 3$, $x_2 = 0$, which is outside
region II. Such a point is known as a virtual point. It
can be easily seen from the original nonlinear equations
that the origin is the only singular point in the plane.

We shall extrapolate the isoclines constructed in region I
to region II. Let us note that since the right-hand sides
of eqs. (1) are continuous, the isoclines must also be
continuous.

In any linear region isoclines constitute a family of
straight lines extending from a singular point, virtual or
real. Using the above property we extend the isocline
$M = -1$ into region II by drawing the line passing through
point P and the virtual singular point (see Fig. 3). This
method allows us to extend each isocline into region II.

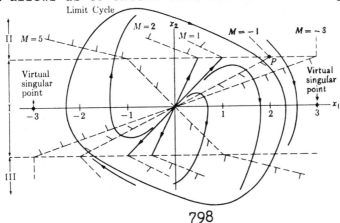

Fig.3

Region III:

Equations (1) become

$$\dot{x}_1 = x_2$$

$$\dot{x}_2 = -2x_1 - 3x_2 - 6$$

We have another virtual singular point at (-3, 0). In this region the construction procedure is the same as in region II. The completed family of isoclines and a representative set of trajectories is shown in Fig. 3.

APPLICATION TO NETWORKS AND SYSTEMS

● **PROBLEM** 14-11

The block diagram of the system is shown in fig. 1

Obtain a phase portrait of the system if the input to the system is an impulse function shown in fig. 2

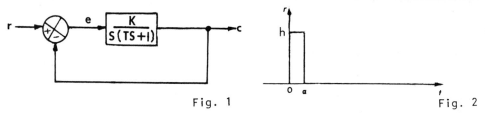

Fig. 1   Fig. 2

Solution: The impulse function can be considered as a sum of two step functions.

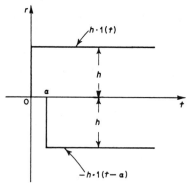

Fig. 3

The equations for the system are

$$T\ddot{e} + \dot{e} + Ke = 0 \qquad \text{for} \quad 0 < t < \alpha$$

$$T\ddot{e} + \dot{e} + Ke = -\infty \qquad \text{for} \quad t = \alpha$$

$$T\ddot{e} + \dot{e} + Ke = 0 \qquad \text{for} \quad t > \alpha$$

For simplicity we shall assume that the poles of the closed-loop transfer function are complex conjugates and lie in the left-half plane.

799

The initial conditions for the error signal are e(0) = h,
ė(0) = 0, since the system is initially at rest.

The trajectory on the e-ė plane starts at A, its path converges to the stable focus (0, 0); for t = α the trajectory
reaches B, see Fig. 4. The trajectory jumps then to C,
where ė = -∞ since the negative step input -h1(t - α).
From C the trajectory jumps to D and then converges to the
origin.

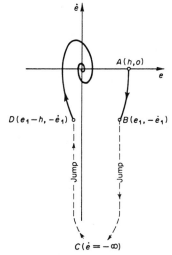

Fig. 4

The system equations in terms of the output signal become

$$T\ddot{c} + \dot{c} + Kc = Kh \quad \text{for} \quad 0 < t < \alpha$$

$$T\ddot{c} + \dot{c} + Kc = 0 \quad \text{for} \quad \alpha < t$$

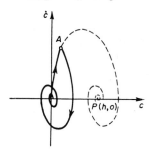

Fig. 5

The initial conditions are c(0) = ċ(0) = 0, thus the
trajectory starts in the c-ċ plane from the origin, see
Fig. 5.

● PROBLEM 14-12

The block diagram of a nonlinear control system is shown in
Fig. 1.

Fig. 2 shows the input-output characteristic curve of the
nonlinear element.

The system does not have any time delay in switching from
the on to the off condition.

For the system described above, determine the step-response and the ramp-response behavior. Take the following constants of the system: $T = 1$, $K = 4$, $e_0 = 0.1$, $e_1 = 0.2$, $M_0 = 0.2$.

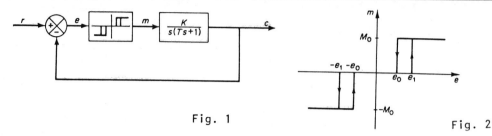

Fig. 1   Fig. 2

Solution: We shall write the equation of the system

$$T\ddot{e} + \dot{e} + Km = T\ddot{r} + \dot{r} \tag{1}$$

For $\dot{e} > 0$, we obtain

$$
\begin{aligned}
m &= M_0 && \text{for} && e > e_1 \\
m &= 0 && \text{for} && e_1 > e > -e_0 && \quad (2)\\
m &= -M_0 && \text{for} && e < -e_0
\end{aligned}
$$

For $\dot{e} < 0$, we obtain

$$
\begin{aligned}
m &= M_0 && \text{for} && e > e_0 \\
m &= 0 && \text{for} && e_0 > e > -e_1 && \quad (3)\\
m &= -M && \text{for} && e < -e_1
\end{aligned}
$$

Based on eqs. (1), (2) and (3) we get for a unit-step input the trajectory shown in Fig. 3. We see that system output oscillation continues indefinitely since the response shows a limit cycle at steady state.

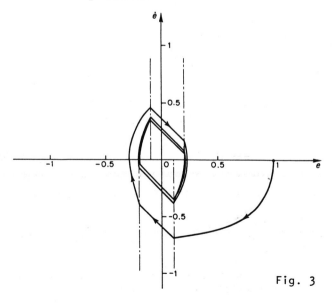

Fig. 3

For a ramp input r(t) = Vt the system equation is

$$T\ddot{e} + \dot{e} + Km = V \qquad (4)$$

The response to the ramp input is determined for three cases:

V > KM$_0$

Fig. 4 shows a phase-plane portrait for V = 1.2, we have

1.2 = V > KM$_0$ = 4 × 0.2 = 0.8

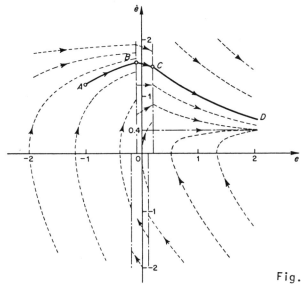

Fig. 4

The trajectory is asymptotic to the line $\dot{e}$ = 0.4 and the error approaches infinity as t → ∞.

V < KM$_0$

The trajectory for this case for V = 0.4 is shown in Fig. 5.

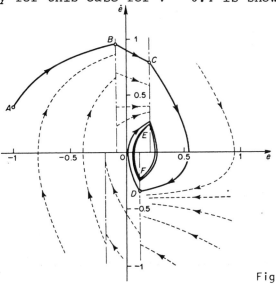

Fig. 5

802

For this case, output oscillation continues indefinitely.
The trajectory starting at point A follows the path
ABCDEF and converges to a limit cycle.

$$V = KM_0$$

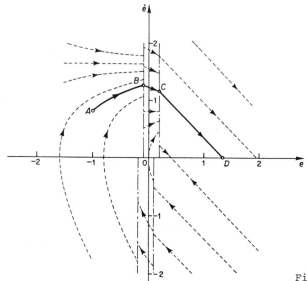

Fig.6

The trajectory for this case for V = 0.8 is shown in Fig. 6.
The trajectory starting at A follows ABCD and converges
to D.

At steady state, the system will have a steady-state error
equal to $\overline{OD}$.

● **PROBLEM 14-13**

The system of two parallel current-carrying conductors of
length 1 is shown in Fig. 1

The lower conductor is fixed, the upper is attached to springs
with total spring constant k. The total force f on the
movable conductor is

$$f(x) = -kx + \frac{\mu_0 li^2}{2\pi(d - x)} \tag{1}$$

We neglected fringing effects.

μ_0 is the permeability of free space.

Defining momentum as $p = m\dot{x}$, eliminate t from the system
equations. Integrate, and use the result to draw a phase
portrait for the system. Use k = m = 1, d = 10,

$$\frac{\mu_0 li^2}{2\pi} = 16.$$

803

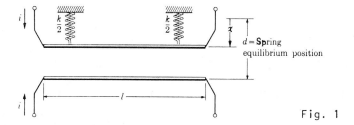

Fig. 1

Solution: From Newton's law, we have

$$m\ddot{x} - f(x) = 0 \tag{2}$$

Since

$$p = m\dot{x} \tag{3}$$

we get

$$\dot{p} = f(x) \tag{4}$$

Combining eqs. (3) and (4) we obtain

$$\frac{\dot{p}}{\dot{x}} = \frac{m}{p}f(x)$$

or $\quad \frac{pdp}{m} = f(x)\,dx \tag{5}$

Substituting for $f(x)$, eq. (1) gives

$$\frac{pdp}{m} = \left[-kx + \frac{\mu_0 li^2}{2\pi(d-x)}\right]dx \tag{6}$$

integrating we obtain

$$\frac{p^2}{2m} + \frac{1}{2}kx^2 + \frac{\mu_0 li^2}{2\pi}\ln(d-x) = h \tag{7}$$

Eq. (7) gives the total energy of the system as a sum of $\frac{p^2}{2m}$, which is kinetic energy and the potential energy $U(x)$ represented by the remaining terms of eq. (7). We shall use the numerical values of the parameters to construct a phase portrait.

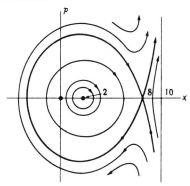

Fig. 2

804

For these values we find from eq. (1) the singular points $(x, p) = (2, 0)$ which is a vortex point and $(x, p) = (8, 0)$ which is a saddle point. To draw the phase portrait we take different values of h and then plot the locus of points on the phase plane which satisfy eq. (7) for a given value of h.

The phase portrait for this system is shown in Fig. 2.

● **PROBLEM** 14-14

The network is shown in Fig. 1.

x_1 and x_2 represent the charges on C_1 and C_2. Write the state equations in terms of x_1 and x_2. Find a pair of normal modes. Sketch a phase portrait for the system. What changes when you diagonalize the system and repeat?

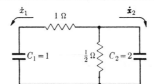

Fig. 1

Solution: $\dot{x}_1$ and $\dot{x}_2$ are the currents through the capacitors and $\frac{x_1}{C_1}$, $\frac{x_2}{C_2}$ are the voltages across them. Then

$$\dot{x}_1 = \frac{x_2}{2} - x_1$$

$$\dot{x}_2 = -\dot{x}_1 - \frac{\frac{x_2}{2}}{\frac{1}{2}} \tag{1}$$

represent KCL for the two nodes. Rearranging gives

$$\begin{bmatrix} \dot{x}_1 \\ \dot{x}_2 \end{bmatrix} = \begin{bmatrix} -1 & \frac{1}{2} \\ 1 & -\frac{3}{2} \end{bmatrix} \begin{bmatrix} x_1 \\ x_2 \end{bmatrix} \tag{2}$$

The eigenvalues are given by

$$\Delta(\lambda) = (\lambda + 1)(\lambda + \frac{3}{2}) - \frac{1}{2} = \lambda^2 + \frac{5}{2}\lambda + 1$$

$$= (\lambda + 2)(\lambda + \frac{1}{2}) \tag{3}$$

The eigenvectors are found by solving

$$[\lambda I - A]p = 0 \tag{4}$$

or

$$(\lambda + 1)p_1 - \frac{1}{2}p_2 = 0$$

$$-p_1 + (\lambda + \frac{3}{2})p_2 = 0 \tag{5}$$

Choosing $p_1 = 1$ gives $p_2 = 2(\lambda + 1)$ so the eigenvectors are

$$p_1 = \begin{bmatrix} 1 \\ -2 \end{bmatrix} \quad \text{and} \quad p_2 = \begin{bmatrix} 1 \\ 1 \end{bmatrix}. \tag{6}$$

So a pair of normal modes is

$$X_1(t) = \begin{bmatrix} 1 \\ -2 \end{bmatrix} \varepsilon^{-2t}, \quad X_2(t) = \begin{bmatrix} 1 \\ 1 \end{bmatrix} \varepsilon^{-\frac{t}{2}} \tag{7}$$

Fig. 2 shows a phase portrait of the system.

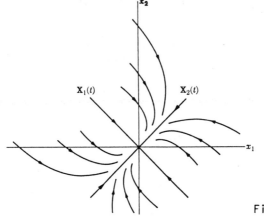

Fig. 2

The four straightline trajectories correspond to the normal modes $X_1(t)$ and $X_2(t)$.

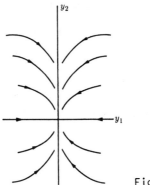

Fig. 3

For the system in diagonal form the eigenvectors are

$$x_1 = \begin{bmatrix} c_1 \\ 0 \end{bmatrix}, \quad x_2 = \begin{bmatrix} 0 \\ c_2 \end{bmatrix} \tag{8}$$

Therefore the normal modes (when real) correspond to the coordinate axes and the phase portrait takes "canonical" form. We often use a diagonalizing transformation x = Py

The transformation

$$\begin{bmatrix} x_1 \\ x_2 \end{bmatrix} = \begin{bmatrix} 1 & 1 \\ -2 & 1 \end{bmatrix} \begin{bmatrix} y_1 \\ y_2 \end{bmatrix}$$

(9)

modifies the phase portrait to the form shown in Fig. 3. A nonsingular linear transformation as in eq. (9) does not change the eigenvalues.

● **PROBLEM** 14-15

A control system with saturation nonlinearity is shown in Fig. 1.

The input-output characteristic curve of the saturation nonlinearity is shown in Fig. 2.

Assume that the system is initially at rest. Construct the trajectories in the phase plane when the system is subjected to a step input r(t) = R. The constants are T = 1, K = 4, e_0 = 0.2, M_0 = 0.2.

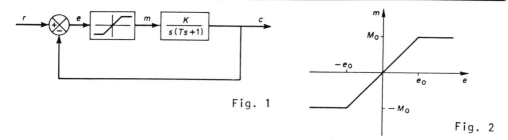

Fig. 1

Fig. 2

Solution: From Fig. 2 we obtain

$$m = e \qquad \text{for} \qquad |e| \leq e_0$$

$$m = M_0 \qquad \text{for} \qquad e > e_0$$

(1)

$$m = -M_0 \qquad \text{for} \qquad e < -e_0$$

The equation for the system is

$$T\ddot{e} + \dot{e} + Km = T\ddot{r} + \dot{r}$$

(2)

For step input:

$$\ddot{r} = \dot{r} = 0 \quad \text{for} \quad t > 0, \text{ thus}$$

$$T\ddot{e} + \dot{e} + Km = 0$$

(3)

For the linear operation of the system

$$T\ddot{e} + \dot{e} + Ke = 0$$

(4)

807

The singular point (0, 0) is either a stable node or a stable focus. For the nonlinear operation we have

$$T\ddot{e} + \dot{e} + KM_0 = 0 \quad \text{for} \quad e > e_0$$
$$T\ddot{e} + \dot{e} - KM_0 = 0 \quad \text{for} \quad e < -e_0 \tag{5}$$

We shall define

$$\frac{d\dot{e}}{de} = \alpha \tag{6}$$

thus

$$\dot{e} = \frac{\dfrac{-KM_0}{T}}{\alpha + \dfrac{1}{T}} \quad \text{for} \quad e > e_0 \tag{7}$$

$$\dot{e} = \frac{\dfrac{KM_0}{T}}{\alpha + \dfrac{1}{T}} \quad \text{for} \quad e < -e_0 \tag{8}$$

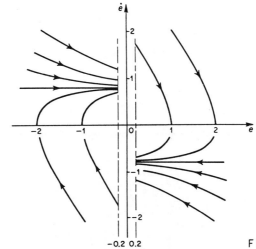

Fig. 3

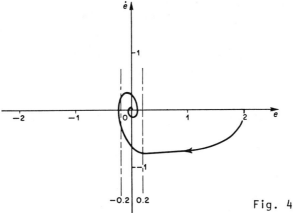

Fig. 4

808

From eq. (7) we see that for $e > e_0$, all trajectories are asymptotic to the line

$$\dot{e} = -KM_0,$$

and from eq. (8), for $e < -e_0$, are asymptotic to

$$\dot{e} = KM_0.$$

Fig. 3 shows a phase plane portrait of the system.

Fig. 4 shows the trajectory when the system is subjected to a step input of magnitude 2.

● **PROBLEM** 14-16

The system is shown in Fig. 1.

Write the state and output equations for the system. Draw plots of x_2 vs. x_1, for different initial conditions. Graph the state variables and outputs as functions of time for

a) $x_{10} = 2$ $x_{20} = 1$

b) $x_{10} = -3$ $x_{20} = 2$

The function sgn(x) is defined as

$$\text{sgn}(x) = \begin{cases} 1 & \text{for} \quad x \geq 0 \\ -1 & \text{for} \quad x < 0 \end{cases}$$

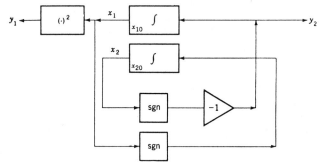

Fig. 1

Solution: The state variables are the integrator outputs, as shown in Fig. 1. The state and output equations are

$$\frac{dx_1}{dt} = -\text{sgn } x_2 \tag{1}$$

$$\frac{dx_2}{dt} = \text{sgn } x_1 \tag{2}$$

$$y_1 = x_1^2 \tag{3}$$

$$y_2 = -\text{sgn } x_2 \tag{4}$$

We shall investigate the behavior of x_1 and x_2 in the state space. Each point of that space corresponds to a state x_1, x_2 of the system.

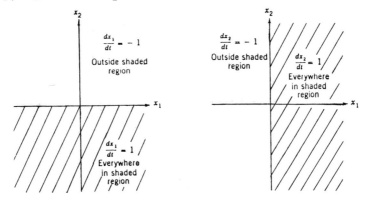

Fig. 2   Fig. 3

Fig. 2 shows $\frac{dx_1}{dt}$ and Fig. 3 shows $\frac{dx_2}{dt}$ for the states of the system.

We see that $\frac{dx_1}{dt}$ and $\frac{dx_2}{dt}$ take on one of the two values $-1, +1$, thus $\frac{dx_2}{dt}$ will be either -1 or 1.

$\frac{dx_2}{dx_1}$ takes on value $+1$ in the second and fourth quadrants and value -1 in the first and third. Fig. 4 shows $\frac{dx_2}{dx_1}$, the arrows indicate the direction in which x_1 and x_2 change.

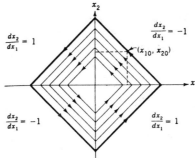

Fig. 4

The point moves along its trajectory in the direction indicated by arrows at a constant speed

$$\sqrt{\left(\frac{dx_1}{dt}\right)^2 + \left(\frac{dx_2}{dt}\right)^2} = \sqrt{1+1} = \sqrt{2}$$

For the initial conditions

$$x_1(0) = x_{10}$$

$$x_2(0) = x_{20}$$

(5)

we obtain trajectory starting from that point. (See Fig. 4.)

The waveforms of $x_1(t)$ and $x_2(t)$ are shown in Fig. 5.

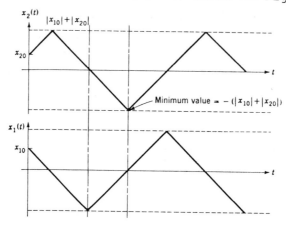

Fig. 5

Although there is no external input to the system, nevertheless it does not reach a steady-state, but instead shows periodic behavior. This is only possible with non-linear systems. Such periodic behavior, which appears as a closed curve in the state-space plots, is called a limit-cycle.

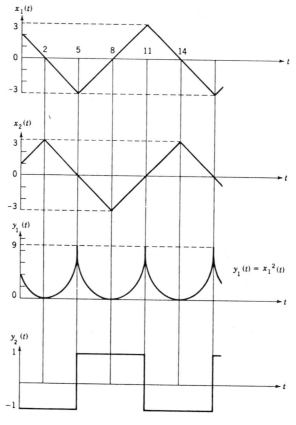

Fig.6

From Fig. 4 we see that the maximum value that x_1 and x_2 reach is $|x_{10}| + |x_{20}|$ and the minimum value is

$-|x_{10}| - |x_{20}|.$

The output waveforms are functions of the state variables and can be obtained after the $x_1(t)$ and $x_2(t)$ waveforms have been established. Fig. 6 shows the state variables and outputs for $x_{10} = 2$ and $x_{20} = 1$. The response for $x_{10} = -3$, $x_{20} = 2$ is shown in Fig. 7.

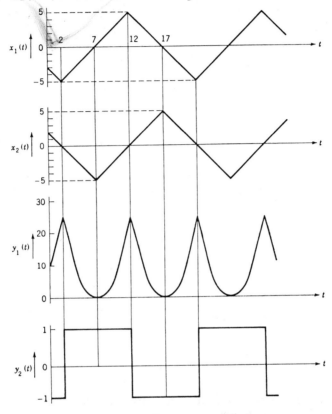

Fig.7

Inspecting Fig. 6 and Fig. 7 we see that:

1. $x_1(t)$ and $x_2(t)$ start at their initial values x_{10} and x_{20}.

2. Thereafter, $dx_1/dt = -\text{sgn } x_2$ and $dx_2/dt = \text{sgn } x_1$. [Note that 1 and 2 above "say" that x_1 and x_2 are solutions to the given D.E.'s (1) and (2).]

3. Output $y_1(t)$ is $x_1(t)$ squared, as dictated by Eq. (3).

4. Output $y_2(t)$ is equal to dx_1/dt and to $-\text{sgn } x_2$.

5. The maximum value of either $x_1(t)$ or $x_2(t)$ is $|x_{10}| + |x_{20}|$ and the minimum value is the negative of the maximum value.

6. The period is $4(|x_{10}| + |x_{20}|)$.

7. When either one of $x_1(t)$, $x_2(t)$ is a max(min), the other crosses zero.

Fig. 1 shows the network representing a tunnel diode with some associated capacitance and inductance.

The normal-form equations for this network are

$$\dot{q} = -i_2 - f(e_1) \qquad e_1 = \frac{q}{c}$$

$$\dot{\lambda} = e_1 - Ri_2 - V \qquad i_2 = \frac{\lambda}{L} \qquad (1)$$

where $f(e_1)$ represents the tunnel diode branch relation. Approximate $f(e_1)$ in a piecewise linear fashion as shown in Fig. 2.

Assume that

$$L = C = 1, \qquad R = 0.1, \qquad V = 2 \qquad (2)$$

and make the identification

$$x_1 = q = e_1, \qquad x_2 = -\lambda = -i_2 \qquad (3)$$

to obtain

$$\dot{x}_1 = x_2 - f(x_1)$$

$$(4)$$

$$\dot{x}_2 = -x_1 + (2 - 0.1x_2)$$

Use the generalized Liénard method to construct the phase portrait of the system.

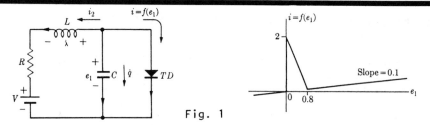

Fig. 1

Fig. 2

Solution: The generalized method applies to cases where the isocline equation is given by

$$M = \frac{\dot{x}_2}{\dot{x}_1} = \frac{-x_1 + \Phi_2(x_2)}{x_2 - \Phi_1(x_1)} \qquad (5)$$

The procedure is as follows:

Plot the two L-characteristics $x_1 = \Phi_2(x_2)$ and $x_2 = \Phi_1(x_1)$ corresponding respectively to M = 0 and M = ∞. The intersections are the singular points. Then at a given point P, the trajectory is approximated by a circular arc centered at S, radius SP as shown in Fig. 3.

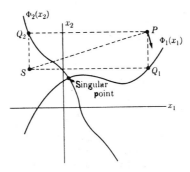

Fig. 3

In our case, $\Phi_1(x_1) = f(x_1)$ and $\Phi_2(x_2) = 2 - 0.1 x_2$.

Fig. 4 shows a phase portrait of the system. Dashed lines represent the two L - characteristics. There is one singular point, which is a stable spiral point and two concentric limit cycles. The outer limit cycle "s" is orbitally asymptotically stable, the inner "u" is unstable.

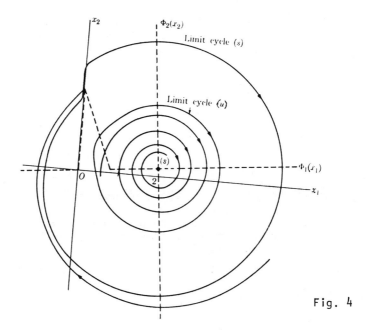

Fig. 4

We conclude from the phase portrait that if the system is in the initial state within the unstable limit cycle, it will tend asymptotically toward the stable singular point. If the initial state is outside the unstable limit cycle, the trajectory will tend toward the stable limit cycle, thus the system will oscillate.

814

Fig. 1 shows two systems. The electrical network is defined by

$$L\ddot{q} - \phi_e(\dot{q}) + Sq = 0 \tag{1}$$

and the mechanical system by

$$M\ddot{x} - \phi_m(\dot{x}) + kx = 0 \tag{2}$$

The functions $-\phi_e$ and $-\phi_m$ represent respectively the characteristic of the gas tube and the friction forces in the mechanical system. Each of these has a nature of a damping force. Both systems behave in a similar way. Use the method of Liénard to draw the phase portrait for the two equivalent systems. Let $V_d = 100$, $V_m = 50$, $L = 4$, $S = 4$, $e_c(0) = 1000$ and $i(0) = 0$.

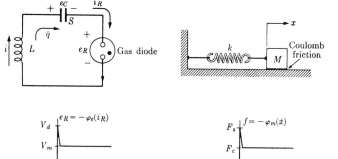

Fig. 1

Solution: We shall follow the procedure:

a) On the phase plane, plot the locus of points M = 0, the zero slope isocline. From equations:

$$\dot{x}_1 = x_2, \qquad \dot{x}_2 = -x_1 + \phi(x_2) \tag{3}$$

$$M = \frac{\dot{x}_2}{\dot{x}_1} = \frac{-x_1 + \phi(x_2)}{x_2} \tag{4}$$

we find that for M = 0

$$x_1 = \phi(x_2) \tag{5}$$

The above equation is the L-characteristic.

b) To find the slope of a trajectory passing through any point P, we project the point horizontally to the L-characteristic (point Q), then from point Q project vertically to the x_1-axis (point S). Swing an arc through P with center at S. This arc will be tangent to the trajectory at point P. A short segment of arc is extended to a neighboring point P'. Then a new center S' is located and the process is repeated. The shorter the segments of arc, the more accurate the trajectory. Note that all intersections of the L-characteristic with the x_1-axis are singular points.

We assign states in eq. (1) as follows:

$$x_1 = \sqrt{S}q = \frac{e_c}{\sqrt{S}}, \qquad x_2 = \sqrt{L}\dot{q} = \sqrt{L}i \qquad (6)$$

The normalized L-characteristic is

$$\phi(x_2) = \frac{1}{\sqrt{S}} \cdot \phi_e(\frac{x_2}{\sqrt{L}}) \qquad (7)$$

Fig. 2 shows the Liénard construction.

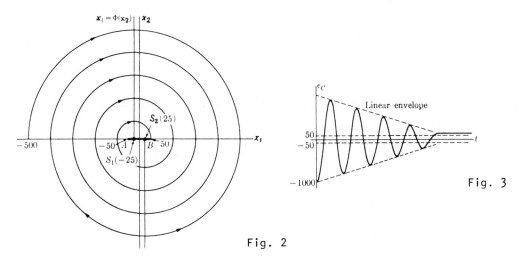

Fig. 2

Fig. 3

Since the L-characteristic consists of only vertical segments, any trajectory must consist of only arcs of circles centered at S_1 or S_2. The spiral diminishes since each time the trajectory crosses the x_1-axis, the radius of arc decreases by 50 units. In our case line AB belongs to the L-characteristic, therefore, when the trajectory crosses a point on this line the arc radius must be zero. Thus the trajectory becomes stationary. Line AB is a continuum of nonisolated singular points.

CHAPTER 15

NONLINEAR SYSTEMS

NONLINEAR SYSTEMS

● PROBLEM 15-1

The system is described by the equation

$$\dot{x} = a(t)x + u(t) \qquad (1)$$

where

$$a(t) = \frac{-t}{1 + t^2} \qquad (2)$$

Determine the stability of the system

a) for the homogeneous case $u(t) = 0$,

b) for the case $u(t) = 1 \quad t \geq 0$.

Solution: a) Separating the variables we find a homogeneous solution. Eq. (1) can be written as

$$\frac{dx}{dt} = -\frac{t}{1 + t^2} x \qquad (3)$$

Then

$$\int \frac{dx}{x} = \int - \frac{t\,dt}{1 + t^2} \qquad (4)$$

so

$$\ln x = -\frac{1}{2} \ln(1 + t^2) + C \qquad (5)$$

and

$$\phi(t) = \frac{1}{\sqrt{1 + t^2}} \qquad (6)$$

where

$$x(t) = \phi(t)x(0) \qquad (7)$$

The notation $\phi(t)$ is used to represent the state transis-

tion matrix. Since $\lim\limits_{t\to\infty}||\phi(t)|| = 0$, the homogeneous
system is asymptotically stable.

b) Applying an input u(t) = 1 over the interval [0,∞)
with an initial condition x(0) = 0, we test the complete
system for input-output stability. We form the impulse
response matrix

$$H(t,s) = [\phi(t)][\phi(s)]^{-1} \tag{8}$$

$$H(t,s) = \sqrt{\frac{1 + s^2}{1 + t^2}} \tag{9}$$

Substituting into

$$x(t) = \phi(t)x(0) + \int_0^t H(t,s)u(s)\,ds \tag{10}$$

gives

$$x(t) = \int_0^t \sqrt{\frac{1 + s^2}{1 + t^2}}\,ds \tag{11}$$

or

$$x(t) = \frac{1}{2}[t + \frac{1}{\sqrt{1 + t^2}} \ln (t + \sqrt{1 + t^2}) \tag{12}$$

$\lim\limits_{t\to\infty} x(t) = \infty$, therefore the system is not input-output
stable.

● **PROBLEM** 15-2

The figure shows a tracking station that measures the range
r to an earth satellite, the azimuth angle α and the eleva-
tion angle β.

1) Find the nonlinear equations which relate the satellite's
   relative position [x_1, x_2, x_3] to the measured quantities.

2) Derive the linear equations which relate small perturbations in satellite location to small perturbations
   in the measurements.

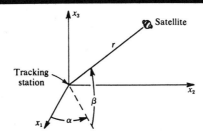

Solution:

1) We have the following equations relating x_1, x_2, x_3 to
   r, α, β.

$$r = \sqrt{x_1^2 + x_2^2 + x_3^2}$$

$$\alpha = \tan^{-1}\left(\frac{x_2}{x_1}\right)$$

$$\beta = \tan^{-1}\left(\frac{x_3}{\sqrt{x_1^2 + x_2^2}}\right)$$

2) Let

$$x(t) = x_n(t) + \delta x(t)$$

$$r(t) = r_n(t) + \delta r(t)$$

$$\alpha(t) = \alpha_n(t) + \delta \alpha(t)$$

$$\beta(t) = \beta_n(t) + \delta \beta(t)$$

The Taylor series expansion gives

$$\delta r = \sum_{i=1}^{3} \frac{\partial r}{\partial x_i} \delta x_i = \frac{1}{r_n} x_n^T \delta x_n \tag{1}$$

$$\delta \alpha = \sum_{i=1}^{3} \frac{\partial \alpha}{\partial x_i} \delta x_i = \frac{1}{x_{1n}^2 + x_{2n}^2}[-x_{2n},\ x_{1n},\ 0]\delta x \tag{2}$$

$$\delta \beta = \sum_{i=1}^{3} \frac{\partial \beta}{\partial x_i} \delta x_i = \frac{1}{r_n^2 \sqrt{x_{1n}^2 + x_{2n}^2}}[-x_{1n}x_{3n},\ -x_{2n}x_{3n}, \\ x_{1n}^2 + x_{2n}^2]\delta x \tag{3}$$

Taking $\delta y = [\delta r \quad \delta \alpha \quad \delta \beta]$ we can express the above results as

$$\delta y(t) = A(t)\delta x(t)$$

where A is a 3 × 3 matrix.

● **PROBLEM 15-3**

The block diagram of a control system with a saturation non-linearity is shown in the figure.

The state equations of the system are

$$\dot{x}_1 = f_1 = x_2 \tag{1}$$

$$\dot{x}_2 = f_2 = u \tag{2}$$

The input-output relation of the saturation nonlinearity is represented by

$$u = (1 - e^{-k|x_1|})\, \text{sgn}\ x_1 \tag{3}$$

where

$$\text{sgn}\ x_1 = \begin{cases} +1 & x_1 \geq 0 \\ -1 & x_1 < 0 \end{cases} \tag{4}$$

Linearize the system for small signals.

Solution: Based on a Taylor series expansion, with high-order terms disregarded, we have

$$\Delta\dot{x}_1 = \frac{\partial f_1}{\partial x_2}\Delta x_2 = \Delta x_2 \tag{5}$$

$$\Delta\dot{x}_2 = \frac{\partial f_2}{\partial x_1}\Delta x_1 = Ke^{-K|x_{01}|}\Delta x_1 \tag{6}$$

where x_{01} denotes a nominal value of x_1. The last two equations are linear and are valid only for small signals. We write them in the matrix form

$$\begin{bmatrix} \Delta\dot{x}_1 \\ \Delta\dot{x}_2 \end{bmatrix} = \begin{bmatrix} 0 & 1 \\ a & 0 \end{bmatrix} \begin{bmatrix} \Delta x_1 \\ \Delta x_2 \end{bmatrix} \tag{7}$$

where

$$a = Ke^{-K|x_{01}|} = \text{constant} \tag{8}$$

To analyze the results of the linearization, let us choose $x_{01} = 0$, then eq. (6) becomes

$$\Delta\dot{x}_2 = K\Delta x_1 \tag{9}$$

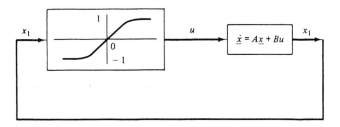

We see that the linearized model is equivalent to the one with a linear amplifier with a constant gain K. For large x_{01}, the nominal operating point lies on the saturated portion of the nonlinearity and a = 0. Thus small changes in x_1 will cause no changes in $\Delta\dot{x}_2$.

● **PROBLEM 15-4**

Fig. 1 shows the nonlinear system containing a Clegg integrator. Determine its frequency response.

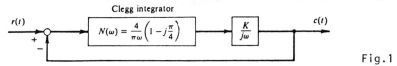

Fig.1

Solution: Since the DF for the Clegg integrator depends on frequency alone, we shall solve the linear problem to arrive at a closed-form solution for $\frac{C}{R}(j\omega)$.

Thus

$$\frac{C}{R}(j\omega) = \frac{N(\omega)\frac{K}{j\omega}}{1 + N(\omega)\frac{K}{j\omega}} \tag{1}$$

820

$$= \frac{1 + j\left(\frac{4}{\pi}\right)}{1 - \frac{\omega^2}{K} + j\left(\frac{4}{\pi}\right)} \tag{2}$$

$$= \frac{\sqrt{1 + \left(\frac{4}{\pi}\right)^2}}{\sqrt{\left(1 - \frac{\omega^2}{K}\right)^2 + \left(\frac{4}{\pi}\right)^2}} \exp\left[-j \tan^{-1}\left(\frac{\frac{4\omega^2}{\pi K}}{1 - \frac{\omega^2}{K} + \left(\frac{4}{\pi}\right)^2}\right)\right] \tag{3}$$

Fig. 2 shows the magnitude and phase angle.

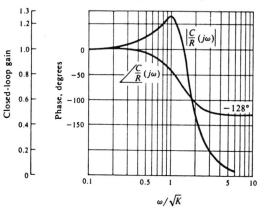

Fig.2

● **PROBLEM** 15-5

For the system shown in the figure, determine the range of a, the parameter of the digital lead compensator, for which the 3, 3 limit cycle mode is impossible.

$$L(s) = \frac{K}{s(\tau s + 1)}, \quad \tau = T_s \tag{1}$$

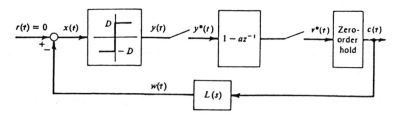

Solution: Let us denote $y(mT_s) = y_m$ and use the periodically repeating sequence of nonlinearity output samples

$$y_m = D, D, D, -D, -D, -D \text{ for } m = 0, 1, 2 \ldots 5$$

and

$$y_{m+6} = y_m \tag{2}$$

For the system linear part we have the following sampled transfer function.

$$\frac{X(z)}{Y(z)} = -K\left[\frac{(1 - az^{-1})(1 - z^{-1})}{s^2(\tau s + 1)}\right]^*$$

$$= -KT_s\left[\frac{z - b - (\frac{\tau}{T_s})(1 - b)(z - 1)}{(z - 1)(z - b)}\right]\frac{z - a}{z} \qquad (3)$$

where

$$b = e^{-\frac{T_s}{\tau}} \qquad (4)$$

For $\frac{T_s}{\tau} = 1$, eq. (3) becomes

$$x_m - 1.368x_{m-1} + 0.368x_{m-2}$$

$$= -KT_s[0.368y_{m-1} + (0.264 - 0.368a)y_{m-2} - 0.264ay_{m-3}] \qquad (5)$$

which is the difference equation. The y_m are given for all m. Using eq. (2) which holds for the x_m as well as for the y_m we take in eq. (5), m = 0, 1, 2, 3, 4, 5 and obtain the system of equations for the six unknowns x_0 to x_5.

$$x_0 - 1.368x_5 + 0.368x_4 = -KDT_s(-0.632 + 0.632a)$$

$$x_1 - 1.368x_0 + 0.368x_5 = -KDT_s(0.104 + 0.632a)$$

$$x_2 - 1.368x_1 + 0.368x_0 = -KDT_s(0.632 - 0.104a)$$

$$x_3 - 1.368x_2 + 0.368x_1 = -KDT_s(0.632 - 0.632a) \qquad (6)$$

$$x_4 - 1.368x_3 + 0.368x_2 = -KDT_s(-0.104 - 0.632a)$$

$$x_5 - 1.368x_4 + 0.368x_3 = -KDT_s(-0.632 + 0.104a)$$

The solution of the system is not unique, because only five equations are linearly independent. We shall express x_0, x_1, x_2, x_3, x_4 in terms of x_5, thus

$$x_0 = x_5 - KDT_s(-0.837 + 0.557a) > 0$$

$$x_1 = x_5 - KDT_s(-1.041 + 1.393a) > 0$$

$$x_2 = x_5 - KDT_s(-0.484 + 1.599a) > 0$$

$$x_3 = x_5 - KDT_s(0.354 + 1.043a) < 0$$

$$x_4 = x_5 - KDT_s(0.558 + 0.204a) < 0$$

822

Note that the inequalities on the right are the conditions required to sustain the postulated Y_m sequence. We determine the boundaries between admissible and inadmissible values of x_5 as functions of a. The conditions on x_5 and x_2 are incompatible for a > 0.303. The conditions on x_0 and x_3 are incompatible for a < -2.45. Thus

$$-2.45 < a < 0.303$$

is the range of a for which the 3, 3 limit cycle mode is possible in this system.

● **PROBLEM** 15-6

Fig. 1 shows the block diagram of a positional servo-mechanism in which backlash is present in the gear train. The linear part of the system $G_s(j\omega)$ consists of the blocks between the error detector and the motor shaft. The transfer function of the linear portion of the system is

$$\frac{\theta_m}{\theta_\epsilon}(j\omega) = G_s(j\omega) = \frac{K_a K_m}{j\omega(1 + j\omega\tau_m)} \tag{1}$$

where

K = amplifier gain

τ_m = time constant of the motor

K_m = velocity gain constant of the motor

Fig. 2 shows the describing function G_n for backlash. The reciprocal of the gear ratio is the proportionality constant associated with the describing function. Assume the following numerical values:

$$K_a = 10, \quad K_m = 0.1, \quad \tau_m = 10, \quad N = 1,$$

the overall backlash, defined as 2A, is 0.05 rad.

Investigate the stability of the system.

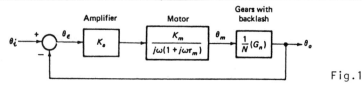

Fig.1

Solution: Using

$$G_s(j\omega) = \frac{1}{j\omega(1 + j\omega 10)} = -\frac{1}{G_n} \tag{2}$$

we determine the stability of the servomechanism. Fig. 3 shows the polar diagram of the $G_s(j\omega)$ and $-\dfrac{1}{G_n}$ character-

istics. The two loci intersect at approximately $\omega = 0.28\dfrac{\text{rad}}{\text{s}}$

and $E_m = 0.09$ rad. Thus the system will oscillate at a frequency and amplitude determined by the point of inter-section. Assuming that the amplitude of the oscillation increases beyond that determined by the point of inter-section, we evaluate the stablity of oscillation. The operating point shifts to the outer or stable side of the $G_s(j\omega)$ curve, and the amplitude of the oscillation decreases,

moving the operating point back to the original intersection.

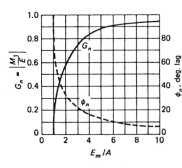

Fig.2

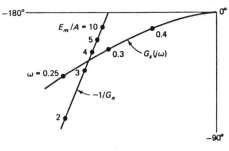

Fig.3

A small decrease in amplitude shifts the operating point to the unstable side of $G_s(j\omega)$. The amplitude tends to

build up and that moves the operating point back to the original intersection. Thus the sustained oscillation is stable and the steady-state operating condition is a limit cycle of amplitude and frequency defined by the point of intersection.

● **PROBLEM** 15-7

A nonlinear time-varying plant is described by

$$\begin{bmatrix} \dot{x}_1 \\ \dot{x}_2 \end{bmatrix} = \begin{bmatrix} 0 & 1 \\ -b & -a(t)x_2 \end{bmatrix}\begin{bmatrix} x_1 \\ x_2 \end{bmatrix} + \begin{bmatrix} 0 \\ 1 \end{bmatrix}[u]$$

where b is a positive constant and $a(t)$ is time-varying. The reference model equation is

$$\begin{bmatrix} \dot{x}_{d1} \\ \dot{x}_{d2} \end{bmatrix} = \begin{bmatrix} 0 & 1 \\ -\omega_n^2 & -2\zeta\omega_n \end{bmatrix}\begin{bmatrix} x_{d1} \\ x_{d2} \end{bmatrix} + \begin{bmatrix} 0 \\ \omega_n^2 \end{bmatrix}[v] \qquad (1)$$

Design a nonlinear controller which will give stable operation of the system.

Solution: We define the error vector by

$$e = x_d - x$$

824

and a Lyapunov function by

$$v(e) = e'Pe$$

where P is a positive-definite real symmetric matrix. Then by using

$$\dot{v}(e) = \dot{e}'Pe + e'P\dot{e} \qquad (e' = e \text{ transpose})$$

and

$$\dot{x}_d = Ax_d + Bv$$

$$\dot{x} = f(x, \dot{u}, t)$$

Then

$$\dot{e} = \dot{x}_d - \dot{x} = Ax_d + Bv - f(x, \dot{u}, t)$$

from

$$e = x_d - x$$

$$x_d = e + x$$

Thus

$$\dot{e} = Ae + Ax + Bv - f(x, u, t)$$

and

$$\dot{v}(e) = [e'A' + x'A' - f'(x, u, t) + v'B']Pe$$
$$\qquad + e'P[Ae + Ax - f(x, u, t) + Bv]$$

Then

$$\dot{v}(e) = e'(A'P + PA)e + 2M$$

where

$$M = e'P[Ax - f(x, u, t) + Bv]$$

We choose the matrix Q to be positive definite

$$Q = \begin{bmatrix} q_{11} & 0 \\ 0 & q_{22} \end{bmatrix}$$

From eq. (1) we find the matrices A and B. Since

$$A'P + PA = -Q$$

we obtain

$$\dot{V}(e) = -(q_{11}e_1^2 + q_{22}e_2^2) + 2M$$

where

$$M = [e_1 \quad e_2] \begin{bmatrix} P_{11} & P_{12} \\ P_{12} & P_{22} \end{bmatrix} \left\{ \begin{bmatrix} 0 & 1 \\ -\omega_n^2 & -2\zeta\omega_n \end{bmatrix} \begin{bmatrix} x_1 \\ x_2 \end{bmatrix} \right.$$
$$\left. - \begin{bmatrix} 0 & 1 \\ -b & -a(t)x_2 \end{bmatrix} \begin{bmatrix} x_1 \\ x_2 \end{bmatrix} - \begin{bmatrix} 0 \\ u \end{bmatrix} + \begin{bmatrix} 0 \\ \omega_n^2 v \end{bmatrix} \right\}$$

$$= (e_1 p_{12} + e_2 p_{22})[-(\omega_n^2 - b)x_1 - 2\zeta\omega_n x_2 + a(t)x_2^2$$
$$+ \omega_n^2 v - u]$$

Choosing u such that

$$u = -(\omega_n^2 - b)x_1 - 2\zeta\omega_n x_2 + \omega_n^2 v + a_m x_2^2 \text{sgn}(e_1 p_{12}$$

$$+ e_2 p_{22}) \qquad (2)$$

where

$$a_m = \max|a(t)|$$

We have for M

$$M = (e_1 p_{12} + e_2 p_{22})[a(t) - a_m \text{sgn}(e_1 p_{12} + e_2 p_{22})]x_2^2$$

where M is nonpositive.

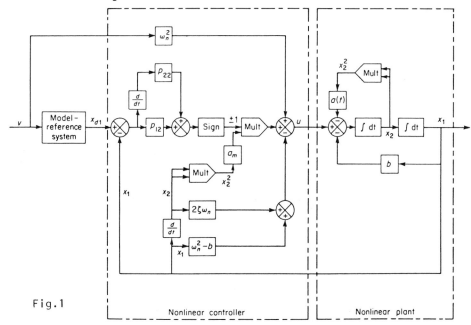

Fig.1

Nonlinear controller   Nonlinear plant

For the control function u given by eq. (2), the equilibrium state e = 0 is asymptotically stable in the large. Therefore, eq. (2) gives a nonlinear control law that yields an asymptotically stable operation. Fig. 1 shows the block diagram for the control system.

● **PROBLEM** 15-8

Investigate the possibility of limit cycle oscillations in the system shown in Fig. 1. Use the table below. Consider time dimensioned in seconds.

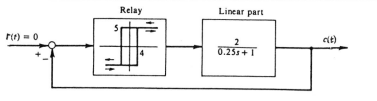

Fig.1

$L_1(s)$	$\sum\limits_{k=1,3,5,\ldots}^{\infty} \text{Re}\,[L_1(jk\omega_0)]$	$\sum\limits_{k=1,3,5,\ldots}^{\infty} \dfrac{1}{k}\,\text{Im}\,[L_1(jk\omega_0)]$
$\dfrac{1}{s+a}$	$\dfrac{\pi}{4\omega_0}\tanh\dfrac{\pi a}{2\omega_0}$	$-\dfrac{\pi}{4a}\tanh\dfrac{\pi a}{2\omega_0}$
$\dfrac{s+b}{(s+a)^2}$	$\dfrac{\pi}{8\omega_0\cosh^2(\pi a/2\omega_0)}$ $\times\left[\sinh\left(\dfrac{\pi a}{\omega_0}\right)-\dfrac{\pi}{\omega_0}(b-a)\right]$	$-\dfrac{\pi}{8a^2\cosh^2(\pi a/2\omega_0)}$ $\times\left[b\sinh\left(\dfrac{\pi a}{\omega_0}\right)-\dfrac{\pi a}{\omega_0}(b-a)\right]$
$\dfrac{(s+b_1)(s+b_2)}{(s+a)^3}$	$\dfrac{\pi}{16\omega_0\cosh^2(\pi a/2\omega_0)}$ $\times\left[2\sinh\left(\dfrac{\pi a}{\omega_0}\right)-\dfrac{\pi^2}{\omega_0^2}(a-b_1)\right.$ $\times(a-b_2)\tanh\left(\dfrac{\pi a}{2\omega_0}\right)$ $\left.+\dfrac{2\pi}{\omega_0}(2a-b_1-b_2)\right]$	$-\dfrac{\pi}{16a^3\cosh^2(\pi a/2\omega_0)}$ $\times\left[2b_1b_2\sinh\left(\dfrac{\pi a}{\omega_0}\right)-\pi^2(a-b_1)\right.$ $\times(a-b_2)\tanh\left(\dfrac{\pi a}{2\omega_0}\right)$ $\left.+2\pi\dfrac{a}{\omega_0}(a^2-b_1b_2)\right]$

<u>Solution:</u> We have, for the system

$$L(\infty) = 0,$$

thus

$$L_1(s) = L(s).$$

From the table we have

$$\text{Re}\,[T(j\omega_0)] = \frac{2\pi}{\omega_0}\tanh\frac{2\pi}{\omega_0} \tag{1}$$

and

$$\text{Im}\,[T(j\omega_0)] = -\frac{\pi}{2}\tanh\frac{2\pi}{\omega_0} \tag{2}$$

The conditions for sustained oscillation are

$$\frac{2\pi}{\omega_0}\tanh\frac{2\pi}{\omega_0} < \frac{2\pi}{\omega_0} \quad\text{or}\quad \tanh\frac{2\pi}{\omega_0} < 1 \tag{3}$$

and

$$-\frac{\pi}{2}\tanh\frac{2\pi}{\omega_0} = -\frac{\pi}{5} \quad\text{or}\quad \tanh\frac{2\pi}{\omega_0} = 0.4 \tag{4}$$

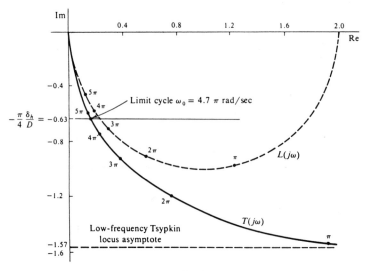

Fig.2

We see that a limit cycle is possible and its period is

$$T_0 = \frac{2\pi}{\omega_0} = \tanh^{-1} 0.4 = 0.424 \text{ sec} \qquad (5)$$

From Fig. 2 we have

$$\frac{d}{d\omega}(\text{Im}T) > 0 \qquad \text{at} \qquad \omega = \omega_0.$$

Thus the limit cycle is stable.

● **PROBLEM** 15-9

Consider an equation for a system

$$\ddot{x} + \omega^2 x + \beta x^3 = 0 \qquad (1)$$

It can represent an electrical network consisting of a linear capacitance and nonlinear inductance. For $\beta > 0$, the force characteristic $\omega^2 x + \beta x^3$ is a hard restoring force; for $\beta < 0$ it is a soft restoring force, see Fig. 1.

Locate singular points for $\beta > 0$ and $\beta < 0$ (points where the restoring force is zero). For oscillatory motion find x(t) in terms of elliptic integrals for $\beta < 0$ and $\beta > 0$.

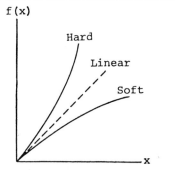

Fig. 1

Solution: If $\beta > 0$, then the only singular point of this system is the origin and the phase trajectories will all be closed. If $\beta < 0$, the restoring force becomes negative for sufficiently large x, and two other singular points appear. They are saddle points, whose coordinates are

$$x = \pm\sqrt{\frac{-\omega^2}{\beta}} \qquad (2)$$

The phase portrait, in this case is more complicated, it is comprised of open and closed trajectories. An energy integral for the system is given by the sum of kinetic energy and the work done by the force. The kinetic energy is

$$\frac{1}{2}mv^2 = \frac{1}{2}(\dot{x})^2 \qquad (3)$$

The work done by the force is

$$\int f(x)\,dx = \int (\omega^2 x + \beta x^3)\,dx = \frac{\omega^2 x^2}{2} + \frac{\beta x^4}{4} \qquad (4)$$

828

Thus, the energy in the system is

$$\frac{1}{2}(\dot{x})^2 + \frac{\omega^2 x^2}{2} + \frac{\beta x^4}{4} = h \qquad (5)$$

A second integration with $u = x$, gives

$$\int_{x_0}^{x} \frac{du}{(2h - \omega^2 u^2 - \frac{\beta u^4}{2})^{\frac{1}{2}}} = t - t_0 \qquad (6)$$

We shall investigate two separate cases.

1) $\beta < 0$

Let us consider only the oscillatory motion of the system, thus the maximum value of $|x|$ must be less than $\sqrt{\frac{-\omega^2}{\beta}}$. The initial conditions are $x(0) = 0$ and $\dot{x}(0)$ is such that the peak value of the oscillation is A. Substituting into the energy integral we find

$$\frac{1}{2}[\dot{x}(0)]^2 = h = \frac{\omega^2 A^2}{2} + \frac{\beta A^2}{4} \qquad (7)$$

From eq. (7) we see that $u^2 = A^2$ must be a root of the polynomial in eq. (6). Thus, it may be factored into the form

$$2h - \omega^2 u^2 - \frac{\beta u^4}{2} = \frac{\beta}{2}(A^2 - u^2)(B^2 + u^2) \qquad (8)$$

where

$$B^2 = A^2 + \frac{2\omega^2}{\beta} \qquad (9)$$

Changing the variable

$$x = A \sin \phi \qquad (10)$$

we transform eq. (6) into the form of an elliptic integral.

Substituting eq. (10) into eq. (6) and using the factorization of eq. (8) we find

$$t = \frac{k(-\frac{2}{\beta})^{\frac{1}{2}}}{A} \int_{0}^{\phi} \frac{du}{\sqrt{1 - k^2 \sin^2 u}} \qquad (11)$$

where

$$k = \frac{1}{(-1 - \frac{2\omega^2}{\beta A^2})^{\frac{1}{2}}} \qquad (12)$$

From eq. (10) and eq. (11) we find the solution for $x(t)$ in terms of the elliptic sine function as

$$x = A sn \left[\frac{A(-\frac{\beta}{2})^{\frac{1}{2}} t}{k} \right] \qquad (13)$$

Setting $\phi = \frac{\pi}{2}$ in eq. (11), we find the period of oscillation T.

$$T = \frac{4k \left(-\frac{2}{\beta} \right)^{\frac{1}{2}}}{A} K(k) \tag{14}$$

where

$$K(k) = \int_0^{\frac{\pi}{2}} \frac{du}{\sqrt{1 - k^2 \sin^2 u}} \tag{15}$$

2) For $\beta > 0$

We choose $x(0) = A$ and $\dot{x}(0) = 0$. Following the procedure for $\beta < 0$ we find

$$x = A cn \left[\frac{A \left(\frac{\beta}{2} \right)^{\frac{1}{2}} t}{k} \right] \tag{16}$$

where

$$k = \frac{1}{\left[2 \left(1 + \frac{\omega^2}{\beta A^2} \right) \right]^{\frac{1}{2}}} \tag{17}$$

The period of oscillation is

$$T = \frac{4k \left(\frac{2}{\beta} \right)^{\frac{1}{2}}}{A} K(k) \tag{18}$$

● **PROBLEM** 15-10

Consider the cubic characteristic

$x^3 = y.$

For an input of the form

$x = A \cos \omega t + B \cos (\gamma \omega t + \theta)$

determine the TSIDF$_s$ for non-harmonically and harmonically related input sinusoids.

Solution: We shall use direct expansion to determine output harmonic content. We obtain

$$y = [A \cos \omega t + B \cos (\gamma \omega t + \theta)]^3$$

$$= A^3 \cos^3 \omega t + 3A^2 B \cos^2 \omega t \cos (\gamma \omega t + \theta) \tag{1}$$

$$+ 3AB^2 \cos \omega t \cos^2 (\gamma \omega t + \theta)$$

$$+ B^3 \cos^3 (\gamma \omega t + \theta)$$

830

$$= (\frac{3A^3}{4} + \frac{3AB^2}{2}) \cos \omega t + \frac{A^3}{4} \cos 3\omega t$$

$$+ (\frac{3A^2B}{2} + \frac{3B^3}{4}) \cos (\gamma \omega t + \theta)$$

$$+ \frac{3A^2B}{4} \left\{ \cos [(\gamma + 2)\omega t + \theta] \right.$$

$$+ \cos [(\gamma - 2)\omega t + \theta] \Big\}$$

$$+ \frac{3AB^2}{4} \left\{ \cos [(2\gamma + 1)\omega t + 2\theta] \right.$$

$$+ \cos [(2\gamma - 1)\omega t + 2\theta] \Big\}$$

$$+ \frac{B^3}{4} \cos (3\gamma \omega t + 3\theta) \tag{2}$$

We have only single-frequency components in the last fully expanded equation. For different values of γ, we have several distinct cases.

Case 1: Harmonically related input simusoids ($\gamma = \frac{1}{3}$):

In eq. (2), the output terms of interest are

$$y = (\frac{3A^3}{4} + \frac{3AB^2}{2}) \cos \omega t + (\frac{3A^2B}{2} + \frac{3B^3}{4}) \cos (\frac{1}{3}\omega t + \theta)$$

$$+ \frac{3AB^2}{4} \cos (-\frac{1}{3}\omega t + 2\theta) + \frac{B^3}{4} \cos (\omega t + 3\theta)$$

$$+ \text{ terms at other frequencies.}$$

The phasor representation gives

$$N_A (A, B, \frac{1}{3}, \theta) = \frac{(\frac{3}{4}A^3 + \frac{3}{2}AB^2) e^{j\omega t} + (\frac{1}{4}B^3) e^{j(\omega t + 3\theta)}}{A e^{j\omega t}}$$

$$= \frac{3}{4}A^2 + \frac{3}{2}B^2 + \frac{1}{4}\frac{B^3}{A}e^{j3\theta} \tag{3}$$

$$N_B (A, B, \frac{1}{3}, \theta) = \frac{(\frac{3}{2}A^2B + \frac{3}{4}B^3) e^{j[\frac{\omega}{3}t+\theta]} + \frac{3}{4}AB^2 e^{j[\frac{\omega}{3}t-2\theta]}}{B e^{j[\frac{\omega}{3}t + \theta]}}$$

$$= \frac{3}{2}A^2 + \frac{3}{4}B^2 + \frac{3}{4}ABe^{-j3\theta} \tag{4}$$

Thus, the TSIDFs are both complex and dependent upon θ.

Case 2: Harmonically related input sinusoids ($\gamma = 3$):

This case is identical with Case 1. The only difference being that the roles of the higher- and the lower - frequency input terms are reversed.

Case 3: Non-harmonically-related input sinusoids:

The first and third terms of eq. (2) owe the only output terms of the same frequency as the input sinusoids. Terming $N_A(A,B)$ and $N_B(A,B)$, the TSIDFs for input frequencies ω and $\gamma\omega$, respectively gives

$$N_A(A,B) = \frac{\frac{3}{4}A^3 + \frac{3}{2}AB^2}{A} = \frac{3}{4}A^2 + \frac{3}{2}B^2 \tag{5}$$

$$N_B(A,B) = \frac{\frac{3}{2}A^2B + \frac{3}{4}B^3}{B} = \frac{3}{4}B^2 + \frac{3}{2}A^2 \tag{6}$$

The expressions are real, independent of γ and independent of θ.

● **PROBLEM** 15-11

For the nonlinear block whose input-output characteristic is defined by Fig. 1, determine the describing function. This characteristic approximates that of a saturating amplifier.

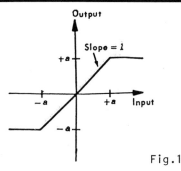

Fig.1

Solution: We shall assume a sinusoidal input of amplitude A. For $A \leq a$, the response is sinusoidal and linear, and for $A > a$, is nonsinusoidal and nonlinear. In case of $A > a$, we calculate the ratio of the fundamental output component obtained by Fourier methods to the input. The response to a large amplitude sinusoidal input is shown in Fig. 2. The peaks of the output are at ±a.

The Fourier components of output consist only of odd harmonic sine terms, due to the output symmetry. The harmonic coefficients are defined by

$$m_n = \frac{4}{\pi} \int_0^{\frac{\pi}{2}} y(t) \sin n\omega t \, d\omega t \tag{1}$$

n = 1, 3, 5, 7. . .

832

Substituting $y(t) = A \sin \omega t$ for $0 < t < t_1$ and $y(t) = a$ for $t_1 < t < \frac{\pi}{2\omega}$ we have

$$\frac{m_n}{A} = \frac{2}{\pi}\left[\frac{\sin(n-1)\omega t_1}{n-1} - \frac{\sin(n+1)\omega t_1}{n+1}\right] + \frac{4a}{n\pi A}\cos n\omega t_1 \quad (2)$$

t_1 is given by

$$t_1 = \frac{1}{\omega} \text{ arc } \sin \frac{a}{A} \quad (3)$$

Substituting we get

$$\frac{m_n}{A} = \frac{2}{n\pi}\left[\frac{\sin(n-1)\omega t_1}{n-1} + \frac{\sin(n+1)\omega t_1}{n+1}\right] \quad (4)$$

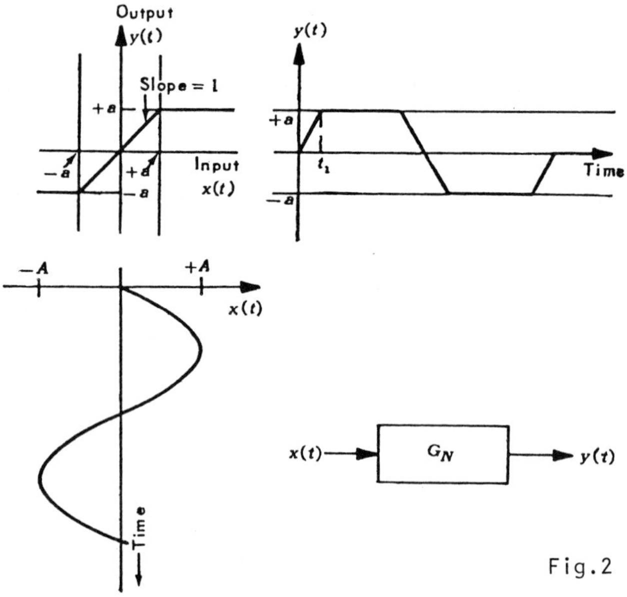

Fig.2

The describing function, corresponding to the fundamental output component, is given by

$$\frac{m_1}{A} = \frac{2}{\pi}(\omega t_1 + \frac{\sin 2\omega t_1}{2}) \quad (5)$$

and the higher harmonic components are

$$\frac{m_3}{A} = \frac{2}{3\pi}\left(\frac{\sin 2\omega t_1}{2} + \frac{\sin 4\omega t_1}{4}\right) \quad (6)$$

$$\frac{m_5}{A} = \frac{2}{5\pi}\left(\frac{\sin 4\omega t_1}{4} + \frac{\sin 6\omega t_1}{6}\right) \quad (7)$$

The plots are shown in Fig. 3.

The higher harmonics have relatively small amplitudes. For $\frac{A}{a} < 1$ (the linear region), $\frac{m_1}{A} = 1$ and the higher harmonics are zero. The describing function is a real

833

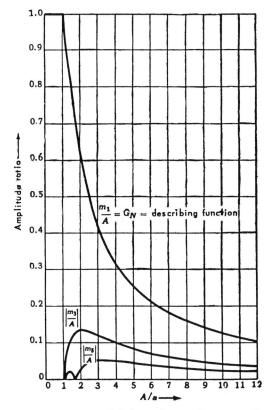

Fig.3

quantity with zero phase shift, dependent only on $\frac{A}{a}$ because the output fundamental is in phase with the input sinusoid.

● **PROBLEM** 15-12

Consider the threshold nonlinearity which occurs in devices which are unresponsive to small signals. In such devices, input must exceed a certain minimum level before an output response is developed. The threshold nonlinearity function is shown in Fig. 1. Determine its descrbing function.

Solution: To find the describing function we use an input of A sin ωt. The describing function is the fundamental of the output (that is the coefficient of sin ωt in the Fourier expansion of the output or the combination of the coefficients of sin ωt and cos ωt if the latter is non-zero). Fig. 1 shows the input and output. Because of the symmetry of the output, the Fourier series contains only odd sine terms. It is given by

$$y(t) = \sum_{n=1}^{\infty} m_n \sin n\omega t \qquad (1)$$

The Fourier coefficients are given by

834

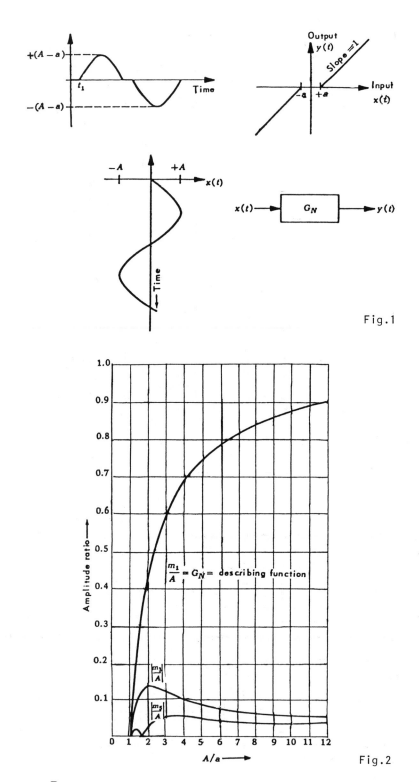

Fig.1

Fig.2

$$m_n = \frac{4}{\pi} \int_0^{\frac{\pi}{2}} y(t) \sin n\omega t \, d\omega t \qquad n = 1, 3, 5, 7 \qquad (2)$$

The function y(t) is defined in intervals

$$y(t) = \begin{cases} 0 & \text{for } 0 < \omega t < \omega t_1 \\ A \sin(\omega t - a) & \text{for } \omega t_1 < \omega t < \frac{\pi}{2} \end{cases} \quad (3)$$

where

$$\sin \omega t_1 = \frac{a}{A}$$

Substituting $y(t)$ into eq. (2)

$$\frac{m_n}{A} = -\frac{2}{n\pi}\left[\frac{\sin(n-1)\omega t_1}{n-1} + \frac{\sin(n+1)\omega t_1}{n+1}\right] \quad (4)$$

The describing function is

$$G_N = \frac{m_1}{A} = \frac{2}{\pi}\left(\frac{\pi}{2} - \omega t_1 - \frac{\sin 2\omega t_1}{2}\right) \quad (5)$$

Fig. 2 shows the describing function and the third and fifth harmonic components. Note that the magnitudes of the harmonics are the same as for saturation. But the fundamental output at low levels increases from zero at $\frac{A}{a} = 1$, so that output distorition for threshold level signals is high and the describing function approximation is not correct. The describing function for threshold is a real number, has zero phase shift and is independent of frequency.

● **PROBLEM** 15-13

An anolog computer is connected as shown to provide a highly sinusoidal amplitude-stablized oscillation at a fixed frequency.

In the drawing π indicates product. The describing equations are

$$\dot{x}_1 = x_1 g(r) - x_2, \quad g(r) = r^2(A - r)$$

$$\dot{x}_2 = x_2 g(r) + x_1, \quad r = \sqrt{x_1^2 + x_2^2} \quad (1)$$

We can write system (1) in terms of the polar coordinates r and θ.

$$\dot{r} = r^3(A - r)$$

$$\dot{\theta} = 1 \quad (2)$$

where

$$r = \sqrt{x_1^2 + x_2^2}$$

$$\theta = \tan^{-1}\left(\frac{x_2}{x_1}\right) \quad (3)$$

For the initial conditions

$$r(0) = A$$

$$\theta(0) = \psi \tag{4}$$

the solution of eq. (2) is

$$r^\circ(t) = A$$

$$\theta^\circ(t) = t + \psi \tag{5}$$

which corresponds to a sinusoid of amplitude A and radian frequency 1. We shall investigate the small changes in initial conditions. We assume that the variations

$$\delta r = r - r^\circ(t) \text{ and } \delta\theta = \theta - \theta^\circ(t)$$

from the original solution will also be small.

Linearize the system by considering small changes in the initial conditions.

Find the effect of a small error in amplitude and a small error in phase.

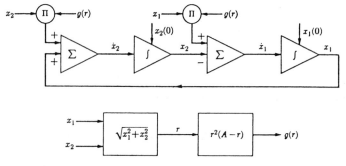

Solution: Differentiating eq. (2) with respect to r and θ we obtain

$$\begin{bmatrix} \dot{y}_r \\ \dot{y}_\theta \end{bmatrix} = \begin{bmatrix} 3r^2A - 4r^3 & 0 \\ 0 & 0 \end{bmatrix} \begin{bmatrix} y_r \\ y_\theta \end{bmatrix} \tag{6}$$

Substituting the generating eq. (5) into eq. (6) we have

$$\begin{bmatrix} \dot{y}_r \\ \dot{y}_\theta \end{bmatrix} = \begin{bmatrix} -A^3 & 0 \\ 0 & 0 \end{bmatrix} \begin{bmatrix} y_r \\ y_\theta \end{bmatrix} \tag{7}$$

$$y_r(0) = \delta r(0)$$

$$y_\theta(0) = \delta\theta(0)$$

This is the equation of first variation evaluated about

the solution of interest. From eq. (7) we determine the effect of a small error in initial amplitude. Let

$$\delta r(0) = \Delta, \quad \delta\theta(0) = 0,$$

thus

$$\delta r(t) \approx y_r(t) = \Delta\epsilon^{-A^3 t} \qquad (8)$$

From eq. (8) we see that the small errors in amplitude decay (approximately) exponentially with time constant A^{-3}, the phase θ is not effected. Eq. (7) indicates that $\dot{y}_\theta = 0$, and therefore $\delta\theta(t) = \delta\theta(0)$. This makes sense, since the initial phase is not part of the system, whereas the amplitude is. That is, A is wired in, but ψ is not. Eq. (5) says that $\theta(t) = t + \theta(0)$, thus it is expected that $\theta(t) + \delta\theta = t + \theta(0) + \delta\theta(0)$.

● **PROBLEM** 15-14

For the system shown in Fig. 1, determine qualitatively the character of the harmonically forced response.

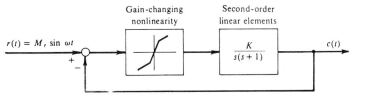

Fig.1

Solution: Note that for the system nonlinearity, the DF is non-phase-shifting. We have $L_1(s) = L_3(s) = 1$ and

$$L_2(s) = \frac{K}{s(s+1)} \qquad (1)$$

thus

$$\rho_1 = 1 \qquad \theta_1 = 0 \qquad (2)$$

$$\rho_2 = \frac{K}{\omega\sqrt{1 + \omega^2}} \qquad \theta_2 = -\frac{\pi}{2} - \tan^{-1}\omega$$

From Fig. 2 we can draw a group of ellipses corresponding to various input frequencies. For a single value of M_r, the result of this construction is shown in Fig. 3. All ellipses pass through the point $(M_r, 0)$, because $\rho_1 = 1$.

Note that $A_{\rho N}$ is the amplitude of the nonlinearity output fundamental. For increasing A, the plot of $A_{\rho N}$ is like a gain-changing function. The rough plot of $A_{\rho N}(A)$ can be made from the limiting results:

$$A_{\rho N} \to m_1 A \qquad \text{as } A \to 0$$

$$A_{\rho N} \to m_2 A \qquad \text{as } A \to \infty \qquad (3)$$

838

The shape of $A_{\rho N}$ resembles the nonlinear characteristic itself with the sharp corners rounded off. We find the magnitude of ρ_N on a point-by-point basis. At $\omega = \omega_5$, we get $A = A_5$ and $\rho_N(A_5)$ is known. For some values of ω, this process is unique, other values yield multivalued solutions, see the multiple intersections in Fig. 3. Amplitude and phase plots of $\frac{C}{R}(j\omega)$ can be built, utilizing all values of ρ_N found graphically. For the particular M_r of Fig. 3 the solution is shown in Fig. 4. Note that these plots are dicontinuous, with jumps occuring at ω_3 and ω_4.

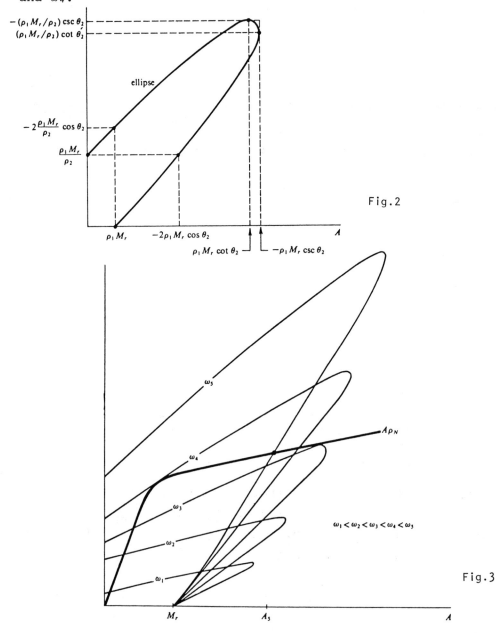

Fig.2

Fig.3

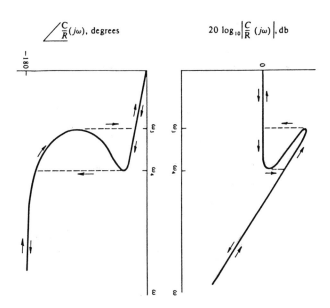

$$\underline{/\dfrac{C}{R}}(j\omega),\ \text{degrees} \qquad\qquad 20\log_{10}\left|\dfrac{C}{R}(j\omega)\right|,\ \text{db}$$

Fig.4

• **PROBLEM** 15-15

Fig. 1 shows the network with the linear inductor, with inductance L, the linear capacitor with capacitance C and the resistive branches 1 and 2 with the branch relations

$$i_1 = f(e_1),\quad e_2 = V + Ri_2 \qquad\qquad (1)$$

Fig. 2 shows the function $f(e_1)$.

The normal-form equations for this network may be written in the form

$$\dot{q} = -f(e_1) - i_2,\qquad e_1 = \frac{q}{c} \qquad\qquad (2)$$

$$\dot{\lambda} = e_1 - Ri_2 - V,\qquad i_2 = \frac{\lambda}{L}$$

The construction of Fig. 2 locates graphically the singular point of eq. (2).

For the selected value of R, there is only one singular point.

We can use this arrangement to measure the complete tunnel-diode characteristic. The curve $f(e_1)$ can be plotted by varying V and waiting for the steady state. But for this, the singular point must be asymptotically stable, otherwise the system never comes to rest.

Let G_0 be the incremental conductance of the negative-conductance characteristic evaluated at the singular point e^0. Determine constraints on R, L, C and G_0 so that the system will be asymptotically stable.

Solution: We write eq. (2) as

$$\dot{q} = -\frac{G_0}{C} q - \frac{1}{L}\lambda$$

$$\dot{\lambda} = \frac{1}{C} q - \frac{R}{L}\lambda - V \tag{3}$$

Then the variational equation is

$$\begin{bmatrix} \delta\dot{q} \\ \delta\dot{\lambda} \end{bmatrix} = \begin{bmatrix} \frac{\partial\dot{q}}{\partial q} & \frac{\partial\dot{q}}{\partial\lambda} \\ \frac{\partial\dot{\lambda}}{\partial q} & \frac{\partial\dot{\lambda}}{\partial\lambda} \end{bmatrix} \begin{bmatrix} \delta q \\ \delta\lambda \end{bmatrix} + 0\left(\left\| \begin{matrix} \delta q \\ \delta\lambda \end{matrix} \right\|^2 \right)$$

$$\approx \begin{bmatrix} -\frac{G_0}{C} & -\frac{1}{L} \\ \frac{1}{C} & -\frac{R}{L} \end{bmatrix} \begin{bmatrix} \delta q \\ \delta\lambda \end{bmatrix} \tag{4}$$

If $K(t) > 0$, the notation $f = 0(K^n)$ means that there exists a finite M such that $\lim_{t \to \infty} |f(t)| \leq MK^n$. That is, f is "on the order of K^n". In eq. (4), we disregard terms proportional to the squares of errors. The characteristic equation of this system is

$$s^2 + \left(\frac{G_0}{C} + \frac{R}{L}\right)s + \frac{1}{LC}(1 + RG_0) = 0 \tag{5}$$

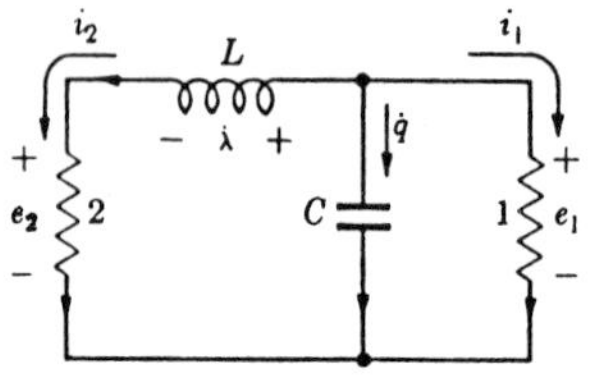

Fig.1 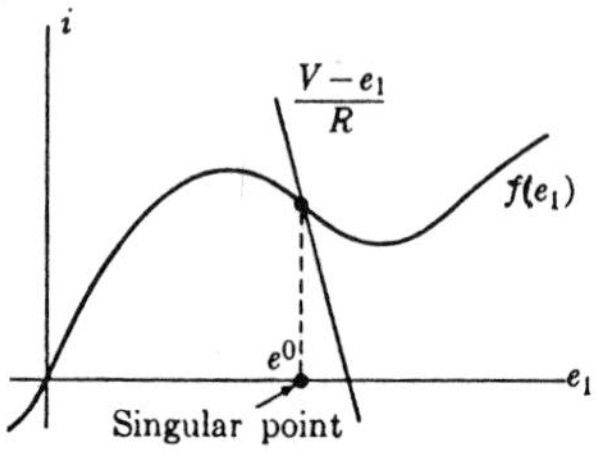   Fig.2

When the roots of eq. (5) are in the left half-plane, the system is asymptotically stable. Thus a sufficient condition for asymptotic stability is

$$\left(\frac{G_0}{C} + \frac{R}{L}\right) > 0, \quad (1 + RG_0) > 0 \tag{6}$$

From the construction of Fig. 2 we see that $|RG_0| < 1$, thus the second condition is fulfilled whenever there is only one singular point. From the first condition we see that stability depends upon the ratio $\frac{L}{C}$. This method of measurement is not feasible when the ratio is too large and the two conditions may be contradictory.

From eq. (6) we find the necessary condition for stability of this configuration

$$\frac{L}{C} \leq \frac{1}{G_0^2} \tag{7}$$

Fig. 1 shows a network used to amplify an impressed signal. A signal source, represented by the current source $I \sin \omega_0 t$ together with a parallel interval source conductance G_s is driving a load whose conductance is G_L. (See Fig. 2)

The source transfers the maximum average power to the load when $G_L = G_s$, and

$$P_{max} = \frac{I^2}{8G_s} \tag{1}$$

Connecting a parametric amplifier, we obtain power amplification. A variable capacitor in the network has been separated into a fixed part and a variable part. The network parameters are

$$C = C_0 + C_1 \sin 2\omega_0 t \qquad 0 < C_1 < C_0 \tag{2}$$

$$\omega_0 = \frac{1}{\sqrt{LC_0}} \tag{3}$$

When the resonance of the combination G_s, G_L, L, C_0 is sufficiently sharp, the network acts as a short-circuit for all frequency components of $e(t)$ other than ω_0. Thus, we can represent $e(t)$ as

$$e(t) = E \sin(\omega_0 t + \theta) \tag{4}$$

in the steady state.

What is the resulting power amplification?

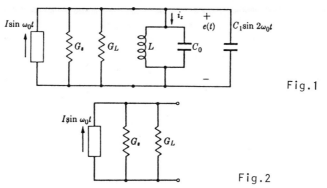

Fig.1

Fig.2

Solution: The Kirchhoff's current law for the network is

$$I \sin \omega_0 t = (G_s + G_L)e(t) + \frac{d}{dt}(eC_1 \sin 2\omega_0 t) + i_z \tag{5}$$

or

$$I \sin \omega_0 t = (G_s + G_L)E \sin(\omega_0 t + \theta) + \frac{C_1 E \omega_0}{2} \tag{6}$$

$$\times\ [3\ \sin(3\omega_0 t + \theta) - \sin(\omega_0 t - \theta)] + i_z$$

where the current through the variable capacitor is $\frac{d}{dt}(ce)$ and not $c\frac{de}{dt}$. Eq. (6) must be satisfied for all t, thus

$$i_z = \frac{-3C_1 E\omega_0}{2}\ \sin\ (3\omega_0 t + \theta) \tag{7}$$

Setting $\theta = 0$ we compute the amplitude of $e(t)$.

$$E = \frac{I}{G_s + G_L - \frac{1}{2}C_1\omega_0} \tag{8}$$

We conclude from eq. (8) that the parametric amplifier acts like a negative conductance $-G_a$ of value

$$-G_a = -\frac{1}{2}C_1\omega_0 \tag{9}$$

The negative conductance increases the maximum power available to the load to

$$P_{max} = \frac{I^2}{8(G_s - G_a)} \tag{10}$$

Comparing eqs. (10) and (1) we see that the parametric amplifier achieves a power amplification K of

$$K = \frac{G_s}{G_s - G_a} \tag{11}$$

For $G_a \to G_s$, we have $K \to \infty$.

● **PROBLEM** 15-17

Consider a system of an optical-beam-riding antitank missile that has its position control loop closed by a human operator. The on-off rate control loop uses a 400-cps carrier rate gyro feedback signal and associated amplification, demodulation and compensation as shown in the figure.

The transfer function characterizing the airframe dynamics from δ to ψ is

$$\frac{\psi}{\delta}(s) = \frac{50(s + 1)}{s(s^2 + 2s + 25)} \tag{1}$$

Find the amplitude of missile control-surface deflection due to a 1-volt demodulator output noise at the carrier frequency. Assume that the in-flight missile trajectory oscillates about the optical tracker-target line of sight, producing a signal $e = 25 \sin 6t$ volts.

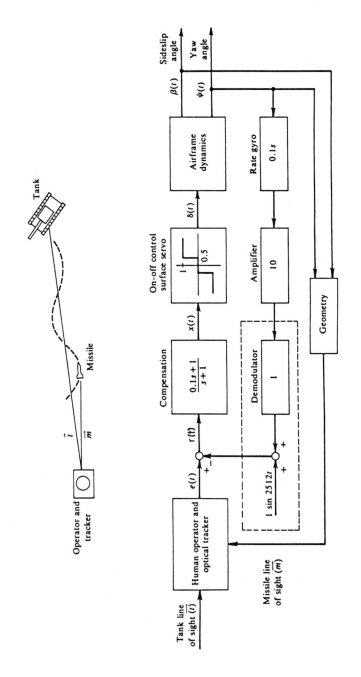

<u>Solution</u>   We model the nonlinearity input

$$x \approx A \sin 6t + B \sin 2,512t \tag{2}$$

where the $\omega = 6$ component is due to the trajectory oscillation and the $\omega = 2512 = 2\pi(400)$ component comes from the system.

We compute the amplitude A of the low-frequency nonlinearity input component. That is, we want

$$\left|\frac{x}{e}(s = j6)\right|.$$

The feedback gain is

$$H(s) = (1)(10)(0.1s) = s \tag{3}$$

disregarding the high frequency component. Also

$$r(s) = e(s) - H(s)\psi(s)$$

or

$$r(s) = e(s) - H(s)\frac{\psi}{\delta}(s)\frac{\delta}{x}(s)\frac{x}{r}(s)r(s) \tag{4}$$

We solve eq. (4) for $r(s)$, thus

$$r(s) = \frac{e(s)}{1 + H(s)\frac{\psi}{\delta}(s)\frac{\delta}{x}(s)\frac{x}{r}(s)} \tag{5}$$

and then

$$\frac{x}{e}(s) = \frac{x}{r}(s)\frac{r}{e}(s) = \frac{\frac{x}{r}(s)}{1 + H(s)\frac{\psi}{\delta}(s)\frac{\delta}{x}(s)\frac{x}{r}(s)} \tag{6}$$

Finally, since $e = 25 \sin 6t$, we multiply eq. (6) by 25 in order to get the amplitude of the $\sin 6t$ term of x. From the figure we substitute the transfer functions into eq. (6). This yields

$$A = \left|\frac{\frac{0.1s + 1}{s + 1}25}{1 + \frac{0.1s + 1}{s + 1}N(A)\frac{50(s + 1)}{s^2 + 2s + 25}}\right|_{s=j6} \tag{7}$$

where the approximation to $N_A(A,B)$ is the DF for the on-off element with dead zone

$$\tilde{N}_A(A,B) = \frac{4}{\pi A}\sqrt{1 - \left(\frac{0.5}{A}\right)^2} \tag{8}$$

Solving, we find $A = 2.5$.

We calculate the incremental-input describing function gain in approximation to

$N_B(A,B)$.

$$\tilde{N}_B(A,B) = \frac{4}{\pi A}\sqrt{1 - \left(\frac{0.5}{A}\right)^2} + \frac{A}{2}\frac{d}{dA}\left[\frac{4}{\pi A}\sqrt{1 - \left(\frac{0.5}{A}\right)^2}\right]$$

$$= \frac{2}{\pi A}\sqrt{1 - \left(\frac{0.5}{A}\right)^2} + \frac{0.5}{\pi A^3\sqrt{1 - \left(\frac{0.5}{A}\right)^2}} \qquad (9)$$

Approximating, we find

$$\tilde{N}_B(A,B)\bigg|_{A=2.5} = 0.26 \qquad (10)$$

Thus the system seen by the sinusoidal noise voltage has been linearized. The required control-surface noise-deflection amplitude δ_n is

$$\delta_n \approx 1 \times \left|\frac{j(0.1)(2512) + 1}{j2512 + 1}\right| \times 0.26 = 0.026 \text{ rad} \qquad (11)$$

The above result is interesting in view of the fact that δ_n would be zero if e were zero.

● **PROBLEM** 15-18

The figure shows the three-node network. The node voltages, branch voltages, branch currents and resistor constraints are shown. Find equations for the node voltages e_1 and e_2 in terms of the system parameters R and I. If I = 129A and R = 1Ω, find the branch voltages and element currents.

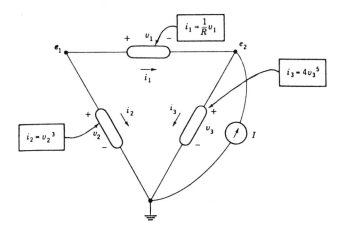

Solution: Step 1 of the mode-voltage method yields the following relations, which satisfy KVL:

$$v_1 = e_1 - e_2 \tag{1}$$

$$v_2 = e_1 \tag{2}$$

$$v_3 = e_2 \tag{3}$$

Step 2 yields

$$i_1 = \frac{1}{R}v_1 = \frac{1}{R}(e_1 - e_2) \tag{4}$$

$$i_2 = v_2^3 = e_1^3 \tag{5}$$

$$i_3 = 4v_3^5 = 4e_2^5 \tag{6}$$

All the element constraints, together with the current I of the current source, are satisfied.

Step 3 leads to the following KCL relations:

$$-i_1 - i_2 = 0 \tag{7}$$

$$i_1 - i_3 + I = 0 \tag{8}$$

We can rewrite them in the form

$$-\frac{1}{R}(e_1 - e_2) - e_1^3 = 0 \tag{9}$$

$$-\frac{1}{R}(e_1 - e_2) - 4e_2^5 + I = 0 \tag{10}$$

The obtained results relate the two unknown node voltages to the known current source value I, taking into account KCL, KVL and the element constraints.

For $I = 129$ and $R = 1$ we obtain

$$e_1 - e_2 + e_1^3 = 0 \tag{11}$$

$$e_1 - e_2 - 4e_2^5 = -129 \tag{12}$$

Solving eqs. (11) and (12) we get

$$e_1 = 1, \; e_2 = 2$$

The branch voltages are

$$v_1 = 1 - 2 = -1$$

$$v_2 = 1$$

$$v_3 = 2$$

and the element currents

$$i_1 = \frac{-1}{1} = -1$$

$$i_2 = (1)^3 = 1$$

$$i_3 = 4(2^5) = 128$$

Fig. 1 shows a simple network which is the key to the conversion of alternating to direct current. The source waveform is given by

$$v_s(t) = 100 \sin 2\pi t \tag{1}$$

and is shown in Fig. 2.

Fig. 3 shows the diode characteristic with a forward conductance $G_f = 1V$ and a back conductance $G_b = 0.01V$.
Find the current $i(t)$ for this network.

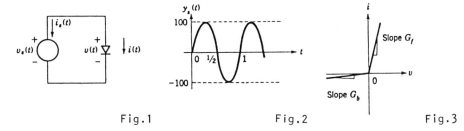

Fig.1   Fig.2   Fig.3

<u>Solution</u>: The topological constraints (KVL and KCL) and the element constraints (diode and source) are

$$V_s(t) = V(t) \tag{2}$$

$$i_s(t) = -i(t) \tag{3}$$

$$i(t) = \begin{cases} G_f V(t) & V \geq 0 \\ G_b V(t) & V < 0 \end{cases} \tag{4}$$

We have for $V_s(t) \geq 0$

$$i(t) = v(t) = v_s(t) = 100 \sin 2\pi t \tag{5}$$

for $V_s(t) < 0$

$$i(t) = 0.01 v(t) = 0.01 v_s(t) = \sin 2\pi t \tag{6}$$

The answer is:

For $n \leq t \leq n + \frac{1}{2}$

$$i(t) = 100 \sin 2\pi t \tag{7}$$

for $n + \frac{1}{2} \leq t \leq n + 1$

$$i(t) = \sin 2\pi t$$

where n = 0, 1, 2, . . .

The results are shown in Fig. 4.

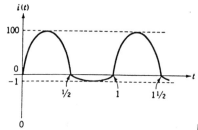

Fig.4

We conclude that the alternating source voltage which is equally positive and negative has given rise, through the diode, to a current which is almost completely positive. As the diode approaches the ideal characteristic, i.e., the ratio $\dfrac{G_f}{G_b}$ increases, the current waveform of Fig. 4 approaches half of the time a positive sinusoid and the rest of the time the value zero.

Fig. 1 shows the sampled system with a limiter. The linear part includes a zero-order hold and an integrator. Test the stability of a limit cycle mode of period $T = 4T_s$, where $y(mT_s) = 1, a, -1, -a$ for $m = 0, 1, 2, 3,$ $0 \le a \le 1.$

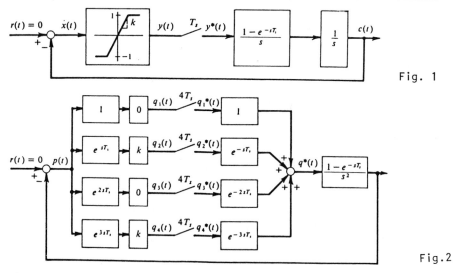

Fig. 1

Fig.2

Solution: We shall give two solutions to this problem.

First Solution: We shall consider the full form of the equivalent sampling system for perturbation analysis as shown in Fig. 2. We evaluate the slope of the non-linearity at the sampling points, thus

849

$$\frac{dy}{dx}[x(mT_s)] = 0, \ k, \ 0, \ k.$$

The Laplace transform of $q_i(t)$ is $Q_i(s)$. We shall find the characteristic equation of this system.

$$Q_2 = -Q_2^* \frac{1 - e^{-sT_s}}{s^2}k - Q_4^* \frac{1 - e^{-sT_s}}{s^2}ke^{-2sT_s} \tag{1}$$

$$Q_4 = -Q_2^* \frac{1 - e^{-sT_s}}{s^2}ke^{2sT_s} - Q_4^* \frac{1 - e^{-sT_s}}{s^2}k \tag{}$$

Let us denote

$$F_1^* = k\left(\frac{1 - e^{-\frac{ST}{4}}}{s^2}\right)^*$$

$$F_2^* = k\left(\frac{1 - e^{-\frac{ST}{4}}}{s^2}e^{-\frac{ST}{2}}\right)^* \tag{2}$$

$$F_3^* = k\left(\frac{1 - e^{-\frac{ST}{4}}}{s^2}e^{\frac{ST}{2}}\right)^*$$

We obtain

$$(1 + F_1^*)Q_2^* + F_2^*Q_4^* = 0$$
$$F_3^*Q_2^* + (1 + F_1^*)Q_4^* = 0 \tag{3}$$

where the z transform with respect to the sampling period $T = 4T_s$ is indicated by the star. We find F_1^*, F_2^*, F_3^*.

$$F_1^* = \frac{kT_s}{z - 1} \tag{4}$$

$$F_2^* = \frac{kT_s}{z - 1} \tag{5}$$

$$F_3^* = \frac{kT_s z}{z - 1} \tag{6}$$

Equating to zero the determinant of the coefficient matrix is eq. (3), we obtain the characteristic equation

$$(1 + F_1^*)^2 - F_2^*F_3^* = 0$$

Using eqs. (4), (5), (6) we obtain

$$z = (1 - kT_s)^2 \tag{7}$$

For kT_s in the range

$$0 < kT_s < 2$$

the closed-loop pole takes values between 0 and 1. This is the range of stability for the linearly perturbed system.

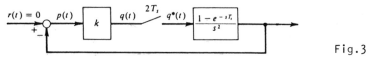

Fig.3

Second solution: Since the limit cycle under consideration is an unbiased even-period mode and the nonlinearity is odd, we can simplify the analysis of this problem. We need only the first half of the parallel paths (see Fig. 2) if the period of the samplers is changed to $2T_s$. We are left with only one path with a nonzero gain, which is the second path with a one-unit predictor and a one-unit delay. In case of a single path, a predictor and corresponding delay have no effect on system stability, therefore they can be ignored. Fig. 3 shows the simplified form of the perturbed system. This is a single-path linear sampled-data system.

This system is formed from the original nonlinear system by replacing the limiter by its gain in the linear region and changing the sampling period from T_s to $2T_s$. The closed-loop root of this system is located at

$$Z = 1 - kT_s \qquad (8)$$

and the stable range is

$$0 < kT_s < 2.$$

● **PROBLEM** 15-21

The drawing shows a satellite of mass m moving in the gravitational field of a body of mass M. The larger body is fixed, since $M \gg m$. Since the orbit remains in a plane, two coordinates r and θ and their derivatives $\dot{r}$ and $\dot{\theta}$ define the state of the system.

The kinetic coenergy and energy are equal for the system, and are given by

$$I'(r, \dot{r}, \dot{\theta}) = I(r, \dot{r}, \dot{\theta}) = \tfrac{1}{2}m(\dot{r}^2 + r^2\dot{\theta}^2) \qquad (1)$$

The potential energy is

$$U(r) = \frac{-\gamma Mm}{r} \qquad (2)$$

γ is a universal gravitational constant.

Write Lagrange's equations for the system. Eliminate time from the equations, and find the quation of the orbit. What is its shape?

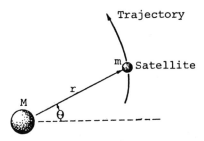

Solution: Lagrange's equation is

$$\frac{d}{dt}\left(\frac{\partial L}{\partial \dot{x}}\right) - \frac{\partial L}{\partial x} = -\frac{\partial R}{\partial \dot{x}} \tag{3}$$

where L is the Lagrangian

$$L = I' - U \tag{4}$$

and R is a dissipation function for the system. We have two state variables r and θ, each yielding an equation of the form of eq. (3). No energy is lost, so R = 0. The Lagrangian is

$$L = I' - U = \frac{1}{2}m(\dot{r}^2 + r^2\dot{\theta}^2) + \frac{\gamma Mm}{r} \tag{5}$$

Then

$$\frac{\partial L}{\partial \dot{r}} = m\dot{r}, \quad \frac{\partial L}{\partial r} = mr\dot{\theta}^2 - \frac{\gamma Mm}{r^2},$$

$$\frac{d}{dt}\frac{\partial L}{\partial \dot{r}} = m\ddot{r}, \quad \frac{\partial L}{\partial \dot{\theta}} = mr^2\dot{\theta}, \quad \frac{\partial L}{\partial \theta} = 0 \tag{6}$$

Substituting into eq. (3) we find the Lagrange's equations for the system.

$$m\ddot{r} - mr\dot{\theta}^2 + \frac{\gamma Mm}{r^2} = 0 \tag{7}$$

$$\frac{d}{dt}(mr^2\dot{\theta}) = 0 \tag{8}$$

Eq. (8) expresses the conservation of angular momentum, since $mr^2\dot{\theta}$ is the angular momentum of the system. From eqs. (1) and (2) we find the energy integral for the system

$$H(r, \dot{r}, \dot{\theta}) = I + U$$

$$= \frac{1}{2}m(\dot{r}^2 + r^2\dot{\theta}^2) - \frac{\gamma Mm}{r} = h \tag{9}$$

To find the orbit we shall eliminate t from the equations and solve for r(θ).

From eq. (8) we get

$$dt = \frac{mr^2}{p}d\theta \tag{10}$$

where P is the angular momentum for the system, P = const. Substituting eq. (10) into eq. (9) we find

$$\frac{1}{2}m\left[\left(\frac{dr}{d\theta}\right)^2\frac{P^2}{m^2r^4} + \frac{P^2}{m^2r^2}\right] - \frac{\gamma Mm}{r} = h \tag{11}$$

or, separating the variables

$$\frac{dr}{r\left[\frac{2mh}{P^2}r^2 + \frac{2\gamma Mm^2}{P^2}r - 1\right]^{\frac{1}{2}}} = d\theta \tag{12}$$

Integrating, we transform eq. (12) into

$$r = \frac{K}{1 - \varepsilon \cos(\theta - \theta_1)} \tag{13}$$

The parameters in eq. (13) can be expressed in terms of the initial conditions r_0, θ_0, $\dot{r}_0$, $\dot{\theta}_0$.

$$K = \frac{P^2}{\gamma Mm^2} \tag{14}$$

$$\varepsilon = 1 + \frac{2hP^2}{\gamma^2 M^2 m^3} \tag{15}$$

$$\theta_1 = \theta_0 - \cos^{-1}(1 - \frac{K}{\varepsilon}) \tag{16}$$

$$h = \frac{1}{2}m(\dot{r}_0 + r_0^2\dot{\theta}_0^2) - \frac{\gamma Mm}{r_0} \tag{17}$$

$$P = mr_0^2\dot{\theta}_0 \tag{18}$$

Note that eq. (13) is the general equation for a conic section in polar form. No matter what initial energy and angular momentum has the satellite, its orbit must be either a straight line, a circle, ellipse, parabola or hyperbola. The shape of the orbit is determined by the value of ε.

$\varepsilon = 0$ circle

$0 < \varepsilon < 1$ ellipse

$\varepsilon = 1$ parabola

$\varepsilon > 1$ hyperbola

● **PROBLEM** 15-22

One of the problems, where the DIDF model input waveform arises, is the determination of the near-circular-orbit period of an earth satellite. The coupled radial and tangential force equations describing that motion are

$$\ddot{r} - r\dot{\psi}^2 = -\frac{\mu}{r^2} \tag{1}$$

$$r\ddot{\psi} + 2\dot{r}\dot{\psi} = 0 \tag{2}$$

<u>Solution</u>: We shall rewrite eq. (2) in the form

$$\left(\frac{1}{r}\right)\frac{d}{dt}(r^2\dot{\psi}) = 0 \tag{3}$$

and we can write

$$r^2\dot{\psi} = h \tag{4}$$

r and ψ are shown in the figure and h is the constant specific angular momentum of the orbit.

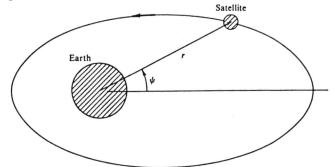

Using eq. (3), we eliminate $\dot{\psi}$ from eq. (1),

$$\ddot{r} - \frac{h^2}{r^3} = -\frac{\mu}{r^2} \tag{5}$$

where μ is a constant defining the specific gravitational force. We shall try to find an approximate solution to this equation. Let

$$r = R + \delta_R \cos \psi, \ \psi = \omega t \tag{6}$$

where R and δ_R are constants. Substituting eq. (6) into eq. (5) gives

$$-\omega^2\delta_R \cos\psi - \frac{h^2}{R^3}\left(1 + \frac{\delta_R}{R}\cos\psi\right)^{-3}$$

$$= -\frac{\mu}{R^2}\left(1 + \frac{\delta R}{R}\cos\psi\right)^{-2} \tag{7}$$

Since the orbit is near-circular we have $\frac{\delta R}{R} \ll 1$. Therefore, we can approximate the nonlinear terms in eq. (7) by appropriate first-order expansions. Thus

$$-\omega^2\delta_R\cos\psi - \frac{h^2}{R^3}\left(1 - \frac{3\delta_R}{R}\cos\psi\right) \approx \frac{-\mu}{R^2}\left(1 - \frac{2\delta_R}{R}\cos\psi\right) \tag{8}$$

The integral part of the DF analysis is the procedure of balancing harmonics. We obtain

$$R = \frac{h^2}{\mu}, \ \omega^2 = \frac{3h^2}{R^4} - \frac{2\mu}{R^3} \tag{9}$$

Thus, for the orbit period T, we find

$$T = \frac{2\pi}{\omega} = 2\pi\frac{h^3}{\mu^2} \tag{10}$$

The exact result, where orbit eccentricity $= \frac{\delta R}{R}$ is

$$T_{exact} = 2\pi\frac{h^3}{\mu^2}\left[1 - \left(\frac{\delta R}{R}\right)^2\right]^{-\frac{3}{2}} \tag{11}$$

Note that the harmonic linearization method is the easy way to obtain the approximate solution.

● **PROBLEM** 15-23

The set of N cities is given with the system of roads which connects every two cities. t_{ij} represents the time of travel along the connecting road from city i to city j. Assume that t_{ij} is not proportional to the distance between the cities and $t_{ij} \neq t_{ji}$. Starting from any of the cities, find the route to the Nth city which requires the minimum amount of time.

Solution: Since there are only a finite number of paths possible for the N cities, we can use the numerical method and compute all possible times. Unfortunately, for large N, it is practically impossible. We shall use the dynamic programming to solve this problem.

Let us define

$I^0(i, N) =$ the time required to travel from the ith city to the Nth city by an optimal path.

$I^0(i, N) = I^0_i$

where

$$I^0(N, N) = I^0_N = 0 \tag{1}$$

Note that, if we start at the ith city and go directly to the jth city, the remainder of the route, j to N must be chosen so as to minimize the remaining time, thus

$$I^0_i = \min_{j \neq i}\left\{t_{ij} + I^0_j\right\} \quad (i = 1, 2, \ldots N - 1) \tag{2}$$

In this dynamic programming recurrence relationship, we have I^0_i on both sides of the equation. That means that the trip from city j to N may involve passage through city i, but the path from city i is what the left-hand side of eq. (2) involves. Thus I^0_i is implied in I^0_j. To preclude a path back to itself, we can write

$$I^0_i = \min_{i < j}\left\{t_{ij} + I^0_j\right\}$$

The unknown function occurs on both sides of the recurrence relationship, thus it cannot be used in its explicit form.

We shall use the iterative procedure and rewrite the recurrence relationship in the form

$$I_i^{(k)} = \min_{j \neq i} [t_{ij} + I_j^{(k-1)}] \tag{3}$$

$$I_N^{(k)} = 0 \quad (i = 1, 2, \ldots N - 1)$$

where (k) denotes the interative step, and $I_i^{(0)}$ represents the initial estimates in the interation. The sequence $I_i^{(k)}$ will converge to a unique I_i^0 in monotonic fashion.

Let us take, as a first slection of the initial estimates

$$I_i^{(0)} = t_{iN}$$

$$I_N^{(0)} = 0 \quad \text{for} \quad i = 1, 2, \ldots N - 1$$

which represents the fact, that we go directly from city i to city N. Using eq. (3) we generate the optimal paths with one intermediate city between i and N, then two intermediate cities and so on. This method assumes an initial set of times which are not optimal, but are physically realizable, and then proceeds to improve these times. It yields, at each stage, a nonoptimal solution to the original problem. We may select, alternatively, as initial estimates the set of equations

$$I_i^{(0)} = \min_{i \neq j} t_{ij}, \quad I_N^{(0)} = 0, \quad i = 1, 2, \ldots N - 1$$

This method evaluates the best path of length k starting at city i and then improves the same by increasing k. Here, each step in the interation gives an optimal solution to a problem shorter than the original one.

● **PROBLEM** 15-24

Consider a system for inertial navigation. For simplicity, assume there is only one dimension of motion. A vehicle carries the system consisting of a platform with a mounted accelerometer and a gyroscope. The accelerometer is fixed rigidly to the platform. The gyroscope and platform are mounted in servo-driven gimbals in such a way that they may rotate with respect to each other and with respect to the vehicle, which moves in the θ-direction. The initial portion of the vehicle is $\theta = 0$. The gyroscope axis is aligned parallel to the z-axis and remains in that orientation through the action of servo-driven gimbals, which isolate the gyroscope from the vehicle's reference frame.

The accelerometer senses any linear acceleration in a direction parallel to the plane of the platform and delivers the data to the computer, which commands the platform servos. These servos should keep the platform aligned perpendicularly to the "local vertical," that is to the direction of the gravity vector. The acceleration is given by

$$a = (g - \dot{\theta}^2 R) \sin(\theta - \tilde{\theta}) + R\ddot{\theta} \cos (\theta - \tilde{\theta}) \qquad (1)$$

where

g = the gravitational acceleration

θ = angular position of vehicle

$\tilde{\theta}$ = angle between platform and x-axis

R = distance to the center of earth R is constant.

The perfect orientation of the platform is $\theta = \tilde{\theta}$, where $\tilde{\theta}$ is measurable in the vehicle and indicates the angular distance from the starting point. The platform servos produce an angular acceleration $\ddot{\theta}$ to keep the orientation of the platform as close as possible to $\theta = \tilde{\theta}$.

$$\ddot{\tilde{\theta}} = \frac{1}{R}[(g - \dot{\theta}^2 R)\sin(\theta - \tilde{\theta}) + R\ddot{\theta} \cos(\theta - \tilde{\theta})] \qquad (2)$$

The initial conditions are

$$\tilde{\theta}(0) = \theta(0), \quad \dot{\tilde{\theta}}(0) = \dot{\theta}(0) \qquad (3)$$

With initial conditions, the solution of eq. (2) is

$$\tilde{\theta}(t) = \theta(t) \qquad (4)$$

Thus, the system gives an exact indication of position. We shall derive the variational equations to determine the errors in $\tilde{\theta}$ due to various causes.

Find the errors resulting from initial misalignment, and from additive noise in $\ddot{\tilde{\theta}}$.

Solution: Let

$$\tilde{\theta} = \begin{bmatrix} \tilde{\theta}_1 \\ \tilde{\theta}_2 \end{bmatrix} = \begin{bmatrix} \tilde{\theta} \\ \dot{\tilde{\theta}} \end{bmatrix} \qquad (5)$$

The additive noise term is $pn(t)$. Eq. (2) can be written

$$\dot{\tilde{\theta}}_1 = \tilde{\theta}_2$$

$$\dot{\tilde{\theta}}_2 = \frac{1}{R}\Big\{(g - \dot{\theta}^2(t)R)\sin[\theta(t) - \tilde{\theta}_1]$$

$$+ R\ddot{\theta}(t)\cos[\theta(t) - \tilde{\theta}_1]\Big\} + pn(t). \qquad (6)$$

857

For the ideal performance, we have

$$\tilde{\theta}_1(0) = \theta(0)$$

$$\tilde{\theta}_2(0) = \dot{\theta}(0) \tag{7}$$

$$p = 0$$

and the solution of eq. (6) is

$$\tilde{\theta}^0(t) = \begin{bmatrix} \tilde{\theta}_1^0(t) \\ \tilde{\theta}_2^0(t) \end{bmatrix} = \begin{bmatrix} \theta(t) \\ \dot{\theta}(t) \end{bmatrix} \tag{8}$$

Eq. (8) is the generating solution for the variational equation. An error $\delta\theta(t)$ will occur, if there is error $\delta\theta(0)$ in the initial alignment or some additive disturbance $\delta p = p \neq 0$. Differentiate eq. (6) with respect to $\tilde{\theta}_1$ and $\tilde{\theta}_2$. Then

$$\frac{\partial}{\partial\tilde{\theta}_1}(\dot{\tilde{\theta}}_1) = 0, \quad \frac{\partial}{\partial\tilde{\theta}_2}(\dot{\tilde{\theta}}_1) = 1$$

$$\frac{\partial}{\partial\tilde{\theta}_1}(\dot{\tilde{\theta}}_2) = -(\frac{g}{R} - \dot{\theta}^2)\cos(\theta - \tilde{\theta}_1) + R\ddot{\theta}\sin(\theta - \tilde{\theta}_1) \tag{9}$$

$$\frac{\partial}{\partial\tilde{\theta}_2}(\dot{\tilde{\theta}}_2) = 0$$

Substituting eq. (8) into eq. (9) (i.e. $\theta = \tilde{\theta}_1$) we get

$$\dot{y} = \begin{bmatrix} 0 & 1 \\ -(\frac{g}{R} - \dot{\theta}^2) & 0 \end{bmatrix} y + \begin{bmatrix} 0 \\ 1 \end{bmatrix} pn(t), \quad y(0) = \delta\tilde{\theta}(0) \tag{10}$$

Eq. (10) is the variational equation for the system. Since $\dot{\theta}^2(t) \ll \frac{g}{R}$, we shall drop the term $\dot{\theta}^2$. Let us first examine the effect of a small initial error in alignment

$$\delta\tilde{\theta}(0) = \begin{bmatrix} \Delta \\ 0 \end{bmatrix} \tag{11}$$

Then, rewiritng eq. (10) in the Laplace domain

$$sy_1 - \Delta = y_2, \quad sy_2 = -\frac{g}{R}y_1 \tag{12}$$

Solving for y_1 and y_2 we get

$$y_1 = \Delta\frac{s}{s^2 + \frac{g}{R}}, \quad y_2 = -\Delta\frac{\frac{g}{R}}{s^2 + \frac{g}{R}} \tag{13}$$

Inverting gives

$$y(t) = \delta\tilde{\theta}(t) \approx \begin{bmatrix} \Delta\cos\omega t \\ -\Delta\omega\sin\omega t \end{bmatrix} \tag{14}$$

where $\omega = \sqrt{\dfrac{g}{R}}$.

The system oscillates with period $\dfrac{2\pi}{\omega} \approx 84.4$ min, if R is the radius of earth. To investigate the effects of noise alone, we set

$$\delta\tilde{\theta}(0) = 0, \; p \neq 0 \tag{15}$$

The transition matrix is

$$\phi(s) = [sI - A]^{-1}$$

$$= \begin{bmatrix} s & -1 \\ \dfrac{g}{R} & s \end{bmatrix}^{-1} = \begin{bmatrix} s & -1 \\ \omega^2 & s \end{bmatrix}^{-1} \tag{16}$$

$$= \dfrac{1}{s^2 + \omega^2} \begin{bmatrix} s & 1 \\ -\omega^2 & s \end{bmatrix} \tag{17}$$

Taking the inverse Laplace transforms

$$\phi(t) = \begin{bmatrix} \cos \omega t & \dfrac{1}{\omega} \sin \omega t \\ -\omega \sin \omega t & \cos \omega t \end{bmatrix} \tag{18}$$

and

$$\phi(t,s) = \phi(t - s) = \begin{bmatrix} \cos \omega(t - s) & \dfrac{1}{\omega} \sin \omega(t - s) \\ -\omega \sin \omega(t - s) & \cos \omega(t - s) \end{bmatrix} \tag{19}$$

Then with $\delta\tilde{\theta}(0) = 0$, the error due to the noise pn(t) is

$$\delta\tilde{\theta}(t) = \int_0^t \phi(t,s) B \, pn(s) \, ds$$

where

$$B = \begin{bmatrix} 0 \\ 1 \end{bmatrix}.$$

Thus

$$\delta\tilde{\theta}(t) \approx \dfrac{P}{\omega} \begin{bmatrix} \displaystyle\int_0^t \sin \omega(t - s) n(s) \, ds \\[2em] \omega \displaystyle\int_0^t \cos \omega(t - s) n(s) \, ds \end{bmatrix}$$

The variational system is not input-output stable and the bounded noise n(t) will lead to unbounded behavior of $\delta\tilde{\theta}(t)$.

DESCRIBING FUNCTIONS

Find a corresponding y(x), for the DF data $N(A) = \frac{4D}{\pi A}$

Use the approximation

$$y(A) \approx \frac{3A}{2} \sum_{i=0}^{\infty} \left(-\frac{1}{2} \right)^i N \left(\frac{A}{2^i} \right) \tag{1}$$

Solution: We must ensure that the series expansion for y(A) converges. Let us consider the replacement of eq. (1) by

$$y(A) \approx \frac{3A}{2} \lim_{\varepsilon \to 1} \sum_{i=0}^{\infty} \left(-\frac{\varepsilon}{2} \right)^i N \left(\frac{A}{2^i} \right) \tag{2}$$

Inserting the given DF data, we obtain

$$y(A) \approx \frac{3A}{2} \lim_{\varepsilon \to 1} \sum_{i=0}^{\infty} \left(-\frac{\varepsilon}{2} \right)^i \frac{4D}{\pi A} (2^i) \tag{3}$$

$$= \frac{6D}{\pi} \lim_{\varepsilon \to 1} \sum_{i=0}^{\infty} (-\varepsilon)^i = \frac{6D}{\pi} \lim_{\varepsilon \to 1} \frac{1}{1 + \varepsilon}$$

$$= \frac{3D}{\pi} \tag{4}$$

Note that without the special form chosen, the resulting infinite series would have been the alternating series 1 - 1 + 1 - 1 . . . etc. ε approaches 1 from below. For this odd single-valued nonlinearity, the characteristic is described by

$$y(x) = \begin{cases} \dfrac{3D}{\pi} & \text{for} \quad x \geq 0 \\[2ex] -\dfrac{3D}{\pi} & \text{for} \quad x < 0 \end{cases} \tag{5}$$

It is an ideal-relay characteristic. The inversion is thus in error by 4.5 percent, since the exact function is y = D.

For the piecewise-linear nonlinearity containing dead zone, a linear band, and saturation, derive the DF.

Solution: We can write y as

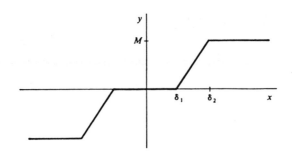

$$y = \begin{cases} 0 & \text{for} \quad 0 < x \leq \delta_1 \\ \dfrac{M}{\delta_2 - \delta_1}(x - \delta_1) & \delta_1 \leq x \leq \delta_2 \\ M & \delta_2 < x \end{cases}$$

Then, in terms of the saturation function $f(\frac{\delta}{A})$ we obtain

$$N(A) = \frac{M}{\delta_2 - \delta_1}[f(\frac{\delta_2}{A}) - f(\frac{\delta_1}{A})]$$

For the different ranges, we have

$$N(A) = 0 \qquad\qquad\qquad \text{for } 0 < A \leq \delta_1$$

$$N(A) = \frac{M}{\delta_2 - \delta_1}[1 - f(\frac{\delta_1}{A})] \qquad \text{for } \delta_1 < A \leq \delta_2$$

$$N(A) = \frac{M}{\delta_2 - \delta_1}[f(\frac{\delta_2}{A}) - f(\frac{\delta_1}{A})] \quad \text{for } \delta_2 < A.$$

● **PROBLEM 15-27**

The relay system is shown in the figure. Find the steady-state following error produced by the system when the input is

a) a step of magnitude R

b) a ramp of magnitude Rt.

Assume a first-order prefilter with time constant τ.

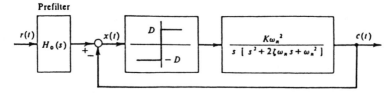

Solution:

a) Since the steady-rate prefilter output is constant, R, the following error of the overall system is equal to the constant-input following error of the limit cycling loop. We require a zero-average-value relay output for the limit cycling loop to be in steady state. This condition can be satisfied only when

861

B = 0. In such case there is no steady-state following error. It can be explained by the open-loop integration in L(s).

b) For the ramp input, the steady-state prefilter output is another ramp, $R(t - \tau)$, displaying a following error R_τ. For the limit cycling loop to be in steady state, an average relay output equal to $\frac{R}{K}$ is required. Using the exact expressions for N_B and N_A, we get

$$\frac{R}{K} = BN_B(A,B) = \frac{2D}{\pi}\sin^{-1}\frac{B}{A} \tag{1}$$

or

$$\frac{B}{A} = \sin\frac{\pi R}{2DK}$$

since $\omega_0 = \omega_n$

$$\frac{4D}{\pi A}\sqrt{1 - \left(\frac{B}{A}\right)^2}\,\frac{K}{2\zeta\omega_n} = 1 \tag{2}$$

Eq. (2) is the limit cycle magnitude condition. Solving eqs. (1) and (2) we get for the limit cycle amplitude

$$A = \frac{2DK}{\pi\zeta\omega_n}\cos\frac{\pi R}{2DK} \tag{3}$$

The input following error is

$$B = \frac{DK}{\pi\zeta\omega_n}\sin\frac{\pi R}{DK} \tag{4}$$

Thus the ramp following error of the overall system is

$$\text{Ramp following error} = R\tau + \frac{DK}{\pi\zeta\omega_n}\sin\frac{\pi R}{DK} \tag{5}$$

where $DK \geq R$. Note that it is impossible to command any faster response from this system.

● **PROBLEM 15-28**

Using the approximation

$$N(A) \approx \frac{2}{3A}[y(A) + y(\tfrac{A}{2})] \tag{1}$$

a) find the DF for the odd polynomial characteristic

$$y(x) = c_1 x + c_2 x|x| + c_3 x^3 + c_4 x^3|x| + c_5 x^5 \tag{2}$$

b) find the DF for an ideal-relay characteristic

$$y(x) = D\frac{x}{|x|} \tag{3}$$

Solution:

a) Substituting eq. (2) into eq. (1) we get

$$N(A) \approx c_1 + \frac{5}{6}c_2A + \frac{3}{4}c_3A^2 + \frac{17}{24}c_4A^3 + \frac{11}{16}c_5A^4 \quad (4)$$

The exact DF IS

$$N(A) = c_1 + \frac{8}{3\pi}c_2A + \frac{3}{4}c_3A^2 + \frac{32}{15\pi}c_4A^3 + \frac{5}{8}c_5A^4 \quad (5)$$

Note that the terms in C_1 and C_3 are exact. The terms in C_2 and C_4 are in error by less than 4 percent. The term in C_5 is in error by 10 percent.

b) Substituting eq. (3) into eq. (1) gives

$$N(A) \approx \frac{4D}{3A} \quad (6)$$

and the exact result is

$$N(A) = \frac{4D}{\pi A} \quad (7)$$

In using the approximate DF calculation only a 4.5 percent error occurs in the results.

● **PROBLEM** 15-29

The idealized representation of saturation is shown in Fig. 1. The response to a sinusoidal input is shown in Fig. 2. Find the describing function for the non-linearity.

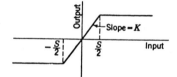

Fig.1

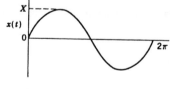

Fig.2

Solution: The input is $x = X \sin \omega t$ and for $X < \frac{S}{2}$, the output is directly proportional to the input. For $X > \frac{S}{2}$, the output reaches the value $\frac{KS}{2}$ and remains constant.

863

For $0 < \omega t < \sin^{-1} \dfrac{S}{2X}$ (1)

the output is proportional to the input. We evaluate the coefficients of the Fourier series from the integral

$$Y_k = \frac{4}{\pi} \int_0^{\frac{\pi}{2}} y(t) \sin k\omega t \; d(\omega t)$$

$$= \frac{4}{\pi} \int_0^{\beta} K \sin \omega t \sin k\omega t \; d(\omega t) + \frac{4}{\pi} \int_{\beta}^{\frac{\pi}{2}} \frac{KS}{2} \sin k\omega t \; d(\omega t) \quad (2)$$

where

$$\beta = \sin^{-1} \frac{S}{2X} \quad (3)$$

The fundamental is given by

$$Y_1 = KX \left(\frac{2\beta}{\pi} + \frac{\sin 2\beta}{\pi} \right) \quad (4)$$

and the nondimensionalized describing function is

$$\frac{N}{K} = \frac{Y_1}{KX} = \frac{2\beta}{\pi} + \frac{\sin 2\beta}{\pi} \quad (5)$$

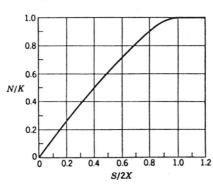

Fig.3

Fig. 3 shows the curve of this function plotted against $\dfrac{S}{2X}$. The value of the describing function is unity for $\dfrac{S}{2X} > 1$.

● **PROBLEM** 15-30

The idealized representation of a dead-zone characteristic is shown in Fig. 1. Find the describing function for the non-linearity.

Solution: The sinusoidal input is $x(t) = X \sin \omega t$. For $X < \dfrac{d}{2}$, there is no output. Fig. 2 shows the output for $X > \dfrac{d}{2}$. When x is smaller than $\dfrac{d}{2}$, the output is zero. For $0 < \omega t < \beta$, where

864

$$\beta = \sin^{-1} \frac{d}{2x} \tag{1}$$

the output is zero.

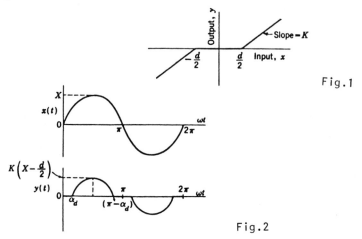

Fig.1

Fig.2

The Fourier series has only odd sine terms, because the output is an odd function and there is symmetry of the half cycles. Due to the symmetry of the output, the coefficients of the Fourier series can be evaluated by taking four times the integral over one-quarter of a cycle:

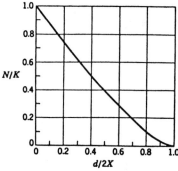

Fig.3

$$Y_k = \frac{4}{\pi} \int_0^{\frac{\pi}{2}} y(t) \sin k\omega t \, d(\omega t) \qquad k = 1, 3, 5, \ldots \tag{2}$$

For $x > \frac{d}{2}$, the output of the nonlinearity is

$$y(t) = K(X \sin \omega t - \frac{d}{2}) \tag{3}$$

and the coefficient of the fundamental is

$$Y_1 = \frac{4}{\pi} \int_\beta^{\frac{\pi}{2}} K(X \sin \omega t - \frac{d}{2}) \sin \omega t \, d(\omega t)$$

$$= KX(1 - \frac{2\beta}{\pi} - \frac{\sin 2\beta}{\pi}) \tag{4}$$

865

We use the nondimensionalized ratio of describing function divided by K to eliminate the slope K.

$$\frac{N}{K} = \frac{Y_1}{KX} = 1 - \frac{2\beta}{\pi} - \frac{\sin 2\beta}{\pi} \qquad (5)$$

Fig. 3 shows eq. (5) plotted as a function of $\frac{d}{2X}$.

For $\frac{d}{2X} > 1$, the output is zero and the value of the describing function is zero.

● **PROBLEM** 15-31

For the on-off nonlinearity with dead zone shown in Fig. 1, obtain the describing function.

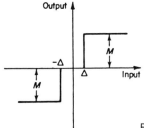

Fig.1

Solution: Fig. 2 shows the input and output waveforms for an element with the given nonlinearity.

For $0 \leq \omega t \leq \pi$, the output of the nonlinear element is

$$y(t) = 0 \quad \text{for } 0 < t < t_1$$

$$= M \quad \text{for } t_1 < t < \frac{\pi}{\omega} - t_1$$

$$= 0 \quad \text{for } \frac{\pi}{\omega} - t_1 < t < \pi$$

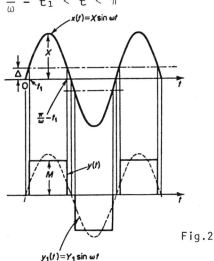

Fig.2

Since the output waveform is an odd function, the fundamental harmonic component of the output y(t) is given by

866

$$y_1(t) = Y_1 \sin \omega t$$

where

$$Y_1 = \frac{1}{\pi} \int_0^{2\pi} y(t) \sin \omega t \, d(\omega t)$$

$$= \frac{4}{\pi} \int_0^{\frac{\pi}{2}} y(t) \sin \omega t \, d(\omega t)$$

$$= \frac{4}{\pi} \int_{\omega t_1}^{\frac{\pi}{2}} M \sin \omega t \, d(\omega t) = \frac{4M}{\pi} \cos \omega t_1$$

We have

$$\sin \omega t_1 = \frac{\Delta}{X}$$

and

$$\cos \omega t_1 = \sqrt{1 - \left(\frac{\Delta}{X}\right)^2}$$

Thus

$$y_1(t) = \frac{4M}{\pi} \sqrt{1 - \left(\frac{\Delta}{X}\right)^2} \sin \omega t$$

The describing function for an on-off element with dead zone is

$$N = \frac{4M}{\pi X} \sqrt{1 - \left(\frac{\Delta}{X}\right)^2}$$

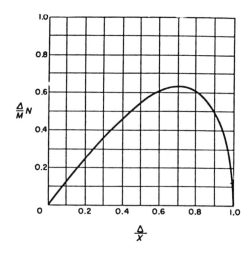

Fig.3

In this case the describing function is a function of the input amplitude X. The plot of $\frac{\Delta N}{M}$ vs. $\frac{\Delta}{X}$ is shown in Fig. 3.

Fig. 1 shows the on-off controller.

Find the DF for this controller.

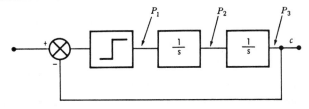

Fig.1

Solution: Let us assume that a sinusoidal signal of amplitude A enters the nonlinear element. Fig. 2 shows the output signal which is a square wave of amplitude u_{max}.

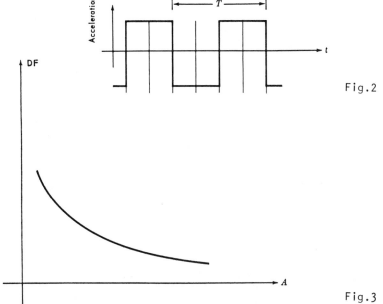

Fig.2

Fig.3

Note that the fundamental component of this wave is in phase with the input. It's amplitude is

$$\frac{4}{\pi}u_{max}.$$

We have

$$C_1 = \frac{4}{\pi}u_{max}$$

$$\phi_1 = 0$$

and obtain for DF

$$DF = \frac{4u_{max}}{\pi A}.$$

DF or the "equivalent harmonic gain" depends strongly upon the signal level A. In this example the output is the same for any magnitude of the input. The equivalent gain varies from 0 to ∞ as shown in Fig. 3.

The control system with saturation nonlinearity is shown in Fig. 1, where G(s) is a minimum phase transfer function. A plot of the $-\frac{1}{N}$ locus and the $G(j\omega)$ locus are shown in Fig. 2.

We see that the $-\frac{1}{N}$ locus starts from the -1 point on the negative real axis and moves to $-\infty$. N is a function only of the amplitude of the input $x(t) = X \sin \omega t$ and the $G(j\omega)$ locus is a function only of ω. How can the plot be used to determine the presence of limit cycles?

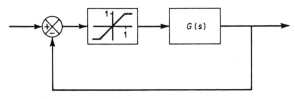

Fig.1

Solution: The intersection of the two loci corresponds to a stable limit cycle. From the $-\frac{1}{N}$ locus, we find the amplitude of the limit cycle as $X = X_1$. From the $G(j\omega)$ locus, we find the frequency $\omega = \omega_1$. Since there is no reference input, the output of this system at steady state shows a sustained oscillation with amplitude equal to X_1 and frequency equal to ω_1. When the gain of the transfer function, G(s) is decreased so that the $-\frac{1}{N}$ locus and the $G(j\omega)$ locus do not intersect, the system becomes stable.

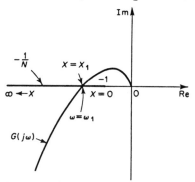

Fig.2

Any oscillation in the system output will die out. No sustained oscillation will exist at steady state, because the $-\frac{1}{N}$ locus is to the left of the $G(j\omega)$ locus, or the $G(j\omega)$ locus does not enclose the $-\frac{1}{N}$ locus.

Consider a system with a nonlinearity in the forward gain, given by a describing function G_N. If the linear portion is $G(j\omega)$, then the system is unstable if

$$1 + G(j\omega)G_N = 0$$

or

$$G(j\omega) = -\frac{1}{G_N} \qquad (1)$$

By means of a sketch in the complex plane, show qualitatively how this may be used to check stability of a saturating servomechanism. The transfer function is

$$G(j\omega) = \frac{Kv}{j\omega(1 + j\omega T_1)(1 + j\omega T_2)} \qquad (2)$$

The describing function for saturation is

$$G_N = \frac{2}{\pi}\left(\text{arc sin }\frac{a}{A} + \frac{a}{A}\sqrt{1 - \left(\frac{a}{A}\right)^2}\right) \qquad (3)$$

where A is input amplitude and a is the saturation level.

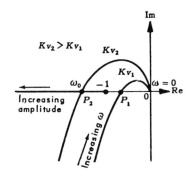

Solution: We shall start with plotting $G(j\omega)$ and $-\frac{1}{G_N}$ on polar coordinates. The effect of G_N varying from unity for $\frac{A}{a} \le 1$ to zero as $\frac{A}{a} = \infty$ is shown in the figure.

The plot of $-\frac{1}{G_N}$ varies from -1 at small signals to $-\infty$ for large signals. The transfer functions are sketched in for two values of Kv. At P_2 instability occurs, which is the intersection between the transfer function characteristic and the negative reciprocal of the describing function.

The amplitude of the resultant oscillation may be determined from the value of $-\frac{1}{G_N}$ at P_2.

The oscillation occurs at the value of $\omega = \omega_0$ of $G(j\omega)$ at the point of intersection.

For the relay control system shown in Fig. 1, find the DIDF linearized equivalent system. The prefilter has been chosen such that for all expected r(t), the amplitude-ratio condition at x(t) is satisfied.

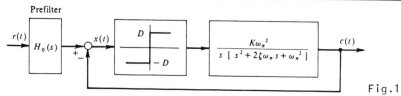

Fig.1

Solution: Assuming the relay to act as a non-phase-shifting gain, we derive the uncalibrated root locus of the limit cycling loop, as shown in Fig. 2. The point at which the locus crosses the jω axis determines the limit cycle frequency.

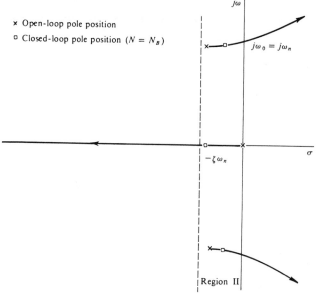

Fig.2

From the characteristic equation, using the DF methods, we determine the limit cycle amplitude and frequency.

$$s^3 + 2\zeta\omega_n s^2 + \omega_n^2 s + N_A K \omega_n^2 = 0 \tag{1}$$

and

$$\omega_n = \omega_0 \tag{2}$$

$$A = \frac{2DK}{\pi\zeta\omega_n} \tag{3}$$

The limit cycle DIDF has been taken as

$$N_A \approx \frac{4D}{\pi A} \tag{4}$$

which is the value of $N_A(A,B,\omega)$ valid to 5 percent for all $\frac{B}{A} < \frac{1}{3}$.

871

Using the value of A (eq. (3)) and the signal DIDF

$$N_B \approx \frac{2D}{\pi A} \tag{5}$$

computed again using the fact that $\frac{B}{A} < \frac{1}{3}$, gives the characteristic equation of the linearized limit cycling loop.

$$s^3 + 2\zeta\omega_n s^2 + \omega_n^2 s + \zeta\omega_n^3 = 0 \tag{6}$$

Eq. (6) has one real and two complex-conjugate roots, indicated by the squares on the three root-locus branches in Fig. 2. We find, for small ζ, that the three roots are approximately located within region II.

$$s_{1,2,3} \approx -\zeta\omega_n, \; -\frac{\zeta\omega_n}{2} \pm j\omega_n\sqrt{1 - \frac{\zeta^2}{4}} \tag{7}$$

Including the prefilter transfer function we have

$$\frac{C}{R}(s) \approx H_0(s) \frac{\zeta\omega_n^3}{(s + \zeta\omega_n)(s^2 + \zeta\omega_n s + \omega_n^2)} \tag{8}$$

The system is adaptive with respect to the parameters K and D. They do not appear in the input-output transfer function and the input-output dynamics do not change with changing K and D.

● **PROBLEM 15-36**

For the backlash element shown in Fig. 1, compute the DF.

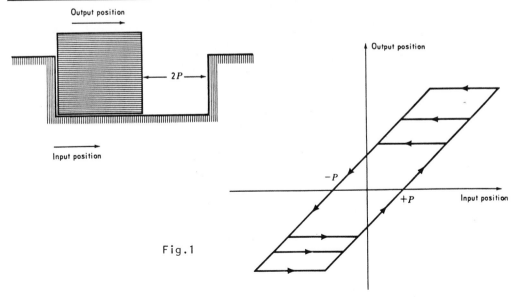

Fig.1

Solution: Note that backlash occurs to a certain extent in all mechanical gear trains. Its action is characterized

by the "play" 2P. The actual output wave form resulting
from a sinusoidal input is shown in Fig. 2.

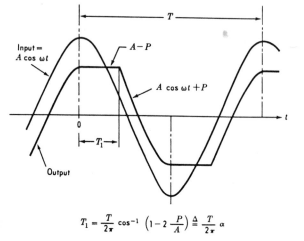

$$T_1 = \frac{T}{2\pi} \cos^{-1}\left(1 - 2\frac{P}{A}\right) \triangleq \frac{T}{2\pi}\,\alpha$$

Fig.2

Input amplitude A exceeds P. We compute the cosine and
sine components of the fundamental harmonic component of
the output.

$$A_1 = \frac{4}{T}\left[\int_0^{T_1}(A - P)\cos \omega t\, dt + \int_{T_1}^{\frac{T}{2}}(A\cos \omega t + P)\cos \omega t\, dt\right]$$

$$B_1 = \frac{4}{T}\left[\int_0^{T_1}(A - P)\sin \omega t\, dt + \int_{T_1}^{\frac{T}{2}}(A\cos \omega t + P)\sin \omega t\, dt\right]$$

where $\omega = \frac{2\pi}{T}$

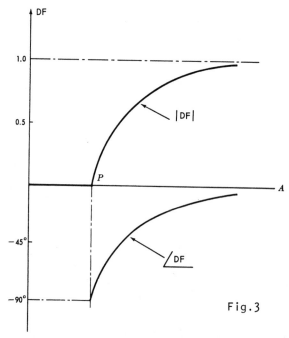

Fig.3

Integrating, we obtain

$$A_1 = A\left[1 - \frac{\alpha}{\pi} + \frac{2}{\pi}(1 - 2\frac{P}{A})\sin\alpha - \frac{1}{2\pi}\sin 2\alpha\right]$$

$$B_1 = A\left[\frac{1}{\pi} - \frac{2}{\pi}(1 - 2\frac{P}{A})\cos\alpha + \frac{1}{\pi}\cos 2\alpha\right]$$

where

$$\alpha = \cos^{-1}\left[1 - 2\frac{P}{A}\right]$$

The DF is given by

$$DF = \begin{cases} 0 & \text{for} \quad A < P \\ \dfrac{C_1}{A}\, e^{j\phi_1} & \text{for} \quad A > P \end{cases}$$

where

$$C_1 = \sqrt{A_1^2 + B_1^2}$$

$$\phi_1 = -\tan^{-1}\frac{B_1}{A_1}$$

Fig. 3 shows the DF (gain and phase) plotted versus input amplitude A. It's interesting to note that the backlash element will introduce a phase lag of as much as 90°.

PHASE PLANE

● **PROBLEM** 15-37

For the on-off servo system shown in Fig. 1, obtain a phase plane portrait.

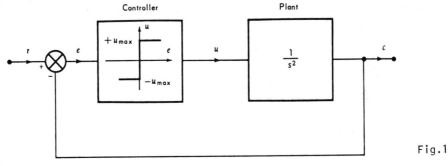

Fig.1

<u>Solution</u>: A set of equations that define the on-off servo are

$$\dot{x}_1 = x_2 \tag{1}$$

$$\dot{x}_2 = \frac{r - x_1}{|r - x_1|}u_{max}$$

From this we get

874

$$\frac{dx_2}{dx_1} = \frac{r^0 - x_1}{|r^0 - x_1|}\frac{u_{max}}{x_2}$$

We separate the variables and integrate

$$\int_{x_2(0)}^{x_2} x_2\,dx_2 = \frac{r^0 - x_1}{|r^0 - x_1|}u_{max}\int_{x_1(0)}^{x_1} dx_1 \tag{2}$$

The solution is

$$\frac{1}{2}[x_2^2 - x_2^2(0)] = \frac{r^0 - x_1}{|r^0 - x_1|}u_{max}[x_1 - x_1(0)] \tag{3}$$

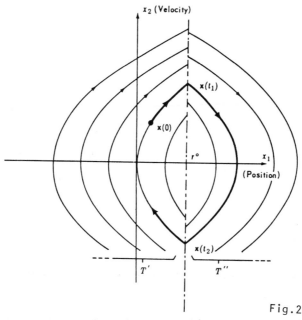

Fig.2

Equation (3) is the trajectory equation. Fig. 2 shows the phase trajectories for the system, they consist of two families of parabolas:

$$x_2^2 = 2u_{max}\,x_1 + const \qquad \text{for } x_1 < r^0$$

$$x_2^2 = -2u_{max}\,x_1 + const \qquad \text{for } x_1 > r^0 \tag{4}$$

They are denoted by T' and T" respectively.

The constants in the eqs. (4) can be determined from the initial states $x_1(0)$ and $x_2(0)$. Let us assume that initially we are located in the initial state indicated by X(0) in the figure. Since $x_1(0)$, is less than r^0, the state changes along the T' trajectory that passes through the point in question. This trajectory is marked boldface. To determine if we move to the left or to the right we use eq.(1)

$$\frac{dx_1}{dt} = x_2$$

Since dt and x_2 are positive, dx_1 must be positive and the motion along the trajectory is to the right. The arrows indicate the changes of the state of the system. After t_1 sec., we reach the state $x(t_1)$. Now x_1 is not less than r^0, thus the relay switches and we transfer to a T" trajectory.

We follow this trajectory until the relay switches again at the time t_2, and we again transfer to a T' trajectory. After T sec. the system returns to the original state $x(0)$. The periodic oscillation will continue indefinitely, unless we change the reference input T^0. The relay system is "limit cycling." This phenomenon of the limit-cycle oscillation is a typical nonlinear phenomenon that is common in nonlinear control systems.

● **PROBLEM** 15-38

For the feedback control system shown in Fig. 1 with a nonlinear controller, obtain the system response to a step input.

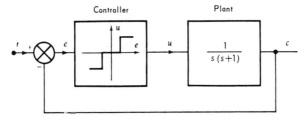

Fig.1

Solution: First we establish a mathematical model for the system. The plant transfer function $G_p(s)$ is

$$G_p(s) = \frac{C(s)}{U(s)} = \frac{1}{s^2 + s}$$

We have

$$\ddot{c} + \dot{c} = u \tag{1}$$

For the nonlinear controller we have

$$u = N(e) = N(r^0 - c) \tag{2}$$

where the nonlinear function $N(e)$ is defined by

$$N(e) = \begin{cases} +1 & \text{for} & e > 1 \\ 0 & \text{for} & -1 < e < 1 \\ -1 & \text{for} & e < -1 \end{cases}$$

Now we can choose c and $\dot{c}$ as components of the state vector x. It is more convenient to work with e and $\dot{e}$, thus

$$x = \begin{bmatrix} x_1 \\ x_2 \end{bmatrix} = \begin{bmatrix} e \\ \dot{e} \end{bmatrix} = \begin{bmatrix} r^0 - c \\ -\dot{c} \end{bmatrix} \tag{3}$$

876

Combining eqs. (1) and (3) we get the nonlinear system of equations.

$$\dot{x}_1 = x_2$$

$$\dot{x}_2 = -x_2 - N(x_1)$$

To obtain the isoclines, we form the ratio

$$\frac{dx_2}{dx_1} = S = \frac{-[x_2 + N(x_1)]}{x_2}$$

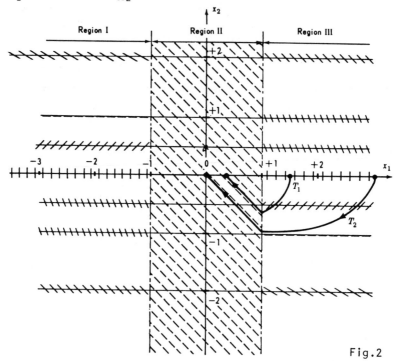

Fig.2

Fig. 2 shows the $x_1 x_2$ plane divided into three regions, each corresponding to the three different states of the nonlinear element N.

Region I is defined by

$$e = x_1 < -1$$

thus

$$N(e) = N(x_1) = -1$$

From eq.(2) we get

$$S = -\frac{x_2 - 1}{x_2}$$

and the isoclines are indepedent of x_1. Therefore they are horizontal lines. For example the locus for $S = 1$ is

$$1 = -\frac{x_2 - 1}{x_2}$$

or

$$x_2 = \frac{1}{2}$$

Fig. 2 shows some representative isoclines and the short line segments along the isoclines indicate the slope associated with these isoclines.

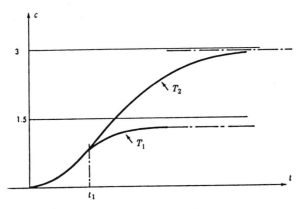

Fig.3

Region II is defined by $-1 < e < 1$ and represents the dead zone of the nonlinear element. Thus

$$N(e) = N(x_1) = 0$$

From eq. (3) we have

$$S = -\frac{x_2 + 0}{x_2} = -1$$

All the trajectories in this region have a slope -1.

Region III: We have $e = x_1 > 1$

and

$$N(e) = N(x_1) = 1.$$

Therefore

$$S = -\frac{x_2 + 1}{x_2}$$

The isoclines are horizontal lines. We can sketch now the phase trajectories. Fig. 3 shows the response picture in the time domain, the trajectories T_1 and T_2 are shown. They correspond to step inputs r^0 of magnitudes 1.5 and 3 respectively.

● **PROBLEM** 15-39

The phenomenon of synchronization in a large class of quasiharmonic oscillators can be fairly well described by the equation

$$\ddot{x} - \varepsilon(1 - x^2)\dot{x} + x = A \sin \omega t \tag{1}$$

It has a uniqe asymptotically orbitally stable limit cycle, whose frequency is close to 1. It is necessary sometimes to synchronize such a system to a frequency close to, but not identical to, its own natural frequency. This can be done by exciting the system with a signal of the desired frequency ω, as described by eq. (1). Detemine the conditions under which such synchronization can take place. Put eq. (1) into the form

$$\dot{x} = y, \quad \dot{y} = -x + \mu f(\theta, x, y) \tag{2}$$

by setting $\theta = \omega t$ and

$$y = \frac{dx}{d\theta}$$

Assume solutions of the form

$$x(\theta) = X \cos\theta + Y \sin\theta, \quad y(\theta) = -X \sin\theta + Y \cos\theta \tag{3}$$

Locate the singular points, and determine their asymptotic stability.

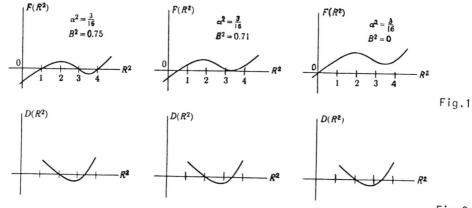

Fig.1

Fig.2

Solution: Making the substitution, we have

$$y = \frac{dx}{d\theta} = \frac{1}{\omega}\frac{dx}{dt}, \quad \frac{d^2x}{d\theta^2} = \frac{dy}{d\theta} = \frac{1}{\omega^2}\frac{d^2x}{dt^2} \tag{4}$$

so eq. (1) becomes

$$\omega^2\dot{y} - \varepsilon(1 - x^2)\omega y + x = A \sin\theta \tag{5}$$

Rearranged, the normal form equations are

$$\frac{dx}{d\theta} = y, \quad \frac{dy}{d\theta} = -x + \mu[\alpha x + (1 - x^2)y + B \sin\theta] \tag{6}$$

where

$$\mu = \frac{\varepsilon}{\omega}, \quad \mu\alpha = \frac{\omega^2 - 1}{\omega^2}, \quad \mu B = \frac{A}{\omega^2} \tag{7}$$

Where α represents the detuning, it reflects the difference between the excitation frequency ω and 1.

Taking into account eq. (3) we get

$$\frac{dX}{d\theta} = \frac{\mu}{2\pi} P(X,Y) = \frac{\mu}{2}[X(1 - \frac{R^2}{4}) - \alpha Y - B]$$

$$\frac{dY}{d\theta} = \frac{\mu}{2\pi} Q(X,Y) = \frac{\mu}{2}[\alpha X + Y(1 - \frac{R^2}{4})]$$

(8)

where

$$R^2 = X^2 + Y^2$$

P and Q are given by

$$P(X,Y) = -\int_0^{2\pi} \sin\theta \; f(\theta, x, y) d\theta$$

$$Q(X,Y) = \int_0^{2\pi} \cos\theta \; f(\theta, x, y) d\theta$$

(9)

Eq. (3) is used to substitute for x and y in the integrals, $f(\theta,x,y)$ is defined by eqs. (2) and (6) and the integration is carried out.

The singular points are given by

$$\frac{dX}{d\theta} = 0, \; \frac{dY}{d\theta} = 0$$

or

$$X(1 - \frac{R^2}{4}) - \alpha Y = B, \; \alpha X + Y(1 - \frac{R^2}{4}) = 0 \quad (10)$$

Taking the sum of the squares of both eqs. (10) we determine R^2 for each singular point as a positive solution of the equation

$$F(R^2) = R^2[(1 - \frac{R^2}{4})^2 + \alpha^2] - B^2 = 0 \quad (11)$$

The X- and Y-coordinates of the singular points are

$$X = \frac{(1 - \frac{R^2}{4})B}{(1 - \frac{R^2}{4})^2 + \alpha^2}, \; Y = \frac{-\alpha B}{(1 - \frac{R^2}{4})^2 + \alpha^2} \quad (12)$$

where R^2 is a solution of eq. (11). Fig. 1 shows the plot of $F(R^2)$ for three different values of B.

We find the variational equation, to examine the behavior of eqs. (8) in the vicinity of each singular point

880

$$\begin{bmatrix} \dfrac{du_x}{d\theta} \\[2mm] \dfrac{du_y}{d\theta} \end{bmatrix} = \dfrac{\mu}{2\pi}\left[\dfrac{\partial\,(P,Q)}{\partial\,(X,Y)}\right]\begin{bmatrix} u_x \\[2mm] u_y \end{bmatrix}$$

$$= \dfrac{\mu}{2}\begin{bmatrix} (1 - \dfrac{R^2}{4} - \dfrac{X^2}{2}) & (-\alpha - \dfrac{XY}{2}) \\[4mm] (\alpha - \dfrac{XY}{2}) & (1 - \dfrac{R^2}{4} - \dfrac{Y^2}{2}) \end{bmatrix}\begin{bmatrix} u_x \\[4mm] u_y \end{bmatrix} \qquad (13)$$

From eq. (13) we find that a singular point is asymptotically stable if

$$\mathrm{tr}\left[\dfrac{\partial\,(P,\ Q)}{\partial\,(X,\ Y)}\right] = \pi(2 - R^2) < 0 \qquad (14)$$

and

$$\dfrac{1}{\pi^2}\det\left[\dfrac{\partial\,(P,Q)}{\partial\,(X,Y)}\right] = D(R^2) = (1 - \dfrac{3}{4}R^2)(1 - \dfrac{R^2}{4}) + \alpha^2 > 0 \qquad (15)$$

We evaluate the above quantities at the singular point in question.

The singular point becomes unstable, when one or both of these inequalities is reversed. Comparing eqs. (15) and (11) we find that

$$D(R^2) = \dfrac{dF(R^2)}{dR^2} \qquad (16)$$

Fig. 2 shows the plot of this function. From this figure we see that condition eq. (15) for stability is always fulfilled when eq. (11) has only one positive root, and it is fulfilled for the roots of the smallest and largest radii when eq. (11) has three positive roots. From eq. (14) we see that any singular point located at a radius $R < \sqrt{2}$ is unstable. Fig. 3 illustrates the information contained in eqs. (11), (14) and (15).

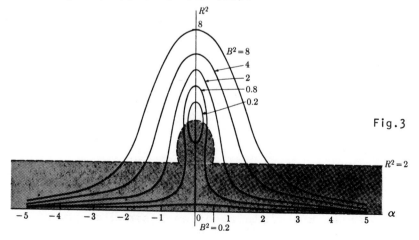

Fig.3

881

The curves represent the locus of solutions R^2 of eq. (11) as a function of α, with forcing amplitude B as a parameter and each is a frequency-response curve. They represent the squared amplitude R^2 of the quasiharmonic solution as a function of the frequency of the excitation. For sufficiently small values of B, the corresponding curves are multivalued. Eqs. (14) and (15) determine the region in which any singular point is unstable. In the Fig. 3 this region is indicated by the shaded area.

When, for a given pair of values of α and B, multiple singular points exist, then only one of them is stable. For any fixed value of the forcing amplitude B, there exist values of α, beyond which no singular point can be stable. The explanation is that the existence of a stable singular point of eq. (8) corresponds to a stable synchronized almost sinusoidal solution of eq. (6).

LIMIT CYCLE

● **PROBLEM** 15-40

The system is described by

$$\dot{x}_1 = x_1 g(r) - x_2, \quad g(r) = r^2 (A - r)$$

$$\dot{x}_2 = x_1 + x_2 g(r), \quad r = \sqrt{x_1^2 + x_2^2} \tag{1}$$

and has a unique closed trajectory

$$C : x^0(t) = \begin{bmatrix} A \cos t \\ A \sin t \end{bmatrix}, \quad x^0(t) = x^0(t + 2\pi) \tag{2}$$

This system represents an oscillator stabilized at an amplitude $A > 0$. Show that the trajectory C is asymptotically orbitally stable.

(Loosely, that implies that trajectories near C approach it as $t \to \infty$).

Solution: Theory says that $n - 1$ of the n characteristic multipliers of the equations of first variation must be inside the unit circle (the other one will be on it) for the limit cycle to be asymptotically orbitally stable. In the polar coordinates we have

$$r = \sqrt{x_1^2 + x_2^2}, \quad \theta = \tan^{-1} \frac{x_2}{x_1} \tag{3}$$

Differentiating, we obtain

$$\dot{r} = \frac{x_1 \dot{x}_1 + x_2 \dot{x}_2}{\sqrt{x_1^2 + x_2^2}}, \quad \dot{\theta} = \frac{x_1 \dot{x}_2 - \dot{x}_1 x_2}{x_1^2 + x_2^2} \tag{4}$$

Substituting eq. (1) into eq. (4) we get

882

$$\dot{r} = \frac{x_1^3 g - x_1 x_2 + x_2 x_1 + x_2^2 g}{\sqrt{x_1^2 + x_2^2}} = g\sqrt{x_1^2 + x_2^2} = r^3(A - r) \qquad (5)$$

$$\dot{\theta} = \frac{x_1^2 + x_1 x_2 g - x_1 x_2 g + x_2^2}{x_1^2 + x_2^2} = 1$$

The equation of first variation is

$$\begin{bmatrix} \delta\dot{r} \\ \delta\dot{\theta} \end{bmatrix} = \begin{bmatrix} \dfrac{\partial\dot{r}}{\partial r} & \dfrac{\partial\dot{r}}{\partial\theta} \\ \dfrac{\partial\dot{\theta}}{\partial r} & \dfrac{\partial\dot{\theta}}{\partial\theta} \end{bmatrix} \begin{bmatrix} \delta r(0) \\ \delta\theta(0) \end{bmatrix} \qquad (6)$$

and

$$\begin{bmatrix} \delta\dot{r} \\ \delta\dot{\theta} \end{bmatrix} = \begin{bmatrix} 3Ar^2 - 4r^3 & 0 \\ 0 & 0 \end{bmatrix} \begin{bmatrix} \delta r(0) \\ \delta\theta(0) \end{bmatrix} \qquad (7)$$

From eq. (2), we rewrite the trajectory C in polar coordinates.

$$r = \sqrt{x_1^2 + x_2^2} = \sqrt{(A\cos t)^2 + (A\sin t)^2} = A$$

$$\theta = \tan^{-1}\frac{x_2}{x_1} = \tan^{-1}\frac{A\sin t}{A\cos t} = t \qquad (8)$$

Substituting eq. (8) into eq. (7) we have

$$\begin{bmatrix} \delta\dot{r} \\ \delta\dot{\theta} \end{bmatrix} = \begin{bmatrix} -A^3 & 0 \\ 0 & 0 \end{bmatrix} \begin{bmatrix} \delta r(0) \\ \delta\theta(0) \end{bmatrix} \qquad (9)$$

Then the eigenvalues of the characteristic equation of eq. (9) are $\lambda_1 = -A^3$ and $\lambda_2 = 0$, and the characteristic multipliers are e^{-A^3} and 1. Since e^{-A^3} is inside the unit circle, C is asymptotically orbitally stable.

● **PROBLEM** 15-41

For the nonlinear time-lag differential equation

$$\dot{x}(t) + \delta x(t - T) = \varepsilon x^3(t - T) \qquad (1)$$

determine the limit cycle amplitude and frequency.

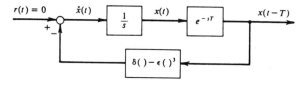

Solution: The figure shows the block diagram of this

883

system. The time-lag effect is represented by the factor e^{-sT}, the Laplace transform of an ideal delay.

The equation for a limit cycle is

$$\frac{1}{j\omega_o} e^{-j\omega_o T} (\delta - \varepsilon \frac{3}{4} A_o^2) = -1 \qquad (2)$$

Note that the DF for a cubing operator,

$$N(S) = \frac{3}{4} A^2$$

has been used.

The angle and magnitude conditions give

$$-\frac{\pi}{2} - \omega_o T = -\pi - 2n\pi \qquad (3)$$

We thus find, that

$$\omega_o = (2n + \frac{1}{2}) \frac{\pi}{T} \qquad \text{for } n = 0, 1, 2, \ldots \qquad (4)$$

and

$$\frac{1}{\omega_o} (1) (\delta - \varepsilon \frac{3}{4} A_o^2) = 1 \qquad (5)$$

Solving the equations for A_o, we obtain

$$A_o = \sqrt{\frac{4}{3\varepsilon} \left[\delta - (2n + \frac{1}{2}) \frac{\pi}{T} \right]} \qquad (6)$$

Thus for $\delta > \frac{\pi}{2T}$, a limit cycle is predicted.

● **PROBLEM** 15-42

Fig. 1 shows the system with a limit cycle with the frequency of oscillation at 5.9 $\frac{rad}{sec}$, as shown in Fig. 2.

We want to decrease the frequency of the limit cycle to $4\frac{rad}{sec}$. Assuming that the nonlinear element is fixed, determine the necessary change in the gain of G(s).

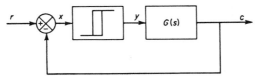

Fig.1

Solution: We have Im $\{G(j\ 5.9)\} = \alpha$, where α is the

884

location of the $-\frac{1}{N}$ line. We multiply $G(j\omega)$ by a constant K so that

$$K \text{ Im } \{G(j4)\} = \alpha$$

Dividing and solving for K we get

$$K = \frac{\text{Im}\{G(j5.9)\}}{\text{Im}\{G(j4)\}} \tag{1}$$

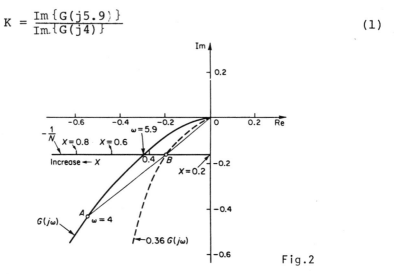

Fig.2

But from Fig. 2, by similar triangles, this means that K must be $\frac{\overline{OB}}{\overline{OA}}$, where A and B are as shown.

We find

$$\frac{\overline{OB}}{\overline{OA}} = 0.36$$

thus, if the value of the gain of G(s) is decreased to 36% of the original value, the frequency of the new limit cycle will become $4 \frac{\text{rad}}{\text{sec}}$.

● **PROBLEM** 15-43

Fig. 1 shows the block diagram of a temperature control system. Investigate the limit cycle behavior of the system.

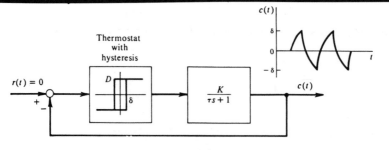

Fig.1

Solution: We shall use the DF method to solve the problem. Fig. 2 shows the graphical construction. We conclude that no oscillation takes place. The DF result is wrong, since a limit cycle must occur in this on-off system if DK > δ . It is interesting to find out why the DF formulation leads to incorrect conclusion.

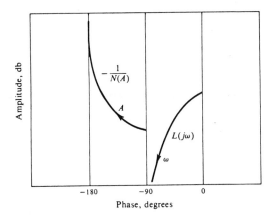

Fig.2

We shall consider the exact limit cycle waveform, shown in Fig. 1, which consists of matched single-exponential segments. There are two slope discontinuities in the wave form during each limit cycle period. The failure of the DF method is due to the waveform harmonic content. Note,that the third-to first-harmonic amplitude ratio is 1/3 at the nonlinearity output. Each of the harmonics is attenuated by the linear elements by the factor

$$\frac{K}{\sqrt{1 + n^2 \omega_o \tau^2}}$$

where n is the harmonic number.

At the input to the nonlinearity, we have

$$\left|\frac{A_3}{A_1}\right|_x = \frac{1}{3}\sqrt{\frac{1 + \omega_o^2\tau^2}{1 + 9\omega_o^2\tau^2}} \tag{1}$$

and the following limiting inequality holds.

$$0.111 < \left|\frac{A_3}{A_1}\right|_3 < 0.333 \tag{2}$$

Thus, at the input to the nonlinearity, the third-harmonic amplitude is somewhere between 11.1 and 33.3 percent of the fundamental. Due to the large amplitude at x, the unaccounted for phase shift of the third harmonic is sufficient to distort the original input zero-crossing assumptions, and thus negate the DF calculation.

886

Determine all possible limit cycles of the relay control system shown in the figure , using the Routh-Hurwitz test.

Find the limit cycles displayed by the relay control system analytically using the equations:

$$n_p(A_o) \text{Re} \{L(j\omega_o)\} - n_q(A_o) \text{Im} \{L(j\omega_o)\} = -1 \quad (1)$$

$$n_p(A_o) \text{Im} \{L(j\omega_o)\} + n_q(A_o) \text{Re} \{L(j\omega_o)\} = 0 \quad (2)$$

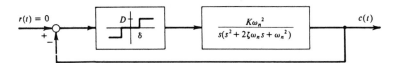

ROUTH-HURWITZ
LIMIT CYCLE DETERMINATION

Characteristic equation	Limit cycle equations
$s^2 + a_1 s + a_0 = 0$	$a_1 = 0$ $\omega_0^2 = a_0$
$s^3 + a_2 s^2 + a_1 s + a_0 = 0$	$a_1 a_2 - a_0 = 0$ $\omega_0^2 = a_1$
$s^4 + a_3 s^3 + a_2 s^2 + a_1 s + a_0 = 0$	$a_1 a_2 a_3 - a_1^2 - a_0 a_3^2 = 0$ $\omega_0^2 = \dfrac{a_1}{a_3}$

Solution: For a relay with dead zone, the DF is

$$n_p = \frac{4D}{\pi A} \sqrt{1 - \left(\frac{\delta}{A}\right)^2} \quad , \qquad n_q = 0 \quad (3)$$

The characteristic equation $1 + N(A)L(S) = 0$ is representable as

$$s^3 + 2\zeta\omega_n s^2 + \omega_n^2 s + \frac{4\omega_n^2 KD}{\pi A} \sqrt{1 - \left(\frac{\delta}{A}\right)^2} = 0 \quad (4)$$

The coefficients are

$$a_2 = 2\zeta\omega_n, \; a_1 = \omega_n^2, \; a_o = \frac{4\omega_n^2 KD}{\pi A} \sqrt{1 - \left(\frac{\delta}{A}\right)^2} \quad (5)$$

From the table, we find that the limit cycle frequency is exactly the undamped natural frequency of the second-order part of L(s),

$$\omega_o = \sqrt{a_1} \quad (6)$$

and

$$(2\zeta\omega_n)(\omega_n^2) - \frac{4\omega_n^2 KD}{\pi A_o} \sqrt{1 - \left(\frac{\delta}{A_o}\right)^2} = 0 \quad (7)$$

Solving eq. (7) we get

$$A_o = \frac{\sqrt{2}\ KD}{\pi\zeta\omega_n} \sqrt{1 \pm \sqrt{1 - \left(\frac{\pi\zeta\omega_n\delta}{KD}\right)^2}} \qquad (8)$$

Note that, since A_o must be real, δ must be in the range

$$\delta > \frac{KD}{\pi\zeta\omega n} \qquad \text{condition for no limit cycle} \qquad (9)$$

When the above condition is not met, two limit cycles are possible, corresponding to two signs under the radical in the solution for A_o. Of these two limit cycle states, only one is stable. Analytically, we have that the real and inaginary parts of $L(j\omega)$ are

$$Re[L(j\omega)] = \frac{-2K\zeta\omega_n^3}{(\omega_n^2 - \omega^2)^2 + (2\zeta\omega\omega_n)^2} \qquad (10)$$

$$Im[L(j\omega)] = \frac{- K\omega_n^2 (\omega_n^2 - \omega^2)}{\omega[(\omega_n^2 - \omega^2) + (2\zeta\omega\omega_n)^2]}$$

From eq. (2) we have

$$Im[L(j\omega_o)] = 0 \qquad (11)$$

Thus the limit cycle frequency is equal to the natural frequency of the second-order part of $L(j\omega)$.

$$\omega_o = \omega_n \qquad (12)$$

Substituting eq. (12) into eq. (1) we get

$$\frac{K}{2\zeta\omega_n} \frac{4D}{\pi A_o} \sqrt{1 - (\frac{\delta}{A_o})^2} = 1 \qquad (13)$$

Solving we obtain

$$A_o = \frac{\sqrt{2}\ KD}{\pi\zeta\omega_n} \sqrt{1 \pm \sqrt{1 - \left(\frac{\pi\zeta\omega_n\delta}{KD}\right)^2}} \qquad (14)$$

Determine the stability of each of the limit cycles

$$A_o = \frac{\sqrt{2}\ KD}{\pi \zeta \omega_n} \sqrt{1 \pm \sqrt{1 - \left(\frac{\pi \zeta \omega_n \delta}{KD}\right)^2}}, \qquad \delta \geq 0 \qquad (1)$$

for the system with

$$L(s) = \frac{K\omega_n^2}{s(s^2 + 2\zeta\omega_n s + \omega_n^2)} \qquad (2)$$

$$n_p = \frac{4D}{\pi A} \sqrt{1 - (\frac{\delta}{A})^2}, \qquad n_q = 0 \qquad (3)$$

Solution: The characteristic equation of $1 + N(A)L(s)$ is

$$s^3 + 2\zeta\omega_n s^2 + \omega_n^2 s + \frac{4KD\omega_n^2}{\pi A} \sqrt{1 - (\frac{\delta}{A})^2} = 0 \qquad (4)$$

Substituting $s = j\omega$, we get

$$- 2\zeta\omega_n \omega^2 + \frac{4DK\omega_n^2}{\pi A} \sqrt{1 - (\frac{\delta}{A})^2} + j(\omega\omega_n^2 - \omega^3) = 0 \qquad (5)$$

The real and imaginary parts of eq. (5) are U and V, respectively.

Stability requires that

$$\frac{\partial U}{\partial A} \frac{\partial V}{\partial \omega} - \frac{\partial U}{\partial \omega} \frac{\partial V}{\partial A} > 0 \qquad (6)$$

Taking the required four partial derivatives and evaluating them at (A_o, ω_o) gives

$$\left.\frac{\partial U}{\partial A}\right|_{A_o, W_o} = - \frac{4DK\omega_n^2}{\pi A_o^2} \left[\frac{1 - 2(\frac{\delta}{A_o})^2}{\sqrt{1 - (\frac{\delta}{A_o})^2}}\right]$$

$$\left.\frac{\partial V}{\partial A}\right|_{A_o, \omega_o} = - 4\zeta\omega_n^2 \qquad (7)$$

$$\left.\frac{\partial V}{\partial A}\right|_{A_o, \omega_o} = 0$$

$$\left.\frac{\partial V}{\partial \omega}\right|_{A_o, \omega_o} = - 2\omega_n^2$$

Substituting eq. (7) into eq. (6) yields

889

$$1 - 2 \left(\frac{\delta}{A_o}\right)^2 > 0 \qquad (8)$$

or

$$A_o > \sqrt{2}\,\delta \qquad (9)$$

which is the requirement for a stable oscillation.

Eq. (1) is real only in the range

$$\delta < \frac{KD}{\pi\zeta\omega_n} \qquad (10)$$

In that range, the larger A_o is greater than $\sqrt{2}\,\delta$, and the smaller is less than $\sqrt{2}\,\delta$. Therefore, only the larger amplitude limit-cycle is stable.

STATE REPRESENTATION, POPOV, LIAPUNOV

● **PROBLEM** 15-46

Consider the system shown in the figure. The controller is a nonlinear on-off type controller. It is also called a maximum effort or bang-bang type. This block diagram can represent the altitude-control system of a space capsule. We assume that the reaction-jet moment u is applied in a bang-bang fashion. Obtain a state space representation of this system.

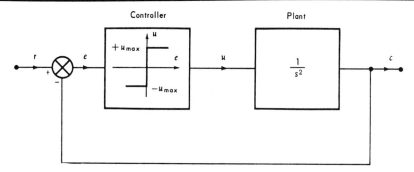

Solution: The plant differential equation is

$$\ddot{c} = u$$

We introduce the state vector

$$x = \begin{bmatrix} x_1 \\ x_2 \end{bmatrix}$$

where

$$x_1 = c$$

$$x_2 = \dot{c}$$

and rewrite the plant equations in the form

$$\dot{x}_1 = x_2$$

$$\dot{x}_2 = u \tag{1}$$

where

$$c = x_1$$

From the equation

$$u = \frac{e}{|e|} u_{max} = \frac{r - c}{|r - c|} u_{max} = \frac{r - x_1}{|r - x_1|} u_{max} \tag{2}$$

We determine the maximum control force $\pm u_{max}$ which depends upon the sign of the error $e = r - c$.

In this case we have the nonlinear controller equation q - scalar. Combining eq. (1) and eq. (2) and eliminating u we get the closed-loop system

$$\dot{x}_1 = x_2$$

$$\dot{x}_2 = \frac{r - x_1}{|r - x_1|} u_{max}$$

● **PROBLEM 15-47**

Fig. 1 shows the block diagram of a feedback control system that has a saturation element in the forward path. A position control system using a motor as an actuator is one of the examples. The saturation is in the power amplitier of the motor controller, with K representing the gain of the linear portion of the amplifier characterisitc. The linear transfer function of the system is

$$G(s) = \frac{1}{s(s + 1)(s + 2)} \tag{1}$$

Find the maximum value of K for which the system is stable

a)   without saturation,

b)   with saturation.

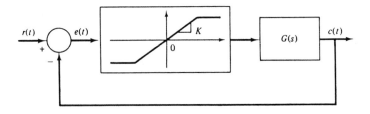

Fig.1

Solution:

a)   Without saturation, the characteristic equation is

$$s(s + 1)(s + 2) + K = S^3 + 3s^2 + 2s + K \qquad (2)$$

We form the Routh array

s^3	1	2
s^2	3	K
s^1	$2 - \dfrac{K}{3}$	0
s^0	K	0

From row S^0, we require $K > 0$, and from row s^1 we require $K < 6$, so $K_{max} = 6$.

b) With saturation, we will use the Popov criterion. It states that if two lines through the origin with slopes k_1 and k_2 (positive) completely enclose the nonlinearity, a sufficient condition for stability is that the Nyquist plot of the linear portion not intersect the circle with center on the real axis that passes through points $-\dfrac{1}{k_1}$ and $-\dfrac{1}{k_2}$. If $k_1 = 0$, the circle degenerates to a vertical line at $-\dfrac{1}{k_2}$. In our case, $k_1 = 0$ and $k_2 = K$.

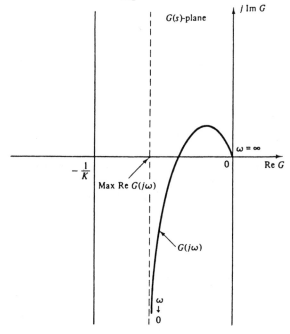

Fig.2

For the saturation nonlinearity, the result of the Popov criterion is independent of the level of saturation. Fig. 2 shows the Nyquist locus of $G(j\omega)$. For stability of the closed-loop system, the $G(j\omega)$ locus must not intersect the vertical line, which passes through the point ($-\dfrac{1}{K}$, j0). Finding the maximum magnitude of the real part of $G(j\omega)$ we determine the maximum value of K.

The real part of $G(j\omega)$ is

$$\text{Re } G(j\omega) = \frac{-3}{9\omega^2 + (2 - \omega^2)^2} \tag{3}$$

The frequency at which Re $G(j\omega)$ is a maximum is determined from

$$\frac{d}{d\omega} [\text{Re } G(j\omega)] = \frac{\omega[-18 + 4(2 - \omega^2)]}{[9\omega^2 + (2 - \omega^2)^2]^2} = 0 \tag{4}$$

thus

$$\omega = 0$$

We get

$$\text{Max Re } G(j\omega) = -\frac{3}{4} \tag{5}$$

The closed-loop system is stable for

$$K < \frac{4}{3} \tag{6}$$

which is less than the critical value of $K = 6$ for the linear system.

● **PROBLEM** 15-48

Using the Lyapunov's direct method, determine the stability of the origin $x = 0$ for the system described by

$$\dot{x}_1 = x_2 - ax_1 (x_1^2 + x_2^2)$$
$$\dot{x}_2 = -x_1 - ax_2 (x_1^2 + x_2^2) \tag{1}$$

Solution: In the direct method we take a function $V(x)$. If

(1) $V(0) = 0$ and $V(x) > 0$ for all $x \neq 0$,

(2) $\dot{V}(x) < 0$ for all $x \neq 0$,

(3) $V(x) \to \infty$ as $||x|| = \sqrt{x_1^2 + x_2^2 + \ldots} \to \infty$,

then the origin is globally asymptotically stable. Usually $V(x)$ is chosen as a quadratic function of the state variables, and represents in a way the energy of the system. By condition (1) the energy is positive, V is positive definite. By condition (2) the energy is always decreasing, and by condition (3), $V(x)$ increases with $||x||$.

Let us take as a trial Lyapunov function

$$V(x) = C_1 x_1^2 + C_2 x_2^2$$

where c_1, c_2 are positive constants. $V(x)$ is positive definite and $V(x) \to \infty$ as $||x|| \to \infty$. The time derivative is

893

$$\dot{V}(x) = 2c_1 x_1 \dot{x}_1 + 2c_2 x_2 \dot{x}_2 \qquad (2)$$

Substituting eq. (1) we get

$$\dot{V}(x) = 2c_1 x_1 [x_2 - ax_1(x_1^2 + x_2^2)] + 2c_2 x_2 [-x_1 - ax_2(x_1^2 + x_2^2)] \quad (3)$$

Ther term $x_1 x_2$ is cancelled for $c_1 = c_2$, thus

$$\dot{V}(x) = -2ac_1(x_1^2 + x_2^2)^2$$

For $a > 0$, $\dot{V}(x)$ is negative definite and the origin is globally asymptotically stable.

CHAPTER 16

OPTIMIZATION

The transfer function of a linear process is

$$\frac{C(s)}{U(s)} = \frac{10}{s(s + 1)(s + 2)} \tag{1}$$

Design a feedback controller with state feedback so that the eigenvalues of the closed-loop system are at -2, -1 + j, -1 - j. Note that the transfer function does not have common poles and zeros, thus the process is completely state controllable.

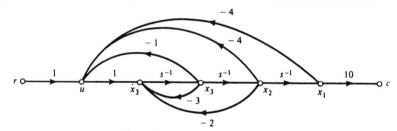

Solution: The state equations for eq. (1) can be written using direct decomposition.

$$\begin{bmatrix} \dot{x}_1 \\ \dot{x}_2 \\ \dot{x}_3 \end{bmatrix} = \begin{bmatrix} 0 & 1 & 0 \\ 0 & 0 & 1 \\ 0 & -2 & -3 \end{bmatrix} \begin{bmatrix} x_1 \\ x_2 \\ x_3 \end{bmatrix} + \begin{bmatrix} 0 \\ 0 \\ 1 \end{bmatrix} [u] \tag{2}$$

For the third-order system the feedback matrix G is in the form

$$G = [g_1 \quad g_2 \quad g_3] \tag{3}$$

The characteristic equation of the closed-loop system is

$$\lambda^3 + (3 + g_3)\lambda^2 + (2 + g_2)\lambda + g_1 = 0 \tag{4}$$

For the eigenvalues -2, $-1 + j$, $-1 - j$ the character-istic equation is

$$\lambda^3 + 4\lambda^2 + 6\lambda + 4 = 0 \tag{5}$$

Comparing the coefficients in eqs. (4) and (5) we obtain

$$g_1 = 4, \quad g_2 = 4, \quad g_3 = 1$$

$$G = [4, 4, 1] \tag{6}$$

The state diagram of the overall system is shown below.

● **PROBLEM 16-2**

For the control system shown in the figure, determine ζ that will minimize the following performance index

$$J = \int_0^\infty x'(t) Q x(t) dt$$

when the system is subjected to a unit step disturbance.

Assume that the system is initially at rest and

$$x = \begin{bmatrix} x_1 \\ x_2 \end{bmatrix} = \begin{bmatrix} x \\ \dot{x} \end{bmatrix}, \qquad Q = \begin{bmatrix} 1 & 0 \\ 0 & \mu \end{bmatrix} \qquad (\mu > 0)$$

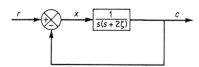

Solution: We obtain from the figure the following equation for the system.

$$\ddot{c} + 2\zeta\dot{c} + c = r$$

The initial conditions are equal to zero and $x = r - c$, $r(t) = 1(t)$, thus

$$\ddot{x} + 2\zeta\dot{x} + x = 0 \qquad (t > 0)$$

The state-space representation of the last equation is

$$\begin{bmatrix} \dot{x}_1 \\ \dot{x}_2 \end{bmatrix} = \begin{bmatrix} 0 & 1 \\ -1 & -2\zeta \end{bmatrix} \begin{bmatrix} x_1 \\ x_2 \end{bmatrix}, \quad \begin{bmatrix} x_1(0) \\ x_2(0) \end{bmatrix} = \begin{bmatrix} 1 \\ 0 \end{bmatrix}$$

or

$$\dot{x} = Ax$$

where

$$A = \begin{bmatrix} 0 & 1 \\ -1 & -2\zeta \end{bmatrix}$$

A is a stable matrix and the value of J is determined from

$$J = x'(0)Px(0)$$

where

$$A'P + PA = -Q.$$

From the last equation we find P.

$$\begin{bmatrix} 0 & -1 \\ 1 & -2\zeta \end{bmatrix} \begin{bmatrix} p_{11} & p_{12} \\ p_{12} & p_{22} \end{bmatrix} + \begin{bmatrix} p_{11} & p_{12} \\ p_{12} & p_{22} \end{bmatrix} \begin{bmatrix} 0 & 1 \\ -1 & -2\zeta \end{bmatrix} = \begin{bmatrix} -1 & 0 \\ 0 & -\mu \end{bmatrix}$$

or

$$-2p_{12} = -1$$

$$p_{11} - 2\zeta p_{12} - p_{22} = 0$$

$$2p_{12} - 4\zeta p_{22} = -\mu$$

Solving these equations we obtain

$$P = \begin{bmatrix} p_{11} & p_{12} \\ p_{12} & p_{22} \end{bmatrix} = \begin{bmatrix} \zeta + \dfrac{1 + \mu}{4\zeta} & \dfrac{1}{2} \\ \dfrac{1}{2} & \dfrac{1 + \mu}{4\zeta} \end{bmatrix}$$

We compute the performance index

$$J = x'(0)Px(0)$$

$$= \left(\zeta + \frac{1 + \mu}{4\zeta} \right) x_1^2(0) + x_1(0)x_2(0) + \frac{1 + \mu}{4\zeta} x_2^2(0)$$

where $x_1(0) = 1$ and $x_2(0) = 0$, thus

$$J = \zeta + \frac{1 + \mu}{4\zeta}$$

To minimize J, we set $\dfrac{\partial J}{\partial \zeta} = 0$, thus

$$\frac{\partial J}{\partial \zeta} = 1 - \frac{1 + \mu}{4\zeta^2} = 0$$

Solving for ζ we find

$$\zeta = \frac{\sqrt{1 + \mu}}{2}$$

For $\zeta = \dfrac{\sqrt{1 + \mu}}{2}$, J reaches its minimal value, since $\dfrac{\partial^2 J}{\partial \zeta^2} > 0$
for $\zeta = \dfrac{\sqrt{1 + \mu}}{2}$

The second-order system is described by

$$\ddot{x}_1 + 2\zeta \dot{x}_1 + x_1 = 0, \quad \zeta > 0 \tag{1}$$

The performance index is defined by

$$V(x,\zeta) = \int_0^\infty x'(t)Qx(t)\,dt$$

where

$$x = \begin{bmatrix} x_1 \\ x_2 \end{bmatrix} \qquad Q = \begin{bmatrix} 1 & 0 \\ 0 & \mu \end{bmatrix} \qquad \mu \geq 0$$

Determine the value of ζ that minimizes the performance index.

Solution: We write eq. (1) in the form

$$\begin{bmatrix} \dot{x}_1 \\ \dot{x}_2 \end{bmatrix} = \begin{bmatrix} 0 & 1 \\ -1 & -2\zeta \end{bmatrix} \begin{bmatrix} x_1 \\ x_2 \end{bmatrix}$$

or

$$\dot{x} = Ax$$

where

$$A = \begin{bmatrix} 0 & 1 \\ -1 & -2\zeta \end{bmatrix}$$

Q is a positive semidefinite real symmetric matrix, thus the Liapunov function can be chosen as:

$$V(x) = x'Px$$

$$\dot{V}(x) = -x'Qx$$

We determine the positive definite real symmetric matrix P from

$$A'P + PA = -Q$$

Solving for P we get

$$P(\zeta) = \begin{bmatrix} \zeta + \dfrac{1 + \mu}{4\zeta} & \dfrac{1}{2} \\[2ex] \dfrac{1}{2} & \dfrac{1 + \mu}{4\zeta} \end{bmatrix}$$

Since $V(x,\zeta) = x'(0)Px(0)$ we have $V(x,\zeta) = \zeta x_1^2(0) + \left[\dfrac{1 + \mu}{4\zeta}\right][x_1^2(0) + x_2^2(0)] + x_1(0)x_2(0)$

The value of $\zeta(\mu)$ which minimizes V when $x_2(0) = 0$ is found from

$$\zeta_{opt}(\mu) = \frac{\sqrt{1 + \mu}}{2}$$

and

$$\zeta_{opt}(0) = \frac{1}{2}$$

Note that this type of optimization leads to a system that is too oscillatory. Choosing $\mu = 1$ we get

$$\zeta_{opt}(1) = \frac{1}{\sqrt{2}} = 0.707$$

The value of $\zeta = 0.707$ is satisfactory. We conclude that

$$Q = \begin{bmatrix} 1 & 0 \\ 0 & 1 \end{bmatrix}$$

or minimizing

$$\int_0^\infty (x_1^2 + x_2^2)\,dt$$

is a better error criterion than

$$Q = \begin{bmatrix} 1 & 0 \\ 0 & 0 \end{bmatrix}$$

or minimizing

$$\int_0^\infty x_1^2\,dt.$$

● PROBLEM 16-4

The control system is described by

$$\dot{x} = Ax + Bu \tag{1}$$

where

$$A = \begin{bmatrix} 0 & 1 \\ 0 & 0 \end{bmatrix}, \quad B = \begin{bmatrix} 0 \\ 1 \end{bmatrix}, \quad x(0) = \begin{bmatrix} c \\ 0 \end{bmatrix}$$

Determine the constants k_1 and k_2 in the linear control law

$$u = -Kx = -k_1 x_1 - k_2 x_2 \tag{2}$$

that minimizes the performance index

$$J = \int_0^\infty x'x\,dt$$

The undamped natural frequency is 2 $\frac{rad}{s}$.

Solution: Substituting eq. (2) into eq. (1) we obtain

$$\dot{x} = Ax - BKx$$

In the matrix form

$$\begin{bmatrix} \dot{x}_1 \\ \dot{x}_2 \end{bmatrix} = \begin{bmatrix} 0 & 1 \\ 0 & 0 \end{bmatrix} \begin{bmatrix} x_1 \\ x_2 \end{bmatrix} + \begin{bmatrix} 0 \\ 1 \end{bmatrix} [-k_1 x_1 - k_2 x_2]$$

$$= \begin{bmatrix} 0 & 1 \\ -k_1 & -k_2 \end{bmatrix} \begin{bmatrix} x_1 \\ x_2 \end{bmatrix} \qquad (3)$$

Thus

$$A - BK = \begin{bmatrix} 0 & 1 \\ -k_1 & -k_2 \end{bmatrix}$$

Eliminating x_2 from eq. (3) we obtain

$$\ddot{x}_1 + k_2 \dot{x}_1 + k_1 x_1 = 0$$

$$k_1 = 4$$

since the undamped natural frequency is $2 \frac{\text{rad}}{\text{sec}}$.

We have

$$A - BK = \begin{bmatrix} 0 & 1 \\ -4 & -k_2 \end{bmatrix}$$

For $k_2 > 0$, A - BK is a stable matrix. To determine the value of k_2 that minimizes J

$$J = \int_0^\infty x'x\,dt$$

we proceed as follows:

Assume the indefinite integral is

$$\int x'x\,dt = -x'Px$$

Taking derivatives on both sides yields

$$-x'x = \dot{x}'Px + x'P\dot{x}$$

Now

$$\dot{x} = (A - BK)x \quad \text{and} \quad \dot{x}' = x'(A - BK)'$$

Substituting

$$-x'x = x'(A - BK)'Px + x'P(A - BK)x$$

$$= x'[(A - BK)'P + P(A - BK)]x$$

900

$$= x'[-I]x$$

P is determined from the relationship

$$(A - BK)'P + P(A - BK) = -I$$

The performance index is

$$J = \int_0^\infty x'x \, dt = (-x'Px)\Big|_0^\infty$$

$$= -x'Px\Big|_\infty + x'Px\Big|_0$$

For the integral to converge, x'x must approach 0 as $t \to \infty$. Since $x'x = x_1^2 + x_2^2$, this is equivalent to

$$\lim_{t\to\infty} x_1 = 0, \quad \lim_{t\to\infty} x_2 = 0$$

Therefore

$$x'Px\Big|_\infty = 0$$

and

$$J = x'Px\Big|_0 = x'(0)Px(0)$$

We determine P from

$$(A - BK)'P + P(A - BK) = -I$$

in the matrix form

$$\begin{bmatrix} 0 & -4 \\ 1 & -k_2 \end{bmatrix} \begin{bmatrix} p_{11} & p_{12} \\ p_{12} & p_{22} \end{bmatrix} + \begin{bmatrix} p_{11} & p_{12} \\ p_{12} & p_{22} \end{bmatrix} \begin{bmatrix} 0 & 1 \\ -4 & -k_2 \end{bmatrix} = \begin{bmatrix} -1 & 0 \\ 0 & -1 \end{bmatrix}$$

Solving for the matrix P we get

$$P = \begin{bmatrix} p_{11} & p_{12} \\ p_{12} & p_{22} \end{bmatrix} = \begin{bmatrix} \dfrac{5}{2k_2} + \dfrac{k_2}{8} & \dfrac{1}{8} \\ \\ \dfrac{1}{8} & \dfrac{5}{8k_2} \end{bmatrix}$$

The performance index is given by

$$J = x'(0)Px(0)$$

$$= [c \ 0] \begin{bmatrix} p_{11} & p_{12} \\ p_{12} & p_{22} \end{bmatrix} \begin{bmatrix} c \\ 0 \end{bmatrix} = p_{11}c^2$$

$$= \left(\frac{5}{2k_2} + \frac{k_2}{8} \right) c^2$$

To minimize J we set $\dfrac{\partial J}{\partial k_2} = 0$, thus

$$\frac{\partial J}{\partial k_2} = \left(\frac{-5}{2k_2^2} + \frac{1}{8} \right) c^2 = 0$$

Solving, we get $k_2 = \sqrt{20}$.

For $k_2 = \sqrt{20}$, we have $\dfrac{\partial^2 J}{\partial k_2^2} > 0$, thus for $k_2 = \sqrt{20}$, J reaches its minimum value

$$J_{min} = \frac{\sqrt{5}}{2}c^2$$

The control law of the designed system is

$$u = -4x_1 - \sqrt{20}x_2$$

● **PROBLEM** 16-5

Consider the functional

$$F = \int_0^1 Ldt \qquad L = \frac{x^2}{2} \tag{1}$$

Find a function $x(t)$ which passes through $x(0) = \dfrac{7}{4}$ and $x(1) = \dfrac{5}{4}$, and which minimizes the functional such that

$$|\dot{x}| \leq 1 \tag{2}$$

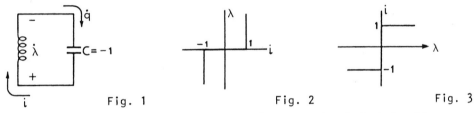

Fig. 1   Fig. 2   Fig. 3

Solution: To include the inequality constraint into the functional F let us define a new Lagrangian

$$L'(x,\dot{x}) = \Delta(1 - |\dot{x}|) + \frac{x^2}{2} \tag{3}$$

where Δ is defined as

$$\Delta(x) = \begin{cases} \dfrac{1}{2\mu}x^2 & x < 0 \\ 0 & x \geq 0 \end{cases} \tag{4}$$

and μ is a small, positive constant.

The modified functional is

$$F' = \int_0^1 L'dt \tag{5}$$

For $x \Rightarrow q$, we find that this is equivalent to a network composed of a -1-farad capacitor and a nonlinear inductor with the magnetic coenergy functon

902

$$W'_m(\dot{q}) = \Delta(1 - |\dot{q}|) \tag{6}$$

Fig. 1 shows the network and Fig. 2 the inductor branch relation $\lambda(i)$.

The Euler-Lagrange equation for the functional F' is

$$\frac{d}{dt}\left(\frac{\partial L'}{\partial \dot{x}}\right) - \frac{\partial L'}{\partial x} = 0 \tag{7}$$

or

$$\frac{d}{dt}[\lambda(i)] - q = 0 \tag{8}$$

where

$$i = \dot{q}.$$

Eq. (8) is the KVL equation for the network and all extremals must satisfy it. The form of the inductor characteristic requires that $\dot{q}$ remain at the extreme values ±1 for all time except at the points where λ changes sign. We shall rewrite eq. (8) in the form

$$\dot{q} = i(\lambda) \tag{9}$$

$$\dot{\lambda} = q$$

Fig. 3 shows $i(\lambda)$.

To obtain the complete solution to the minimization problem we shall select the appropriate initial charge $q(0) = \frac{7}{4}$ and a value of initial flux linkage $\lambda(0)$, which produces an extremal passing through the point $q(1) = \frac{5}{4}$. We find

$$\lambda(0) = -\frac{33}{32} \tag{10}$$

Using these initial conditions we find the resultant waveforms $q(t)$ and $\lambda(t)$. See Fig. 4. The circled numbers indicate corresponding points on the three curves.

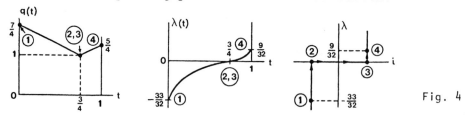

Fig. 4

The initial value $\lambda(0)$ determines the time at which the inductor current switches from -1 to +1.

Choosing the initial condition properly, we can obtain any final value which satisfies the inequality

$$|q(1) - q(0)| \leq 1 \tag{11}$$

Due to the velocity constraints, boundary conditions which do not satisfy eq. (11) represent problems with no solution.

The system is described by

$$\dot{x} = \begin{bmatrix} -3 & 2 \\ -1 & -1 \end{bmatrix} x + \begin{bmatrix} 1 & -1 \\ 0 & 1 \end{bmatrix} \begin{bmatrix} u_1(t) \\ u_2(t) \end{bmatrix} \tag{1}$$

with the control input $u(t)$ limited by $u^T(t)u(t) \leq 1$ for all t. Find the feedback control law which minimizes $\dot{V}(x)$.

Solution: Let P be a positive definite symmetric matrix, and choose

$$V(x) = x^T Px.$$

P satisfies $A^T P + PA = -I$.

Solving we get

$$P = \frac{1}{40} \begin{bmatrix} 7 & -1 \\ -1 & 18 \end{bmatrix} \tag{2}$$

The derivative of $V(x)$ is

$$\dot{V}(x) = \dot{x}^T Px + x^T P\dot{x}$$

$$= [Ax + Bu]^T Px + x^T P[Ax + Bu] \tag{3}$$

$$= x^T A^T Px + u^T B^T Px + x^T PAx + x^T PBu$$

$$= x^T [A^T P + PA]x + u^T B^T Px + [u^T B^T Px]^T$$

$$= -x^T x + 2u^T B^T Px$$

Eq. (1) was used to substitute for $\dot{x}$ in eq. (3). We also used the equality

$$A^T P + PA = -I.$$

Note that $u^T B^T Px$ is symmetric since its size is $|x|$. The control $u(t)$ is selected to minimize $u^T B^T Px$ subject to the restriction placed upon $u(t)$. $u(t)$ is parallel to $B^T Px$ but with the opposite sign, it has its largest possible magnitude. Thus

$$u(t) = -\frac{B^T Px(t)}{||B^T Px(t)||} \tag{4}$$

Substituting the expression for the matrices B and P we get

$$u(t) = \frac{\begin{bmatrix} -7x_1(t) + x_2(t) \\ 8x_1(t) - 19x_2(t) \end{bmatrix}}{(113x_1^2 - 318x_1x_2 + 362x_2^2)^{\frac{1}{2}}} \tag{5}$$

The transfer function of the dc motor is

$$\frac{\Omega(s)}{V(s)} = \frac{K'}{s + a} \tag{1}$$

The motor is initially at rest, $\omega(0) = 0$. Find the input $v(t)$ which minimizes J

$$J = \int_0^T v^2(\tau)\,d\tau \tag{2}$$

and gives an angular velocity $\omega(T) = 100$ at $t = T$.

Solution: We use the system weighting function to write the input-output relationship. Since

$$W(t,0) = L^{-1}\left\{\frac{K'}{s + a}\right\} = K'e^{-at},$$

the weighting function is $W(t,\tau) = K'e^{-a(t - \tau)}$. We get

$$\omega(T) = \int_0^T W(T,\tau)v(\tau)\,d\tau \tag{3}$$

in the form of an inner product

$$100 = <W(T,\tau),\ v(\tau)>$$

thus the Cauchy-Schwarz inequality gives

$$100 = |<W(T,\tau),\ v(\tau)>| \leq ||W(T,\tau)|| \cdot ||v(\tau)|| \tag{4}$$

When the equality holds we obtain the minimum value of $||v(\tau)||$. Thus

$$||v(\tau)|| = \frac{100}{||W(T,\tau)||} = \frac{100}{\left\{\int_0^T [K'e^{-a(T-\tau)}]^2\,d\tau\right\}^{\frac{1}{2}}}$$

$$= \frac{100\sqrt{2a}}{K'[1 - e^{-2aT}]^{\frac{1}{2}}} \tag{5}$$

The equality holds only when $v(t)$ and $W(T,t)$ are linearly dependent, that means for some α

$$v(t) = \alpha W(T,t)$$

We get

$$||v|| = |\alpha|\,||w|| = \frac{100}{||w||}$$

and

$$|\alpha| = \frac{100}{||w||^2} \quad \text{and} \quad v_{optimal}(t) = \frac{100}{||w||^2}W(T,t)$$

$$= \frac{200ae^{-a(T-t)}}{K'[1 - e^{-2aT}]} \tag{6}$$

Given the system

$$\frac{dx_1}{dt} = m \tag{1}$$

$$\frac{dx_2}{dt} = x_1^2 + \rho m^2$$

with $x_1(t_0) = x_{10}$,

minimize the following performance index

$$J(m) = \int_{t_0}^{t_f} (x_1^2 + \rho m^2) dt$$

Solution: The Hamiltonian for the system is

$$H = \lambda_1 m + \lambda_2 (x_1^2 + \rho m^2)$$

The adjoint equations are

$$\frac{d\lambda_1}{dt} = -2x_1 \lambda_2$$

$$\frac{d\lambda_2}{dt} = 0$$

and the final condition is

$$\lambda(t_f) = \begin{pmatrix} 0 \\ 1 \end{pmatrix}$$

which gives

$$\lambda_2(t) = 1$$

We obtain

$$\frac{\partial H}{\partial m} = \lambda_1 + 2\rho m \tag{2}$$

$$\frac{d\lambda_1}{dt} = -2x_1 \tag{3}$$

Let us choose the simple control

$$m(t) = \frac{-x_{10}}{t_f - t_0} \qquad t_0 \le t \le t_f \tag{4}$$

Integrating eq. (1) forward we obtain

$$x_1(t) = x_{10} \left(\frac{t_f - t}{t_f - t_0} \right)$$

This result and the condition $\lambda_1(t_f) = 0$ allows us
to integrate eq. (3) backward, thus

$$\lambda_1(t) = \frac{x_{10}(t_f - t)^2}{t_f - t_0}$$

From eq. (2)

$$\frac{\partial H}{\partial m} = \frac{x_{10}}{t_f - t_0}[(t_f - t)^2 - 2\rho]$$

Let us take $G(t) = 1$ for distance measure. The direction of steepest descent is given by

$$\delta_1 m = m_1(t) - m_2(t) = -W(t)\frac{\partial H_1}{\partial m}$$

and

$$\delta m = -\alpha\frac{x_{10}}{t_f - t_0}[(t_f - t)^2 - 2\rho]$$

In this case $\alpha > 0$ is proportional to the distance represented by the move δm. From eq. (4) we get

$$m^{(2)}(t) = -\frac{x_{10}}{t_f - t_0}\left\{1 + \alpha[(t_f - t)^2 - 2\rho]\right\}$$

We know that $m(t_f) = 0$, thus we choose a distance given by

$$\alpha = \tfrac{1}{2}\rho$$

We get the new m

$$m^{(2)}(t) = \frac{-x_{10}(t_f - t)^2}{2\rho(t_f - t_0)}$$

which satisfies the correct final condition. We obtain the optimal value m*(t) as

$$m^*(t) = \frac{-x_{10}}{\sqrt{\rho}}\frac{\sinh[(t_f - t)/\sqrt{\rho}]}{\cosh[\dfrac{t_f - t_0}{\sqrt{\rho}}]}$$

● **PROBLEM** 16-9

Given the system

$$\frac{dx_1}{dt} = -x_1 + m$$

minimize

$$J(m) = \int_{t_0}^{t_f}[x_1^2(t) + \rho m^2(t)]dt$$

Solution: The matrices describing this problem are

$$A = -1 \qquad P = 0$$

$$B = 1 \qquad Q = 2$$

$$G = 1 \qquad R = 2\rho$$

$$H = 0 \qquad r(t) = 0$$

where

$$x(t) = Ax + Bm$$

$$c = Gx + Hm$$

$$J(m) = \tfrac{1}{2}x^T P x + \tfrac{1}{2}\int_{t_0}^{t_f} [x^T Q x + m^T R m]\, dt$$

From

$$U = BRB^T$$

$$V = G^T Q G$$

we get

$$U = \frac{1}{2\rho}$$

$$V = 2$$

$$\mu(t) = 0$$

From

$$\frac{dK}{dt} = -(KA + A^T K) - V + KUK$$

and the Riccati equation we get

$$\frac{dk_1}{dt} = 2k_1 - 2 + \frac{k_1^2}{2\rho}$$

with final condition

$$k_1(t_f) = 0$$

Direct integration leads to

$$t_f - t = \int_{k_1}^{0} \frac{d\zeta}{2\zeta - 2 + \frac{\zeta^2}{2\rho}}$$

We obtain the optimal control

$$k_1 = \frac{2}{1 + \alpha \coth \alpha (t_f - t)}$$

where

$$\alpha = \sqrt{1 + \rho^{-1}}$$

908

Assume the control system to be

$$u(t) = -Kx(t)$$

For the control system shown in Fig. 1 design the optimal feedback gain matrix K such that the following performance index is minimized:

$$J = \int_0^\infty (x'Qx + u'u)\,dt$$

where

$$Q = \begin{bmatrix} 1 & 0 \\ 0 & \mu \end{bmatrix} \qquad (\mu \geq 0)$$

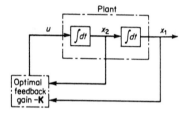

Fig. 1

Solution: The equation for the plant is

$$\begin{bmatrix} \dot{x}_1 \\ \dot{x}_2 \end{bmatrix} = \begin{bmatrix} 0 & 1 \\ 0 & 0 \end{bmatrix} \begin{bmatrix} x_1 \\ x_2 \end{bmatrix} + \begin{bmatrix} 0 \\ 1 \end{bmatrix} [u]$$

We have

$$Q = \begin{bmatrix} 1 & 0 \\ 0 & \sqrt{\mu} \end{bmatrix} \begin{bmatrix} 1 & 0 \\ 0 & \sqrt{\mu} \end{bmatrix} = S'S$$

then

$$[S' \mid A'S'] = \begin{bmatrix} 1 & 0 & 0 & 0 \\ 0 & \sqrt{\mu} & 1 & 0 \end{bmatrix}$$

The rank of $[S' \mid A'S']$ is found to be 2. Thus A-BK is a stable matrix, and the Liapunov approach gives the correct result. We shall use the reduced-matrix Riccati equation in the design of the optimal control system.

In a more general form, the performance index is

$$J = \int_0^\infty (x'Qx + U'RU)\,dt \qquad\qquad (1)$$

where Q and R are positive definite and symmetric. Since U = -KX,

$$J = \int_0^\infty [x'Qx + x'K'RKx]dt = \int_0^\infty x'[Q + K'RK]x\,dt \qquad (2)$$

Define a matrix P such that P is positive definite and symmetric, and such that

$$\int (x'Qx + U'RU)dt = -x'Px \qquad (3)$$

Taking derivatives and using eq. (2) we get

$$x'[Q + K'RK]x = -\dot{x}'Px - x'P\dot{x} \qquad (4)$$

Now

$$\dot{x} = (A - BK)x,$$

so

$$\dot{x}' = x'(A - BK)'.$$

Substituting into eq. (4)

$$x'[Q + K'RK]x + x'(A - BK)'Px + x'P(A - BK)x = 0 \qquad (5)$$

or

$$x'[Q + K'RK + (A - BK)'P + P(A - BK)]x = 0 \qquad (6)$$

Therefore

$$Q + K'RK + (A - BK)'P + P(A - BK) = 0 \qquad (7)$$

or

$$Q + K'RK + A'P - K'B'P + PA - PBK = 0 \qquad (8)$$

It can be shown that for minimizing J, the optimal K is

$$K = R^{-1}B'P \qquad (9)$$

Substituting into eq. (8), we get (using $P' = P$, $(R^{-1})' = R^{-1}$)

$$Q + PBR^{-1}RR^{-1}B'P + A'P - PBR^{-1}B'P$$

$$+ PA - PBR^{-1}B'P = 0 \qquad (10)$$

Simplifying

$$A'P + PA - PBR^{-1}B'P + Q = 0 \qquad (11)$$

or

$$\begin{bmatrix} 0 & 0 \\ 1 & 0 \end{bmatrix}\begin{bmatrix} p_{11} & p_{12} \\ p_{12} & p_{22} \end{bmatrix} + \begin{bmatrix} p_{11} & p_{12} \\ p_{12} & p_{22} \end{bmatrix}\begin{bmatrix} 0 & 1 \\ 0 & 0 \end{bmatrix}$$

$$- \begin{bmatrix} p_{11} & p_{12} \\ p_{12} & p_{22} \end{bmatrix}\begin{bmatrix} 0 \\ 1 \end{bmatrix}[1][0\ 1]\begin{bmatrix} p_{11} & p_{12} \\ p_{12} & p_{22} \end{bmatrix} + \begin{bmatrix} 1 & 0 \\ 0 & \mu \end{bmatrix}$$

$$= \begin{bmatrix} 0 & 0 \\ 0 & 0 \end{bmatrix}$$

Simplifying we get

$$\begin{bmatrix} 0 & 0 \\ p_{11} & p_{12} \end{bmatrix} + \begin{bmatrix} 0 & p_{11} \\ 0 & p_{12} \end{bmatrix} - \begin{bmatrix} p_{12}^2 & p_{12}p_{22} \\ p_{12}p_{22} & p_{22}^2 \end{bmatrix}$$

$$+ \begin{bmatrix} 1 & 0 \\ 0 & \mu \end{bmatrix} = \begin{bmatrix} 0 & 0 \\ 0 & 0 \end{bmatrix}$$

We obtain the set of equations:

$$1 - p_{12}^2 = 0$$

$$p_{11} - p_{12}p_{22} = 0$$

$$\mu + 2p_{12} - p_{22}^2 = 0$$

Solving the equations and requiring P to be positive definite we get

$$P = \begin{bmatrix} p_{11} & p_{12} \\ p_{12} & p_{22} \end{bmatrix} = \begin{bmatrix} \sqrt{\mu + 2} & 1 \\ 1 & \sqrt{\mu + 2} \end{bmatrix}$$

The optimal feedback gain matrix is

$$K = R^{-1}B'P = [1][0 \; 1]\begin{bmatrix} p_{11} & p_{12} \\ p_{12} & p_{22} \end{bmatrix}$$

$$= [p_{12} \quad p_{22}] = [1 \; \sqrt{\mu + 2}]$$

and the optimal control signal

$$u = -Kx = -x_1 - \sqrt{\mu + 2}x_2$$

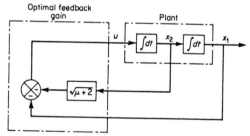

Fig. 2

The control law given by eq. (1) gives an optimal result for any initial state under the given performance index. The block diagram for this system is shown in Fig. 2.

● **PROBLEM** 16-11

Given the system described by

$$\begin{bmatrix} \dot{x}_1 \\ \dot{x}_2 \end{bmatrix} = \begin{bmatrix} 0 & 1 \\ 0 & 0 \end{bmatrix}\begin{bmatrix} x_1 \\ x_2 \end{bmatrix} + \begin{bmatrix} 0 \\ 1 \end{bmatrix}[u]$$

determine the function u that minimizes the performance index

$$J = \int_0^\infty (x'Qx + u'u)\,dt \qquad Q = \begin{bmatrix} 1 & 0 \\ 0 & \mu \end{bmatrix}$$

Solution: Let us define

$$V = x'Px$$

where

$$P = \begin{bmatrix} p_{11} & p_{12} \\ p_{12} & p_{22} \end{bmatrix}$$

Thus

$$V = p_{11}x_1^2 + 2p_{12}x_1x_2 + p_{22}x_2^2$$

and

$$\dot{V} = 2p_{11}x_1x_2 + 2p_{12}x_2^2 + (2p_{12}x_1 + 2p_{22}x_2)u$$

Then

$$H(x, u) = \frac{dV}{dt} + L(x, u) = \frac{dV}{dt} + x'Qx + U'U$$

$$= 2p_{11}x_1x_2 + 2p_{12}x_2^2 + (2p_{12}x_1 + 2p_{22}x_2)u$$

$$+ x_1^2 + \mu x_2^2 + u^2 \tag{1}$$

We differentiate $H(x,u)$ with respect to u and set the resulting equation equal to zero. Note that the optimal control signal $u(t)$ minimizes $H(x,u)$.

$$\frac{\partial H}{\partial u} = 2p_{12}x_1 + 2p_{22}x_2 + 2u = 0$$

Solving, we obtain

$$u = -p_{12}x_1 - p_{22}x_2 \tag{2}$$

Substituting eq. (2) into eq. (1) and setting the result equal to zero, we obtain

$$x_1^2(1 - p_{12}^2) + x_1x_2(2p_{11} - 2p_{12}p_{22})$$

$$+ x_2^2(\mu + 2p_{12} - p_{22}^2) = 0$$

Since the last equation holds for any x_1 and x_2 we have

$$1 - p_{12}^2 = 0$$

$$p_{11} - p_{12}p_{22} = 0$$

$$\mu + 2p_{12} - p_{22}^2 = 0$$

Solving the equations and requiring that P be positive definite we get

$$p_{11} = \sqrt{\mu + 2}, \; p_{12} = 1, \; p_{22} = \sqrt{\mu + 2}$$

The optimal control law is

$$u = -x_1 - \sqrt{\mu + 2x_2}$$

and the system state equation becomes

$$\begin{bmatrix} \dot{x}_1 \\ \dot{x}_2 \end{bmatrix} = \begin{bmatrix} 0 & 1 \\ 0 & 0 \end{bmatrix} \begin{bmatrix} x_1 \\ x_2 \end{bmatrix} + \begin{bmatrix} 0 \\ 1 \end{bmatrix} [-x_1 - \sqrt{\mu + 2x_2}]$$

or

$$\begin{bmatrix} \dot{x}_1 \\ \dot{x}_2 \end{bmatrix} = \begin{bmatrix} 0 & 1 \\ -1 & -\sqrt{\mu + 2} \end{bmatrix} \begin{bmatrix} x_1 \\ x_2 \end{bmatrix}$$

Since the coefficient matrix in the last equation is a stable one, the origin of the system is asymptotically stable.

● **PROBLEM** 16-12

The system is described by

$$\ddot{x} + 2\zeta\dot{x} + x = 0 \qquad 0 < \zeta \le 0.9 \qquad (1)$$

Find the value of ζ which minimizes the performance index J.

$$J = \int_0^\infty [x^2(t) + \dot{x}^2(t)] dt$$

Solution: Solving eq. (1) we get

$$x(t) = k_1 e^{\lambda_1 t} + k_2 e^{\lambda_2 t}$$

where

$$\lambda_1 = -\zeta + \sqrt{\zeta^2 - 1} \qquad \lambda_2 = -\zeta - \sqrt{\zeta^2 - 1}$$

The parameters k_1 and k_2 are determined from the initial conditions. Thus

$$\dot{x}(t) = k_1\lambda_1 e^{\lambda_1 t} + k_2\lambda_2 e^{\lambda_2 t}$$

We have

$$x^2(t) = k_1^2 e^{2\lambda_1 t} + 2k_1 k_2 e^{(\lambda_1 + \lambda_2)t} + k_2^2 e^{2\lambda_2 t}$$

and

$$\dot{x}^2(t) = k_1^2\lambda_1^2 e^{2\lambda_1 t} + 2k_1 k_2 \lambda_1 \lambda_2 e^{(\lambda_1 + \lambda_2)t} + k_2^2\lambda_2^2 e^{2\lambda_2 t}$$

We obtain for the performance index

$$J = \int_0^\infty [x^2(t) + \dot{x}^2(t)] dt = \int_0^\infty [k_1^2(1 + \lambda_1^2)e^{2\lambda_1 t}$$

$$+ 2k_1 k_2 (1 + \lambda_1\lambda_2)e^{(\lambda_1 + \lambda_2)t}$$

$$+ k_2^2(1 + \lambda_2^2)e^{2\lambda_2 t}] dt.$$

The real parts of λ_1 and λ_2 are negative. Integrating and evaluating at $t = \infty$ and $t = 0$ we get

$$J = - [\frac{k_1^2 (1 + \lambda_1^2)}{2\lambda_1} + \frac{2k_1 k_2 (1 + \lambda_1 \lambda_2)}{\lambda_1 + \lambda_2} + \frac{k_2^2 (1 + \lambda_2^2)}{2\lambda_2}]$$

Using

$$\lambda_1 + \lambda_2 = -2\zeta, \quad \lambda_1 \lambda_2 = 1$$

we can rewrite the expression for J.

$$J = \zeta (k_1^2 + k_2^2) + \frac{2k_1 k_2}{\zeta} = \zeta (k_1 + k_2)^2 - 2k_1 k_2 (\zeta - \frac{1}{\zeta}) \quad (2)$$

The initial conditions are

$$x(0) = c_1 = k_1 + k_2 \qquad \dot{x}(0) = c_2 = k_1 \lambda_1 + k_2 \lambda_2$$

Since $0 < \zeta \le 0.9$, $\lambda_1 \ne \lambda_2$ and k_1 and k_2 are obtained as

$$k_1 = \frac{c_1 \lambda_2 - c_2}{\lambda_2 - \lambda_1}, \quad k_2 = \frac{c_2 - \lambda_1 c_1}{\lambda_2 - \lambda_1}$$

Substituting the values of k_1 and k_2 into eq. (2) we obtain

$$J = \zeta c_1^2 + c_1 c_2 + \frac{c_1^2 + c_2^2}{2\zeta} \quad (3)$$

Differentiating with respect to ζ and taking $\frac{\partial J}{\partial \zeta} = 0$ we obtain

$$\frac{\partial J}{\partial \zeta} = c_1^2 - \frac{c_1^2 + c_2^2}{2\zeta^2} = 0$$

or

$$\zeta = \sqrt{\frac{c_1^2 + c_2^2}{2c_1^2}} \quad (4)$$

Eq. (4) gives the optimal value of ζ, since $\frac{\partial^2 J}{\partial \zeta^2} > 0$.

Since $\zeta \le 0.9$ we get

$$\sqrt{\frac{c_1^2 + c_2^2}{2c_1^2}} \le 0.9$$

or

$$c_2^2 \le 0.62 c_1^2$$

For $c_2 = 0$ we get

$$\zeta = \frac{1}{\sqrt{2}}$$

as the optimal value of ζ. For $c_1 = 1$ and $c_2 = 0$ we have

$$\min J = \zeta + \frac{1}{2\zeta} = 1.414$$

For $c_2^2 > 0.62 c_1^2$, eq. (3) becomes

$$\zeta = \sqrt{\frac{c_1^2 + c_2^2}{2c_1^2}} > \sqrt{\frac{1}{2} + \frac{0.62}{2}} = 0.9$$

914

Since $0 < \zeta \le 0.9$, eq. (3) will not give the optimal value of ζ.

We shall define

$$c_2^2 = (0.62 + 2h)c_1^2, \quad h > 0$$

Eq. (2) becomes

$$J = (\zeta + \frac{0.81 + h}{\zeta})c_1^2 + c_1 c_2$$

Minimizing I with respect to ζ, where

$$I = \zeta + \frac{0.81 + h}{\zeta}$$

we find the optimal value of ζ which minimizes J for given values of c_1 and c_2. For $\zeta^2 = 0.81 + h$, the value of I is minimum. This value of ζ cannot be accepted because $0 < \zeta \le 0.9$.

The value of I increases as ζ decreases from $\sqrt{0.81 + h}$, the optimal value of ζ when $c_2^2 > 0.62c_1^2$ is 0.9.

Thus, the optimal value of ζ is

$$\zeta = \sqrt{\frac{c_1^2 + c_2^2}{2c_1^2}} \quad \text{for} \quad c_2^2 \le 0.62c_1^2$$

$$= 0.9 \quad \text{for} \quad c_2^2 > 0.62c_1^2$$

● **PROBLEM** 16-13

A linear time-invariant system is described by the state equation

$$\dot{x}(t) = -2x(t) + u(t) \tag{1}$$

with the initial condition $x(t_0)$. Determine the optimal control $u^0(t)$ such that

$$J = \frac{1}{2}x^2(t_f) + \frac{1}{2}\int_{t_0}^{t_f} [x^2(t) + u^2(t)]dt \tag{2}$$

reaches its minimal value.

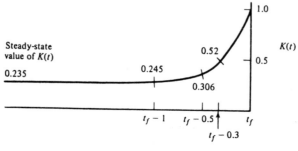

Fig. 1

Solution: Eq. (1) is of the form

$$\dot{x} = Ax + B \tag{3}$$

and eq. (2) is of the form

$$J = \frac{1}{2}x'Px + \frac{1}{2}\int_{t_0}^{t_f}[x'Qx + U'RU]dt \tag{4}$$

We define matrix M as

$$M = \begin{bmatrix} A & -BR^{-1}B' \\ -Q & -A' \end{bmatrix} \tag{5}$$

Let

$$w = \begin{bmatrix} w_{11} & w_{12} \\ w_{21} & w_{22} \end{bmatrix}$$

be a matrix such that

$$T = w^{-1}Mw = \begin{bmatrix} -\Lambda & 0 \\ 0 & \Lambda \end{bmatrix} \tag{6}$$

where Λ is a diagonal matrix containing the eigenvalues of M with positive real part. Define

$$H(\tau) = -e^{-\Lambda t}(w_{22} - Pw_{12})^{-1}(w_{21} - Pw_{11})e^{-\Lambda t} \tag{7}$$

and

$$K(t_f - \tau) = [w_{21} + w_{22}H(\tau)][w_{11} + w_{12}H(\tau)]^{-1} \tag{8}$$

The optimal control is

$$U^0 = -R^{-1}B'K(t)x(t) \tag{9}$$

Let us assume that the final time t_f is finite.
From eqs. (3) and (4) we have

$$A = -2, \ B = 1, \ P = 1, \ Q = 1, \ R = 1.$$

Using eq. (5) we find the matrix M.

$$M = \begin{bmatrix} A & -BR^{-1}B' \\ -Q & -A' \end{bmatrix} = \begin{bmatrix} -2 & -1 \\ -1 & 2 \end{bmatrix}$$

The eigenvalues of M are $\lambda_1 = -\sqrt{5}$, $\lambda_2 = \sqrt{5}$, and the eigenvectors are

$$p_1 = \begin{bmatrix} 1 \\ 0.235 \end{bmatrix} \quad p_2 = \begin{bmatrix} 1 \\ -4.235 \end{bmatrix} \tag{10}$$

916

We get

$$W = \begin{bmatrix} w_{11} & w_{12} \\ w_{21} & w_{22} \end{bmatrix} = \begin{bmatrix} 1 & 1 \\ 0.235 & -4.235 \end{bmatrix} \tag{11}$$

The eigenvectors are arranged in W in such a way that they correspond to the requirement of the matrix T in eq. (6). Using the results, we obtain

$$H(\tau) = -0.146e^{-2\sqrt{5}\tau} \tag{12}$$

Substituting $H(\tau)$ in eq. (8) we find the time-varying Riccati gain.

$$K(t_f - \tau) = \frac{0.235 + 0.6189e^{-2\sqrt{5}\tau}}{1 - 0.146e^{-2\sqrt{5}\tau}} \tag{13}$$

When $t = t_f$, $\tau = 0$, $K(t_f) = P = 1$.
A plot of $K(t)$ as a function of t is shown in Fig. 1.

From eq. (9), setting $t \to \infty$, we obtain the steady-state solution of $K(t)$. In the case of $t_f = \infty$, since the system described in eq. (1) is completely controllable, the optimal solution exists. The constant Riccati gain K is solved from eq. (9) by setting $\tau = \infty$. Thus $K = 0.235$ which is the steady-state value of $K(t)$.
For the finite-time problem the optimal control is

$$u^0(t) = -R^{-1}B'K(t)x(t) \tag{14}$$

$$= -K(t)x(t) \qquad t \le t_f$$

and for the infinite-time problem, it is

$$u^0(t) = -0.235x(t) \tag{15}$$

● PROBLEM 16-14

For the system described by

$$\frac{d^2c}{dt^2} + 2\zeta\frac{dc}{dt} + c = m(t)$$

minimize the performance index

$$J(m) = \frac{1}{2}\int_{t_0}^{\infty} [c^2(t) + \rho m^2(t)]dt$$

with the initial condition

$$c(t_0) = c_0, \quad \frac{dc(t_0)}{dt} = 0$$

Solution: Since the output variable $c(t)$ is defined as a deviation around the final steady-state, the condition m = 0 is required to hold the process of the desired steady state $r(t) = 0$. Let the state variables be

$$x_1 = c, \quad x_2 = \frac{dc}{dt}$$

The problem is described by

$$x = Ax(t) + Bm(t)$$

$$c(t) = Gx(t) + Hm(t)$$

$$J(m) = \frac{1}{2} \int_{t_0}^{\infty} [c^T(t)Qc(t) + m^T(t)Rm(t)]dt$$

where

$$A = \begin{pmatrix} 0 & 1 \\ -1 & -2\zeta \end{pmatrix} \qquad B = \begin{pmatrix} 0 \\ 1 \end{pmatrix}$$

$$R = \rho \qquad Q = 1 \qquad G = (1 \; 0) \qquad H = 0$$

Using

$$U(t) = B(t)R^{-1}(t)B^T(t)$$

$$V(t) = G^T(t)Q(t)G(t)$$

we get

$$U = \frac{1}{\rho}\begin{pmatrix} 0 & 0 \\ 0 & 1 \end{pmatrix} \qquad V = \begin{pmatrix} 1 & 0 \\ 0 & 0 \end{pmatrix}$$

Then

$$-(K_s A + A^T K_s) - V + K_s U K_s = 0$$

becomes

$$2k_{12} - 1 + \frac{1}{\rho}k_{12}^2 = 0$$

$$2\zeta k_{12} + k_{22} - k_{11} + \frac{1}{\rho}k_{12}k_{22} = 0 \qquad (1)$$

$$4\zeta k_{22} - 2k_{12} + \frac{1}{\rho}k_{22}^2 = 0$$

The matrix K_s is symmetric, $k_{ij} = k_{ji}$.

$$K_s = \begin{pmatrix} k_{11} & k_{12} \\ k_{12} & k_{22} \end{pmatrix}$$

918

K_s must be positive definite, thus

$$k_{11} > 0, \quad k_{22} > 0, \quad k_{11}k_{22} > k_{12}^2$$

From eq. (1) and restricting conditions, we find

$$k_{11} = \rho[\alpha\sqrt{4\zeta^2 + 2(\alpha - 1)} - 2\zeta]$$

$$k_{12} = \rho(\alpha - 1)$$

$$k_{22} = \rho[\sqrt{4\zeta^2 + 2(\alpha - 1)} - 2\zeta]$$

where

$$\alpha = \sqrt{1 + \rho^{-1}}$$

From

$$m^*(t) = -R^{-1}(t)B^T(t)[k(t)x(t) - u(t)]$$

where

$$\mu(t) = 0,$$

we find the optimal control.

$$m(t) = -\frac{1}{\rho}[k_{12}x_1(t) + k_{22}x_2(t)]$$

$$= -[(\alpha - 1)x_1 + (\sqrt{4\zeta^2 + 2(\alpha - 1)} - 2\zeta)x_2]$$

$$= -\left[(\alpha - 1)c + (\sqrt{4\zeta^2 + 2(\alpha - 1)} - 2\zeta)\frac{dc}{dt}\right]$$

The figure shows the block diagram for the optimal control.

In this example the feedback control element is unrealizable and has to be approximated, for example by

$$\frac{\left(\dfrac{k_{22}}{\rho}\right)s + \dfrac{k_{12}}{\rho}}{\gamma s + 1}$$

where γ is small. The characteristic equation of this closed loop is

$$s^2 + s\sqrt{4\zeta^2 + 2(\alpha - 1)} + \alpha = 0$$

thus the closed loop has natural frequency ω_n,

$$\omega_n = \sqrt{\alpha}$$

and damping factor ζ_c,

$$\zeta_c = \frac{1}{2}\sqrt{\frac{4\zeta^2 + 2(\alpha - 1)}{\alpha}}$$

The scalar system is described by

$$x(k+1) = x(k) + u(k)$$

with the boundary conditions $x(0) = 0$, $x(3) = 3$.

Find the controls $u(0)$, $u(1)$, $u(2)$ which minimize

$$J = \sum_{k=0}^{2} \{u^2(k) + \Delta t_k^2\}$$

Solution: The terminal point is $x(3)$, let $g(x(k))$ be the minimum cost from $x(k)$ to the terminal point, thus

$$g(x(3)) = 0$$

Then

$$g(x(2)) = \min_{u(2)} \{\text{cost from } x(2) \text{ to the terminal point}\}$$

$$= \min_{u(2)} \{u^2(2) + \Delta t_2^2 + g(x(3))\}$$

The restriction is $x(3) = 3$.

Using $x(3) = x(2) + u(2)$ we find

$$u(2) = 3 - x(2)$$

For any $x(2)$, there is a uniquely required $u(2)$.

Let $\Delta t_k = 1$.

Then

$$g(x(2)) = [3-x(2)]^2 + 1$$

and

$$g(x(1)) = \min_{u(1)} \{u^2(1) + 1 + g(x(2))\}$$

From $x(2) = x(1) + u(1)$ we get

$$g(x(1)) = \min_{u(1)} \{u^2(1) + 2 + [3-x(1) - u(1)]^2\}$$

To minimize $u(1)$ we set

$$\frac{\partial}{\partial u(1)} \{u^2(1) + 2 + [3-x(1) - u(1)]^2\} = 0$$

Solving, we get

$$u(1) = \frac{3 - x(1)}{2}$$

and

$$g(x(1)) = \left[3-x(1) - \frac{3-x(1)}{2}\right]^2 + 2 + \left[\frac{3-x(1)}{2}\right]^2$$

We use this result together with

$$x(1) = x(0) + u(0)$$

in

$$g(x(0)) = \min_{u(0)} \{u^2(0) + 1 + g(x(1))\}$$

We get

$$u(0) = \frac{3 - x(0)}{3} = 1$$

The difference equation gives $x(1) = 1$.

We get

$$u(1) = 1$$

and

$$x(2) = x(1) + u(1) = 2$$

and

$$u(2) = 3 - x(2) = 1.$$

Finally, we obtain

$$x(3) = x(2) + u(2) = 3.$$

In tx space, the sequence $x(0)$, $x(1)$, $x(2)$, $x(3)$ lies on a straight line.

● **PROBLEM 16-16**

For the system described by

$$\frac{dx_1}{dt} = -x_1 + m \qquad (1)$$

find the control that drives the system from an initial state $x_1(t_0) = x_{10}$ to the origin $x_1(t_f) = 0$ in minimum time $(t_f - t_0)$, with the constraint

$$\int_{t_0}^{t_f} m^2(t) dt = \gamma > 0 \qquad (2)$$

Solution: In this case we have to minimize the final time. From

$$J(m) = \int_{t_0}^{t_f} F(x,m)\,dt + G[x(t_f)]$$

we write the performance criterion

$$J(m) = \int_{t_0}^{t_f} dt = t_f - t_0$$

We define

$$J_p(m) = \int_{t_0}^{t_f} (1 + pm^2)\,dt$$

We obtain a solution to the original problem using the Lagrange multiplier principle.

We define the additional state variable

$$x_2 = \int_{t_0}^{t} (1 + pm^2)\,dt$$

and get

$$\frac{dx_2}{dt} = 1 + pm^2$$

$$S(x(t_f)) = x_2(t_f)$$

Therefore

$$H = \lambda_1(-x_1 + m) + \lambda_2(1 + pm^2)$$

$$\frac{d\lambda_1}{dt} = \lambda_1$$

$$\frac{d\lambda_2}{dt} = 0$$

$$\lambda_2(t_f) = \frac{\partial S}{\partial x_2(t_f)} = 1$$

and we get

$$\lambda_2(t) = 1$$

Differentiation of H yields

$$\frac{\partial H}{\partial m} = \lambda_1 + 2pm = 0$$

thus

$$m = -\frac{\lambda_1}{2p}$$

Solving the differential equation for λ_1 gives

$$\lambda_1 = \lambda_{10} e^{(t-t_0)}$$

We can use the condition $H^* = 0$ to determine the free initial condition λ_{10}, because the final time t_f is unspecified.

$$H = \lambda_{10}\, e^{(t-t_0)} \left[-x_1 - \frac{\lambda_{10}}{2p}\, e^{(t-t_0)} \right]$$

$$+ 1 + \frac{\lambda_{10}^2}{4p}\, e^{2(t-t_0)} = 0$$

Setting $\quad t = t_0 \quad$ we get

$$\frac{1}{4p}\, \lambda_{10}^2 + x_{10}\, \lambda_{10} - 1 = 0$$

Solving this equation we get

$$\lambda_{10} = -2p\, x_{10}(1\pm\alpha)$$

where

$$\alpha = \sqrt{1 + \frac{1}{px_{10}^2}}$$

Thus

$$m(t) = x_{10}(1\pm\alpha)\, e^{(t-t_0)}$$

To satisfy the final condition on x_1, we use the minus sign, thus

$$m(t) = -x_{10}(\alpha-1)\, e^{(t-t_0)} \tag{3}$$

Combining eqs. (1) and (3) we obtain

$$\frac{dx_1}{dt} + x_1 = -x_{10}(\alpha-1)\, e^{(t-t_0)}$$

Solving this equation and using the initial condition we have

$$x_1(t) = x_{10}[\cosh(t-t_0) - \alpha\sinh(t-t_0)] \tag{4}$$

Substituting eq. (4) with the value of λ_{10} into the expression for H we get

$$H = 0$$

From the condition $x_1(t_f) = 0$ we find the final time

t_f ,

$$t_f - t_0 = \coth^{-1}\alpha \tag{5}$$

We choose p that satisfies eq. (2). Using eq. (3) we obtain

$$\int_{t_0}^{t_f} m^2(t)\,dt = x_{10}^2(\alpha-1)^2 \int_{t_0}^{t_f} e^{2(t-t_0)}\,dt = \gamma$$

or

$$x_{10}^2(\alpha-1)^2[e^{2(t_f-t_0)} - 1] = 2\gamma$$

We see that p depends on x_{10} and γ; it satisfies the relation

$$x_{10}^2(\alpha-1)^2[e^{(2\coth^{-1}\alpha)} - 1] = 2\gamma \tag{6}$$

Solving eq. (6) and taking into account the initial condition x_{10} and constraint γ, we obtain

$$p = \frac{1}{x_{10}^2(\alpha^2-1)}$$

Eq. (5) gives the minimum time to reach the origin subject to the given constraint. Eq. (3) gives the optimal control and eq. (4) the optimal response.

● PROBLEM 16-17

For the system

$$\frac{dx_1}{dt} = m$$

determine m(t) that minimizes

$$J(m) = \int_{t_0}^{t_f} e^{\sigma(t-t_0)}(x_1^2 + \rho m^2)\,dt$$

with t_0 fixed and t_f not fixed.

TABLE

$t - t_0$	$x_1(t)/x_{10}, \sigma = 3$	$x_1(t)/x_{10}, \sigma = 0$
0	1.000	1.000
0.1	0.682	.928
0.2	0.470	.865
0.5	0.175	.730
1.0	0.089	.647

Solution: We define

$$x_2(t) = \int_{t_0}^{t} e^{\sigma(\tau - t_0)} (x_1^2 + \rho m^2) \, d\tau$$

Taking the time derivative, we obtain

$$\frac{dx_2(t)}{dt} = e^{\sigma(t - t_0)} [x_1^2 + \rho m^2]$$

The system is nonautonomous, therefore we must transform it to an autonomous system. If our system is described by

$$\dot{x} = f(x, m, t)$$

and our performance index is described by

$$J(m) = \int_{t_0}^{t_f} F(x, m, t) \, dt + G[x(t_f)]$$

we can define an augmented state vector

$$x^a = \begin{pmatrix} x_0 \\ x \\ x_{n+1} \end{pmatrix}$$

where

$$x_0 = t$$

$$x_{n+1} = \int_{t_0}^{t_f} F(x, m, x_0) \, dt$$

Then

$$\dot{x}^a = f^a(x^a, m)$$

$$f^a = \begin{pmatrix} 1 \\ f \\ F \end{pmatrix}$$

is an autonomous system.

Using the above transformation we obtain

$$\frac{dx^a}{dt} = f^a(x, m)$$

925

with

$$f = \begin{pmatrix} 1 \\ m \\ e^{\sigma(x_0 - t_0)} (x_1^2 + \rho m^2) \end{pmatrix}$$

which is an autonomous system.

The Hamiltonian is

$$H = \lambda_0 + \lambda_1 m + \lambda_2 e^{\sigma(x_0 - t_0)} (x_1^2 + \rho m^2)$$

with

$$x_0 = t.$$

For λ_2, we have $\dfrac{d\lambda_2}{dt} = 0$ and $\lambda_2(t_f) = 1$ so that we find $\lambda_2 = 1$.

From the minimum principle

$$\frac{\partial H}{\partial m} = \lambda_1 + 2\rho e^{\sigma(t - t_0)} m = 0$$

we find the optimal control

$$m = - \frac{\lambda_1}{2\rho} e^{-\sigma(t - t_0)}$$

We obtain the system

$$\frac{dx_1}{dt} = \frac{-\lambda_1}{2\rho} e^{-\sigma(t - t_0)}$$

$$\frac{d\lambda_1}{dt} = -2e^{\sigma(t - t_0)} x_1$$

with

$$x_1(t_0) = x_{10} \quad , \quad \lambda_1(t_f) = 0 \ .$$

Solving the above system we obtain

$$\frac{x_1^*(t)}{x_{10}} = \frac{\alpha_1 e^{-\alpha_2(t_f - t)} - \alpha_2 e^{-\alpha_1(t_f - t)}}{\alpha_1 e^{-\alpha_2(t_f - t_0)} - \alpha_2 e^{-\alpha_1(t_f - t_0)}}$$

$$\frac{m^*(t)}{x_{10}} = - \frac{1}{\rho} \frac{e^{-\alpha_2(t_f - t)} - e^{-\alpha_1(t_f - t)}}{\alpha_1 e^{-\alpha_2(t_f - t_0)} - \alpha_2 e^{-\alpha_1(t_f - t_0)}}$$

where

$$\alpha_1 = \frac{-\sigma + \sqrt{\sigma^2 + \dfrac{4}{\rho}}}{2} \quad , \quad \alpha_2 = \frac{-\sigma - \sqrt{\sigma^2 + \dfrac{4}{\rho}}}{2}$$

The table shows the results for $\rho = 1$, $t_f = 1$ and for $\sigma = 3$, $\sigma = 0$.

Using the maximum principle, investigate the optimal control of a process. The dynamics of the system is represented by

$$x_{n+1} = ax_n + u_n \tag{1}$$

where x_n is the state of the system at the nth sampling instant and u_n is the control signal at the nth instant. For an N-stage process, the total performance index is

$$S = \sum_{n=0}^{N-1} \{x_n^2 + \lambda u_n^2\} + x_N^2 = S_0 + S_1 + \ldots + S_n + \ldots S_N \tag{2}$$

Since $u_N = 0$, the cost of termination of the system is

$$S_N = x_N^2 \; .$$

Determine the optimum control sequence when the initial state is

$$x(0) = 0.$$

Solution: We introduce the function $f_N(x_0)$ such that

$$f_N(x_0) = \{S_0 + f_{N-1}(ax_0 + u_0)\} \tag{3}$$

is minimum with respect to u_0, $N = 1,2, \ldots$
Rewriting for the last stage, we obtain

$$f_1(x_{N-1}) = \min_{u_{N-1}} \{[x_{N-1}^2 + \lambda u_{N-1}^2] + f_N(ax_{N-1} + u_{N-1})\}$$

$$= \min_{u_{N-1}} \{[x_{N-1}^2 + \lambda u_{N-1}^2] + f_N(x_N)\}$$

$$= \min_{u_{N-1}} \{[x_{N-1}^2 + \lambda u_{N-1}^2] + (ax_{N-1} + u_{N-1})^2\} \tag{4}$$

Taking the partial derivative with respect to u_{N-1} we compute the minimum for the first-order process. Thus

$$\frac{\partial}{\partial u_{N-1}} \{[x_{N-1}^2 + \lambda u_{N-1}^2] + (ax_{N-1} + u_{N-1})^2\}$$

$$= 2\lambda u_{N-1} + 2(ax_{N-1} + u_{N-1}) = 0 \tag{5}$$

For the final stage the optimal control law is

$$u_{N-1} = -\frac{a}{1+\lambda} x_{N-1} \tag{6}$$

Eq. (6) is the optimal control law for the process, because the system is time-invariant. In general

$$u_n = -\frac{a}{1+\lambda} x_n \tag{7}$$

The optimal control law is a negative feedback function. With this optimal control law, the minimal total performance index or cost is proportional to x_0^2.

● **PROBLEM** 16-19

For the system

$$\frac{C(s)}{R(s)} = \frac{1}{s^2 + 2\zeta s + 1} \qquad (\zeta \geq 1)$$

$$E(s) = L[e(t)] = R(s) - C(s)$$

compute the following performance indexes:

$$\int_0^\infty |e(t)|dt, \quad \int_0^\infty t|e(t)|dt, \quad \int_0^\infty t[|e(t)| + |\dot{e}(t)|]dt$$

The system is initially at rest and is subjected to a unit-step input.

<u>Solution:</u> $R(s) = \frac{1}{s}$ for a unit-step input, and

$$E(s) = \frac{s^2+2\zeta s}{s^2+2\zeta s+1} \cdot \frac{1}{s}$$

$$= \frac{-\zeta + \sqrt{\zeta^2-1}}{2\sqrt{\zeta^2-1}} \cdot \frac{1}{s + \zeta + \sqrt{\zeta^2-1}} + \frac{\zeta + \sqrt{\zeta^2-1}}{2\sqrt{\zeta^2-1}} \cdot \frac{1}{s + \zeta - \sqrt{\zeta^2-1}}$$

Note that for $\zeta \geq 1$ the system does not overshoot and $|e(t)| = e(t)$ for all $t \geq 0$.

We shall compute now the performance indexes:

1) $\int_0^\infty |e(t)|dt = \int_0^\infty e(t)dt = \lim_{s \to 0} E(s)$

928

① $J=\int_0^\infty e^2\, dt$ ④ $J=\int_0^\infty |e|\, dt$

② $J=\int_0^\infty (e^2+\dot e^2)\, dt$ ⑤ $J=\int_0^\infty t|e|\, dt$

③ $J=\int_0^\infty te^2\, dt$ ⑥ $J=\int_0^\infty t(|e|+|\dot e|)\, dt$

$$= \frac{-\zeta + \sqrt{\zeta^2-1}}{2\sqrt{\zeta^2-1}} \cdot \frac{1}{\zeta + \sqrt{\zeta^2-1}} + \frac{\zeta + \sqrt{\zeta^2-1}}{2\sqrt{\zeta^2-1}} \cdot \frac{1}{\zeta - \sqrt{\zeta^2-1}}$$

$$= 2\zeta \qquad (\zeta \geq 1)$$

2) $\displaystyle\int_0^\infty t\,|e(t)|\,dt = \int_0^\infty te(t)\,dt = \lim_{s\to 0}\left[-\frac{d}{ds}E(s)\right]$

$$= \lim_{s\to 0}\left[\frac{-\zeta + \sqrt{\zeta^2-1}}{2\sqrt{\zeta^2-1}} \cdot \frac{1}{(s+\zeta+\sqrt{\zeta^2-1})^2} + \frac{\zeta + \sqrt{\zeta^2-1}}{2\sqrt{\zeta^2-1}} \cdot \frac{1}{(s+\zeta-\sqrt{\zeta^2-1})^2}\right]$$

$$= 4\zeta^2-1 \qquad (\zeta \geq 1)$$

3) Since $\dot e(t) < 0$ and $|\dot e(t)| = -\dot e(t)$ for $t \geq 0$ we have

$$\int_0^\infty t(|e(t)| + |\dot e(t)|)\,dt = \int_0^\infty [te(t) - t\dot e(t)]\,dt$$

$$= \lim_{s\to 0}\left[-\frac{d}{ds}E(s) + \frac{d}{ds}sE(s)\right] = \lim_{s\to 0}\left[(-1+s)\frac{d}{ds}E(s) + E(s)\right]$$

$$= 4\zeta^2-1 + 2\zeta \qquad (\zeta \geq 1)$$

Below are computed some other performance indexes.

929

$$\int_0^\infty e^2(t)\,dt = \zeta + \frac{1}{4\zeta} \qquad (\zeta > 0)$$

$$\int_0^\infty te^2(t)\,dt = \zeta^2 + \frac{1}{8\zeta^2} \qquad (\zeta > 0)$$

$$\int_0^\infty \left[e^2(t) + \dot{e}^2(t) \right] dt = \zeta + \frac{1}{2\zeta} \qquad (\zeta > 0)$$

The figure shows plots of the six computed performance indexes as functions of ζ.

● **PROBLEM** 16-20

The equation that describes the control system shown in Fig. 1 is

$$\dot{x} = Ax + Bu \qquad (1)$$

where

$$x = \begin{bmatrix} x_1 \\ x_2 \end{bmatrix} , \qquad A = \begin{bmatrix} 0 & 1 \\ 0 & 0 \end{bmatrix} , \qquad B = \begin{bmatrix} 0 \\ 1 \end{bmatrix}$$

The control constraint is to be

$$u = -Kx = -k_1 x_1 - k_2 x_2 \qquad (2)$$

Find k_1 and k_2 so that the performance index

$$J = \int_0^\infty (x'x + u'u)\,dt$$

is minimized.

Fig. 1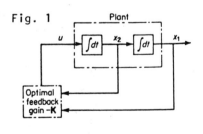

Fig. 2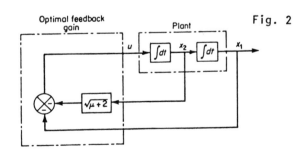

Solution: Substituting eq. (2) into eq. (1) we obtain

$$\dot{x} = Ax - BKx = (A-BK)x = \begin{bmatrix} 0 & 1 \\ -k_1 & -k_2 \end{bmatrix} x$$

If k_1 and k_2 are positive constants, then A–BK is a stable matrix and $x(\infty) = 0$. Thus the performance index can be written

$$J = \int_0^\infty (x'x + x'K'Kx)\,dt = \int_0^\infty x'(I + K'K)\,x\,dt$$

$$= x'(0)\,P\,x(0)$$

We determine P from

$$(A-BK)'\,P + P(A-BK) = -(I + K'K)$$

or

$$\begin{bmatrix} 0 & -k_1 \\ 1 & -k_2 \end{bmatrix} \begin{bmatrix} p_{11} & p_{12} \\ p_{12} & p_{22} \end{bmatrix} + \begin{bmatrix} p_{11} & p_{12} \\ p_{12} & p_{22} \end{bmatrix} \begin{bmatrix} 0 & 1 \\ -k_1 & -k_2 \end{bmatrix}$$

$$= - \begin{bmatrix} 1 & 0 \\ 0 & 1 \end{bmatrix} - \begin{bmatrix} k_1^2 & k_1 k_2 \\ k_1 k_2 & k_2^2 \end{bmatrix}$$

We can rewrite the above equation in the form of three equations.

$$-2k_1 p_{12} = -1 - k_1^2$$

$$p_{11} - k_2 p_{12} - k_1 p_{22} = -k_1 k_2$$

$$2p_{12} - 2k_2 p_{22} = -1 - k_2^2$$

Solving the equations we obtain

$$P = \begin{bmatrix} p_{11} & p_{12} \\ \\ p_{12} & p_{22} \end{bmatrix} = \begin{bmatrix} \frac{1}{2}\left(\frac{k_2}{k_1} + \frac{k_1}{k_2}\right) + \frac{k_1}{2k_2}\left(\frac{1}{k_1} + k_1\right) & \frac{1}{2}\left(\frac{1}{k_1} + k_1\right) \\ \\ \frac{1}{2}\left(\frac{1}{k_1} + k_1\right) & \frac{1}{2}\left(\frac{1}{k_2} + k_2\right) + \frac{1}{2k_2}\left(\frac{1}{k_1} + k_1\right) \end{bmatrix}$$

We have

$$J = x'(0)\,P\,x(0) = x_1^2(0)\left[\frac{1}{2}\left(\frac{k_2}{k_1} + \frac{k_1}{k_2}\right) + \frac{k_1}{2k_2}\left(\frac{1}{k_1} + k_1\right)\right]$$

$$+ \left(\frac{1}{k_1} + k_1 \right) x_1(0) \; x_2(0) \; + \; x_2^2(0) \left[\frac{1}{2} \left(\frac{1}{k_2} + k_2 \right) + \frac{1}{2k_2} \left(\frac{1}{k_1} + k_1 \right) \right].$$

Setting $\frac{\partial J}{\partial k_1} = 0$ and $\frac{\partial J}{\partial k_2} = 0$, we minimize J.

$$\frac{\partial J}{\partial k_1} = \left[\frac{1}{2} \left(- \frac{k_2}{k_1^2} + \frac{1}{k_2} \right) + \frac{k_1}{k_2} \right] x_1^2(0) + \left(\frac{-1}{k_1^2} + 1 \right) x_1(0) x_2(0)$$

$$+ \left[\frac{1}{2k_2} \left(- \frac{1}{k_1^2} + 1 \right) \right] x_2^2(0) = 0$$

$$\frac{\partial J}{\partial k_2} = \left[\frac{1}{2} \left(\frac{1}{k_1} - \frac{k_1}{k_2^2} r \right) + \frac{-k_1}{2k_2^2} r \left(\frac{1}{k_1} + k_1 \right) \right] x_1^2(0) +$$

$$\left[\frac{1}{2} \left(\frac{-1}{k_2^2} + 1 \right) - \frac{1}{2k_2^2} \left(\frac{1}{k_1} + k_1 \right) \right] x_2^2(0) = 0 .$$

The value of J becomes minimum for any given initial conditions $x_1(0)$ and $x_2(0)$, when

$$k_1 = 1, \qquad k_2 = \sqrt{3}$$

For the optimal control law, we have

$$K = [k_1 \; k_2] = [1 \; \sqrt{3}]$$

Fig. 2 shows the block diagram of this optimal control system with $\mu = 1$.

● **PROBLEM** 16-21

Fig. 1 shows the linear digital process described by the state equation

$$x(k+1) = Ax(k) + Bu(k) \tag{1}$$

where

$$A = \begin{bmatrix} 0.5 & 0 \\ 0 & 0.2 \end{bmatrix}, \qquad B = \begin{bmatrix} 1 \\ 1 \end{bmatrix}$$

Minimize the performance index

$$\Delta V(x) = V[x(k+1)] - V[x(k)] \tag{2}$$

finding the optimal control $u^\circ(k)$.

Assume that

$$Q = I, \quad V(x) = x'(k)Px(k)$$

and

P is the solution of

$$-Q = A'PA - P \qquad (3)$$

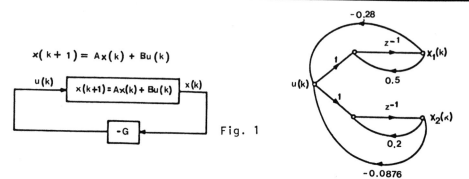

$$x(k+1) = Ax(k) + Bu(k)$$

Fig. 1

Fig. 2

Solution: The eigenvalues of A are 0.5 and 0.2, thus the process represented by eq. (1) is asymptotically stable.

We write P as

$$P = \begin{bmatrix} P_{11} & P_{12} \\ \\ P_{12} & P_{22} \end{bmatrix} \qquad (4)$$

Substituting eq. (4) into eq. (3) and solving for the elements of P we find

$$P_{12} = 0$$

$$P_{11} = \frac{1}{1-(0.5)^2} = 1.333$$

$$P_{22} = \frac{1}{1-(0.2)^2}$$

Matrix P is positive definite.

The optimal control which minimizes the performance index of eq. (2) is given by

$$u^\circ(k) = -(B'PB)^{-1} B'PA \, x(k)$$

The optimal feedback gain is

$$G = (B'PB)^{-1} B'PA = [0.28 \quad 0.0876]$$

and the optimal control is

$$u^\circ(k) = -0.28 \, x_1(k) - 0.0876 \, x_2(k)$$

Fig. 2 shows the state diagram of the closed-loop system

933

with the eigenvalues of the optimal closed-loop system at approximately 0 and 0.3324.

A two-stage discrete-time system is described by

$$x(k+1) = x(k) + u(k)$$

where

$$x(0) = 10.$$

Using dynamic programming find u(0) and u(1) which minimize

$$J = [x(2)-20]^2 + \sum_{k=0}^{1} [x^2(k) + u^2(k)]$$

There are no restrictions on u(k).

Solution:Since g[x(k)] is defined as the minimum cost from state x(k) at time k to some final state x(2), we have

$$g[x(2)] = [x(2)-20]^2$$

$$g[x(1)] = \min_{u(1)} \left\{ x^2(1) + u^2(1) + g[x(2)] \right\}$$

$$= \min_{u(1)} \left\{ x^2(1) + u^2(1) + [x(1) + u(1)-20]^2 \right\}.$$

Differentiating with respect to u(1) and equating to zero gives

$$u(1) = \frac{20 - x(1)}{2}$$

and

$$g[x(1)] = x^2(1) + \frac{[20-x(1)]^2}{2}$$

Substituting into

$$g[x(0)] = \min_{u(0)} \{x^2(0) + u^2(0) + g[x(1)]\}$$

and setting

$$\frac{\partial\{\ \}}{\partial u(0)} = 0$$

gives

$$u(0) = \frac{20 - 3x(0)}{5}$$

We have x(0) = 10, u(0) = -2

and from the difference equation we find

$x(1) = 8$

Using this, we find

$u(1) = 6$

We find

$x(2) = x(1) + u(1) = 14$

and the minimum cost is

$g[x(0)] = 240.$

The performance index is given by

$$J = \frac{1}{2} \sum_{k=0}^{10} [x^2(k) + 2u^2(k)] \tag{1}$$

Find the optimal control $u°(k)$, $k = 0,1,2 \ldots 10$ that minimizes the performance index, subject to the equality constraint. The initial state is $x(0) = 1$ and the final state is $x(11) = 0$.

<u>Solution</u>: We define the Hamiltonian, from

$H[x(k), u(k), p(k+1), k]$

$= F[x(k), u(k), k] + < p(k+1), f[x(k), u(k), k] >$

We get

$H[x(k), u(k), p(k+1)] = \frac{1}{2} [x^2(k) + 2u^2(k)]$

$\qquad + p(k+1)[x(k) + 2u(k)]$

From

$$\frac{\partial H°(k)}{\partial x°(k)} = p°(k)$$

and

$$\frac{\partial H°(k)}{\partial p°(k+1)} = x°(k+1)$$

We obtain the canonical state equations,

$p°(k+1) - p°(k) = -x°(k)$

$x°(k+1) - x°(k) = 2u°(k)$

From

$$\frac{\partial H°(k)}{\partial u°(k)} = 0$$

we find the optimal control

935

$$u^\circ(k) = -p^\circ(k+1)$$

The transversality condition

$$\frac{\partial G^\circ(N)}{\partial x^\circ(N)} = p^\circ(N)$$

is not needed, because the endpoint x(11) is fixed.

The solutions of

$$p^\circ(k+1) - p^\circ(k) = -x^\circ(k)$$

$$x^\circ(k+1) - x^\circ(k) = 2u^\circ(k)$$

$$u^\circ(k) = -p^\circ(k+1)$$

are

$$x^\circ(k) = 0.289[2.732-2p^\circ(0)](3.732)^k \tag{2}$$

$$+ 0.289[0.732 + 2p^\circ(0)](0.268)^k$$

$$p^\circ(k) = [-0.289 + 0.211p^\circ(0)](3.732)^k$$

$$+ [0.289 + 0.789p^\circ(0)](0.268)^k \tag{3}$$

Substitution of $x^\circ(11) = 0$ in eq. (2) yields

$$p^\circ(0) = 1.366.$$

Thus

$$x^\circ(k) = (0.268)^k$$

$$u^\circ(k) = -2.732(0.268)^{k+1}$$

If x(11) is a free-end condition, we add a terminal cost to the performance index, and obtain

$$J = \frac{1}{2}x^2(11) + \frac{1}{2}\sum_{k=0}^{10}[x^2(k) + 2u^2(k)]$$

The transversality condition of eq. (1) gives

$$x^\circ(11) = p^\circ(11)$$

Equating eqs. (2) and (3) at k=11 we obtain

$$p^\circ(0) = 1.366$$

The same results are obtained for $x^\circ(k)$ and $u^\circ(k)$.

The first-order digital process is described by

$$x(k+1) = x(k) + u(k)$$

Minimize the performance index.

$$J = \frac{1}{2} \sum_{k=0}^{\infty} u^2(k)$$

Solution: We have

$\phi = 1$, $\theta = 1$, $\hat{Q} = 0$, $M = 0$ and $\hat{R} = 1$.

The optimal control which minimizes J is given by

$$u^\circ(k) = -(\hat{R} + \theta'K\theta)^{-1} (\theta'K\phi + M') \, x^\circ(k)$$

which can be reduced to

$$u^\circ(k) = \frac{-K}{1+K} \, x^\circ(k)$$

K is the scalar constant Riccati gain, which is the solution of the Riccati equation

$$K = \phi'K\phi + \hat{Q} - (\phi'K\theta+M)(\hat{R}+\theta'K\theta)^{-1}(\theta'K\phi+M')$$

or

$$K = K - \frac{K^2}{1+K}$$

Solving the equation we get K=0. Thus $u^\circ(k) = 0$ for all k and the closed-loop system is not asymptotically stable.

Because $\hat{Q} = 0$, the optimal linear regulator design does not lead to an asymptotically stable system and the state variable is not observed by the performance index.

● **PROBLEM** 16-25

A single-lag linear system is represented by the differential equation

$$\dot{x}_1 = -x_1 + u \tag{1}$$

with the initial condition

$$x_1(0) = 1$$

Bring the output x_1 to zero at some final time t_f

$$x_1(t_f) = 0 \quad \text{for } t = t_f \tag{2}$$

Find the optimal control signal u to minimize the

performance index

$$S = \int_0^{t_f} (x_1^2 + u^2) dt \tag{3}$$

Solution: We assume that there is no constraint on the magnitude of u. We have the following set of differential equations:

$$\dot{x}_0 = f_0(x_1, u) = x_1^2 + u^2 \tag{4}$$

$$\dot{x}_1 = f_1(x_1, u) = -x_1 + u \tag{5}$$

The Hamiltonian, for $p_0 = -1$, is

$$H = p_0 f_0 + p_1 f_1 = -(x_1^2 + u^2) + p_1(-x_1 + u) \tag{6}$$

To maximize H, we take the derivative of H with respect to u, thus

$$\frac{\partial H}{\partial u} = -2u + p_1 = 0 \tag{7}$$

Solving, we get

$$u = \frac{p_1}{2} \tag{8}$$

and

$$H_{max} = -x_1^2 + u^2 - 2x_1 u \tag{9}$$

To apply the optimal control law $u = \frac{p_1}{2}$, we have to express u in terms of x_1 and t. We have, from the canonical equations,

$$\dot{x}_1 = \frac{\partial H}{\partial p_1} = f_1(x_1, u) = -x_1 + u \tag{10}$$

$$\dot{p}_1 = -\frac{\partial H}{\partial x_1} = 2x_1 + p_1 \tag{11}$$

Substituting eq. (8) into eq. (11) we get

$$\dot{x}_1 = -x_1 + u \tag{12}$$

$$\dot{u} = x_1 + u \tag{13}$$

In the matrix form

$$\begin{bmatrix} \dot{x}_1 \\ \dot{u} \end{bmatrix} = \begin{bmatrix} -1 & 1 \\ 1 & 1 \end{bmatrix} \begin{bmatrix} x_1 \\ u \end{bmatrix} \tag{14}$$

From

$$\det[A - \lambda I] = (-1-\lambda)(1-\lambda) - 1$$

$$= (\lambda^2 - 2) = (\lambda + \sqrt{2})(\lambda - \sqrt{2}) = 0 \qquad (15)$$

we obtain the characteristic roots.

Solving for x_1 and u we have

$$x_1(t) = c_1 e^{\sqrt{2}\,t} + c_2 e^{-\sqrt{2}\,t} \qquad (16)$$

$$u(t) = (1 + \sqrt{2})c_1 e^{\sqrt{2}\,t} + (1 - \sqrt{2})c_2 e^{-\sqrt{2}\,t} \qquad (17)$$

From the initial condition $x_1(0) = 1$, we find the constants c_1 and c_2 .

$$x_1(0) = c_1 + c_2 = 1 \qquad (18)$$

Let us choose $c_1 = 0$ and $c_2 = 1$, then

$$x_1(t) = e^{-\sqrt{2}\,t} \qquad (19)$$

$$u(t) = (1 - \sqrt{2})e^{-\sqrt{2}\,t} = -0.414 e^{-\sqrt{2}\,t} \qquad (20)$$

The final time t_f is equal to infinity, and the maximum value of the Hamiltonian is equal to zero at $t_0 = 0$.

Substituting eqs. (19) and (20) into (9) we obtain

$$H_{max} = -x_1^2(0) + u^2(0) - 2x_1(0)u(0)$$

$$= -1 + (1 - 2\sqrt{2} + 2) - 2(1 - \sqrt{2}) = 0$$

Thus, the necessary condition is satisfied.

● **PROBLEM 16-26**

Describe the network shown in the figure, by a Hamiltonian H and a set of canonical equations. The branch relations are

$$q_1 = e_1 , \quad q_2 = 2e_2, \quad q_3 = 3e_3$$

$$\lambda_4 = (i_4)^{\frac{1}{3}} \qquad \lambda_5 = 5i_5 \qquad (1)$$

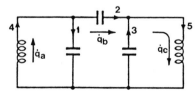

Solution: We choose the loop charges q_M as the generalized

939

and
coordinates, where

$$q_M = \begin{bmatrix} q_a \\ q_b \\ q_c \end{bmatrix}, \qquad
\begin{aligned}
q_a &= \int_0^t i_4(u)\,du \\[6pt]
q_b &= \int_0^t i_2(u)\,du \\[6pt]
q_c &= \int_0^t i_5(u)\,du
\end{aligned} \qquad (2)$$

$$q_c(0) = \begin{bmatrix} q_1(0) \\ q_2(0) \\ q_3(0) \end{bmatrix}$$

The network Lagrangian is
$$L = W'_m(B_L^t \dot{q}_m) - W_e\left[B_C^t q_m + q_c(0)\right] \qquad (3)$$
where
$$W'_m(B_L^t \dot{q}_M) = \frac{3}{4}(\dot{q}_a)^{\frac{4}{3}} + \frac{5}{2}(\dot{q}_c)^2 \qquad (4)$$

$$W_e[B_C^t q_M + q_c(0)] = \frac{[q_a - q_b + q_1(0)]^2}{2} + \frac{[q_b + q_2(0)]^2}{4}$$
$$+ \frac{[q_c - q_b + q_3(0)]^2}{6} \qquad (5)$$

Note that B_C and B_L are submatrices of the fundamental loop matrix for the network.

We define the generalized momenta as

$$p_a = \lambda_a = \frac{\partial L}{\partial \dot{q}_a} = \lambda_4 = (\dot{q}_a)^{\frac{1}{3}}$$

$$p_b = \lambda_b = \frac{\partial L}{\partial \dot{q}_b} = 0 \qquad (6)$$

$$p_c = \lambda_c = \frac{\partial L}{\partial \dot{q}_c} = \lambda_5 = 5\dot{q}_c$$

The Hamiltonian is

$$H(q_a, q_b, q_c, \lambda_a, \lambda_c) = \lambda_a \dot{q}_a + \lambda_c \dot{q}_c - L$$

$$= \frac{1}{4} \lambda_a^4 + \frac{1}{10} \lambda_c^2 + W_e [B_c^t q_M + q_c(0)] \tag{7}$$

and represents the total energy stored in the network. Since $\lambda_b = 0$, the Hamiltonian is a function of five and not six variables. The canonical equations are

$$\dot{q}_a = \frac{\partial H}{\partial \lambda_a} = \lambda_a^3$$

$$\dot{q}_c = \frac{\partial H}{\partial \lambda_c} = \frac{\lambda c}{5} \tag{8}$$

$$\dot{\lambda}_a = -\frac{\partial H}{\partial q_a} = -q_a + q_b - q_1(0)$$

$$-\frac{\partial H}{\partial q_b} = 0 = q_a - q_b + q_1(0) - \frac{q_b + q_2(0)}{2} + \frac{q_c - q_b + q_3(0)}{3}$$

$$\dot{\lambda}_c = -\frac{\partial H}{\partial q_c} = \frac{-q_c + q_b - q_3(0)}{3}$$

● **PROBLEM 16-27**

For the system described by

$$x(k+1) = x(k) + u(k)$$

with $x(0) = x_0$

find the optimal control $u^o(k)$, $k = 0, 1, 2 \ldots 9$ such that the performance index

$$J_{10} = \frac{1}{2} [10x^2(10)] + \frac{1}{2} \sum_{k=0}^{9} [x^2(k) + u^2(k)]$$

is minimized.

Solution: Substitution of the system parameters into

$$V = \begin{bmatrix} \Omega^{-1} & \Omega' \hat{\theta} \hat{R}^{-1} \theta' \\ \Gamma \Omega^{-1} & \Omega' + \Gamma \Omega^{-1} \hat{\theta} \hat{R}^{-1} \theta' \end{bmatrix}$$

where

$$\Omega = \phi - \theta \hat{R}^{-1} M'$$

$$\Gamma = \hat{Q} - M \hat{R}^{-1} M'$$

941

yields

$$V = \begin{bmatrix} 1 & 1 \\ 1 & 2 \end{bmatrix}$$

The eigenvalues of V are

$$\lambda_1 = 2.618 \qquad \text{and} \qquad \lambda_2 = \frac{1}{\lambda_1} = 0.382.$$

We form the matrix

$$W^{-1}VW = \begin{bmatrix} \Lambda & 0 \\ 0 & \Lambda^{-1} \end{bmatrix} = \begin{bmatrix} 2.618 & 0 \\ 0 & 0.382 \end{bmatrix}$$

The columns of W are the eigenvectors of V because W is a similarity transformation. Thus

$$W = \begin{bmatrix} W_{11} & W_{12} \\ W_{21} & W_{22} \end{bmatrix} = \begin{bmatrix} 1 & 1 \\ 1.618 & -0.618 \end{bmatrix}$$

From

$$U = -(W_{22} - SW_{12})^{-1}(W_{21} - SW_{11})$$

we have

$$U = -(-0.618 - 10)^{-1}(1.618 - 10) = -0.789$$

From

$$H(k) = \Lambda^{-k}U\Lambda^{-k}$$

we get

$$H(k) = -0.789(0.382)^{2k}$$

The Riccati gain of the finite-time problem is

$$K(N-k) = [1.618 - 0.618(-0.789)(0.382)^{2k}]$$

$$\times [1 - 0.789(0.382)^{2k}]^{-1}$$

or

$$K(i) = [1.618 + 0.488(0.382)^{2(N-i)}]$$

$$\times [1 + 0.789(0.382)^{2(N-i)}]^{-1}$$

For N = ∞, the constant Riccati gain is given by

942

$$K = \lim_{k \to \infty} K(N-k) = W_{21}W_{11}^{-1} = 1.618$$

Optimal control is given by

$$U^\circ(i) = -G(i)x^\circ(i)$$

$$G(i) = [\hat{R} + \theta'K(i+1)\theta]^{-1}[M' + \theta'K(i+1)\phi]$$

$$K(i) = \hat{Q} + \phi'K(i+1)\phi - [M' + \theta'K(i+1)\phi]'G(i)$$

$$K(N) = S$$

For $N = \infty$, G and K are constant.

● **PROBLEM** 16-28

Given the digital process

$$x(k+1) = x(k) + u(k) \tag{1}$$

with $x(0) = x_0$, find $u^\circ(k)$, $k = 0,1, \ldots 9$

such that the following performance index is minimized.

$$J_{10} = \frac{1}{2}[10x^2(10)] + \frac{1}{2}\sum_{k=0}^{9}[x^2(k) + u^2(k)]$$

Solution: We find that

$$S = 10, \qquad \hat{Q} = 1, \qquad \hat{R} = 1, \qquad M = 0, \qquad \phi = \theta = 1.$$

Substituting these parameters into

$$K(i) = \hat{Q} + \phi'K(i+1) - [M' + \theta'K(i+1)\phi]'G(i)$$

yields the Riccati equation

$$K(i) = 1 + K(i+1) - K(i+1)G(i) \tag{2}$$

where

$$G(i) = \frac{K(i+1)}{1 + K(i+1)} \tag{3}$$

The optimal control is

$$u^\circ(i) = -G(i)x^\circ(i)$$

Starting with the boundary condition $K(10) = S = 10$, we solve recursively eqs. (2) and (3). The results are shown below.

i	K(i)	G(i)
0	1.6180	0.6180
1	1.6180	0.6180

2	1.6180	0.6180
3	1.6180	0.6180
4	1.6180	0.6180
5	1.6182	0.6182
6	1.6188	0.6188
7	1.6236	0.6236
8	1.6562	0.6562
9	1.9091	0.9091
10	10	0

Substituting the optimal control into eq. (1), we find the optimal trajectory of x(k).

$$x°(k) = \prod_{i=0}^{k-1} [1 - G(i)]x_0$$

for k = 0,1, ... 10.

● **PROBLEM** 16-29

The matrix Riccati equation arises frequently in the solution of optimal control problems. The matrix Riccati equation is

$$\dot{X} = A(t)X + XB(t) + XC(t)X + D(t)$$

where X is an n × n matrix and A(t), B(t), C(t) and D(t) are n × n matrices. The matrices A(t), B(t), C(t) and D(t) are absolutely integrable as functions of t in the interval $t_0 \leq t \leq t_1$.

Show that the matrix Riccati equation can be solved by using the two simultaneous linear matrix differential equations,

$$\dot{Y} = -B(t)Y - C(t)Z$$

(1)

$$\dot{Z} = D(t)Y + A(t)Z$$

Solution: If the matrix Y is nonsingular in the interval $t_0 \leq t \leq t_1$, then the solution to the matrix Riccati equation can be given by

$$X(t) = Z(t)Y^{-1}(t) \qquad t_0 \leq t \leq t_1$$

Since

$$\dot{X} = \dot{Z}Y^{-1} + Z\dot{Y}^{-1}$$

$$\dot{Y}^{-1} = -Y^{-1}\dot{Y}Y^{-1}$$

we get

$$\dot{X} = [D(t)Y + A(t)Z]Y^{-1}$$

$$+ Z(-Y^{-1})[-B(t)Y - C(t)Z]Y^{-1} = D(t) + A(t)X$$

$$+ XB(t) + XC(t)X.$$

It is difficult to solve eq. (1) for Y and Z when the coefficient matrices in the matrix Riccati equation are time-varying.

If matrices $A(t)$, $B(t)$, $C(t)$ and $D(t)$ are constant we can obtain a formal solution for $Y(t)$ and $Z(t)$. In such case, we choose $t_0 = 0$ and define $W(t)$ and G as follows:

$$W(t) = \begin{bmatrix} Y(t) \\ \hline Z(t) \end{bmatrix} \qquad G = \begin{bmatrix} -B & -C \\ \hline D & A \end{bmatrix}$$

Equation (1) can be represented by

$$\dot{W}(t) = GW(t)$$

whose solution is

$$W(t) = e^{Gt} W(0).$$

If e^{Gt} is obtained as a square matrix, we get $Y(t)$ and $Z(t)$. If $Y(t)$ is nonsingular, then the solution $X(t)$ to the time-invariant matrix Riccati equation is given by

$$X(t) = Z(t) Y^{-1}(t)$$

● **PROBLEM 16-30**

The system is described by

$$\dot{x} = Ax \qquad , \quad x(0) = c$$

$$x = [x_1, \ldots x_n]$$

Assume that all the eigenvalues of A have negative real parts. The performance index is given by

$$J = \int_0^\infty x*Qxdt$$

where Q is a constant positive semidefinite Hermitian or

real symmetric matrix. Evaluate the performance index.

Suppose that the system equation is

$$
\begin{bmatrix} \dot{x}_1 \\ \dot{x}_2 \\ \dot{x}_3 \end{bmatrix} = \begin{bmatrix} 0 & 1 & 0 \\ 0 & 0 & 1 \\ -1 & -2 & -a \end{bmatrix} \begin{bmatrix} x_1 \\ x_2 \\ x_3 \end{bmatrix}
$$

and the initial condition is

$$
x(0) = c = \begin{bmatrix} x_1(0) \\ 0 \\ 0 \end{bmatrix}
$$

Determine the value of the parameter $a(a > 0)$ that minimizes the performance index J.

$$
J = \int_0^\infty x'x \, dt
$$

Solution: Let $x*Qx = -\dfrac{d}{dt}(x*Px)$

where P is a constant positive definite Hermitian or real symmetric matrix.

We have

$$
x*Qx = -\dot{x}*Px - x*P\dot{x} = -x*A*Px - x*PAx
$$

Hence

$$
A*P + PA = -Q
$$

From the Liapunov theorem we know that if A is stable, then for a given Q there exists a P that satisfies this equation. A is stable because the eigenvalues of A have negative real parts.

We evaluate J from

$$
J = \int_0^\infty x*Qxdt = -x*Px \Big|_0^\infty
$$

$$
= -x*(\infty) \, P \, x(\infty) + x*(0) \, P \, x(0)
$$

Since all the eigenvalues of A have negative real parts,

$x(\infty) \rightarrow 0.$

Thus, we have

$$J = x^*(0) \, P \, x(0) = c^* \, Pc$$

From

$$A'P + PA = -I$$

we determine P.

Since P is symmetric, we have

$$\begin{bmatrix} 0 & 0 & -1 \\ 1 & 0 & -2 \\ 0 & 1 & -a \end{bmatrix} \begin{bmatrix} p_{11} & p_{12} & p_{13} \\ p_{12} & p_{22} & p_{23} \\ p_{13} & p_{23} & p_{33} \end{bmatrix} + \begin{bmatrix} p_{11} & p_{12} & p_{13} \\ p_{12} & p_{22} & p_{23} \\ p_{13} & p_{23} & p_{33} \end{bmatrix}$$

$$\begin{bmatrix} 0 & 1 & 0 \\ 0 & 0 & 1 \\ -1 & -2 & -a \end{bmatrix} = \begin{bmatrix} -1 & 0 & 0 \\ 0 & -1 & 0 \\ 0 & 0 & -1 \end{bmatrix}$$

For J we have

$$J = c'Pc = \begin{bmatrix} x_1(0) & 0 & 0 \end{bmatrix} \begin{bmatrix} p_{11} & p_{12} & p_{13} \\ p_{12} & p_{22} & p_{23} \\ p_{13} & p_{23} & p_{33} \end{bmatrix} \begin{bmatrix} x_1(0) \\ 0 \\ 0 \end{bmatrix}$$

$$= p_{11} \, x_1^2(0)$$

To determine J we need only p_{11}. We find

$$p_{11} = \frac{a^2 + 5a - 1}{4a - 2}$$

Thus

$$J = \frac{a^2 + 5a - 1}{4a - 2} \, x_1^2(0)$$

From the condition

$$\frac{\partial J}{\partial a} = 0$$

we find

$$\frac{4a^2 - 4a - 6}{(4a - 2)^2} = 0$$

Solving, we obtain

$a_1 = 1.823$

$a_2 = -0.823$

For $a = 1.823$, $\dfrac{\partial^2 J}{\partial a^2} > 0$. Thus $a = 1.823$ minimizes J.

The designed system satisfies Liapunov's theorem for asymptotic stability, thus the eigenvalues of A have negative real parts.

● **PROBLEM** 16-31

For the digital process described by

$$x(k+1) = \phi \; x(k) + \theta \; u(k) \qquad (1)$$

where

$$\phi = \begin{bmatrix} 0 & 1 \\ -1 & 1 \end{bmatrix} , \qquad \theta = \begin{bmatrix} 0 \\ 1 \end{bmatrix} ,$$

$$x(0) = \begin{bmatrix} 1 \\ 1 \end{bmatrix}$$

find the optimal control u(k), k = 0,1,2, ... 7, minimizing the performance index

$$J_8 = \sum_{k=0}^{7} [x_1^2(k) + u^2(k)]$$

Solution: For this problem we have

$$M = 0, \qquad \hat{R} = 2 \qquad \text{and}$$

$$\hat{Q} = \begin{bmatrix} 2 & 0 \\ 0 & 0 \end{bmatrix} , \qquad S = \begin{bmatrix} 0 & 0 \\ 0 & 0 \end{bmatrix}$$

Substituting these parameters to

$$K(i) = \hat{Q} + \phi'K(i+1)\phi - [M' + \theta'K(i+1)\phi]'G(i)$$

we find

948

$$K(i) = \begin{bmatrix} 2 & 0 \\ 0 & 0 \end{bmatrix} + \begin{bmatrix} 0 & -1 \\ 1 & 1 \end{bmatrix} K(i+1) \left[\begin{bmatrix} 0 & 1 \\ -1 & 1 \end{bmatrix} - \begin{bmatrix} 0 \\ 1 \end{bmatrix} G(i) \right] \qquad (2)$$

The optimal control is

$$u^\circ(i) = -G(i)x^\circ(i) \qquad\qquad (3)$$

From

$$G(i) = [\hat{R} + \theta'K(i+1)\theta]^{-1}[M' + \theta'K(i+1)\phi]$$

we obtain

$$G(i) = \left[2 + [0\ 1]K(i+1) \begin{bmatrix} 0 \\ 1 \end{bmatrix} \right]^{-1} [0\ 1]K(i+1) \begin{bmatrix} 0 & 1 \\ -1 & 1 \end{bmatrix} \qquad (4)$$

Beginning with the boundary condition

$$K(8) = S = \begin{bmatrix} 0 & 0 \\ 0 & 0 \end{bmatrix}$$

we solve eqs. (2) and (4).

$$K(7) = \begin{bmatrix} 2 & 0 \\ 0 & 0 \end{bmatrix} \quad , \qquad G(7) = [0\ 0]$$

$$K(6) = \begin{bmatrix} 2 & 0 \\ 0 & 2 \end{bmatrix} \quad , \qquad G(6) = [0\ 0]$$

$$K(5) = \begin{bmatrix} 3 & -1 \\ -1 & 3 \end{bmatrix} \quad , \qquad G(5) = [-0.5\ 0.5]$$

$$K(4) = \begin{bmatrix} 3.2 & -0.8 \\ -0.8 & 3.2 \end{bmatrix} \quad , \qquad G(4) = [-0.6\ \ 0.4]$$

$$K(3) = \begin{bmatrix} 3.23 & -0.922 \\ -0.922 & 3.69 \end{bmatrix} \quad , \qquad G(3) = [-0.615\ \ 0.462]$$

$$K(2) = \begin{bmatrix} 3.297 & -0.973 \\ -0.973 & 3.729 \end{bmatrix}, \qquad G(2) = [-0.651 \quad 0.481]$$

$$K(1) = \begin{bmatrix} 3.301 & -0.962 \\ -0.962 & 3.75 \end{bmatrix}, \qquad G(1) = [-0.652 \quad 0.485]$$

$$K(0) = \begin{bmatrix} 3.305 & -0.97 \\ -0.97 & 3.777 \end{bmatrix}, \qquad G(0) = [-0.653 \quad 0.486]$$

Substituting the feedback gains into eqs. (1) and (3) we compute the optimal control and the optimal trajectories. The results are shown below.

i	$u^\circ(i)$	$x_1^\circ(i)$	$x_2^\circ(i)$
0	0.1678	1	1
1	0.5708	1	0.1678
2	0.235	0.1678	−0.2614
3	−0.071	−0.2614	−0.1942
4	−0.115	−0.1942	−0.0038
5	0.0358	−0.0038	0.0754
6	0	0.0754	0.115
7	0	0.115	0.0396
8	0	0.0396	−0.0754

We can show that for large values of N the Riccati gain approaches the steady-state solution,

$$K = \begin{bmatrix} 3.308 & -0.972 \\ -0.972 & 3.780 \end{bmatrix}$$

The constant optimal control is

$$G = [-0.654 \quad 0.486]$$

● **PROBLEM** 16-32

The state equations of a second-order process are given by

$$\dot{x}(t) = Ax(t) + Bu(t) \qquad (1)$$

where

$$A = \begin{bmatrix} 0 & 0 \\ 0 & 0 \end{bmatrix}, \quad B = \begin{bmatrix} 1 & 0 \\ 0 & 1 \end{bmatrix}$$

Find the optimal sampled-data control $u(t) = u(kT)$, $k = 0,1,2, \ldots$ such that the performance index

$$J = \frac{1}{2} \int_0^\infty \left[[x_1(t) - x_2(t)]^2 + x_1^2(t) + x_2^2(t) + u_1^2(t) \right.$$
$$\left. + u_2^2(t) \right] dt$$

is minimized.

<hr>

Solution: We use the following equations

$$J = \frac{1}{2} < x(t_f), Sx(t_f) > + \frac{1}{2} \int_0^{t_f} [< x(t), Q(t) >$$

$$+ < u(t), R \ u(t) >] dt$$

where Q and S are symmetric positive semidefinite matrices $(n \times n)$, and R is a symmetric positive definite matrix $(p \times p)$.

The following weighting matrices are defined

$$Q = \begin{bmatrix} 2 & -1 \\ -1 & 2 \end{bmatrix}, \quad R = \begin{bmatrix} 1 & 0 \\ 0 & 1 \end{bmatrix}$$

The linear continuous-data regulator problem has an asymptotically stable solution for $T = 0$, because the process in eq. (1) is controllable and stabilizable and the pair $[A,D]$ is observable, where $DD' = Q$. The coefficient matrix of the sampled-data process is

$$\phi(T) = e^{AT} = I$$

and

$$\theta(T) = \int_0^T \phi(\lambda) B d\lambda = \begin{bmatrix} T & 0 \\ 0 & T \end{bmatrix}$$

We see that the sampled-data system

$$x[(k+1)T] = \phi(T)x(kT) + \theta(T)u(kT) \qquad (2)$$

is completely controllable.

Since $\hat{R}$ as given by eq. (1) is nonsingular, the pair $[\phi, D]$ is observable for any $DD' = Q$. Therefore the system in eq. (2) has an asymptotically stable solution for the optimal linear digital regulator problem. In most cases the optimal linear digital regulator problem can be solved by formulating the discrete Riccati equation,

$$K(k) - Q + M\hat{R}^{-1}M^{-1}$$

$$= (\phi' - M\hat{R}^{-1}\theta')K(k+1)[I + \theta\hat{R}^{-1}\theta'K(k+1)]^{-1}(\phi - \theta\hat{R}^{-1}M')$$

or

$$K(k) = \phi'K(k+1)\phi + \hat{Q} - [\theta'K(k+1)\phi + M']'$$

$$\times [\hat{R} + \theta'K(k+1)\theta]^{-1}[\theta'K(k+1)\phi + M']$$

and solve for the Riccati gain $K(T)$.

The optimal control is given by

$$u^\circ(k) = - (\hat{R} + \theta'K\theta)^{-1}(\theta'K\phi + M')\, x^\circ(k)$$

In our case the sampling period T is not specified and it is difficult to solve the Riccati equation.

To solve the problem we shall use the sampling sensitivity method. We first solve the optimal linear continuous-data regulator problem. For the continuous-data regulator, the Riccati equation is

$$K(0)A + A'K(0) + Q = K(0)BR^{-1}B'K(0)$$

where $\quad K(0) = \lim_{T \to 0} K(T)$

or after substitution of the system parameters

$$K^2(0) = Q = \begin{bmatrix} 2 & -1 \\ -1 & 2 \end{bmatrix}$$

Thus

$$K(0) = \frac{1}{2} \begin{bmatrix} 1 + \sqrt{3} & 1 - \sqrt{3} \\ 1 - \sqrt{3} & 1 + \sqrt{3} \end{bmatrix}$$

Using

$$G(0) = \lim_{T \to 0} G(T) = R^{-1}B'K(0)$$

we determine the optimal feedback gain

$$G(0) = R^{-1}B'K(0) = K(0) \qquad (3)$$

Note that the first-order sampling period sensitivity of $K(T)$ is always zero. To find the second-order sensitivity $\dfrac{\partial^2 K(T)}{\partial T^2}$

we solve Π from

$$[A' - K(0)BR^{-1}B'][K(0) - \Pi] + [K(0) - \Pi]$$

$$\times \ [A - BR^{-1}B'K(0)] + Q = 0$$

which becomes

$$-2K^2(0) + 2K(0)\Pi + Q = 0$$

Thus

$$\Pi = \frac{1}{4} \begin{bmatrix} 1 + \sqrt{3} & 1 - \sqrt{3} \\ \\ 1 - \sqrt{3} & 1 + \sqrt{3} \end{bmatrix}$$

Substituting Π into

$$\frac{\partial^2 K(T)}{\partial T^2} = [A' - K(0)BR^{-1}B'] \frac{\Pi}{6} [A - BR^{-1}B'K(0)]$$

we obtain

$$\frac{\partial^2 K(T)}{\partial T^2} = \frac{1}{6} K(0)\Pi K(0) = \frac{1}{24} \begin{bmatrix} 1 + 3\sqrt{3} & 1 - 3\sqrt{3} \\ \\ 1 - 3\sqrt{3} & 1 + 3\sqrt{3} \end{bmatrix}$$

The second-order approximation of $K(T)$ is

$$K(T) \cong K(0) + \frac{T^2}{2} \frac{\partial^2 K(T)}{\partial T^2} = \begin{bmatrix} e_1 + e_2 & e_1 - e_2 \\ \\ e_1 - e_2 & e_1 + e_2 \end{bmatrix} \qquad (4)$$

where

$$e_1 = \frac{1}{2} + \frac{T^2}{48}$$

$$e_2 = \frac{\sqrt{3}}{2} + \frac{\sqrt{3}\ T^2}{16}$$

Solving the Riccati equation

$$\frac{\phi'K(T)\phi - K(T) + \hat{Q}}{T} = \left[\frac{M + \phi'K(T)\theta}{T}\right]\left[\frac{\hat{R} + \theta'K(T)\theta}{T}\right]^{-1}\left[\frac{\theta'K(T)\phi + M'}{T}\right]$$

we find the exact Riccati gain

$$K(T) = \begin{bmatrix} d_1 + d_2 & d_1 - d_2 \\ d_1 - d_2 & d_1 + d_2 \end{bmatrix} \tag{5}$$

where

$$d_1 = \frac{1}{2}\left[1 + \frac{T^2}{12}\right]^{\frac{1}{2}}$$

$$d_2 = \frac{\sqrt{3}}{2} + \frac{\sqrt{3}\ T^2}{8}$$

Substituting the approximation for $K(T)$ in eq. (4) and the result into

$$G(T) = \left[\frac{\hat{R} + \theta'K(T)\theta}{T}\right]^{-1}\left[\frac{\theta'K(T)\phi + M'}{T}\right] \tag{6}$$

we obtain

$$G(T) \cong \frac{1}{2}\begin{bmatrix} f_1 + f_2 & f_1 - f_2 \\ f_1 - f_2 & f_1 + f_2 \end{bmatrix}$$

where

$$f_1 = \frac{24 + 12T + T^2}{24 + 24T + 8T^2 + T^3}$$

$$f_2 = \frac{8\sqrt{3} + 12T + \sqrt{3}\ T^2}{8 + 8\sqrt{3}\ T + 8T^2 + \sqrt{3}\ T^3}$$

Substituting the exact expression of K(T) in eq. (5) into eq. (6) we get the exact feedback gain matrix

$$
G(T) = \frac{1}{2}
\begin{bmatrix}
g_1 + g_2 & g_1 - g_2 \\
\\
g_1 - g_2 & g_1 + g_2
\end{bmatrix}
$$

where

$$
g_1 = \frac{\left[1 + \frac{T^2}{12}\right]^{\frac{1}{2}} + \frac{T}{2}}{1 + T\left[1 + \frac{T^2}{12}\right]^{\frac{1}{2}} + \frac{T^2}{3}}
$$

and

$$
g_2 = \frac{\sqrt{3}\left[1 + \frac{T^2}{4}\right]^{\frac{1}{2}} + \frac{3T}{2}}{1 + \sqrt{3}\,T\left[1 + \frac{T^2}{4}\right]^{\frac{1}{2}} + T^2} \quad .
$$

● **PROBLEM 16-33**

The block diagram of a linear system with a controller is shown below,

where

$$
G(s) = \frac{10}{s^2} \tag{1}
$$

and the input transform is

$$
R(s) = \frac{0.5}{s} \tag{2}
$$

For the system described, determine the optimal control $U_o(s)$ and the optimal closed-loop transfer function

$M_o(s)$ such that

(1) $J_e = \displaystyle\int_0^\infty e^2(t)\,dt = \text{minimum}$ (3)

955

(2) $J_u = \int_0^\infty u^2(t)\,dt \le 2.5$ (4)

We introduce the notation of spectral factorization representing a function as

$$E(s) = Y(s)\ Y(-s) = Y\ \overline{Y} \tag{5}$$

where Y has poles and zeros only in the left-half plane, and $\overline{Y}$ only in the right. We denote

$$Y = \{E(s)\}_+ \tag{6}$$

and

$$\overline{Y} = \{E(s)\}_- \tag{7}$$

Further, we introduce notation

$$A = [E(s)]_+ \tag{8}$$

and

$$B = [E(s)]_- \tag{9}$$

A is that part of the partial-fraction expansion of E(s) containing terms with poles in the left-half plane, B in the right.

The optimal control is given by

$$U_{o_k} = \frac{D}{\{k^2 D\overline{D} + N\overline{N}\}_+} \left[\frac{\overline{N}R}{\{k^2 D\overline{D} + N\overline{N}\}_-} \right]_+ \tag{10}$$

where N and D are respectively the numerator and denominator of G(s). After finding U_{o_k} in terms of k, k is determined from eq. (2) using Table 1 to evaluate the integral, which by Parseval's theorem is

$$J_u = \int_0^\infty u^2(t)\,dt = \frac{1}{2\pi j} \int_{-j\infty}^{j\infty} u(s)u(-s)\,ds \tag{11}$$

956

Table 1   Tabulation of the Definite Integral

$$J_n = \frac{1}{2\pi j} \int_{-j\infty}^{j\infty} \frac{N(s)N(-s)}{D(s)D(-s)}\,ds$$

$$N(s) = N_{n-1}s^{n-1} + N_{n-2}s^{n-2} + \ldots + N_1 s + N_0$$

$$D(s) = D_n s^n + D_{n-1}s^{n-1} + \ldots + D_1 s + D_0$$

$$J_1 = \frac{N_0^2}{2D_0 D_1}$$

$$J_2 = \frac{N_1^2 D_0 + N_0^2 D_2}{2D_0 D_1 D_2}$$

$$J_3 = \frac{N_2^2 D_0 D_1 + (N_1^2 - 2N_0 N_2)D_0 D_3 + N_0^2 D_2 D_3}{2D_0 D_3(-D_0 D_3 + D_1 D_2)}$$

$$J_4 = \frac{N_3^2(-D_0^2 D_3 + D_0 D_1 D_2) + (N_2^2 - 2N_1 N_3)D_0 D_1 D_4 + (N_1^2 - 2N_0 N_2)D_0 D_3 D_4 + N_0^2(-D_1 D_4^2 + D_2 D_3 D_4)}{2D_0 D_4(-D_0 D_3^2 - D_1^2 D_4 + D_1 D_2 D_3)}$$

Solution: For N = 10 and D = s^2 we have

$$\{k^2 D\overline{D} + N\overline{N}\}^+ = \{k^2 s^4 + 100\}^+ \tag{12}$$

We perform the spectral factorization of the last equation.

We write

$$k^2 s^4 + 100 = (a_2 s^2 + a_1 s + a_0)(a_2 s^2 - a_1 s + a_0)$$

$$= a_2^2 s^4 - (a_1^2 - 2a_2 a_0)s^2 + a_0^2 \tag{13}$$

Equating the corresponding coefficients on both sides of eq. (10)

$$a_2^2 = k^2 \tag{14}$$

$$a_2 = k \tag{15}$$

$$a_1^2 - 2a_2 a_0 = 0 \tag{16}$$

$$a_0^2 = 100 \tag{17}$$

or

$$a_0 = 10 \tag{18}$$

Substituting the results of a_0 and a_2 into eq. (13) we have

$$a_1 = \sqrt{20k} \tag{19}$$

Thus

$$\{k^2 D\overline{D} + N\overline{N}\}^+ = (a_2 s^2 + a_1 s + a_0)$$

$$= ks^2 + \sqrt{20k}\, s + 10 \tag{20}$$

and

$$\{k^2 D\bar{D} + N\bar{N}\}^- = ks^2 - \sqrt{20k}\, s + 10 \qquad (21)$$

Equation (10) gives

$$\left[\frac{\bar{N}R}{\{k^2 D\bar{D} + N\bar{N}\}^-} \right]_+ = \left[\frac{10\, \frac{0.5}{s}}{ks^2 - \sqrt{20k}\, s + 10} \right]_+ = \frac{0.5}{s} \qquad (22)$$

and

$$U_{o_k} = \frac{s^2}{ks^2 + \sqrt{20k}\, s + 10}\, \frac{0.5}{s} = \frac{0.5s}{ks^2 + \sqrt{20k}\, s + 10} \qquad (23)$$

Substituting U_{o_k} into eq. (11) for U, we obtain

$$J_u = \frac{1}{2\pi j} \int_{-j\infty}^{j\infty} \frac{0.5s}{ks^2 + \sqrt{20k}\, s + 10} \cdot \frac{-0.5s}{ks^2 - \sqrt{20k}\, s + 10}\, ds \qquad (24)$$

From Table 1 we find

$$J_u = \frac{N_1^2 D_0 + N_0^2 D_2}{2 D_0 D_1 D_2} \qquad (25)$$

where

$$N_1 = 0.5,\ N_0 = 0,\quad D_0 = 10,\quad D_1 = \sqrt{20k},\quad D_2 = k$$

Thus

$$J_u = \frac{0.125}{k\,\sqrt{20k}} = 2.5 \qquad (26)$$

and

$$k = 0.05$$

We have the optimal control and the optimal closed-loop transfer function respectively,

$$U_o(s) = \frac{0.5s}{0.05s^2 + s + 10} \qquad (27)$$

$$M_o(s) = \frac{G(s)}{R(s)}\, U_o(s) = \frac{10}{0.05s^2 + s + 10} = \frac{200}{s^2 + 20s + 200} \qquad (28)$$

958

The performance index is given by

$$J = \frac{1}{2} \sum_{k=0}^{10} [x^2(k) + 2u^2(k)]$$

Find the optimal control $u°(k)$, $k = 0,1,2 \ldots 10$ that minimizes this performance index, subject to the equality constraint

$$x(k+1) = x(k) + 2u(k) \qquad (1)$$

Consider two cases

1) The initial state is $x(0) = 1$ and the final state is $x(11) = 0$.

2) The initial state is $x(0) = 1$ and the final state $x(11)$ is free.

Solution: 1) The adjoined performance index is

$$J_c = \sum_{k=0}^{10} F_c[x(k), u(k)]$$

where

$$F_c[x(k),u(k)] = \frac{1}{2} [x^2(k) + 2u^2(k)] + \lambda(k+1)[x(k+1) - x(k)$$

$$- 2u(k)] = F_c(k)$$

From

$$\frac{\partial F_c°(k)}{\partial x°(k)} + \frac{\partial F_c°(k-1)}{\partial x°(k)} = 0$$

we determine the discrete Euler-Lagrange equation

$$\lambda°(k+1) - \lambda°(k) - x°(k) = 0 \qquad (2)$$

From

$$\frac{\partial F_c°(k)}{\partial \lambda_i°(k+1)} = 0 \qquad\qquad i = 1,2, \ldots n$$

we get

$$\frac{\partial F_c°(k)}{\partial \lambda°(k+1)} = x°(k+1) - x°(k) - 2u°(k) = 0 \qquad (3)$$

Eq. (3) is equality constraint or the state equation in eq. (1) under the optimal condition.

959

From

$$\frac{\partial F_c^{\ o}(k)}{\partial u_j^{o}(k)} = 0 \qquad\qquad j = 1,2, \ldots p$$

we determine the optimal control

$$\frac{\partial F_c^{\ o}(k)}{\partial u^{o}(k)} = 2u^{o}(k) - 2\lambda^{o}(k+1) = 0$$

Thus

$$u^{o}(k) = \lambda^{o}(k+1) \tag{4}$$

Substituting we obtain two simultaneous difference equations:

$$\lambda^{o}(k+1) - \lambda^{o}(k) - x^{o}(k) = 0$$

$$x^{o}(k+1) - 2\lambda^{o}(k+1) - x^{o}(k) = 0$$

where

$$x(0) = 1 \quad , \quad x(11) = 0.$$

We can solve this set of equations using the z-transform method or the state transition method, thus

$$x^{o}(k) = 0.289[2.732 + 2\lambda^{o}(0)](3.732)^{k}$$

$$\quad + 0.289[0.732 - 2\lambda^{o}(0)](0.268)^{k} \tag{5}$$

$$\lambda^{o}(k) = [0.289 + 0.211\lambda^{o}(0)](3.732)^{k}$$

$$\quad + [-0.289 + 0.789\lambda^{o}(0)](0.268)^{k} \tag{6}$$

Substituting $x^{o}(11) = 0$ into eq. (5) we find the initial value of $\lambda^{o}(k)$.

$$\lambda^{o}(0) = -1.366$$

The optimal trajectory of $x^{o}(k)$ is described by

$$x^{o}(k) = (0.268)^{k} \tag{7}$$

and the optimal control is

$$u^{o}(k) = -2.732(0.268)^{k+1}$$

$$\quad = -0.732x^{o}(k) \tag{8}$$

2) For x(0) = 1 and x(11) free, we use the transversality condition

$$\left. \frac{\partial F_c^{\,\circ}(k-1)}{\partial x^{\circ}(k)} \right|_{k=11} = 0$$

We obtain

$$\lambda^{\circ}(11) = 0 \qquad\qquad (9)$$

Substituting eq. (9) into eq. (6) we get

$$\lambda^{\circ}(0) = -1.366$$

and the same results as in eq. (7) and eq. (8)

● **PROBLEM** 16-35

The digital system is described by

$$x(k+1) = Ax(k) + Bu(k) \qquad\qquad (1)$$

where

$$A = \begin{bmatrix} 0 & 1 \\ -0.5 & -0.2 \end{bmatrix}, \quad B = \begin{bmatrix} 0 \\ 1 \end{bmatrix}$$

a) For this system find the constant state feedback gain matrix G such that u(k) = -Gx(k) will bring any initial state x(0) to x(N) = 0, for N = 2. Determine the optimal control u°(k) for k = 0,1 and the optimal state trajectory x°(k) when

$$x(0) = \begin{bmatrix} 1 \\ 1 \end{bmatrix}$$

and

$$x^{\circ}(2) = 0.$$

b) For the control subject to amplitude constraint

$|u(k)| \leq 1$, determine the region of controllable states

of x(0) in the state plane for $N \leq 2$, such that x(N) = 0.

c) Find the optimal control u°(k) which will bring the initial state

$$x(0) = \begin{bmatrix} 1 \\ \\ 1 \end{bmatrix}$$

to $x(2) = 0$ and satisfy

$$J = \frac{1}{2} \sum_{k=0}^{1} u^2(k) = \text{minimum} \qquad (2)$$

Determine the optimal trajectory $x°(k)$ and the optimal value of J.

d) Determine the region of controllable states of $x(0)$ in the state plane for $N = 2$ such that $x(N) = 0$, and $J \leq 1$.

Where J is given by eq. (2), solve the problem for $J \leq 0.25$.

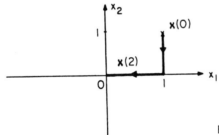

Fig. 1

Solution: a) Let us rewrite eq. (1) as

$$x(k+1) = (A - BG) \ x(k)$$

The solution is

$$x(k) = (A - BG)^k x(0).$$

Since $x(2) = 0$, then $(A - BG)^2 = 0$.

Now $G = [g_1 \ g_2]$, so

$$(A - BG)^2 = \begin{bmatrix} 0 & 1 \\ -0.5 - g_1 & -0.2 - g_2 \end{bmatrix} \begin{bmatrix} 0 & 1 \\ -0.5 - g_1 & -0.2 - g_2 \end{bmatrix}$$

$$= \begin{bmatrix} -0.5 - g_1 & -0.2 - g_2 \\ (0.2 + g_2)(0.5 + g_1) & (0.2 + g_2)^2 - (0.5 + g_1) \end{bmatrix}$$

$$= 0$$

Thus

$$g_1 = -0.5 \text{ and } g_2 = -0.2$$

and we have

$$u^0(k) = [0.5 \quad 0.2]x(k).$$

The optimal state feedback gain matrix is

$$G = [-0.5 \quad -0.2].$$

We get, for the given initial state

$$u^0(0) = 0.7 \qquad x^0(1) = \begin{bmatrix} 1 \\ 0 \end{bmatrix}$$

$$u^0(1) = 0.5 \qquad x^2(2) = \begin{bmatrix} 0 \\ 0 \end{bmatrix}$$

Fig. 1 shows the optimal state trajectory.

b) For the amplitude constraint

$$|u(k)| \le 1 \text{ and for } N = 2, \text{ the state transition}$$
equation is

$$x(2) = A^2 x(0) + ABu(0) + Bu(1) = 0$$

Solving for $x(0)$ we get

$$x(0) = \begin{bmatrix} 2 & -0.8 \\ 0 & 2 \end{bmatrix} \begin{bmatrix} u(0) \\ u(1) \end{bmatrix} \tag{3}$$

Substituting the four possible combinations of $u(k) = +1$ and -1 into eq. (3), we find the vertices of the region of controllable states.

$$u(0) = u(1) = 1 \qquad x(0) = \begin{bmatrix} 1.2 \\ 2 \end{bmatrix}$$

$$u(0) = u(1) = -1 \qquad x(0) = \begin{bmatrix} -1.2 \\ 2 \end{bmatrix}$$

$$u(0) = -u(1) = 1 \qquad x(0) = \begin{bmatrix} 2.8 \\ -2 \end{bmatrix}$$

$$-u(0) = u(1) = 1 \qquad x(0) = \begin{bmatrix} -2.8 \\ 2 \end{bmatrix}$$

Fig. 2 shows the region of controllable states.

963

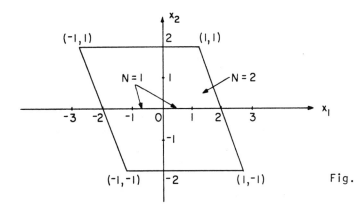

Fig. 2

c)   For the performance index in eq. (2), R = 1.   Using

$$W = \sum_{k=0}^{N-1} A^{-k-1}BR^{-1}B'(A^{-k-1})'$$

we determine the matrix W for N = 2.

$$W = \begin{bmatrix} 4.64 & -1.6 \\ -1.6 & 4 \end{bmatrix}$$

The optimal control is given by

$$u^0(k) = -R^{-1}B'(A^{-k-1})'W^{-1}x(0)$$

and

$$u^0(k) = -B'(A^{-k-1})'W^{-1}x(0)$$

We get

$$u^0(0) = 0.7 \quad \text{and} \quad u^0(1) = 0.5$$

The optimal performance index is

$$J^0 = \frac{1}{2}x'(0)W^{-1}x(0) = 0.37$$

d)   We have for $J \leq 1$

$$\frac{1}{2}x'(0)W^{-1}x(0) \leq 1$$

or

$$0.125x_1^2(0) + 0.1x_1(0)x_2(0) + 0.145x_2^2(0) \leq 1$$

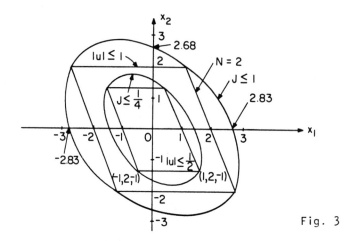

Fig. 3

The last expression describes an ellipse in the state plane.
Fig. 3 shows the regions of controllable states for J < 1
and J < 0.25.

CHAPTER 17

DIGITAL CONTROL SYSTEMS

DESIGN-CONTROLLER

● **PROBLEM** 17-1

Fig. 1 shows the digital control system with cascade digital controller.

The transfer function of the digital controller is

$$G_C(z) = \frac{(1 - 0.2z^{-1})(1 - 0.1z^{-1})}{(1 - 0.5z^{-1})(1 - 0.8z^{-1})}, \quad T = 1 \text{ sec}$$

For Fig. 2 find H(s).

Fig. 2 shows the digital control system with continuous-data feedback controller.

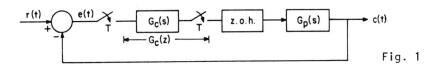

Fig. 1

Solution: H(s) can be realized by an RC network, since $G_C(z)$ has an equal number of poles and zeros, and all the zeros are simple and lie inside the unit circle $|z| = 1$.

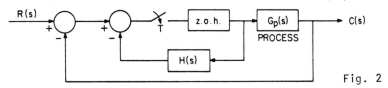

Fig. 2

H(s) will be placed in the system as shown in Fig. 2. Comparing the systems, we see that

$$G_C(z)h_o = \frac{h_o}{1 + h_o H(s)}$$

Solving for H(s)

$$H(s) = \frac{1 - G_c(z)}{h_o G_c(z)}$$

Now, a zero-order-hold is represented by

$$h_o = \frac{1 - e^{-Ts}}{s}$$

Substituting gives

$$\frac{H(s)}{s} = \frac{1}{1 - e^{-Ts}} \frac{1 - G_c(z)}{G_c(z)}$$

Taking the z-transforms

$$Z\left[\frac{H(s)}{s}\right] = \frac{1}{1 - z^{-1}} \frac{1 - G_c(z)}{G_c(z)}$$

Thus

$$Z\left[\frac{H(s)}{s}\right] = \frac{-z^{-1} + 0.38z^{-2}}{(1 - z^{-1})(1 - 0.2z^{-1})(1 - 0.1z^{-1})}$$

$$= \frac{-0.86z}{z - 1} - \frac{2.25z}{z - 0.2} + \frac{3.11z}{z - 0.1}$$

Taking the inverse transforms gives

$$z^{-1}Z\left[\frac{H(s)}{s}\right] = -0.86 - 2.25(0.2)^K + 3.11(0.1)^K$$

so the continuous time representation is

$$z^{-1}Z\left[\frac{H(s)}{s}\right] = -0.86 - 2.25(0.2)^{\frac{t}{T}} + 3.11(0.1)^{\frac{t}{T}}$$

$$= -0.86 - 2.25e^{+\ln 0.2} + 3.11e^{+\ln 0.1}$$

since T = 1.

Finding the corresponding Laplace transforms we have

$$\frac{H(s)}{s} = \frac{-0.86}{s} - \frac{2.25}{s + 1.61} + \frac{3.11}{s + 2.3}$$

or

$$H(s) = \frac{-3.53s - 3.18}{(s + 1.61)(s + 2.3)}$$

967

The controlled process of a digital control system is given by

$$G(z) = \frac{0.005z^{-1}(1 - 0.9z^{-1})}{(1 - z^{-1})(1 - 0.905z^{-1})}$$

Determine the effect of a weighting factor, c on the system.

TABLE 1

Input function	N	M(z)
Step input $u_s(t)$	1	$1 - (1 - z^{-1})F(z)$
Ramp input $tu_s(t)$	2	$1 - (1 - z^{-1})^2 F(z)$
Parabolic input $t^2 u_s(t)$	3	$1 - (1 - z^{-1})^3 F(z)$

Solution: From Table 1 we have for a unit-ramp input the closed-loop transfer function given by the deadbeat-response

$$M(z) = 2z^{-1} - z^{-2}$$

Figs. 1 and 2 show the unit-step response and the unit-ramp response of the designed system, c = 0.

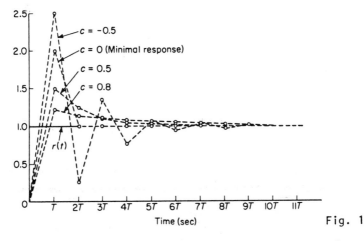

Fig. 1

From the drawings we see that the ramp response reaches the input in two sampling instants and the step response has a peak overshoot of 100 percent.

$$1 - M_W(z) = \frac{1 - M(z)}{1 - cz^{-1}} = \frac{(1 - z^{-1})^2}{1 - cz^{-1}}$$

or

$$M_W(z) = \frac{(2 - c)z^{-1} - z^{-2}}{1 - cz^{-1}}$$

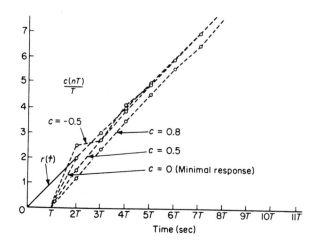

Fig. 2

We plotted the unit-step and unit-ramp responses of the system at the sampling instants for several values of c (where $|c| \leq 1$), to show the effects of the weighting factor c. For negative values of c the overshoot of the step response is increased from the deadbeat design. For $c = 0.8$, the peak overshoot in the step response is 20 percent. We see that the best choice for c is 0.5.

● **PROBLEM** 17-3

The system is described by

$$x[(k + 1)T] = Ax(kT) + Bu(kT)$$

where

$$A = \begin{bmatrix} 0 & 1 \\ -1 & -2 \end{bmatrix} \qquad B = \begin{bmatrix} 0 \\ 1 \end{bmatrix}$$

and the state feedback is

$$u(kT) = -Gx(kT)$$

$$G = [g_1 \ g_2]$$

Examine the effect of the feedback gain matrix $G = [g_1 \ g_2]$.

Solution: Assuming that $g_2 = 0$, we get the following characteristic equation of the closed-loop system.

$$|zI - A + BG| = z^2 + (g_1 + 2)z + (1 - g_1) = 0$$

The two eigenvalues of the closed-loop system cannot be arbitrarily assigned, since there is only one parameter in g_1. We can rewrite the equation

$$1 + \frac{g_1(z - 1)}{z^2 + 2z + 1} = 0$$

Fig. 1 shows the root loci of the equation based on the pole-zero configuration of

$$\frac{z - 1}{z^2 + 2z + 1}$$

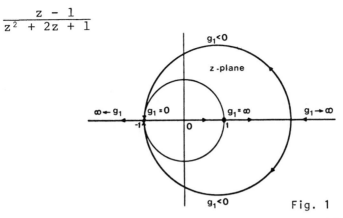

Fig. 1

For negative values of g_1 both roots are outside the unit circle and for positive values of g_1 one root stays always to the left of the -1 point in the z-plane. We see that with only $x_1(kT)$ available for feedback the system cannot be stabilized for any value of g_1. In case, when only $x_2(kT)$ is available for feedback we have

$$G = [0 \quad g_2]$$

and the characteristic equation is

$$|zI - A + BG| = z^2 + (g_2 + 2)z + (1 + g_2) = 0$$

or

$$1 + \frac{g_2(z + 1)}{z^2 + 2z + 1} = 0$$

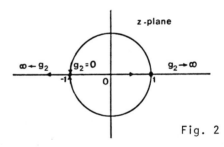

Fig. 2

Fig. 2 shows the root loci of this equation. The open-loop pole at $z = -1$ cannot be moved. The other root moves from $z = \infty$ to $z = -\infty$ on the real axis in the z-plane as g_2 varies between $-\infty$ and ∞. Thus for the single-input system the system cannot be stabilized. Changing B to

$$B = \begin{bmatrix} 1 & 0 \\ 0 & 1 \end{bmatrix}$$

the feedback matrix becomes

$$G = \begin{bmatrix} g_{11} & g_{12} \\ g_{21} & g_{22} \end{bmatrix}$$

and

$$|zI - A + BG| = z^2 + (2 + g_{11} + g_{22})z + g_{11}(2 + g_{22})$$

$$+ (1 - g_{12})(1 + g_{21}) = 0$$

When $x_1(kT)$ is not available for feedback we have

$$g_{11} = g_{21} = 0$$

$$z^2 + (2 + g_{22})z + (1 - g_{12}) = 0$$

The two eigenvalues of A - BG can be arbitrarily assigned since we have two independent parameters in g_{12} and g_{22}. For $g_{12} = g_{22} = 0$ we get

$$z^2 + (2 + g_{11})z + 2g_{11} + g_{21} + 1 = 0$$

and again the eigenvalues can be arbitrarily assigned.

● **PROBLEM 17-4**

The state diagram of a digital control process which is subject to state feedback is shown in Fig. 1. The eigenvalues of the closed-loop system are 0.5 + j0.5 and 0.5 - j0.5.

The coefficient matrices are

$$A = \begin{bmatrix} 0 & 1 \\ -0.368 & 1.368 \end{bmatrix} \qquad B = \begin{bmatrix} 0 \\ 1 \end{bmatrix}$$

$$D = [0.264 \quad 0.368] \qquad E = 0$$

and the feedback gain matrix is

$$G = [0.132 \quad 0.368].$$

The equivalent feedback dynamic controller which approximates the state feedback is

$$H(z) = K\frac{z + \alpha_1}{z + \beta_1} \cong K(1 + d_1 z^{-1})$$

Obtain the response of the system.

<u>Solution:</u> Using

$$P = K[1d_1] = G\begin{bmatrix} D \\ D(A - BG)^{-1} \end{bmatrix}^{-1}$$

$$= [0.132 \quad 0.368 \begin{bmatrix} 0.264 & 0.368 \\ 0.896 & -0.528 \end{bmatrix}^{-1}$$

971

$$= 0.8514[1 \qquad -0.1216$$

we have

$$K = 0.8514 \qquad d_1 = -0.1216$$

We want the dynamic controller to be physically realizable, thus $\beta_1 \ll d_1$. Let $\beta_1 = 0.0005$, then

$$\alpha_1 = d_1 + \beta_1 = -0.1211$$

The transfer function of the feedback controller is

$$H(z) = 0.8514\frac{z - 0.1211}{z - 0.0005}$$

Fig. 2 shows the block diagram of the closed-loop system with the dynamic controller.

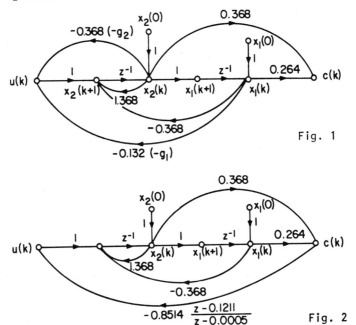

Fig. 1

Fig. 2

We shall compare the characteristics and the performance of both systems shown in Figs. 1 and 2.

For the system with the state feedback we have

$$|zI - A + BG| = z^2 - z + 0.5 = 0$$

and the roots of the characteristic equation are $z_1 = 0.5 + j0.5$, $z_2 = 0.5 - j0.5$ as required. For the system with the dynamic controller feeding back from the output we have

$$z^3 - 1.05202z^2 + 0.557398z - 0.027678 = 0$$

with the roots

$$z_1 = 0.0548 \qquad z_2 = 0.4986 + j0.5043$$

972

$$z_3 = 0.4986 - j0.5043$$

We see that the dominant characteristic roots are close to $0.5 \pm j0.5$ even though the use of the dynamic controller in place of the state feedback has made the overall system a third-order system. From Fig. 1, for the initial state

$$x_1(0) = 1, \qquad x_2(0) = 0$$

we have

$$C(z) = \frac{0.264z^2 - 0.448z}{z^2 - z + 0.5} \, x_1(0)$$

The response of $c(k)$ is shown in Fig. 3.

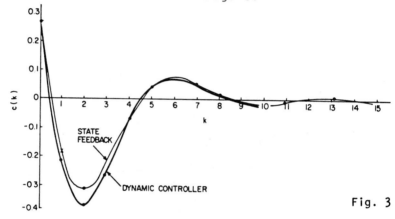

Fig. 3

Using Fig. 2 we find the z-transform of the output of the system with the dynamic controller for the initial states $x_1(0) = 1$, $x_2(0) = 0$

$$C(z) = \frac{0.264z^3 - 0.4967z^2 + 0.000248z}{z^3 - 1.05202z^2 + 0.5573z - 0.02767} x_1(0)$$

From Fig. 3 we conclude that the responses for both systems are very close. The system with the dynamic controller gives a higher overshoot.

● **PROBLEM 17-5**

The second-order digital system is represented by the state equations

$$x(k + 1) = Ax(k) + Bu(k)$$

where

$$A = \begin{bmatrix} 1 & -1 \\ 0 & 1 \end{bmatrix} \qquad B = \begin{bmatrix} 1 \\ 1 \end{bmatrix}$$

For the eigenvalues of the closed-loop system $\lambda_1 = 0.4$ and $\lambda_2 = 0.6$, find the feedback gain matrix G with the state feedback $u(k) = -Gx(k)$. The state feedback stabilizes the system by moving the eigenvalues of A, $\lambda = 1$, 1 inside the unit circle.

973

Solution: Let us transform the system into the phase-variable canonical form. From

$$S = [B, AB, A^2B, \ldots A^{n-1}B]$$

we get the controllability matrix

$$S = [B \quad AB] = \begin{bmatrix} 1 & 0 \\ 1 & 1 \end{bmatrix}$$

which is nonsingular. Then

$$M_1 = [0 \quad 1]S^{-1} = [-1 \quad 1]$$

$$M = \begin{bmatrix} M_1 \\ M_1 A \end{bmatrix} = \begin{bmatrix} -1 & 1 \\ -1 & 2 \end{bmatrix}$$

In canonical form

$$y(k + 1) = A_1 y(k) + B_1 u(k)$$

where

$$A_1 = M A M^{-1} = \begin{bmatrix} 0 & 1 \\ -1 & 2 \end{bmatrix}$$

$$B_1 = M B = \begin{bmatrix} 0 \\ 1 \end{bmatrix}$$

The characteristic equation of the transformed closed-loop system is

$$|\lambda I - A_1 + B_1 G_1| = 0$$

where

$$G_1 = [g_1^* \quad g_2^*]$$

is the feedback matrix.

We can write

$$\lambda^2 + (g_2^* - 2)\lambda + (g_1^* + 1) = 0$$

Equating the corresponding coefficients we get

$$G_1 = [g_1^* \quad g_2^*] = [-0.76 \quad 1]$$

From

$$G = G_1 M = [-0.76 \quad 1]\begin{bmatrix} -1 & 1 \\ -1 & 2 \end{bmatrix} = [-0.24 \quad 1.24]$$

we find the feedback matrix of the original system.

974

The eigenvalues of A - BG are $\lambda_1 = 0.4$ and $\lambda_2 = 0.6$, and the state feedback control that produces the desired eigenvalues is

$$u(k) = [0.24 \quad -1.24]x(k)$$

The transfer function of a digital controller is given by

$$\frac{E_2(z)}{E_1(z)} = G_c(z) = \frac{5(1 + 0.25z^{-1})}{(1 - 0.5z^{-1})(1 - 0.1z^{-1})} \qquad (1)$$

Realize the transfer function by various methods.

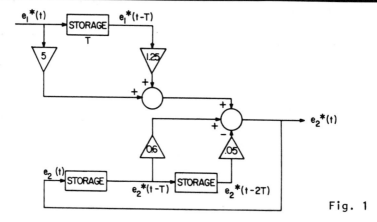

Fig. 1

A direct digital program of the transfer function $G_c(z)$

Solution:

1) Direct Digital Programming

Eq. (1) can be written in the form

$$E_2(z)(1 - 0.6z^{-1} + 0.05z^{-2})$$

$$= E_1(z) \cdot 5 \cdot (1 + 0.25z^{-1})$$

Taking the inverse z-transform and solving for $e_2^*(t)$ we have

$$e_2^*(t) = 5e_1^*(t) + 1.25e_1^*(t - T) + 0.6e_2^*(t - T)$$

$$- 0.05e_2^*(t - 2T) \qquad (2)$$

where T is the sampling period.

Fig. 1 shows the block diagram of the direct digital programming of eq. (1) according to eq. (2).

Applying the direct decomposition to eq. (1) we get

$$E_2(z) = (5 + 1.25z^{-1})X(z)$$

975

$$X(z) = E_1(z) + 0.6z^{-1}X(z) - 0.05z^{-2}X(z)$$

Fig. 2 shows the digital program that realizes the last two equations.

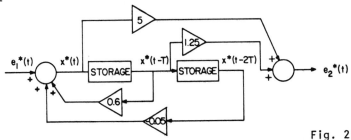

Fig. 2

A direct digital program of the transfer function $G_c(z)$

2) Cascade Digital Programming

We can write eq. (1) in the form

$$\frac{E_2(z)}{E_1(z)} = \frac{1 + 0.25z^{-1}}{1 - 0.5z^{-1}} \cdot \frac{5}{1 - 0.1z^{-1}} \qquad (3)$$

Based on equation (3) we obtain the digital program shown in Fig. 3.

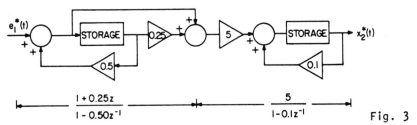

Fig. 3

A cascade digital program of the transfer function $G_c(z)$

3) Parallel Digital Programming

Using partial-fraction expansion we can write eq. (1) in the form

$$\frac{E_2(z)}{E_1(z)} = \frac{9.375}{1 - 0.5z^{-1}} - \frac{4.375}{1 - 0.1z^{-1}} \qquad (4)$$

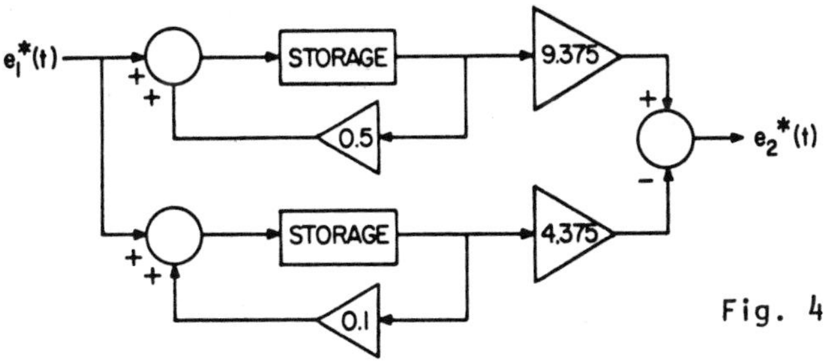

Fig. 4

A parallel digital program of the transfer function $G_c(z)$

This transfer function is realized by the parallel connection of two first-order programs, see Fig. 4.

Consider a DC motor, whose block diagram is shown in Fig. 1.

This motor is used as a control system which controls a pure inertia load in such a way that any non-zero initial values in the armature current i_a and the motor-load velocity ω are reduced to zero as quickly as possible. Neglect the armature inductance of the motor and assume the armature resistance $R_a = 1$ ohm.

$K_a = 0.345 \dfrac{Nm}{amp}$ torque constant

$B = 0.25$ kg-m   viscous friction coefficient

$K_b = 0.345 \dfrac{Nm}{amp}$ back emf constant

$J = 1.41 \times 10^{-3} kgm^2$ motor and load inertia.

Find the feedback gain matrix that makes the closed-loop eigenvalues $\lambda_1 = \lambda_2 = 0$. The feedback matrix is $G = [g_1 \ g_2]$ and the control is sample-and-hold, so that $T = 0.005$ sec

$u(t) = -Gx(kT)$

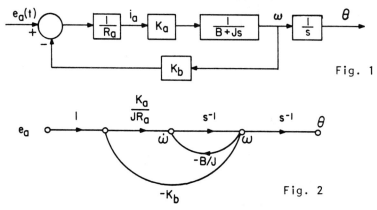

Fig. 1

Fig. 2

Solution: The state diagram of the dc motor is shown in Fig. 2.

The state variables are $\theta(t)$ and $\omega(t)$, and the state equations are

$x(t) = Ax(t) + Bu(t)$

where

$x(t) = \begin{bmatrix} \theta(t) \\ \omega(t) \end{bmatrix}$

$u(t) = e_a(t)$

977

$$A = \begin{bmatrix} 0 & 1 \\ 0 & -\dfrac{BR_a + K_aK_b}{JR_a} \end{bmatrix} \qquad B = \begin{bmatrix} 0 \\ \dfrac{K_a}{JR_a} \end{bmatrix}$$

Substituting the system parameters we obtain

$$A = \begin{bmatrix} 0 & 1 \\ 0 & -267.19 \end{bmatrix} \qquad B = \begin{bmatrix} 0 \\ 244.68 \end{bmatrix}$$

Deriving the control through sample-and-hold we can write

$$u(kT) = -Gx(kT)$$

$$G = [g_1 \quad g_2]$$

and

$$T = 0.005 \text{ sec.}$$

We have to find g_1 and g_2 such that the eigenvalues of the closed-loop digital control system are $\lambda_1 = \lambda_2 = 0$.

Note that the eigenvalues of the closed-loop digital control system are the eigenvalues of $\phi(T) - \phi(T)G$, where

$$\phi(T) = e^{AT}$$

and

$$\theta(T) = \int_0^T \phi(\lambda)Bd\lambda$$

For $T = 0.005$ sec the state transition matrix $\phi(T)$ is

$$\phi(T) = L^{-1}[(sI - A)^{-1}]\Big|_{t=T=0.005} = \begin{bmatrix} 1 & 0.00276 \\ 0 & 0.263 \end{bmatrix}$$

$$\theta(T) = L^{-1}[(sI - A)^{-1}Bs^{-1}]\Big|_{t=T=0.005} = \begin{bmatrix} 0.00205 \\ 0.675 \end{bmatrix}$$

We write the state equations in the difference equation form

$$x[(k + 1)T] = \phi(T)x(kT) + \theta(T)u(kT)$$

The state feedback control is given by

$$u(kT) = -Gx(kT)$$

The characteristic equation of the open-loop system, or of $\phi(T)$ is

$$\Delta_0(z) = |zI - \phi(T)| = \begin{vmatrix} z - 1 & -0.00276 \\ 0 & z - 0.263 \end{vmatrix}$$

$$= z^2 - 1.263z + 0.263 = 0$$

978

The eigenvalues of $\phi(T) - \theta(T)G$ can be arbitrarily placed by state feedback, because the pair $[\phi(T), \theta(T)]$ is completely controllable. Let the desired closed-loop characteristic equation be

$$\Delta_c(z) = z^2$$

Using

$$k(z) = Adj[zI - \phi(T)] \cdot \theta(T) = \begin{bmatrix} 0.00205z + 0.00132 \\ 0.675(z - 1) \end{bmatrix}$$

From

$$G = -[\Delta_{01} \quad \Delta_{02}]K^{-1}$$

we determine the feedback matrix G.

$$\Delta_{01} = \Delta_0(z)\Big|_{z=0} = 0.263$$

$$\Delta_{02} = \frac{d}{dz}\Delta_0(z)\Big|_{z=0} = (2z - 1.263)\Big|_{z=0} = -1.263$$

$$K = [k_1 \quad k_2]$$

where

$$k_1 = Adj[zI - \phi(T)] \cdot \theta(T)\Big|_{z=0} = \begin{bmatrix} 0.001323 \\ -0.675 \end{bmatrix}$$

$$k_2 = \frac{d}{dz}Adj[zI - \phi(T)] \cdot \theta(T)\Big|_{z=0} = \begin{bmatrix} 0.00205 \\ 0.675 \end{bmatrix}$$

Substituting the results into the expression for G we get

$$G = [296.3 \quad 0.970]$$

We can write the state equation of the closed-loop system as

$$\begin{bmatrix} x_1[(k + 1)T] \\ x_2[(k + 1)T] \end{bmatrix} = \begin{bmatrix} 0.395 & 0.00078 \\ -200 & -0.392 \end{bmatrix}\begin{bmatrix} x_1(kT) \\ x_2(kT) \end{bmatrix}$$

Solving this equation in the z-domain we get

$$X(z) = \begin{bmatrix} z^{-1} + 0.395z^{-2} & 0.00078z^{-2} \\ -200z^{-2} & z^{-1} - 0.392z^{-2} \end{bmatrix}x(0)$$

We see that the responses will become zero after two sampling periods for any initial conditions for $x_1(kT)$ and $x_2(kT)$.

STATE-DISCRETE

Consider the nonlinear, time-varying system

$$x(k + 3) + x(k + 2)x(k + 1) + \alpha\sin(\omega k)x^2(k) + x(k)$$

$$= u(k) - 3u(k + 1) \tag{1}$$

Draw a simulation diagram and express the system in state space form.

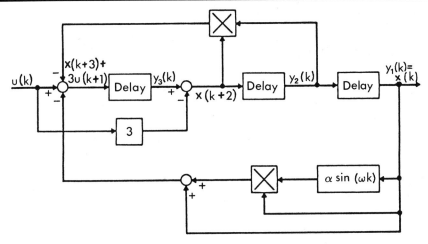

Solution: To draw the simulation diagram let us rewrite the equation as

$$x(k + 3) + 3u(k + 1) = u(k) - x(k + 2)x(k + 1)$$

$$- \alpha\sin(\omega k)x^2(k) - x(k) \tag{2}$$

Note that, the signal at the node labeled $x(k + 3) +$ $3u(k + 1)$ is delayed once, giving $x(k + 2) + 3u(k)$ and at that point $3u(k)$ is subtracted leaving $x(k + 2)$.

The states, labeled as shown, are

$$y_1(k) = x(k) \tag{3}$$

$$y_2(k) = x(k + 1) \tag{4}$$

$$y_3(k) = x(k + 2) + 3u(k) \tag{5}$$

The state equations are

$$y_1(k + 1) = y_2(k) \tag{6}$$

$$y_2(k + 1) = y_3(k) - 3u(k) \tag{7}$$

$$y_3(k + 1) = -y_2(k)y_3(k) + 3y_2(k)u(k)$$

$$- \alpha\sin(\omega k)y_1^2(k) - y_1(k) + u(k) \tag{8}$$

with the output equation

$$x(k) = [1 \ 0 \ 0]y(k). \tag{9}$$

A homogeneous discrete-time system is described by

$$x(k + 1) = Ax(k)$$

1) Show that for a nontrivial steady-state solution to exist, the matrix A must have unity as an eigenvalue.

2) Find a 2 x 2 nondiagonal, symmetric matrix with this property and compute the steady-state solution.

Solution:

1) In case of a constant steady-state solution, for sufficiently large k we have

$$x(k + 1) = x(k)$$

Let us denote it x_c. We have

$$x_c = Ax_c$$

which is the eigenvalue equation

$$Ax_e = \lambda x_e \quad \text{with} \quad \lambda = 1.$$

That proves part 1.

2) For

$$A = \begin{bmatrix} a_{11} & a_{12} \\ a_{12} & a_{22} \end{bmatrix}$$

we have

$$|A - I\lambda| = \lambda^2 - (a_{11} + a_{22})\lambda + (a_{11}a_{22} - a_{12}^2) = 0$$

The roots are

$$\lambda_{1,2} = \frac{1}{2}(a_{11} + a_{22}) \pm \sqrt{[\frac{1}{2}(a_{11} + a_{22})]^2 - (a_{11}a_{22} - a_{12}^2)}$$

Let us set $a_{11} = a_{22} = 2$, then

$$\lambda_{1,2} = 2 \pm \sqrt{4 - 4 + a_{12}^2}$$

For $\lambda = 1$ we have $a_{12} = 1$. Thus

$$A = \begin{bmatrix} 2 & 1 \\ 1 & 2 \end{bmatrix}$$

and the steady-state solution x_c is the eigenvector associated with the root $\lambda = 1$,

$$adj[A - I\lambda]\Big|_{\lambda=1} = \begin{bmatrix} 2 - \lambda & -1 \\ -1 & 2 - \lambda \end{bmatrix}\Big|_{\lambda=1} = \begin{bmatrix} 1 & -1 \\ -1 & 1 \end{bmatrix}$$

The steady-state solution will be proportional to $x_c = \begin{bmatrix} 1 \\ -1 \end{bmatrix}$

Consider a system with two inputs $u_1(k)$ and $u_2(k)$ and three outputs $y_1(k)$, $y_2(k)$, $y_3(k)$, whose input-output difference equations are

$$y_1(k + 3) + 6[y_1(k + 2) - y_3(k + 2)] + 2y_1(k + 1)$$

$$+ y_2(k + 1) + y_1(k) - 2y_3(k)$$

$$= u_1(k) + u_2(k + 1) \qquad (1)$$

$$y_2(k + 2) + 3y_2(k + 1) - y_1(k + 1) + 5y_2(k) + y_3(k)$$

$$= u_1(k) + u_2(k) + u_3(k) \qquad (2)$$

$$y_3(k + 1) + 2y_3(k) - y_2(k) = u_3(k) - u_2(k) + 7u_3(k + 1) \quad (3)$$

For this system draw a simulation diagram, select state variables and write the matrix state equations.

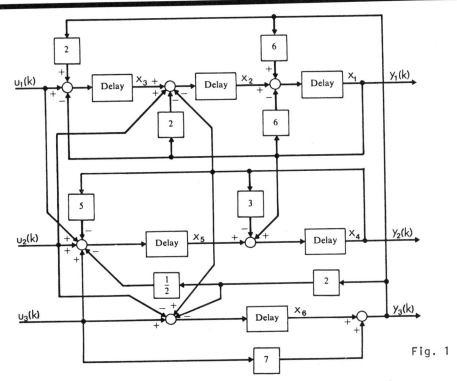

Fig. 1

Solution: In eq. (1) let us delay each term three times, thus

$$y_1(k) = D\{-6[y_1(k) - y_3(k)] + D[u_2(k) - 2y_1(k)$$

$$- y_2(k) + D[u_1(k) - y_1(k) + 2y_3(k)]]\} \qquad (4)$$

In equation (2) we delay each term twice

$$y_2(k) = D\{-3y_2(k) + y_1(k) + D[u_1(k) + u_2(k) + u_3(k)$$

$$- 5y_2(k) - y_3(k)]\} \qquad (5)$$

982

In equation (3) we delay once

$$y_3(k) = 7u_3(k) + D\{u_3(k) - u_2(k) - 2y_3(k) + y_2(k)\} \tag{6}$$

(see figure 1) Simulation diagram.

Using the diagram and denoting $x_1 \ldots x_6$ as indicated we can write

$$x(k+1) = \begin{bmatrix} -6 & 1 & 0 & 0 & 0 & 6 \\ -2 & 0 & 1 & -1 & 0 & 0 \\ -1 & 0 & 0 & 0 & 0 & 2 \\ 1 & 0 & 0 & -3 & 1 & 0 \\ 0 & 0 & 0 & -5 & 0 & -1 \\ 0 & 0 & 0 & 1 & 0 & -2 \end{bmatrix} x(k) + \begin{bmatrix} 0 & 0 & 42 \\ 0 & 1 & 0 \\ 1 & 0 & 14 \\ 0 & 0 & 0 \\ 1 & 1 & -6 \\ 0 & -1 & -13 \end{bmatrix} \begin{bmatrix} u_1(k) \\ u_2(k) \\ u_3(k) \end{bmatrix} \tag{7}$$

and

$$y(k) = \begin{bmatrix} 1 & 0 & 0 & 0 & 0 & 0 \\ 0 & 0 & 0 & 1 & 0 & 0 \\ 0 & 0 & 0 & 0 & 0 & 1 \end{bmatrix} x(k) + \begin{bmatrix} 0 & 0 & 0 \\ 0 & 0 & 0 \\ 0 & 0 & 7 \end{bmatrix} \begin{bmatrix} u_1(k) \\ u_2(k) \\ u_3(k) \end{bmatrix} \tag{8}$$

DIGITAL OBSERVER

● **PROBLEM** 17-11

The state equations describing the digital process are

$$x(k+1) = Ax(k) + Bu(k) \tag{1}$$

where

$$A = \begin{bmatrix} 0 & 1 \\ -1 & 1 \end{bmatrix} \qquad B = \begin{bmatrix} 0 \\ 1 \end{bmatrix}$$

The output equation is

$$c(k) = Dx(k)$$

$$D = \begin{bmatrix} 2 & 0 \end{bmatrix}$$

Design a digital observer that observes the states $x_1(k)$ and $x_2(k)$ from the output $c(k)$.

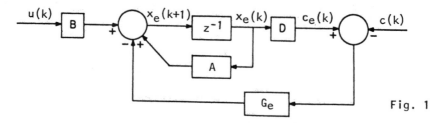

Fig. 1

Solution: The block diagram of the digital observer is shown in Fig. 1.

The characteristic equation is

$$|\lambda I - A + G_e D| = 0$$

or

$$\lambda^2 + (2g_{e1} - 1)\lambda + 1 + 2g_{e2} - 2g_{e1} = 0 \qquad (2)$$

$$G_e = \begin{bmatrix} g_{e1} \\ g_{e2} \end{bmatrix}$$

We will try to design the observer that has a deadbeat response such that $x_e(k)$ reaches $x(k)$ in one sampling period. The characteristic equation for a deadbeat response is

$$\lambda^2 = 0$$

From eq. 2, $g_{e1} = 0.5$ and $g_{e2} = 0$. The coefficient matrix for the closed-loop observer is

$$A - G_e D = \begin{bmatrix} -1 & 1 \\ -1 & 1 \end{bmatrix}$$

The state equations of the observer, for $u(k) = 0$, are

$$x_e(k + 1) = \begin{bmatrix} -1 & 1 \\ -1 & 1 \end{bmatrix} x_e(k) + \begin{bmatrix} 1 & 0 \\ 0 & 0 \end{bmatrix} x(k) \qquad (3)$$

Assuming arbitrary initial states of the system and the observer

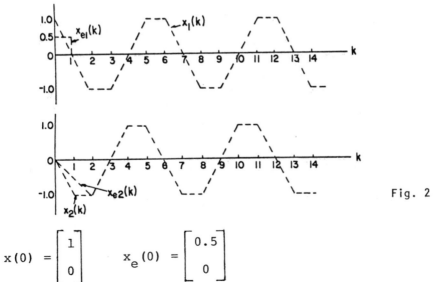

Fig. 2

$$x(0) = \begin{bmatrix} 1 \\ 0 \end{bmatrix} \qquad x_e(0) = \begin{bmatrix} 0.5 \\ 0 \end{bmatrix}$$

we solve equations (1) and (3). Fig. 2 shows the true states, $x_1(k)$, $x_2(k)$ and the observed states $x_{e1}(k)$ and $x_{e2}(k)$.

The heating system shown in the figure is to be controlled by a sampled data system.

The temperature T_2 of the liquid leaving tank 2 must be controlled by the temperature of the liquid in tank 1, T_1. Use $\frac{\rho V_1}{\omega} = 1$, $\frac{\rho V_2}{\omega} = \frac{1}{2}$, $\omega c_p = 2$ and

T_0 - temperature of liquid entering tank 1.

T_1 - temperature of liquid of tank 1.

T_2 - temperature of liquid of tank 2.

ρ - density of water

V_1, V_2 - volume of water in tanks 1 and 2, $V_1 + V_2 = $ const.

ω - flow rate of water

q - heat input

C_p - heat capacity of water.

For the system described above design a compensator $D(z)$ which will provide a response to a step change in set point that rises to and remains at the desired final value in two sampling instants.

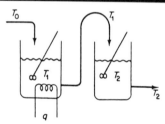

Solution: The equations describing the system are

$$\frac{\rho V_1}{\omega} \frac{d}{dt} T_1 + T_1 = T_0 + \frac{q(t)}{\omega c_p}$$

$$\frac{\rho V_2}{\omega} \frac{d}{dt} T_2 + T_2 = T_1$$

We shall define state variables as

$$x_1 = T_2 - T_{2s}$$

$$x_2 = 2(T_1 - T_{1s})$$

$$m = q - q_s$$

where s denotes the steady state values. We obtain

$$\frac{d}{dt}x_1 = -2x_1 + x_2$$

$$\frac{d}{dt}x_2 = -x_2 + m$$

or

$$\begin{bmatrix} \dot{x}_1 \\ \dot{x}_2 \end{bmatrix} = \begin{bmatrix} -2 & 1 \\ 0 & -1 \end{bmatrix}\begin{bmatrix} x_1 \\ x_2 \end{bmatrix} + \begin{bmatrix} 0 \\ 1 \end{bmatrix} m$$

$$A = \begin{pmatrix} -2 & 1 \\ 0 & -1 \end{pmatrix}$$

Then

$$(sI - A) = \begin{pmatrix} s + 2 & -1 \\ 0 & s + 1 \end{pmatrix}$$

and

$$(sI - A)^{-1} = \frac{1}{(s + 1)(s + 2)}\begin{pmatrix} s + 1 & 1 \\ 0 & s + 2 \end{pmatrix}$$

$$= \begin{pmatrix} \dfrac{1}{s + 2} & \dfrac{1}{(s + 1)(s + 2)} \\ 0 & \dfrac{1}{s + 1} \end{pmatrix}$$

and

$$L^{-1}[(sI - A)^{-1}] = \Phi(t) = \begin{pmatrix} e^{-2t} & e^{-t} - e^{-2t} \\ 0 & e^{-t} \end{pmatrix}$$

Let

$$\alpha = e^{-T}$$

then

$$\Phi(T) = \alpha\begin{pmatrix} \alpha & 1 - \alpha \\ 0 & 1 \end{pmatrix} = e^{AT}$$

We have

$$D(T) = \int_0^T e^{A\tau} B d\tau = \int_0^T \begin{pmatrix} e^{-2\tau} & e^{-\tau} - e^{-2\tau} \\ 0 & e^{-\tau} \end{pmatrix}\begin{pmatrix} 0 \\ 1 \end{pmatrix} d\tau$$

$$= (1 - \alpha)\begin{pmatrix} \dfrac{1 - \alpha}{2} \\ 1 \end{pmatrix} \tag{1}$$

986

Then

$$\Phi(-T)D(T) = \frac{1}{\alpha}\begin{pmatrix} \dfrac{1}{\alpha} & 1 - \dfrac{1}{\alpha} \\ 0 & 1 \end{pmatrix}\begin{pmatrix} \dfrac{(1-\alpha)^2}{2} \\ 1 - \alpha \end{pmatrix} = \begin{pmatrix} \dfrac{-(1-\alpha)^2}{2\alpha^2} \\ \dfrac{1-\alpha}{\alpha} \end{pmatrix}$$

and

$$\Phi(-2T)D(T) = \frac{1}{\alpha^2}\begin{pmatrix} \dfrac{1}{\alpha^2} & 1 - \dfrac{1}{\alpha^2} \\ 0 & 1 \end{pmatrix}\begin{pmatrix} \dfrac{(1-\alpha)^2}{2} \\ 1 - \alpha \end{pmatrix}$$

$$= \begin{pmatrix} \dfrac{-(1-\alpha)^2(1+2\alpha)}{2\alpha^4} \\ \dfrac{1-\alpha}{\alpha^2} \end{pmatrix}$$

Then

$$-S = (\Phi(-T)D(T) \quad \Phi(-2T)D(T))$$

$$= \begin{pmatrix} \dfrac{-(1-\alpha)^2}{2\alpha^2} & \dfrac{-(1-\alpha)^2(1+2\alpha)}{2\alpha^4} \\ \dfrac{1-\alpha}{\alpha} & \dfrac{1-\alpha}{\alpha^2} \end{pmatrix}$$

Then

$$-S^{-1} = \frac{1}{|-S|}\begin{bmatrix} \dfrac{1-\alpha}{\alpha^2} & \dfrac{(1-\alpha)^2(1+2\alpha)}{2\alpha^4} \\ \dfrac{-(1-\alpha)}{\alpha} & \dfrac{-(1-\alpha)^2}{2\alpha^2} \end{bmatrix}$$

$$|-S| = \frac{-(1-\alpha)^3(1+\alpha)}{2\alpha^5}$$

We obtain

$$S^{-1} = \begin{pmatrix} \dfrac{-2\alpha^3}{(1-\alpha)^2(1+\alpha)} & \dfrac{-(1+2\alpha)\alpha}{1-\alpha^2} \\ \dfrac{2\alpha^4}{(1-\alpha)^2(1+\alpha)} & \dfrac{\alpha^3}{1-\alpha^2} \end{pmatrix}$$

We assume the system to be initially at rest with the following initial condition

$$x(0) = \begin{pmatrix} 1 \\ 2 \end{pmatrix}$$

The feedback control law is

$$m(t_k) = (S^{-1})_{\text{row } 1} \, x(t_k)$$

$$= \frac{-2\alpha^3}{(1-\alpha)^2(1+\alpha)}x_1(t_k) - \frac{(1+2\alpha)\alpha}{1-\alpha^2}x_2(t_k)$$

$$m(0) = \frac{-2\alpha^3}{(1-\alpha)^2(1+\alpha)} - \frac{2(1+2\alpha)\alpha}{1-\alpha^2} = \frac{-2\alpha(1+\alpha-\alpha^2)}{(1-\alpha)^2(1+\alpha)} \qquad (2)$$

Then from

$$x(t_{k+1}) = \Phi(T)x(t_k) + D(T)m(t_k)$$

and eqs. (1) and (2) we get

$$x_1(T) = \frac{\alpha}{1+\alpha} \qquad (3)$$

The temperature deviation $T_2 - T_{2s}$ in the second tank is the output variable.

$$C_1(t) = x_1(t) = T_2 - T_{2s} \text{ is the}$$

deviation about the final steady state. Since the control is time optimal

$$x_1(iT) = T_2(iT) - T_{2s} = 0 \qquad i \geq 2$$

We have for the response, in terms of an output variable $c(t)$ defined about the initial steady state

$$c(0) = 0$$

$$c(T) = 1 - \frac{\alpha}{1+\alpha} = \frac{1}{1+\alpha}$$

$$c(iT) = 1 \qquad\qquad i \geq 2$$

Taking the Z-transform we obtain

$$C(z) = \frac{(z+\alpha)}{(1+\alpha)z(z-1)}$$

The input is a unit-step change, thus

$$R(z) = \frac{z}{z-1}$$

and the desired transmission ratio is

$$T(z) = \frac{C(z)}{R(z)} = \frac{(z+\alpha)}{(1+\alpha)z^2}$$

From the state equations we calculate the transfer function

$$\frac{dx_1}{dt} = -2x_1 + x_2$$

$$\frac{dx_2}{dt} = -x_2 + m$$

Setting $x_1 = c$ we can write

$$\frac{d^2c}{dt^2} + \frac{3dc}{dt} + 2c = m$$

and

$$G_p(s) = \frac{C(s)}{M(s)} = \frac{1}{(s + 1)(s + 2)}$$

Inverting we obtain

$$g(t) = e^{-t} - e^{-2t}$$

Applying

$$g^1(t_k - t_o) = \int_{(k-1)T}^{kT} g(\theta)d\theta$$

we get

$$g^1(t_k - t_o) = \alpha^{k-1}(1 - \alpha) - \frac{\alpha^{2(k-1)}}{2}(1 - \alpha^2)$$

which is the delta-function response. Taking the Z-transform we have

$$Z\{g^1(t_k - t_o)\} = \frac{(1 - \alpha)^2(z + \alpha)}{2(z - \alpha^2)(z - \alpha)}$$

which is used in

$$D(z) = \frac{T(z)}{G_p(z)[1 - T(z)]}$$

as $G_p(z)$

The digital compensator is

$$D(z) = \frac{2(z - \alpha^2)(z - \alpha)}{(1 - \alpha)^2(1 + \alpha)(z + \frac{\alpha}{1 + \alpha})(z - 1)} = \frac{M(z)}{E(z)} \tag{4}$$

Rewriting eq. (4) in the form

$$m(t_k) = \frac{2}{(1 + \alpha)^2(1 + \alpha)}[e(t_k) + \alpha(1 - \alpha)e(t_{k-1}) - \alpha^3 e(t_{k-2})]$$

$$+ \frac{1}{1 + \alpha}[m(t_{k-1}) + \alpha m(t_{k-2})]$$

with

$$e(t_k) = r(t_k) - c(t_k)$$

we may obtain the time-optimal control by measuring only the output variable. Note that $m(t_k)$ is defined relative to an initial value of zero.

The state equation of the system is

$$x(k + 1) = Ax(k) + Bu(k) \qquad (1)$$

where

$$A = \begin{bmatrix} 0 & 1 \\ -1 & 1 \end{bmatrix} \qquad B = \begin{bmatrix} 0 \\ 1 \end{bmatrix}$$

The output equation is

$$c(k) = Dx(k) = [2 \quad 0]x(k)$$

Design a first-order observer for the system.

k	0	1	2	3	4	5
$x_1(k)$	1	0	-1	-1	0	1
$x_{e1}(k)$	1	0	-1	-1	0	1
$x_2(k)$	0	-1	-1	0	1	1
$x_{e2}(k)$	1.25	-1	-1	0	1	1
$\bar{w}_e(k)$	0.50	-2	0	2	2	0
$c(k)$	2	0	-2	-2	0	2

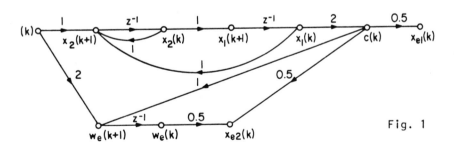

Fig. 1

Solution: The matrix P for the transformation of A into the adjoint phase-variable canonical form is given by

$$P = \begin{bmatrix} a_1 & 1 \\ 1 & 0 \end{bmatrix} \begin{bmatrix} D \\ DA \end{bmatrix}$$

For the second-order case the matrix A_2 is given by

$$A_2 = QA_1Q^{-1} = \begin{bmatrix} 0 & 0 & \cdots & -\alpha_{n-1} & -\alpha_1\alpha_{n-1} - a_n + a_1\alpha_{n-1} \\ 1 & 0 & \cdots & -\alpha_{n-2} & -\alpha_1\alpha_{n-2} - a_{n-1} + a_1\alpha_{n-2} \\ 0 & 1 & \cdots & -\alpha_{n-3} & -\alpha_1\alpha_{n-3} - a_{n-2} + a_1\alpha_{n-3} \\ & & \cdots & & \\ 0 & 0 & \cdots & -\alpha_1 & -\alpha_1^2 - a_2 + a_1\alpha_1 \\ 0 & 0 & \cdots & 1 & -a_1 + \alpha_1 \end{bmatrix}$$

$$A_2 = \begin{bmatrix} -\alpha_1 & -\alpha_1^2 - a_2 + a_1\alpha_1 \\ 1 & -a_1 + \alpha_1 \end{bmatrix}$$

a_1 and a_2 are the coefficients of the characterisitic equation of A,

$$|\lambda I - A| = \lambda^2 - \lambda + 1 = 0$$

Thus $a_1 = -1$ and $a_2 = 1$.

We identify the coefficient α_1 with the characteristic euqation of $\bar{A}_2$ which is

$$\lambda + \alpha_1 = 0$$

In the transformed domain the reduced-order observer is described by

$$\bar{w}_e(k + 1) = \bar{A}_2\bar{w}_e(k) + E_2c(k) + \bar{B}_2u(k) \tag{2}$$

$$w_e(k) = [\bar{w}_e(k) \quad c(k)]'$$

$$\bar{A}_2 = -\alpha_1$$

$$E_2 = -\alpha_1^2 - a_2 + a_1\alpha_1 = -\alpha_1^2 - \alpha_1 - 1$$

$$\bar{B}_2 = 2$$

From eq. (2), setting $\alpha_1 = 0$ we find a deadbeat response for the error of the observer

$$\bar{w}_e(k + 1) = -c(k) + 2u(k) \tag{3}$$

Fig. 1 shows a state diagram for the system and the first-order observer.

It's easy to show that $x_e(k)$ reaches $x(k)$ for $k \geq 1$, for arbitrary initial states $x(0)$ and $\bar{w}_e(0)$. The table shows

the responses of the system for $x(0) = \begin{bmatrix} 1 \\ 0 \end{bmatrix}$, $\overline{w}_e(0) = 0.5$ and
$u(k) = 0$ for all k. Let us consider the closed-loop system
$u(k) = -Gx_e(k)$. The feedback gain matrix G is given by

$$G = [-0.654 \qquad 0.486]$$

Substituting the control and

$$w_e(k) = [\overline{w}_e(k) \qquad c(k)]^1 \qquad \text{and}$$

$$x_e(k) = (QP)^{-1} w_e(k) \qquad (4)$$

into eq. (3) we get

$$\overline{w}_e(k + 1) = -c(k) - 2Gx_e(k)$$

$$= -2x_1(k) + [1.308 \qquad -0.972] \begin{bmatrix} x_1(k) \\ 0.5\overline{w}_e(k) + x_1(k) \end{bmatrix}$$

or

$$\overline{w}_e(k + 1) = -1.664x_1(k) - 0.486\overline{w}_e(k)$$

We have the state equations of the digital system described
by eq. (1), conditioned with the state feedback as

$$x_1(k + 1) = x_2(k) \qquad (5)$$

$$x_2(k + 1) = -0.832x_1(k) + x_2(k) - 0.243\overline{w}_e(k)$$

The state diagram of the feedback system and its first-
order observer are shown in Fig. 2.

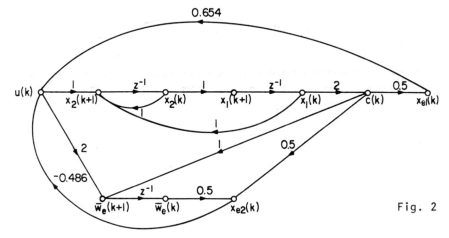

Fig. 2

We get the responses of $x(k)$ and $\overline{w}_e(k)$, $k = 1,2 \ldots$
solving eqs. (4) and (5), for the initial states $x(0) = [1 \ 0]'$ and $\overline{w}_e(0) = 0.5$. Next, we determine the observed
states $x_e(k)$ from

$$x_e(k) = (QP)^{-1}w_e(k) = \begin{bmatrix} x_1(k) \\ 0.5\overline{w}_e(k) + x_1(k) \end{bmatrix}$$

Note that $x_2(k)$ is the only state observed through the first-order observer, $x_{e1}(k)$ is directly $x_1(k)$.

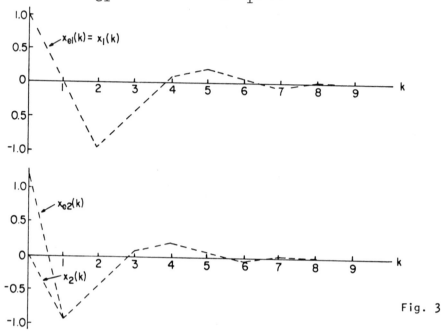

Fig. 3

The time responses of $x(k)$ and $x_e(k)$ for the closed-loop system are shown in Fig. 3. Comparing the results we conclude that the reduced-order observer causes the closed-loop system to have a higher overshoot.

MICROPROCESSOR CONTROL

● **PROBLEM** 17-14

The digital control system is described by the state equations

$$x(k + 1) = Ax(k) + Bu(k) \qquad (1)$$

where

$$A = \begin{bmatrix} 0 & 1 \\ -1 & -2 \end{bmatrix} \qquad B = \begin{bmatrix} 0 \\ 1 \end{bmatrix}$$

with $u(k)$ undergoing the amplitude quantization with quantization level q. Quantize the system.

Solution: The state transition equation of eq. (1) is

$$x(N) = A^N x(0) + \sum_{k=0}^{N-1} A^{N-k-1} Bu(k) \qquad (2)$$

where

$$A^N = z^{-1}\{z(zI - A)^{-1}\} = \begin{bmatrix} (-1)^{N+1}(N-1) & (-1)^{N+1}N \\ (-1)^N N & (-1)^N(N+1) \end{bmatrix}$$

The vector $A^{N-k-1}B$ can be expressed as a linear combination of B and AB. The pair [B AB] is completely controllable. We can write

$$A^{N-k-1}B = (-1)^{N-k-2}[(N-k-2)B + (N-k-1)AB].$$

We can write eq. (2) as

$$x(N) = A^N x(0) + n_1 qB + n_2 qAB$$
$$= A^N x(0) + n_1 q\begin{bmatrix} 0 \\ 1 \end{bmatrix} + n_2 q\begin{bmatrix} 1 \\ -2 \end{bmatrix} \qquad (3)$$

q is the quantization level and n_1, n_2 are integers. We see that for any sampling instant k = N, $x_1(N)$ and $x(N)$ can assume a discrete set of values only. The drawing shows the realizable states of $x_1(k)$ and $x_2(k)$.

For any given initial state $x(0)$, $x_1(k)$ and $x_2(k)$ can reach the states that are spaced at intervals of q in the x_1 vs. x_2 plane.

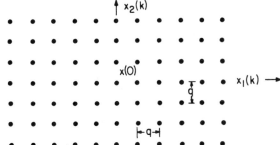

● PROBLEM 17-15

Consider a first-order digital controller with the transfer function

$$D(z) = \frac{C(z)}{U(z)} = \frac{1 + bz^{-1}}{1 + az^{-1}} \qquad (a < 1)$$

Note that usually a digital controller in a control system is implemented digitally in such a way that the round-off of digital data is modeled by amplitude quantization. Examine the effects of quantization.

<u>Solution</u>: Fig. 1 shows a state diagram of the controller and Fig. 2 a model with quantizers positioned at proper locations.

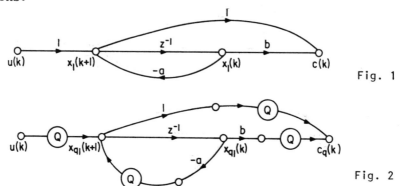

Fig. 1

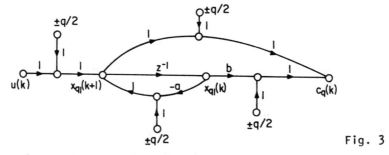

Fig. 2

We replace quantizers by a branch with unity gain and an external source with signal magnitude of $\pm \frac{q}{2}$. The state diagram of this system is shown in Fig. 3.

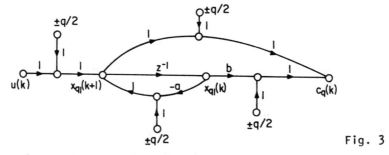

Fig. 3

We assume the same level of quantization for all four quantizers. From Fig. 1 we get the dynamic equations of the controller

$$x_1(k + 1) = -ax_1(k) + u(k) \tag{1}$$

$$c(k) = (b - a)x_1(k) + u(k) \tag{2}$$

For the system with quantizers, the dynamic equations are

$$x_{q1}(k + 1) = -ax_{q1}(k) + u(k) \pm q \tag{3}$$

$$c_q(k) = (b - a)x_{q1}(k) + u(k) \pm 2q \tag{4}$$

Let $e_x(k)$ be the least-upper bound of the quantization error in the state variable $x_1(k)$, we have

$$e_x(k) = x_1(k) - x_{q1}(k)$$

From eqs. (1) and (3) we obtain

$$e_x(k + 1) = -ae_x(k) \pm q$$

whose solution is

$$e_x(N) = (-a)^N e_x(0) + \sum_{k=0}^{N-1} (-a)^{N-k-1}(\pm q)$$

The magnitude of the steady-state value of $e_x(N)$ is

$$\left| \lim_{N \to \infty} e_x(N) \right| = \left| \lim_{N \to \infty} \sum_{k=0}^{N-1} (-a)^{N-k-1}(\pm q) \right| = \frac{q}{1 + a}$$

From eqs. (2) and (4) we get the least-upper bound of the quantization error in the output

$$e_c(N) = c(N) - c_q(N) = (b - a)e_x(N) \pm 2q$$

and the magnitude of the steady-state error bound is

$$\left| \lim_{N \to \infty} e_c(N) \right| = \frac{(b - a)q}{1 + a} + 2q = \frac{2 + a + b}{1 + a} q$$

Taking the z-transforms of $x_1(k)$ and $x_{q1}(k)$ as functions of the equivalent signal sources, we obtain

$$E_x(z) = X_1(z) - X_{q1}(z) = \frac{-z^{-1}}{1 + az^{-1}} \frac{\pm qz}{z - 1}$$

We neglected the initial states, and obtained

$$\left| \lim_{N \to \infty} e_x(N) \right| = \left| \lim_{z \to 1} (1 - z^{-1}) E_x(z) \right| = \frac{q}{1 + a}$$

Evaluating the z-transform of the output, we obtain the least-upper bound of the quantization error in the output

$$E_c(z) = C(z) - C_q(z) = \left[\frac{1 + bz^{-1}}{1 + az^{-1}} + 1 \right] \frac{\pm qz}{z - 1}$$

Thus

$$\left| \lim_{N \to \infty} e_c(N) \right| = \left| \lim_{z \to 1} (1 - z^{-1}) E_c(z) \right|$$

$$= \left| \lim_{z \to 1} \left[\frac{1 + bz^{-1}}{1 + az^{-1}} + 1 \right] q \right| = \frac{2 + a + b}{1 + a} q$$

A digital system is described by

$$x(k + 1) = Ax(k) + Bu(k)$$

where

$$A = \begin{bmatrix} 0 & 1 \\ 0 & 0 \end{bmatrix} \qquad B = \begin{bmatrix} 0 \\ 1 \end{bmatrix}$$

$$u(k) = -Gx(k)$$

$$G = [g_1 \quad g_2]$$

Examine the effect of quantization on the feedback gain matrix.

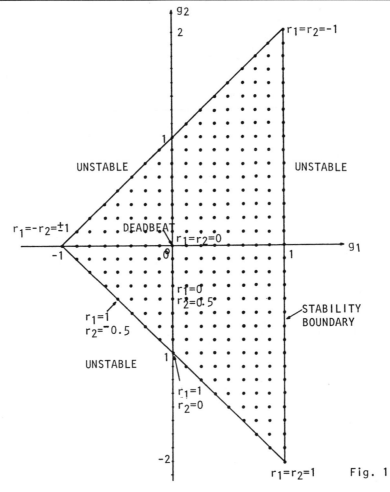

Fig. 1

Solution: The characteristic equation of the closed-loop system is

$$|zI - A + BG| = z^2 + g_2 z + g_1 = (z - r_1)(z - r_2) = 0 \qquad (1)$$

where r_1 and r_2 are the closed-loop eigenvalues. We have

$$G = [g_1 \quad g_2] = [r_1r_2 \quad -(r_1 + r_2)]$$

The values g_1 and g_2 are quantized with quantization level q. Fig. 1 shows the quantization of feedback gains g_1 and g_2 in the $g_1 - g_2$ parameter plane. The region of stable operations is shown when the feedback gains g_1 and g_2 are subject to fixed-point data representation with a 3-bit word-length. The quantization level is

$$q = 2^{-3} = \frac{1}{8}.$$

The dots indicate the realizable values of g_1 and g_2.

To investigate the realizable closed-loop poles due to the quantization of g_1 and g_2 let us represent the roots of eq. (1) in polar form

$$z_1 = re^{j\theta}$$

$$z_2 = re^{-j\theta}$$

The characteristic equation can be written

$$z^2 - 2r \cos \theta z + r^2 = 0$$

We have

$$r = \pm\sqrt{g_1}$$

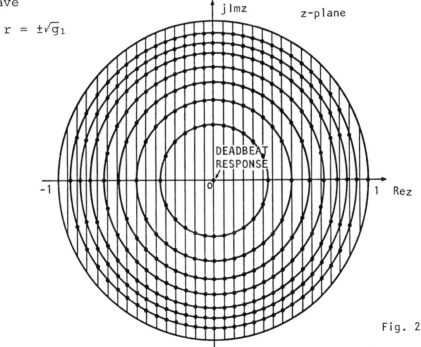

Fig. 2

Note that quantizing g_1 with a quantization level q locates the roots on concentric circles in the z-plane. The circles are centered at the origin, with radii equal to 0, $\sqrt{q}$, $\sqrt{2q}$

. . . Fig. 2 shows the circles for $q = \frac{1}{8}$.

The real parts of the roots are $r \cos \theta$, therefore quantizing $g_2 (= -2r \cos \theta)$ is equivalent to limiting the real parts of the roots to a finite set of numbers. Denoting the real part of the roots by μ, we have

$$\mu = r \cos \theta = -\frac{g_2}{2}$$

The intersections of the circles and the vertical lines represent the realizable poles due to the prescribed quantization of g_1 and g_2.

INDEX